PENGUIN REFERENCE
The Penguin Dictionary of Science

Michael Clugston was born in Lurgan, Northern Ireland, in 1950 and educated at Epsom College and Wadham College, Oxford, where he was the Major Scholar in Chemistry in 1969. During the four-year Chemistry course he won university prizes both for Finals and for research done during the fourth year. He then stayed on for another two years to research for a D.Phil. in theoretical chemistry, under the guidance of Peter Atkins. He spent a year as a Postdoctoral Fellow at Harvard before returning to the UK to get married. He held an SRC Fellowship at Cambridge, as well as a Research Fellowship at St Edmund's College, Cambridge. In 1978 he went into full-time school mastering at Tonbridge School, where he has been ever since. He has published textbooks for the Advanced-level market, *Principles of Physical Chemistry* and *Chemistry: Principles and Applications* (with Peter Atkins), as well as several sets of computer programs. With the huge changes that have occurred at Advanced level, he has written a completely new textbook, *Advanced Chemistry*, which was published in 2000 by Oxford University Press. He was shortlisted for the Salters Prize for Chemistry teachers in the same year.

The Penguin Dictionary of
SCIENCE

Editor: M. J. Clugston
Author team:
N. J. Lord
B. T. Meatyard
J. A. Scarfe and
J. R. C. Whyte

PENGUIN BOOKS

PENGUIN BOOKS

Published by the Penguin Group
Penguin Books Ltd, 80 Strand, London WC2R 0RL, England
Penguin Group (USA) Inc., 375 Hudson Street, New York, New York 10014, USA
Penguin Group (Canada), 90 Eglinton Avenue East, Suite 700, Toronto, Ontario, Canada M4P 2Y3
(a division of Pearson Penguin Canada Inc.)
Penguin Ireland, 25 St Stephen's Green, Dublin 2, Ireland (a division of Penguin Books Ltd)
Penguin Group (Australia), 250 Camberwell Road, Camberwell, Victoria 3124, Australia
(a division of Pearson Australia Group Pty Ltd)
Penguin Books India Pvt Ltd, 11 Community Centre, Panchsheel Park, New Delhi – 110 017, India
Penguin Group (NZ), 67 Apollo Drive, Rosedale, North Shore 0632, New Zealand
(a division of Pearson New Zealand Ltd)
Penguin Books (South Africa) (Pty) Ltd, 24 Sturdee Avenue, Rosebank, Johannesburg 2196, South Africa

Penguin Books Ltd, Registered Offices: 80 Strand, London WC2R 0RL, England

www.penguin.com

First published as the *New Penguin Dictionary of Science* 1998
Revised and updated second edition first published 2004
This revised and updated third edition first published 2009
1

Set in ITC Stone Sans and ITC Stone Serif
Typeset by Data Standards Ltd, Frome, Somerset
Printed in England by Clays Ltd, St Ives plc

ISBN: 978-0-141-03796-7

www.greenpenguin.co.uk

Mixed Sources
Product group from well-managed
forests and other controlled sources
www.fsc.org Cert no. SA-COC-1592
FSC © 1996 Forest Stewardship Council

Penguin Books is committed to a sustainable future
for our business, our readers and our planet.
The book in your hands is made from paper
certified by the Forest Stewardship Council.

Contents

When writing about science, I have a parrot perched on my shoulder who cries out every so often in his raucous voice, 'Can this not be said more simply?' The authors of this dictionary have clearly followed my recipe.

But there remains an intrinsic difficulty about conveying the meaning of scientific terms to non-scientists, comparable to that of explaining musicianship to someone who has never learnt to play an instrument, because science is best learnt by doing it and is hard to pick up from books. Some of the entries in this dictionary, though clear to the scientifically literate, may therefore convey little meaning to, say, a barrister trying to understand a brief on the supposed carcinogenic effects of over-head electric power lines.

On the other hand, those trained predominantly in one scientific discipline who are looking up terms in either their own field or another will find the definitions in these pages concise, rigorous and lucid. A biochemist wondering what is meant by 'd orbitals' in 'transition metals' will find them neatly defined and illustrated; a math-ematician wanting to know about the genetics of Down's syndrome will be well informed in seven lines of text. In some areas the dictionary has developed into an encyclopaedia: for example, the entry about delocalized chemical bonds would pass muster in a textbook. In trying to follow the literature in my own field, the bio-medical sciences, I am faced with a proliferation of acronyms. No sooner have I familiarized myself with one, it seems that two more appear, like Hercules decapi-tating the hydra, the many-headed monster which grew two more heads for every one he cut off. The authors have included a helpful sprinkling of these necessary evils.

The dictionary will be useful to students of all ages from about fifteen years upwards. As a schoolboy taught physics by a teacher whose experiments chronically misfired, I hoped that when I got to university I would learn to understand – in the words of Goethe's Faust, 'Was die Welt im Innersten zusammenhält' – the essence of the forces holding the world together. Gravity perplexed me: how could the Sun attract the Earth when it was 93 million miles away? It took me a long time to realize that our understanding did not reach beyond measuring the force that makes the apple fall and the one that makes the Earth circle the Sun. When Newton found these two forces to have the same origin, he put them into mathematical formulae which define gravity but do not explain why all bodies attract each other. Relativity refined Newton's theories, but brought understanding no nearer.

In a similar vein, the great physicist Richard Feynman recalled playing trains when he was little. When he had loaded small balls on his goods wagons, he asked his father why they rolled backwards when he started the train and forwards when he stopped it. His father replied that it was due to inertia, but that this was merely a name to describe what we cannot understand. This dictionary defines, but does not seek to explore the deepest layers of meaning.

Some sociologists of science now teach that scientific laws are subjective, relative and ephemeral, and, like the laws of Marxism, are upheld by cliques in order to establish and maintain their power. If that were so, then much of this dictionary would soon become as redundant as the Central Committee of the Communist Party. In fact, anyone trying to revise its definitions will have a hard time, and the editors of a new edition, due perhaps in 2025, may look forward to finding little to cross out, but much important new material to add.

Max Perutz, Cambridge, 1998

When I was asked to write from scratch the *New Penguin Dictionary of Science*, my reaction was a mixture of pleasure and trepidation. I was delighted and flattered to be asked to edit the new book, to replace a best-seller which had been continuously in print for over fifty years. Such has been the pace of development during that fifty years that today no single scientist could cover all areas of science to the depth of knowledge required. So my major challenge was to gather together a strong author team.

The aim was to cover all the areas which received detailed coverage in previous editions: mathematics, physics and chemistry. The coverage of human biology, biochemistry, molecular biology and genetics has been significantly increased. The other main area of expansion has been in the provision of diagrams. The seventh edition of the previous work had a total of fifty diagrams; this book has over three hundred.

The myriad of terms used in modern science needed to be explained at a level that can be understood by an Advanced level student, an undergraduate in one of the sciences, the interested lay person and the professional scientist who is curious about ideas in other disciplines. With such a diverse target audience in mind, I wanted to create a team of authors who were highly expert in their own subjects and yet had the gift of communicating those ideas simply.

I started by asking Nick Lord, my friend and colleague (at Tonbridge School) to do the maths entries. Nick is a gifted mathematician, so I knew his definitions would be rigorous. His professional experience as a teacher served well to prevent definitions from becoming too pedantic. I am especially grateful to Nick for his constant encouragement during the long gestation period of the book.

For the physics entries, I was able to persuade Julian Scarfe, a former student of mine, to take on the task. Julian enjoyed great success at Cambridge, excelling in chemistry, before deciding on a career in physics, started off with a Ph.D. at Cambridge. Julian very kindly delayed his entry into full-time employment long enough to write the bulk of the physics entries. It has been encouraging how often we agreed who was more suited to start off the definition of terms that straddle the physics/chemistry divide!

My own training as a physical chemist meant that the biology entries were always going to be the most difficult. Ever since Barry Meatyard became involved in the project, I have been grateful for his enthusiasm for suggesting new headwords and for his skill in writing definitions that are clear and accurate; his teaching experience and research connections at Warwick University's Institute of Education have contributed to the quality of his entries. It has been a great bonus that Barry has been so committed to increasing the coverage of the biological sciences.

For the entries on computing, I relied at the start of the project on another former student, James Whyte. Having won the top First in genetics at Cambridge, James was also able in the final stages of the project to help with those entries with a strong genetics bias. The final stages of the project coincided with the rapid development of the Internet; for such entries it is sensible to look no further than an expert current student. I am very grateful to Alastair Maw for providing the Internet entries.

The chemistry entries I have written myself. For *all* entries the whole author team collaborated magnificently to improve entries written by individual members.

I would like to thank a number of other people who have contributed to the first edition of the *New Penguin Dictionary of Science*. Paul Todd has commented on some of the physics entries and has confirmed a few of the more obscure facts we quote. Mark Farmer and Jared Prakash have helped with transfer of the text between computer platforms. Jean Cook has tracked down various books and articles.

I have been especially grateful for help with the entries on the biological side from a number of different people. Myles Jackson was highly influential in the early development of these entries, before he returned permanently to the United States. I have also appreciated the advice of a friend from Cambridge days, Chris Edge, whose expertise in biochemistry was invaluable. Paul Ridd and Ed Southern gave valuable feedback on the whole of an early draft of these entries. Their helpful comments were greatly appreciated.

Professor Southern generously agreed to cast his eye over the molecular biology entries when I met him during a stay at Trinity College, Oxford. During the course of writing the text, I have been fortunate to enjoy Schoolmaster Fellowships at two Oxford Colleges (Trinity and Merton) and two Cambridge Colleges (Sidney Sussex and Gonville and Caius). My own Colleges (Wadham College, Oxford, and St Edmund's College, Cambridge) generously provided accommodation during the final checking stages.

Our text would be significantly less good than it is were it not for the exceptional professional calibre of John Woodruff and Eddie Mizzi. They have been diligent well beyond the call of duty and have worked tirelessly to improve the efforts of the author team. We owe them a deep debt of gratitude. Eddie would like me to echo his thanks for the help he received from the Publishers of *ChemDraw*. John would like to thank CSW Informatics Ltd, Oxford, for providing extra office facilities.

As is inevitable in a project of this magnitude, there are bound to be errors and omissions. The author team would be genuinely interested to hear of any factual errors, which should be addressed to the Editor, care of Penguin. For example, we have identified the origin of about four hundred eponymous laws, effects, reactions, and so on. We have been stumped on a few (most notably Devarda's alloy and Norton's theorem); help with positive identification would be most welcome.

Finally and most significantly, I am particularly grateful for the help and support I have received from my wife, Corinne. She has been understanding and long-suffering during the writing of the book. She has encouraged me when the going got tough; I would have found it very hard to complete the task without her support. My children, Anna and John, have seen less of their father than they should have. While thanking my family, I must record my great debt of gratitude for the unfailing love and encouragement of my mother, Nessie Clugston, who died just before publication and to whose memory I dedicate this book.

Mike Clugston, Tonbridge, 1998

Preface to the Second Edition

The second edition is dedicated to the memory of Max Perutz OM who discovered, in full molecular detail, how we breathe. It was in his lab that the structure of DNA was worked out exactly fifty years ago.

In addition to the author team, I have been grateful for the helpful advice provided by several colleagues at Tonbridge School, notably Paul Pattenden, Ian Pinkstone and Andrew Worrall.

I was also grateful to receive suggestions for corrections to the text from R. StJ. Richards from Aberdeenshire, Scotland, and Mr W. Young from Queensland, Australia. Thanks to them a couple of errors have been corrected.

Mike Clugston, Tonbridge, 2003

Preface to the Second Edition

Preface to the Third Edition

During the creation of the third edition, every entry has been carefully reconsidered. Some entries have been updated: for example, the term 'planet' was redefined in 2006. A good number of entries have been expanded while yet others have been rewritten in an attempt to make them even clearer. Furthermore, a major emphasis for this edition has been exhaustive checking of all factual details. For the first time, Internet sources have been extensively used for cross-checking. On the rare occasions where differences were found, further checks were made until a definitive source was located.

As always, I have been grateful to the author team for their help. I was particularly grateful that James Whyte has been willing to give us the benefit of his incisive comments, despite having particularly heavy time commitments at the moment. Nick Lord carefully read the whole of the text and made many helpful suggestions, improving the appearance of the equations in particular. I have been especially grateful to my Tonbridge colleague Paul Ridd for the extensive help he gave in the production of this edition. He has commented on a large number of the biology entries and his wise advice on changes has been most welcome.

I was very grateful to the Master and Fellows of St Catharine's College, Cambridge for a Schoolmaster Fellowship during the final stages of the preparation of the third edition. I was grateful for suggestions made by Sir Alan Battersby, Luciano da F. Costa, Philip Oliver and Peter Wothers.

Mike Clugston, Tonbridge, 2009

How to Use this Book

The headwords are listed in alphabetical order according to the letter-by-letter convention, in which spaces between headwords containing two or more words are ignored. Thus, for example, **acetyl CoA** follows **acetylcholine**. Prefixes, such as numbers in the names of chemical compounds, are ignored for the purposes of alphabetization. Greek letters will be found at the beginning of the letter section corresponding to their English equivalent; thus α is located at the beginning of the A section.

Extensive cross-referencing is used to help the reader navigate the dictionary. A cross-reference is indicated by an arrowhead (➤). Terms that may not be familiar to the less knowledgeable reader are cross-referenced to indicate where explanations may be found. Cross-referencing is also used to identify other entries where more detail on a similar topic may be found; words with contrasting meanings are denoted by the word 'Compare' in front of the arrowhead. A final use of cross-referencing is to point to related entries which could be used to locate the present headword in a broader context; this type of cross-reference is indicated by a double arrowhead (➤➤).

Bold type is used within an entry for definitions of terms related to the headword. Thus, for example, there are no separate entries for **spherical aberration** or **chromatic aberration**, which are dealt with in the **aberration** entry and picked out there in bold. Bold type is also used for related terms which do not contain the headword itself. For example, **cilia** and **flagella** are defined under **undulipodium**, but they are included as headwords, each entry consisting simply of a cross-reference to **undulipodium**.

Italic is used for terms or parts of terms, particularly in mathematical, biological and chemical notation and nomenclature, which are customarily italicized. It is also used for foreign terms and, occasionally, for emphasis.

Abbreviations are given in a single form only. Thus the abbreviation used here for nuclear magnetic resonance is NMR, but the variants N.M.R., nmr and n.m.r. will also be encountered in the scientific literature. For simplicity we have given one such form only, being guided in our choice by prevailing practice or recognized authorities on usage.

If a headword has significantly different meanings, the entry is divided into numbered sub-entries. If these separate meanings are associated with different branches of science, the branch is indicated by an abbreviation at the beginning of the sub-entry. Where such headwords are cross-referenced, the relevant number is appended to the cross-reference in brackets if the sense referred to is not the first.

For eponymous entries, the full names and dates of the scientist(s) concerned are included in the entry. If someone is commonly known by a forename other than their first, other forename(s) are enclosed in brackets. For a sequence of headwords named after the same person, full details are given only in the first. The various headwords including the name Einstein, for example, are all named after Albert Einstein. Occasionally the picture is not so simple: the surname Thomson, for example, is shared by J.J. Thomson, discoverer of the electron, and William Thomson, better known by his title of Baron Kelvin.

THE PENGUIN
DICTIONARY OF
SCIENCE

a- A prefix, for words of Greek origin, meaning 'not', as in 'asymmetric'.

a Symbol for ➤acceleration.

A 1 Symbol for the unit ➤ampere.
2 Symbol for the purine base ➤adenine.

Å Symbol for the unit ➤angstrom.

A$_r$ Symbol for ➤relative atomic mass.

α The Greek letter ➤alpha. In physics it occurs most commonly in the phrase ➤alpha decay and related terms. It is also used as a ➤locant in chemistry.

AAS Abbr. for ➤atomic absorption spectroscopy.

A band One of the transverse bands making up the repeating striated pattern of cardiac and skeletal ➤muscle. The 'A' stands for 'anisotropic'.

abdomen ➤thorax.

Abelian group Another name for a ➤commutative group. Named after Niels Henrik Abel (1802–29).

aberration The distortion of an image, such as that of a star, by a ➤lens or mirror (see the diagram). **Spherical aberration** results from the inability of a lens or mirror with spherical surfaces to focus peripheral rays (away from the ➤optical axis) at the same point as paraxial rays (near the optical axis). **Coma** is the formation by a lens or mirror of a slightly off-axis, fan-shaped image for incident rays from an off-axis object. In **astigmatism**, a lens or mirror has different focal lengths across different diameters, and focuses off-axis rays on two lines instead of a single point. **Chromatic**

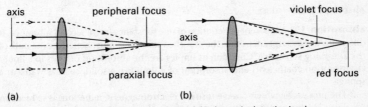

aberration (a) Spherical aberration and (b) chromatic aberration in a lens.

aberration is a result of the material from which a lens is made having different refractive indices for different wavelengths, giving coloured edges to an image formed by white light; it does not arise with mirrors.

abiotic factor Any factor, feature or process, not derived from or caused by living organisms, that might affect an organism in its environment. Examples are hours of daylight, climate and rock type. Compare ➤biotic factor.

abortion The expulsion or removal of the ➤foetus from the uterus before the end of the normal gestation period.

ABO system ➤blood groups.

ABS A thermoplastic ➤copolymer made from acrylonitrile, butadiene and styrene. It has a low density but is very hard, and is used for telephones, suitcases and car body panels.

abscissa Another name for 'x coordinate' (➤➤Cartesian coordinates).

absolute Indicating a quantity measured in normal physical units as opposed to being a ratio. Compare ➤relative. Exceptions to this principle are noted in the entries that follow.

absolute alcohol Pure ➤ethanol. This is hard to achieve because the ➤azeotrope formed by ethanol and water contains only 96% ethanol.

absolute configuration ➤(R–S) system.

absolute expansivity The volume ➤expansivity of a liquid without reference to its containing vessel. Compare ➤apparent expansivity.

absolute humidity The mass of water vapour in the atmosphere per unit volume of air. Compare ➤relative humidity.

absolute temperature ➤thermodynamic temperature, which is the preferred term.

absolute value That positive ➤real number equal to a given number x, except possibly for sign; written $|x|$.

absolute zero The zero of ➤thermodynamic temperature. A system at absolute zero has no thermal energy, thus a quantum-mechanical system is normally in its ground state. Absolute zero is not attainable in practice, though temperatures of 10^{-3} K are routinely achieved in low-temperature physics, and temperatures below 10^{-6} K have been reached.

absorbed dose ➤dose.

absorption 1 The apparent penetration of one substance into a second substance. It differs from ➤adsorption in that the absorbed substance must penetrate the bulk of the absorbing substance. Common examples are the absorption of a gas in a liquid (➤absorption coefficient) and absorption of a liquid in a solid, such as water in a sponge.

2 The process by which a wave (usually electromagnetic radiation) is reduced in intensity by interaction with matter. Absorption of electromagnetic radiation usually involves the excitation of atoms or molecules into higher energy states. ➤➤absorption spectroscopy.

absorption coefficient 1 The maximum volume of a gas at ➤standard temperature and pressure that can be absorbed by a given volume of solvent. The two gases with the largest absorption coefficients in water are sulfur dioxide and ammonia.

2 ➤Beer's law, Lambert's laws.

absorption edge In ➤absorption spectroscopy, a sharp rise in absorption observed as the frequency of the incident radiation changes. Absorption edges occur when the frequency corresponds to a photon energy that is just sufficient to excite an electron from a deep ➤energy level of an atom or ion to create a photoelectron (➤photoelectric effect). ➤➤X-ray absorption spectroscopy.

absorption spectroscopy A form of ➤spectroscopy in which the intensity of electromagnetic radiation is reduced on passing through the sample. Compare ➤emission spectroscopy. The frequencies absorbed give rise to the **absorption spectrum**. For atoms the spectrum shows **absorption lines**, whereas for a molecule there are **absorption bands**. Important examples are ➤infrared spectroscopy and ➤nuclear magnetic resonance.

a.c. Abbr. for ➤alternating current.

Ac Symbol for the element ➤actinium.

acceleration Symbol a; unit m s^{-2}. The rate of change of ➤velocity with respect to time. Thus acceleration is a ➤vector quantity. Note that a decrease in speed (a **deceleration**) falls under this general definition of acceleration, contrary to the common usage of the word.

acceleration of free fall (acceleration due to gravity) Symbol g. The magnitude of the acceleration of a body that is acted upon only by the Earth's ➤gravitational field at the Earth's surface. Because of the equivalence of ➤gravitational mass and ➤inertial mass, it is also the strength of the Earth's gravitational field. It varies with position on the Earth; its value is about 9.8 m s^{-2}.

accelerator (particle accelerator) A device, typically very large, such as a ➤linear accelerator, ➤cyclotron or ➤synchrotron, designed to accelerate charged particles to very high speeds and energies. Accelerators are used in ➤particle physics to force collisions between high-energy particles, which may then interact to form new and different particles.

accelerometer A device for measuring ➤acceleration.

acceptor 1 (Chem.) ➤pi acid.

2 (Phys.) In a semiconductor, an impurity (such as boron, aluminium or gallium in silicon) that accepts an electron and therefore creates a hole. ➤➤doping; compare ➤donor.

accessory pigment One of two main classes of coloured compounds associated with ➤chlorophyll in order to form photosystems for the absorption of light in ➤photosynthesis. **Carotenes** are red, orange or yellow, whereas **phycobilins** are blue or red.

accommodation A ➤reflex action that enables the focal length of the lens of the ➤eye to be changed to bring objects into focus. Near objects require the lens to become more convex, whereas distant objects require a less convex lens. The change in shape of the lens is facilitated by its elastic properties and the action of the ciliary muscles.

accretion disk ➤nova, Seyfert galaxy.

accumulator A secondary cell (➤cell (2)), of which the most common type is the ➤lead–acid battery.

accuracy How close a measurement is to the true value of the quantity measured. Compare ➤precision.

acetal An organic molecule formed from an ➤aldehyde by replacing the C=O group with a C(OR)$_2$ group, where R is any alkyl group. The archetypal example, $CH_3CH(OCH_2CH_3)_2$, was called acetal but is now systematically called 1,1-diethoxyethane; this is a liquid used as a solvent, in perfumes for example. Cyclic acetals, made using ethane-1,2-diol and removed using acidic hydrolysis, are used as ➤protecting groups.

acetaldehyde The older, but still common, name for ➤ethanal.

acetamide ➤ethanamide.

acetate A salt or ester of ➤ethanoic acid (acetic acid).

acetic acid The older, but still common, name for ➤ethanoic acid.

acetic anhydride ➤ethanoic anhydride.

acetoacetic ester ➤ethyl 3-oxobutanoate.

acetone An older name for ➤propanone. Acetone is formed by heating calcium ethanoate (calcium acetate). However, it has three carbon atoms rather than the two in ethanoic acid. The ➤IUPAC names make clear this difference, which is disguised by the traditional names.

acetophenone ➤phenylethanone.

acetylating agent ➤ethanoylating agent.

acetylation ➤ethanoylation.

acetyl chloride ➤ethanoyl chloride.

acetylcholine $CH_3COOCH_2CH_2N^+(CH_3)_3$ A ➤neurotransmitter substance released into the synaptic gap at some (**cholinergic**) nerve endings and neuromuscular junctions. Its function is to initiate a ➤nerve impulse in an adjoining neurone or to initiate contraction in a ➤muscle fibre. Excessive accumulation of acetylcholine in the synapse is prevented by the release of the hydrolytic enzyme **acetylcholinesterase**.

acetyl coenzyme A (acetyl CoA) An intermediate in many key metabolic processes, particularly in the transfer of the products of ➤glycolysis to the ➤Krebs cycle and in fatty acid metabolism. It is derived from pantothenic acid.

acetylene ➤ethyne.

acetylsalicylic acid ➤aspirin.

achromatic lens A lens system that is free from chromatic ➤aberration.

acid A substance that reacts with a ➤base to form a salt and water only. An archetypal reaction is therefore the neutralization of an ➤alkali:

$$H_3O^+(aq) + OH^-(aq) \rightarrow 2H_2O(l).$$

This reaction can, however, be viewed in two alternative ways, which give rise to the two fundamental, more sophisticated, definitions of an acid. The reaction can be seen as a transfer of a proton (H^+ ion) from the ➤oxonium ion to the hydroxide ion. This leads to the ➤Brønsted–Lowry theory of acids and bases (➤➤conjugate acid and base). The extent of proton donation depends on whether the acid is a ➤strong acid or a ➤weak acid. The acidity of the solution can be quantified using ➤pH, and measured approximately using ➤indicators; for example, litmus paper turns red in acids.

An alternative, and equally valid, view regards the reaction above as the attachment of the proton to the lone pair of the hydroxide ion. This leads to the ➤Lewis theory of acids and bases.

acid amide ➤amide.

acid anhydride A molecule related to a carboxylic acid by the loss of water, hence 'anhydride'. An important example is ethanoic anhydride, $(CH_3CO)_2O$, which is a colourless liquid used as an ➤ethanoylating agent. It reacts less vigorously than ethanoyl chloride, which is useful in the industrial manufacture of aspirin.

acid–base indicator ➤indicator.

acid dissociation constant ➤acid ionization constant.

acid dye ➤dye.

acid halide ➤acyl halide.

acidic Having the properties of an ➤acid. An **acidic solution** has a ➤pH less than 7 at $25\,°C$.

acidification Addition of an acid to a solution.

acid ionization constant (acid dissociation constant) The equilibrium constant describing the extent of ionization of a ➤weak acid which is, by definition, only partially ionized. The chemical ➤equilibrium set up when a weak acid dissolves in water is $HA(aq) + H_2O(l) \rightleftharpoons H_3O^+(aq) + A^-(aq)$. The acid ionization constant is $K_a = [H_3O^+][A^-]/[HA]$; pK_a is defined like ➤pH, viz.

$pK_a = -\log_{10}(K_a/\text{mol dm}^{-3})$.

acid rain Rain-water with a lower ➤pH than would occur naturally. Rain-water is naturally acidic as it contains dissolved carbon dioxide. Acidification caused by pollution has markedly lowered the pH of rain taken from several sites over the past 50 years. For example, one site in Scandinavia had rainfall which averaged pH 5.4 in 1956 and 4.3 in 1974. Such acidification results mainly from the release into the atmosphere of the oxides of nonmetals. These tend to be acidic gases, in particular sulfur dioxide, SO_2, which is released when coal containing sulfur impurities is burnt, and nitrogen dioxide, NO_2, from car exhausts. The consequences for the environment of this increased acidity are severe: trees suffer considerable damage and lakes become more acidic, which threatens fish stocks.

acid salt A salt in which only some of the replaceable hydrogen atoms of an acid have been replaced. A common example is sodium hydrogencarbonate (bicarbonate).

aclinic line A line on the surface of the Earth joining points of zero ➤magnetic inclination.

acoustics The study of sound and acoustic waves.

acoustic wave A ➤longitudinal wave of vibration in a medium, possibly sound but not necessarily audible to the human ear. Acoustic waves are the generalization of sound to a wider range of frequencies, in the same way that electromagnetic waves are the generalization of light.

acoustoelectronics The use of acoustics in electronic devices. In particular, the low speed of sound compared with light allows acoustic ➤delay lines to be constructed easily.

acquired characteristic ➤Lamarckism.

acquired immune deficiency syndrome ➤AIDS.

acrosome A compartment that contains digestive enzymes, found at the head of a sperm. On contact with an egg the enzymes are released, facilitating the entry of the sperm into the egg.

acrylamide (polyacrylamide) An inert medium used in the form of a gel in ➤electrophoresis to separate large molecules such as ➤proteins and ➤nucleic acids.

acrylic A member of a class of ➤thermoplastic ➤copolymers made from acrylonitrile (propenenitrile, CH_2=CHCN) which are transparent. They can be made as wool-like fibres or resins.

acrylonitrile ➤propenenitrile.

ACTH Abbr. for ➤adrenocorticotrophic hormone.

actin A protein that makes up the very thin filaments of ➤muscle cells. Changes in their interaction with thicker filaments composed of ➤myosin cause muscle contraction.

actinide (actinoid) One of the 14 elements following actinium in the ➤periodic table; they are the analogues of the ➤lanthanides in the row above. Their chemistry is dominated by their radioactivity. The nuclear properties of two actinides, uranium and plutonium, make them of pre-eminent importance as ➤fissile nuclides. ➤➤Appendix table 5.

actinium Symbol Ac. The element with atomic number 89 and relative atomic mass 227, which is below lanthanum in the periodic table and is the archetypal ➤actinide element.

actinoid ➤actinide.

action (action integral) Symbol I or S; unit J s. A concept in mechanics defined for a system between two specified times t_1 and t_2:

$$I = \int_{t_1}^{t_2} L(q, \dot{q})\, \mathrm{d}t,$$

where L is the ➤Lagrangian, which is a function of both the coordinates q and their rates of change $\dot{q} = \mathrm{d}q/\mathrm{d}t$. See also ➤Hamilton's principle.

action potential The state of polarization of the membrane of a nerve cell that allows a ➤nerve impulse to propagate.

action spectrum For a photochemical reaction, a plot of the rate of reaction against wavelength. It is applied particularly to biological processes that require or are affected by light, such as ➤photosynthesis or the initiation of flowering in some plants in response to day length.

activated charcoal A specially prepared form of charcoal that has a high adsorption, used, for example, in gas masks.

activated complex The cluster of atoms in the region close to the transition state (➤activation energy).

activation energy Symbol E_a. The minimum energy that is necessary before a chemical reaction can occur; the activation energy provides the energy needed to break (at least partially) some bonds before others can be formed. For a ➤bimolecular reaction mechanism not only must the two species collide but they must also possess the activation energy for the reaction before the collision will result in a reaction. The activation energy is shown graphically in the diagram; the ➤transition state corresponds to the maximum of the curve. A ➤catalyst works

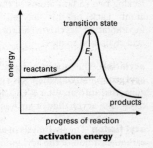

activation energy

by providing an alternative pathway with a lower activation energy. ➤➤Arrhenius equation.

active component A component in an electric circuit that introduces power (like an amplifier), or has a directional effect (like a diode).

active mass An obsolete term, which was used in the ➤law of mass action, that is essentially synonymous with concentration.

active site ➤enzyme.

active transport The process in living cells in which a solute (such as Na^+ and Ca^{2+}) is transported across a membrane from a region of low to high solute concentration (➤diffusion). Energy is needed for this process to occur, and is usually provided by ➤ATP. The ➤sodium pump, which maintains differential concentrations of sodium ions on either side of the membrane of, for example, nerve cells, relies on active transport. Compare ➤passive transport.

activity 1 (Chem.) The effective concentration of a species, which should be used in place of the concentration in all equilibrium expressions. For example, technically the pH is the negative logarithm of the aqueous oxonium ion activity, although in almost all applications this is approximated by its concentration. The activity for a solution is directly proportional to the concentration, the constant of proportionality being the **activity coefficient**. ➤Debye–Hückel theory. The corresponding term for gases is the ➤fugacity.
   2 (Phys.) The rate of disintegration (events per unit time) of a nuclide by ➤radioactive decay.

activity series One of the alternative names for the ➤electrochemical series.

acute angle An ➤angle that is less than $90°$.

acyclic Not having a ring structure.

acyclovir Tradename Zovirax. The first successful **antiviral** agent, used, for example, to cure diseases caused by the ➤herpes virus, such as cold sores on the lips. Its structure (see the diagram) resembles the structure of ➤guanosine, crucially lacking two carbon atoms of the ribose ring and one oxygen atom. Lack of this oxygen atom means that a virus incorporating acyclovir cannot continue to build a nucleic acid chain and so dies.

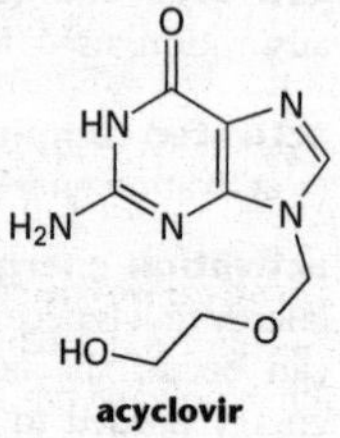

acyclovir

acylation The process of replacing a hydrogen atom with an **acyl group**, —COR, where R is a general alkyl group. For example, an alcohol R′OH is converted into an ester R′OCOR. The process is achieved by an **acylating agent**, often an ➤acyl halide or an ➤acid anhydride.

acyl fission The hydrolysis of an ester where the bond broken is that between the carbonyl carbon of the acyl group and the oxygen atom, as shown in the diagram.

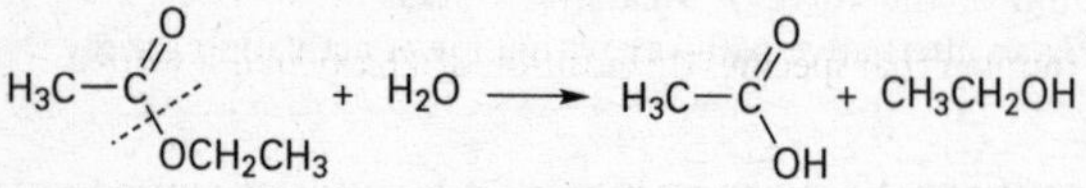

$$H_3C-C\overset{O}{\underset{OCH_2CH_3}{}} + H_2O \longrightarrow H_3C-C\overset{O}{\underset{OH}{}} + CH_3CH_2OH$$

acyl fission

acyl halide (acid halide) A molecule that has an acyl group attached to a halogen atom, to form the structure RCOHal, which is frequently used for ➤acylation. The most common examples, **acyl chlorides**, are made from carboxylic acids using a ➤chlorinating agent. An important example is ➤ethanoyl chloride, CH_3COCl.

Ada A high-level computer language used for real-time control, particularly in military applications such as missile control systems. Named after Lord Byron's daughter Augusta Ada Lovelace (1815–52).

adamantane $C_{10}H_{16}$ A fused tricyclic hydrocarbon (see the diagram). It is a colourless solid. The derivative 1-adamantanamine hydrochloride is useful against viral infections, preventing the virus from penetrating the host cell.

adamantane

adaptation Any characteristic that increases an organism's chance of survival in a particular habitat.

adatom An atom on a surface after ➤adsorption.

ADC Abbr. for ➤analogue-to-digital converter.

addition The mathematical operation in which the sum of two numbers, vectors, matrices, polynomials, etc., is found.

addition compound An alternative name for an ➤adduct.

addition formulae Formulae such as

$\sin(A + B) = \sin A \cos B + \cos A \sin B,$
$\cos(A + B) = \cos A \cos B - \sin A \sin B,$
$\tan(A + B) = (\tan A + \tan B)/(1 - \tan A \tan B),$

which relate ➤trigonometric functions of the sum of two or more angles to the trigonometric functions of the individual angles.

addition polymer A ➤polymer made by the addition of monomers together, the mechanism for which makes it a chain polymer. The most important examples are the ➤polyalkenes, the archetypal example of which is ➤polyethylene. Compare ➤condensation polymer.

addition reaction A reaction in which one molecule is made from two. The most common example is the ➤electrophilic addition reactions of alkenes. Carbonyl compounds undergo ➤nucleophilic addition reactions. An important cyclic addition reaction is the ➤Diels–Alder reaction. An addition reaction followed by an elimination reaction is called a ➤condensation reaction.

additive primaries ➤primary colours.

address A number that specifies the position of a particular ➤word in a computer's memory.

adduct (addition compound) The result of two chemical species adding together. A common example is the reaction between a Lewis acid and a Lewis base, as in the adduct between boron trifluoride and ammonia: $NH_3(g) + BF_3(g) \rightarrow H_3N.BF_3(s)$. The adduct in this case is a white solid, held together by ➤coordinate bonding.

adenine Symbol A. A ➤purine base found in ➤DNA and ➤RNA. ➤➤base pairing; genetic code.

adenosine A ➤nucleoside consisting of the base ➤adenine covalently bonded to the sugar ➤ribose (or ➤deoxyribose in **deoxyadenosine**). The phosphate esters adenosine monophosphate (➤AMP), adenosine diphosphate and adenosine triphosphate (➤ATP) are biochemically important.

adenosine diphosphate ➤ADP.

adenosine monophosphate ➤AMP.

adenosine triphosphate ➤ATP.

adenovirus One of a group of double-stranded DNA-containing ➤viruses associated with a variety of mammalian respiratory infections, including the common cold and some types of tumour. Adenoviruses are icosahedral (20-sided) in shape and are of interest since non-pathogenic forms have been implicated as possible ➤vectors of DNA in the treatment of genetic conditions such as cystic fibrosis. Tumour-forming adenoviruses code for a protein which disrupts the host cell's RNA synthesis, turning the cell into a tumour cell.

ADH Abbr. for antidiuretic hormone (➤vasopressin).

adiabatic Describing a process that occurs without the transfer of heat.

adiabatic demagnetization A method of achieving very low temperatures by use of a magnetic field on paramagnetic salts. The magnetic field is switched on in an ➤isothermal step, magnetizing the material. The material is then isolated from its surroundings and the magnetic field slowly decreased to zero in an ➤adiabatic step.

adipic acid ➤hexanedioic acid.

adipose tissue A body tissue comprising large fat-filled cells. Adipose tissue is found beneath the skin and around internal organs such as blood vessels, the kidneys and the heart. The fat stored can act as an energy reserve, and also provides heat insulation (as in the blubber of whales). This is especially important for mammals that hibernate.

admittance Symbol Y; unit Ω^{-1}. The reciprocal of the ➤impedance of a component or circuit.

ADP Abbr. for adenosine diphosphate (➤ATP).

adrenal cortex The ➤endocrine gland forming the outer shell of each ➤adrenal gland. It secretes various ➤corticosteroid hormones together with small amounts of the sex hormones, androgens and oestrogens. The adrenal cortex is associated with the regulation of water re-absorption by the ➤kidneys and other homeostatic mechanisms (➤homeostasis).

adrenal glands A pair of ➤endocrine glands situated above the kidneys. The outer ➤adrenal cortex surrounds the central ➤adrenal medulla.

adrenaline (US **epinephrine)** A ➤hormone secreted by the ➤adrenal medulla and by parts of the peripheral ➤nervous system in response to a requirement for rapid action by the body. Adrenaline increases the heart rate and diverts blood from the gut to the muscles, thereby preparing the body for increased activity. This is what causes the 'butterflies in the stomach' feeling in a stressful situation. Adrenaline is sometimes referred to as the 'fight or flight' hormone since it prepares the body for either of these options. An injection of adrenaline is the most effective treatment for life-threatening ➤anaphylaxis. The spelling **Adrenalin** (with a capital 'A') is a tradename for a commercial brand of adrenaline.

adrenaline

adrenal medulla The ➤endocrine gland forming the inner core of each ➤adrenal gland. The adrenal medulla produces ➤adrenaline.

adrenocorticotrophic hormone (ACTH) The ➤hormone produced in response to stress or fright by the anterior ➤pituitary gland, which stimulates the ➤adrenal cortex. ACTH is used in the treatment of the symptoms of ➤asthma and rheumatic diseases.

adsorption The attachment of an **adsorbate** to the surface of an **adsorbent**. A typical example is the adsorption of a gas onto a solid. There are two main ways in which the adsorption can occur. An actual chemical bond can form in the process of **chemisorption**; an example is the W—H bond formed when hydrogen gas adsorbs onto tungsten metal (W). The alternative is **physisorption**, in which the adsorbed species is held to the surface only by much weaker ➤van der Waals forces. Adsorption is especially important to the function of ➤heterogeneous catalysts. Compare ➤absorption. ➤➤chromatography.

advection The movement of an airmass, particularly of atmospheric moisture, when horizontal air currents flow, or the transfer of some quantity, such as heat, by such movements. **Advection fog** forms when warm moist air (for example, from the sea) blows over a cold land mass and cools below its ➤dew point.

aerial (antenna) The part of a radio transmitter from which electromagnetic waves emanate, or the part of a receiver that detects them and transforms them into an electrical signal.

aerobic respiration An oxygen-requiring process in organisms, which releases energy from organic molecules. Oxygen is needed as an acceptor for H^+ ions released during the oxidation of reduced ➤coenzymes produced in a series of enzyme-driven reactions in the ➤Krebs cycle. Such reactions occur inside the ➤mitochondria. All eukaryotic cells, which possess mitochondria, can carry out aerobic respiration. Prokaryotes lack mitochondria, but those that possess mesosomes can respire aerobically. The complete oxidation of one molecule of glucose via aerobic respiration can yield up to 38 molecules of ➤ATP. Compare ➤anaerobic respiration.

aerodynamics The study of forces on a body (particularly an ➤aerofoil) moving through the air, and the air currents around it.

aerofoil (US **airfoil)** A shape designed to produce, when in a moving current of air, a force perpendicular to the direction of motion (➤lift) which is much greater than the force parallel to the direction of motion (➤drag). Aerofoils are used for aeroplane wings and control surfaces, as well as to create downforce on racing cars to increase grip on corners.

aerosol A dispersion of a solid or a liquid in a gas. Smoke and fog are everyday examples of aerosols. ➤colloid.

AES Abbr. for Auger electron spectroscopy; ➤Auger effect.

aestivation A physiological state of dormancy during a period of the year which is too hot or dry for the normal maintenance of life. It occurs in desert-living fish (such as the African lungfish) and amphibians (such as the spadefoot toad). Compare ➤hibernation.

aether ➤ether (2).

AFM Abbr. for ➤atomic force microscope.

Ag Symbol for the element ➤silver, from its Latin name *argentum*.

agar A ➤polysaccharide extract, derived from some species of red algae, which forms a gel as a 0.5–3.0% solution. It is used extensively in the food and pharmaceutical industries as a thickening and gelling agent to stabilize, for example, processed cheese and ointments. It is also used in culture media as a gelling agent for nutrient solutions on which bacteria or other cells can be grown. Such a culture preparation in a Petri dish is referred to as an **agar plate**.

agarose A ➤polysaccharide sub-fraction of agar gel used as a medium in some column ➤chromatography techniques and in ➤electrophoresis to separate DNA fragments, for example. In this latter application it is sometimes preferred to ➤acrylamide due to its low toxicity.

Agrobacterium tumefaciens A ➤bacterium infecting plants and causing 'crown gall' disease. Non-virulent strains of the bacterium are extensively used in the ➤genetic engineering of plants. The bacteria carry a range of plasmids, and together the bacteria and plasmids provide a convenient natural mechanism for inserting DNA into plant cells. This is a rare example of a system for the natural transfer of genes between a bacterial ➤plasmid and a ➤eukaryotic ➤genome.

AI 1 (Comput.) Abbr. for ➤artificial intelligence.
   2 (Biol.) Abbr. for artificial ➤insemination.

AIDS (acquired immune deficiency syndrome) A variety of opportunistic infections resulting from the disabling of the body's ➤immune system after infection with ➤HIV. AIDS involves a range of otherwise rare conditions including Kaposi's sarcoma, a rare skin cancer, and respiratory infections, especially pneumocystis pneumonia. The development of AIDS almost always results in death.

air The mixture of gases that comprises the atmosphere of the Earth. It does not have a fixed composition, as is typical of mixtures. The dominant three gases, however, are always nitrogen (about 78%), oxygen (about 21%) and argon (about 1%). Other common constituents are the other noble gases, carbon dioxide and water vapour (the percentage of which depends on the ➤relative humidity). Components present in air in unnaturally high quantity, such as sulfur dioxide, nitrogen oxides and ozone, cause air ➤pollution. ➤➤acid rain.

airfoil US form of ➤aerofoil.

Airy rings ➤Fraunhofer diffraction. Named after George Biddell Airy (1801–92).

Al Symbol for the element ➤aluminium.

Ala Abbr. for ➤alanine.

alabaster Naturally occurring hydrated calcium sulfate, $CaSO_4 \cdot 2H_2O$.

alanine (Ala) A common ➤amino acid. ➤➤Appendix table 7.

albedo The proportion of ➤radiant flux or ➤luminous flux incident on a body that is reflected by it.

albinism A lack of pigmentation in an organism, caused by a ➤recessive gene.

albumin (serum albumin) One of the main groups of ➤protein in ➤blood plasma.

alcohol One of the class of organic molecules containing an alkyl group connected to a hydroxyl group, ROH. The archetypal example ➤ethanol, CH_3CH_2OH, is also commonly referred to simply as 'alcohol'. A ➤dihydric alcohol has two hydroxyl groups; ➤➤polyhydric alcohol. The physical properties of alcohols depend on the ➤hydrogen bonding between the hydroxyl groups of neighbouring molecules; for example, their boiling points are significantly higher than those of the corresponding isomeric ethers. They react with ➤chlorinating agents to form chloroalkanes, and with acids to form esters. They can be dehydrated by concentrated sulfuric acid to ➤alkenes. The oxidation product of an alcohol depends on the structure of the alkyl group. A **primary alcohol** has two hydrogen atoms on the neighbouring carbon atom (next to the hydroxyl group): primary alcohols are oxidized to aldehydes and thence to carboxylic acids. A **secondary alcohol** has only one hydrogen atom on the neighbouring carbon atom: secondary alcohols are oxidized to ketones. A **tertiary alcohol** has no hydrogen atoms on the neighbouring carbon atom: tertiary alcohols resist oxidation (see the diagram). Ethanol is by far the most important alcohol; it is used in alcoholic drinks, as an industrial solvent and in the manufacture of ethanoic acid. ➤Methanol is used as a fuel. Higher alcohols are used to manufacture detergents and plasticizers.

alcohol (a) Primary, (b) secondary and (c) tertiary alcohols.

aldehyde One of the class of organic molecules containing most usually an alkyl group (or an aryl group or a hydrogen) attached to a carbonyl group, RCHO. The most important example is ➤ethanal, CH_3CHO. They are formed by oxidation of a primary alcohol and are themselves oxidized to carboxylic acids; CH_3CH_2OH is oxidized to CH_3CHO which is oxidized to CH_3COOH. Their easy oxidation allows them to be distinguished from ➤ketones by ➤Fehling's solution or ➤Tollens' reagent. Like ketones they react with ➤DNP to give characteristically orange derivatives.

aldohexose A ➤monosaccharide having an aldehyde structure at the carbonyl carbon in the open-chain form and with six carbon atoms. An example is ➤glucose.

aldol condensation An important ➤base-catalysed condensation reaction used to make carbon–carbon bonds. The archetypal reaction is that of ethanal, with sodium hydroxide providing the base catalysis. One molecule loses a proton to form the ➤carbanion $[CH_2CHO]^-$, which then attacks another ethanal molecule to give (after protonation) $CH_3CH(OH)CH_2CHO$, an **aldol**, because of the aldehyde (—CHO) and alcohol (—OH) groups it contains. Dehydration then forms $CH_3CH{=}CHCHO$.

aldose A ➤monosaccharide with an aldehyde group in the open-chain form. An example is ➤glucose.

algae (singular **alga**) An umbrella term for ➤plants and ➤protoctistans comprising aquatic organisms found in both fresh- and salt-water habitats. The term has no direct taxonomic significance. Algae range from macroscopic seaweeds such as giant kelp, which frequently exceeds 30 m in length, to microscopic filamentous and single-celled forms such as *Spirogyra* and *Chlorella*.

algebra Historically, that branch of mathematics concerned with the solution of equations and associated processes in which letters representing numbers are manipulated by rules mimicking those of arithmetic. More generally, modern or abstract algebra is concerned with the study of ➤groups (2), ➤vector spaces, or indeed any situation in which there is a set of elements together with one or more rules for combining them satisfying certain fixed axioms.

algebraic number A ➤real or ➤complex number that is a ➤root of a ➤polynomial equation with ➤integer coefficients. For example, $\sqrt{2} + i$ is an algebraic number because it is a root of $x^4 - 2x^2 + 9 = 0$.

Algol Abbr. for algorithmic language. An early high-level computer language conceptually important in the development of other languages.

algorithm A formal sequence of instructions, given in an unambiguous and logical language, that may be followed in order to achieve a particular purpose, such as sorting a set of numbers in a computer's memory.

alicyclic Describing an organic compound that contains a ring, but not an ➤aromatic ring. An example is cyclohexane. They tend to react in similar ways to the alkanes rather than the aromatic compounds like benzene. The two smallest alicyclic compounds (cyclopropane and cyclobutane) do react more vigorously; the ring can be opened, by hydrogenation for example.

aliphatic Describing an open-chain organic compound. Common examples include alkanes, alkenes, alcohols, aldehydes and ketones. Compare ➤alicyclic.

aliquot A small sample taken from a solution.

alizarin An orange-red dye, originally extracted from the root of the madder plant.

alkadiene ➤alkene.

alkali A water-soluble ➤base. The central species present in an **alkaline solution** is the hydroxide ion, $OH^-(aq)$. An alkaline solution has a pH greater than 7 and turns litmus paper blue.

alkali metal A member of Group 1 in the periodic table: one of the six elements lithium, sodium, potassium, rubidium, caesium and francium. They get their name from the fact that their hydroxides are alkaline in solution: sodium hydroxide and potassium hydroxide are the two most important alkalis both in the laboratory and in industry. The elements themselves are soft metals which can be cut with a knife. They are very strong reductants, having very negative ➤standard electrode potentials, reacting violently with water to form hydrogen gas. The reactivity increases down the group. Their oxidation number in all their compounds is +1, as in sodium hydroxide, NaOH. The oxides are bases which react with water to form alkaline solutions. These solutions contain the hydroxide ion and react with acids to form salts. ➤Sodium chloride is known as common salt, or just salt, and gives its name to the whole class of substances. It is a typical ionic compound. ➤Lithium shows some anomalous behaviour, as is typical of the first member of a group.

alkaline ➤alkali.

alkaline earth metal A member of Group 2 in the periodic table: one of the six elements beryllium, magnesium, calcium, strontium, barium and radium. Technically only the last three elements have alkaline oxides, but the name is used generally to describe the complete group. The elements are reductants, especially ➤magnesium. Their oxidation number in all their compounds is +2, as in calcium oxide, CaO. The oxides are high-melting solids of increasing solubility in water, because the ➤lattice energy of the oxide falls as the ions get larger down the group. The halides are generally fairly ionic solids; those of beryllium, however, are noticeably more covalent as the charge density of the beryllium ion is high.

alkaloid One of a diverse group of nitrogen-containing compounds synthesized by plants and fungi. Many are pharmacologically active; some are extremely important drugs, whereas others are highly toxic. Alkaloids include **atropine**, **cocaine** and **morphine**, and these and their derivatives are used as pain-killers and anaesthetics (➤narcotic drug). **Caffeine** is a common alkaloid, found in tea and coffee, which has a stimulatory effect. **Strychnine** is highly toxic, and was formerly used in pest control.

alkanal An ➤aliphatic ➤aldehyde.

alkane (paraffin) A member of the ➤homologous series of ➤saturated hydrocarbons of formula C_nH_{2n+2}. Their names (see the table) are used as the basis for naming many other compounds. They are found in crude oil and natural gas. The first four members are gases and the next eleven liquids; the others are waxy solids.

alkane The first ten alkanes.

Name	Molecular formula
Methane	CH_4
Ethane	C_2H_6
Propane	C_3H_8
Butane	C_4H_{10}
Pentane	C_5H_{12}
Hexane	C_6H_{14}
Heptane	C_7H_{16}
Octane	C_8H_{18}
Nonane	C_9H_{20}
Decane	$C_{10}H_{22}$

Their most characteristic chemical feature is a general lack of reactivity (which explains their old name of 'paraffin', from the Latin *parvum affinitas*, little affinity). They do, though, undergo ➤halogenation to form halogenoalkanes. They also combust readily and many of our most important fuels are alkanes, such as ➤methane or ➤butane, or mixtures of alkanes such as ➤gasoline.

alkanol An ➤aliphatic ➤alcohol.

alkene (olefin) A member of the ➤homologous series of ➤unsaturated hydrocarbons of formula C_nH_{2n}, the most characteristic feature of which is a carbon–carbon double bond. (A similar compound with two double bonds, such as ➤butadiene, is called an **alkadiene**.) Their characteristic chemical reactions are ➤electrophilic addition reactions, such as that with bromine water. This provides a test for distinguishing an alkene from an alkane: an alkene decolourizes bromine water whereas an alkane does not. They undergo addition polymerization reactions to form the ➤polyalkenes which are particularly important polymers.

alkoxide An anion formed from an alcohol. For example, sodium reacts with ➤ethanol to form sodium ethoxide, $Na^+CH_3CH_2O^-$.

alkyd resin A resinous product of the ➤condensation reaction between a ➤polyhydric alcohol and an acid, such as ➤phthalic acid. They are used in paints, surface coatings and adhesives.

alkylarene A molecule containing an alkyl group attached to a benzene ring. An example is methylbenzene, $C_6H_5CH_3$. They are often made by the ➤Friedel–Crafts reaction.

alkylation The introduction of an ➤alkyl group into a molecule. An important example is the ➤Friedel–Crafts reaction.

alkyl group Symbol R. A group formed from an alkane by removal of a single hydrogen atom. **Methyl** ($-CH_3$) and **ethyl** ($-CH_2CH_3$) are the first two examples. ➤➤butyl for details of possible isomers.

alkyl halide ➤halogenoalkane.

alkyl radical A ➤radical formed from an alkane by removal of a single hydrogen atom. The **methyl radical** ($CH_3\cdot$) is the simplest example.

alkyne (acetylene) A member of the ➤homologous series of ➤unsaturated hydrocarbons of formula C_nH_{2n-2}, the most characteristic feature of which is a carbon–carbon triple bond. The first and most important example is ➤ethyne (acetylene), C_2H_2. An alkyne undergoes electrophilic addition reactions, as does an ➤alkene.

allele Different and distinctive (by virtue of their products or effects) sequences of genetic material that occupy the same ➤locus on each of a pair of ➤homologous chromosomes. Classically, an allele is one of the forms in which a ➤gene can exist. In a diploid cell there are normally two alleles for a given gene, which may be the same (homozygous) or different (heterozygous). ➤➤dominant and recessive.

allene Propadiene, $CH_2=C=CH_2$, or one of its derivatives.

allergy A physiological response occurring when an ➤antigen makes contact with ➤mast cells of the immune system. This contact triggers the release of histamine, which induces local inflammation and produces symptoms of asthma, hay fever or the swelling of insect bites. Common allergies are to grass and other pollen, and to microscopic mites found in house dust and bedding.

allograft ➤graft.

allopatric Describing species that occur in geographically isolated populations or in distinctly separated zones within the same area. Allopatric speciation (➤evolution) refers to new species arising under such conditions. Compare ➤sympatric.

allosteric enzyme An enzyme whose activity can be altered by modulator molecules acting at a site other than the substrate binding site. The regulation of the production of many substances by sequential biochemical pathways in cells is achieved by the end-product molecule binding to the allosteric site of an enzyme acting on a precursor substance. The presence of the product in this site alters the conformation of the substrate binding site so it cannot accept its substrate, thus shutting off the supply of materials to make the product.

allotropy The existence of two or more alternative forms of the same element in the same physical state. For example, carbon has three solid **allotropes**: ➤diamond, ➤graphite and fullerite (➤buckminsterfullerene). (Solid allotropes can also be called polymorphs.) There are important examples of allotropes in the gas phase too, such as dioxygen, O_2, and ➤ozone, O_3. ➤➤monotropy; enantiotropy.

alloy A combination of two or more metals. Most commonly, the combination can be treated as though the two metals form a mixture, though some compound formation does occur (➤Hume–Rothery rules). Alloying is used to improve the physical properties of the resulting metal: the alloy is often stronger or harder than either pure metal. Important examples of alloys are brass and bronze (➤➤solder). **Alloy steels** contain iron, the nonmetal carbon and at least one other metal. A typical stainless steel contains iron, carbon, 18% (by mass) chromium and 8% nickel.

allyl alcohol (prop-2-en-1-ol) $CH_2{=}CHCH_2OH$ A colourless liquid used in the production of synthetic resins.

Alnico A series of iron alloys containing aluminium, nickel and cobalt, used for making magnets.

alpha A prefix used in a number of different ways, especially in biochemistry. For sugars such as ➤glucose, it describes the isomer with the hydroxyl group on carbon atom 1 in the 'axial' position (see diagram at glucose). For amino acids, it describes the position of the carboxyl group relative to the amine group (**alpha amino acids** have the two groups on the same carbon atom). For proteins, it describes the most common helical structure present (see the diagram of the **alpha helix** overleaf).

The prefix is also used in astronomy to indicate, usually, the brightest star in a constellation, for example *Alpha Centauri*.

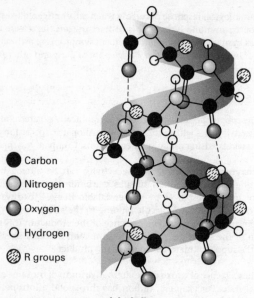

alpha helix

alpha decay The mode of radioactive decay that consists of the emission of ^{4}He nuclei (known in this context as **alpha particles**, with the emission known as **alpha radiation**). Examples are ^{238}U $\rightarrow$ ^{234}Th + ^{4}He and ^{226}Ra $\rightarrow$ ^{222}Rn + ^{4}He. The daughter nuclide has two fewer protons and two fewer neutrons than the parent nuclide. Alpha particles typically come from massive nuclei, which emit this particularly stable, small combination of nucleons. ➤radioactivity.

alpha-iron An allotrope of iron stable up to 910 °C. ➤iron–carbon phase diagram.

alpha particle ➤alpha decay.

alpha radiation ➤alpha decay.

alternating current (a.c.) Current in an electric circuit that changes direction periodically. The variation of current and voltage is usually in the form of a ➤sine curve. As an adjective, 'a.c.' refers to devices and components designed for use with alternating current.

alternation of generations A pattern of life cycle, particularly well illustrated in mosses and ferns, in which a sexual phase, the ➤gametophyte (producing gametes), alternates with an asexual phase, the ➤sporophyte (producing spores). The life cycle of the flowering plants (➤Magnoliopsida) derives from this pattern, but the gametophytes are reduced to the pollen grain (male) and a group of cells in the ovule (female), from which the embryo is derived.

alternator A ➤generator that turns mechanical energy into electrical energy in the form of an alternating current by electromagnetic ➤induction. The magnetic field required is produced by d.c. electromagnets. Alternators are used to supply electric power in conventional motor vehicles, but the same principle in more sophisticated form is used in many power stations.

altimeter An instrument, usually a ➤barometer, for measuring ➤altitude or height. The altimeter uses the variation of pressure with altitude in the ➤international standard atmosphere to convert the pressure into the height above a chosen datum (either mean sea level, or a chosen reference point such as an airport).

altitude **1** Vertical height above mean sea level.

2 (Math.) The perpendicular distance from the ➤vertex of a triangle to the opposite side or from the apex of a ➤pyramid to its base.

3 (Astron.) The angular elevation of a celestial body above the ➤horizon.

altruism In animal behaviour, an act or activity that appears to benefit another individual, usually a closely related family member. For example, when the nest or young of some ground-nesting birds are threatened by a predator, the parents will attempt to attract the attention of the predator by feigning injury. Such behaviour is now recognized by evolutionary biologists as promoting the survival of certain combinations of genes. Thus the parent bird might be promoting the survival of copies of its genes in its offspring, despite putting itself at risk.

ALU Abbr. for ➤arithmetic and logic unit.

alum A generic name for a hydrated double sulfate of two metals, one with oxidation number +1 and the other with oxidation number +3. The archetypal example is **potash alum**, $K_2SO_4 \cdot Al_2(SO_4)_3 \cdot 24H_2O$, which exists as colourless octahedral crystals. Other common examples are ammonium alum, with NH_4^+ in place of K^+, and chrome alum, with Cr^{3+} in place of Al^{3+}. They all tend to crystallize in the same crystal structure, so alums are ➤isomorphous.

alumina An alternative, very common, name for **aluminium oxide**, Al_2O_3, a white, hard solid with a high melting point, useful for a number of catalytic processes such as ➤cracking. It is found in nature as ➤corundum and, in hydrated form, as ➤bauxite. Impure forms are also common, such as the chromium-containing ➤ruby. Hydrated aluminium oxide, precipitated from an aluminium ion solution by adding sodium hydroxide, reacts with both acids and alkalis. This key ➤amphoteric feature of aluminium oxide is used in the industrial manufacture of pure alumina by a process called the **Bayer process**.

aluminate An anion containing aluminium. Simple examples include the mineral ➤spinel and the **tetrahydroxoaluminate** ion present in alkaline solutions of alumina, with the formula $[Al(OH)_4]^-$. ➤Cryolite is a fluorine-containing aluminate.

aluminium (US **aluminum**) Symbol Al. The element with atomic number 13 and relative atomic mass 26.98, which is in Group 13. It is a metal which in the last hundred years has become second only to iron in importance. Although it is the most abundant metal in the Earth's crust, it is very hard to extract, so up until 1886 it was a highly prized metal available only to the very rich. Then the ➤Hall–Héroult process brought down the price of the element drastically. Its twin advantages over iron are its low density and its resistance to corrosion. Drink cans, airframes, engine heads, alloy wheels, ships' hulls and power lines can all be made from aluminium or its alloys. The oxidation number for aluminium in all its compounds is +3. Aluminium oxide, Al_2O_3, is commonly called ➤alumina. The **aluminium halides**, of general formula $AlHal_3$, are an important series of compounds with interesting bonding, which is intermediate in character between ionic and covalent. Aluminium fluoride is an ionic solid melting at over 1000 °C. Aluminium chloride, however, is found in two forms. The hydrated form is a crystalline solid which dissolves in water, as is typical of ionic compounds. The anhydrous form, on the other hand, decomposes at 180 °C and is better considered as covalent. Aluminium chloride is an important Lewis acid catalyst in the ➤Friedel–Crafts reaction. Another important organic reaction involving an aluminium compound is polymerization using ➤Ziegler–Natta catalysts, which uses triethylaluminium. Aluminium sulfate, $Al_2(SO_4)_3$, is widely used in water treatment for coagulating colloids.

aluminosilicate One of a class of aluminium- and silicon-containing solids which have a rich structural chemistry. Clays, micas and feldspars are all aluminosilicates. Industrially important aluminosilicates include the ➤zeolites, which have an open structure.

aluminum US spelling of ➤aluminium.

alveolus (plural **alveoli)** ➤lung.

AM Abbr. for amplitude ➤modulation.

Am Symbol for the element ➤americium.

amalgam An alloy in which one of the metals is mercury. Amalgams are used for fillings in dentistry, for example.

ambident nucleophile A ➤nucleophile that can attack from either of two atoms. An important example is the cyanide ion, CN^-, which can bond to an ➤electrophile using either the carbon or nitrogen atom to form nitriles, RCN, or isonitriles, RNC, respectively.

americium Symbol Am. The element with atomic number 95 and with most stable isotope of nucleon number 243, which is an ➤actinide. Its chemistry is dominated by its radioactivity. The most common aqueous ion is Am^{3+}; the most stable oxide, AmO_2, has a higher oxidation number.

Ames test A test for the tendency of a substance to induce ➤mutation and ➤cancer. The substance to be tested is added to a culture of approximately 10^9 bacteria, a mutant strain of *Salmonella* that requires histidine as a nutrient. After incubating the chemical with the bacteria, some bacteria revert to producing histidine. The greater the reversion rate, the more carcinogenic the chemical (➤carcinogen). Named after Bruce Nathan Ames (b. 1928).

amide An organic compound containing the **amide group**, the $—CONH_2$ group. They are white crystalline solids such as ➤ethanamide, CH_3CONH_2. They can be made by reacting an ➤acyl halide with ammonia. A **substituted amide** has one of the hydrogen atoms on the amide group replaced by another group, as in the product of

➤ethanoylation of ethanamide, $CH_3CONHCOCH_3$. The resulting —CONH— structure is characteristic of bonding between amino acids (➤peptide bond).

amine An organic compound containing a nitrogen atom bonded to at least one alkyl (or aryl) group. A **primary amine** has the formula RNH_2, a **secondary amine** $RR´NH$, and a **tertiary amine** $RR´R´´N$. The group —NH_2 is often called the **amino group**. Aromatic amines also exist, such as phenylamine, $C_6H_5NH_2$, which are important for the production of ➤diazonium salts. They are all bases which form salts with hydrochloric acid, for example. In medical terminology such a salt is often called a **hydrochloride**, as for an ➤alkaloid such as atropine.

amino acid A compound in which an amino group and a carboxylic acid group are both present. The most important examples are the α-amino acids, in which the two groups are bonded to the same carbon atom, to give the general formula $RCH(NH_2)COOH$. Other possibilities do exist, however, as in the neurotransmitter ➤GABA. The amino acid exists in a variety of ionic forms as the pH is varied. At very low pH, the amine group is protonated to form —NH_3^+. At very high pH, the carboxylic acid group is deprotonated to form —CO_2^-. At the ➤isoelectric point, the amino acid exists as a ➤zwitterion. The twenty most important α-amino acids are listed in Appendix table 7. Amino acids can be polymerized to form ➤proteins. The **amino acid sequence**, the order in which the amino acids are arranged in a ➤polypeptide chain, can be found using ➤Sanger's reagent. ➤➤dansyl; Edman degradation.

4-aminobenzoic acid (*p*-aminobenzoic acid) ➤PABA.

ammeter An instrument for measuring electric current. Most are ➤galvanometers with a ➤shunt, but some a.c. ammeters sense the magnetic field produced by the current (so the circuit does not have to be broken), and others work by measuring the heating effect of the current (**thermoammeter**). ➤➤moving-coil ammeter; moving-magnet ammeter.

ammine complex A complex with ammonia as the ligand. An example is the hexaamminenickel(II) ion, $[Ni(NH_3)_6]^{2+}$. The most famous ammine complex, that of copper, has four ammine ligands and two water ligands; its formula is $[Cu(NH_3)_4(H_2O)_2]^{2+}$.

ammonia The hydride of nitrogen, NH_3. It is a colourless gas with a pungent smell, which condenses to a colourless liquid, with useful solvent properties. Liquid ammonia has a higher boiling point than expected, because the molecules are held together by ➤hydrogen bonding. It is the chemical manufactured in the highest amount (in terms of moles); the dominant manufacturing process is the ➤Haber–Bosch process. In the laboratory, it is made by heating an ammonium salt with an alkali.

ammonia

The lone pair of electrons on the nitrogen atom (see the diagram) makes the molecule susceptible to protonation: it is a weak ➤Brønsted base. For the equilibrium established in water, see ➤base ionization constant. Thus a simple test for ammonia is that the gas turns moist litmus paper blue. Ammonia reacts with hydrochloric acid to form ammonium chloride:

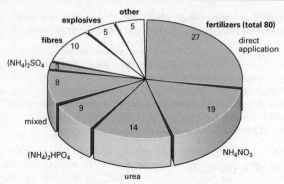

ammonia Pie chart showing the uses of ammonia.

$$NH_3(g) + HCl(g) \rightarrow NH_4Cl(s).$$

The characteristic white fumes of ammonium chloride, produced when ammonia comes into contact with the top of a concentrated hydrochloric acid bottle, constitute a second test for ammonia. (However, an ➤amine gives similar results, and so the absence of amines must also be confirmed.) More generally, the lone pair makes ammonia a ➤Lewis base; this explains both its reaction with Brønsted acids and its ability to form an ➤ammine complex with metal ions. The other main aspect of ammonia's reactivity is its strength as a reductant. For example, in the presence of a catalyst ammonia can be oxidized by oxygen to nitrogen oxide, the first step in the ➤Ostwald process. Ammonia has numerous uses, as shown in the diagram, which are dominated by its importance as a fertilizer.

ammoniacal A solution in aqueous ammonia. An important example is **ammoniacal silver nitrate** used in ➤Tollens' reagent.

ammonia clock ➤atomic clock.

ammonia-soda process ➤Solvay process.

ammonium hydroxide A traditional description for the species present in aqueous ammonia; see ➤base ionization constant for a better description.

ammonium salts Salts formed between an acid and ammonia that contain the **ammonium ion**, NH_4^+. The ammonium ion is formed by protonation of the lone pair on the nitrogen atom of ammonia (see the diagram). Ammonium salts tend to be white crystalline solids. Examples are **ammonium carbonate**, $(NH_4)_2CO_3$, used in smelling salts, **ammonium chloride**, NH_4Cl, used in the ➤Leclanché cell, **ammonium nitrate**, NH_4NO_3, used in fertilizers and explosives such as amatol, and **ammonium sulfate**, $(NH_4)_2SO_4$, another fertilizer.

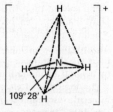

ammonium salts contain the ammonium ion.

amniocentesis A prenatal diagnosis in which cells removed from the **amniotic fluid** around a foetus can be grown in culture to reveal any chromosomal abnormalities such as ➤Down's syndrome.

amoeboid A type of cell locomotion in which a change of state of regions of the ➤cytoplasm between a viscous gel form and a more liquid sol form results in a flow of the cell's contents, producing movement. It is characteristic of *Amoeba*, a genus of protoctistans, though it also occurs in a wide range of other cells including certain ➤leucocytes in humans.

amorphous Without crystalline form. Amber is an example of an amorphous substance, as is a ➤glass.

amount of substance ➤mole.

AMP Abbr. for adenosine monophosphate, a ➤nucleotide component of DNA and RNA. In its cyclic form (**cyclic AMP**) it is a key metabolite in many ➤hormone control systems, particularly operating on membranes. AMP is also a precursor of ➤coenzymes such as ➤FAD and ➤NAD$^+$, and a regulator of a range of metabolic processes in, for example, *E. coli*.

ampere Symbol A. The SI unit of ➤current. It is defined as the current, carried by two parallel, infinitely long, thin conductors, 1 m apart in a vacuum, that causes a force of 2×10^{-7} N m^{-1} between the conductors. Named after André Marie Ampère (1775–1836).

ampere-hour A convenient unit of electric charge equal to 3600 C, the charge passed when a current of 1 A flows for 1 h.

Ampère's (circuital) theorem The ➤magnetic field strength H along any closed loop L around a conductor carrying current I is given by

$$\oint_L \boldsymbol{H} \cdot \mathrm{d}\boldsymbol{l} = I,$$

or in a vacuum

$$\oint_L \boldsymbol{B} \cdot \mathrm{d}\boldsymbol{l} = \mu_0 I,$$

where $\boldsymbol{B}$ is the ➤magnetic flux density. I is the total current through the loop, so it is the current carried by a conductor multiplied by the number of times the conductor passes through L. This is the integral form of the fourth of ➤Maxwell's equations in the absence of alternating electric fields.

amphibian A member of the tetrapod (four-limbed) vertebrate class **Amphibia**. Typically amphibians are animals with smooth, scale-less skins, eggs without shells and aquatic larvae with external gills. Modern amphibians comprise three orders: the frogs and toads (**Anura**), the newts and salamanders (**Urodela**), and the legless caecilians (**Apoda**), although some early ancestors of the amphibians were heavily armoured with bony plates in their skin.

amphibole A class of silicate minerals which includes ➤asbestos.

amphiprotic A solvent, such as water, that can both provide protons and react with them.

amphoteric An oxide or hydroxide that can react with both acids and alkalis. Common amphoteric hydroxides include hydrated ➤alumina, zinc hydroxide and lead(II) hydroxide. ➤Metalloids tend to have amphoteric oxides.

amplifier An electronic device for increasing a current or voltage by a constant factor (the ➤gain). It requires an external source of power. Amplifiers vary greatly in size: typical ➤operational amplifiers can be incorporated on an integrated circuit and consist of one or more ➤transistors, while ➤power amplifiers can be used to produce currents large enough to power huge loudspeakers from very small signals. ➤➤pre-amplifier.

amplitude The largest magnitude of a quantity that varies periodically. For example, if $F(t) = F_0 \sin \omega t$, then the amplitude of F is F_0. For quantities that are not sinusoidal, it is the maximum absolute difference of the quantity from its equilibrium value or other chosen datum.

amplitude modulation (AM) ➤modulation.

amu Abbr. for ➤atomic mass unit.

amylase One of a group of ➤enzymes that digest ➤starch and other polymeric ➤carbohydrates by hydrolysing the bonds between adjacent monomeric units. Amylases are found in germinating seeds (as in the malting of barley for beer and whisky production), in the saliva of mammals, and in the secretions of the ➤pancreas and glands of the ➤duodenum.

amyl group One of a set of isomeric five-carbon alkyl groups of formula C_5H_{11}—. The *n*-amyl group is $CH_3(CH_2)_4$—. The *iso*-amyl group is $(CH_3)_2CH(CH_2)_2$—. The important solvent *iso*-**amyl alcohol**, now called 3-methylbutan-1-ol, forms with ethanoic acid an ester which smells strongly of pear-drops.

amylopectin ➤starch.

amylose ➤starch.

anabolic steroid A synthetic ➤androgen that selectively enhances the growth of skeletal muscle. Anabolic steroids are used to stimulate the growth of muscle in commercial meat production and (illegally) by athletes to enhance performance.

anabolism Those components of ➤metabolism in a living cell in which complex structures are synthesized from simpler ones. The synthesis of carbohydrates from water and carbon dioxide during photosynthesis is a good example of an **anabolic process**. Compare ➤catabolism.

anaemia (anemia) An abnormal reduction in the number or haemoglobin content of ➤erythrocytes in the blood. The condition may be caused by a deficiency of minerals, particularly iron, in the diet, or be symptomatic of other medical conditions, such as stomach ulcers, heavy menstrual periods or pregnancy. **Pernicious anaemia**, the incidence of which increases with age, is caused by the reduced ability of the body to absorb vitamin B_{12}. Symptoms of anaemia include pale skin and lack of energy. ➤sickle cell anaemia.

anaerobic respiration The breakdown of food materials such as glucose to yield energy in the form of ➤ATP in the absence of oxygen. Anaerobic respiration in ➤yeasts produces ethanol as a waste product, and this is the basis of all alcoholic drink manufacture. Anaerobic respiration can also occur in animal cells when the supply of oxygen to actively working tissues (such as ➤muscles during strenuous exercise) is limited. Under such conditions, lactic acid accumulates in the cells and produces the characteristic feeling of tiredness in the muscles. The lactic acid is removed when the oxygen supply is restored and the tissue returns to rest. ➤fermentation; glycolysis. Compare ➤aerobic respiration.

Analar reagent A reagent of higher than usual purity.

analgesic Capable of reducing pain (adjective); a pain-killer (noun).

analogue (US analog) Describing a system, particularly an electronic device, that uses a continuous physical quantity (e.g. ➤potential difference) to represent information. Compare ➤digital.

analogue computer A computer that uses electrical signals and components as direct physical analogues of the variables it is modelling. These are continuous, in contrast to digital computers, in which discrete electrical pulses are logically but not physically equivalent to the values being modelled.

analogue-to-digital converter (ADC) A device for converting an analogue signal, such as a voltage from a sensor, into a digital signal for processing by a computer. Compare ➤digital-to-analogue converter.

analysis **1** (Chem.) The determination of the species present in a sample. ➤Qualitative analysis simply identifies the species present, with no details of the amounts of each. Examples include the many forms of ➤spectroscopy: for example ➤atomic absorption spectroscopy, which can identify the elements present in a complex mixture, or ➤infrared spectroscopy, which can identify specific molecules from their characteristic fingerprint region. ➤Quantitative analysis identifies not only the species present but also their abundances in the sample. ➤Titration is commonly used in quantitative analysis.
    2 (Math.) The branch of mathematics concerned with the theory behind limiting operations, such as those used in ➤calculus and in dealing with infinite ➤series, together with their generalizations (e.g. to ➤metric spaces and ➤Hilbert spaces).

analyte The substance being analysed.

analytic function A ➤complex function f that has a ➤derivative f' at each point of the region in which it is defined:

$$f'(z) = \lim_{h \to 0}[f(z + h) - f(z)]/h, \text{ with } h \text{ complex.}$$

There is a rich theory of analytic functions: the ➤real and ➤imaginary parts of an analytic function are ➤harmonic functions which satisfy the ➤Cauchy–Riemann equations; ➤integration of analytic functions leads to results such as ➤Cauchy's theorem, and an analytic function gives rise to a ➤conformal transformation.

anaphase The phase of cell division in which the ➤chromosomes move apart to form new nuclei. The movement is generated by the contraction of fibres of the spindle. ➤meiosis; mitosis.

anaphylaxis An abnormally severe immune response to a drug such as penicillin, a food such as peanuts or an insect bite, resulting in either a localized reaction or a more general one with lowering of blood pressure and breathing difficulties. **Anaphylactic shock** can be life-threatening.

anastigmatic lens A ➤lens with different radii of curvature in different directions (the surface of a spheroid rather than a sphere) that is designed to correct astigmatism (➤aberration).

anatomy The study of the structure of an organism, particularly the arrangement of internal ➤organs.

AND A logic operation (➤truth table).

androgen A male sex ➤hormone, such as testosterone, secreted by the testes. Androgens are responsible for the maintenance of sperm production, muscular development and aspects of sexual behaviour. Small amounts of androgens are also found in females, and are responsible in part for stimulating the growth of pubic hair.

anemometer An instrument for measuring wind velocity.

aneroid barometer ➤barometer.

Angiospermophyta ➤Magnoliopsida.

angle The shape, or measure of the shape, formed by two line segments that meet at a common point. The size of the angle is the amount of rotation needed to superimpose one line segment on the other; by convention anticlockwise rotations are positive. Angles are measured in ➤degrees, ➤grades or ➤radians according to whether 360, 400 or 2π units are assigned to one revolution.

angstrom Symbol Å. An obsolescent unit of length, used in atomic physics and chemistry, equal to 10^{-10} m; it is being superseded by the nanometre. Named after Anders Jonas Ångström (1814–74). (Note that the name of the unit has no accents.)

angular acceleration Unit s^{-2}. The rate of change of ➤angular velocity.

angular displacement The angle through which something has rotated or the change in an angle. Note that angular displacement, unlike linear displacement, is not a vector (though angular velocity is), because addition of rotations is not ➤commutative.

angular distance 1 The angle subtended by a displacement or line at the point of observation.

2 (Astron.) The size of an object (**angular diameter**) or the separation between two objects (**angular separation**) expressed as an angle on the ➤celestial sphere, commonly measured in arc ➤seconds.

angular frequency Symbol ω; unit s^{-1}. The ➤frequency of an oscillation multiplied by 2π. For sinusoidal oscillations, it is a useful quantity because it appears directly as an argument in, for example, $F = F_0 \sin \omega t$, without cumbersome factors of 2π.

angular momentum Symbol L; unit J s. For a single particle, the vector quantity $r \times p$, where p is the momentum of the particle and r is its displacement from a chosen origin. For a composite system like a ➤rigid body the angular momentum is the sum of the angular momenta of the body's constituent particles, also given by $L = I \times \omega$, where I is the moment of inertia of the body and ω its ➤angular velocity. The angular momentum of an isolated system is conserved. The rate of change of angular momentum of a body is equal to the total ➤couple applied to it. The angular momentum vector and angular velocity vector are not parallel unless they fall along the ➤principal axis of the rigid body. For this reason, differentiating $L = I \times \omega$ to obtain an equation of motion $G = dL/dt = Ia$, where a is the angular acceleration (by analogy with $F = dp/dt = ma$), is applicable only for rotation about a principal axis (➤Euler's equations). In quantum mechanics, angular momentum takes on a more fundamental role: it is quantized in integral and half-integral multiples of $h/2\pi$, where h is the ➤Planck constant. Astronomical bodies possess angular momentum by reason of their rotation about their axis and of their revolution in orbit around another body.

angular momentum quantum number In quantum mechanics, the quantum number corresponding to ➤angular momentum. It is quantized in integral and half-integral multiples of $h/2\pi$, where h is the ➤Planck constant. ➤➤spin angular momentum quantum number; orbital angular momentum quantum number.

angular velocity Symbol ω; unit s^{-1}. The rate of change of angular displacement, often of a rotating or revolving body. Angular velocity is a ➤vector quantity, with the vector pointing perpendicular to the plane of rotation (conventionally according to the orientation of a right-hand screw).

anharmonic An oscillation that is not ➤harmonic. The vibration of molecules can often be understood in terms of ➤simple harmonic motion with a small anharmonic correction term.

anhydride ➤acid anhydride.

anhydrite Naturally occurring ➤calcium sulfate, $CaSO_4$.

anhydrous Without water. It is commonly used to describe crystals that do not have ➤water of crystallization. Thus anhydrous copper(II) sulfate is $CuSO_4$, whereas hydrated copper(II) sulfate is $CuSO_4 \cdot 5H_2O$.

aniline Traditional name for ➤phenylamine.

animal A member of the ➤kingdom **Animalia**. Traditionally, the kingdom Animalia contains multicellular ➤eukaryotic ➤heterotrophs which typically feed by ingesting complex organic food into a specialized internal cavity in the body. Some parasites such as tapeworms are exceptions to this, and feed by absorption through the surface of the whole body. More recently, the manner in which certain gene clusters (*Hox* genes) are expressed has been suggested as a defining character for the kingdom.

anion The ion that moves towards the ➤anode in ➤electrolysis; as the anode is positively charged in electrolysis, anions are negative ions. Common anions include Cl^-, OH^- and ➤oxoanions. Compare ➤cation.

anisotropic Not ➤isotropic. A system (such as a crystal rather than a homogeneous medium) that has different properties in different directions is anisotropic. For example, the conductivity of ➤graphite is 5000 times greater parallel to the layers than perpendicular to the layers.

annealing **1** (Materials) A heat treatment applied to a metal, alloy, ceramic or glass to relieve ➤strain. Annealing may involve heating, holding at a fixed temperature for a certain period, then a slow, controlled cooling. The treatment softens metals, making them less brittle and easier to work.
   2 (Biol.) ➤denature.

annelid A member of the animal ➤phylum **Annelida**, comprising the segmented worms and including classes represented by earthworms, leeches and the marine ragworms.

annihilation The process by which a particle and its ➤antiparticle cease to exist at the same time. For example, an electron and a positron can annihilate, producing a pair of photons.

annual variation The change in ➤magnetic declination over the course of a year, caused by a change in the magnetic field of the Earth.

annular eclipse ➤eclipse.

annulene A class of monocyclic hydrocarbons with ➤conjugated double bonds. [18]Annulene (see the diagram) is a brown-red crystalline solid, which can exist in *cis*- and *trans*-forms (➤*cis–trans* isomerism).

annulus The region between two ➤concentric circles.

anode **1** (Chem.) The electrode at which ➤oxidation occurs. In an electrolytic cell, the anode is positively charged; the ➤anions are attracted towards the anode and oxidized by loss of

[18]annulene

electrons. In an electrochemical cell, on the other hand, the anode is negatively charged, as the oxidation process leaves an excess of electrons at the electrode. Compare ➤cathode.
   2 (Phys.) An obsolete name for the collector in a ➤thermionic valve.

anodizing A process for increasing the width of the protective layer of an oxide coating, especially on a piece of aluminium, by making it the anode in an electrolytic cell usually containing sulfuric acid. The oxygen released at the anode oxidizes the metal, and forms the thicker coating which improves the resistance of the metal to corrosion. **Anodized aluminium** is used, for example, as exterior cladding on the Pompidou Centre in Paris.

anomaly A parameter used to describe the position of a planet within its ➤orbit. It is the angle, shown in the diagram, between the major axis of the ellipse and a line joining the planet to the Sun, measured in the direction of the planet's motion.

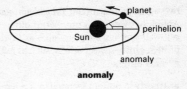

anomaly

ANS Abbr. for autonomic ➤nervous system.

ANSI Acronym for American National Standards Institute.

antagonistic A term applied to pairs of ➤muscles whose action allows movement at a joint, as in a limb. Examples are the biceps and triceps in the upper arm: contraction of the biceps and relaxation of the triceps causes the arm to flex (bend) at the elbow; relaxation of the biceps and contraction of the triceps extends (straightens) the elbow joint. All skeletal movement is brought about by such pairs of antagonistic muscles.

antenna **1** (plural **antennas**) (Phys.) ➤aerial.
  2 (plural **antennae**) (Biol.) One of a pair of appendages on the head of an ➤arthropod, usually associated with the tactile senses. Some insects have very sensitive antennae which can detect a ➤pheromone at great distances.

anterior Describing the position of a structure in the body of an animal that lies closest to the surface that is at the front during normal locomotion. The anterior vena cava, for example, is at the front of the heart of a quadruped. Compare ➤posterior.

anther The terminal part of a stamen of a flowering plant (➤flower), containing four chambers in which pollen grains are formed.

anthocyanin A group of water-soluble, glycosidic ➤flavonoid pigments found in plants, particularly in the ➤vacuoles of flowers, fruits and leaves. The colour of anthocyanins depends on ➤pH, but they are commonly responsible for the red, purple and blue pigmentations of petals.

Anthophyta ➤Magnoliopsida.

anthracene A tricyclic ➤aromatic hydrocarbon (see the diagram). It exists as white crystals and is used to make dyes.

anthracene

anthraquinone A dye which is a derivative (see the diagram) of anthracene.

anthrax A contagious and often fatal disease of animals, including humans, caused by the bacterium *Bacillus anthracis*. It is acquired by the inhalation of spores or through the skin. Symptoms include haemorrhage in the lungs and convulsions. Anthrax was the first disease for which the causative agent was identified and the bacterium was first isolated in pure culture by Robert Koch in 1876. Five years later Louis Pasteur developed a vaccine against anthrax.

anthraquinone

antibiotic A biochemical substance, produced particularly by mould fungi, employed in combating infection, especially by bacteria. ➤Penicillin, produced by strains of the mould *Penicillium notatum*, is the most widely used antibiotic. Antibiotics work in a number of different ways: some, such as Ampicillin, inhibit cell wall synthesis; others, such as Erythromycin or Tetracycline, inhibit bacterial protein synthesis; yet others, such as Actinomycin, inhibit bacterial DNA synthesis. ➤sulfa drugs.

antibody A protein secreted by a ➤B cell as part of the body's ➤immune response. Antibodies are produced in response to foreign substances entering the body and react with such invaders, rendering them harmless. Antibody reactions are highly specific: a useful analogy is that of a key (the foreign substance or ➤antigen) fitting into a lock (the antibody). Lymphocytes produce a wide variety of highly specific antibodies; the reason lies in the ➤V segment of the genes coding for the proteins concerned. **Monoclonal antibodies** are particularly pure, and are derived from cells called **hybridomas** which are the result of fusing antibody-secreting cells with tumour cells. Hybridomas can be cultured indefinitely and screened so that each ➤cell line produces only one kind of antibody. Monoclonal antibodies have a wide range of applications as tools for diagnosis and immunization, and for tracing molecules used in biological experiments. They can also be used to target radioactive or cytotoxic agents (those poisonous to cells) to specific cells in the treatment of cancers.

antibonding orbital ➤molecular orbital.

anticodon A triplet of unpaired ➤nucleotide residues at the end of a loop of a tRNA molecule. This triplet of varying sequence can base-pair to a complementary triplet (the ➤codon) in an mRNA molecule, thus directing the addition of a specific amino acid during ➤protein synthesis.

anticyclone A region of high atmospheric pressure. Anticyclones are associated with stable weather conditions; in cooler months they can produce fog and low cloud. Compare ➤depression.

antidiuretic hormone (ADH) ➤vasopressin.

antiferromagnetism The alignment of ➤magnetic dipole moments on atoms or ions in a crystal (typically antiparallel moments on adjacent ions) such that the ➤magnetic susceptibility is almost zero. Antiferromagnetism is exhibited, for example, by manganese(II) oxide. Like ferromagnets (➤ferromagnetism), antiferromagnets are ordered only below a critical temperature, the ➤Néel temperature.

antifluorite structure An important crystal structure that closely resembles the ➤fluorite structure, but with the positions of the positive and negative ions reversed. The crystals of the alkali metal oxides, such as Na_2O, have the antifluorite structure.

antifreeze A substance that depresses the freezing point of water and thus helps to prevent ice from forming. A well-known antifreeze is ethane-1,2-diol.

antigen Any substance (usually a protein or protein–polysaccharide complex) that, when introduced into an animal with a functioning ➤immune system, elicits a specific ➤immune response by binding to the corresponding ➤antibody. ➤➤allergy.

antihistamine Any substance that inhibits the effects of ➤histamine. Antihistamines are used to reduce the severity of the symptoms of allergic reactions such as insect bites and hay fever (➤allergy).

antiknock agent ➤tetraethyllead.

antilogarithm The number of which a given number is its ➤logarithm in a fixed ➤base (3). Thus, if $x = \log_b y$, then $y = \text{antilog}_b x$, a statement equivalent to $y = b^x$.

antimony Symbol Sb (from *stibium*). The element with atomic number 51 and relative atomic mass 121.8, which is in Group 15. Compounds of the element have been known for millennia. **Stibnite**, antimony(III) sulfide, Sb_2S_3, was traditionally used for cosmetic eye shadow. The element itself, a blue-white brittle ➤metalloid, is unfamiliar. Its main use is in alloys such as pewter, an alloy with tin. It forms compounds mainly with oxidation number +3, as in the sulfide above, the oxide, Sb_2O_3, or the ➤tartrate salt with potassium better known as tartar emetic. There is a small number of compounds with oxidation number +5, as in antimony(v) fluoride, SbF_5.

antinode A point, line or plane where the displacement (positive or negative) is a local maximum in a ➤standing wave. Compare ➤node.

antioxidant A substance added to prevent oxidation, in rubber or food for example.

antiparallel vectors ➤Vectors that are ➤parallel but in opposite directions.

antiparticle A particle with the same mass, spin and lifetime as a specified particle, but with opposite charge, magnetic moment and other internal ➤quantum numbers (like ➤strangeness). For example, the ➤positron is the antiparticle of the electron, and vice versa. **Antimatter** is composed of antiparticles.

antipyretic Capable of reducing fever.

antiquark The ➤antiparticle of a ➤quark. A ➤meson is composed of a quark and an antiquark; an antiproton is composed of the antiquarks $\bar{u}$, $\bar{u}$ and $\bar{d}$.

antisense gene A ➤gene in which the DNA sequence is the exact opposite (according to the rules of ➤base pairing) of the normal form of the gene. Antisense genes are used in ➤molecular biology to block the function of the normal gene. The antisense gene produces an mRNA whose sequence is complementary to that of the normal mRNA and thus binds to it. When this occurs neither mRNA can be translated into protein (➤protein synthesis). This is the principle behind genetically modified tomatoes, which do not soften once they are ripe. Fruit normally softens because it produces the enzyme polygalacturonase, which breaks down pectins in ➤cell walls. By inserting an antisense gene, the mRNA coding for the enzyme is blocked and the fruit stays firm.

antiserum ➤Serum containing a specific ➤antibody which will bind to a particular ➤antigen. Antiserum is available for a wide range of medical applications, including treatment of ➤Rhesus factor problems and snake bites.

antiviral Any chemical that is effective against viral infection. It took much longer for the first agent, ➤acyclovir, to be discovered compared with the first antibiotics, but the pace of development is quickening. For example, oseltamivir (Tamiflu) is useful in the treatment of influenza virus. As a class, antivirals are not curative but often act by inactivating the enzymes needed for viral replication.

apatite A mineral composed of the decayed and compressed remains of bones, with chemical composition $Ca_5(PO_4)_3X$, where X is F, Cl or OH. Tooth enamel consists mainly of apatite. Hydroxyapatite is converted to the less soluble and harder fluorapatite by using fluoride toothpaste. Apatites constitute the most important ores of phosphorus and are abundant in North Africa.

aperture An opening in an instrument through which light (or other electromagnetic radiation) may propagate. The shape of an aperture determines any ➤diffraction pattern exhibited by the light. In some optical instruments, notably the camera, the diameter of the aperture is variable and determines the total intensity of light admitted.

aperture synthesis The simulation of a large aperture in radio astronomy by using the rotation of the Earth to sweep an aerial across an arc in space. Thus a two-dimensional radio ➤interferometer can be constructed as a single line of aerials.

aphelion The point in the orbit of a planet or other body orbiting the Sun at which it is farthest from the Sun. Compare ➤perihelion.

apical dominance A phenomenon whereby the presence of an intact terminal bud on the shoot of a plant suppresses the growth of lateral shoots. Removal of the apical bud causes lateral buds to develop and produce branches. This is exploited by horticulturalists in clipping hedges to encourage dense growth. The process is controlled by plant growth substances, particularly ➤auxin.

aplanatic Describing a lens or mirror that has no spherical ➤aberration or coma.

apoenzyme The protein component of an enzyme. ➤➤holoenzyme.

apogee The point in the orbit of the Moon or an artificial satellite orbiting the Earth at which it is farthest from the Earth. Compare ➤perigee.

apparent depth The depth, as apparent to an observer, of an object in a fluid of ➤refractive index greater than 1 (see the diagram). The apparent depth is the real depth divided by the refractive index of the fluid.

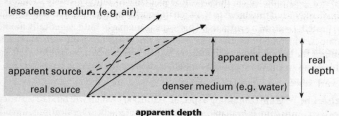

less dense medium (e.g. air)

apparent source

real source

apparent depth

denser medium (e.g. water)

real depth

apparent depth

apparent expansivity The volume ➤expansivity of a liquid found by reference to a measuring container that also expands with increasing temperature. It is the difference between the ➤absolute expansivity and the expansivity of the container.

apparent power In a circuit dissipating power, the product of the ➤root mean square voltage and the root mean square current. It must be multiplied by the ➤loss factor to give the true power dissipated.

apparent solar time The time of day reckoned by the position of the Sun in the sky. It is therefore the time that would be indicated by a sundial. Compare ➤mean solar time; ➤➤equation of time.

Appleton layer (F layer) The uppermost layer of the ➤ionosphere, between altitudes of 150 and 400 km. It reflects high-frequency radio waves. Named after Edward Victor Appleton (1892–1965).

application A software package that is suitable for a particular task, the most common being word processing, spreadsheet and database applications.

aprotic ➤dipolar aprotic solvent.

aq Label (abbr. for 'aqueous') used in chemical equations to denote that a chemical species is dissolved in water.

aqua complex (aquo complex) A complex with water molecules as the ligands. An example is the hexaaquairon(II) ion, $[Fe(H_2O)_6]^{2+}$.

Aquadag Tradename for a colloidal suspension of graphite used as an electrically conducting coating.

aquamarine A blue-green variety of ➤beryl.

aqua regia Latin for 'royal water'. A mixture of concentrated nitric and hydrochloric acids used to dissolve gold and other noble metals. Nitrosyl chloride, NOCl, produced by reaction between the acids, acts as the oxidant, while the hydrochloric acid produces chloride ions to complex the resulting metal ion.

aqueous Dissolved in water (from the Latin *aqua*).

aquo complex ➤aqua complex.

Ar Symbol for the element ➤argon.

arachnid A member of the class **Arachnida** of the phylum **Chelicerata** within the superphylum **Panarthropoda**. Arachnids include the spiders and scorpions. Typically arachnids have their bodies divided into two parts in which the segments are fused to give an anterior **prosoma** and a posterior **opisthosoma** with a pair of claw-like **chelicerae** and four pairs of legs.

aragonite A crystalline polymorph of calcium carbonate, $CaCO_3$. Compare ➤calcite.

aramid fibre A polyamide fibre incorporating an aromatic group. ➤➤Kevlar; Nomex.

arc 1 A portion of the ➤circumference of a circle; used more generally for a portion of any curve.
    2 A prefix denoting an ➤inverse trigonometric or ➤inverse hyperbolic function.

Archaean See table at ➤era.

Archimedes' principle The ➤buoyancy force on an object immersed in a fluid is equal to the weight of the fluid displaced by the body. Named after Archimedes (*c.* 287–212 BC).

arc length The length of a ➤curve. The arc length of a ➤plane curve is given by the ➤integral

$$\int \sqrt{1 + (\mathrm{d}y/\mathrm{d}x)^2}\, \mathrm{d}x$$

between limits of integration corresponding to the endpoints of the arc.

arc second ➤second.

area A measure of the extent of a two-dimensional region. There are many ➤mensuration formulae for the areas of standard shapes; in general, a ➤double integral is required.

arene An ➤aromatic hydrocarbon. The most important example is ➤benzene.

Arg Abbr. for ➤arginine.

Argand diagram A way of representing a ➤complex number in which the complex number $z = x + iy$ is represented by the point (x, y). In this context, the x axis is called the **real axis**, and the y axis the **imaginary axis** (see the diagram). If ➤polar coordinates (r, θ) are used, corresponding to the polar form of z, then r is called the ➤modulus and θ the ➤argument of z. The addition of complex numbers is represented by adding vectors in the Argand diagram, and multiplication by i causes rotation through 90°. Named after Jean Robert Argand (1768–1822).

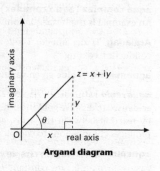

Argand diagram

arginine (Arg) A common ➤amino acid. ➤➤Appendix table 7.

argon Symbol Ar. The element with atomic number 18 and relative atomic mass 39.95, which is in Group 18. It is the most abundant ➤noble gas, comprising about 1% of the atmosphere, from which it is separated by cooling and subsequent fractional distillation of liquid air. It is completely unreactive, forming no stable compounds with any other element. It is used where an inert atmosphere is required, as in welding.

argument 1 The angle that the line representing a ➤complex number in the ➤Argand diagram makes with the positive x axis.
 2 Another name for the ➤independent variable of a ➤function.

arithmetic and logic unit (ALU) The part of the ➤central processing unit of a computer that performs arithmetic (such as additions and multiplications) and logic operations (such as ➤AND and ➤EOR) on binary numbers.

arithmetic mean The arithmetic mean of numbers $x_1, x_2, \ldots, x_n$ is the ➤average $(x_1 + x_2 + \ldots + x_n)/n$, often denoted $\bar{x}$. The **weighted arithmetic mean** with respect to fixed weights $w_1, w_2, \ldots, w_n$ is

$$\sum_{i=1}^{n} w_i x_i \Big/ \sum_{i=1}^{n} w_i.$$

arithmetic progression A ➤sequence in which the difference between successive terms is constant, e.g. 2, 6, 10, 14…, the numbers of the elements in the s, p, d, f ➤blocks of the periodic table. Compare ➤geometric progression.

armature 1 An alternative name for the ➤rotor of an electric motor or generator.
 2 The part of a device based on electromagnetic induction, such as a ➤relay, that closes a ➤magnetic circuit.
 3 ➤keeper.

aromatic Describing an organic compound containing at least one ➤benzene ring, or one benzene ring with another nonmetal such as nitrogen substituted for a carbon atom and its attached hydrogen (a **heterocyclic aromatic** compound, such as ➤pyridine). All aromatic compounds, like the archetypal benzene molecule itself, benefit from stabilization of the electronic structure of the molecule by ➤delocalization. Aromatic compounds tend to react by ➤electrophilic substitution. Compare ➤aliphatic.

aromaticity The characteristic property shown by an aromatic compound.

array A series of items of data arranged in a useful way for manipulation by a computer. The **dimension** of the array is the number of parameters necessary to identify a specific item.

Arrhenius equation An equation that predicts the dependence of the ➤rate constant of a reaction on temperature. An **Arrhenius plot** is a graph of the natural logarithm of the rate constant k against the reciprocal of the thermodynamic temperature T:

$$\ln k = \ln A - E_a/RT.$$

The first term involves the **Arrhenius constant** A, which is a constant for a particular reaction. The second term involves the ➤activation energy E_a (also a constant for a particular reaction) and the ➤gas constant R. The activation energy for a reaction can be found from the gradient of the resulting straight line, which is $-E_a/R$. Named after Svante August Arrhenius (1859–1927).

arsenic Symbol As. The element with atomic number 33 and relative atomic mass 74.92, which is a member of Group 15. Its most common ores are realgar, As_4S_4, and orpiment, As_2S_3. Arsenic is a ➤metalloid. It forms compounds with oxidation numbers +3 and +5. Thus it has two oxides: arsenic(III) oxide (formerly called **arsenious oxide**), As_2O_3, a white amorphous powder used in pigments, and arsenic(v) oxide (formerly called **arsenic oxide**), As_2O_5, which is also a white amorphous solid. Arsenic compounds such as As_2O_3 are used as poisons. Surprisingly, the arsenic-containing compound Salvarsan was the first successful chemical for curing a specific disease, syphilis.

arsine AsH_3 The hydride of arsenic, which is a very poisonous colourless gas with a garlic odour. It is used to make an ➤n-type semiconductor, as it readily decomposes to form arsenic. ➤➤Marsh's test.

artery A thick-walled elastic vessel that carries blood away from the heart under high pressure. Arterial blood, except that in the pulmonary artery, is rich in oxygen and has a characteristic bright red colour.

arthropod A member of the largest ➤phylum (now regarded by taxonomists as a superphylum) in the animal kingdom, **Panarthropoda** (formerly Arthropoda). Arthropods have a body cavity formed by blood spaces (**haemocoel**), segmented bodies with jointed appendages on some or all of the segments, and an exoskeleton containing ➤chitin. Arthropods include ➤crustaceans, ➤arachnids and ➤insects.

artificial intelligence (AI) The modelling by computers of human processes which appear to involve intelligence. Game playing and the understanding of human (as opposed to computer) language are favoured areas of research, and perceptual tasks such as vision are also included in the realm of AI. Robotics is much concerned with the practical application of AI.

aryl group A group derived from an ➤aromatic compound by the removal of a hydrogen atom in the same way that an alkyl group is derived from an alkane. The simplest aryl group is that derived from benzene, C_6H_5—.

As Symbol for the element ➤arsenic.

asbestos A fibrous silicate mineral known since antiquity for its ability to resist heat. Recently it has become recognized as a health hazard, particularly in the form called **blue asbestos**. It causes **asbestosis**, a respiratory disease, and mesothelioma, a type of cancer.

ASCII Acronym for American standard code for information interchange. A standard seven-bit code (➤bit), which therefore has 2^7, or 128, members. These include the alphanumeric characters and control characters such as carriage return.

ascorbic acid ➤vitamin C.

asexual reproduction Reproduction by one of a number of processes that all result in the production of offspring genetically identical to the parent. The processes never involve the fusion of ➤gametes. Asexual reproduction is common in single-celled organisms and in plants, for which artificial asexual reproduction techniques such as taking cuttings and budding are commercially important in the horticultural industry. Any offspring resulting from such techniques constitutes a ➤clone. Asexual reproduction in animals is restricted to the lower phyla such as Cnidaria. Compare ➤sexual reproduction.

Asn Abbr. for ➤asparagine.

Asp Abbr. for ➤aspartic acid.

asparagine (Asn) A common ➤amino acid. ➤➤Appendix table 7.

aspartic acid (Asp) A common ➤amino acid. ➤➤Appendix table 7.

aspect ratio The ratio of the width of an ➤aerofoil (perpendicular to the airflow and the lift) to its length (parallel to the airflow). For example, gliders (sailplanes) have a high aspect ratio wing.

aspirin (acetylsalicylic acid) A white solid that is a widely used drug (see the diagram); it is ➤antipyretic and ➤analgesic. It inhibits the formation of ➤prostaglandins. It has gradually lost its pre-eminence as a pain-killer to ➤paracetamol.

COOH

OCOCH₃

aspirin

assembler A computer program that converts ➤assembly language programs into ➤machine code.

assembly language A low-level computer language which is effectively equivalent to machine code, but in a format more readily understood by humans. For instance, numeric machine code instructions are represented in assembly language as verbal instructions, and ➤registers may be given names.

assimilation Typically any biochemical process that involves the synthesis of more complex compounds from simpler ones. Thus carbon assimilation relates to the processes associated with the synthesis of carbon compounds such as ➤carbohydrates from the initial products of ➤photosynthesis. Nitrogen assimilation includes the synthesis of ➤proteins.

association 1 (Chem.) The aggregation of molecules together. In the most common example, a ➤dimer is formed, as when two NO_2 molecules associate to form N_2O_4. Association makes the relative formula mass higher and this makes ➤colligative properties smaller than expected on the basis of no association.

2 (Biol.) In ecology, an assemblage of species that are commonly found together in a particular set of environmental conditions.

associative A property of a ➤binary operation $*$ defined on a ➤set such that, for all a, b, c in the set, $a * (b * c)$ is the same as $(a * b) * c$. For example, on the set of ➤integers, addition is associative but subtraction is nonassociative, since $3 - (2 - 1)$ is not equal to $(3 - 2) - 1$.

astatine Symbol At. The element with atomic number 85 and most stable isotope 210, which is the ➤halogen with the highest mass. Although its chemistry is dominated by its radioactivity, it is very similar to the other halogens, forming the astatide ion, At^-, for example.

asteroid (minor planet) A small rocky or metallic body orbiting the Sun. Most are in the **asteroid belt** between Mars and Jupiter. About 20 000 asteroids have been observed, the largest of which has a radius of only 457 km. There may be a million larger than 1 km, below which they extend in size down to dust particles. The asteroids are believed to be the remnants of planetesimals, the precursors of the planets of the Solar System, that failed to accrete into a planet.

asthma A condition characterized by severe contraction of the smooth muscle of the airway, and by clogging of the air passage by mucus. Asthmatic attacks are frequently caused by an ➤allergy.

astigmatism ➤aberration.

astrometric binary ➤binary star.

astrometry The branch of astronomy concerning the measurement of the positions of celestial bodies and their movements.

astronomical telescope ➤telescope.

astronomical unit Symbol AU. A unit of length used in astronomy, equal to the mean distance between the Earth and the Sun (149 597 871 km).

astronomy The study of celestial bodies, including the Sun, the stars and the planets.

astrophysics The physics associated with celestial bodies, particularly the formation and evolution of stars.

Asx Abbr. for either asparagine or aspartic acid.

asymmetric Not possessing symmetry.

asymmetric carbon atom ➤chiral.

asymptote A straight line that a curve approaches as the distance of points on the curve from the origin tends to infinity. For example, the curve $y = x/(x - 2)$ has asymptotes $y = 1$ and $x = 2$. ➤➤conic.

asymptotic freedom The name given to the observed decrease in the strength of interaction between ➤quarks as they are brought closer. As they approach zero separation, they are virtually free.

At Symbol for the element ➤astatine.

atactic polymer A ➤polymer in which the groups off the main chain are arranged at random. Compare ➤isotactic polymer and ➤syndiotactic polymer (where there is a diagram). ➤➤Ziegler–Natta catalysts.

-ate A suffix used to indicate an ➤anion. In modern notation, the end of the element's name is changed to -ate and then the oxidation number is placed in brackets, as in manganate(VII). In organic chemistry it indicates the anion formed from a carboxylic acid, as in ethanoate, $CH_3CO_2^-$. In older literature, when the oxidation number of an element was omitted, it was used to indicate the higher of two oxidation numbers: compare ➤-ite. Thus 'sulfate' is the older name for 'sulfate(VI)', and 'sulfite' is the older name for 'sulfate(IV)'.

athermanous Not permitting the passage of heat radiation. It is the equivalent for heat radiation of 'opaque' for light. Compare ➤diathermanous.

atherosclerosis A disease of the heart and arteries characterized by the deposition of inelastic material on the walls of these vessels ('hardening of the arteries'). A range of factors, including genetic effects and high levels of ➤cholesterol, has been implicated in the various forms of the disease, which is a major cause of death in Western populations.

atm Symbol for the unit ➤atmosphere (2).

atmosphere 1 The gaseous environment surrounding the surface of the Earth (or, in general, any other celestial body). The atmosphere is categorized by layers (see the diagram), with the ➤ionosphere as a separate label applied to the upper mesosphere and thermosphere. The variations of temperature and pressure in the atmosphere (caused mostly by solar heating) result in what we know as weather. ➤➤air; international standard atmosphere.

2 Symbol atm. A convenient unit of pressure equal to 101 325 Pa. It is equal to atmospheric pressure in the ➤international standard atmosphere at zero altitude (mean sea level).

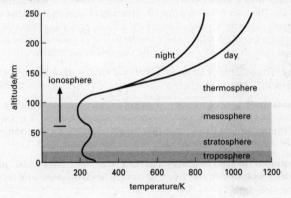

atmosphere 1 Variation of temperature with altitude in the Earth's atmosphere.

atmospheric pressure The pressure exerted by the atmosphere. Atmospheric pressure varies with altitude, reaching half its sea-level value at about 5 km.

atmospherics Electromagnetic radiation caused by phenomena in the atmosphere, particularly lightning, resulting in the static that interferes with radio reception.

atom The smallest part of an element that has an independent existence. The atom consists of a minute central ➤nucleus surrounded by electrons. The number of electrons in a neutral atom is equal to the number of protons in the nucleus. The electrons are arranged in a particular way (➤electronic structure).

atomic absorption spectroscopy (AAS) An analytical technique for identifying elements by using ➤absorption spectroscopy. A sample is burnt in a cool flame, and the sample's absorption of incident light from a heated cathode of the element of interest is monitored. A modern commercial AAS instrument can detect concentrations as low as one part per billion, and analyse for up to 20 elements at the same time.

atomic bomb ➤nuclear weapon.

atomic clock A very accurate clock that uses the precise energy difference (and therefore the frequency of photons at resonance) between ➤energy levels in an atom or molecule. The **ammonia clock** uses energy levels of the ammonia molecule, while the **caesium clock** is based on the difference between levels in the ➤hyperfine structure of the caesium atom (which forms the basis for the SI definition of the ➤second).

atomic emission spectroscopy An analytical technique for identifying elements using ➤emission spectroscopy, based on analysing the frequency of electromagnetic radiation emitted from the element. See ➤hydrogen atom emission spectrum and ➤sodium spectrum for further details.

atomic energy ➤nuclear fission; nuclear fusion.

atomic force microscope (AFM) An instrument for mapping solid surfaces at the atomic scale. The force between a fine tip (typically of diamond) and the surface is kept constant as the tip is scanned across the surface, producing contours of equal force. The principle of operation is very similar to that of the ➤scanning tunnelling microscope, though the latter uses electron currents rather than forces.

atomicity The number of atoms in a chemical species. Thus water, H_2O, has an atomicity of 3.

atomic mass unit Symbol amu or u. A unit of atomic mass, used in atomic physics in preference to true mass for convenience. One atom of carbon-12 has an atomic mass of 12 amu. The unit is defined so that the atomic mass of a pure isotope is very close to the number of protons and neutrons in the nucleus.

atomic number (proton number) Symbol Z (from the German *Zahl*, 'number'). The number of protons in the nucleus of an atom. It labels the position of the element in the ➤periodic table.

atomic orbital Simplistically, a region of space around the nucleus of an atom that is occupied by electrons bound within the atom; more rigorously, a solution of the ➤Schrödinger equation for an atom. The solutions are labelled by quantum numbers (➤principal quantum number; orbital angular momentum quantum number; magnetic quantum number). The resulting notation for orbitals is explained under ➤sodium spectrum. The filling of particular orbitals gives rise to the ➤electronic structure of an atom. ➤➤pairing of electrons.

atomic physics The physics of the interaction of the atomic ➤nucleus with the electrons surrounding it.

atomic radius ➤radius (atomic and ionic).

atomic spectroscopy ➤Spectroscopy involving atoms. ➤➤atomic absorption spectroscopy; atomic emission spectroscopy.

atomic theory The idea that all matter is made up of atoms. The theory originated with the philosophical speculations of the ancient Greeks, but no real supporting evidence was found until the early 19th century, when the concept was reintroduced by John Dalton. Dalton used the atomic theory to explain the laws of chemical combination, such as the law of ➤constant composition, the law of ➤multiple proportions and the law of ➤reciprocal proportions. The detailed structure of the atom was discovered in the 20th century. The ➤nucleus was identified in 1911, and the details of ➤electronic structure were worked out in the 1920s using ➤quantum theory.

atomic volume The ➤molar mass of an element divided by its density. In 1870 Lothar Meyer plotted the atomic volume of each element against its relative atomic mass. The graph rose and fell at periodic intervals, an early indication of the usefulness of a ➤periodic table.

atomic weight An obsolete name for ➤relative atomic mass.

ATP Abbr. for **adenosine triphosphate**, the predominant supplier of metabolic energy in living cells. ATP is a ➤nucleoside triphosphate consisting of the base adenine, the sugar ribose and three phosphate groups (see the diagram). ATP supplies the chemical energy to drive endergonic reactions (➤endergonic process), perform mechanical work, provide heat and even produce light. ATP releases energy by hydrolysing to **adenosine diphosphate** (ADP) via a phosphate cleavage. In the presence of an appropriate ATPase ➤enzyme, the bond connecting the third phosphate group is broken, releasing energy which may be used to drive a closely coupled reaction. ATP is one of a family of nucleoside phosphates that have a range of important metabolic functions in cells. ➤aerobic respiration; anaerobic respiration.

ATP

atrium (plural atria) ➤heart.

atropine An ➤alkaloid derived from, in particular, deadly nightshade (*Atropa belladonna*), which affects the nervous control of the heart rate.

attenuated A term applied particularly to pathogenic (disease-causing) microorganisms that have been treated to reduce their virulence. Such organisms are useful in, for example, the production of ➤antiserum without inducing disease symptoms.

attenuation The reduction of the intensity of a wave as it passes through a medium that absorbs or scatters it.

atto- A prefix in ➤SI units meaning 10^{-18} base units.

Au Symbol for the element ➤gold, from its Latin name *aurum*.

AU Symbol for ➤astronomical unit.

audiofrequency The frequency range of sound waves, conventionally taken as 30 Hz to 20 kHz. An ➤acoustic wave outside this frequency range is inaudible to the human ear.

aufbau principle (building-up principle) The principle that the ➤orbital that fills first is the one with the lowest energy. In atoms, the order is 1s, 2s, 2p, 3s, 3p, 4s and 3d,... ➤electronic structure. (*Aufbau* is German for 'building up'.)

Auger effect The emission of an electron from an ion in an excited state due to the previous emission of an electron (see the diagram). The energies of the electrons ejected depend only on the levels and atoms involved, and **Auger electron spectroscopy** is often used to determine the chemical composition of surfaces. Named after Pierre Victor Auger (1899–1993).

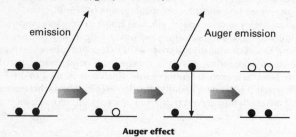

emission   Auger emission

Auger effect

auric Traditional name for gold(III).

aurora Lights in the upper atmosphere visible in the sky at high northern (**aurora borealis**) or southern (**aurora australis**) latitudes. They are caused by high-energy particles in the ➤solar wind, trapped by the Earth's ➤magnetosphere, interacting with atoms and molecules in the atmosphere.

aurous Traditional name for gold(I).

austenite A phase in the ➤iron–carbon phase diagram.

autocatalysis Catalysis by a product of a reaction, which then speeds up the reaction. An example occurs in the iodination of propanone:

$$I_2(aq) + CH_3COCH_3(aq) \rightarrow CH_3COCH_2I(aq) + HI(aq).$$

The reaction is acid-catalysed and an acid is being made in the reaction. Autocatalysis can have a remarkable effect on chemical reactions, as in the oscillatory behaviour of the Zhabotinskii–Belousov reaction.

autoclave A laboratory device used to sterilize equipment and culture media, particularly those used for growing fungi and bacteria. The material to be sterilized is heated under pressure so that temperatures in excess of 100 °C can be achieved without liquids boiling.

autoimmune diseases Diseases that result from the inability in some humans to differentiate between 'self' and 'not-self', leading to the production of an auto-antibody (an ➤antibody against oneself). Such diseases include Hashimoto's disease of the thyroid, rheumatoid arthritis and Crohn's disease.

autoionization The ionization of an atom in two steps. First, the atom is excited by an external source, such as an electron beam or electromagnetic radiation, promoting an electron into a higher energy level (see the diagram). The electron then falls back into a lower energy level, providing the energy to eject a second electron from a shallower level, thus ionizing the atom.

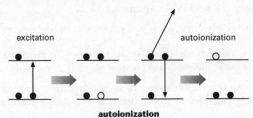

autoionization

autonomic nervous system ➤nervous system.

autosome A chromosome other than a ➤sex chromosome.

autotroph Literally 'self-feeder'. An autotrophic organism is one that can obtain energy directly from sources other than ingested or absorbed food. Autotrophic processes include ➤photosynthesis and ➤chemosynthesis. Compare ➤heterotroph.

auxin (indole-acetic acid, IAA) One of a group of plant growth substances produced by actively growing tissues. Auxin helps to promote the division of cells in plant tissues grown in culture, in leaf and fruit fall, and in the initiation of flowering and sex expression. Auxin has also been isolated from a wide range of natural sources, including human urine. The synthetic auxin ➤2,4-D is a powerful selective weed-killer.

auxochrome An atom or group of atoms that shifts and intensifies the absorption of a ➤chromophore, usually by increasing the extent of the ➤delocalization in the dye molecule.

avalanche 1 The creation of many ions originating from a single charged particle passing through a gas in a strong electric field. It explains the principle behind a ➤Geiger counter.
2 ➤avalanche breakdown.

avalanche breakdown The ➤breakdown of a p–n junction in strong electric fields. The carriers gain sufficient energy in the field to undergo ionizing collisions and create new electron–hole pairs. The new carriers go on to create more ions, until the process multiplies to the extent of breakdown. Compare ➤Zener breakdown.

average A single number which, in some sense, is representative of a ➤sample or set of data. Examples include the ➤arithmetic mean, ➤median (2) and ➤mode.

Aves The vertebrate class comprising the birds. Birds are thought to be descended from a group of theropod dinosaurs. Characteristic features include: ➤homoio-thermy, cleidoic (shelled) eggs, hollow bones, feathers, forelimbs modified to form wings, a lack of teeth (although fossil types such as *Archaeopteryx* have teeth), a keel-like breastbone and a **furcula** (wishbone). There are approximately 9800 species of birds, 12% of which are currently declining or facing extinction.

Avogadro constant (Avogadro's number) Symbol L (derived from the ➤Loschmidt constant). The number of species per ➤mole of a substance. The accepted value is 6.022×10^{23} mol^{-1}. Named after (Lorenzo Romano) Amedeo (Carlo) Avogadro (1776–1856). ➤➤Appendix table 2.

Avogadro's principle Equal volumes of gas, at the same temperature and pressure, contain the same number of molecules. This law follows from the ➤ideal gas equation and is obeyed, approximately, by real gases. At room temperature and pressure, the ➤molar volume of a gas is approximately 24 dm^3 mol^{-1}.

avoirdupois An obsolescent system of weight (strictly, ➤mass) used chiefly in English-speaking countries. Its main units are the ➤ounce, ➤pound and ➤ton.

axenic culture A pure culture of an organism. The term is usually applied to fungi and bacteria growing on artificial media under laboratory conditions where only one species is present.

axial vector (pseudovector) A ➤vector that does not change direction under an ➤inversion. For example, ➤angular momentum is an axial vector because $L = r \times p = (-r) \times (-p)$.

axil ➤leaf.

axiom A mathematical statement that is accepted as true without proof, usually because it is self-evident or postulated as true.

axis (plural **axes) 1** One of the reference lines used in a ➤coordinate system.
  2 A distinguished line associated with a geometric figure, such as an ➤axis of symmetry.

axis of revolution A line about which a curve or area is rotated through 360° to form a surface or solid.

axis of rotation The line about which an astronomical body, such as the Earth, rotates. It intersects the body's surface at the two poles.

axis of symmetry A line about which a geometric figure is symmetrical in the sense that, for every point P in the figure, there is an image point P' such that the axis is the perpendicular bisector of PP'.

axon Part of a ➤neurone that carries an impulse away from the cell body of a nerve cell. Some human axons may be over a metre in length, but are only about 20 μm across. Giant axons up to 1 mm in diameter are found in squid, and these have been used to elucidate the nature of the ➤nerve impulse.

azeotrope (constant boiling mixture) When a mixture of liquids deviates from ➤Raoult's law, there may be a maximum or minimum in the boiling point/composition graph: the **azeotropic mixture** boils without changing composition, as if

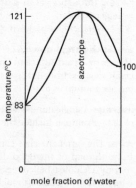

azeotrope Formation of a nitric acid/water azeotrope.

it were a pure substance. Nitric acid and water form a **maximum boiling azeotrope** at 68% nitric acid (by mass) and 121 °C (see the diagram opposite). Ethanol and water form a **minimum boiling azeotrope** at 96% ethanol (by mass) and 78.2 °C. Further separation is impossible by fractional distillation, thus anhydrous ethanol has to be prepared by chemical drying, for example with calcium oxide.

azide A salt of **hydrazoic acid**, HN_3, containing the azide ion N_3^-. An example is sodium azide, NaN_3.

azimuth (azimuthal angle) A parameter that measures angle around a cylinder, or something that can be mapped on to a cylinder. For example, the ➤longitude of a point on the surface of the Earth is its azimuth. ➤➤celestial sphere.

azimuthal quantum number An alternative name for ➤orbital angular momentum quantum number.

azo compound ➤diazo compound.

azo coupling The formation of a ➤diazo compound.

azo dyes ➤diazo dyes. *Azote* is an obsolete French name for ➤nitrogen. This explains the 'az' in many names associated with nitrogen.

AZT (azidodideoxythymidine, zidovudine) Tradename Retrovir. A ➤nucleoside analogue approved for the treatment of ➤HIV, which acts by inhibiting ➤reverse transcriptase, needed for viral replication. AZT is commonly used in combination with other nucleoside analogues.

B

B Symbol for the element ➤boron.

B Symbol for ➤magnetic flux density.

Ba Symbol for the element ➤barium.

bacillus One of a number of types of ➤bacteria having a rod-shaped cell. It is also the generic name of some of these, for example *Bacillus subtilis*.

back bonding The additional movement of electron density from a Lewis acid back to a Lewis base when some complexes form, complementing the bond from the Lewis base to the Lewis acid (➤Lewis theory). An important example occurs in the carbonyl complexes of transition metals. The carbonyl group donates its electron pair to form a sigma bond. At the same time, it acts as a ➤pi acid, accepting electron density from the transition metal.

back cross A cross (mating) between a parent and one of its offspring. Such crosses are used to ascertain the ➤genotype of an individual offspring. A cross performed between such offspring and an organism of known genotype that is not one of its parents is termed a **test cross**.

back e.m.f. An e.m.f. (➤electromotive force), generated by electromagnetic ➤induction, that opposes the flow of current in a circuit.

background The contribution to the signal collected as data in an experiment (typically spectroscopy) that comes from sources with no significant features in the region of interest. The background is usually estimated as a simple function and subtracted from the data collected in order to leave the significant data.

background radiation ➤cosmic background radiation.

bacteria (singular **bacterium**) A major and widespread group of single-celled ➤prokaryotic organisms. Bacterial cells are surrounded by a cell wall different in structure to those of ➤eukaryotic organisms. Bacteria are important as agents of many infections and diseases, such as cholera, and also play an important part in the recycling of materials in nature as a result of decomposition of organic materials and the fixation of nitrogen (➤nitrogen cycle). The genetic mechanisms of many bacteria such as *E. coli* have been extensively studied, and they are routinely used in ➤genetic engineering as tools for manipulating DNA. Some bacteria are used commercially for industrial processes such as the conversion of ethanol into ethanoic acid in the manufacture of vinegar.

bacteriophage (phage) A ➤virus that infects bacteria. Bacteriophages destroy bacterial cells, and this destruction shows up as clear areas called ➤plaques in a colony of bacteria growing on nutrient jelly. Many phages can be used in genetic engineering applications as a ➤cloning vector to insert segments of DNA into other bacterial cells and to amplify the numbers of specific fragments of DNA. The structure of the DNA of several phages is known in detail. For example, the sequence of the 48 502 base pairs of the entire length of DNA found in lambda phage is known. This is now used as a standard laboratory source of DNA.

bacterium The singular of ➤bacteria.

Bakelite An obsolete tradename for a ➤phenol–formaldehyde resin. It was named after Leo Hendrik Baekeland (1863–1944), who first discovered this class of polymers.

baking soda (bicarbonate of soda) A traditional name for sodium hydrogencarbonate, $NaHCO_3$, used in baking powders.

balance An instrument for comparing two weights (and hence two masses).

ballistic galvanometer An instrument for measuring the total flow of charge. It measures the ➤impulse imparted by magnetic effects rather than the force (as in a standard moving-coil galvanometer), thus it effectively integrates (➤integration) the current with respect to time.

ballistic missile A missile that moves freely under gravity after an initial ➤impulse is given to it.

Balmer series The series of lines in the visible region of the ➤hydrogen atom emission spectrum. The lines are caused by an electron falling from an excited state to the second quantum level, labelled $n = 2$ (where n is the ➤principal quantum number). Johann Jakob Balmer (1825–98) related the inverse of the wavelength λ of the lines to the inverse square of the principal quantum number of the upper level:

$$\frac{1}{\lambda} = R_H \left(\frac{1}{4} - \frac{1}{n^2} \right)$$

where R_H is the Rydberg constant. A similar equation, the ➤Rydberg–Ritz equation, accounts for all the lines in all the series of the hydrogen spectrum.

banana bond A descriptive name for the shape of the ➤three-centre bond present in molecules such as diborane, B_2H_6.

band gap ➤energy gap.

band-pass filter A ➤filter that allows only signals within a particular frequency range to pass.

band spectrum A spectrum produced by molecules as opposed to atoms (which show line spectra). The frequencies are much less well-defined than for atoms, because the molecules can also be excited into other vibrational and rotational energy levels. This gives rise to a set of closely spaced lines which are often not resolved and instead merge into a band.

band structure The dependence of the energy of electrons in a solid on their ➤crystal momentum. ➤➤energy band.

bandwidth The range of frequencies of a ➤sinusoidal signal that can be sent along a particular channel. In combination with the ➤signal-to-noise ratio, the bandwidth determines the rate at which information can be transmitted. Optical fibres have a bandwidth of the order of 1 GHz, compared with 4 kHz for conventional telephone lines.

bar A unit of pressure: 1 bar = 10^5 Pa = 10^5 N m^{-2}. It is a very convenient unit for atmospheric pressure, and weather maps give pressures in millibars (a pressure of 1000 mbar is the same as a pressure of 1 bar).

Barfoed's test A chemical test for ➤monosaccharide reducing sugars in solution. Aqueous copper(II) ethanoate and ethanoic acid is added to the solution to be tested and boiled. A red precipitate of copper(I) oxide indicates a positive result. Named after Christen Thomsen Barfoed (1815–89).

barium Symbol Ba. The element with atomic number 56 and relative atomic mass 137.3, which is a member of Group 2. It is the densest of the common elements of the group and this explains its name which comes from the Greek for 'heavy'. Its main ores are barytes, $BaSO_4$, and witherite, $BaCO_3$. The metal is extracted by electrolysis of its fused chloride, as it is too reactive to be made in any other way. Its chemistry is dominated by a unique ➤oxidation number of +2, as in the oxide BaO, the chloride $BaCl_2$ and the titanate $BaTiO_3$, a piezoelectric crystal discussed under ➤titanium. Barium peroxide, BaO_2, is bonded $Ba^{2+}O_2{}^{2-}$. The hydroxide $Ba(OH)_2$, called **caustic baryta**, is a strong alkali. Probably barium's most important use is in medical radiography, when barium sulfate is administered as a **barium meal**. This coats the upper part of the digestive tract, making it visible to X-rays because barium scatters X-rays effectively.

Barkhausen effect The step-like increase in the magnetization of a ferromagnetic material as magnetic domains suddenly snap into alignment (➤ferromagnetism). Named after Heinrich Georg Barkhausen (1881–1956).

barn Symbol b. A unit of ➤scattering cross-section used in nuclear physics, equal to 10^{-28} m^2. The term is a whimsical likening of the very small to the very large.

barometer An instrument for measuring ➤pressure. An **aneroid barometer** measures the expansion or contraction of a sealed capsule, and is calibrated for pressure by using ➤Boyle's law.

Barr body The compact X chromosome of females. Female cells have evolved a mechanism for permanently inactivating one of the ➤sex chromosomes, called the Barr body. This chromosome is condensed into ➤chromatin and can be seen in the optical microscope during ➤meiosis. The presence of a Barr body has been used as a test to confirm the sex of athletes. Named after Murray Llewellyn Barr (1908–95).

baryon A subatomic particle that consists of three ➤quarks. Examples are the proton (uud, i.e. two up quarks and a down quark), the neutron (udd), the lambda (uds, i.e. up, down and strange) and the sigma (uus, uds or dds depending on charge). Baryons and ➤mesons together make up the ➤hadrons. Baryons are hadrons with half-integral spin, whereas mesons have integral spin. The name baryon comes from the Greek for 'heavy'.

baryon number A ➤quantum number B defined so that all baryons have $B = +1$, and all **antibaryons** have $B = -1$; for all other particles $B = 0$. The total baryon number is conserved in all particle interactions.

baryta An old name for barium oxide, BaO.

basal metabolic rate (BMR) The rate at which the chemical processes of the body must function in order to sustain life. The BMR provides a clinical test based on oxygen consumption which determines the rate of the body's ➤metabolism under standard conditions (physical and mental rest) and can be used, for example, to establish the effectiveness of training programmes for athletes.

basalt A fine-grained ➤igneous rock. It is composed predominantly of calcium-rich plagioclase ➤feldspar, together with the ➤silicate minerals pyroxene and olivine.

base 1 (Chem.) A substance that reacts with an acid to form a salt and water only. There are two more general definitions of the word 'base': ➤Lewis base and ➤Brønsted base. In the classic acid–base reaction of a hydrogen ion, H^+, with a hydroxide ion, OH^-, they focus on different aspects of the same process. The Lewis theory sees the hydroxide ion as the supplier of the electron pair used to make the bond, whereas the Brønsted theory focuses instead on the hydroxide ion as a proton acceptor. Compare ➤acid.
  2 (Math.) That number, ➤powers (2) of which are used to represent whole numbers in a place-value number system. Thus the base used in everyday calculations is 10 because, for example, 325 means $3 \times 10^2 + 2 \times 10^1 + 5 \times 10^0$. The name of a number system derives from the base, as in **binary** (base 2), **octal** (base 8), **decimal** (base 10), **hexadecimal** (base 16), and so on.
  3 (Math.) That number, ➤powers (2) of which are used to define ➤logarithms. Thus y is the logarithm of x to base b, written $y = \log_b x$, if $x = b^y$.
  4 (Math.) A side of a ➤polygon or ➤face of a solid that has been singled out for some purpose, such as defining an ➤altitude (2).
  5 (Phys.) ➤bipolar junction transistor.
  6 (Biol.) ➤nitrogenous base.

base-catalysed condensation reaction A very important class of reactions for making carbon–carbon bonds in organic synthesis. The essential features are the removal of a proton from one molecule, catalysed by the base, followed by attack of the carbanion formed at another molecule. An important example is the ➤aldol condensation. ➤➤Knoevenagel reaction.

base ionization constant (base dissociation constant) Symbol K_b. The equilibrium constant describing the species present in the solution of a base. For the base ammonia, for example, the aqueous solution contains an equilibrium mixture:

$$NH_3(aq) + H_2O(l) \rightleftharpoons NH_4^+(aq) + OH^-(aq),$$

for which the base ionization constant K_b is

$$K_b = [NH_4^+][OH^-]/[NH_3],$$

where the square brackets represent the concentration of the species.

base metal A metal, such as lead, that is quite easily oxidized and therefore tarnishes readily. Compare ➤noble metal.

base pairing The linking by ➤hydrogen bonding of pairs of nitrogenous bases to make the double-stranded form of ➤DNA. In base pairing, a ➤purine is always base-paired with a ➤pyrimidine. Thus adenine always pairs with (is **complementary** to) thymine, and guanine always pairs with cytosine by mutual hydrogen bonding (see the diagram). Such specificity of base pairing allows for accurate ➤DNA replication during cell division. This is an important natural phenomenon in living cells, and is also the basis of the manipulation and cloning (➤clone) of DNA in ➤genetic engineering.

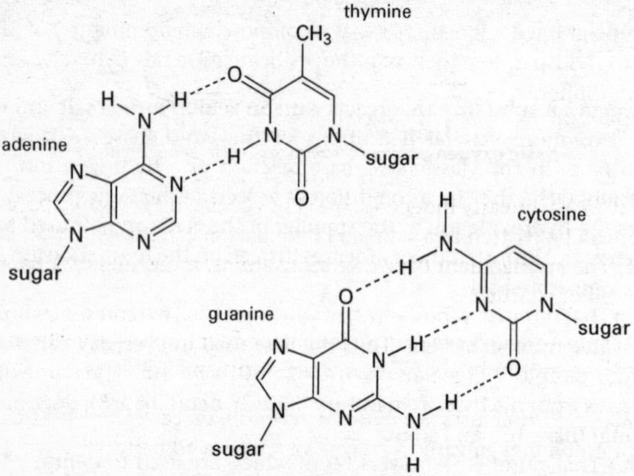

base pairing

base unit (fundamental unit) A ➤unit whose size is arbitrarily defined independently of other units. Compare ➤derived unit; ➤➤SI units; Appendix table 1.

basic Having the properties of a ➤base.

BASIC Acronym for beginner's all-purpose symbolic instruction code. A high-level computer language, designed as an aid to teaching programming.

basic oxygen process The most commonly used process for converting iron into steel. Oxygen is passed into molten iron through a water-cooled lance (see the diagram opposite) to oxidize the excess carbon in the impure iron ('pig iron'), produced in a ➤blast furnace, to carbon dioxide. The 'basic' refers to the basic lining, which reacts with acidic impurities to form a slag. The vessel can be tilted to allow the pig iron to be added, and the molten steel and slag to be removed.

basic salt A salt resulting from the partial neutralization of a base. It contains both hydroxide ions and another negative ion, as in $Pb(OH)_2 \cdot PbCl_2$.

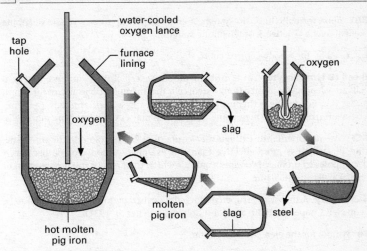

basic oxygen process Furnace and cycle of operation.

basis **1** (Math.) A ►linearly independent set of vectors in a ►vector space such that every vector can be written as a ►linear combination of vectors in the set.

2 (Chem.) The arrangement of two or more atoms, molecules or ions within the ►primitive cell of a lattice.

basophil A type of ►leucocyte with an irregular (polymorphic) nucleus that releases ►histamine and ►serotonin, triggering reactions to an ►allergy.

battery A device that uses a chemical reaction to generate electricity; it is a collection of individual galvanic ►cells (2). ►►lead–acid battery.

baud rate In communications, the number of ►bits per second that can be transmitted over a ►channel.

bauxite The most important ore of aluminium, named after the town of Les Baux in France, where the first major source was discovered. The formula for the ore is $Al_2O_3 \cdot xH_2O$, where x lies between 1 and 3. Aluminium is extracted from bauxite by the ►Hall–Héroult process.

Bayesian statistics and probability A particular way of thinking about probabilities in which the probability is interpreted primarily as a number that indicates the degree of rational belief in the truth of a proposition. For a die, the Bayesian method assigns a probability of 1/6 to the event that the die lands with the 4 facing up; in the absence of further information, this would be true for both an unbiased *and* a biased die. The value of 1/6 is clearly *unlikely* to be the frequency ratio of 4s in a large number of trials, if the die is in fact biased.

Bayesians treat the parameters of statistical distributions in hypothesis testing as random variables themselves, and hence assign probability distributions (rather than ►confidence intervals) to unknown parameters. Named after Thomas Bayes (1702–61).

BBS Abbr. for bulletin board system. A computer set up to receive and distribute ➤email over a ➤network or telephone system.

b.c.c. Abbr. for ➤body-centred cubic.

B cell (B lymphocyte) An ➤antibody-producing cell of the ➤immune system. B cells are ➤leucocytes formed from ➤stem cells that mature in bone marrow or lymph nodes (i.e. not in the ➤thymus gland). B cells express their antibody (➤immunoglobulin) on their cell membrane. Some B cells may develop into ➤memory cells.

BCF An important fire extinguisher containing bromochlorodifluoromethane, $CBrClF_2$, which is more effective than, for example, carbon dioxide because it decomposes to produce bromine ➤radicals which then reduce the proliferation of chain carriers in the flame.

BCS theory A theory of ➤superconductivity devised by John Bardeen (1908–91), Leon Neil Cooper (b. 1930) and John Robert Schrieffer (b. 1931).

Be Symbol for the element ➤beryllium.

bearing A way of specifying the directions of displacements which is much used in navigation. The bearing of B from A is the ➤angle between AB and AN, where N is any point due north of A. Bearings are conventionally given in three-figure form such as $063°$ or $216°$, with clockwise rotations counted as positive.

beats The oscillations observed at half the difference in frequency between two superposed signals. If $x = \sin(\omega_1 t)$ and $y = \sin(\omega_2 t)$, then

$$x + y = 2 \sin[\tfrac{1}{2}(\omega_1 + \omega_2)t] \cos[\tfrac{1}{2}(\omega_1 - \omega_2)t].$$

Beaufort scale A nautical scale of wind speed. The scale runs from 0 (calm), through 7 (moderate gale), to 12 (hurricane) and beyond. The relationship with speed is not linear, the **Beaufort number** being proportional to the wind speed to the power 2/3. Named after Francis Beaufort (1774–1857).

Beckmann rearrangement The conversion of the ➤oxime of a ➤ketone into an ➤amide. A Beckmann rearrangement is used to prepare the monomer of nylon-6 (see the diagram). Named after Ernst Otto Beckmann (1853–1923).

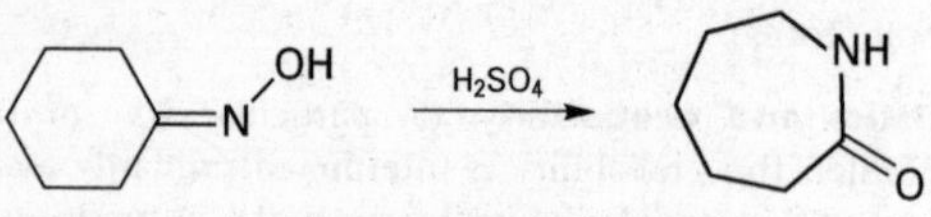

Beckmann rearrangement

Beckmann thermometer A mercury ➤thermometer with a large bulb, which enables it to be very sensitive to temperature differences over a small range. It typically measures to $0.01°C$.

becquerel Symbol Bq. The SI unit of radioactivity: one becquerel corresponds to one disintegration event per second. Named after (Antoine) Henri Becquerel (1852–1908), the discoverer of radioactivity.

Beer's law The intensity of light passing through a solution decreases as $e^{-\varepsilon cd}$, where d is the distance travelled through the solution, c is the concentration of the solution and ε is the **extinction coefficient**. Named after August Beer (1825–63). ➤Lambert's laws.

bel A unit of power ratio. The commonly used derivative is the ➤decibel.

benchmark A test of the speed of a computer, consisting of a set of calculations and memory storage instructions supposedly representative of the operations that are carried out by a typical program. Running the same test on different machines allows an objective comparison between them.

bending moment For a cross-section of a beam or similar rigid object, the total moment acting on the object on one side of the cross-section (also equal to the total moment on the other side if the beam is in static equilibrium).

Benedict's solution (Benedict's reagent) An alkaline solution of Cu^{2+} ions complexed with citrate ions. If a solution being tested contains a reducing sugar, the Cu^{2+} ions are reduced to a red precipitate of copper(I) oxide. Some formulations of Benedict's solution can be used to estimate the concentration of reducing sugar quantitatively. Named after Stanley Rossiter Benedict (1884–1936). ➤Barfoed's test; Fehling's solution.

benthic Describing animals that live on or near the sea bed or lake bed, often attached to it, in particular deep-sea organisms. Compare ➤pelagic.

(a)

benzaldehyde (benzenecarbaldehyde) C_6H_5CHO The derivative of ➤benzene with an ➤aldehyde group attached to the ring. It is a colourless oil smelling of almonds, which is used both in perfumes and flavourings.

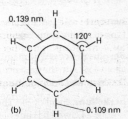
(b)

benzene C_6H_6 The archetypal ➤aromatic organic compound. It exists as a colourless liquid. The key feature of the **benzene ring** is the ➤delocalization of the double bonds (see the diagram), which is why the older ➤Kekulé structure does not describe the bonding so well. This ring is such a stable structure that it is almost always reformed on reaction, so the most common reaction of benzene and other aromatic compounds is ➤electrophilic substitution.

benzenecarbaldehyde ➤benzaldehyde.

benzenecarboxylate ➤benzoate.

benzenecarboxylic acid ➤benzoic acid.

benzenediol ➤dihydroxybenzene.

benzene hexachloride (BHC) $C_6H_6Cl_6$ The molecule formed by adding one chlorine atom to

(c)

benzene (a) Conventional symbol for a benzene ring. (b) Configuration of atoms, and bond lengths. (c) Delocalization of electrons gives rise to high electron densities above and below the ring.

each carbon atom in a benzene ring, which makes it no longer aromatic. It is a colourless liquid, important as the first effective ➤insecticide.

benzenesulfonic acid $C_6H_5SO_2OH$ A derivative of ➤benzene, which exists as a colourless oil. It is formed when benzene reacts with concentrated sulfuric acid. The parent molecule itself is much less important than its derivatives with alkyl groups in the ➤*para* position of the ring, which are the archetypal ➤detergents.

benzenoid Structurally related to ➤benzene.

benzoate A salt or ester of benzoic acid. Sodium benzoate is used as a food preservative.

benzoic acid (benzenecarboxylic acid) C_6H_5COOH The derivative of ➤benzene with a ➤carboxylic acid group attached to the ring. It is a white crystalline solid, whose main use is as a food preservative.

benzophenone ➤diphenylmethanone.

benzoylation The ➤acylation reaction in which the acyl group added is the benzoyl group, C_6H_5CO—, on reaction with a reagent such as **benzoyl chloride**, C_6H_5COCl.

benzpyrene (benzopyrene) $C_{20}H_{12}$ A polycyclic aromatic hydrocarbon (see the diagram). Benzpyrene is extremely ➤carcinogenic and is one of the pollutants in tobacco smoke most responsible for causing lung cancer in smokers.

benzpyrene

benzyl group The $C_6H_5CH_2$— group.

benzyne C_6H_4 A reactive intermediate (see the diagram), formed during some organic reactions.

benzyne

Bergius process The manufacture of fuel by the liquefaction of coal at about 450 °C in the presence of hydrogen at about 200 atm. Named after Friedrich Karl Rudolph Bergius (1884–1949).

berkelium Symbol Bk. The element with atomic number 97 with longest lived isotope 247, which is one of the transuranium elements. Its chemistry is dominated by its radioactivity. Named after Berkeley, the university where many of the transuranium elements were discovered.

Bernoulli's theorem A theorem concerning steady-state flow in an incompressible fluid of zero ➤viscosity. The sum of the pressure, the gravitational potential energy per unit volume and the kinetic energy per unit volume is constant along a ➤streamline. This is usually expressed as

$$p + \rho gh + \tfrac{1}{2}\rho v^2 = \text{constant}.$$

Named after Daniel Bernoulli (1700–82).

Berthollide compound ➤nonstoichiometric compound. Named after Claude Louis Berthollet (1748–1822).

beryl $Be_2Al_2Si_6O_{18}$ A natural beryllium-containing aluminosilicate which is the chief ore of beryllium. **Emerald** is a gem-quality variant with a characteristic green colour; **aquamarine** is a blue-green variety.

beryllium Symbol Be. The element with atomic number 4 and relative atomic mass 9.012, which is the Group 2 element with lowest mass. It occurs in beryl, from which, after a number of processes, it is extracted by ➤electrolysis: electrolysis has to be used as beryllium is a very reactive metal. The metal itself is grey and hard. Beryllium is able to reduce alkalis, because of the formation of the tetrahydroxoberyllate ion, $[Be(OH)_4]^{2-}$, due to the high ➤charge density of the beryllium ion. It forms an oxide **beryllia**, BeO, which is a ➤refractory white solid used in the exhaust nozzles of rockets, where it can withstand the high temperatures without melting. Other compounds include beryllium chloride, $BeCl_2$, a white crystalline solid. Its structure (see the diagram) shows considerable covalent character, as expected from the high charge density of the beryllium ion. Note that the coordination number is 4, as is common in beryllium compounds. The uses of beryllium and its compounds are limited as it is extremely toxic to humans, possibly because the smaller beryllium ion replaces magnesium in vital biochemical molecules.

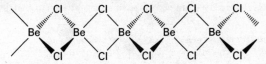

beryllium $BeCl_2$ as an extended chain.

Bessel function One of a family of ➤special functions that occur as solutions of Bessel's equation,

$$x^2 \frac{d^2 y}{dx^2} + x \frac{dy}{dx} + (x^2 - n^2)y = 0,$$

which arises when solving the ➤wave equation in ➤cylindrical coordinates. Named after Friedrich Wilhelm Bessel (1784–1846).

Bessemer process The conversion of impure iron into steel by pouring molten cast iron into a **Bessemer converter**, a large egg-shaped vessel, and using a stream of air to oxidize most of the carbon in the cast iron. It has been largely superseded by the ➤basic oxygen process. Named after Henry Bessemer (1813–98).

beta A prefix used in a number of different ways, especially in biochemistry. For sugars such as ➤glucose, it describes the isomer with the hydroxyl group on carbon atom 1 in the 'equatorial' position (see diagram at glucose). For amino acids, it describes the position of the carboxyl group relative to the amine group (**beta amino acids** have the two groups on neighbouring carbon atoms). For proteins, it describes a common structure present (see the diagram of the **beta pleated sheet** overleaf).

The prefix is also used in astronomy to indicate, usually, the second brightest star in a constellation, for example *Beta Persei*.

beta-adrenergic receptor A type of receptor for ➤adrenaline and its derivative noradrenaline that responds to specific drugs.

beta blocker A clinically applied substance that inhibits ➤beta-adrenergic receptors in the parasympathetic ➤nervous system, causing a reduction in heart rate. Beta blockers thus produce a calming effect and are used to reduce stress.

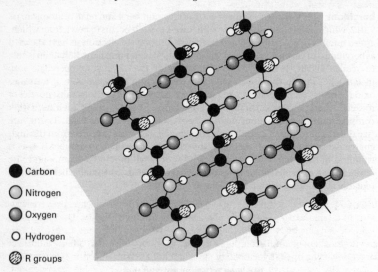

● Carbon
○ Nitrogen
◐ Oxygen
○ Hydrogen
⊘ R groups

beta pleated sheet

beta decay The mode of radioactive decay that consists of the emission of electrons (known in this context as **beta particles**, with the emission known as **beta radiation**). This is always accompanied by the emission of an antineutrino, for example $^{77}Ge \rightarrow {}^{77}As + e^- + \bar{v}_e$. Because of the nature of beta particles, in beta decay there is a negligible change in mass but the element with one more proton is produced. Processes in which a ➤positron (and a ➤neutrino) are emitted are also usually called beta decay. Beta decay is governed by the ➤weak interaction.

beta emitter An unstable nucleus that undergoes ➤beta decay.

beta-iron A term occasionally used for iron between its Curie temperature (➤Curie–Weiss law), 768 °C, and its transition temperature to gamma-iron, 910 °C. ➤➤iron–carbon phase diagram.

betatron An electron ➤accelerator that uses an oscillating magnetic field to produce a circular stream of high-energy electrons.

B **field** ➤magnetic flux density.

Bh Symbol for the element ➤bohrium.

BHC Abbr. for ➤benzene hexachloride.

bi- A prefix meaning two or double, as in ➤bicyclic.

Bi Symbol for the element ➤bismuth.

bias A d.c. voltage applied to a semiconductor device, such as a transistor, to set its ➤operating point.

biaxial crystal ➤birefringence.

bicarbonate A traditional name for a ➤hydrogencarbonate; an ➤acid salt of carbonic acid in which only one of the hydrogen atoms has been replaced. The most famous example is **bicarbonate of soda**, sodium hydrogencarbonate, $NaHCO_3$, which is used in baking powder.

biconcave lens A lens that is ➤concave on both surfaces, typically a diverging lens.

biconvex lens A lens that is ➤convex on both surfaces, typically a converging (magnifying) lens.

bicyclic A compound having two rings. An example is ➤naphthalene.

bidentate ligand A ➤ligand that can bind to a central metal ion by using the electron pairs from two donor atoms within the ligand. An important example is ➤1,2-diaminoethane. Such a bidentate ligand can form a ➤chelate.

big bang The supposed event, about 14 billion years ago, in which the Universe was born in a rapid expansion from a condition of ultra-high density. The Universe's ultimate fate depends on whether its initial energy is sufficient to exceed the gravitational energy of its components, analogous to the orbit of a body around a larger one being open (hyperbola) or closed (ellipse) depending on the body's total energy. The big bang theory is favoured by the presently available observational evidence (particularly the ➤cosmic background radiation). The name was a derisory label given to the theory by the proponents of the competing ➤steady state theory. ⯮Hubble constant.

bile A yellow-green fluid, composed of salts, ➤cholesterol, lecithin (a ➤phospholipid) and other organic metabolites and trace metals, which is produced by the ➤liver and stored in the gall bladder before secretion into the duodenum (the first part of the small intestine). The main function of bile is to aid the digestion of a ➤lipid by emulsifying it, thus producing a larger surface area for the action of lipase enzymes.

billion A thousand million. (It was formerly used in the UK to mean a million million.) In science it is desirable to avoid confusion by using the prefix ➤giga- or the notation 10^9, as appropriate.

bimetallic strip A strip composed of two metals with different linear ➤expansivity welded together in such a way that the strip bends when the temperature changes and usually makes or breaks a contact, for example in a ➤thermostat.

bimolecular Any step in a reaction mechanism that involves two species (*not* necessarily molecules, however). An important example is the bimolecular ➤nucleophilic substitution mechanism S_N2.

bimorph cell A strip composed of two ➤piezoelectric materials, designed to bend when a potential difference is applied, or to produce a potential difference when bent. They are used, for example, in some types of microphone.

binary Number ➤base 2 used in digital computers, since this requires only the representation of the numbers 0 and 1.

binary compound A compound containing only *two* elements. Water, H_2O, is a binary compound, but sodium hydroxide, NaOH, is not.

binary operation A rule for assigning to any two ➤elements (2) of a ➤set another element of that set. For example, addition is a binary operation on the set of ➤integers.

binary star A system that consists of two stars (components) bound together by gravitational attraction. The components can be detected by eye (➤visual binary), by their Doppler shifts (**spectroscopic binary**), by changes in brightness as they eclipse each other (**eclipsing binary**) or by inference from their motion (**astrometric binary**). About half the nearest hundred stars are binary stars or groups of more components (**multiple stars**), so binary stars should not be regarded as unusual. ➤➤cluster; double star.

binding energy **1** The energy required to raise an electron in a solid to the ➤Fermi energy. The ➤ionization energy equals the binding energy plus the ➤work function. In photoelectron spectroscopy, binding energies are used to characterize specific elements.
2 The potential energy (from the ➤strong interaction) that holds the atomic nucleus together. It is the energy required to split a nucleus into its component nucleons. It can be evaluated as the difference between the total energy of the nucleus and the sum of the ➤rest masses of its components.

binomial coefficient The ➤coefficient $\binom{n}{k} = {}^nC_k = n!/k!(n-k)!$ that appears in the ➤binomial theorem.

binomial distribution A ➤discrete random variable X taking values $0, 1, \ldots, n$ with ➤probabilities $P(X = k) = \binom{n}{k}p^k q^{n-k}$, where $\binom{n}{k}$ is a ➤binomial coefficient and $q = 1 - p$. It gives the number of 'successes' in n independent repetitions of something that can only happen in two ways – success with probability p and failure with probability q. The binomial distribution has ➤mean np and ➤variance npq, and for large values of n may be approximated by a ➤normal distribution having these ➤parameters (3). For large n and small p, it can also be approximated by a ➤Poisson distribution with mean np.

binomial system ➤Linnaean system.

binomial theorem For whole numbers n,

$$(a+b)^n = a^n + \ldots + \binom{n}{k}a^{n-k}b^k + \ldots + b^n,$$

where the coefficient of $a^{n-k}b^k$ is the ➤binomial coefficient $\binom{n}{k}$. For other values of n, the binomial expansion is an ➤infinite power series:

$$(1+x)^n = 1 + nx + n(n-1)x^2/2! + \ldots + n(n-1)\ldots(n-r+1)x^r/r! + \ldots,$$

which converges for values of x between -1 and 1.

bioaccumulation The progressive concentration of a non-biodegradable substance in the bodies of successive organisms in a ➤food chain. It is of particular significance when, for example, pesticide toxins are accumulated in the bodies of carnivores, often with fatal effects. If a field of grain is treated with a pesticide such that each seed receives the equivalent of one arbitrary unit of toxin, and fieldmice eat 10 grains a day, each mouse will have a daily intake of 10 units. If a kestrel eats 10 such fieldmice it will receive a dose of 100 units. The deaths of many birds of prey and the appearance of ➤DDT in human food chains led to the restriction of its use as an insecticide.

bioassay The use of a measurable biological effect to determine the concentration of a substance applied to produce the effect. ➤➤bioluminescence.

biochemical oxygen demand ➤biological oxygen demand.

biochemistry The branches of chemistry and biology associated with the study of the chemistry of living systems and their products.

biodegradable A substance that is capable of being broken down into its constituents by the natural activities of living organisms. For example, the organic components of sewage can be broken down into simple inorganic substances by the activities of various bacteria and fungi. **Non-biodegradable** materials persist in the environment, and those that are toxic or interfere with living processes in other ways (such as nylon fishing lines ensnaring seabirds) are serious causes of ➤pollution.

biodiversity The variety among living organisms from all sources, including terrestrial, marine and other aquatic ➤ecosystems of which they are part. The term includes diversity within species, between species and of ecosystems. Ultimately, all biodiversity is determined by ➤genes. Biodiversity is an important concept in planning for conservation of natural resources and was a central theme of the United Nations Conference on the Environment held in Rio de Janeiro in 1992.

bioengineering The application of engineering principles to biological systems. The design and development of artificial limbs and heart valves are examples. The term is also frequently used as a synonym for ➤genetic engineering.

biofuel A fuel that is produced from a natural organic product. For example, **biodiesel** is formed by a **transesterification** reaction in which a vegetable oil (such as oilseed rape, soya or palm oil) reacts with methanol, under acid or base catalysis, to form a mixture of the methyl ➤esters of whatever long-chain carboxylic acids were present in the oil. Rudolf Diesel, who originally designed his engine to work on peanut oil, said in 1912 that 'The use of vegetable oils for engine fuels may seem insignificant today, but such oils may become in the course of time as important as the petroleum and coal tar products of the present time.'

Concerns have been raised that **first-generation biofuels** that turn food into fuel exacerbate world-wide food shortages: **second-generation biofuels** use only the residual non-edible parts of the crop such as stems and husks. A Royal Society report in 2008 warned that each biofuel needs to be assessed individually over its *whole* cycle from planting to combustion, that widespread use of biofuels would have implications for land use (unintended consequences may outweigh any benefit) and that global economic and social impacts must be fully considered.

biological clock (biorhythm, endogenous rhythm) A process or series of events in a living organism that is repeated at regular intervals of time. Biorhythms may have diurnal (**circadian**) periodicity or may operate over longer time-scales. They may be controlled internally, as is the ➤menstrual cycle, or modulated by environmental stimuli such as day length, tidal cycles and temperature.

biological control A method of pest control in which predators or diseases of a pest are used rather than chemicals to reduce its numbers. An example commonly used in the production of commercial glasshouse crops is the control of whitefly by the parasitic wasp *Encarsia*. The adult wasps lay eggs on the larval whitefly, and the wasp larvae hatch and feed off the developing whitefly, thereby killing it. Biological control can only reduce the effect of a pest rather than completely eradicate it, since the predator requires a certain number of prey in order for it to survive. Recent developments in biological control involve the use of ➤pheromones.

biological (biochemical) oxygen demand (deficit) (BOD) An index used in a test for the level of ➤pollution in water samples, particularly c·ganic pollution by sewage. The basis of the test is to measure the amount of oxyg·n consumed by a known volume of the sample in a given time. Bacteria in the sample carry out ➤respiration using the organic matter as an energy source and consume oxygen. The BOD is thus an indirect measure of the organic matter present that is supporting bacterial populations; clean water usually has a low BOD.

biology The study of living organisms. Modern biology weaves together a wide range of scientific disciplines including the physical and chemical sciences, mathematics and computing.

bioluminescence The production of light by many groups of organisms, including fungi, bacteria, insects such as glow-worms and fireflies, marine microorganisms (that are responsible for phosphorescence) and some fish. Bioluminescence is used, particularly by fireflies, glow-worms and some deep-sea organisms, as a signalling system and possibly as a defence mechanism. Light production is under the control of an enzyme, **luciferase**, which acts on luciferin. The process requires ➤ATP, and luciferase systems are used as a ➤bioassay for ATP.

biomass The total mass of material in an organism or group of organisms, particularly those at a given trophic level in a ➤food chain. The measurement of biomass is important in determining the overall efficiency of ecological and agricultural systems.

biome In ecology, a major complex of ➤ecosystems determined by a particular set of climatic conditions and comprising a characteristic set of organisms. Tropical forests, coral reefs and deserts are examples. The exact composition of species depends on the geographical context, but typically assemblages of species in biomes on different continents have a similar ➤adaptation. Thus plants growing in desert biomes throughout the world tend to have fleshy leaves with thick cuticles for water storage and conservation.

biophysics The study of the physics of biological structures.

bioreactor ➤fermenter.

biorhythm ➤biological clock.

biosphere An all-encompassing term in ecology used to define that part of the Earth's crust, oceans and atmosphere that is inhabited by living organisms.

biosynthesis The manufacture by a biological system of a particular product. Thus the manufacture of ➤proteins by ➤cells is referred to as **protein biosynthesis.**

biotechnology The application of biological systems and processes to the production of substances of commercial, medical or nutritional significance.

Biot–Fourier equation (heat equation) The equation for the variation of temperature T with position and time in a conductor with ➤thermal conductivity κ, ➤specific heat capacity c and density ρ:

$$\frac{\partial T}{\partial t} = \frac{\kappa}{c\rho}\nabla^2 T.$$

It is a second-order partial differential equation identical in form to the ➤diffusion equation. Named after Jean-Baptiste Biot (1774–1862) and (Jean-Baptiste) Joseph Fourier (1768–1830).

biotic factor Any factor, feature or process due to, or derived from, living organisms that might affect any other organism in its environment. Examples are predators and the shade afforded by trees in a forest. Compare ➤abiotic factor.

biotin One of the B-complex ➤vitamins.

Biot–Savart law The law relating the magnetic field as a function of position r in the vicinity of a conductor to the current in the conductor:

$$B = \int \frac{\mu_0}{4\pi} \frac{\mathrm{d}I \times \hat{r}}{r^2},$$

where the integral is along the conductor, $\mathrm{d}I$ is an element of current and μ_0 is the ➤permeability of free space. It is the general solution to the fourth of ➤Maxwell's equations in the same way that ➤Coulomb's law is the general solution to the first of Maxwell's equations. Named after J-B. Biot and Félix Savart (1791–1841).

bipolar junction transistor (bipolar transistor) A transistor based on two ➤p–n junctions, either as an n–p–n or a p–n–p triple layer, which constitute the **emitter, base** and **collector.** The main conventional current flow is from the collector to the emitter, and it depends not only on the potential difference applied from the collector to the emitter but also (more sensitively at the normal ➤operating point) on the current flowing through the base. Thus the bipolar transistor is typically pictured as a current amplifier.

birefringence (double refraction) The property of a material (typically a crystal) that exhibits different refractive indices for light polarized in different directions. Birefringent crystals are **uniaxial** (when two directions are equivalent) or **biaxial** (when all three directions are distinct). Each distinct direction is called an **optic axis.** When plane-polarized light passes through a birefringent crystal with its plane of polarization parallel to an optic axis, it remains plane-polarized. If, however, it consists of components along two distinct optic axes, a phase delay is introduced between the components and the light becomes elliptically polarized (or circularly polarized in certain circumstances). Another consequence of the difference in refractive indices is the appearance of two refracted rays, hence the alternative name. ➤polarization.

bis- A prefix indicating that two identical groups or ligands are present, as in bisbenzenechromium.

bisect To divide a line, angle, area, etc., into two equal parts.

bismuth Symbol Bi. The element with atomic number 83 and relative atomic mass 209.0, which is the element in Group 15 with the largest mass. It is the only metal in the group; the element immediately above, antimony, is a ➤metalloid. The metal is extracted by reduction of its oxide ore **bismite**, Bi_2O_3, with carbon. The metal is used to form alloys (with tin, lead and cadmium) which have low melting points, and so can be used to trigger sprinkler systems in the event of a fire. Bismuth compounds have two main oxidation numbers. The higher is +5, as for nitrogen, most commonly found in the bismuthate(v) ion. **Sodium bismuthate(v)**, $NaBiO_3$, is such a strong oxidant that it can be used to test for manganese in solution as it oxidizes the manganese(II) ion to the characteristic purple colour of the manganate(VII) ion, itself normally a powerful oxidant. The lower is +3, as in bismuth(III) chloride, $BiCl_3$, a white solid which is partially hydrolysed in solution, forming an equilibrium mixture with **bismuthyl chloride**, $BiOCl$.

bistable ➤flip-flop.

bit Acronym for binary digit. An n-bit ➤word can therefore represent the integers from 0 to $2^n - 1$.

bitmap A graphics display on a computer in which the screen display corresponds ➤pixel by pixel with bits held in the computer's memory.

bitumen ➤petroleum.

biuret test A chemical test to detect the presence of proteins in a solution. **Biuret reagent** consists of copper(II) sulfate, ➤Rochelle salt and sodium hydroxide; the test is based on the reaction between Cu^{2+} ions and two adjacent ➤peptide bonds. A violet colour is a positive test.

Bk Symbol for the element ➤berkelium.

black-body radiation Electromagnetic radiation that is emitted by an idealized object, a **black body**, that absorbs all incident radiation; the black body re-emits a spectrum dependent only on its surface temperature. By analogy with a gas in the same circumstances, black-body radiation can be characterized by an internal energy per unit volume (energy density) ε, ➤radiation pressure p and temperature T. The radiation pressure satisfies $p = \frac{1}{3}\varepsilon$, and from this one can derive the temperature dependence of the energy density as $\varepsilon \propto T^4$. **Wien's law** states that the wavelength λ_{max} at which there is maximum emission is given by $\lambda_{max} \propto T^{-1}$.

Max Planck subsequently deduced that the full spectrum of black-body radiation is given by ➤Planck's law of radiation. Albert Einstein proposed that monochromatic radiation consists of a number of independent energy quanta (now known as photons) of energy hf corresponding to oscillators of frequency f. We now identify these as bosons, and applying ➤Bose–Einstein statistics to their occupation of energy levels, with a ➤density of states proportional to f^2, yields Planck's law.

black hole A celestial body that is so dense that even photons inside its ➤event horizon are prevented from escaping by gravitational attraction towards its centre. ➤General relativity predicts that the radius to which a star must contract to form a black hole is given by $2GM/c^2$ (the **Schwartzschild radius**), which for the Sun is a mere 3 km. Although some unusual phenomena occur near black holes, they are not the cosmological all-devouring monsters portrayed in popular science fiction. ➤➤Hawking radiation.

black lead ➤graphite.

blast furnace The apparatus used to extract iron from its ores (see the diagram). The charge fed in at the top of the blast furnace consists of an iron ore (such as haematite, Fe_2O_3), limestone, $CaCO_3$, and coke, C. A blast of hot air at about $900\,°C$ is fed in at the bottom of the furnace through openings called **tuyères**. At the bottom of the

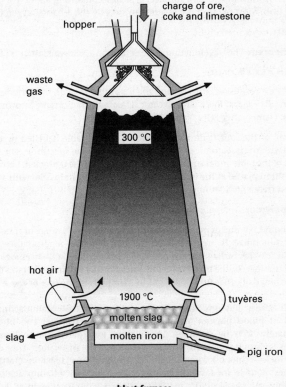

blast furnace

furnace, the temperature reaches about 1900 °C. At that temperature carbon can reduce haematite to iron, but this is not the most important reduction process: carbon monoxide is the dominant reductant. Carbon monoxide is formed by reduction by carbon of carbon dioxide, formed, for example, by decomposition of limestone:

$$CaCO_3(s) \rightarrow CaO(s) + CO_2(g),$$
$$CO_2(g) + C(s) \rightarrow 2CO(g).$$

The carbon monoxide can reduce the iron ore to iron at temperatures below 800 °C, which occurs higher up in the furnace:

$$Fe_2O_3(s) + 3CO(g) \rightarrow 2Fe(l) + 3CO_2(g).$$

Any acidic impurities in the iron ore, such as sand, are removed by reacting with calcium oxide (from the limestone's decomposition) to form ➤slag, which floats on the molten metal:

$$CaO(s) + SiO_2(s) \rightarrow CaSiO_3(l).$$

The molten iron from the blast furnace, typically containing about 4% carbon, is called pig iron. Most pig iron is converted to steel by the ➤basic oxygen process.

blind spot ➤eye.

Bloch's theorem The wavefunction ψ of electrons in a ➤Bravais lattice is of the form

$$\psi_k(r) = \exp(ik.r)u_k(r)$$

where k is the ➤reciprocal lattice vector, and u is a function with the periodicity of the lattice: $u(r + R) = u(r)$ if R is a lattice vector. The wavefunctions are known as **Bloch functions**. Named after Felix Bloch (1905–83).

block One of the four main areas in the ➤periodic table, labelled by the letter associated with the last atomic orbital to be filled (➤sodium spectrum). The **s block** is on the left of the table and the **p block** on the right, with the **d block** in between. The **f block** is often placed at the bottom of the table as it contains elements which are studied less frequently. ➤➤Appendix table 5.

block copolymer ➤copolymer.

blood A liquid ➤tissue in many animal groups, with a wide range of transport and metabolic functions. In many groups the blood carries a respiratory pigment responsible for the carriage of oxygen and carbon dioxide (➤haemoglobin); in others, such as insects, the blood carries little respiratory gas. In vertebrates the blood has a wide range of other functions, and in mammals forms part of the ➤immune system.

An adult human has approximately 5 to 6 litres of blood. Mammalian blood is composed of **blood plasma** (approximately 55% by volume) and **blood cells** (approximately 45% by volume). The plasma is a pale, straw-coloured fluid containing approximately 10% dissolved materials in water. The dissolved materials include blood proteins such as fibrinogen to assist in clotting, and ➤antibodies; soluble food substances such as ➤glucose, ➤amino acids and oil droplets; ions such as Na^+, Cl^- and HCO_3^-; waste substances such as ➤urea and carbon dioxide; and ➤hormones. **Blood serum** is blood plasma with the protein component removed. Also present may be traces of other substances derived from diet or lifestyle, such as alcohol or drugs. These and other dissolved substances can be detected by various diagnostic tests. Different types of blood cell are given in the table below.

In vertebrates, blood is pumped around the body by the ➤heart, which generates pressure to circulate blood to the body organs via an ➤ artery. Blood flows from the arteries into thin-walled, narrow vessels called ➤capillaries where exchange of materials with body organs takes place. Blood returns to the heart in a ➤vein.

blood Different types of blood cell.

Type	Number per mm^3	Function
Erythrocyte	5 000 000	Carriage of oxygen and carbon dioxide
Leucocytes	7000	Defence: actively seek and engulf bacteria,
Phagocyte		particularly in wounds
Lymphocyte	2500	Antibody-producing cells
Thrombocyte	250 000	Cell fragments which assist in blood clotting

blood–brain barrier A complex group of barriers and transport systems that control the types of substance entering the extracellular space of the brain from the blood.

blood clot An amorphous structure resulting from the precipitation of certain proteins in the blood. Blood clots form at the site of external wounds to prevent blood loss and the entry of ➤pathogens, but may also form internally where they may block blood vessels, causing a **thrombosis**. Blood clots are the cause of 'strokes' if they occur in the brain, and of heart attacks if they occur in the coronary arteries which supply the heart with oxygen. ➤➤fibrin.

blood groups A series of systems for classifying blood types based on a series of ➤antigen responses. The immunologist Karl Landsteiner was responsible for elucidating the **ABO system**, which is widely used for typing blood for transfusions and can also be used in forensic applications. The ABO system relies on the relationship between two antigens occurring on the surface of ➤erythrocytes and their ➤antibodies found in the plasma. The presence of an antigen on the surface of the erythrocytes is matched by an absence of the antibody to that antigen in the plasma. The possession of these antigens is genetically determined, and particular combinations produce four blood groups designated A, B, AB and O (group O is designated the 'universal donor'). The ➤alleles concerned are designated I_A, I_B and I_O.

The practical significance of blood groups is that only blood of a certain type can be transfused into the body to replace blood lost as a result of an accident or during surgery. Failure to match the blood groups in such circumstances can be fatal (see the table overleaf). The antigens are termed A and B, and the antibodies anti-A and anti-B. Thus an individual of blood group A has A antigens on the cells and lacks anti-A in the plasma, but possesses anti-B. If blood of type B were to be transfused into such an

blood groups The ABO system.

Genetic basis

Group	Genotype
A	$I_A I_A$, $I_A I_O$
B	$I_B I_B$, $I_B I_O$
AB	$I_A I_B$
O	$I_O I_O$

Antigen/antibody basis

Group	Antigen on erythrocytes	Antibody in plasma
A	A	anti-B
B	B	anti-A
AB	A and B	none
O	none	anti-A and anti-B

Compatibility of blood of different groups during transfusion:
Y = compatible (no agglutination); N = incompatible (agglutination).
Group O is referred to as the 'universal donor'.

	Donor group			
	A	B	AB	O
Recipient group				
A	Y	N	N	Y
B	N	Y	N	Y
AB	Y	Y	Y	Y
O	N	N	N	Y

individual, the anti-B antibodies would react with the B antigen on the donor cells to produce an agglutination response which causes the blood to clot.

blood pressure The pressure of blood in the vessels of an animal's vascular system, particularly in humans the pressure in the arteries. Blood pressure is highest at **systole**, normally around 120 mm Hg, and lowest at **diastole**, around 80 mm Hg (➤heart). For clinical and diagnostic purposes, these systolic and diastolic values are expressed in the form 120/80. The maintenance of an appropriate blood pressure is an example of homeostasis, and is achieved by the vasomotor centre of the autonomic ➤nervous system which controls cardiac output. Blood pressure depends on a number of factors including exercise, diet and lifestyle, and can be used diagnostically to identify individuals at risk from heart disease and strokes.

blooming The deposition on the surface of a lens of a layer of material that has a lower ➤refractive index than the lens, so as to minimize reflections.

blotting ➤DNA blotting; northern blotting; western blotting.

blue-green algae ➤cyanobacteria.

blue vitriol Traditional name for hydrated copper(II) sulfate, $CuSO_4 \cdot 5H_2O$.

B lymphocyte ➤B cell.

BMR Abbr. for ➤basal metabolic rate.

BOD Abbr. for ➤biological oxygen demand.

body-centred cubic (b.c.c.) A crystal structure (see the diagram) that has a lattice point at the centre of each cube, in addition to eight at the vertices. This structure is not as tightly packed as the close-packed structures (➤close packing). The Group 1 elements crystallize with the b.c.c. structure. Compare ➤caesium chloride structure.

Bohr frequency condition ➤electronic transition. Named after Niels Henrik David Bohr (1885–1962).

body-centered
cubic

bohrium Symbol Bh. The element with atomic number 107.

Bohr magneton ➤magneton; Appendix table 2.

Bohr model The earliest attempt at explaining the detailed structure of the atom; Bohr accepted Rutherford's model of a ➤nucleus at the centre of an atom and then tried to explain how the electrons in an atom were distributed around the nucleus. He considered that in the simplest picture the electrons would orbit the atom just as the planets orbit the Sun. In order to be in circular motion, any object must experience a centripetal force, which causes a centripetal acceleration. This centripetal force is provided by the electrostatic attraction between the positively charged nucleus and the negatively charged electrons. Bohr suggested that only certain stable **orbits** existed, which had integral values for the orbital angular momentum of the electron around the nucleus; that is, he proposed that the orbital angular momentum was quantized. He could use the model to explain the observed ➤hydrogen atom emission spectrum. The Bohr model was flawed because electrons are charged, whereas planets are not. By the laws of electromagnetism, any accelerating charged object should emit electromagnetic radiation. An orbiting electron in the Bohr model would therefore rapidly lose energy by radiation and spiral into the nucleus. To advance our knowledge of atoms beyond this simple model required a complete revolution in the way we interpret matter (➤de Broglie's equation; quantum theory; Schrödinger equation).

Bohr radius Symbol a_0. A convenient unit of length in atomic physics, equal to the radius of the lowest-energy orbit in the Bohr model of the hydrogen atom. In terms of ➤fundamental constants, $a_0 = \varepsilon_0 h^2 / \pi m_e e^2$, and is approximately 5.29×10^{-11} m. ➤➤Appendix table 2.

boiling point (boiling temperature) The temperature at which a liquid is in dynamic equilibrium with its vapour. At the boiling point, the liquid's ➤vapour pressure equals the atmospheric pressure. At the **normal boiling point** the atmospheric pressure is 1 atm. If the liquid is heated in an open vessel, its temperature will remain constant at the boiling point as all of the liquid evaporates to form a gas.

bolometer A device for detecting low intensities of radiation.

Boltzmann constant Symbol k or k_B. A ➤fundamental constant relating energy to temperature; its value is $1.38 \times 10^{-23}\,J\,K^{-1}$. It appears in the ➤Boltzmann distribution, and in Boltzmann's expression for ➤entropy. It also appears in conjunction with temperature in almost every application of ➤statistical mechanics. Named after Ludwig Edward Boltzmann (1844–1906). ➤➤Appendix table 2.

Boltzmann distribution (canonical distribution) The distribution giving the probability p_j of the occupation of energy level j (with energy E_j) for a system at temperature T (strictly, in thermal contact with a heat reservoir also at temperature T). It is given by

$$p_j = \frac{\exp(-E_j/kT)}{\sum_i \exp(-E_i/kT)}$$

where k is the ➤Boltzmann constant and the sum is over all states i. The numerator is often used as a weighting factor (the **Boltzmann factor**) for the occupation of energy levels. The denominator (the ➤partition function) need not necessarily be calculated explicitly as it simply ➤normalizes the distribution. The Boltzmann distribution is a key idea in statistical mechanics, and arises from maximizing the ➤entropy of the probability distribution with the constraint that the mean value of the energy is fixed. The concept of the Boltzmann distribution is slightly different from that of ➤Fermi–Dirac statistics and ➤Bose–Einstein statistics in that it deals with the energy of a system as a whole, not with the occupation numbers of individual levels.

bomb calorimeter An instrument used to measure the ➤standard enthalpy change of combustion of a substance by combusting it in oxygen at about 25 atm pressure in a strong, closed, thick-walled vessel immersed in a water-bath. The mass of sample burnt and the temperature rise in the calorimeter are measured. The calorimeter is calibrated using benzoic acid.

bond A link between two or more atoms. ➤ionic bonding; ➤covalent bonding; ➤hydrogen bonding.

bond angle The angle between two covalent bonds (➤covalent bonding). ➤➤valence-shell electron-pair repulsion theory.

bond dissociation energy The energy (usually measured per mole) required to break a specified bond.

bond energy The average energy (usually measured per mole) required to break a bond A—B. If the bond only occurs in a single molecule, as for the H—Cl or H—H bonds, the value is exact, and is equal to the ➤bond dissociation energy of the particular bond. When the bond can occur in a wide range of molecules, as for C—H, the average value may well be substantially different from the bond dissociation energy for a particular bond. Bond energies can, however, be used to make approximate calculations of the energy change in a particular reaction.

bond enthalpy Often used synonymously with 'bond energy'. The subtle difference is explained under ➤enthalpy.

bonding orbital ➤molecular orbital.

bond length The length of a particular covalent bond. Typical lengths are about 0.1–0.2 nm.

bond pair ➤valence-shell electron-pair repulsion theory.

bone A composite tissue in vertebrates, consisting of protein fibres (➤collagen), bone cells (osteocytes and osteoblasts) and salts, including calcium phosphate. The various bones comprise the vertebrate ➤skeleton.

bone marrow A vascular, cellular substance in the central cavity of some bones. Bone marrow is of two types. **Red bone marrow** is found in vertebrae, ribs and the bones of the pelvic girdle, and contains myeloid tissue responsible for the formation of several types of cell in the blood that function in the immune system. Red bone marrow has been a target for transplants to cure some forms of leukaemia and for genetic manipulation in ➤gene therapy. In mature animals the marrow of the long bones ceases to produce blood cells and becomes **white bone marrow**, associated with fat metabolism.

Boolean algebra An algebraic system in which elements (with values true or false) are combined by operations ∧ (AND), ∨ (OR) and ´ (NOT) subject to certain natural axioms. Although originally devised by George Boole (1815–64) for the analysis of logical arguments, it has found extensive use in the design of the logic circuitry used in computers.

borane One of a collection of binary compounds of boron and hydrogen, which form interesting structures. The simplest is **diborane**, B_2H_6, which shows one key feature of the collection, the presence of delocalized ➤three-centre bonds. More than two dozen larger boranes are known with shapes ranging from delicate spiders' webs to untidy birds' nests. ➤➤hydroboration.

borax (disodium tetraborate) An important ore of boron which occurs in large quantities in California, Tibet and Turkey. Its formula is best written as $Na_2[B_4O_5(OH)_4]\cdot 8H_2O$. It is used in the **borax bead test**, in which a crystal of borax is melted to a bead in a platinum loop, placed in the substance to be identified and heated in a flame. The colour of the bead is characteristic of the metal: chromium, for example, turns green, and cobalt turns blue.

boric acid (boracic acid) H_3BO_3 or $B(OH)_3$ A white crystalline solid used as a mild antiseptic and in large quantities to make borosilicate glass. It is a very weak acid, with its acidity arising from an unusual equilibrium:

$$B(OH)_3(aq) + 2H_2O(l) \rightleftharpoons H_3O^+(aq) + [B(OH)_4]^-(aq).$$

boride A binary compound of boron with a less electronegative element. An example is the refractory titanium boride, TiB_2.

Born–Haber cycle A particularly important thermodynamic cycle that calculates the standard enthalpy change for the synthesis of an ionic compound in terms of a series of five simple steps. Consider the example of sodium chloride (see the diagram). Adding together the atomization enthalpy of sodium, the ionization enthalpy of sodium, the atomization enthalpy of chlorine, the electron-gain enthalpy of chlorine and the lattice formation enthalpy of sodium chloride gives the standard enthalpy change of formation of sodium chloride. Named after Max Born (1882–1970) and Fritz Haber (1868–1934).

Born interpretation The square of the ➤modulus of the ➤wavefunction gives the probability density of finding an electron in an infinitesimal volume. The Born interpretation was historically important as it allowed a physical picture to be developed from the ➤Schrödinger equation. Named after M. Born.

boron Symbol B. The element with atomic number 5 and relative atomic mass 10.81, which is the element in Group 13 with the lowest atomic number. Unusually, the relative atomic mass of commercial samples is slowly increasing with time as the nuclear industry extracts the isotope boron-10 for its neutron absorption properties. The element itself is a very high melting solid with several allotropes all of which have a giant covalent structure, the central motif of which is the B_{12} icosahedron (see the diagram). It is a nonmetal, in clear distinction to aluminium, the element immediately below it in Group 13. Boric oxide, B_2O_3, is a transparent crystalline solid used to make ➤borosilicate glass. The ➤boron trihalides are important Lewis acids. Boron carbide, $B_{13}C_2$, is very hard and is used to make bullet-proof vests and armour plating.

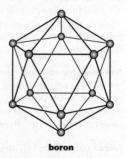

Born-Haber cycle for sodium chloride.

Diagram labels:
$Na^+(g) + e^-(g) + Cl(g)$
121
$Na^+(g) + e^-(g) + \frac{1}{2}Cl_2(g)$
498   351
$Na^+(g) + Cl^-(g)$
$Na(g) + \frac{1}{2}Cl_2(g)$
108
$Na(s) + \frac{1}{2}Cl_2(g)$
411   787
$NaCl(s)$

boron

boron trihalides Compounds, such as BF_3 and BCl_3, which are covalently bonded: BF_3 is a colourless gas at room temperature whereas BCl_3 is a colourless liquid, reflecting the higher ➤dispersion forces caused by having more electrons in the molecule. They are used frequently as ➤Lewis acids.

borosilicate glass A ➤glass, with a giant covalent structure containing silicate (SiO_4) and borate (BO_3) units linked together, which can withstand changes in temperature much more easily than other glasses and which is therefore much used for laboratory glassware.

Bose condensation (Bose–Einstein condensation) A state in which (below a threshold temperature) the ground state of a system of bosons has nonzero

➤occupation number and the ➤chemical potential is zero. Under such conditions, several thousand atoms 'condense' to form a single entity. Although predicted in the early years of ➤quantum theory, it was observed experimentally for the first time in 1995 in a gas of rubidium atoms at 0.17 μK. Bose condensation, which has been called the fifth state of matter (the fourth being plasma), is important in the theory of ➤superfluids. Named after Satyendra Nath Bose (1894–1974) and Albert Einstein (1879–1955).

Bose–Einstein statistics The expression for the ➤occupation numbers n_i of a set of quantum states with energies ε_i occupied by ➤bosons:

$$n_i = \frac{1}{\exp[(\varepsilon_i - \mu)/kT] - 1}$$

where k is the Boltzmann constant, T is the thermodynamic temperature and μ is the ➤chemical potential (always negative). Compare ➤Fermi–Dirac statistics. The idea was put forward initially by S. N. Bose, who, fearing that his paper might be rejected, sent it to A. Einstein for checking.

boson A particle with integral ➤spin. Examples include the ➤photon, the ➤phonon, ➤gluons and the W and Z bosons. ➤➤Bose–Einstein statistics; identical particles. Named after S. N. Bose. Compare ➤fermion ➤➤ standard model.

botany The scientific study of plants. Botany is a very wide subject encompassing a range of disciplines such as genetics, physiology, ecology, pathology and biochemistry. In recognition of this, the synonymous term 'plant sciences' is now commonly used by universities and other academic institutions.

bottom quark The fifth ➤quark of the ➤standard model to be discovered, in the form of the ➤upsilon particle.

botulism One of a variety of forms of poisoning caused by a toxin from the bacterium *Clostridium botulinum*. All forms of botulism are potentially fatal and are treated as emergencies. Food-borne botulism can occur when inappropriately processed food is sealed in the absence of oxygen, such as in canning, since the bacterial cells proliferate under anaerobic conditions.

boundary conditions Conditions imposed on the solution of a ➤differential equation which are sufficient to determine the arbitrary constants that occur in the ➤general solution.

boundary layer The layer of a flowing fluid closest to a boundary, such as the surface of an ➤aerofoil. It behaves differently from the fluid further away from the boundary because of forces between the boundary and the fluid, and its thickness is proportional to the ➤kinematic viscosity of the fluid.

bound state A ➤quantum state (usually of an electron within an atom or ion) with less energy than the potential energy at infinity. Bound states have discrete ➤energy levels. Compare ➤free electron.

Bourdon gauge A simple pressure gauge that uses the distortion of a tube sealed at one end to sense the pressure applied to the other. Named after Eugène Bourdon (1808–84).

bovine somatotrophin ➤growth hormone.

bovine spongiform encephalopathy ➤BSE.

Boyle's law For a fixed mass of gas at constant temperature, the volume of the gas is inversely proportional to the pressure. This is only strictly true for ideal gases (it can be derived from the ➤ideal gas equation), but it remains a close approximation for real gases. It is one of the earliest scientific laws, and remains useful to this day. Named after Robert Boyle (1627–91).

Br Symbol for the element ➤bromine.

Brackett series ➤Rydberg–Ritz equation. Named after Frederick Sumner Brackett (1896–1980).

Bragg's law Reflection of X-rays from a ➤lattice plane in a crystal occurs at a **Bragg angle** θ (the angle between the plane and the incident beam of X-rays) satisfying the equation

$$n\lambda = 2d \sin \theta$$

where d is the distance between successive lattice planes, λ is the wavelength of the X-rays, and n is an integer (the **order of reflection**). Named after (William) Lawrence Bragg (1890–1971).

brain The enlarged anterior part of the ➤nervous system in vertebrates; it is continuous with the spinal cord. The brain has distinct regions and layers (see the diagram), each associated with the reception and processing of specific stimuli received from sense organs. The brain coordinates such information and generates ➤nerve impulses in appropriate neurones to produce appropriate responses. The

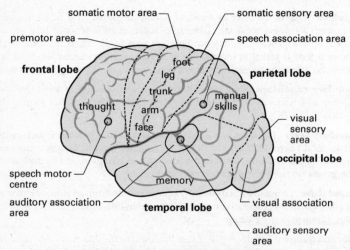

brain Anatomical divisions of the human left cerebral hemisphere (indicated in bold) and areas associated with particular functions.

brain is also responsible for memory, and in a human has areas which seem to control aesthetic and artistic activity and logical thought. The term 'brain' is also applied to the concentrations of nervous tissue at the anterior end of animals of other groups such as earthworms. The human brain is divided into three main areas: the **forebrain (prosencephalon)**, **midbrain (mesencephalon)** and **hindbrain (rhombencephalon)**. The forebrain consists of an **olfactory lobe** (providing the sense of smell), the ➤**cerebrum**, which is the centre for memory, hearing, sight, speech, voluntary muscle action, thinking and reasoning (the largest part of the brain in humans, see the diagram), and the **hypothalamus**, which contains receptors associated with ➤homeostasis and with the ➤pituitary gland. The outer layer of the brain consists mainly of the axons of nerve cells (**white matter**), whereas the inner layers contain the cell bodies (**grey matter**). However, in the cerebrum of more advanced vertebrates the grey matter is found at the surface. The midbrain is mainly an optic centre. The hindbrain consists of the **cerebellum**, which is the centre for coordination of balance and integration of muscle action, and the **medulla oblongata** (brain stem), which merges with the spinal cord and provides the centre of reflexes involving the ➤eye, breathing and heart rate. The dysfunction of certain reflexes seated here is the basis of the establishment of ➤brain death.

brain death A clinical diagnosis based on the functioning of certain ➤reflexes known to be located in the base of the brain, the brain stem: if these reflexes can repeatedly be shown to be non-functional over a given period of time, the brain is considered to be dead. The determination of brain death is critical in the definition of death in, for example, accident victims whose organs may be used in transplants.

branched-chain molecule A molecule in which the main chain has at least one side chain off it. An example is methylpropane, an isomer of the straight-chain molecule butane.

brass An alloy of copper (between 55 and 90%) with zinc and small quantities of other metals.

Bravais lattice An infinite regular array of lattice points arranged so that the array looks the same when viewed in the same direction from any lattice point. The name is also given to the set of lattice vectors joining these points, that is vectors R such that $R = n_1 a_1 + n_2 a_2 + n_3 a_3$ for any combination of integers n_1, n_2 and n_3, where a_1, a_2 and a_3 are the **primitive vectors** of the lattice. There are only fourteen distinct symmetry

---

Bravais lattice The symmetry of the different Bravais lattices can be expressed in terms of the sides and angles in the unit cell. The three sides are a, b and c. The angle α is that between sides b and c; similarly for the angles β and γ.

Cubic	$a = b = c$	$\alpha = \beta = \gamma = 90°$
Tetragonal	$a = b$, c different	$\alpha = \beta = \gamma = 90°$
Orthorhombic	a, b, c all different	$\alpha = \beta = \gamma = 90°$
Rhombohedral	$a = b = c$	$\alpha = \beta = \gamma$ but not equal to $90°$
Hexagonal	$a = b$, c different	$\alpha = \beta = 90°$, $\gamma = 120°$
Monoclinic	a, b, c all different	$\alpha = \gamma = 90°$, β different
Triclinic	a, b, c all different	α, β, γ all different and not equal to $90°$

Cubic

Tetragonal

Orthorhombic

Rhombohedral

Hexagonal

Monoclinic

Triclinic

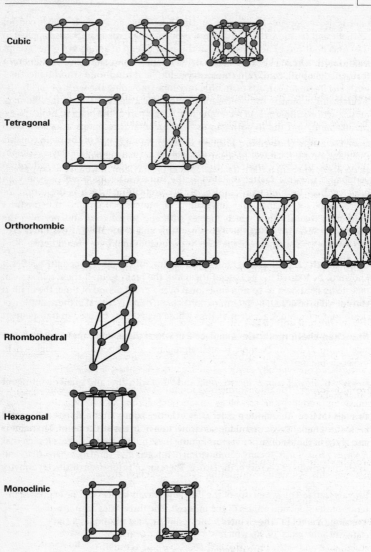

➤space groups that a Bravais lattice can have, and the lattices are characterized in this way, also being divided into seven **crystal systems** by their ➤point group (see the diagram opposite). Named after Auguste Bravais (1811–63).

breakdown A sudden change in a material from an ➤insulator to a ➤conductor. ➤➤avalanche breakdown; Zener breakdown.

Bremsstrahlung The radiation emitted by a charged particle as it is slowed down (it accelerates, albeit negatively, and must therefore radiate), usually by an electric field. The name comes from the German for 'braking radiation'.

Brewster angle (polarizing angle) The angle of reflection θ from a surface at which elliptically polarized incident light becomes plane-polarized. It is given by $\tan \theta = n_2/n_1$, where n_1 is the refractive index of the medium in which the ray travels and n_2 the refractive index of the medium on the other side of the surface. The reflection coefficient for light polarized in the incident plane is zero at this angle. Named after David Brewster (1781–1868).

bridge **1** (Phys.) The generic name for a circuit, such as a ➤Wheatstone bridge, in which a quantity is measured by detecting a null current between two arms of a circuit.

   2 (Comput.) A connection between two computer networks. The result appears to the user as a larger, single network, in which the bridge is not apparent.

brine ➤chlor-alkali industry.

British thermal unit (BTU) An obsolete unit of energy, equal to the heat required to increase the temperature of one pound of water by one degree Fahrenheit.

brittle Describing a material that can be fractured without appreciable deformation. Ionic crystals are typically brittle, as are ceramics. Compare ➤malleable.

bromide ion Br^- The anion formed by bromine. It is found in solid bromides such as potassium bromide, KBr, as well as in aqueous solution. The test for the aqueous bromide ion is to add acidified silver nitrate and look for the appearance of a cream precipitate, AgBr. Silver bromide was heavily used as the photosensitive layer in photographic emulsions.

bromination A reaction in which bromine replaces hydrogen in a molecule. A useful **brominating agent** for carbon atoms adjacent to carbon–carbon double bonds is *N*-bromosuccinimide (see the diagram).

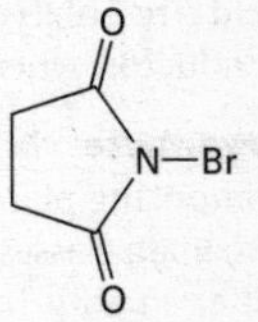

bromination The brominating agent *N*-bromosuccinimide.

bromine Symbol Br. The element with atomic number 35 and relative atomic mass 79.90, which is a ➤halogen. It exists as a red-brown liquid with a brown toxic vapour. The liquid contains Br_2 molecules, formed by ➤covalent bonding between two bromine atoms. The most important source of bromine is seawater, which contains bromide ions, Br^-, at a sufficiently high concentration for their commercial extraction by reaction with chlorine. In this reaction, chlorine is showing its greater strength as an oxidant. Bromine, in turn, is a stronger oxidant than iodine and can displace iodine from the iodide ion. In inorganic chemistry bromine occurs most often as ➤bromide ion, although some covalent compounds do exist, such as phosphorus tribromide, PBr_3. Organic compounds containing bromine are common, such as bromoform (tribromomethane, $CHBr_3$), the fire extinguisher ➤BCF and the refrigerant bromoethane, CH_3CH_2Br. ➤➤Grignard reaction.

bromothymol blue An ➤indicator which changes colour close to neutral ➤pH.

bronchus (plural **bronchi**) ➤lung.

Brønsted acid A chemical species that is a proton donor. ➤➤conjugate base. Named after Johannes Nicolaus Brønsted (1879–1947).

Brønsted base A chemical species that is a proton acceptor. ➤➤conjugate acid.

Brønsted–Lowry theory (of acids and bases) A theory that describes acids and bases in terms of the transfer of a proton from the acid to the base. This is the traditional view, which can be further elaborated by calculation of the ➤pH of the resulting solution, which depends on the concentration of the oxonium ion. Common acids such as hydrochloric acid, nitric acid and sulfuric acid are all Brønsted acids. The theory has the disadvantage that it cannot be applied to any reaction in which protons are not involved. For these the ➤Lewis theory offers a useful alternative description of the species. Named after J. N. Brønsted and (Thomas) Martin Lowry (1874–1936). ➤➤acid; base.

bronze An alloy of copper and tin together with small quantities of other metals.

brown dwarf A body contracted directly from the ➤interstellar medium (like a ➤star) but too small for the normal nuclear reactions that occur within a star to be initiated. In a star with less than 8% of the Sun's mass, gravitational compression does not raise the temperature high enough to allow full hydrogen burning.

Brownian motion The rapid random motion of particles, such as smoke particles in a gas, as they are given random impulses by the smaller gas molecules. Such motion is an example of a ➤stochastic process. Named after Robert Brown (1773–1858).

brown-ring test ➤nitrate ion.

brush discharge A luminous discharge from a conductor in air when the electric field is not high enough to produce a spark. It tends to occur around pointed parts of a conductor, where the field is locally higher than elsewhere.

bryophyte A member of the phylum **Bryophyta** within the plant kingdom, comprising plants that lack a ➤vascular system and reproduce by spores. There are two main classes: the mosses (class Musci) and the liverworts (class Hepaticae), which differ mainly in the details of how the spore-bearing structure develops. Most bryophytes prefer damp shady conditions, but they are also often the dominant plants on damp moorlands, and some species are highly drought-tolerant.

BSE Abbr. for bovine spongiform encephalopathy, a neurological and ultimately fatal disease of cattle (commonly called 'mad cow disease'). The symptoms include uncoordinated muscle control and other behavioural changes associated with lesions in the brain. BSE was first identified in 1986 and is one of a number of ➤prion diseases that occur in several mammals, including humans. Currently interest lies in increasing evidence of a link between BSE and a variant of ➤Creutzfeldt–Jakob disease in humans through contaminated meat products in the food chain.

BTU Abbr. for ➤British thermal unit.

bubble cap ➤fractional distillation.

bubble chamber A device for detecting ➤ionizing radiation by observing the bubbles formed when the liquid in the chamber (typically hydrogen) is vaporized by the radiation. Compare ➤cloud chamber.

bubblejet (inkjet) A form of computer printing technology that relies upon heating a bubble of ink within a nozzle. Boiling of the ink increases the pressure in the nozzle, and fires the bubble onto the paper. Bubblejet technology made available a cheaper alternative to laser printers.

Buchner funnel A type of filter funnel, often made from porcelain, that has a circular base with perforated holes through which air can be sucked to speed up the filtration. Named after Eduard Buchner (1860–1917).

buckminsterfullerene The third allotrope of carbon (➤allotropy), with the formula C_{60}, which was first isolated as a solid (**fullerite**) in 1990. The other two allotropes of carbon, diamond and graphite, have been known for millennia. For their discovery of buckminsterfullerene, Harry Kroto, Bob Curl and Rick Smalley won the 1996 Nobel prize for Chemistry. It is named after (Richard) Buckminster Fuller (1895–1983), whose geodesic sphere has a form similar to that of the molecule. The polyhedral 'cage' structure is reminiscent of a soccer ball, and has gained the nickname '**buckyball**'.

buffer A temporary memory store in a computer or ➤peripheral, used to increase the efficiency of data transfer. If the peripheral cannot accept data as quickly as the computer provides them, the data are placed in a buffer so that, providing the buffer is not full, the computer need not wait for the peripheral to catch up.

buffer solution A solution that resists changes in ➤pH. Buffer solutions can occur either at acidic or basic pH. An **acidic buffer solution** consists of a solution of a weak acid and its conjugate base, for example ethanoic acid and sodium ethanoate. A **basic buffer solution** consists of a solution of a weak base and its conjugate acid, for example ammonia and ammonium chloride.

Buffer action depends on the fact that the pH varies slowly when a small quantity of ➤strong acid or ➤strong base is added to the solution. To see this qualitatively, consider the equilibrium occurring for a weak acid:

$$HA(aq) + H_2O(l) \rightleftharpoons H_3O^+(aq) + A^-(aq).$$

By ➤Le Chatelier's principle, adding hydrogen ions from a strong acid will shift the equilibrium to the left, thus removing A^- ions and forming HA. The ➤Henderson–Hasselbalch equation explains buffer action quantitatively.

Buffer solutions are used in experiments where a nearly constant pH is required, such as those involving ➤electrophoresis. They are also used to calibrate pH meters. Buffer solutions are particularly important in biochemistry, where the two most common buffer systems are the carbonate buffer H_2CO_3/HCO_3^- and the phosphate buffer $H_2PO_4^-/HPO_4^{2-}$. For example, blood must be maintained around pH 7.4, whereas saliva must be maintained around pH 6.6.

bug A local error in a computer program, as distinct from an error in system design.

building-up principle ➤aufbau principle.

bulk density The mass per unit volume occupied of a material that is not perfectly homogeneous (like a powder) because it consists of lumps with air in between. Bulk density includes the volume of the air gaps in the reckoning.

bulk modulus Symbol K. The ratio of the pressure applied to a material to the resultant fractional change in volume it undergoes. ➤➤compressibility.

bumping Violent boiling. It can be reduced by **anti-bumping granules**, such as pieces of porous pot, which provide surfaces on which liquid can vaporize.

Bunsen burner The traditional burner used in laboratories that combusts a gas (usually methane) in a controllable air supply. Named after Robert Wilhelm Eberhard Bunsen (1811–99).

buoyancy The force on a body in a fluid caused by the pressure difference between the bottom of the body and the top of the body. ➤Archimedes' principle.

burette (US **buret**) ➤titration.

Burkitt's lymphoma A type of cancer believed to be caused by a herpes-like virus, the Epstein–Barr virus (EBV). The cancer cells of most of those affected contain a specific translocation between chromosomes 8 and 14. Named after Denis Parsons Burkitt (1911–93).

bus In a computer, a bus is a parallel connection between two parts of the system. A full-width bus can carry a complete ➤word at once, as well as control signals. This allows much faster transfer of data around the system than a serial connection, which can only carry one bit at a time.

busbar A solid rod or rail designed as a conductor for high currents.

butadiene (buta-1,3-diene) $CH_2{=}CHCH{=}CH_2$ One of the most important ➤dienes, a derivative of which (➤isoprene) is the monomer of rubber. It is a colourless gas used for the production of the copolymers ➤ABS and ➤SBR. ➤➤Diels–Alder reaction.

butanal $CH_3CH_2CH_2CHO$ The straight-chain ➤aldehyde with four carbon atoms, which exists as a colourless liquid.

butane $CH_3CH_2CH_2CH_3$ The straight-chain alkane with four carbon atoms, which exists as a colourless gas, and which can be readily compressed to form **Butagas**. It is also a component of Calor gas used for cooking. Compare ➤methylpropane.

butanedioic acid (succinic acid) $HOOCCH_2CH_2COOH$ The dicarboxylic acid with four carbon atoms. It is a colourless crystalline solid. ➤➤Krebs cycle.

butanoic acid (butyric acid) $CH_3CH_2CH_2COOH$ The carboxylic acid with four carbon atoms. It is a colourless liquid which smells of rancid butter, whereas butanoate esters are used for flavourings.

butanol $C_4H_{10}O$ One of two isomeric alcohols. **Butan-1-ol** has the formula $CH_3CH_2CH_2CH_2OH$ whereas **butan-2-ol** is $CH_3CH_2CH(OH)CH_3$. Both are colourless liquids used as solvents. Their oxidation products are different. ➤alcohols.

butanone $CH_3CH_2COCH_3$ The ➤ketone with four carbon atoms, which exists as a colourless liquid used as a solvent.

butene C_4H_8 One of two isomeric alkenes, both colourless gases. **But-1-ene** is $CH_2=CHCH_2CH_3$ whereas **but-2-ene** is $CH_3CH=CHCH_3$ (which has ➤*cis* and *trans* isomers).

butenedioic acid $HOOCCH=CHCOOH$ This molecule has two isomers, both of which are colourless crystalline solids. The *cis* isomer was called **maleic acid** and the *trans* isomer was called **fumaric acid**. The latter is involved in the ➤Krebs cycle.

butyl group C_4H_9— In older literature this had a prefix specifying which of four isomers was described: *n-* was $CH_3CH_2CH_2CH_2$—, *sec-* was $CH_3CH_2CH(CH_3)$—, *iso-* was $(CH_3)_2CHCH_2$— and *tert-* was $(CH_3)_3C$—.

butyl rubber The generic name for copolymers of methylpropene that can undergo ➤vulcanization. It is resistant to atmospheric ozone and so can be used out of doors.

butyric acid ➤butanoic acid.

bypass capacitor A capacitor placed in a circuit in parallel with a component, or set of components, to act as a ➤shunt to allow direct current to reach the component while bypassing the alternating current through the capacitor itself.

byte An 8-bit ➤word, often the basic unit of storage in a digital computer. ➤➤bit.

C

c **1** Symbol for the prefix centi-.
2 (Chem.) Symbol for the crystalline state.

c **1** Symbol for ➤concentration.
2 Symbol for the ➤speed of light in a vacuum.

C **1** Symbol for the element ➤carbon.
2 Symbol for the unit ➤coulomb.
3 (Biol.) Symbol for the pyrimidine base ➤cytosine.
4 (Comput.) A high-level computer language, which nevertheless has some elements of similarity to ➤assembly language. C++ is a more modern, ➤object-oriented version which concentrates on modular program design.

C **1** Symbol for ➤capacitance.
2 Symbol for ➤heat capacity.

ℂ Symbol for the set of ➤complex numbers.

Ca Symbol for the element ➤calcium.

cache High-speed memory which acts as a ➤buffer between the ➤CPU and the slower main memory of a computer. The aim is to predict, on the basis of previous storage and retrieval operations, which parts of main memory a program will access, and to transfer these parts into the cache so that they may be accessed faster.

cadmium Symbol Cd. The element with atomic number 48 and relative atomic mass 112.4, which lies immediately below zinc in Group 12 and hence in the last column of the ➤d-block elements. Like zinc, it does not share the common characteristics of the ➤transition metals and should not be regarded as one. The element itself is a white lustrous toxic metal, which is reasonably reactive in water. Its major use is in ➤nickel–cadmium batteries. It is also used in control rods for nuclear reactors as its nucleus absorbs neutrons well. The most important oxidation number for cadmium is +2, as in the oxide CdO, which is a solid, whose colour ranges from red to black because of lattice defects. Both **cadmium chloride**, $CdCl_2$, and **cadmium iodide**, CdI_2, are white crystalline solids which have (different) layer structures, which are characteristic of many other layered solids. The sulfide CdS is a natural pigment called **cadmium yellow**.

caesium (US **cesium**) Symbol Cs. The element with atomic number 55 and relative atomic mass 132.9, which is in Group 1. Like sodium and potassium, it is a vigorously reactive reductant. The trend down the group is to greater reactivity, so caesium is even more reactive than potassium. When placed in a glass vessel full of water, its reaction with water is so violent that the glass vessel shatters. Its high reactivity is put to good use as a ➤getter. In common with the other elements of Group 1, the oxidation number in all its compounds is +1, as in caesium chloride CsCl (➤caesium chloride structure). The oxide, Cs_2O, is an orange crystalline solid. All caesium salts are soluble so a ➤flame test is needed to detect the **caesium ion**, Cs^+, which gives a blue flame.

caesium chloride structure A crystal structure (see the diagram) adopted by caesium bromide and caesium iodide as well as the chloride. The structure consists of two interpenetrating simple cubic arrays, which contrasts with the much more important ➤rock-salt structure. This structure is *not* correctly described as body-centred cubic because the ion in the centre of the cube is of the opposite charge to the ones at the edges of the cube.

Cs^+ ⬤ Cl^- ◯

caesium chloride structure

caesium clock ➤atomic clock.

caffeine ➤alkaloid.

cage compound ➤clathrate.

Cahn–Ingold–Prelog system ➤(R–S) system. Named after Robert Sidney Cahn (1899–1981), Christopher Kelk Ingold (1893–1970), and Vladimir Prelog (1906–98).

calcite A crystalline polymorph of calcium carbonate, $CaCO_3$. Calcite shows ➤birefringence. ➤➤aragonite.

calcium Symbol Ca. The element with atomic number 20 and relative atomic mass 40.08, which is in Group 2. It is one of the two most important elements in the group (with magnesium). First, in biochemistry calcium is well known to be essential for bones and teeth (➤apatite). Second, many of its compounds (➤calcium carbonate; calcium oxide) are important in nature. Third, several of its compounds are used in huge quantities by industry (➤calcium carbonate). The element itself is a white solid which is a vigorous reductant, and hence tarnishes rapidly. It reacts with water to form calcium hydroxide, $Ca(OH)_2$.

Its only oxidation number in compounds is +2, as in calcium oxide, CaO, or calcium chloride, $CaCl_2$. Both calcium fluoride, CaF_2, and calcium titanate, $CaTiO_3$, have important crystal structures (➤fluorite structure; ➤perovskite structure). The calcium ion, Ca^{2+}, is a major cause of ➤hard water.

calcium carbide ➤calcium dicarbide.

calcium carbonate $CaCO_3$ A white solid which is a very important compound both in nature and in industry. It exists in two crystalline forms, calcite and aragonite. It is the main constituent of a huge range of structures in nature, as in **chalk**, ➤limestone hills and coral reefs. It is also the main constituent of **marble**. Calcium carbonate is used in industry in the ➤blast furnace and in the ➤Solvay process, both of these uses depending on its thermal decomposition to calcium oxide and carbon dioxide:

$$CaCO_3(s) \rightarrow CaO(s) + CO_2(g).$$

calcium chloride $CaCl_2$ A white solid which is used in the laboratory as a drying agent, especially for ammonia (where the more common concentrated sulfuric acid would react with the gas rather than dry it). The molten salt is electrolysed to extract calcium metal.

calcium dicarbide (calcium carbide) CaC_2 A solid, containing the **dicarbide ion** C_2^{2-}, produced commercially by reaction between coke and calcium oxide at high temperatures. It reacts with water to form ➤ethyne, C_2H_2. This pair of reactions is potentially a bridge between inorganic chemicals and organic ones and they may come into their own in the future, when readily available sources of organic compounds, such as natural gas and crude oil, are exhausted. Calcium dicarbide can also react with nitrogen and so provide artificial nitrogen fixation.

calcium hydrogencarbonate (calcium bicarbonate)   ➤hardness of water.

calcium hydroxide $Ca(OH)_2$ The white powder formed when an equivalent quantity of water is added to calcium oxide is called ➤slaked lime; when more water has been added, the colourless solution is called ➤limewater. Slaked lime is used in agriculture for raising the pH of acidic soils. Limewater is used to test for ➤carbon dioxide.

calcium oxide (lime, quicklime)   CaO A white crystalline solid (with the ➤rock-salt structure) made in huge quantities in industry (by thermal decomposition of ➤calcium carbonate), where it is the most common base. A slurry of lime can be used to react with sulfur dioxide to desulfurize the waste gas from a coal-powered power station.

calcium sulfate $CaSO_4$ A common ore of calcium occurring both as the anhydrous salt, **anhydrite**, and the hydrated salt, $CaSO_4 \cdot 2H_2O$ in ➤gypsum and ➤**alabaster**. ➤➤hardness of water.

calculus   Arguably the most important branch of mathematics, developed independently by Isaac Newton (1642–1727) and Gottfried Wilhelm Leibniz (1646–1716) and consisting of two fundamental processes, ➤differentiation and ➤integration, which gives rise to **differential calculus** and **integral calculus**. Differentiation is concerned with finding the instantaneous ➤rates of change of ➤functions and hence the ➤gradients of ➤tangents (2) to graphs. Integration deals with the reverse process of finding functions from their rates of change and is used to find ➤areas, ➤volumes, ➤arc lengths, ➤centres of mass, ➤moments of inertia, etc. It is only a slight exaggeration to say that modern physical science would be inconceivable without calculus: in any situation dealing with change, differentiation in the form of ➤differential equations is the tool *par excellence* to summarize the process of change, and integration the tool to convert 'rate' information about a quantity to information about its actual magnitude.

calculus of variations   A branch of mathematics concerned with finding ➤functions that minimize or maximize certain quantities associated with them. A classic early problem was the brachistochrone problem: find the curve between two points in a vertical plane along which a particle would move freely under gravity in the shortest possible time. Here, typically, the time taken for a curve $y = f(x)$, with $y' = df(x)/dx$ is given by an integral of the form $\int_a^b F(x, y, y')\,dx$ and the minimizing curve (in this case a ➤cycloid) satisfies **Euler's equation** $\partial F/\partial y = d/dx(\partial F/\partial y')$.

calibration The process of setting up a measuring instrument to correspond with a standard. Typically, a value is associated with a number of arbitrary graduations on the scale of the instrument, and readings between these graduations are interpolated.

californium Symbol Cf. The element with atomic number 98 and most stable isotope 251, which is an ➤actinide. As such, its chemistry is dominated by its radioactivity. Californium is the heaviest element available in gram quantities. Its compounds normally have ➤oxidation number +3, as in the light green bromide, $CfBr_3$, but there are some compounds with oxidation number +2, as in $CfBr_2$.

callus An undifferentiated mass of dividing cells, particularly in plant tissue culture. The term is also applied to the 'wound tissue' produced when woody plants are damaged, as when a branch falls from a tree, and in medicine to describe an early stage in the formation of bone regenerating after a fracture.

calomel electrode (calomel half-cell) A standard ➤half-cell containing **calomel** (mercury(I) chloride, Hg_2Cl_2). It is frequently used for measuring standard electrode potentials, as the half-cell with zero potential (by definition), the ➤standard hydrogen electrode, is awkward to use. The **saturated calomel electrode** has a standard electrode potential of +0.246 V at 25 °C.

calorie Symbol cal. A unit of energy in the ➤c.g.s. system of units, equal to the heat required (about 4.2 J) to increase the temperature of one gram of water by one degree Celsius. Physiologists use the name **Calorie** (with a capital 'C') for what should be the kilocalorie (1000 cal) when referring to the energy content of food.

calorimeter An instrument for measuring the heat added to a system, usually as a result of a chemical reaction. ➤➤bomb calorimeter.

Calvin cycle ➤dark reaction. Named after Melvin Calvin (1911–97).

cambium (plural **cambia)** A ➤meristem giving rise to parallel rows or plates of cells in plants. The **vascular cambium** gives rise to subsequent annual layers of ➤xylem and ➤phloem, and forms the characteristic rings found in the wood of trees.

Cambrian See table at ➤era.

camera A device used in photography for recording images. For still images, it consists of a lens system, a photographic film (or alternative means of recording electronically) and a system for controlling the exposure of the film by timing the movement of a shutter. For moving images, the device must either record onto moving film (a **cine camera**) or use an electronic means of encoding and storing the images (a **television camera**, **digital camera** or **video camera**).

camphor A white crystalline solid (see the diagram) with a characteristic smell used in the manufacture of celluloid (➤cellulose nitrate).

camphor

Canada balsam A yellowish liquid with a ➤refractive index close to that of glass, and hence used for mounting specimens for examination under a microscope.

cancer A growth caused by uncontrolled or abnormal cell division, or the resulting condition experienced by the host. Cancer cells grow in an uncontrolled and abnormal way to form a so-called **neoplasm**; if the cells clump together this may result in a **tumour**. Benign tumours remain stationary and do not invade other tissues. Those from which cells are shed and migrate into other tissues and organs are malignant. Cancers are classified according to the type of tissue they affect. They include **carcinomas** (cancer of epithelial tissues, e.g. the lung), **sarcomas** (cancer of connective tissues, e.g. myelomas of bone) and **lymphomas** (cancer of the lymph or blood, e.g. leukaemia). Cancers have a wide range of causes: ➤carcinogens in the environment, radiation and genetic and dietary factors; some ➤viruses have been implicated. ➤➤oncogene; *ras* gene.

candela Symbol cd. The SI unit of ➤luminous intensity. It is the luminous intensity of a source of frequency 5.4×10^{10} Hz whose radiant intensity is 1/683 W per ➤steradian.

Cannizzaro reaction A ➤disproportionation reaction in which two molecules of an aldehyde form one alcohol and one carboxylate ion, as the reaction is base-catalysed. For example, the Cannizzaro reaction for benzaldehyde, C_6H_5CHO, in NaOH forms $C_6H_5CH_2OH$ and $Na^+C_6H_5CO_2^-$. Named after Stanislao Cannizzaro (1826–1910), who was the first to argue cogently for the existence of molecules.

canonical distribution ➤Boltzmann distribution.

canonical form One of two or more **resonance structures** proposed for a chemical species (e.g. see the diagram at ➤Kekulé structure for benzene). Each canonical form necessarily involves *two-centre*, two-electron bonds. This model has been superseded by structures based on ➤delocalization.

capacitance Symbol C; unit ➤farad F. It is the charge Q stored on a ➤capacitor per unit potential difference V between its plates, so that $Q = CV$.

capacitor A component in an electrical circuit designed to have a large ➤capacitance. The simplest such device is a pair of parallel plates (area A), a distance D apart, with air in the gap between them. This has capacitance $C = \varepsilon_0 A/D$, where ε_0 is the ➤permittivity of free space. If a material of ➤permittivity ε replaces the air gap the capacitance is given by $C = \varepsilon A/D$; thus capacitors typically use ➤dielectrics with a high permittivity. ➤➤electrolytic capacitor.

capacitor microphone A microphone that measures the change in ➤capacitance between two plates caused by small changes in their separation, and thus detects sound waves incident on the plates and causing them to vibrate.

capillary The smallest of the blood vessels in the body. Capillaries are only about 10 μm in diameter, and have walls only one cell thick to facilitate the exchange of materials between the ➤blood and surrounding tissues. Blood flows into capillaries from ➤arteries, and capillaries join up to form ➤veins.

capillary action (capillarity) The tendency for a liquid to rise up a narrow vertical cavity (**capillary tube**) because the intermolecular forces between the liquid and the tube are stronger than the corresponding forces within the liquid. ≫surface tension.

capric acid Traditional name for decanoic acid, $C_9H_{19}COOH$.

caproic acid Traditional name for hexanoic acid, $C_5H_{11}COOH$.

caprolactam The cyclic ≻lactam from caproic acid; the monomer from which ≻nylon-6 is made. See ≻Beckmann rearrangement for its synthesis.

caprylic acid Traditional name for octanoic acid, $C_7H_{15}COOH$.

carat A unit of mass, used for precious stones, equal to 0.2 g. It is also the name given to a scale used to measure the purity of gold, 24 carat gold being pure gold.

carbamide ≻urea.

carbanion An ≻anion in which a carbon atom carries most of the negative charge, as in R_3C^-. Common sources of carbanions include ≻base-catalysed condensations and especially the ≻Grignard reaction. Carbanions attack ≻carbonyl groups to form new carbon–carbon bonds. Compare ≻carbocation.

carbene A class of very reactive species with carbon having only two bonds, as in the simple CH_2 species. The two electrons on the carbon atom can have either paired electrons, a **singlet carbene**, or unpaired electrons, a **triplet carbene**. Their most characteristic reactions are ≻insertion reactions.

carbenium ion ≻carbocation.

carbide A compound containing carbon as the more electronegative atom, as in silicon carbide, SiC, and tungsten carbide, WC, both of which are hard materials useful for making cutting tools. Many carbides of the ≻transition metals are ≻interstitial, such as the important carbide ≻cementite, Fe_3C, found in steels. ≫calcium dicarbide.

carbocation A ≻cation in which carbon carries most of the positive charge, as in $(CH_3)_3C^+$. (The traditional name for such a species was 'carbonium ion'. The correct alternative 'carbenium ion' is also commonly used.) Compare ≻carbanion.

Carbocations are important intermediates in organic chemistry. They are formed, for example, in the S_N1 ≻nucleophilic substitution reaction and are also involved in some ≻elimination reactions. **Tertiary carbocations**, with three alkyl groups attached to the positive carbon atom as in $(CH_3)_3C^+$, are more stable than **secondary carbocations** such as $(CH_3)_2CH^+$, which are in turn more stable than **primary carbocations** such as $CH_3CH_2^+$. So the primary carbocation $CH_3CH_2CH_2^+$ can undergo a ≻rearrangement into the more stable $(CH_3)_2CH^+$.

carbohydrase An ≻enzyme that hydrolyses ≻carbohydrates.

carbohydrates An abundant class of biological molecules composed of carbon, with hydrogen and oxygen in a ratio of about two hydrogens and one oxygen per one carbon. Carbohydrates occur as ≻monosaccharides (simple sugars such as ≻glucose), ≻disaccharides (such as ≻sucrose), ≻oligosaccharides and ≻polysaccharides (such as ≻starch).

carbolic acid Obsolete name for ►phenol.

carbon Symbol C. The element with atomic number 6 and relative atomic mass 12.01, which is at the top of Group 14. As carbon-based life forms, we have assigned carbon the most important place in chemistry: the study of carbon constitutes one of the three major branches of the subject, ►organic chemistry.

The element itself has three ►allotropes. Two have been familiar since prehistory, ►graphite and ►diamond. The third allotrope, fullerite (►buckminsterfullerene), was discovered in 1990. Several of the apparently amorphous forms of carbon, such as charcoal, have a microscopic structure like graphite's. Carbon is a nonmetal; its most unusual physical property is that graphite is a conductor, a fact that leads to its use as an electrode material for electrolysis.

The most important ►oxidation number for carbon is +4, as in ►carbon dioxide, in which two oxygen atoms form two bonds each to the carbon atom. Although the concept of oxidation number is much less useful in organic chemistry, the vast majority of organic compounds also have carbon forming four covalent bonds with other atoms. ►Carbon dioxide, CO_2, is acidic but ►carbon monoxide, CO, is noticeably less so. Both oxides are of considerable importance both in everyday life and in industry.

There is a whole range of halides, generally called ►halogenoalkanes. Tetrachloromethane and trichloromethane have been important solvents, although their use is declining as they are suspected as carcinogens. ►►carbon cycle; photosynthesis; radiocarbon dating.

carbonate A salt of carbonic acid, which contains the **carbonate ion** CO_3^{2-}. An important example is ►calcium carbonate.

carbon cycle **1** (carbon–nitrogen cycle) (Astron.) A sequence of nuclear reactions that forms the basis for the production of energy in stars. The net result is the formation of a ^{4}He nucleus (and two positrons) from four protons, but the cycle proceeds via ^{12}C, ^{13}N, ^{13}C, ^{14}N, ^{15}O and ^{15}N (see the diagram opposite). It is the predominant mechanism for energy generation in stars greater than twice the Sun's mass. ►►proton–proton reaction.

2 (Biol.) A sequence of events occurring in nature by which carbon in the form of carbon dioxide is removed from the atmosphere and incorporated into carbon compounds during ►photosynthesis; subsequently this carbon is returned to the atmosphere as a by-product of ►respiration (see the diagram opposite). There may be many intermediate stages as carbon compounds pass through ►food chains and are recycled via the decomposition of dead organisms and excretory products by ►bacteria.

carbon dating ►radiocarbon dating.

carbon dioxide CO_2 The oxide of carbon (see the diagram) with ►oxidation number +4, which exists as a colourless gas. It undergoes ►sublimation at $-78\,°C$ to form ►dry ice. The gas is
$$O=C=O$$
carbon dioxide
nontoxic which, together with its conveniently low sublimation temperature, accounts for its use both as a refrigerant and as 'smoke' at pop concerts. Its density, which is greater than that of air, together with its resistance to combustion, also makes it a useful fire extinguisher.

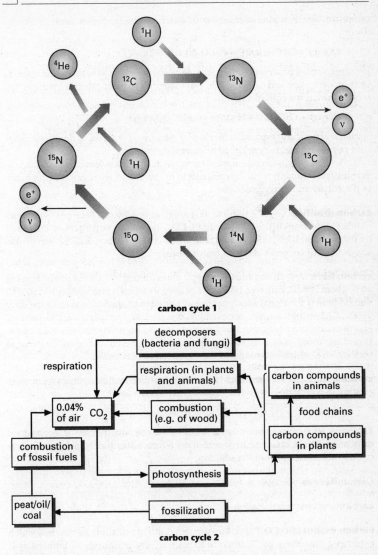

carbon cycle 1

carbon cycle 2

The gas is reasonably soluble in water, forming an equilibrium mixture with carbonic acid:

$$CO_2(g) + H_2O(l) \rightleftharpoons H_2CO_3(aq).$$

The test for carbon dioxide stems partly from this acidic nature. Limewater, calcium hydroxide, $Ca(OH)_2(aq)$ is used to react with the gas. Reaction produces calcium carbonate, which is almost insoluble in water, so the limewater turns 'milky' or 'cloudy':

$$CO_2(g) + Ca(OH)_2(aq) \rightarrow CaCO_3(s) + H_2O(l).$$

However, further passage of carbon dioxide causes the solution to turn clear again, as calcium carbonate and carbonic acid react together to form calcium hydrogencarbonate, which is a colourless solution:

$$CO_2(g) + CaCO_3(s) + H_2O(l) \rightarrow Ca(HCO_3)_2(aq).$$

Similar reactions are responsible for causing ➤hardness of water, as limestone reacts with rain-water naturally acidified by carbon dioxide.

The level of carbon dioxide in the atmosphere has been a recent cause for concern, as rising levels (mainly from the combustion of fossil fuels) are responsible for much of the enhanced ➤greenhouse effect.

carbon disulfide CS_2 The sulfur analogue of carbon dioxide. Because of the larger number of electrons in CS_2 compared with CO_2, the ➤dispersion forces are larger and it exists as a colourless, smelly liquid rather than as a gas. Carbon disulfide was used as a solvent until concerns about its toxicity arose.

carbon fibre Very strong and lightweight fibres formed by heating acrylic fibres in air to about 300 °C, followed by further heating in an inert atmosphere to 1500 °C. Carbon fibre is a material which has found a number of applications since the early 1980s. Carbon fibre brakes, for example, are used in Formula 1 cars to produce exceptionally short braking distances: their use on high-performance road cars is becoming more common as engineers have improved the cold-temperature performance, which had hindered their application previously.

carbonic acid H_2CO_3 A weak acid formed when ➤carbon dioxide dissolves in water existing in equilibrium with the ➤hydrogencarbonate ion:

$$H_2CO_3(aq) + H_2O(l) \rightleftharpoons H_3O^+(aq) + HCO_3^-(aq).$$

The pure acid cannot be isolated, however, as carbon dioxide is evolved on distillation. The acid is the parent acid of the ➤carbonates and the acid salts are the hydrogencarbonates. ➤➤buffer solution.

Carboniferous See table at ➤era.

carbonium ion ➤carbocation.

carbon monoxide CO The oxide of carbon with ➤oxidation number +2, which exists as a colourless gas. Carbon monoxide is very poisonous to humans as it combines with ➤haemoglobin to form **carboxyhaemoglobin**, which starves the body of oxygen causing asphyxiation. Carbon monoxide is produced by incomplete combustion of fossil fuels, especially in petrol engines.

Carbon monoxide is a strong reductant. It burns in oxygen with a blue flame to form carbon dioxide:

$$2CO(g) + O_2(g) \rightarrow 2CO_2(g).$$

In industry its power as a reductant is used in the ►blast furnace for the extraction of iron.

carbon–nitrogen cycle ►carbon cycle (1).

carbon tetrachloride ('carbon tet') ►tetrachloromethane.

carbonyl A complex containing a metal ion coordinated to carbon monoxide ligands. An important example is tetracarbonylnickel(0), $Ni(CO)_4$, used in the ►Mond process. Note the unusual oxidation number of zero for the metal in this compound. ►►eighteen-electron rule.

carbonyl chloride ►phosgene.

carbonyl group The group CO (see the diagram) that is characteristic of aldehydes and ketones, which are therefore collectively called **carbonyl compounds.**

Carborundum A tradename for ►silicon carbide.

carboxyhaemoglobin ►carbon monoxide.

carboxylate The anion formed from a ►carboxylic acid. The anion is delocalized and so should be written RCO_2^- in contrast to the acid RCOOH, as the two oxygen atoms are indistinguishable in the anion and distinguishable in the parent acid (►delocalization). A simple example is the ethanoate ion, $CH_3CO_2^-$.

carboxyl group The group —COOH, which is present in a number of compounds, such as the ►carboxylic acids and ►amino acids.

carboxylic acid An organic molecule formed from an ►alkyl group bonded to a ►carboxyl group, and which therefore has a general formula RCOOH. They are weak acids in solution, forming an equilibrium mixture with the carboxylate ion:

$$RCOOH(aq) + H_2O(l) \rightleftharpoons H_3O^+(aq) + RCO_2^-(aq).$$

An important example is ►ethanoic acid. A **dicarboxylic acid** contains two carboxyl groups and a **tricarboxylic acid** contains three carboxyl groups; they are especially important in nature (►Krebs cycle).

carcinogen A substance or agent that causes ►cancer, such as tars in tobacco smoke, or asbestos.

carcinoma ►cancer.

cardiac Of or relating to the heart. A **cardiac arrest**, for example, is a heart attack.

cardinal number For a finite ►set, the number of ►elements in the set. The cardinal number of an infinite set is assigned by putting it into ►one-to-one correspondence with a standard set of known cardinality such as the ►integers (**cardinality** $\aleph_0$, 'aleph-nought') or the ►real numbers.

cardioid ►epicycloid.

cardiovascular Of or relating to the heart and blood system, and usually (particularly in athletic training programmes) taken to include the lungs and respiratory system.

caries Decay (necrosis) of teeth (➤tooth) caused by bacteria in the mouth that metabolize carbohydrates to produce acid, which dissolves tooth enamel. The characteristic pain (toothache) associated with severe caries is due to the increased pressure of gases in the pulp cavity produced by the decay process.

carnallite Natural potassium magnesium chloride, $KCl \cdot MgCl_2 \cdot 6H_2O$, which is an important ore of both metals.

carnivore An animal that feeds largely or exclusively on other animals. The term is used most frequently for members of the order **Carnivora**, which includes flesh-eating mammals such as dogs, cats, bears and seals. These carnivores are secondary consumers in ➤food chains. **Carnivorous plants** such as the Venus fly-trap feed on insects and other small animals. Compare ➤herbivore; ➤omnivore.

Carnot cycle A sequence of reversible ➤changes of state in a ➤heat engine working between two heat reservoirs (see the diagram). The heat input (Q_1) is at a higher (constant) temperature (T_1) than the heat output (Q_2), which occurs at temperature T_2, with a reversible ➤adiabatic step in between. The work W done by the engine equals $Q_1 - Q_2$; its efficiency, W/Q_1, equals $(T_1 - T_2)/T_1$. It is typically considered to be based on a gas, and it is primarily a hypothetical construction rather than a practical heat engine. Named after (Nicolas Léonard) Sadi Carnot (1796–1832). ➤Carnot's principle.

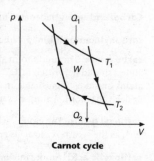

Carnot cycle

Carnot's principle No heat engine can operate with higher efficiency between two heat reservoirs at specified temperatures than is achieved in a ➤Carnot cycle. A more efficient engine could be shown to violate the second law of ➤thermodynamics. Named after N. L. S. Carnot.

carotene (carotenoid) One of a number of organic molecules comprising rings of carbon atoms linked to chains in which there are ➤conjugated bonds. They are red, yellow or orange lipid-soluble pigments which are responsible for the pigmentation of plant organs such as ripe tomatoes and peppers (➤accessory pigment). Some pigments function in the capture of light in ➤photosynthesis, transferring energy to chlorophyll *a*. *β*-carotene (see the diagram) is the precursor of vitamin A, which in turn oxidizes to retinal, the pigment involved in vision.

β-**carotene**

carotid artery One of a pair of arteries (one on each side of the neck) in vertebrates supplying the head and brain with oxygenated blood.

carrier 1 (Biol.) An organism that contains an ➤allele for a particular characteristic but does not express it. The term is particularly applied to individuals that harbour alleles for genetic disorders such as phenylketonuria, a hereditary enzyme deficiency. Such alleles are ➤recessive, and are not expressed in a ➤heterozygous individual offspring carrying a ➤dominant allele. However, the recessive allele can be passed to the offspring. Thus if two carriers mate there is a chance (1 in 4) that an individual offspring will be ➤homozygous and express the disease symptoms. ➤➤haemophilia; sex-linkage.

   2 (Phys.) A particle that carries electric charge in a conductor. In a semiconductor, carriers may be ➤holes or electrons. In a gas or ionic solution, carriers may be ions.

   3 (Phys.) ➤modulation.

carrier gas ➤chromatography.

carrier wave The ➤radio frequency wave used as the carrier for transmitting information by ➤modulation.

Cartesian coordinates A system for specifying the location of a point in a plane (or space) by means of its distances from a fixed origin along two (or three) fixed, mutually perpendicular axes (➤axis). In the plane, the axes are labelled x and y with coordinates written (x, y). In space, the axes are labelled x, y and z with coordinates written (x, y, z); by convention, a right-handed choice of axes is made: one for which the positive x, y, z directions line up with the respective positions of the thumb, forefinger and middle finger of the right hand (see the diagram). Named after René Descartes (1596–1650).

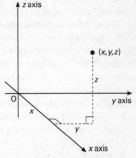

Cartesian coordinates

cartography The study and production of maps and ➤map projections.

cascade Any process that proceeds by a series of similar steps, with the output of one step feeding the input of the next. For example, ➤fractional distillation is a cascade; the name **cascade process** is given to the separation of isotopes in a series of steps.

Cassegrain telescope ➤telescope.

Cassinian curve (oval) The ➤locus of a point the product of whose distances from two fixed points is constant, a^2. A **lemniscate** is the special case where the separation of the fixed points is $2a$; in ➤polar coordinates its equation is $r^2 = 2a^2 \cos 2\theta$. Named after Giovanni Domenico Cassini (1625–1712).

Cassini division The conspicuous gap, about 4700 km wide, between the two main components of the rings of Saturn. Named after G. D. Cassini, who discovered it in 1675.

cassiterite (tinstone) SnO_2 The principal ore of ➤tin.

cast iron An iron alloy with a carbon content of about 2–4% by mass. It is usually made by melting ►pig iron to remove some impurities, and cast into moulds.

catabolism Those components of ►metabolism in a living cell in which complex molecules are broken down into simpler ones. The breaking down of carbohydrates into carbon dioxide and water during ►aerobic respiration is a good example of a **catabolic process**. Compare ►anabolism.

catalysis An increase in the rate of a chemical reaction caused by a ►catalyst.

catalyst A substance that increases the rate of a chemical reaction and is recoverable at the end unchanged in chemical composition or mass. The catalyst increases the rate of a reaction by providing an alternative pathway for the reaction that has a lower ►activation energy. Catalysts involved in biochemical processes are called ►enzymes. Catalysts are classified as: **homogeneous catalysts**, which are in the same phase as the reactants, and **heterogeneous catalysts**, which are in a different phase from the reactants. An example of a homogeneous catalyst is iron(III) ions in the reaction between iodide ions and peroxodisulfate ions. An example of a heterogeneous catalyst is the iron used in the Haber–Bosch process for ammonia; this illustrates the common situation of a solid catalysing the reaction between two gases.

Catalysts do not affect the equilibrium position of a reaction. So acid can catalyse both the formation of an ester and its hydrolysis. Small quantities of catalyst are usually sufficient to achieve a huge increase in rate. For example, just 2 g of platinum (along with rhodium) is the active ingredient in a ►catalytic converter. One molecule of the enzyme triosephosphate isomerase can catalyse the reaction of 400 000 molecules s^{-1} (this being limited by the rate of diffusion to the enzyme). Catalysts can be physically altered. The platinum/rhodium gauze used as a catalyst in the ►Ostwald process is roughened over time.

The catalyst may need a support. This acts partly to provide simple physical support and partly to increase the surface area. An example is the silica support for the phosphoric acid catalyst in the ►direct hydration of ethene. The catalyst may work better if a ►promoter is used. Catalysts may be poisoned and hence rendered ineffective by another chemical (►catalytic converter). Catalysts, especially enzymes, may be very specific. For example, urease catalyses the hydrolysis of urea, NH_2CONH_2, but not the hydrolysis of the isoelectronic molecule CH_3CONH_2. ►►autocatalysis.

catalytic converter ('cat') An apparatus designed to reduce levels of all three main pollutants emitted from the exhaust of a petrol engine (carbon monoxide, NO_x, and unburnt hydrocarbons), primarily by converting carbon monoxide and nitrogen monoxide to carbon dioxide and nitrogen:

$$2CO(g) + 2NO(g) \rightarrow 2CO_2(g) + N_2(g).$$

Up to 90% of the unburnt hydrocarbons are also combusted, so a car with a cat should emit only around 10% of the pollutants produced by a car without one fitted. The lead produced by decomposition of tetraethyllead in leaded petrol would poison the catalyst, so unleaded petrol must be used. One disadvantage is that catalytic converters catalyse other reactions too, including the combustion of small quantities of sulfur impurities in the petrol. Hence we can detect a sulfurous smell from them.

catalytic cracking ➤cracking.

catalytic RNA ➤ribozyme.

catechol Traditional name for ➤*o*-dihydroxybenzene.

catecholamine One of a number of derivatives of the amino acid tyrosine which function as ➤neurotransmitters.

catenary The name of the curve formed by a thin, uniform, flexible chain hanging from two fixed supports. With the y axis vertical, its equation is given by $y = a \cosh(x/a)$ for an appropriate constant a.

catenation The process of joining atoms together to form a chain. Carbon shows by far the greatest tendency to **catenate**, forming millions of catenated molecules. ➤Sulfur too can catenate.

catenoid The ➤surface of revolution formed by rotating a ➤catenary about the x axis.

cathetometer A device for measuring small vertical distances. It consists of a movable telescope on a graduated vertical pillar.

cathode 1 (Chem.) The electrode at which ➤reduction occurs. In an electrolytic cell the cathode is negatively charged, and the ➤cations are attracted towards the cathode and reduced by gain of electrons. In an electrochemical cell, on the other hand, the cathode is positively charged. Compare ➤anode.
   2 (Phys.) An obsolete name for the emitter in a ➤thermionic valve.

cathode ray A stream of electrons, so named because they were first produced from the cathode in ➤gas-discharge tubes.

cathode-ray oscilloscope (CRO) ➤oscilloscope.

cathode-ray tube (CRT) An evacuated tube designed to allow an electron beam from an ➤electron gun at one end to be scanned across a luminous screen at the other by varying electric and magnetic fields. Most televisions and computer monitors have until recently been based on cathode-ray tubes.

cathodic protection ➤rusting.

cation The ion that moves towards the ➤cathode in ➤electrolysis; as the cathode is negatively charged in electrolysis, cations are positive ions. Common cations include H^+, Na^+ and Al^{3+}. Compare ➤anion.

Cauchy–Riemann equations The ➤partial differential equations $\partial u/\partial x = \partial v/\partial y$ and $\partial u/\partial y = -\partial v/\partial x$ satisfied by the ➤real and ➤imaginary parts u and v of an ➤analytic function $f(x + iy) = u(x, y) + iv(x, y)$. Named after Augustin Louis Cauchy (1789–1857) and (Georg Friedrich) Bernhard Riemann (1826–66).

Cauchy–Schwarz inequality The ➤inequality $(\sum a_i b_i)^2 \leqslant (\sum a_i^2)(\sum b_i^2)$ for two sets of numbers $a_1,\ldots, a_n$ and $b_1,\ldots, b_n$; it also refers to the analogous inequality between ➤integrals, $(\int f(x)g(x)\, dx)^2 \leqslant (\int f(x)^2\, dx)(\int g(x)^2\, dx)$. Named after A. L. Cauchy and Hermann Amandus Schwarz (1843–1921).

Cauchy's theorem A fundamental theorem concerning ➤analytic functions which states that the ➤contour integral of an analytic function around any closed curve is always zero.

causality A concept based on the fact that the effects of an event, or information about it, cannot travel faster than the speed of light. Event A can affect event B only if event A is in the **absolute past** of event B (and thus event B is in the **absolute future** of event A). This seems trivial in Galilean spacetime, but when special relativity is taken into account the situation is less straightforward. If the interval between A and B is space-like (each is in the other's **absolute elsewhere**), then for some observers the time-ordering of events A and B is reversed by the ➤Lorentz transformation. For some observers A occurs before B, while for others B occurs before A; therefore one cannot cause the other.

caustic curve The ➤envelope of light rays emitted from a point source (which may be at infinity) after ➤reflection from a curved mirror. The familiar caustic formed on the surface of a mug of tea in strong sunlight is, in fact, a nephroid (➤epicycloid). Caustics may also be formed by ➤refraction at a curved boundary between two media.

caustic potash An old name for ➤potassium hydroxide.

caustic soda An old name for ➤sodium hydroxide.

cavitation The formation of cavities in a fast-moving fluid as a result of a pressure drop, for example downstream of propeller blades.

cc Abbr. for the unit cubic centimetre; the symbol cm^3 is now preferred.

CCD Abbr. for **charge-coupled device**, a semiconductor device in which charge moves through the ➤substrate (2) in discrete units and in a controllable way. CCDs have many uses, for example as memory, as filters or for imaging in solid-state cameras.

c.c.p. Abbr. for cubic close-packed or cubic ➤close packing.

cd Symbol for the unit ➤candela.

Cd Symbol for the element ➤cadmium.

cDNA ➤DNA.

CD-ROM Abbr. for compact disc read-only memory. A read-only backing store using the same physical format as an audio compact disc, but on which any sort of data (text, images, sound, programs) may be held. The capacity of a CD-ROM is considerably greater than that of a ➤floppy disk.

Ce Symbol for the element ➤cerium.

celestial body Generic name for stars, planets and their satellites, and other objects in and between galaxies which might be observed by an astronomer.

celestial equator The ➤great circle (on the ➤celestial sphere) in the plane through the centre of the Earth that is perpendicular to a line joining the ➤celestial poles.

celestial mechanics The study of the motion of stars and other celestial bodies.

celestial meridian The great circle on the ➤celestial sphere in the plane of the ➤celestial poles and of a specified observer.

celestial poles The points at which a line through the rotational axis of the Earth cuts the ➤celestial sphere. The north celestial pole is marked by the bright star Polaris; there is no bright star near the south celestial pole.

celestial sphere A hypothetical sphere, of infinite radius and with the Earth at its centre, onto which all celestial bodies are projected in order to give them angular coordinates. Three systems for coordinates are possible. They all use equivalents of latitude and longitude (which for the Earth are measured relative to the equator and

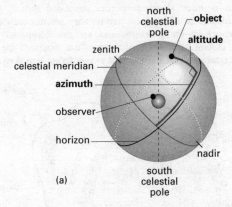

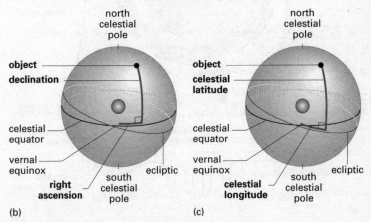

celestial sphere Coordinate systems: (a) horizontal coordinates, (b) equatorial coordinates and (c) ecliptic coordinates.

the prime meridian, respectively). In **horizontal** coordinates, the **altitude** and **azimuth** of a body at a given time are measured relative to the observer's horizon and celestial meridian; in **equatorial** coordinates, the **declination** and **right ascension** are measured relative to the celestial equator and vernal ➤equinox; in **ecliptic** coordinates, the **celestial latitude** and **celestial longitude** are measured relative to the ➤ecliptic and the vernal equinox (see the diagram).

cell 1 (Biol.) The basic structural and functional unit of living things. The cell theory, attributed to Matthias Jakob Schleiden and Theodor Schwann in 1838, states that 'Cells are of universal occurrence and are the basic units of an organism'; in 1858 Rudolf Virchow proposed that all cells come from pre-existing cells. Today the essence of the theory still holds good: cells are considered to be the smallest unit in which life can exist to manifest all the properties associated with it. Although ➤viruses can be regarded as of a biological nature, they are not capable of the independent maintenance of life processes and are not composed of cells. Structurally, all cells consist of a mass of ➤cytoplasm bounded by a ➤cell membrane (see the diagrams). Cells are basically of two types: ➤prokaryotic cells

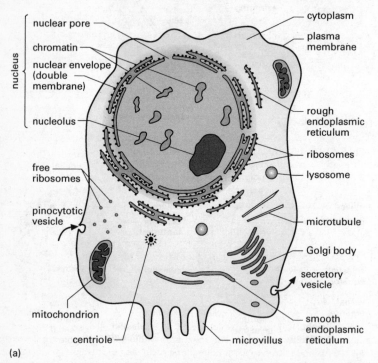

cell 1 Simplified representations of (a) an animal cell and (b) (see opposite) a plant cell as viewed at the resolution of the electron microscope.

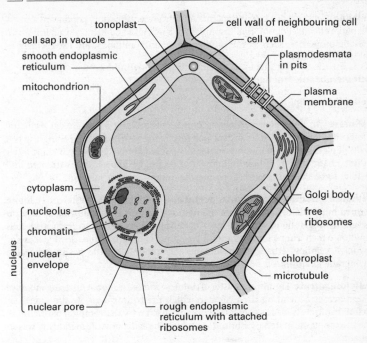

tonoplast

cell sap in vacuole

smooth endoplasmic reticulum

mitochondrion

cell wall of neighbouring cell

cell wall

plasmodesmata in pits

plasma membrane

cytoplasm

nucleus {
nucleolus

chromatin

nuclear envelope

nuclear pore
}

Golgi body

free ribosomes

chloroplast

microtubule

rough endoplasmic reticulum with attached ribosomes

(b)

have no membrane around the ➤nucleus (2) or other ➤organelles; ➤eukaryotic cells have membrane-bound nuclei and organelles. Eukaryotic cells are very diverse in their structure and function and comprise the bodies of ➤animals, ➤plants and ➤fungi.

2 (Phys., Chem.) A device for transforming electrical to chemical energy, or vice versa. An **electrolytic cell** is used in ➤electrolysis to cause chemical change using electricity. A **galvanic cell** does the opposite: chemical reactions within the cell produce an e.m.f. A combination of galvanic cells is a ➤battery. **Primary cells** cannot be recharged. The ➤Daniell cell, for example, uses the spontaneous reaction between zinc and copper(II) ions to produce a current flow around a circuit, which can then be used to power electrical equipment. ➤➤Leclanché cell; fuel cell. **Secondary cells** can be recharged; two important examples are the ➤lead–acid battery and the ➤nickel–cadmium cell.

cell division The process by which a ➤cell replicates. Cell division in ➤prokaryotic cells is the simplest and proceeds by replication of the single DNA strand, followed by simple fission of the cytoplasm caused by invagination of the ➤plasma membrane and ➤cell wall. ➤Eukaryotic cells divide by ➤mitosis; the production of ➤gametes by eukaryotic cells occurs by ➤meiosis.

cell line Cells that can be cultured indefinitely under laboratory conditions and are therefore effectively immortal. Examples are 3T3, HeLa and BHK 21. Cell lines with specific properties, such as the production of an ➤antibody, are used widely in biological research.

cell membrane ➤plasma membrane.

celluloid ➤cellulose nitrate.

cellulose An unbranched ➤polysaccharide consisting of β-(1-4)-linked ➤glucose molecules. It is the most abundant polysaccharide in nature, comprising the cell walls of plants in which it occurs in densely packed fibrils. These also include other polysaccharides such as hemicellulose and pectin. Cellulose confers strength and rigidity to cell walls but is freely permeable to substances in solution.

cellulose ethanoate (cellulose acetate) The ethanoate ➤ester of cellulose formed by ➤ethanoylation. Once the hydroxyl groups are ethanoylated, they no longer hydrogen-bond and the ester dissolves in propanone to form a viscous solution which can be forced through ➤spinnerets to give a fibre called **viscose** or rayon. A cellulose ethanoate membrane is commonly used as a semipermeable membrane in ➤osmosis.

cellulose nitrate The nitrate ➤ester of cellulose formed using a ➤nitrating mixture. Its properties depend on the extent to which the hydroxyl groups are esterified. Cellulose nitrate is used as gun-cotton in explosives and with a camphor plasticizer as the thermoplastic material **celluloid**, on which the early movie film industry was so reliant.

cell wall A rigid or semi-rigid boundary layer of the ➤cells of plants, fungi and bacteria. The cell wall of all cells lies outside the ➤plasma membrane and is therefore extracellular. In plants, cell walls are composed mainly of ➤cellulose with lignin in woody cells, but derivatives of ➤chitin are found in fungi and bacteria. Cell walls impart rigidity and structural integrity to cells but generally play little part in controlling the exit and entry of substances in solution.

Celsius scale A temperature scale whose unit, the degree Celsius (°C), is the same size as the ➤kelvin. However, the origin of the scale is different, with 0 °C equivalent to 273.15 K (actually, ➤absolute zero is defined as 273.16 K below the ➤triple point of water, which is 0.01 °C). On the Celsius scale, 0 °C is the normal ➤melting point of ice, and 100 °C is the normal ➤boiling point of water. In the UK the scale used to be called the **centigrade** scale. Named after Anders Celsius (1701–44), who actually proposed a slightly different scale; the Celsius scale was invented in 1743 by Jean Pierre Christen (1683–1755).

cement A powdery material used in building, usually as an ingredient of concrete or mixed with water and sand to make mortar. Portland cement is made by heating a finely ground mixture of ➤limestone with carefully controlled amounts of ➤clay in a rotary kiln at 1500 °C. The resulting clinker is reground and mixed with 3% ➤gypsum. When mixed with water, it sets in a few hours. The composition of cement is very complicated and is only poorly approximated by the description that it is a calcium aluminosilicate. Worldwide annual production of cement is over 1 billion tonnes; China is the largest producer.

cementite The hard, brittle compound Fe_3C which is present in a large number of steels. ➤iron–carbon phase diagram.

Cenozoic See table at ➤era.

centi- The SI prefix for 10^{-2} base units, as in ➤centimetre.

centigrade ➤Celsius scale.

centimetre (US **centimeter)** Symbol cm. The fundamental unit of length in the c.g.s. system, equal to 10^{-2} m.

central dogma In biology, the notion that genetic information flows in one direction, from ➤DNA via ➤RNA to ➤protein, and not from protein to RNA or from RNA to DNA. It was first proposed by Francis Crick in 1958 before the discovery of ➤reverse transcriptase: Crick was under the impression that 'dogma' meant an idea for which there was no reasonable evidence.

central limit theorem A result of central importance in ➤statistics which states that, for a wide class of ➤populations (2) described by a ➤random variable with ➤mean μ and ➤variance σ^2, the distribution of means of ➤random samples of size n (for large n) is approximately normal with mean μ and variance σ^2/n. This key result is the main reason why the ➤normal distribution occupies such a prominent position in statistics and is important in justifying the procedures of ➤statistical inference. An early version was introduced by Abraham de Moivre and improved upon by Pierre Simon de Laplace.

central nervous system (CNS) ➤nervous system.

central processing unit (CPU) The core part of a computer, consisting of a control unit and the ➤arithmetic and logic unit. The CPU executes a ➤program by fetching instructions from memory, decoding and executing them.

centre of curvature ➤curvature.

centre of gravity The point at which the total ➤weight of a body can be considered to act. In a uniform gravitational field (including, to a good approximation, the gravitational field of the Earth at its surface), this is the same as the ➤centre of mass.

centre of mass (centroid) The point at which the total mass of a body can be considered to lie, for the purpose of studying linear motion. For a body composed of a number of point masses m_i in positions R_i, this is given by:

$$R_0 = \sum_i m_i R_i / \sum_i m_i,$$

where the denominator is simply M, the total mass of the body. For a body with distributed mass that can be expressed as a density $\rho(r)$, the centre of mass is given by:

$$R_0 = \frac{1}{M} \int \rho(r) r \ dV,$$

where the integral is over the extent of the body. All the forces acting on a body can be reduced to a single force acting on the body through the centre of mass, plus a single ➤couple.

centre of pressure The point at which the total force on an ➤aerofoil can be considered to act.

centre of symmetry A point about which a geometric figure is symmetrical in the sense that for every point P in the figure there is an image point P′ such that the centre of symmetry is the midpoint of PP′.

centrifugal force A ➤fictitious force appropriate to a frame of reference that is rotating relative to an ➤inertial frame. To use the rotating frame (such as a frame rotating with the Earth) as if it were inertial, a force is required equal to $m\omega^2r$, where ω is the magnitude of the angular velocity of the frame relative to an inertial frame (such as the frame of the ➤fixed stars) and r is the displacement of the body (mass m) from the centre of rotation. Note that the centrifugal force is always directed away from the centre along r, hence its name. ➤➤Coriolis force.

centrifuge A device for separating out particles from a suspension by rapidly spinning the suspension in a tube. ➤➤density gradient centrifugation.

centriole An ➤organelle associated with ➤microtubule formation in animal and ➤protoctistan cells. Centrioles occur in pairs, usually located at right angles to each other. At least some centrioles have recently been shown to contain DNA, supporting the hypothesis that they may be ancestrally derived from free-living motile bacteria.

centripetal acceleration The acceleration of a body that is moving in a circle (when viewed in an ➤inertial frame). It is equal to $-\omega^2r$, where ω is the magnitude of the angular velocity ω of the body and r is its displacement from the centre of the rotation. It is always directed towards the centre along r, hence its name. Its magnitude can also be written as v^2/r, where v is the speed of the body.

centripetal force The force required to provide the ➤centripetal acceleration that causes a body to move in a circle. Alternatively, in a frame of reference rotating with the body, it may be viewed as the force required to balance the ➤centrifugal force.

centroid ➤centre of mass.

centromere The region of a ➤chromosome by which it attaches to the spindle during ➤meiosis and ➤mitosis.

cephalopod A member of the ➤mollusc class **Cephalopoda**, containing active animals with tentacles, a beak for feeding and well-developed eyes. Examples are squid, cuttlefish and octopus.

Cepheid variable An unstable star that shows periodic changes in ➤magnitude (2), typically over a few days, as a result of pulsation (changes in its radius). Named after the prototype star, *Delta Cephei*.

ceramic A material composed of minute **platelets** that can be easily shaped when wet; when heated to a very high temperature it becomes permanently rigid. A common method of making a ceramic is to add sand (essentially ➤silica) to a ➤clay or a ➤feldspar. The typical characteristics of ceramics, all of which stem from their giant (and complicated) aluminosilicate structure, are high melting point, high strength, good electrical insulation and good thermal insulation. Ceramics have many uses, including crockery and furnace linings. Ceramic tiles are used on the space shuttle to resist the very high temperatures generated by atmospheric friction on re-entry. An exciting new application was promised by the discovery in 1986 of special ceramics that show high-temperature (above liquid nitrogen)➤superconductivity. While they are used for specialist magnets for scientific research, their more widespread commercial applications have been more limited than expected because of their brittle nature.

cerebellum ➤brain.

cerebrospinal fluid The ➤extracellular fluid surrounding the brain and spinal cord that cushions them from physical shock. In particular, it bathes the outer membranes (**meninges**) of the central nervous system, inflammation of which is called **meningitis**.

cerebrum (cerebral hemispheres) Paired outgrowths of the vertebrate forebrain, which function as a control and association centre. It is particularly well developed in the higher mammals, and its surface area and functionality are increased in humans by a high degree of folding. It is dominated by the **cerebral cortex** in which all motor and mental functions are coordinated. Certain areas of the cortex are associated with specialized activities, such as speech and movement of particular limbs (➤brain).

Cerenkov radiation Alternative spelling of ➤Cherenkov radiation.

cerium Symbol Ce. The element with atomic number 58 and relative atomic mass 140.1, which is a ➤lanthanide. It is a grey malleable metal, one use of which is in machine-gun tracer bullets. As usual for lanthanides, the most common ➤oxidation number in its compounds is +3, as in $CeCl_3$, which exists as white crystals. It does, however, show another oxidation number, cerium(IV), which corresponds to the removal of all the electrons in its outer shell, as in **ceria**, cerium(IV) oxide, CeO_2.

CERN The major European particle physics laboratory near Geneva. It has a number of large particle accelerators, including a 450 GeV Super Proton Synchrotron and a Large Electron–Positron (LEP) collider, the largest lepton accelerator in the world with a circumference of 27 km. The even more powerful Large Hadron Collider began operation in 2008. The name comes from the original title of its founding organization, the Conseil Européen pour la Recherche Nucléaire.

cesium US form of ➤caesium.

cetane number A measure of the burning characteristic of a diesel fuel, in a similar way that the ➤octane number measures the burning characteristic of a petrol.

Cf Symbol for the element ➤californium.

CFC Abbr. for ➤chlorofluorocarbon.

c.g.s. A system of units adopted in 1874 by the British Association, based on the centimetre, the gram and the second. It was superseded by the ➤SI units in use in science today.

chain reaction **1** (Chem.) A reaction mechanism in which an overall process can be broken down into a series of steps, in which **chain carriers** are involved. A classic example is the **radical chain reaction** involved in the ➤chlorination of methane. The three main stages are **chain initiation**, in which a chain carrier (here the chlorine ➤radical) is formed:

$$Cl_2(g) \rightarrow 2Cl^{\bullet}(g).$$

The **chain propagation** steps are a pair of linked reactions that destroy and then recreate a chain carrier:

$$CH_4(g) + Cl^{\bullet}(g) \rightarrow CH_3^{\bullet}(g) + HCl(g),$$
$$CH_3^{\bullet}(g) + Cl_2(g) \rightarrow CH_3Cl(g) + Cl^{\bullet}(g).$$

The final **chain termination** step sees the destruction of the chain carriers (both the chlorine and methyl radicals can be removed, giving three termination reactions):

$$CH_3^{\bullet}(g) + Cl^{\bullet}(g) \rightarrow CH_3Cl(g),$$
$$Cl^{\bullet}(g) + Cl^{\bullet}(g) \rightarrow Cl_2(g),$$
$$CH_3^{\bullet}(g) + CH_3^{\bullet}(g) \rightarrow C_2H_6(g).$$

Detection of ethane is good proof of this mechanism, as is the identification of methyl radical by ➤electron spin resonance. ➤Combustion occurs via radical chain reactions.

2 (Phys.) A nuclear reaction that is self-sustaining. The product of one stage of the reaction is one of the reactants of the next, and a chain of reaction steps proceeds sequentially. For the nuclear reaction involving uranium that takes place in a nuclear reactor, it is neutrons from the fission of uranium nuclei that induce further fission. The average number of product neutrons that go on to induce fission determines the way the reaction proceeds: less than one is **subcritical** and the reaction rate will decay; exactly one is **critical** and the reaction rate is sustained; more than one is **supercritical** and the reaction accelerates, possibly out of control.

chain rule A rule for the ➤differentiation of **composite functions**: if y is a function of u, and u is a function of x, then

$$\frac{dy}{dx} = \frac{dy}{du} \cdot \frac{du}{dx}.$$

chair conformation ➤conformation.

chalcogens The generic name for the elements of Group 16; this is much less commonly used than the corresponding name ➤'halogens' for the elements of Group 17.

chalcopyrite (copper pyrites) $CuFeS_2$ A brassy yellow ore, which is the most important source of copper.

chalk Natural ➤calcium carbonate, $CaCO_3$.

Chandrasekhar limit The maximum possible size for a ➤white dwarf (about 1.4 times the Sun's mass). Named after Subrahmanyan Chandrasekhar (1910–95).

change of state (change of phase) A transition between one state of matter and another. ➤Melting is the change of state from solid to liquid, with **freezing** the reverse change. ➤Boiling is the change of state from liquid to gas, with **condensation** the reverse change. ➤Sublimation is the change of state from solid directly to gas or vice versa. During a change of state, the temperature remains constant as the ➤latent heat is transferred.

change of variable A technique of ➤integration in which the substitution $x = x(u)$ is made so that

$$\int f(x)\,dx = \int f(x(u))\frac{dx}{du}\,du,$$

the hope being that the right-hand integral is more tractable than the left. For more than one variable, change of variables involves a change of coordinate system; for example, changing to plane ➤polar coordinates (➤Jacobian):

$$\iint f(x,y)\,dx\,dy = \iint f(r\cos\theta, r\sin\theta)\,r\,dr\,d\theta.$$

channel In communications, a path for data transmission.

chaos theory A branch of applied mathematics based on the fact that seemingly irregular, unpredictable phenomena can arise from the sensitivity of some well-defined ➤differential equations to small changes in their ➤initial conditions. Chaos theory promises applications in many areas, for example in fluid turbulence and in meteorology.

characteristic curve For a ➤semiconductor device, the dependence of one parameter on another, in particular the dependence of current on voltage. For example, characteristic curves for a ➤field-effect transistor (drain current against drain–source voltage, for different gate–source voltages V_{GS}) are shown in the diagram.

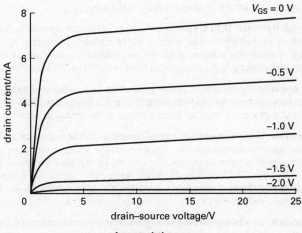

characteristic curve

characteristic equation The equation $\det(A - xI) = 0$ associated with a square $n \times n$ ➤matrix A, where det denotes ➤determinant and I is the $n \times n$ ➤identity (3) matrix. It is a ➤polynomial equation of degree n whose roots are the ➤eigenvalues of A.

characteristic wave A wave polarized along the optic axis (➤birefringence) of a crystal so that the polarization does not change on passing through the crystal.

charcoal Impure amorphous carbon, with a microscopic structure of ➤graphite, made by the destructive distillation of organic material, as in **animal charcoal**, **wood charcoal**, etc. Charcoal is very useful for ➤adsorption of gases. In the laboratory, a charcoal block is sometimes used to reduce metal oxides.

charge ➤electric charge.

charge carrier ➤carrier (2).

charge conjugation A ➤symmetry transformation that reverses the sign of the charge of every charged particle. Thus an electron would become positively charged. ➤➤charge–parity symmetry; charge–parity–time symmetry.

charge-coupled device ➤CCD.

charge density Symbol ρ. The charge per unit volume in a body where the charge is distributed (rather than being located at point charges). Charge density appears in the differential form of the first of ➤Maxwell's equations, and thus determines ➤electric field. The charge density of a cation affects the extent of covalent bonding (➤Fajans' rules). ➤➤electron density.

charge–parity (CP) symmetry The symmetry operation that is a combination of ➤charge conjugation and ➤parity. It appears that the ➤weak interaction, which is not invariant under parity, is not precisely invariant under CP either, and the CP quantum number of neutral kaons (➤K meson) can be changed by the weak interaction. This is known as **charge–parity violation**.

charge–parity–time (CPT) symmetry The symmetry operation that is a combination of ➤charge conjugation, ➤parity and ➤time reversal symmetry. There are fundamental reasons in quantum mechanics (combined as the **CPT theorem**) for believing that all the interactions of nature are invariant under CPT.

charge transfer A mechanism of the production of ➤colour in which electron density passes from one species to another; the purple of the manganate(VII) ion is caused by charge transfer from an oxide ➤ligand to the central manganese ion.

Charles's law The volume of a fixed mass of gas, at constant pressure, is directly proportional to the thermodynamic temperature. The law provides a quantitative statement of the familiar rule of thumb that heating a gas causes it to expand. The law can be derived from the ➤ideal gas equation. Named after Jacques Alexandre César Charles (1746–1823), who once held the world altitude record, which was set in a balloon using hydrogen for lift.

charm quark One of the flavours of ➤quark in the ➤standard model. Charm is paired with ➤strangeness, as up is paired with down.

chelate A species in which an ion is enveloped by complexation (see the diagram) with a chelating agent, which provides at least two atoms with lone pairs that can form ➤coordinate bonds with the metal ion. Examples of chelating agents are ➤1,2-diaminoethane, ➤oxalate ions and, especially, ➤EDTA. The name comes from the Greek for 'crab's claw'.

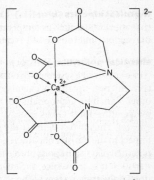

chelate The chelate $[Ca(EDTA)]^{2-}$.

chemical bonding The characteristic phenomenon in chemistry is that atoms bond together; in a chemical reaction some bonds are broken and others made. The crucial clue to the identity of the chemical bond was found in 1916 by Gilbert Newton Lewis, when he identified the electron pair. In his own words, 'the cardinal phenomenon of all chemistry is the electron pair bond'.

Lewis introduced the concept of ➤covalent bonding. This model assumed that the electron pair was *shared* between the atoms being bonded. In the same year, the idea of ➤ionic bonding was also proposed. This took the extreme view that the electron pair was given to one of the atoms bonded rather than being shared: in effect an electron was *transferred* from one atom to another.

It is clear now that the two models complement each other and that there is a smooth transition from one extreme to the other as the ➤electronegativities of the atoms become more different (➤➤coordinate bonding; Fajans' rules). Chemical bonding is sometimes broadened in scope to the forces between molecules (➤intermolecular forces; hydrogen bonding).

chemical change A change in which the chemical nature of the substances involved changes, which distinguishes it from a ➤physical change. When pure ice melts, the substance does not change as both ice and liquid water are composed of H_2O molecules, so melting is a physical change. When metallic sodium is added to water, hydrogen gas is evolved and aqueous sodium hydroxide is formed. The nature of the substances has changed, and so the process is a chemical change.

chemical combination, laws of ➤constant composition, law of; multiple proportions, law of; reciprocal proportions, law of.

chemical equation The expression of a ➤chemical change in terms of the formulae of the reacting substances. For example, the ➤Haber–Bosch process for ammonia can be expressed by the equation

$$N_2(g) + 3H_2(g) \rightarrow 2NH_3(g).$$

All chemical reactions reach chemical ➤equilibrium: the concentrations of the reactants (the species on the left of the equation) and the products (the species on the right of the equation) are related by the ➤equilibrium constant.

chemical formula ➤formula.

chemical kinetics (kinetics) The study of the ➤rates of reaction and the factors that affect them, such as temperature (➤activation energy), concentration (➤order of reaction) and the use of a ➤catalyst.

chemical potential Symbol μ. In thermodynamics, the partial derivative of the ➤free energy of a system with respect to the number of particles (or alternatively for convenience, usually in chemistry, the amount of substance (➤mole)), with the other variables appropriate to that free energy held constant (pressure and temperature for ➤Gibbs energy, volume and temperature for ➤Helmholtz free energy). Thus two systems at the same chemical potential are in equilibrium with respect to the transfer of particles, as the change in free energy is zero. The chemical potential plays an important role in the ➤Bose–Einstein statistics and ➤Fermi–Dirac statistics. It is, in essence, the energy with which a particle can be transferred to or from a quantum system without changing its equilibrium with its surroundings.

chemical reaction A reaction in which ➤chemical change takes place.

chemical shift A change in the frequency (or equivalently the wavelength) of a ➤spectral line caused by a change in the chemical environment of an atom or nucleus. It is of importance in ➤nuclear magnetic resonance.

chemiluminescence The emission of light caused by a chemical reaction. ➤➤bioluminescence.

chemiosmotic coupling hypothesis ➤oxidative phosphorylation.

chemisorption ➤adsorption.

chemistry The study of the reactions between chemical species. There are three main branches of chemistry. **Physical chemistry** seeks to understand the physical basis for chemical change, based on ➤chemical bonding, ➤thermodynamics, and the study of ➤equilibrium and ➤chemical kinetics. **Organic chemistry** studies the behaviour of compounds of carbon; intimately related to this is the study of ➤biochemistry, where the organic compounds are involved in life processes. The third major area of chemistry is **inorganic chemistry**, which studies all the other elements apart from carbon. A final subdivision that straddles the last two areas is **organometallic chemistry**, currently an area of intense research.

chemoorganotroph ➤heterotroph.

chemosynthesis The ability to synthesize organic compounds using energy derived from inorganic reactions, exhibited by certain ➤bacteria. Such bacteria are termed **chemoautotrophs**, and include the nitrifying bacteria of the ➤nitrogen cycle.

chemotherapy The use of chemical substances in the treatment of disease. The term is particularly applied to the treatment of some cancers. ➤Cytotoxic substances are applied to the body in carefully controlled doses which are calculated to limit the growth of cancer cells while doing minimal damage to healthy body cells. However, since the main action of chemotherapeutic agents is to inhibit cell division, inevitably noncancerous cells are also affected and adverse side effects are common. The plant alkaloids vincristine and vinblastine, used in the treatment of childhood leukaemias, and alkylating agents such as chlorambucil, used to treat non-Hodgkin's lymphoma, are examples of substances used for chemotherapy.

Cherenkov radiation Radiation from a charged particle that is moving through a medium (usually with a high refractive index) faster than the speed of light in that medium. This produces an effect analogous to the shock wave produced in air when an aircraft moves faster than the speed of sound. Named after Pavel Alekseyevich Cherenkov (1904–90).

Chile saltpetre Traditional name for sodium nitrate, $NaNO_3$.

chimera (chimaera) An organism derived from the ➤genotypes of two different organisms, resulting from manipulation of early ➤embryos to fuse cells together. Such techniques are used to study the early stages of development. A 'shoat' is the result of hybridization of cells from a sheep and a goat: the resulting animal is goat-like but with the wool of a sheep. The commercial viability of such an animal has yet to be established. The production of chimeras raises a wide range of ethical issues.

chimeric DNA A recombinant ➤DNA molecule containing unrelated genes.

chip A single crystal of semiconductor (usually silicon), on which an ➤integrated circuit is created.

chiral A species that cannot be superimposed on its mirror image. Such a chiral species shows ➤optical activity. The most common source of **chirality** is an **asymmetric carbon atom**, that is, a carbon atom attached to four different groups, as in the amino acid alanine; some inorganic complexes are also chiral (see the diagram). ➤➤(R–S) system.

(a)   (b)

chiral (a) Alanine ➤enantiomers; (b) $[Co(en)_3]^{3+}$ enantiomers.

chi-squared test A widely used statistical procedure to test how well an observed ➤frequency distribution may be fitted by a theoretically predicted frequency distribution. The test involves calculating

$$\chi^2 = \sum (O - E)^2 / E,$$

where O is the observed frequencies and E the expected (predicted) frequencies and using statistical tables to assess the significance of the result obtained.

chitin A linear polymer of the glucose derivative N-acetyl-D–glucosamine (an amino sugar). Chitin is the main structural component of the exoskeletons of arthropods such as insects and crustaceans, and is also found in many fungi.

chloral (trichloroethanal) Trivial name for the molecule CCl_3CHO. **Chloral hydrate**, $CCl_3CH(OH)_2$, is used as a sedative.

chlor–alkali industry   The manufacture of chlorine and sodium hydroxide (respectively tenth and eighth in the top ten of industrial chemicals) by the **electrolysis of brine**.

The two current technologies, the **diaphragm cell** and the **membrane cell**, share several common features. The electrolyte is saturated brine (about 25% aqueous sodium chloride). The chloride ion is oxidized at the ➤anode to chlorine gas:

$$2Cl^- \rightarrow Cl_2 + 2e^-.$$

The titanium anode resists attack by chlorine. Water is reduced at the ➤cathode (made of steel or nickel) to hydrogen gas, also creating hydroxide ions:

$$2H_2O + 2e^- \rightarrow H_2 + 2OH^-.$$

As the products of the electrolysis react with each other, they must be kept apart. This is done in the diaphragm cell by using a porous asbestos diaphragm. The membrane cell uses a polymer membrane that allows cations to pass through. The membrane cell is currently considered to be the more environmentally sound.

chlorate   A salt of chloric acid. If an oxidation number is appended, the relationship to the acid is clear: chlorate(III) is the salt from chloric(III) acid. When no oxidation number is specified, the name is the traditional name for chlorate(V), ClO_3^-. Chlorate(I), ClO^-, is an important ingredient of bleaching powder.

chloric acid   One of a series of oxoacids of chlorine. An ➤oxidation number is appended to identify the specific acid: **chloric(III) acid**, $HClO_2$, is the acid in which the oxidation number of chlorine is +3. When no oxidation number is specified, the name is the traditional name for **chloric(V) acid**, $HClO_3$. The other two oxoacids are $HClO$, **chloric(I) acid** or hypochlorous acid, and $HClO_4$, **chloric(VII) acid** or perchloric acid. The strength of the acids increases very significantly as more oxygen atoms are added, as the extent of delocalization in the anion gets greater. Perchloric acid is a ➤strong acid. In addition it is a very strong oxidant, especially with organic compounds.

chloride   A compound containing the **chloride ion**, the anion formed from chlorine, Cl^-. The single negative charge is typical of ➤halide ions, as chlorine has seven electrons in its outer shell and so reacts to gain just one more in ➤ionic bonding (➤octet rule). The aqueous chloride ion can be detected by adding acidified ➤silver nitrate and observing a white precipitate of silver chloride.

chlorinating agent   A reagent that replaces a ➤hydroxyl group with a chlorine atom. The three most important examples are phosphorus pentachloride, PCl_5, phosphorus trichloride, PCl_3, and ➤sulfur dichloride oxide, SCl_2O.

chlorination   **1** The replacement of a hydrogen atom by a chlorine atom. The chlorination of alkanes is an important reaction in organic chemistry which is a typical radical substitution process occurring by a radical ➤chain reaction.

2 The treatment of water supplies with a chlorine-containing compound, such as sodium chlorate(I), in order to disinfect the water.

chlorine   Symbol Cl. The element with atomic number 17 and relative atomic mass 35.45, which is the most important of the ➤halogens. It is manufactured in the

➤chlor–alkali industry. The element itself is a yellow-green (Greek *chloros*) gas with a choking smell; it was the first substance to be used in chemical warfare during the First World War. The gas contains Cl_2 molecules, formed by covalent bonding between two chlorine atoms. Chlorine is a strong oxidant, oxidizing metals to form their chlorides. For example, sodium, when heated, burns in chlorine gas to form a white solid, sodium chloride:

$$2Na(s) + Cl_2(g) \rightarrow 2NaCl(s).$$

Metals usually form ionic chlorides which dissolve easily in water to form solutions containing the chloride ion, Cl^-; nonmetals usually form covalent chlorides which tend to hydrolyse in water, reacting to form acidic solutions (➤ionic bonding; ➤covalent bonding; ➤hydrolysis). For example, phosphorus pentachloride, PCl_5, reacts violently with water to form phosphoric acid. Chlorine forms a series of oxides, such as the yellow-brown gas Cl_2O, the yellow gas ClO_2 and the colourless oil Cl_2O_7. Chlorine dioxide, ClO_2, is used in large quantities to bleach wood pulp, despite being highly explosive.

chlorine-containing organic compounds The simplest examples are the chloroalkanes, such as **chloroethane**, CH_3CH_2Cl, which is used as a spray to treat sports injuries: it has a conveniently low boiling point of $12\,°C$, so when the skin supplies the latent heat to cause evaporation the skin cools. The chloroalkanes have the typical properties of ➤halogenoalkanes. More than one chlorine atom can be present in the molecule, as in ➤chloroform, $CHCl_3$. Aromatic compounds can also contain chlorine, as in **chlorobenzene**, C_6H_5Cl. Three famous chlorine-containing organic compounds are best known by their initials: the polymer ➤PVC, the insecticide ➤DDT and the disinfectant ➤TCP.

chlorite A traditional name for chlorate(III) ion, ClO_2^-. Compare the name ➤'chlorate' for the oxoanion with the higher oxidation number.

chloroethene (vinyl chloride) $CH_2{=}CHCl$ A colourless gas that is the monomer from which ➤PVC is made.

chlorofluorocarbon (CFC) A ➤halogenoalkane that contains both chlorine and fluorine, such as CCl_2F_2 or CCl_3F. CFCs were heavily used for air-conditioning systems, propellants for cosmetics and insecticides, refrigerants and foaming agents. However, their decomposition when exposed to ultraviolet light from the Sun to form chlorine radicals was recognized as a danger as early as 1974 by F. Sherwood Rowland. Chlorine radicals, produced when these rather unreactive CFCs reach the stratosphere, react with ozone: $Cl^{\bullet} + O_3 \rightarrow ClO^{\bullet} + O_2$. The depletion in the ozone level causes an ➤ozone hole. The search for replacements for CFCs has focused on molecules, such as ➤hydrofluorocarbons, that do not contain chlorine and hence cannot produce harmful chlorine radicals.

chloroform (trichloromethane) $CHCl_3$ A volatile colourless liquid, notorious from early films as the anaesthetic of choice with which to knock out a victim. Its former wide use as a solvent has been drastically reduced because of its proven carcinogenic effects.

chlorophyll One of a group of green pigments in photosynthetic organisms that capture light energy in the first phase (the ➤light reaction) of photosynthesis. Chlorophylls consist of a magnesium-containing ➤porphyrin ring structure linked to different side chains. Chlorophyll *a* is found in all photosynthetic groups. ➤Cyanobacteria have only chlorophyll *a*; higher plants have chlorophyll *b*; the brown algae have chlorophyll *c*. Chlorophylls function with a range of ➤accessory pigments in complexes called ➤photosystems which absorb a range of the visible spectrum and transfer the energy to the primary photochemical process of photosynthesis. ➤➤chloroplast.

chloroplast A ➤membrane-bound ➤organelle containing ➤chlorophyll, found in all ➤eukaryotic photosynthetic organisms (see the diagram). Chloroplasts are discoid in shape, and range from 5 to 10 μm in length; individual cells contain 1 to 40 of them, depending on the cell type. The chloroplast envelope consists of a double membrane, the inner layer of which forms stacks of flat 'thylakoid' ➤lamellae called **grana**. The matrix of the chloroplast is called the **stroma**. The ➤light reaction of photosynthesis occurs on the thylakoids, whereas the ➤dark reaction occurs in the stroma.

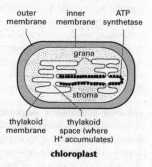

outer membrane · inner membrane · ATP synthetase · grana · stroma · thylakoid membrane · thylakoid space (where H⁺ accumulates)

chloroplast

chlorous acid Traditional name for chloric(III) acid (➤chloric acid).

cholesteric liquid crystal ➤liquid crystal.

cholesterol A steroid ➤lipid derivative forming part of the structure of ➤plasma membranes, particularly in animal cells, and having a wide range of key metabolic functions. Cholesterol is synthesized mainly in the liver, but is also produced in other organs such as the intestine, skin, nervous tissue and the ➤gonads. An adult human can synthesize approximately 800 mg of cholesterol a day, and in addition a substantial amount may be present in the diet. Cholesterol regulates its own biosynthesis via the LDL–cholesterol complex (LDL is a low-density ➤lipoprotein) and is itself a precursor for many important steroid ➤hormones. Control of the levels of cholesterol in the body is of interest since it is implicated in some forms of coronary heart disease and ➤atherosclerosis.

choline $HOCH_2CH_2N^+(CH_3)_3$ An amino alcohol, found in cells, where it is involved in the metabolism of ➤phospholipids, particularly those associated with the membranes of nerve cells. It is a member of the B group of ➤vitamins. ➤➤acetylcholine.

chord A straight line joining two points on a curve.

chordate A member of the animal ➤phylum **Chordata**, containing organisms that have a dorsal supporting rod of tissue (a ➤notochord) at some stage of their life history. All ➤vertebrates are chordates, but the phylum also contains simpler types, such as sea-squirts, in which only the free-swimming larva has a notochord.

chromate(VI) A salt or ester of chromic(VI) acid. The chromate(VI) ion is CrO_4^{2-}, which forms a yellow aqueous solution. ➤dichromate(VI) ion.

chromatic aberration ➤aberration.

chromatid A ➤chromosome that is still attached to its copy (sister chromatid) at the ➤centromere after replication in the early phases of ➤meiosis or ➤mitosis.

chromatin A complex of ➤eukaryotic ➤DNA combined with proteins; the proteins that bind with DNA include ➤histones. The unfolding of the compact form of chromatin (**heterochromatin**) frequently activates ➤transcription.

chromatogram A record obtained by ➤chromatography.

chromatography A separation and identification technique that relies on the different ➤partition coefficients or ➤adsorption characteristics of the components of a mixture for two different phases: the **mobile phase** and the **stationary phase.** Chromatography is particularly effective if the components are coloured, as their separation can then be clearly seen as separate bands of colour; this gives the technique its name. In **paper chromatography** the stationary phase is water held within a piece of paper and the mobile phase is a solvent, often an aqueous solution of ammonia, or an alcohol, or both. It may be used to separate the pigments in chlorophyll. **Thin-layer chromatography** (TLC) uses a thin layer of alumina or silica on a plate as the stationary phase. The detection by TLC of the molecule pregnanediol in urine confirms pregnancy. TLC, using two different solvents in perpendicular directions, is particularly effective for identifying amino acids. **Liquid chromatography** (or **column chromatography**) uses a column, typically containing alumina or silica, down which a solvent runs. A column containing the polymer Sephadex may be used to separate the ➤purines. **High-performance liquid chromatography** (HPLC) is an advanced version of this technique in which solvent is forced through a thin column under pressure. The components are often identified by ultraviolet ➤spectroscopy. HPLC with ➤chiral columns may be used to separate chiral molecules. **Gas–liquid chromatography** (GLC) uses a liquid coated onto a support held in a thin tube folded on itself as the stationary phase. A **carrier gas,** usually nitrogen, is used to carry the sample through the apparatus. The components are often identified by mass spectrometry (MS). This joint technique (GCMS) has been used since 1983 for routine drug testing of athletes. ➤gel filtration.

chrome alum An ➤alum containing Cr^{3+} ions and K^+ ions.

chrome yellow The traditional name for lead(II) chromate, $PbCrO_4$, which is a yellow pigment widely used for road markings.

chromium Symbol Cr. The element with atomic number 24 and relative atomic mass 52.00, which is in the first row of the ➤transition metals. The metal itself is hard, lustrous and resistant to corrosion and so is much used to protect steel objects either by **chrome-plating** or as a constituent of ➤stainless steel. Chromium shows the typical properties of the transition metals. It forms compounds with a number of different oxidation numbers. For example, the two common oxidation numbers in aqueous solution are +3 as in the chromium(III) ion, $Cr^{3+}(aq)$, a green aqueous solution, and +6 as in the dichromate(VI) ion, $Cr_2O_7^{2-}(aq)$, an orange aqueous solution. A common test for reductants is to add acidified dichromate(VI) and observe a colour change from orange to green. There are other oxidation numbers too, such as +4 in the oxide 'chrome dioxide', CrO_2, which was heavily used as a coating for video and audio tape. The more common oxide is chromium(III) oxide, Cr_2O_3, a green crystalline solid which is a rare example of an *inorganic* compound implicated as a carcinogen. (Compounds with the highest oxidation number, **hexavalent chromium**, achieved notoriety for similar reasons in the film *Erin Brockovich*.) Some of the different colours chromium shows have already been mentioned; the aqueous chromium(II) ion, Cr^{2+}(aq), which is only stable in the absence of air, is blue. An interesting organometallic compound is the ➤sandwich compound ➤dibenzenechromium.

chromophore A group in a molecule that causes absorption of electromagnetic radiation, most usually used in the ultraviolet region. The ➤carbonyl group and the diazo group (➤diazo dyes) are good chromophores.

chromosome A thread-like structure in the nucleus of ➤eukaryotic cells composed of ➤DNA and ➤protein. Chromosomes carry the genetic information of the cell in their DNA, organized into linear arrangements of ➤genes. The locations of genes on particular chromosomes and their relative positions along the chromosome can be mapped by classic breeding experiments, or by more recent techniques for probing specific DNA sequences (➤DNA probe). Chromosomes can be observed directly in cells only while they are dividing. Each species has a constant chromosome number; humans have 23 pairs of chromosomes. Chromosome 22 was the first human chromosome to be fully sequenced.

chromosphere The layer of the ➤Sun surrounding the ➤photosphere.

chromyl chloride CrO_2Cl_2 A deep red fuming liquid, which is a strong oxidant.

chrysalis ➤pupa.

chymotrypsin A protein-digesting enzyme that hydrolyses amino-acid chains (➤hydrolysis). In particular, it acts at residues (groups forming part of a larger molecule) such as phenylalanine, tyrosine and methionine. It is commonly used in analysing the primary structure of ➤proteins.

Ci Symbol for the unit ➤curie.

cilium (plural cilia) ➤undulipodium.

cinnabar The principal ore of mercury, mercury(II) sulfide, HgS, which exists as a red crystalline solid.

cinnamic acid (3-phenylpropenoic acid) $C_6H_5CH{=}CHCOOH$ One of two (*cis–trans*) isomers (the *trans* isomer being more important) which are contained in balsams and are used in flavours, perfumes and pharmaceuticals.

circadian ➤biological clock.

circle A plane ➤curve, every point of which is a fixed distance from a fixed point, or the region enclosed by such a curve. The fixed distance r is the ➤radius of the circle; the fixed point is its centre. With the centre as origin, the Cartesian equation of the circle is $x^2 + y^2 = r^2$. Its circumference is $2\pi r$ and the area enclosed is πr^2, where ➤π is one of the fundamental constants of mathematics. Of all closed curves with a given ➤perimeter, the circle encloses the greatest area.

circuit **1** An arrangement of electrical or electronic components designed to be part of a device.

2 A closed loop of electrical components (in the context of **making** or **breaking a circuit**), arranged in such a way that allows current to flow.

circuit board The solid, insulating support for an electronic circuit. ➤➤printed circuit.

circuit-breaker A device that breaks a circuit when a current that exceeds a chosen value flows. Unlike a ➤fuse, it does not break destructively but in a way such that it can be reset to its original state. Typically, the force produced by electromagnetic ➤induction breaks the circuit mechanically.

circular functions Another name for ➤trigonometric functions arising from the fact that the point (cos θ, sin θ) lies on the circle $x^2 + y^2 = 1$.

circularly polarized A state of ➤polarization of light in which the electric field vector rotates as a circle. It is the special case of ➤elliptically polarized light in which the two components are equal. ➤➤plane-polarized.

circular measure The measurement of angles in ➤radians.

circumference Usually, the ➤perimeter of a circle but occasionally also used for the perimeter of other closed curves.

***cis–trans* isomerism (geometric isomerism)** A form of ➤stereoisomerism in which the isomers differ in the positions of the groups relative to a double bond. One isomer has similar groups arranged across (Latin *trans*) the molecule; the other isomer has them on the same side (Latin *cis*), as shown in diagram (a). As a double bond resists rotation, the two isomers are not normally interconverted. *Cis–trans* isomerization can be achieved if a photon is absorbed. Such an event is the primary mechanism of vision in all animal eyes when the molecule 11-*cis*-retinal becomes all-*trans*-retinal. *Cis–trans* isomers can also exist in inorganic complexes; the *cis* isomer *cis*-platin (see diagram (b)) is a potent anticancer drug.

cis-but-2-ene   *trans*-but-2-ene

(a)

cis-platin   *trans*-platin

(b)

***cis–trans* isomerism**

citrate A salt or ester of citric acid.

citric acid (2-hydroxypropane-1,2,3-tricarboxylic acid) A white crystalline solid (see the diagram) present in citrus fruits such as lemons. It is used extensively in soft drinks. ➤➤Krebs cycle.

```
        COOH
        |
        CH2
        |
HO—C—COOH
        |
        CH2
        |
        COOH
```
citric acid

citric acid cycle ➤Krebs cycle.

CJD Abbr. for ➤Creutzfeldt–Jakob disease.

Cl Symbol for the element ➤chlorine.

cladistics ➤taxonomy.

Claisen condensation A ➤base-catalysed condensation reaction between two molecules of an ➤ester to form a β-keto ester, as when ethyl ethanoate, $CH_3COOCH_2CH_3$, forms $CH_3COCH_2COOCH_2CH_3$. Named after Ludwig Claisen (1851–1930).

class A category used in the classification of organisms (➤taxonomy) which contains one or more ➤orders. In turn, classes are grouped together in a ➤phylum.

classical physics Those areas of physics that make no reference to ➤quantum mechanics. The term is used almost exclusively as a contrast to 'quantum physics'.

classification (of organisms) ➤taxonomy. Humans are the species *sapiens* of the genus *Homo* of the order Primate of the class Mammalia of the subphylum Vertebrata of the phylum Chordata of the animal kingdom. ➤Appendix table 8.

clathrate (cage compound) A compound formed by trapping a small molecule in the holes of a solid matrix of some kind. ➤Zeolites are particularly good at imprisoning other molecules, such as benzene. Ice can also form clathrates.

clay A weathered form of aluminosilicate, usually of very complex composition.

Clemmensen reduction The selective reduction of carbonyl compounds (➤carbonyl group) using amalgamated zinc and concentrated hydrochloric acid: the CO group is reduced to CH_2. Named after Erik Christian Clemmensen (1876–1941).

climate change A catch-all term to describe changes to global weather patterns: the main driver of the change is ➤global warming. The term 'climate change' reflects the fact that some parts of the world may, in the short term, benefit from this warming, by experiencing less severe winters for example.

clonal selection theory A theory that explains the action of ➤B cells in producing ➤antibodies. The theory states that a foreign antigen selects the lymphocyte with an antibody of complementary shape by combining with it. This triggers the specific antibody-secreting lymphocyte to multiply, thereby increasing the number of antibodies available to attack that particular antigen. ➤➤immune system.

clone A population of genetically identical organisms or cells derived from an original single organism or cell by ➤asexual reproduction or ➤mitosis. The term is also applied to populations of viruses or DNA molecules copied from original parent viruses or DNA molecules. The ability to clone DNA to produce large amounts of identical fragments is an important technique in ➤genetic engineering. The birth of the first cloned mammal, Dolly the sheep, was announced in 1997 by scientists at the Roslin Institute, Scotland. ➤DNA replication; polymerase chain reaction.

cloning vector A ➤virus, ➤plasmid or artificial chromosome that can be used to insert a DNA fragment of interest into a host cell in order to produce multiple copies of the fragment. Cloning vectors are one of the basic tools of ➤genetic engineering. ➤➤vector (4).

closed shell ➤electronic structure.

close packing Arranging spheres to minimize the space occupied; the resulting two structures are adopted by nearly fifty elements in their crystals. When identical spheres are packed together in one layer, the structure is a regular hexagonal array. Continuing the closest packing in three dimensions, each sphere in the second layer is placed over the gaps in the first layer. A third layer can be added in two ways. One way (see diagram (a)), called **hexagonal close packing** (h.c.p.), places the spheres in the third layer directly above those in the first layer. Magnesium and zinc adopt this structure. Alternatively, the spheres in the third layer could lie over the gaps in the first layer that were not covered by the second layer (see diagram (b)). This is **cubic close packing** (c.c.p.); the structure produced is also frequently called **face-centred cubic** (f.c.c.). Aluminium, copper, silver and gold adopt this structure. ➤➤Bravais lattice.

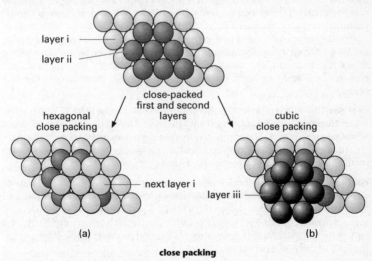

layer i
layer ii

close-packed first and second layers

hexagonal close packing

cubic close packing

next layer i

layer iii

(a)   (b)

close packing

cloud Condensed water vapour or sublimed ice (or a mixture of both) in the atmosphere. Clouds are formed when air that is saturated with water vapour cools. At ground level, clouds are called ➤fog.

cloud chamber A detector for ➤ionizing radiation. Supersaturated vapour condenses at the trail of ions left by the radiation. Compare ➤bubble chamber.

cluster A group of many stars held together by gravitational attraction. The smaller **open clusters** contain from about a dozen to a few thousand stars and are found in the galactic plane. **Globular clusters** contain from tens of thousands to a million or more stars, with the highest density of stars closest to the centre; they are found in a halo surrounding our Galaxy, and in similar haloes round other galaxies.

cluster compound A compound in which groups of transition metal atoms are held together, as in $[Mo_6Cl_8]^{4+}$.

cm Symbol for the unit ➤centimetre.

Cm Symbol for the element ➤curium.

CMOS Abbr. for complementary metal oxide semiconductor; a standard set of ➤logic gates, using complementary n-type and p-type transistors.

Cnidaria ➤coelenterate.

CNS Abbr. for central ➤nervous system.

Co Symbol for the element ➤cobalt.

CoA Abbr. for coenzyme A (➤acetyl coenzyme A).

coal A naturally occurring black or brown solid fuel formed from the arrested bacterial decay of vegetation that grew millions of years ago. The higher 'rank' the coal is, the higher the carbon content is, from about 50% in **peat** through **lignite**, **bituminous** and **carbonaceous**, to **anthracite**, which is about 95% carbon. Carbonization of coal formed coal gas, coal tar and ➤coke. The reserves of coal in the UK far exceed those of petroleum and natural gas, with the potential for lasting several hundred years. The Industrial Revolution was founded on 'King Coal', and coal may well again assume a dominant place in energy needs later in the 21st century.

coal gas The volatile product of carbonization of coal at about 1250 °C, which forms a useful fuel that was used as town gas until the second half of the 20th century, when ➤natural gas took over. The energy density of coal gas is about 18 MJ m^{-3}. The main constituents are about half hydrogen and over a quarter methane, together with smaller quantities of carbon monoxide, alkenes, nitrogen and carbon dioxide.

coal tar The liquid product of carbonization of coal, consisting of a range of components, especially aromatic hydrocarbons, such as benzene and phenol, and ammoniacal liquor. Despite being a black smelly liquid, it was further manufactured into a range of products from soap to perfumes. In the 19th century especially, coal tar was the raw material for a wide range of chemical processes (a role taken over in the 20th century by ➤petroleum).

coaxial Of cylinders, sharing the same ➤axis; thus **coaxial cable** consists of a central wire surrounded by a cylindrical shell of insulator, in turn surrounded by a cylindrical conductor (typically braided).

cobalt Symbol Co. The element with atomic number 27 and relative atomic mass 58.93, which is in the first row of the ➤transition metals. The metal itself is hard, silvery and magnetic, and together with nickel is used in alloys with iron to form permanent magnets (≫Alnico). One isotope, cobalt-60, is frequently used in radiotherapy.

Cobalt shows the typical properties of a transition metal. It can be found with a number of oxidation numbers. The most important is +2, as in aqueous cobalt(II) ion, Co^{2+} (aq), which forms a pink solution, or the green solid oxide CoO. Anhydrous cobalt(II) chloride is blue: the colour change from blue to pink is used in **cobalt chloride paper** to test for water. With water ligands, the cobalt(III) ion is a strong oxidant. When complexed with ammonia, however, the cobalt(III) state forms hundreds of ammine complexes, such as $[Co(NH_3)_6]^{3+}$. Coloured compounds abound; one particularly interesting example is $[Co(NH_3)_4Cl_2]^+$, the *cis* isomer of which is violet and the *trans* isomer green. **Cobalt blue** glass is used in flame tests as it can absorb the yellow light of sodium, allowing the much less intense lilac colour of any potassium present to be seen more clearly. A cobalt carbonyl, $Co_2(CO)_8$, is used as the catalyst in the ➤Fischer–Tropsch synthesis. Vitamin B_{12} is a cobalt-containing organometallic compound.

COBOL Acronym for common business-oriented language, a high-level computer language widely used in commercial applications.

coccus One of a number of species of spherical bacteria. Members of the ➤genus *Staphylococcus* cause respiratory infections, boils and some forms of food poisoning in humans.

codon A sequence of three consecutive ➤nucleotides in a DNA or RNA molecule that forms a unit of genetic information specifying a particular amino acid to be inserted in a particular position in a ➤polypeptide chain. ≫genetic code; protein synthesis; ribosome.

coefficient 1 A number multiplying an algebraic expression.

2 A dimensionless constant of proportionality relating one physical quantity to another.

coefficient of expansion ➤expansivity.

coefficient of friction Symbol μ. The ratio of the frictional force F on a body to the force R between it and a surface (the direction of F being parallel to that surface): $F = \mu R$. Different coefficients of friction apply depending on whether the body is moving, stationary or rolling.

coefficient of restitution ➤Newton's experimental law.

coelenterate A member of the animal ➤phylum **Coelenterata** (or **Cnidaria**), containing animals with a simple radial body pattern composed of two layers of cells separated by a jelly-like matrix and enclosing a single cavity (the **coelenteron**) serving as a ➤gut. Such an arrangement is called **diploblastic.** Coelenterates are mainly marine organisms and include jellyfish, comb jellies and sea anemones, as well as the freshwater hydra. They are characterized by unique cells called **nematocysts** which, when suitably stimulated, can eject a number of different types of barbed threads by which they capture and paralyse their prey.

coenzyme A nonprotein organic molecule that plays an accessory yet necessary role in the catalytic action of an ➤enzyme. Many ➤vitamins function as coenzymes, particularly those of the B group. ➤➤FAD; NAD$^+$; NADP$^+$; thiamine.

coenzyme A (CoA) ➤acetyl coenzyme A.

coenzyme Q A ➤quinone coenzyme functioning in the ➤electron–transport chain in respiration.

coenzyme I ➤NAD$^+$.

coenzyme II ➤NADP$^+$.

coercive force The magnetic field strength required to bring to zero (and subsequently reverse) the remanent magnetic flux density (➤remanence) of a permanent magnet. ➤ hysteresis.

cofactor **1** (Biol.) A nonprotein substance that binds to specific regions of an ➤enzyme, thereby maintaining the shape of the active site or taking part directly in the binding of ➤substrate to the active site. Cofactors include metal ions, ➤coenzymes and ➤prosthetic groups. Some metal cofactors are ➤trace elements.
  2 (Math.) ➤determinant.

coherence Of waves or similar signals, or their source, a measure of how well correlated the waves are between two times (**temporal coherence**) or two nearby points (**spatial coherence**). Coherent waves differ by a constant ➤phase (2). The **coherence time** is the period over which the phase relationship remains nearly constant. Compare ➤incoherent wave.

coke The solid product formed when coal, most particularly the form called **coking coal**, is heated in the absence of air in **coke ovens**. Coke is largely carbon. Its main use is in the ➤blast furnace.

cold-bloodedness ➤poikilothermy.

cold emission Nonthermionic emission of electrons from a surface; it is typically caused by strong electric fields (➤field emission).

collagen A family of highly characteristic fibrous ➤proteins found in all multicellular animals, constituting a quarter of the total protein in mammals. Collagen forms a major part of the protein component of bone and connective tissue.

collective excitation Excited states of a system that occur when what are normally considered individual particles of a system move in a coherent way (as ➤normal modes) so as to produce an oscillation. For example, the electrons in a conductor may oscillate to produce a ➤plasmon, or the particles in a solid may oscillate to produce a ➤phonon. ➤➤second quantization.

collector One of the electrodes of a ➤bipolar junction transistor.

collider A particle ➤accelerator, usually a ➤synchrotron, that is designed to cause collisions between particles at very high energies. ➤CERN.

colligative properties Those properties of a solution that depend solely on the number of solute particles present and not on their nature. Colligative properties can be understood in terms of the ➤entropy changes taking place when a solution is made. A nonvolatile solute will **lower the vapour pressure** and **elevate the boiling point** of a solvent, as the solution is more entropically favoured (due to an entropy of mixing) relative to the vapour than it was in the absence of the solute. A solute that does not dissolve in the solid solvent will **depress the freezing point** of a solvent; again, the solution is more entropically favoured and so forms at a lower temperature. The fourth colligative property is ➤osmosis.

collimator An optical device in which incident light passes through a system of lenses or prisms in order to produce a parallel beam.

collinear Three or more points are collinear if they all lie on a straight line.

colloid A colloid consists of a **dispersed phase** surrounded by a **dispersion medium.** In 1861 Thomas Graham distinguished between substances by noting whether they could penetrate a membrane during ➤dialysis; those that did not he called 'colloids'. A further test for a colloid is that a **colloidal solution** will scatter light, whereas a true solution will be transparent (➤Tyndall effect). Both these properties are explicable by assuming that the radius of the particles is significantly greater than that of ordinary molecules. Colloids range in radius from 1.7×10^{-9} m for colloidal gold up to about 10^{-6} m. Many materials such as proteins, vegetable fibres and rubber are most stable in the **colloidal state.** The major classes of colloid are **solid sols,** suspensions and smokes (where the dispersed phase is a solid); **gels,** ➤emulsions and fogs (where the dispersed phase is a liquid); and **foams** (where the dispersed phase is a gas). In **aerosols** the dispersion medium is a gas.

colon ➤gut.

colony A distinct population of organisms usually occupying a specific location. Groups of nesting seabirds are examples of colonies. The term is also applied to the descendants of a single bacterial cell growing on an agar plate which form a mass visible to the naked eye.

colorimeter A device for measuring ➤colour, in particular the hue, saturation and luminous intensity.

colour The subjective appearance of the wavelength or wavelengths present in a beam of light perceived by the eye. Rather than a continuous sensor of wavelength, the eye perceives colour using three distinct sensors, each of which responds over a range of wavelengths concentrated around red, green or blue-violet. It is these three intensities of response that determine the colour perceived. Two different systems can be used to parametrize colour: **RGB** adds together red, green and blue; **HSL** uses hue, saturation and ➤luminous intensity. **Hue** is the angle around a colour wheel indicating which primary colour predominates; **saturation** describes the purity of the colour (for example, how much red and blue is mixed with a green).

colourblindness The inability of some individuals to distinguish between certain combinations of colours, such as reds and greens. The inability is genetically determined by a ➤recessive ➤allele carried on the X chromosome, and is thus sex-linked (➤sex-linkage).

colour centre An electron state in an ionic crystal that is localized around a ➤defect. Transitions to excited states often occur with energies in the visible spectrum, giving colour to a crystal which would otherwise be transparent.

colour charge A property of ➤quarks that corresponds, for the ➤strong interaction, to electric charge for the electromagnetic interaction. The colour charge consists of three components (labelled red, green and blue, hence the name) because the local symmetry of ➤quantum chromodynamics is ➤SU(3), while that of ➤quantum electrodynamics (the field theory of electromagnetism) is just U(1), requiring only one component of charge.

colour vision The **trichromatic theory** of colour vision relies on there being three different functional types of cone (➤eye) which are sensitive to different wavelengths of light (red, green and blue). Rods and cones are sensitized by light to generate impulses which pass along the optic nerve to the brain, where they are interpreted as images in the visual cortex. The nerve impulses are initiated by the bleaching effect of light on a pigment (**rhodopsin** or **visual purple**).

columbium ➤niobium.

column chromatography ➤chromatography.

coma **1** (Phys.) ➤aberration.
  2 (Astron.) A bright cloud of ions and dust surrounding the nucleus of a comet.
  3 (Med.) A deep state of unconsciousness in which a patient fails to demonstrate response to stimuli such as sound, light and touch. Coma may be induced by drugs or by severe trauma. Comas commonly last from a few days to several weeks. The outcome ranges from death to full recovery, dependent on the degree of pharmacological or traumatic damage, but many patients recovering from coma exhibit on-going neurological symptoms.

combination A selection of objects from a given set in which the order of selection is immaterial. The number of combinations of k objects chosen from n is written $\binom{n}{k}$ or ^nC_k and equal to $n!/k!(n-k)!$ (➤binomial coefficient). Compare ➤permutation.

combining volumes, law of ➤Gay-Lussac's law.

combustion The burning of a substance in oxygen. Combustion of fuels, notably ➤fossil fuels such as natural gas, provides much of the heat needed in everyday life and the vast majority of the energy for movement from place to place, for which the ➤internal combustion engine is most responsible. Combustion of a hydrocarbon produces water and either carbon dioxide (in plentiful air) or carbon monoxide (in a restricted air supply). The mechanism of combustion is most usually a radical ➤chain reaction. **Combustion analysis** measures the mass of water (absorbed by magnesium perchlorate, $Mg(ClO_4)_2$) and carbon dioxide (absorbed by ➤soda-lime) formed on complete combustion from which the carbon and hydrogen content of an organic compound can be determined. For a quantitative measure of the heat released by a particular fuel, see ➤standard enthalpy change of combustion.

comet An interplanetary body centred around a nucleus of dust and ice. When a comet nears the Sun, the Sun's heat vaporizes the ice to produce the ➤coma and ➤tail of the comet. ➤➤Halley's Comet.

commensalism ➤symbiosis.

common ion effect Addition of a substance that contains an ion in common with one of those present in the solution of a sparingly soluble compound causes some of the sparingly soluble compound to precipitate. This occurs because the value of the ➤solubility product is constant.

common salt Sodium chloride, NaCl.

commutative A property of a ➤binary operation $*$ defined on a ➤set such that for all a and b in the set, $a * b$ is the same as $b * a$. For example, on the set of three-dimensional ➤vectors, addition of vectors is commutative but forming the ➤vector product is noncommutative.

commutator **1** (Math.) For a pair of operators (or matrices) A and B, the commutator is defined as:

$$[A, B] = AB - BA.$$

Commutators play an important role in, for example, quantum mechanics, particularly in the ➤Heisenberg uncertainty principle.

2 (Phys.) Part of an electric motor or generator that connects successive segments of the ➤rotor to the ➤stator (or vice versa) to enable a continuous torque to be maintained (and hence current to be generated continuously).

compass A device for sensing magnetic north. The simplest compass is just a floating magnetized needle; more sophisticated systems have mechanisms for keeping them horizontal or are based on sensors with no moving parts. ➤➤gyrocompass.

competitive inhibition A process by which the activity of an ➤enzyme is reduced by another substance. Specifically the substance is a structural analogue of the natural ➤substrate and competes with it for access to the active site. Competitive inhibitors have a wide range of pharmaceutical applications. Compare ➤noncompetitive inhibition. ➤➤sulfa drugs.

compiler A computer program that converts a program written in a ➤high-level language to its ➤low-level language equivalent. This process is completed before the program can be executed, so the resulting compiled code forms a stand-alone program, independent of the high-level code (compare ➤interpreter). The advantage of compilation is that it produces a program which will run faster, but the disadvantage is that the process of compilation must be repeated whenever the program is changed.

complement A group of blood serum proteins made by a number of body cells, particularly in the liver. They are normally inactive but in the presence of invading microorganisms, particularly bacteria, they become activated and help identify the foreign cells as invaders. They thus work in association with the ➤immune system as part of the body's defence mechanism.

complementary angles Two ➤angles that add up to $90°$.

complementary colour The colour of light that must be added to a specified colour in order to produce white light. Thus yellow and blue are complementary. This additive property applies to light, for which colours add naturally, not to paints or filters, for example, for which colours subtract. ➤primary colour.

complementary DNA ➤DNA.

complementary function ➤linear differential equation.

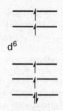

complex
The electronic structure of a high-spin d^6 complex consists of four electrons in the lower of two energy levels and two in the higher. There are thus four unpaired electrons.

complementary (of the bases in DNA) ➤base pairing.

complex (coordination compound) A collection of species held together by ➤coordinate bonding (hence the alternative name). The ➤Lewis base when forming a complex is described as the **ligand**. Complexes can be charged as in **complex ions**, such as $[Fe(H_2O)_6]^{2+}$ or $[Fe(CN)_6]^{4-}$, or uncharged as in carbonyl complexes such as $Fe(CO)_5$.

A complex ion consists of a central metal ion surrounded by a certain number of ligands, each of which donates an electron pair to the central metal ion. Examples of ligands include uncharged molecules with at least one lone pair, such as water in $[Fe(H_2O)_6]^{2+}$ or ammonia in $[Ni(NH_3)_6]^{2+}$, or negatively charged ions, such as cyanide ion in $[Fe(CN)_6]^{4-}$ or chloride ion in $CuCl_4^{2-}$.

Complexes involving d orbitals undergo ligand field splitting (➤ligand field theory). Hence complexes with from four to eight electrons can occur either as **low-spin** or **high-spin** complexes. Low-spin complexes, such as the d^6 ion $[Fe(CN)_6]^{4-}$, have all six of the electrons paired in the lower energy level. High-spin d^6 complexes, such as $[Fe(H_2O)_6]^{2+}$, have only four of the electrons in the lower energy level and two in the upper, producing four unpaired electrons (see the diagram) and hence ➤paramagnetism.

complex conjugate The number $x - iy$, denoted by $\bar{z}$ or z^*, associated with a ➤complex number $z = x + iy$.

complex function A ➤function defined on a region of the ➤complex plane which takes ➤complex number values.

complex number A number of the form $z = x + iy$ where x and y are ➤real numbers (the ➤real and ➤imaginary parts, respectively). Complex numbers may be represented on an ➤Argand diagram and may be written in the alternative polar form $z = r(\cos\theta + i\sin\theta) = re^{i\theta}$ where r and θ are the ➤modulus and ➤argument of z, respectively. Every operation (such as adding, multiplying, dividing, taking roots, exponentiating) that makes sense for real numbers can be extended to complex numbers, although there are surprises: $\sin z = 3$ has solutions; one value of i^i is real; the ➤fundamental theorem of algebra; and Euler's exquisite formula $e^{i\pi} + 1 = 0$ relating five of the most important numbers in mathematics.

complexometric titration The analysis of a solution by formation of a ➤chelate from the metal ions in the solution, most often using ➤EDTA. The hardness of water can be determined by EDTA titration in a slightly basic buffer solution, using the dye ➤eriochrome black T as indicator.

complex plane Alternative name for the ➤Argand diagram.

component **1** (Math.) The components of the ➤vector $x\mathbf{i} + y\mathbf{j} + z\mathbf{k}$ are the vectors $x\mathbf{i}$, $y\mathbf{j}$, $z\mathbf{k}$ along the coordinate axes.
2 (Chem.) One of the substances in a ➤mixture. ➤phase rule.

composite material A combination of two or more different materials with superior mechanical properties than are possessed by either component. A biological example is wood, which is a composite of cellulose fibres cemented together by lignin. Ancient manufactured examples include wattle and daub used to make homes. A more modern example is reinforced concrete, which uses steel rod reinforcements to counteract the weakness of concrete in tension. Glass fibre reinforced plastic (GRP) composites have a high strength-to-weight ratio and are often used for the hulls of boats.

compound A species formed by the chemical combination of two or more elements. The compound iron(II) sulfide has very different properties from a ➤mixture of iron and sulfur. For example, the mixture can be separated by a magnet; the compound cannot. Dilute hydrochloric acid reacts with iron in the mixture to give hydrogen gas; it reacts with the compound to form hydrogen sulfide gas.

compound microscope An optical ➤microscope that consists of more than one lens.

compound pendulum ➤pendulum.

compressibility The reciprocal of the ➤bulk modulus of a material. It is the fractional change in volume per unit pressure applied.

compression ratio In an internal combustion engine, the ratio of the volume of the cylinder before compression to its volume after compression (➤Otto cycle). A compression ratio of about 9 or 10 is typical in a petrol engine.

compressive stress The stress in a material that is being compressed (along the axis of compression). Compare ➤tensile stress.

Compton effect (Compton scattering) The scattering of electromagnetic radiation by a free electron. Classically, the electron is forced to oscillate by the oscillating electromagnetic field, and re-radiates at lower frequency. In the quantum-mechanical picture, in which a photon collides with an electron at rest, the energy of the photon is reduced in the collision (which must conserve momentum), resulting in an increase of wavelength, which is given by $\Delta\lambda = (2h/m_0c)\sin^2(\frac{1}{2}\phi)$, where h is the ➤Planck constant, m_0 is the rest mass of the electron, c is the speed of light and ϕ is the angle through which the photon is scattered. The quantity h/m_0c is known as the **Compton wavelength** of the electron. Named after Arthur Holly Compton (1892–1962).

computer A device capable of input, data processing and output. A stored-program computer, such as those in current use, follows an ►algorithm specified by the ►program that it is executing. The first computers were mechanical, but they have been superseded by electrical devices since their development in the 1950s. ►analogue computer; digital computer.

concave Not ►convex. Thus a concave ►polygon such as a chevron has a 're-entrant' interior angle greater than 180°.

concavo-convex Describing a lens with one concave face and one convex face.

concentration Symbol c. A measure of the quantity of a dissolved substance in a stated quantity of a solution. For example, the concentration of the ions present in bottled water is usually quoted in milligrams per litre. In chemistry, the **molar concentration** of a solution is defined as the amount of substance (in►moles) of the ►solute per unit volume of the solution. The usual units of molar concentration are mol dm^{-3}. ►►molality.

concentric Of circles, having a common centre.

condensation 1 ►change of state.
  2 ►Bose condensation.

condensation polymer A polymer formed by a ►condensation reaction between two compounds each of which has *two* ►functional groups. Polyamides, commonly called nylons, are made from diacyl chlorides together with diamines. 1,2-Diaminohexane and hexanedioyl chloride form nylon-6,6 (see the diagram at ►nylon). The —CONH— link is strongly reminiscent of the peptide link in proteins. One use of nylon is for **Velcro** fastenings. ►►aramid fibre.
  ►Polyesters are made from diacyl chlorides together with dialcohols. Polyethylene terephthalate (**PET**), discovered by John Whinfield in 1941, is the most widely used, for example to make containers for carbonated drinks. PET can be drawn into a filament to produce the most important synthetic fibre, called ►Terylene in the UK and **Dacron** in the US. PET can be used to make strong and yet lightweight plastic film, called **Mylar**, which is used for the backing for audio tapes. ►►polycarbonate.

condensation reaction A reaction in which two molecules join together, with the loss of a small molecule, typically water. The most important synthetically are the ►base-catalysed condensation reactions.

condensed matter Matter in which atoms and molecules interact closely with each other, thus liquids and solids as distinct from gases.

condenser 1 An apparatus for converting a vapour into a liquid. The hot vapour flows along a cooled tube, to which the vapour can transfer its heat, in so doing condensing to liquid. In most condensers, such as the ►Liebig condenser, the tube is cooled with flowing water. ►►reflux.
  2 An obsolete name for a ►capacitor.

condenser lens An optical device that converges light rays from a source of illumination, typically at an object to be viewed by a microscope.

conditional probability The ➤probability that an event occurs assessed in the light of the knowledge that some other event has occurred. The conditional probability of A given B is denoted $p(A|B)$ and defined as $p(A \cap B)/p(B)$ (➤probability for notation). ➤➤Bayesian statistics and probability.

conditioning The process by which an animal can be trained to respond to an otherwise inappropriate stimulus. In a series of classic experiments, Ivan Pavlov observed that dogs would salivate in response to the sound of a bell after a period of 'conditioning' during which they were given food at the same time as the bell was rung. Conditioning forms the basis of much of human learned behaviour, and is exploited by commercial advertising.

conductance The reciprocal of ➤resistance.

conduction The transmission of heat (**thermal conduction**) or electric charge (**electrical conduction**) through a material without any macroscopic motion. Conduction in metals is caused by the movement of electrons. Thermal conduction in ionic solids is mostly due to the transfer of thermal energy as lattice vibrations.

conduction band The highest, partially occupied ➤energy band of a metal or semiconductor. Because the conduction band is not full, electrons can be moved to states of very similar energy by a potential difference; the flow of electrons constitutes an electric current.

conductivity, electrical Symbol σ; unit S m^{-1}. The reciprocal of ➤resistivity. The local current density j at any point in a conductor is given by σE where E is the electric field at that point. In the context of solutions of electrolytes (➤electrolysis), the reciprocal of the resistivity is called the **electrolytic conductivity**, κ. The **molar conductivity** is the electrolytic conductivity divided by the concentration: $\Lambda = \kappa/c$. ➤➤Kohlrausch's law.

conductivity, thermal ➤thermal conductivity.

conductometric titration A titration during which the conductivity of the solution is monitored. It is most useful for titrations between ➤weak acids and weak bases, for which ➤indicators do not allow a good end-point to be seen. The diagram overleaf shows a typical result: the sharp change in slope of the graph allows the ➤equivalence point to be found.

conductor A material that conducts heat or electricity. Although most good conductors of heat are also good conductors of electricity (metals), there are many exceptions (for example sapphire).

cone 1 (Math.) A solid figure formed by joining all points in a plane region (the **base**) by a straight line to a common point (the **apex**) not in the plane of the base. A **right circular cone** has circular base and apex symmetrically positioned above the centre of the base: it has a volume $\frac{1}{3}\pi r^2 h$ and a curved surface area $\pi r l$ where r is the radius of the base, h the perpendicular height of the cone, and l the slant height.
    2 (Biol.) ➤eye; ➤➤colour vision.

confidence interval An ➤interval centred on a ➤sample estimate of a ➤parameter of a ➤population giving a measure of the likely precision of that estimate. A 95%

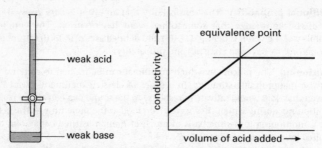

conductometric titration The weak base is only partially ionized, so the conductivity is quite low at the start. As acid is added it reacts with the base to form a salt, which is in effect fully ionized, thus greatly increasing the conductivity of the solution. After the equivalence point has been reached, adding more acid, which is only partially ionized, has hardly any effect on the conductivity.

confidence interval has the property that, for 95% of all random samples, the confidence interval calculated from them will actually contain the true parameter value, which is termed the '**population parameter**'.

configuration 1 (Chem.) ➤(R–S) system.
2 ➤electronic structure.

conformal transformation A ➤transformation (2) that preserves the size and sense of ➤angles between crossing lines or curves.

conformation The detailed three-dimensional shape of a molecule. **Conformers** are different conformations of a molecule which can be interconverted, without breaking any chemical bonds, by the *rotation* of bonds only. Consider the molecule ethane, viewed along the carbon–carbon bond. The energy of the molecule depends slightly on the conformation of the groups; the **staggered** conformation, shown in the diagram at ➤Newman projection, is the most stable. Cyclohexane has carbon–hydrogen bonds in two different conformations, called **chair** and **boat** (see the diagram). At temperatures around −50 °C, the ➤nuclear magnetic resonance signals from the axial and equatorial hydrogens in the chair conformer can be resolved; interconversion between the conformers at higher temperatures causes a coalesced signal.

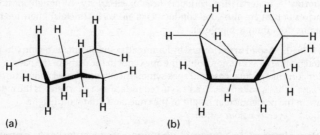

(a)   (b)

conformation (a) Chair and (b) boat conformations of cyclohexane. In the chair conformation, the six emboldened atoms are described as **axial**; the other six are called **equatorial**.

congeners Elements in the same ➤group of the periodic table. Copper, silver and gold are congeners.

congenital Describing a condition that develops in an embryo before birth. It may be the result of an inherited genetic condition such as cystic fibrosis or thalassaemia. Alternatively it can be taken to mean a condition that arises due to the passage of a substance or infective agent across the placenta from mother to embryo. Deformities caused by ➤cytotoxic agents such as thalidomide, or infection with ➤HIV or ➤rubella virus, are examples. ➤➤hereditary.

congruent Two geometric figures are congruent if they have the same shape and size so that they 'look' identical.

conic (conic section) The ➤locus of a point that moves in a plane such that its distance from a fixed point (the **focus**) is a constant multiple (the eccentricity e) of its distance from a fixed line (the **directrix**): see the diagram overleaf. The name 'conic' derives from the fact that they may also be defined as those ➤curves produced by the intersection of a ➤cone and a plane.

A conic is called a ➤circle, ➤ellipse, ➤parabola or ➤hyperbola according to whether $e = 0$, $e < 1$, $e = 1$, $e > 1$, respectively. One of Newton's many triumphs was to prove that the orbit of a body moving under an inverse square law force is necessarily a conic section.

conjugate 1 (Phys.) In optics, describing pairs of locations (points, lines or planes) that share the reversible relationship that if an object is placed at one, the image appears at the other.
2 (Math.) ➤complex conjugate.

conjugate acid The species related to a ➤Brønsted base by the addition of a proton. For example, in the equilibrium

$$NH_3(aq) + H_2O(l) \rightleftharpoons NH_4^+(aq) + OH^-(aq),$$

the ammonium ion NH_4^+ is the conjugate acid of ammonia, NH_3.

conjugate base The species related to a ➤Brønsted acid by the removal of a proton. For example, in the equilibrium shown above, H_2O is acting as an acid; its conjugate base is hydroxide ion, OH^-.

conjugated bonds Two or more double (or triple) bonds separated by just one single bond (see the diagram). The π systems of the two double bonds can then be delocalized. This conjugation affects the absorption frequencies in spectroscopy and allows for the possibility of a ➤Michael reaction. Benzene can be seen as having a conjugated ring system.

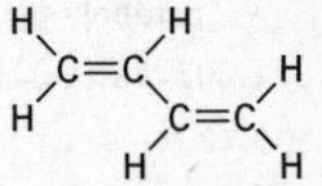

buta-1,3-diene

conjugated bonds
The molecule buta-1,3-diene has conjugated bonds.

conjunction An occasion on which a planet is aligned with or is closest to the direction of the Sun, as viewed from the Earth. Compare ➤opposition.

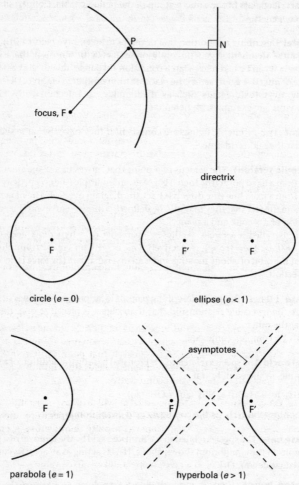

conic The focus–directrix definition of a conic, FP/PN = *e*, and the four types of conic.

conservation law A law stating that the total value of some physical quantity in an isolated system must remain constant. Most conservation laws are ultimately derived from considerations of symmetry.

conservation of angular momentum, law of The total ➤angular momentum of an isolated system, on which no external torques are operating, is constant. This law is a consequence of the invariance of the ➤Hamiltonian to rotations in space.

conservation of charge, law of The total electric charge within an isolated system is constant. This law is ultimately derived from the local ➤gauge symmetry of the electromagnetic interaction.

conservation of energy, law of The total energy of an isolated system is constant. This law is a consequence of the invariance of the ➤Hamiltonian to translations in time (i.e. the interactions of particles depend only on their positions and momenta, not also on the explicit variable of time).

conservation of mass, law of The total mass within an isolated system is constant. Conservation of mass applies only in the limit of low speeds compared with the speed of light. In ➤relativity, the sum of the ➤rest masses of individual particles is not conserved, though the rest mass of an entire system must be (as its energy and momentum are conserved).

conservation of momentum, law of The total momentum of an isolated system, on which no external forces operate, is constant. This law is a consequence of the invariance of the ➤Hamiltonian to translations in space (i.e. the interactions of particles depend only on their positions and momenta relative to each other, not also on the explicit variables of position, which are a consequence of where we choose the origin). It can also be derived from Newton's laws of motion, but in many senses the concept of conservation of momentum is more fundamental than Newton's laws themselves.

conservative field A vector field whose ➤curl is zero everywhere. This has three important consequences: the ➤line integral of the field from one point to another is independent of the route chosen; the line integral of the field around a closed loop is zero; and, most importantly, there exists a scalar ➤potential (2) function whose gradient at any point is the value of the field at that point. An electric field is conservative in the absence of changing magnetic fields (the third of ➤Maxwell's equations). A gravitational field is also conservative.

consolute temperature The temperature at which phase separation of two partially miscible liquids occurs. At a **lower consolute temperature**, increase in temperature forms two phases. At an **upper consolute temperature**, increase in temperature forms a single phase.

constant A quantity which, in a given context, takes a fixed value.

constant acceleration formulae ➤motion under uniform acceleration.

Constantan An alloy of copper and nickel used because its resistance varies little with temperature.

constant boiling mixture ➤azeotrope.

constant composition, law of All pure chemical substances have the same elements combined together in the same ratios by mass. This was one of the great early laws of chemistry which were eventually explained by the ➤atomic theory.

constant heat summation, law of ➤Hess's law.

constellation A named pattern of stars visible to the naked eye. There are 88 officially recognized constellations. ➤zodiac.

consumer Any organism in a ➤food chain that feeds by obtaining energy and materials from other organisms. Thus all animals and ➤saprotrophs such as fungi are consumers. Compare ➤producer.

contact potential The potential difference between two different metals when they are placed in contact with each other. It is essentially the difference between the two ➤work functions of the metals. It depends on temperature, which is the principle behind the ➤thermocouple.

contact process Currently the main manufacturing process for ➤sulfuric acid, accounting for over 150 million tonnes per year worldwide. The first step in the process involves burning molten sulfur in air to form ➤sulfur dioxide:

$$S(l) + O_2(g) \rightarrow SO_2(g).$$

The second oxidation step, which takes sulfur from ➤oxidation number +4 to +6, is an ➤equilibrium:

$$2SO_2(g) + O_2(g) \rightleftharpoons 2SO_3(g).$$

This exothermic reaction needs to be conducted at a compromise temperature of $450\,^{\circ}C$, at which temperature the rate is not too slow and the yield is not too low. The catalyst is vanadium(v) oxide, V_2O_5, promoted by potassium sulfate, K_2SO_4. The sulfur trioxide formed is absorbed in concentrated sulfuric acid and the resulting more concentrated solution diluted back to its previous strength. (The direct addition of water to sulfur trioxide produces a mist which does not condense to form liquid sulfuric acid.)

continental drift The theory that accounts for the present distribution of the Earth's continents by the movement of large land masses with respect to one another over the course of geological time. The German astronomer and geophysicist Alfred Wegener (1880–1930) developed the theory, though it had been proposed in the mid-19th century. He postulated that the modern continents were originally part of a single large land mass which he termed **Pangaea**. This later broke up into two supercontinents: ➤**Laurasia** in the north and ➤**Gondwana** in the south, from which the modern continents are derived. Wegener's main evidence was in the form of the geometric fit between some of the continents, and matching sets of fossils and rock types, for example between Brazil and West Africa. In the 1960s his ideas received support from studies of the spreading of the Atlantic sea floor, and continental drift is fully corroborated by the modern theory of ➤plate tectonics.

continuity equation The equation that must be obeyed when a physical quantity satisfies a ➤conservation law. If ρ is the density of the quantity and v is its velocity field:

$$\nabla \cdot (v\rho) + \frac{\partial \rho}{\partial t} = 0.$$

For electric charge this becomes

$$\nabla \cdot \boldsymbol{j} + \frac{\partial \rho}{\partial t} = 0,$$

where $\boldsymbol{j}$ is the current density and ρ is the charge density.

continuous A ➤function f is continuous at a point a if the ➤limit

$$\lim_{x \to a} f(x)$$

exists and equals $f(a)$. This definition expresses the intuitive idea that the ➤graph (2) of a continuous function should have no breaks in it. It may be generalized successively to ➤metric spaces and to ➤topological spaces.

continuous random variable A ➤random variable that can take any value in a continuum. It is fully described by its **probability density function** f with the property that $\int_a^b f(x) \, dx$ is equal to the probability that the random variable takes values between a and b. Compare ➤discrete random variable.

continuous variation Variation of a character in a population of organisms that grades from one extreme to another and thus cannot be subdivided into distinct classes. Dimensions such as height and hand span in humans are examples. The frequency data for such characters usually follow a ➤normal distribution. Compare ➤discontinuous variation.

continuum A continuous range of values (by contrast with a **discrete** set of values) that can be taken by a physical quantity or random variable. The energies and quantum states available to a ➤free electron form a continuum, while those of an electron bound within an atom do not.

contour integral The ➤line integral of a ➤complex function along a curve in the ➤complex plane. Notation such as $\oint f(z) \, dz$ is sometimes used if the curve is closed.

control rods Rods or tubes, usually of graphite, that are inserted into the core of a ➤nuclear reactor to absorb neutrons, and can thus be used to control the reaction.

convection Movement of a fluid that carries heat from one part of a system to another.

convective cloud Cloud that forms when warm moist air rises in an unstable atmosphere and condenses, as a result of local heating of the surface. Such clouds (cumulus and cumulonimbus) are responsible for showers and thunderstorms, and are characterized by their large vertical development over a relatively small horizontal area.

convergence That property of a sequence or series that converges to a ➤limit. It is not always obvious what the limit of a sequence is: for example the sequence $(1+1/n)^n$ converges to the limit e = 2.718....

convergence limit The lines in different series of an emission spectrum from an atom, such as the ➤Lyman series, get closer together until individual lines can no longer be distinguished at the convergence limit.

convergent evolution The process by which different groups of organisms may evolve similar characters in response to particular environmental requirements. For example, unrelated desert-living plants on different continents may have similar adaptations of fleshy leaves and thick cuticles to conserve water.

converging lens A lens that focuses a parallel beam of light to a point.

converse The deduction of the premises of a ➤theorem from its conclusion. Not all theorems have valid converses. For example, the theorem 'all ➤prime numbers have an even number of ➤factors' is true (they all have precisely two factors), but the converse is not true because 6 has four factors yet is not prime. ➤➤iff.

conversion electron An electron ejected from an atom or ion as a result of ➤internal conversion (2).

convex A geometric figure is convex if, given any two points of the figure, the line joining them lies wholly within the figure.

coolant A fluid used to remove heat from a system such as an engine, usually by forced convection.

Cooper pair A pair of electrons that acts as a single particle in ➤superconductivity. Named after Leon Neil Cooper (b. 1930).

coordinate bonding (dative covalent bonding) A type of ➤covalent bonding in which one of the two atoms bonded supplies *both* the shared electrons. ➤➤Lewis acid; Lewis base; complex.

coordinate geometry (analytic geometry) A branch of mathematics dealing with the systematic use of algebraic methods to solve geometrical problems by expressing the equations of the lines and curves involved in terms of ➤Cartesian coordinates.

coordinate system A system such as ➤Cartesian coordinates or ➤spherical polar coordinates for specifying the location of a point in space. In a given situation, it usually helps to choose a coordinate system adapted to any symmetries involved.

coordination compound An alternative common name for a ➤complex.

coordination number The number of nearest neighbours of a species, either in the solid state or in a complex. The coordination number of structures with ➤close packing is 12. The coordination number of each ion is 8 in the ➤caesium chloride structure and 6 in the ➤rock-salt structure. In an ➤octahedral complex, the coordination number is 6. ➤➤valence-shell electron-pair repulsion theory.

coplanar Lying in the same plane.

copolymer A ➤polymer manufactured from two or more different ➤monomers. **Acrylic fibres**, for example, are made by **copolymerization** of ethene derivatives such as propenenitrile, $CH_2{=}CHCN$. Butadiene forms two important copolymers ➤SBR and ➤ABS.

copper Symbol Cu (from the Latin *cuprum*). The element with atomic number 29 and relative atomic mass 63.55, which is in the first row of the ➤transition metals. The element itself has a characteristic red-brown colour and, despite its low abundance in nature, is much used for wires, as it has a high electrical conductivity. It is also generally unreactive, not reacting with water, for example, and so it is also used in piping (►patina). The relative ease of its extraction explains why it has been known since ancient times. It forms several important alloys, such as ►bronze (with tin), ►brass (with zinc) and ►Monel (with nickel).

Copper shows the typical properties of transition metals. It is found with a number of ►oxidation numbers. The most important is +2, as in the very familiar blue solid hydrated copper(II) sulfate. Other examples include the black solid oxide CuO and the green solid carbonate $CuCO_3$. It also forms compounds with oxidation number +1, as in the brick-red solid oxide Cu_2O, formed on reduction of ►Fehling's solution by an aldehyde. Anhydrous copper(II) sulfate is white and the colour change to blue when water ligands are present is used to test for water. Complexes are very common, including the ►ammine complex. Copper(I) chloride is used as a catalyst in the ►Sandmeyer reaction.

copper(II) sulfate, hydrated $CuSO_4 \cdot 5H_2O$ This is one of the most familiar crystalline substances in the laboratory, as it forms beautiful blue crystals. Its crystal structure is unusually complicated, however, as only four of the water molecules act as ligands round the copper(II) ion, the fifth is hydrogen-bonded.

core 1 (Astron.) The inner portion of a planet or star.
    2 (Phys.) The central part of a ►nuclear reactor in which the reaction takes place.
    3 (Phys.) A material with high ►magnetic permeability inserted into the coil of a device such as a transformer.

core level An ►energy band (in a solid) of lower energy than the ►conduction band and ►valence band. Core levels typically have a very small spread of allowable energy, and are usually treated as discrete energy levels (hence the name).

Coriolis force A ►fictitious force in a rotating frame of reference that acts perpendicular to the velocity of a body and the axis of rotation (►centrifugal force). The Coriolis force can be difficult to visualize. If a ball is thrown outwards from the centre of a turntable, describing a straight line in an ►inertial frame, an observer in a frame rotating with the turntable sees the ball accelerate as if a force were acting perpendicular to its velocity (see the diagram). One manifestation of the Coriolis force is the direction of winds along (not perpendicular to) atmospheric isobars in

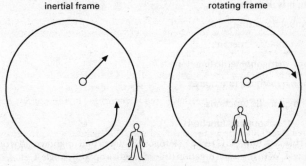

inertial frame   rotating frame

Coriolis force

latitudes away from the equator. The air accelerates until, in a frame rotating with the Earth, the Coriolis force balances the force from the pressure gradient acting perpendicular to the isobars. Named after Gaspard Gustave de Coriolis (1792–1843). ➤➤geostrophic flow.

corollary A ➤theorem that follows easily from the proof of another theorem.

corona 1 The extended outer atmosphere of the ➤Sun.
   2 A halo of colour around the Sun or the Moon as observed from the Earth, caused by diffraction and refraction in the atmosphere.

corona discharge An electrical ➤discharge around a conductor in a gas that occurs when the electric field around the conductor exceeds the value required to ionize the gas, but is insufficient to cause a spark.

correlation coefficient The correlation coefficient ρ between paired observations $(x_1, y_1),\ldots,(x_n, y_n)$ is defined by $\rho = s_{xy}/s_x s_y$, where s_{xy} is the ➤covariance and s_x and s_y are the ➤standard deviations of the x_i and y_i. The value of ρ lies between -1 and 1 and it measures the extent to which a linear relationship exists between the data points with $\rho = 1$ corresponding to a straight line of positive gradient through the points and $\rho = -1$ to one of negative gradient. It is also called **Pearson's product-moment correlation coefficient** after Karl Pearson (1857–1936).

corrosion Any surface deterioration, especially on metals. The most common form of corrosion is ➤rusting, with which it is often taken to be synonymous.

corticosteroid A steroid derived from ➤cholesterol, synthesized in the ➤adrenal cortex. **Cortisol** (hydrocortisone) prevents excessive ➤immune response.

cortisone A steroid ➤hormone secreted by the ➤adrenal glands. It has a range of functions, including the reduction of inflammation, for example at damaged joints such as cartilage injuries in athletes.

corundum Natural aluminium oxide, Al_2O_3. It is the second hardest mineral after diamond. The abrasive emery paper is impure corundum. Some impure forms are more prized than the pure material. Ruby, for example, has a few percent of chromium ions in place of aluminium ions, which imparts the beautiful colour to the stone. Sapphire usually has some pairs of Al^{3+} ions replaced by one Ti^{4+} ion and one Fe^{2+} ion; it is then blue.

cos Symbol for cosine.

cosec Symbol for cosecant.

cosecant ➤trigonometric functions.

cosech ➤hyperbolic functions.

cosh ➤hyperbolic functions.

cosine ➤trigonometric functions.

cosine rule A ➤formula (2) used in ➤trigonometry to find unknown sides or angles of a triangle. With the usual notation (side of length a opposite angle A, etc.), it states: $a^2 = b^2 + c^2 - 2bc\cos A$.

cosmic background radiation (microwave background, cosmic microwave radiation) The almost ➤isotropic radiation that permeates the Universe. Its distribution with frequency corresponds to ➤black-body radiation with a temperature of 2.73 K. It is regarded as evidence for the ➤big bang theory.

cosmic rays High-energy atomic and subatomic particles incident on the Earth from space. Some probably originate in ➤supernova explosions.

cosmid ➤vector (4).

cosmology The study of the structure, origin and dynamics of the Universe. ➤➤big bang; general relativity.

cot Symbol for cotangent.

cotangent ➤trigonometric functions.

coth ➤hyperbolic functions.

cotyledon A leaf derived from the ➤embryo in a ➤seed. Cotyledons usually develop as food reserves. The ➤Magnoliopsida are separated into two groups according to the number of cotyledons in the seed. The Liliidae (➤monocotyledons) have one cotyledon and include the grass, lily and orchid families; the Magnoliidae (➤dicotyledons) have two cotyledons and include the cabbage, rose and sunflower families. ➤➤endosperm.

coulomb Symbol C. The SI derived unit of ➤electric charge, equal to the charge moved when a current of 1 A flows for 1 s. Named after Charles Augustin de Coulomb (1736–1806).

Coulomb's law The force between two point charges q_1 and q_2 is along a line joining the two, and has a magnitude given by:

$$F = \frac{q_1 q_2}{4\pi\varepsilon_0 r^2},$$

where ε_0 is the ➤permittivity of free space. The force is attractive if the charges are of opposite sign, repulsive if they are of the same sign. Coulomb's law is a classic example of an ➤inverse square law; it is a general solution of ➤Gauss's law. Named after Coulomb, although Henry Cavendish had discovered (but not published) the law earlier.

coulometer ➤voltameter.

countable set A ➤set that can be put in ➤one-to-one correspondence with a subset of the set of whole numbers. Thus all finite sets and the set of ➤rational numbers are countable but the set of ➤real numbers is not, according to a famous theorem of Georg Cantor (1845–1918) who pioneered these concepts.

counter A device for counting events, particularly the incidence of individual particles, such as a ➤Geiger counter or ➤scintillation counter.

couple A pair of forces of magnitude F that are equal and opposite but applied at points separated by a distance d perpendicular to the forces. The combined moment of the forces produces a torque Fd on the object on which they act (➤moment of force).

coupling reaction ➤diazo dyes.

covalent bonding In the covalent model of ➤chemical bonding, the atoms are held together by sharing electron pairs. The model was introduced by Lewis in 1916: ➤Lewis structures show the electron pairs explicitly.

Imagine two hydrogen atoms approaching each other. When their ➤atomic orbitals are close enough to **overlap**, there is an increase in the electron density between the atoms. The electrons are now attracted to two nuclei, rather than one: this lowering of energy causes a **covalent bond**. If the two atoms approach closer, the energy rises again, because of repulsion between the two nuclei. There is an optimum distance where the energy is least. So a covalent bond has a specific distance, the **bond length**, and a specific direction, between the nuclei. The shapes of covalent molecules can be predicted by ➤valence-shell electron-pair repulsion theory. The electron pairs are not always shared equally. When the electron pair forming a covalent bond is shared unequally, the atom with the larger ➤electronegativity acquires a partial negative charge, as in $H^{\delta+}$—$Cl^{\delta-}$. A covalent bond that is shared unequally is called a **polar covalent bond**. All bonds between nonidentical atoms are polar to some extent. ➤Ionic bonding is simply an extreme form of polar covalent bonding in which the electron density is heavily distorted towards one of the two atoms.

covalent radius ➤radius (3).

covariance The covariance of paired observations $(x_1, y_1), \ldots, (x_n, y_n)$ is defined by

$$\frac{1}{n} \sum_{i=1}^{n} (x_i - \bar{x})(y_i - \bar{y}),$$

where $\bar{x}$ and $\bar{y}$ are the means of the x_i and y_i.

CP Abbr. for ➤charge–parity (symmetry).

CPT Abbr. for ➤charge–parity–time (symmetry).

CPU Abbr. for ➤central processing unit.

Cr Symbol for the element ➤chromium.

cracking Cracking breaks a large molecule down into smaller molecules. The starting molecule is often an alkane from the fractional distillation of ➤petroleum and the product molecules are smaller alkanes and alkenes, such as

$$C_8H_{18} \rightarrow C_6H_{14} + C_2H_4.$$

Thermal cracking involves heating the alkane to between 800 and 1000 °C, sometimes in the presence of superheated steam. The reaction mechanism involves ➤radicals. The alternative type of cracking is called **catalytic cracking** (or 'cat-cracking'). This does not require such high temperatures, 500 °C being common, but does require a catalyst, such as ➤silica, SiO_2, or ➤alumina, Al_2O_3. The mechanism is less certain but may involve ➤carbocations. The biggest difference is that the carbon skeleton suffers more ➤rearrangement in catalytic cracking. This is put to good use in ➤reforming.

creep The permanent deformation of a material subjected to continuous stresses beyond its ➤elastic limit.

C region (constant region) The ➤amino-acid sequence closest to the carboxyl end of a ➤polypeptide chain comprising an ➤antibody. This region is virtually the same in all antibody molecules and is coded for by the so-called ➤C segments of DNA. ➤➤V segment.

Cretaceous See table at ➤era.

Creutzfeldt–Jakob disease (CJD) A rare and fatal neurological disorder of humans caused by a ➤prion and characterized by porous lesions in the brain. Symptoms include progressive dementia and loss of muscular coordination. Recently links have been suggested between a variant of CJD and ➤BSE (bovine spongiform encephalo-pathy). Named after Hans Gerhard Creutzfeldt (1885–1964) and Alfons Maria Jakob (1884–1931).

critical Generally, an adjective used to describe a parameter for a system that signifies a marked transition from one kind of behaviour to another.

critical angle The smallest angle of incidence of light on the boundary to a medium of lower refractive index at which ➤total internal reflection can occur.

critical damping ➤damped harmonic motion.

critical mass The mass of fissile material above which a nuclear fission reaction can be self-sustaining. Below the critical mass, too many neutrons escape from the material. ➤➤chain reaction (2).

critical point The point on the (three-dimensional $p–V–T$) ➤phase diagram of a substance at which the liquid phase exists at the ➤critical temperature. The corresponding volume and pressure are called the **critical volume** and **critical pressure**. The term 'vapour' should be used only for a substance below its critical point, as it implies that it is possible to form a liquid.

critical temperature The temperature above which a gas cannot be liquefied by compression alone (➤critical point).

critical velocity The maximum velocity in a fluid before ➤turbulent flow develops.

CRO Abbr. for cathode-ray ➤oscilloscope.

cross In genetics or breeding experiments, the action of combining genetic material from two parents to produce offspring or the results of such an action. ➤➤back cross.

crossing over A process occurring during ➤meiosis in which fragments of chromosome are exchanged within a homologous pair. Crossing over produces new combinations of ➤genes, and thus contributes to genetic diversity in a population.

cross-linking A chemical bond linking two chains in a polymer together. It is the cross-linking that is the characteristic feature of thermosetting ➤plastics. Cross-linking is also important in the ►vulcanization of rubber and in biochemistry in holding parts of a protein chain close together.

cross product ►vector product.

cross-section ►scattering cross-section.

crown ether A large ring structure containing ether linkages, as found in, for example, 18-crown-6 (see the diagram). Crown ethers are excellent complexing agents for ions: 12-crown-4 is specifically good at complexing lithium ions.

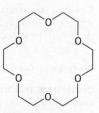

crown ether 18-crown-6

CRT Abbr. for ►cathode-ray tube.

crude oil ►petroleum.

crust The outermost layer of the ►Earth.

crustacean A member of the diverse ►arthropod superclass (►taxonomy) Crustacea, containing animals with a generally poorly defined head, two pairs of antennae, and a tough, often calcified, ►exoskeleton. Examples are crabs, lobsters and woodlice.

cryogenic pump A vacuum pump that removes gas by ►adsorption to very cold surfaces, usually at liquid helium temperature (4.2 K).

cryogenics The study of materials and phenomena at very low temperatures, often at liquid helium temperature (4.2 K) and below.

cryolite Natural sodium hexafluoroaluminate, Na_3AlF_6, used in large quantities in the ►Hall–Héroult process for manufacturing aluminium.

cryometer A thermometer designed for use at temperatures close to absolute zero.

cryostat An experimental vessel or chamber, used in ►cryogenics, that can be cooled to very low temperatures.

crystal A solid whose regular array of particles has definite polyhedral faces meeting at definite angles and showing certain symmetry characteristics (for a more detailed description of the geometric shapes of crystals, see ►crystal structure and ►Bravais lattice). Solids that are not crystalline are called ►amorphous.

crystal field theory An early version of ►ligand field theory, historically important as it enabled John Van Vleck (in 1939) to explain for the first time why complexes are coloured, in terms of ligand field splitting.

crystal lattice The ►lattice underlying a crystal. It may be a ►Bravais lattice if only one type of atom or ion is present, or it may be a lattice with a ►basis (2).

crystalline Consisting of ►crystals. A microcrystalline substance, such as a typical metal, does not look obviously crystalline on the macroscopic level, but under the microscope small crystals can be seen.

crystallization The process of forming ►crystals. ►►fractional crystallization.

crystallography The study of the structure of crystals. Crystallographic techniques can determine the ➤Bravais lattice and the magnitudes of the ➤lattice vectors. ➤➤X-ray crystallography.

crystal momentum A quantity associated with an electron in a ➤crystal lattice. For free electrons it is $h/2\pi$ multiplied by the ➤wave-vector (k) of the electron. In general, k appears in the Bloch function (➤Bloch's theorem) that is the electronic wavefunction. Crystal momentum is similar to the macroscopic idea of ➤momentum, but includes implicitly the electron's interaction with the particles in the lattice. The rate of change of crystal momentum of an electron is equal to the total applied external force on the electron. ➤➤energy band.

crystal oscillator An oscillator of very precise frequency, used in electronic circuits. It is based on a crystal that exhibits the ➤piezoelectric effect.

crystal structure A description of the way in which atoms, ions or molecules pack together to make up a crystal. Identical particles have the space group symmetry of one of the fourteen ➤Bravais lattices, belonging to one of the seven crystal systems (which describe the ➤point group of the lattice). Where more than one species is involved (or where the same species occupies identifiably different positions in the lattice), the Bravais lattice must be supplemented by a ➤basis (2) which describes the position of the two or more species within the primitive unit cell. In total there are 32 point groups and 230 ➤space groups that describe the full range of symmetries of the lattice.

crystal system ➤Bravais lattice.

Cs Symbol for the element ➤caesium.

C segment (constant segment) One of three main kinds of gene that code for an ➤antibody. The C segment codes for the so-called 'constant regions' (➤C regions). ➤➤J segment; V segment.

CS gas o-$ClC_6H_4CHC(CN)_2$ A vapour used as a harassing agent in crowd control. It causes tears, choking and painful breathing.

CT Abbr. for computerized ➤tomography.

Cu Symbol for the element ➤copper, from the Latin *cuprum*.

cube 1 A solid that has six identical square faces with adjacent faces being mutually perpendicular. It is one of the five ➤Platonic solids.
   2 The third ➤power (2) of a number.

cube root ➤root.

cubic 1 A ➤polynomial or polynomial equation of degree 3.
   2 ➤Bravais lattice.

cubic close packing (cubic close-packed, c.c.p.) ➤close packing.

cuboid A solid that has six rectangular faces with adjacent faces being mutually perpendicular.

cultivar A variety of a cultivated plant. There are many cultivars of wheat, for example, each with applications to different products such as pasta or bread. Certain cultivars may be resistant to disease or suit particular growing conditions. It is estimated that there are over 160 000 cultivars of rice.

cumene process A useful industrial synthesis of ➤phenol and ➤propanone. ➤Benzene and ➤propene are reacted in the presence of a phosphoric acid catalyst at 250 °C and 30 atm to form **cumene**, (1-methylethyl)benzene, $C_6H_5C(CH_3)_2H$. This is oxidized in air at 100 °C and 5 atm to its peroxide $C_6H_5C(CH_3)_2OOH$, which decomposes in acid at 70 °C to give phenol, C_6H_5OH, and propanone, CH_3COCH_3.

cuprammonium ion Traditional name for the most well-known ➤ammine complex $[Cu(NH_3)_4(H_2O)_2]^{2+}$, the tetraamminediaquacopper(II) ion. Its aqueous solution has a characteristic dark-blue colour.

cupric Traditional name for copper(II).

cuprous Traditional name for copper(I).

curie Symbol Ci. A unit of radioactivity equal to 3.7×10^{10} disintegrations per second. Named after Pierre Curie (1859–1906). It has been superseded by the corresponding SI unit, the ➤becquerel.

Curie's law The ➤susceptibility of a ➤paramagnetic substance is inversely proportional to its thermodynamic temperature.

Curie–Weiss law The ➤susceptibility of a ferromagnetic substance above its **Curie temperature** (**Curie point**), T_C, is inversely proportional to the thermodynamic temperature minus the Curie temperature:

$$\chi \propto 1/(T - T_C).$$

It is approximately true for many materials exhibiting ➤ferromagnetism. Named after P. Curie and Pierre-Ernest Weiss (1865–1940).

curium Symbol Cm. The element with atomic number 96 and most stable isotope 247, which is one of the ➤actinides. As usual for an actinide, its chemistry is dominated by its radioactivity. Its most common ➤oxidation number is +3, as in the chloride, $CmCl_3$, although compounds with oxidation number +4 are known, such as CmO_2. Named after Marie Curie (1867–1934), the great early experimenter with radioactivity.

curl The curl of a ➤vector field $v = v_1\mathbf{i} + v_2\mathbf{j} + v_3\mathbf{k}$ is the vector field

$$\left(\frac{\partial v_3}{\partial y} - \frac{\partial v_2}{\partial z}\right)\mathbf{i} + \left(\frac{\partial v_1}{\partial z} - \frac{\partial v_3}{\partial x}\right)\mathbf{j} + \left(\frac{\partial v_2}{\partial x} - \frac{\partial v_1}{\partial y}\right)\mathbf{k},$$

written curl v or $\nabla \times v$. ➤Stokes's theorem.

current Generally, the movement of particles through a medium. It is usually taken to mean ➤electric current.

current balance (Kelvin balance) A ➤galvanometer based on the measurement of the force that a current produces by its magnetic field. The force is compared with a known weight.

current density Symbol j; unit A m^{-2}. The current flowing per unit area in a material. It is a vector quantity, and is more useful than a simple scalar measure of current when the current flows in different directions at different points within a material.

curvature A measure of the 'roundness' of a ➤curve, denoted by κ. It is defined at a point P by $d\phi/ds$, where ϕ is the angle that the ➤tangent (2) at P makes with the x axis and s is the ➤arc length from some fixed point on the curve to P. In ➤Cartesian coordinates

$$\kappa = \frac{d^2y}{dx^2} / \left[1 + \left(\frac{dy}{dx} \right)^2 \right]^{3/2}.$$

The ➤radius of curvature $\rho = 1/\kappa$ and the ➤centre of curvature is that point on the ➤normal at P distance ρ from P. These give the centre and radius of the ➤circle of curvature, the circle that most closely 'fits' the curve at P.

curve Any unbroken line in two (**plane curve**) or three (**space curve**) dimensions. A plane curve may be described by a Cartesian equation $y = f(x)$ (with f ➤continuous) or a polar equation or a parametric equation (➤parameter); a space curve is usually described parametrically.

cusp A special type of ➤singularity where two branches of a curve meet and share a common tangent.

cyan- A prefix signifying the presence of a CN group or CN$^-$ ion. The name hides a story: the old name for hydrocyanic acid, HCN, was prussic acid, which in turn indicated that a ➤cyanide ion was essential for producing ➤Prussian blue.

cyanate A salt of cyanic acid that contains the **cyanate ion**, OCN$^-$.

cyanic acid HOCN The oxoacid formed from the ➤pseudohalogen cyanide, which exists as a poisonous liquid.

cyanide ion CN$^-$ The anion formed from ➤hydrocyanic acid, HCN, found in cyanides such as potassium cyanide, a notoriously poisonous solid. The cyanide ion can bind to haemoglobin much more strongly than oxygen can, thus blocking oxygen uptake. Despite this danger, the **cyanide process** uses potassium cyanide to extract gold by forming the complex ion $[Au(CN)_2]^-$ with the gold(I) ion.

cyanobacteria (blue-green algae) A major and ecologically important group of prokaryotes (➤prokaryotic), formerly classified as a group of algae but now considered to be more closely related to ➤bacteria. They are photosynthetic, and some forms are capable of nitrogen fixation (➤nitrogen cycle). Some of the earliest fossils from Precambrian rocks, called stromatolites, are of cells remarkably similar to modern cyanobacteria.

cyanocobalamin ➤vitamin B$_{12}$.

cyanogen $(CN)_2$ The species analogous to the normal diatomic form of the halogens for the ➤pseudohalogen cyanide, which exists as a colourless poisonous gas which smells of bitter almonds.

cyanohydrin An old-fashioned but useful nomenclature indicating the 2-hydroxynitrile formed from a carbonyl compound (➤carbonyl group) by the nucleophilic addition (➤nucleophile) of hydrogen cyanide. Ethanal, CH_3CHO, forms $CH_3CH(OH)CN$.

cycle 1 A period of a ➤periodic system.
   2 A sequence of changes to a system, particularly in thermodynamics, that leaves the system in the same state in which it began. An example is the ➤Born–Haber cycle.

cyclic 1 (Math.) Lying on a ➤circle. A **cyclic quadrilateral** is a ➤convex ➤quadrilateral with vertices lying on a circle.
   2 (Chem.) Having a ring structure. ➤heterocyclic aromatic compound.

cyclic AMP (cyclic adenosine monophosphate) ➤AMP.

cyclic permutation A ➤permutation corresponding to arranging the objects to be permuted in a circle and rotating the circle. For example, 3412 is a cyclic permutation of 1234, whereas 1342 is not.

cyclization The closing of a ring structure during a chemical reaction.

cycloaddition An addition reaction that forms a ring. By far the most important for synthesis is the ➤Diels–Alder reaction.

cycloalkane An alkane that has a ring structure. The three- and four-carbon cycloalkanes, cyclopropane and cyclobutane, react readily as their rings are strained. Cyclopentane and cyclohexane are generally stable. Higher rings become very hard to make. The reactions of cycloalkanes are generally similar to those of the alkanes, with the additional possibility of ring opening reactions, which are common for cyclopropane in particular.

cyclohexane C_6H_{12} The cycloalkane with six carbon atoms, which exists as a colourless flammable liquid. The most stable ➤conformation of the molecule is the chair form (see the diagram at ➤conformation).

cycloid The path traced out by a point on a ➤circle as it rolls along a fixed line. If the line is taken as the x axis, the parametric equations of the cycloid are $x = r(t - \sin t)$ and $y = r(1 - \cos t)$, where r is the ➤radius of the circle. The solution to the brachistochrone problem is a cycloid (➤calculus of variations).

cyclone ➤depression.

cyclopentadiene A cyclic diene (see the diagram) which exists as a colourless liquid with a sweet smell. The **cyclopentadienyl ion**, $C_5H_5^-$, forms a number of important organometallic compounds such as ➤ferrocene.

cyclopentadiene

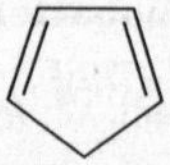

cyclopropane C_3H_6 The cycloalkane with three carbon atoms (see the diagram opposite), which exists as a colourless flammable gas. It was used from the 1920s as an anaesthetic, but, although powerful and safe, it formed dangerously explosive mixtures with air, so it has now been replaced by ►halothane. It is much more reactive than other cycloalkanes and indeed the alkanes, as the molecule has considerable **ring strain**; the bond angles are 60° rather than the preferred carbon–carbon bond angle of 109° 28′ (the tetrahedral angle). So, for example, it reacts with bromine by a ring opening reaction to form $BrCH_2CH_2CH_2Br$.

$$H_2C \!-\!\!-\!\! CH_2$$

cyclopropane

cyclotron A particle ►accelerator based on two semicircular hollow electrodes within which the particles circulate. The particles circle the cyclotron as a magnetic field perpendicular to the circle is applied, together with an alternating electric field between the electrodes. For energies higher than a few megaelectronvolts (MeV), a ►synchrocyclotron or ►synchrotron must be used.

cylinder A solid figure formed by joining by straight lines corresponding points in two ►congruent plane regions identically situated in ►parallel planes. A **right circular cylinder** is formed from two circular regions with the line joining their centres ►perpendicular to each region: it has volume $\pi r^2 h$ and curved surface area $2\pi rh$, where r is the ►radius of each circle and h the distance between their centres.

cylindrical coordinates (cylindrical polar coordinates) A system for specifying the location of a point in space related to ►Cartesian coordinates by using ►polar coordinates in the (x, y) plane and the usual z coordinate.

Cys Abbr. for ►cysteine.

cysteine (Cys) An important ►amino acid. ►►Appendix table 7.

cystine A derivative formed by oxidation of the —SH groups in two ►cysteine molecules, particularly in polypeptides. The resultant structure, a **disulfide bridge**, is important in maintaining the three-dimensional shape of proteins.

cytidine A ►nucleoside consisting of the base ►cytosine covalently bonded to the sugar ►ribose (or ►deoxyribose in **deoxycytidine**).

cyto- Prefix denoting of, or pertaining to, cells.

cytochrome One of a group of iron-containing proteins which function in redox reactions in the ►electron-transport chain of ►oxidative phosphorylation in respiration and ►photophosphorylation in photosynthesis.

cytokine One of several typically soluble proteins which take part in cellular interactions, frequently promoting cell growth and division. ►Interleukin 1 and ►interleukin 2 are cytokines which facilitate communication between cells of the ►immune system.

cytokinin (kinin) One of a group of naturally occurring plant growth substances derived from ►adenine. A cytokinin works in conjunction with ►auxin to promote cell division and differentiation, stimulating, for example, the growth of lateral branches. Cytokinins exert their influence by activating genes and promoting ►protein synthesis. The balance between auxin and cytokinin is a major factor in determining the overall form of a mature plant.

cytology The study of cells and their functions, particularly by use of the ➤microscope.

cytoplasm The semi-fluid contents within the ➤plasma membrane of a cell, excluding the nucleus.

cytoplasmic inheritance The inheritance of genetic factors in DNA found outside the ➤nucleus (2). Such DNA is usually associated with ➤chloroplasts and ➤mitochondria. An example of such inheritance is male sterility in some plants; this is a useful character in plant breeding since it ensures that such plants cannot fertilize themselves.

cytosine Symbol C. A ➤pyrimidine base found in both ➤DNA and ➤RNA. Cytosine binds to guanine by ➤base pairing in double-stranded DNA. ➤➤genetic code.

cytosol The fluid part of the ➤cytoplasm, excluding the membrane-bound ➤organelles.

cytotoxic Toxic or poisonous to cells.

D

D 1 The symbol for ➤ deuterium, the isotope of hydrogen with a nucleon number of 2. The formula D_2O is the formula for ➤ heavy water.

2 Symbol for the unit ➤debye.

3 Symbol for the unit ➤dioptre.

2,4-D Abbr. for 2,4-dichlorophenoxyethanoic acid (see the diagram). It is used extensively as a selective herbicide, killing weeds with broad leaves. It is closely related to ➤2,4,5-T.

2,4-D

D Symbol for diffusion coefficient (➤diffusion equation).

D Symbol for ➤electric displacement.

DAC Acronym for ➤digital-to-analogue converter.

Dacron US name for ➤Terylene.

DALR Abbr. for ➤dry adiabatic lapse rate.

dalton An alternative name for the ➤atomic mass unit. Named after John Dalton (1766–1844).

Dalton's law (of partial pressure) The total pressure of a mixture of ideal gases (➤ideal gas model) – which therefore do not react – is equal to the sum of the pressures each gas would exert if it alone occupied the whole vessel.

damped harmonic motion The motion of an oscillating system that is subject to ➤damping. If the damping is proportional to the velocity of the body, then the equation of motion is:

$$m\frac{d^2x}{dt^2} = -kx - \mu\frac{dx}{dt},$$

where x is the ➤displacement of the body, m is its mass, k is the constant of proportionality of the restoring force (the 'spring constant') and μ is the damping constant. There are three distinct solutions of this second-order ➤differential equation, depending on the constants involved:

$$\mu^2 < 4km, \tag{i}$$

known as **underdamping**. The body oscillates, and the amplitude of the oscillation decreases exponentially with time.

$$\mu^2 = 4km, \tag{ii}$$

known as **critical damping**. The body returns to its equilibrium position at the optimum rate.

$$\mu^2 > 4km,$$ (iii)

known as **overdamping**. The body returns to its equilibrium position, but the excessive damping causes a slower return than in the critical case.

damping A force opposing the motion of a body that is oscillating (or would be oscillating in the absence of damping). It may be a consequence of the environment through which the body is moving (as in air damping) or deliberately introduced to prevent oscillation (as with a ➤dash-pot). Damping causes ➤simple harmonic motion to become ➤damped harmonic motion. The concept of damping is often generalized to the other situations in which a similar equation is found; for example, in electronics a resistance may **damp** an ➤LC circuit.

Daniell cell A primary ➤cell (2) for producing electricity, which relies on the reaction between zinc and copper(II) sulfate to provide the electromotive force. Its standard e.m.f. is 1.10 V. It was one of the earliest cells discovered, by John Frederic Daniell (1790–1845).

dansyl A colloquial shorthand for dimethylaminosulfonyl, a group used in the ➤Edman degradation.

daraf Symbol F^{-1}. A unit of ➤elastance equal to the reciprocal of the ➤farad.

dark field imaging A method of illumination in microscopy. An opaque disc prevents direct transmission of light from the illumination source to the objective, and only objects that bend light rays round the disc by ➤refraction are visible. Such objects, which might be almost transparent and therefore otherwise difficult to detect, appear as bright images on a dark background.

dark matter Matter in the Universe that is inferred to exist by its gravitational effect rather than by direct observation. Up to 90% of the mass of the Milky Way and other ➤galaxies may be composed of dark matter.

dark nebula ➤nebula.

dark reaction (light-independent reaction) A series of reactions occurring in ➤chloroplasts during photosynthesis in which carbon dioxide is converted into carbohydrate. The reactions require ➤ATP and NADPH (➤NADP$^+$) made in the ➤light reaction as the source of energy and reductants needed to bring about the conversion. The dark reaction occurs in the stroma (body) of the chloroplast rather than on the membranes. The chemistry of the process was elucidated by Melvin Calvin, and has been termed the **Calvin cycle.** The key discovery was the role of ribulose bisphosphate (RuBP) as the prime carbon dioxide acceptor molecule. The sequence of reactions initiated by the addition of carbon dioxide to RuBP molecules yields a carbohydrate product, which may be converted into glucose, and regenerates RuBP.

darmstadtium Symbol: Ds. The element with atomic number 110. Named after Darmstadt in Germany, where many of the elements with atomic numbers above 106 were first discovered.

Darwinism The theory of ➤evolution as deduced by Charles Robert Darwin (1809–82) and outlined in his book *On the Origin of Species* (1859). The essence of Darwinism is that evolution is driven by ➤natural selection, a process in which environmental factors favour the survival of organisms that possess certain advantageous features. If such organisms survive long enough to breed, they pass on the ➤genes that control these advantageous features to their offspring. Darwin did not cast his theory in terms of genetics, the precise mechanisms of which were not elucidated until after his death. The modern synthesis of natural selection and genetic processes is termed neo-Darwinism.

dash-pot A device that uses oil or air, usually with a piston, to create a ➤damping force for a part of a mechanical system.

database A computerized system for storing information in a structured, easily accessible form. Typically, a database consists of a number of similar **records**. The database can be interrogated by the user, to search for records that fulfil particular criteria or to extract statistics. Modern databases can be very powerful and process large amounts of data, and form the crucial information system of large organizations.

dative covalent bonding Alternative name for ➤coordinate bonding.

daughter nuclide The ➤nuclide that is the product of the radioactive decay of a **parent nuclide**. For example, $^{234}_{90}\text{Th}$ is the daughter produced by the ➤alpha decay of $^{238}_{92}\text{U}$.

Davy lamp (miner's safety lamp) A source of illumination that enabled miners to work more safely than previously. It used a wire gauze cylinder to conduct much of the heat away from the burning oil so that the temperature outside the gauze did not rise high enough to ignite 'firedamp' (methane) and cause disastrous fires. Named after Humphry Davy (1778–1829).

day For everyday purposes, a unit of time equal to 86 400 s (24 hours). It is the mean value of the ➤solar day. ➤➤equation of time; sidereal day.

db Symbol for the unit ➤decibel.

Db Symbol for the element ➤dubnium.

d-block element An element in the ➤block of the periodic table between the s block and the p block. The most important subset of the d-block elements is the ➤transition metals, those d-block elements that form at least one stable ion with an incompletely filled d subshell.

d.c. Abbr. for direct current. Commonly, d.c. is used in an adjectival sense to describe a circuit designed to be driven by a steady electromotive force.

d.c. amplifier Abbr. for direct coupled (*not* direct current) amplifier. The output of each stage of the amplifier is coupled directly to the input of the next. In fact, this *is* a form of amplifier capable of amplifying direct current.

DDT Abbr. for dichlorodiphenyltrichloroethane (see the diagram overleaf), which is a powerful insecticide. It had an enormous beneficial impact on health in the 20th

century. According to estimates by the World Health
Organization, 5 million lives were saved by its use,
and the lives of 2 billion people improved. However,
this effective agent is now banned in many countries
because, although humans can safely convert DDT to
an excretion product, levels build up in food chains
by ➤bioaccumulation, as non-mammalian species
cannot excrete DDT.

DDT

d–d transition The excitation of an electron from the lower of two energy level groups to the upper. The **d-orbital splitting** into two groups arises from the different shapes of the five d orbitals. Simplistically, three of the five point between the axes and thus, in an octahedral complex, point between the ligands; the other two orbitals point directly at the ligands and so are higher in energy due to repulsion between like charges (➤ligand field theory). d–d transitions are the major source of colour in transition metal complexes; the other is ➤charge transfer.

deadbeat An ➤analogue instrument that has critical damping (➤damped harmonic motion).

dead time The time from when an electrical device, typically a detector such as a ➤Geiger counter, receives a stimulus to the time at which it becomes capable of responding to a second stimulus.

deamination The removal of an amino group (—NH_2) from an organic compound, usually under the influence of a **deaminase** enzyme. In mammals it occurs particularly in the liver enabling the group to be excreted as ➤urea.

de Broglie's equation The equation linking the wave and particle aspects of fundamental particles (➤wave–particle duality). It gives the wavelength λ associated with a particle as

$$\lambda = h/p,$$

where h is the ➤Planck constant and p is the magnitude of the particle's ➤momentum. This equation was important in the historical development of ➤quantum theory. It was first proposed by Louis-Victor Pierre Raymond de Broglie (1892–1987) in his doctoral thesis of 1924.

debye Symbol D. A unit of electric ➤dipole moment approximately equal to 3.336×10^{-30} C m. Named after Peter Joseph William Debye (1884–1966).

Debye–Hückel theory A theory explaining the behaviour of ➤electrolytes; it is based on a simple picture for the ions in the solution, in which each ion is surrounded by an **ionic atmosphere** containing predominantly the ions of opposite charge. The theory applies the ➤Boltzmann distribution of particles at energy levels dictated by the ➤Poisson equation for the electrostatic potential to yield (eventually) an equation that predicts that the logarithm of the ➤activity coefficient γ is directly proportional to the square root of the **ionic strength** I of the solution:

$$\log_{10}\gamma = -|z_+ z_-|AI^{\frac{1}{2}}.$$

Here A is a constant, z_+ and z_- are the charges carried by the positive and negative ions, and the ionic strength is $\frac{1}{2}(c_+ z_+^2 + c_- z_-^2)$ with c the concentration of the ions. This prediction is indeed justified for very dilute solutions (with concentrations of about 10^{-3} mol dm^{-3}). Significant deviations occur for more concentrated solutions. Named after P. J. W. Debye and Erich Armand Arthur Joseph Hückel (1896–1980).

Debye model A model of ➤lattice vibrations that takes into account the quantized nature of the ➤normal modes of a crystal. This results, for example, in a ➤specific heat capacity that is proportional to T^3 at low thermodynamic temperature T, and independent of T at high T (compare ➤Dulong and Petit's law). The temperature that characterizes the transition between these two regimes is called the **Debye temperature**.

Debye–Waller factor A factor taking account of lattice vibrations that appears in the theory of the scattering of particles in a crystal. Named after P. J.W. Debye and Ivar Waller (1898–1991). ➤➤electron diffraction; neutron diffraction.

deca- The SI prefix for a factor of 10. It is rarely used in practice.

decalin A bicyclic hydrocarbon, $C_{10}H_{18}$, used as a solvent. Its structure resembles two cyclohexane rings joined together along one side. It has two isomers (see the diagram), one in which the hydrogens at the join are arranged *trans* and one in which they are *cis*.

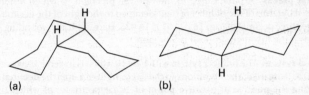

decalin (a) *cis*-decalin and (b) *trans*-decalin.

decane $C_{10}H_{22}$ The alkane with ten carbon atoms. It has 75 isomers, all of which have been isolated in pure form.

decanol One of a collection of isomeric alcohols with ten carbon atoms. They are used, for example, in the preparation of ➤plasticizers and detergents. Decan-1-ol is a colourless liquid with a sweet smell, which is immiscible with water.

decantation The process in which the solution lying above a sediment is poured very gently away from the sediment, which remains in the original vessel.

decay 1 The reduction in a physical quantity over time. ➤exponential decay.
  2 (Phys.) The process of an unstable nuclide spontaneously changing its state by ➤radioactivity. ➤➤alpha decay; beta decay; gamma decay; exponential decay.

decay constant ➤exponential decay.

deceleration ➤retardation. ➤➤acceleration.

deci- A prefix meaning one-tenth of the base unit. It is not officially recognized in SI units, but finds great utility in the volume unit dm^3: 1 dm^3 = 1000 cm^3; 1000 dm^3 = 1 m^3.

decibel Symbol dB. A unit of power ratio which is commonly used in electronics and acoustics. It is a more practical unit for everyday use than the bel (10 dB). The decibel is a dimensionless unit. For a pair of ➤powers P_2 and P_1, the power ratio in decibels is:

$$10 \log_{10}(P_2/P_1).$$

The unit is also applied to **sound pressure levels**: for a pressure p_2 and reference pressure p_1 (conventionally 2×10^{-5} Pa), the sound pressure level in decibels is:

$$20 \log_{10}(p_2/p_1).$$

The difference of 2 in the prefactor of the logarithm in the two cases arises because the power carried by a sound wave is proportional to the square of the pressure. One decibel is about the smallest difference in sound level that the human ear can detect.

deciduous 1 Describing plants that lose their leaves at the end of the growing season (autumn in temperate regions). The majority of temperate flowering trees and shrubs (such as oaks and roses) are deciduous, as are a few conifers such as the larch. Compare ➤evergreen.

2 Those teeth (also called **milk teeth**) belonging to the first of two sets produced by mammals. In humans they are replaced by the permanent teeth from the age of about 7 to 10.

decimal places A phrase used to indicate the precision to which a number is expressed in terms of the number of digits retained to the right of the decimal point. Thus, to three decimal places, 16.024 91 is 16.025. Here, typically, rounding is done upwards if the first digit dropped is 5 or more.

decimal system The number system with base 10 which is in general everyday use. Digits 0 to 9 are used, their positions to the left or right of a dot (the **decimal point**) indicating the positive or negative power of 10 (respectively) of which they are multiples. For example, 316.2 represents

$$3 \times 10^2 + 1 \times 10^1 + 6 \times 10^0 + 2 \times 10^{-1}.$$

declination 1 (Astron.) The ➤angular distance of a ➤celestial body from the ➤celestial equator. Declination is similar in concept to latitude, except that it is defined relative to the ➤celestial sphere rather than the Earth's surface.

2 (Phys.) ➤magnetic declination.

decomposer An organism whose mode of nutrition assists in the process of decay, particularly in the recycling of elements such as carbon and nitrogen in nature. Decomposers are heterotrophic, and include ➤saprotrophs such as ➤bacteria and ➤fungi. However the term is frequently extended to include all organisms in the vitally important 'decomposer food chain' which are responsible for the physical and chemical breakdown of dead organisms. The breakdown of sewage and the making of garden compost are among the activities of decomposers. ➤➤ecosystem.

decomposition The breakdown of one substance into two or more substances. The most common type is **thermal decomposition**, an example of which is heating calcium carbonate which decomposes into calcium oxide and carbon dioxide. **Electrolytic decomposition** uses an electric current to cause decomposition.

decrepitation The crackling of some crystals when heated, caused by loss of water from within the crystals.

defect An imperfection or irregularity in an otherwise perfect crystal, which would have all its constituent particles in exactly the correct positions in every layer of the structure. A **Schottky defect** is a vacant site accompanied by an extra particle at the surface. A **Frenkel defect** is a vacant site accompanied by an **interstitial particle**, one between two normal lattice points. In addition to these **point defects**, there are **line defects** such as ➤dislocations. Defects can cause anomalous properties, such as imparting colour to a colourless crystal. Some defects can be introduced intentionally by irradiation of a crystal.

deficiency disease A disease characterized by pathological symptoms caused by a lack of an essential factor in the diet. Such factors include vitamins (e.g. a lack of vitamin C leads to scurvy, a disease of the gums) and essential minerals, (e.g. a lack of iron causes anaemia). Deficiency diseases also occur in plants, where they are commonly due to mineral deficiencies in the soil (e.g. a lack of iron or magnesium produces chlorotic (yellow) leaves because these elements are required in the synthesis of ➤chlorophyll).

definite integral An integral such as $\int_1^3 x^2 \, dx$ that is evaluated between two given limits. ➤integration. ➤➤Appendix table 3.

definite proportions, law of Alternative name for the law of ➤constant composition.

deflagration An explosion in which the speed of the reaction front is less than the speed of sound in the material, and hence the shock wave precedes the reaction front. Compare ➤detonation.

deformation A mechanically induced change of shape of a material, usually a crystal.

degassing The process of heating a solid surface to release molecules of gas that are adsorbed on the surface. In order to achieve an ultra-high vacuum of less than about 10^{-5} Pa, it is necessary to heat a metal vacuum chamber to temperatures of about 150 °C for a number of hours to degas adsorbed gases from the inner surfaces of the chamber.

degaussing A technique for reducing the permanent magnetization of a ferromagnetic component by passing a decreasing ➤alternating current through a coil surrounding the component. Many computer monitors have a 'degauss' control which employs the technique to demagnetize components within the monitor that may have become permanent magnets and hence distort the display. ➤➤ferromagnetism; magnet, permanent.

degenerate ➤degenerate states.

degenerate gas A gas-like state of ➤identical particles at a temperature sufficiently low and a density sufficiently high for quantum mechanics to be important in determining its properties. The particles obey ➤Bose–Einstein statistics if they are bosons and ➤Fermi–Dirac statistics if fermions. A degenerate electron gas has a

degeneracy pressure in excess of the pressure that would be expected from the ➤kinetic theory of gases for a similar number of particles. Degeneracy pressure plays an important role in the physics of ➤white dwarf and ➤neutron stars.

degenerate semiconductor A semiconductor in which the ➤chemical potential lies within or close to the top of the ➤valence band or the bottom of the ➤conduction band. The position of the chemical potential at an energy with a nonzero ➤density of states provides a higher concentration of ➤carriers (2), and means that the material is essentially metallic.

degenerate states ➤Quantum states of a system that have the same energy. Any ➤linear combination of degenerate states is itself an allowable state, so where degeneracy exists the choice of ➤wavefunctions as a ➤basis (2) is somewhat arbitrary. For example, the five ➤d orbitals of an isolated atom are degenerate. We can define arbitrary x, y and z axes and choose states thus: d_{z^2}, $d_{x^2-y^2}$, d_{xy}, d_{yz}, d_{zx}. If an octahedral ligand field is introduced, the fivefold degeneracy is 'lifted', creating a degenerate triplet and a degenerate doublet at a higher energy. ➤➤ligand field theory.

degradation A reaction in which an organic compound is converted into a simpler compound. ➤➤Edman degradation; Hofmann degradation.

degree **1** A measure of angle such that there are 360 degrees (written $360°$) in a complete revolution.
2 The highest power of a variable occurring in a polynomial. A polynomial of more than one variable such as $x^3 + 3x^2y^2 + y^2$ has (total) degree 4 (from the $3x^2y^2$ term) and is of degree 3 in x and 2 in y.

degree of ionization (degree of dissociation) Symbol α. The fraction of molecules of a ➤weak acid that are ionized. This is related to the ➤acid ionization constant K_a by noting that when the added concentration of weak acid is c, the hydrogen ion and anion concentrations are given by αc and that of the un-ionized acid is $(1 - \alpha)c$:

$$K_a = \alpha^2 c/(1 - \alpha).$$

degrees of freedom Generally, the number of parameters required to define the state of a ➤system, less the number of constraints on the parameters. The 'state' here can mean various things. For example, to define the position of a diatomic molecule of a gas, with bond length L, the position of the centre of mass of the molecule may be specified using three coordinates, together with two angles to determine its orientation. Alternatively, the three coordinates of each atom (a total of 6 parameters) could be specified, together with one constraint (that the separation of the atoms is L). Either way, the result is 5 degrees of freedom. (➤➤equipartition of energy.) If certain parameters of the system are ignored, and only a ➤macrostate is specified, it is still appropriate to examine the number of degrees of freedom, which then depends on the context of the model. For example, a diatomic gas consisting of N molecules has $5N$ true degrees of freedom, but its macrostate is often determined with only two, usually one of temperature, internal energy or entropy, plus either pressure or volume. ➤➤phase rule.

de Haas–van Alphen effect An entirely quantum-mechanical effect that causes the ➤magnetic susceptibility of a conductor to oscillate with an applied magnetic field. This occurs because the angular momentum of the conduction electrons parallel to the field is quantized (➤angular momentum quantum number). The period of the oscillations varies with the cross-sectional area of the ➤Fermi surface perpendicular to the field, and the effect provides a useful technique for mapping out Fermi surfaces. Named after Wander Johannes de Haas (1878–1960) and Pieter Martinus van Alphen (1906–67).

dehydration The removal of the elements of water from a compound, as when ethene, C_2H_4, is produced from ethanol, CH_3CH_2OH. A common **dehydrating agent** is concentrated sulfuric acid, which can dehydrate methanoic acid, $HCOOH$, to produce carbon monoxide, CO.

dehydrogenase An oxidoreductase ➤enzyme that brings about an ➤oxidation reaction in which hydrogen is removed from a ➤substrate. Alcohol dehydrogenases catalyse the conversion of primary ➤alcohols to ➤aldehydes by the removal of hydrogen from the $-CH_2OH$ group of the alcohol.

dehydrogenation Any process in which hydrogen is removed from a molecule, as in the conversion of cyclohexane into benzene. The same catalysts that assist in hydrogenation must also be good at dehydrogenation and so nickel, palladium and platinum are common dehydrogenation catalysts. An enzyme that catalyses such reactions is termed a ➤dehydrogenase.

deionized water Water that has been purified by removal of dissolved ions, usually using an ➤ion exchange resin. The term should not be used interchangeably with ➤'distilled water'.

del (nabla) The differential operator ∇ formally defined by $\nabla = \mathbf{i}\partial/\partial x + \mathbf{j}\partial/\partial y + \mathbf{k}\partial/\partial z$. It can be used to express many of the operations of vector analysis, so that, for example, ➤Laplace's equation may be written $\nabla^2\phi = 0$.

delayed neutron A neutron produced when the product of ➤beta decay is unstable with respect to neutron emission. For example, after the beta-decay process $^{87}_{35}Br \rightarrow ^{87}_{36}Kr + e^-$, the krypton daughter nuclide can subsequently undergo further decay: $^{87}_{36}Kr \rightarrow ^{86}_{36}Kr + n$.

delay line An electrical component that introduces a known delay in the transmission of a signal. This may be achieved using ➤digital electronics, ➤analogue electronics or even, for longer delays, ➤acoustics (by converting the electrical signal into an acoustic one, introducing a delay and converting back again).

deliquescence The process in which a solid absorbs water and subsequently dissolves in it. Pellets of sodium hydroxide are **deliquescent**. If no dissolution occurs, the solid is instead termed ➤hygroscopic.

delocalization The sharing of a single electron pair between more than two atoms. The spreading-out of electron density in a molecule over more than two atoms stabilizes its electronic structure. The simplest covalent bonds are those in which an electron pair binds two atoms together, as in the diatomic chlorine molecule, Cl_2. There is no theoretical reason why just two atoms should be bonded together, and

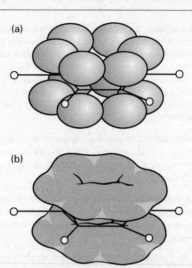

(a)

(b)

delocalization The six p orbitals of the six carbon atoms in the benzene molecule (a) overlap to form three bonding molecular orbitals (π orbitals). The lowest-energy molecular orbital contains one pair of electrons whose electron density is delocalized over all six carbon atoms. This orbital is represented by the two doughnut-shaped regions shown in (b). (The higher-energy π orbitals are not shown, but produce the same average electron density as the lower-energy one.)

when an electron pair binds three or more atoms together it is termed 'delocalized', as it is not localized between a single pair of atoms. An example in inorganic chemistry where this concept is unavoidable is in the structure of the boranes, especially diborane, B_2H_6. A single electron pair bonds both borons and the bridging hydrogen atom, to create a ➤three-centre bond. Delocalization can explain the structure of ➤oxoanions such as carbonate, CO_3^{2-}, in which the double bond extends over all three oxygen atoms. In organic chemistry, delocalization has long been used to explain the particular stability of aromatic compounds such as ➤benzene (see the diagram). It also explains why ethanoic acid is more acidic than ethanol, because delocalization occurs in the resulting ethanoate ion, $CH_3CO_2^-$, but not in the ethoxide ion, $CH_3CH_2O^-$. However, it is much more prevalent than is generally supposed. For example, the lowest-energy orbital in the methane molecule, CH_4, is a five-centre molecular orbital delocalized over all five atoms. The other three orbitals are of higher energy, unlike the common picture of four equal-energy bonds.

delta bond (δ bond) A bond formed by the overlap of d orbitals such that there are four regions of enhanced electron density. Compare the significantly more common ➤pi bond and ➤sigma bond.

delta function ➤Dirac delta function.

delta-iron An allotrope of iron, stable between 1400 °C and its melting point. ➤➤iron–carbon phase diagram.

delta radiation Electrons produced when ➤ionizing radiation interacts with matter.

delta scale A scale for measuring chemical shifts in ➤nuclear magnetic resonance.

demagnetization A reduction of the permanent magnetization of a ferromagnetic material. ➤➤degaussing.

demagnetizing field Symbol H_d. A name given to the difference between the strength of the ➤magnetic field inside and outside a magnetic material. The demagnetizing field results from the ➤magnetization of the material itself.

demodulation The extraction of a signal transmitted on a modulated ➤carrier wave.

de Moivre's theorem The identity $(\cos \theta + i \sin \theta)^n = \cos n\theta + i \sin n\theta$ for ➤complex numbers in polar form. It is true for integer values of n, and for other values of n there is at least one value of the left-hand side that makes it true. Named after Abraham de Moivre (1667–1754).

D enantiomer For an α-amino acid, the ➤enantiomer that has the —COOH, —R and —NH₂ groups arranged clockwise as viewed along the C—H bond. A superior terminology is explained under ➤$(R–S)$ system. Compare ➤L enantiomer.

denature 1 To alter the natural structural state of molecules such as nucleic acids and proteins. Denaturation of ➤DNA, by heating in aqueous solution to near boiling, unwinds the double helix and produces single-stranded forms. Such treatment is a fundamental laboratory procedure for manipulating DNA. When such DNA is allowed to cool, the single strands realign, restoring the original ➤base pairing, and hence the double helix. This process, essentially the reverse of denaturation, is termed **annealing**. The ability to separate and subsequently recombine strands of DNA is used in conjunction with other techniques to produce a ➤DNA fingerprint. Denaturation of proteins, by extremes of pH or heating, interferes with the folding of the amino acid chain. The cooking of egg white (albumin), by which it is converted from a transparent liquid to an opaque white semi-solid, is an everyday example of the denaturing of a protein. Protein denaturation is an irreversible process.
   2 To make a ➤fissile material unsuitable for use in a nuclear weapon by the addition of another isotope.

denatured alcohol US term for ➤methylated spirit.

dendrite A fine branch of a ➤dendron that receives nerve impulses via ➤synapses from other nerve cells. The fine networks of connections provided by dendrites form the basis of all coordination and control processes in animal nervous systems.

dendrochronology A method of dating by counting the annual growth rings of trees to determine the age of structures made of wood and wooden archaeological remains, or to infer climatic histories. The thickness of a tree's annual growth ring depends on the ambient conditions: thicker rings are produced when climatic factors such as rain and temperature increase growth. The pattern over a few decades is therefore characteristic of that period. Starting from living trees, especially bristlecone pines, it is possible to trace the record back thousands of years. Dendrochronology is usually a more accurate method than ➤radiocarbon dating.

dendron The part of a nerve fibre that carries an impulse towards the cell body of a nerve cell. In some sensory cells the dendrons may be over a metre in length. ⤷neurone.

denitrifying bacteria ➤nitrogen cycle.

denominator ➤fraction.

dense A subset Y of a ➤metric space X is dense if every element of X is the limit of some sequence of elements of Y. For example, the set of ➤rational numbers is dense in the set of ➤real numbers.

densitometer An apparatus for measuring the (optical) ➤transmittance or ➤reflectance of a sample. One use is to detect the soundtrack on a cinematic film.

density **1** (Phys.) Symbol ρ; unit kg m^{-3}. The mass of a material per unit volume. **Relative density** is defined with respect to water, which has an **absolute density** of 1000 kg m^{-3} at 4 °C and 1 atm. For example, 1 m^3 of lead, with a relative density of 11.35 at room temperature, has a mass of 11 350 kg.

2 Where a large number of small items have a parameter which itself varies over a larger range, it is sometimes useful to express the number of items per unit interval of the parameter as a density of items as a function of the parameter. For example, we can define population density (the items are people, the parameter is position on the chosen map), charge density (the items are point charges, the parameter is position in space), probability density (➤continuous random variable) and ➤density of states. The common feature of all such definitions is that the interval over which the items are counted must be much larger than the items themselves, but much smaller than the range of the parameter.

density, photographic The logarithm (to base 10) of the opacity (➤opaque) of a photographic sample.

density gradient centrifugation A technique by which large molecules such as proteins and nucleic acids, and subcellular structures such as mitochondria and chloroplasts, may be separated using a ➤centrifuge. The technique relies on the ability to spin solutions of, for example, sucrose or caesium chloride in a tube at high speed to establish a density gradient in the solution. When a sample of, for example, a mixture of proteins to be separated is spun in the tube at high speed, the components migrate down the tube to positions of equivalent density in the solution and are thus separated into bands.

density of states A measure of the number of quantum states per unit energy. In a solid, individual electron states are often very closely spaced in energy. The energies of the states therefore cannot be determined, but it is possible to count the number of states within an infinitesimal energy range, and express this density of states as a function of energy. The density of states at the ➤Fermi energy is a particularly important quantity because it is directly related to the number of electrons that can act as carriers for transport properties such as electrical and thermal conductivity (➤Drude model; solid state physics).

dentine A bone-like material that forms the major part of the structure of a ➤tooth. Dentine is harder than bone and is the major component of ivory.

dentition The number and arrangement of teeth in an animal.

deoxyribonucleic acid ➤DNA.

deoxyribose A ➤pentose ➤monosaccharide (see the diagram) that forms an integral part of the structure of ➤DNA. It is the deoxy derivative of ➤ribose.

D-**deoxyribose** The OH on the right-hand carbon is in the ➤beta position.

deoxythymidine A nucleoside consisting of the base thymine covalently bonded to the sugar deoxyribose. For **deoxyadenosine**, ➤adenosine; for **deoxycytidine**, ➤cytidine; for **deoxyguanosine**, ➤guanosine.

dependent variable ➤variable.

depleted fuel A ➤nuclear fuel with a lower than natural concentration of ➤fissile material, for example after use in a nuclear reactor.

depletion layer A region close to a junction in an ➤inhomogeneous semiconductor in which positive and negative carriers have undergone ➤recombination (2) to create a layer with relatively few carriers. ➤p–n junction.

depolarization A rapid change in the pattern of charge across the membrane of a ➤neurone or a sensory cell which propagates a ➤nerve impulse.

depression A region of low atmospheric pressure. Depressions are usually associated with changeable weather and strong winds. Compare ➤anticyclone.

depression, angle of The angle between the observer's horizontal plane and an object below that plane.

depression of freezing point ➤colligative properties.

deprotonation The process during which a proton, equivalent to the hydrogen ion H^+, is removed from a molecule. Compare ➤protonation.

depth of field In an optical instrument with a lens or lenses, the distance along a direction parallel to the lens axis over which the object may be moved while the image remains in focus. Compare ➤depth of focus.

depth of focus In an optical instrument that forms an image, the distance along a direction parallel to the lens axis over which the image plane may be moved while the image remains in focus. Compare ➤depth of field.

derivative 1 (Math.) ➤differentiation. ➤➤Appendix table 3.

2 (Chem.) A compound closely related to another by substitution of one atom or group for another. For example, methylbenzene, $C_6H_5CH_3$, is a derivative of benzene. Formation of their DNP derivatives can identify aldehydes and ketones (➤2,4-dinitrophenylhydrazine).

derived function ➤differentiation.

derived unit ➤SI units; Appendix table 1.

dermis ➤ skin.

desalination The removal of (common) salt from water, for example to produce drinking water from seawater. A common method uses ➤reverse osmosis.

desiccator An apparatus containing a **desiccant** used for the physical removal of water from a substance. Common desiccants include calcium chloride, silica gel and magnesium perchlorate.

desorption The process of particles, especially gas molecules, detaching from a solid surface (the opposite of ➤adsorption).

destructive distillation The continued distillation of a substance in the absence of air until all the volatile material has been driven off. This often causes some decomposition as well. The destructive distillation of coal produces coal gas and coal tar, and leaves coke.

desulfurization The removal of sulfur, typically from a fuel. This is especially important for the control of pollution, since the sulfur dioxide otherwise produced on combustion is not only poisonous but also contributes to the problem of ➤acid rain.

detector A device designed to register the presence of (and possibly measure some physical quantities of) atomic particles or electromagnetic radiation.

detergent A water-soluble surface-active agent used for cleaning. The classic ➤soap species such as sodium stearate have a ➤hydrophilic end and a ➤hydrophobic one. Synthetic detergents are similar but are derived from petroleum rather than from hydrolysis of an oil or a fat. The molecule acts as a kind of bridge between grease and water. A common anionic detergent is an alkylbenzenesulfonate. The sulfonate group acts as the hydrophilic end, and the benzene ring and alkyl chain (the hydrophobic end) are able to interact strongly with oily materials. The earliest detergents had branched-chain side groups and these turned out to be non-biodegradable. More recent versions have straight chains and are ➤biodegradable.

determinant A number associated with a square ➤matrix A, written det A or $|A|$ and defined by induction as follows. The determinant of a 2×2 matrix is $a_{11}a_{22} - a_{12}a_{21}$; that of a 3×3 matrix is $a_{11}A_{11} - a_{12}A_{12} + a_{13}A_{13}$, where, for example, the **cofactor** A_{12} is the determinant of the 2×2 matrix formed by deleting the row and column of A that contains a_{12}. Higher-order determinants are defined in a similar manner. The determinant det A is equal to the product of the ➤eigenvalues of the matrix A, and A has an inverse if and only if det A is nonzero.

detonation An explosion in which the speed of the reaction front exceeds the speed of sound in a material, and hence the reaction front precedes the shock wave. Compare ➤deflagration.

deuterium Symbol ^2H or D. The isotope of hydrogen with nucleon number 2. Its nucleus is composed of one proton and one neutron. There is very little difference in its chemical behaviour compared with the normal isotope, called protium, which has no neutrons. The major difference is its mass and this can give rise to a kinetic ➤isotope effect. ➤heavy water.

deuteron The nucleus of a ➤deuterium atom, consisting of one proton and one neutron.

Devarda's alloy An alloy containing around 45% aluminium, 50% copper and 5% zinc by mass, used in alkaline solution for reducing nitrates to ammonia, and hence for detecting the nitrate ion. Named after Arturo Devarda (1859–1944).

deviation, angle of The angle through which an incident ray or particle is deflected by ➤refraction or ➤scattering. For a ray of light refracting through two surfaces of a prism, the angle of deviation is minimized when the ray is perpendicular to the symmetry axis of the prism (see the diagram).

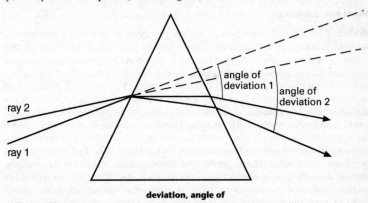

ray 2

ray 1

angle of deviation 1

angle of deviation 2

deviation, angle of

devitrification The crystallization of ➤glass, which in its normal state is ➤amorphous.

Devonian See table at ➤era.

dew Liquid droplets produced by the condensation of a ➤saturated vapour on a cold surface. Frost is produced when dew freezes. If the surface is below the freezing point of the vapour, sublimation occurs instead (➤hoarfrost).

Dewar benzene An isomer of benzene, C_6H_6, which is a bicyclic hydrocarbon (see the diagram). Substituted derivatives have been isolated. Named after James Dewar (1842–1923).

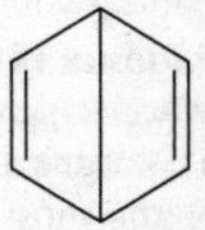

Dewar benzene

Dewar flask   A vessel designed to keep liquids (and sometimes gases) at temperatures different from their surroundings. The container is doublewalled, with the space between the walls evacuated to prevent heat transfer by conduction or convection. The walls are silvered to reduce heat transfer by radiation. The Dewar, known by its tradename of 'Thermos flask' in the UK, finds applications as varied as keeping soup at 325 K on a picnic to keeping liquid nitrogen at 78 K in a laboratory, and, in a more sophisticated form, keeping liquid helium at 4 K.

dew point   The temperature at which a sample of air becomes saturated with water vapour. If the air (with a source of ►nucleation centres, such as dust) is cooled below its dew point, it will condense into water droplets, forming, for example, cloud, fog or dew. ►►relative humidity.

dew-point hygrometer   An instrument for measuring the dew point of air. A polished surface is cooled until water droplets condensing on the surface are detected.

dextrorotatory   Symbol d or (+). A molecule with one ►chiral centre has two enantiomers (►optical activity): the dextrorotatory enantiomer rotates the plane of plane-polarized light to the right (*dexter* is the Latin for 'right'). Compare ►laevorotatory.

dextrose   ►glucose.

diabetes   A clinical condition in which the body is unable to regulate the concentration of solutes, particularly sugar (glucose), in the blood. There are two forms of diabetes, *diabetes mellitus* and *diabetes insipidus*. Blood glucose concentration is under the control of two ►hormones produced by the pancreas; normal blood glucose concentration is about 100 mg per 100 ml of blood. When blood sugar concentration rises (after a meal), ►insulin is released and causes the uptake of glucose from the blood into body cells, particularly in the liver. When blood sugar concentration falls (during exercise) ►glucagon is released, which promotes the release of glucose into the blood from stored glycogen in the liver and muscles. In *diabetes mellitus*, this ►feedback mechanism fails. Too little insulin is produced, or is not used effectively, and blood glucose levels rise dangerously, glucose appearing in the urine. This can be controlled clinically either by the administration of insulin or by 'oral hypoglycemic agents' taken as tablets, depending on the cause. *Diabetes insipidus* is caused by insufficient production of the hormone ►vasopressin (antidiuretic hormone) from the posterior lobe of the ►pituitary gland. This hormone stimulates the reabsorption of water from the distal parts of the kidney tubules. Lack of this hormone causes the production of copious urine which is low in solutes and has no marked taste (hence the *insipidus*).

diagonal matrix   A square ►matrix that has nonzero elements on the ►leading diagonal only.

diagonal relationship   A vague relationship connecting the top element in the early groups of the ►periodic table with the element one period lower and one group to the right. Lithium and magnesium, beryllium and aluminium, boron and silicon are the most important pairs of examples. A simple explanation for beryllium's similarity to aluminium is that the two ions Be^{2+} and Al^{3+} have similar charge densities. A couple of consequences for its chemistry are that beryllium chloride has significant covalent character, dissolving in organic solvents for example, and that beryllium hydroxide is ➤amphoteric, as is aluminium hydroxide (➤alumina).

diakinesis ➤meiosis.

dialysis A biochemical technique by which large molecules such as proteins in solution are separated from smaller species such as salts. The technique is based on the properties of certain membrane structures. The solution to be separated is placed in a cellophane bag of 'Visking tubing' and immersed in tap water. The salts pass through the membrane into the water, leaving a lower salt concentration in the bag. The exact conditions can be varied so that the salt concentration can be reduced to a particular level. Dialysis is frequently used in the purification of proteins. The principle is also used in artificial kidney (dialysis) machines to remove urea from the blood of patients with kidney disease. Blood from a vein is diverted through a system in which it is separated from a specially formulated solution of various salts and glucose by a cellophane membrane. In this way urea can pass out of the blood, leaving appropriate concentrations of salts, glucose and blood proteins for return to the patient. Patients undergoing such treatment are usually dialysed for several hours about twice a week.

diamagnetism An effect that occurs when a ➤magnetic field is applied to a material and the induced ➤magnetization acts to oppose the applied field. The vast majority of materials are weakly **diamagnetic**. Compare ➤ferromagnetism; paramagnetism.

diameter A line segment joining two points on a circle or sphere and passing through its centre, or the length of such a line. ➤➤radius.

1,2-diaminoethane (ethylenediamine) Symbol en. $NH_2CH_2CH_2NH_2$ A colourless liquid used as a very important ➤bidentate ligand, forming strong complexes with many ➤Lewis acids. A typical example is $[Ni(en)_3]^{2+}$. ➤➤chelate.

1,6-diaminohexane ➤nylon-6,6.

diamond One of the allotropes (➤allotropy) of carbon, the crystal structure of which has each carbon atom bonded to four other carbon atoms in a tetrahedral arrangement (see the diagram). Diamond is the hardest known substance, being assigned the maximum value of 10 on the ➤Mohs scale, and is therefore used in special drill bits. It has a high ➤refractive index, which explains its use in jewellery. Artificial diamonds have been made: those manufactured by chemical vapour deposition are now beginning to rival the qualities of the natural material. Graphite is the thermodynamically more stable allotrope, but only by 2.9 kJ mol⁻¹, whereas the activation energy for the interconversion is huge.

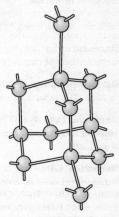

diamorphine (heroin) ➤morphine.   **diamond**

diaphragm **1** (Phys.) A disc, with an aperture at the centre, used to limit the intensity of light passing into an optical instrument. ➤iris.

2 (Biol.) A muscular sheet in mammals separating the thoracic (chest) cavity from the abdominal cavity. Rhythmic contraction and relaxation of the diaphragm produces pressure changes, allowing air to be drawn into and expelled from the lungs during breathing. Spasmodic contraction of the diaphragm produces hiccoughs.

diaphragm cell ➤chlor–alkali industry.

diastase Alternative name for ➤amylase.

diastereomers (diastereoisomers) ➤Stereoisomers that are not ➤enantiomers. For example, an optically active acid can react with an optically active base to form the salt; the (+) acid and (+) base combination is a diastereomer of the (–) acid and (+) base combination. These diastereomers can have slightly different physical properties, enabling them to be separated. Hence the formation of diastereomers is a common method of resolution of ➤racemic mixtures.

diastole ➤heart.

diathermanous Able to transmit radiant heat; the equivalent in heat radiation of transparency in light radiation.

diatom A large and diverse division of photosynthetic ➤protoctistans comprising a single class (**Bacillariophyceae**). The characteristic feature of the group is an extracellular wall consisting of two halves (valves) which contains silica. Diatoms are abundant in fresh and marine waters and are estimated to account for up to one-fifth of global primary ➤productivity. Fossil deposits of diatom remains from the floors of Tertiary and Quaternary lakes are called **diatomite** (➤kieselguhr).

diatomaceous earth (diatomite) ➤kieselguhr.

diatomic molecule A molecule containing two atoms. A **homonuclear diatomic molecule** contains two identical atoms, as in H_2 and Cl_2. A **heteronuclear diatomic molecule** contains two different atoms, as in HCl.

diazepam $C_{16}H_{13}N_2OCl$ A molecule which is the active ingredient in Valium, probably the commonest tranquillizer.

diazo compounds Organic compounds containing two nitrogen atoms bonded together. The most famous examples are ➤diazonium salts.

diazo dyes (azo dyes) Brightly coloured (typically red or orange) dyes made by coupling an aromatic diazonium salt with an alkaline solution of an aromatic amine or a phenol.

diazonium salts Salts containing two nitrogen atoms bonded together, most commonly bonded to an aromatic ring. The archetypal example is benzenediazo-nium chloride, $C_6H_5N_2^+Cl^-$. They are made by ➤diazotization and are used in the production of a range of ➤diazo dyes. They also undergo replacement of the diazo group by a halogen in the ➤Sandmeyer reaction.

diazotization The formation of a diazonium salt by reaction of nitrous acid, HNO_2, with aromatic amines, at temperatures below $5\,^{\circ}C$ (because the diazonium salt decomposes to phenol at about $50\,^{\circ}C$).

dibasic acid (diprotic acid) An acid that can neutralize two moles of sodium hydroxide for every mole of the acid. Partial neutralization forms the ►acid salt. The simplest example is ►sulfuric acid, H_2SO_4. A less obvious one is phosphonic acid, whose formula of H_3PO_3 suggests incorrectly that it is a tribasic acid. One of the hydrogens is in fact bonded to the phosphorus, and so a more informative formula is $(HO)_2HP=O$.

dibenzenechromium An important example both of an organometallic compound and a sandwich compound, which forms brown-black crystals. It consists of a chromium atom sandwiched between two benzene rings (see the diagram).

dibenzenechromium

dibenzoyl peroxide $(C_6H_5CO)_2O_2$ A common initiator for a radical ►chain reaction, as the peroxide is thermally unstable, breaking down into two benzoyl radicals, $C_6H_5CO^{\bullet}$. These can then initiate radical polymerization reactions. Radical addition proceeds in a different manner to ►electrophilic addition; dibenzoyl peroxide causes anti-Markovnikov addition to alkenes (►Markovnikov's rule).

diborane ►borane.

1,2-dibromoethane (ethylene dibromide) $BrCH_2CH_2Br$ A colourless volatile liquid, which was used in large quantities, when leaded petrol was the norm, to promote the formation of volatile lead compounds in the exhaust, notably lead(II) bromide.

dicarboxylic acid A compound containing two carboxylic acid, —COOH, groups. Examples include ethanedioic acid, $(COOH)_2$, tartaric acid, and the amino acid glutamic acid.

dichlorodiphenyltrichloroethane ►DDT.

dichloromethane (methylene dichloride) CH_2Cl_2 A colourless, slightly toxic liquid used as a solvent, especially for paint, and as a refrigerant.

dichroism The ability of a uniaxial crystal (►birefringence) to absorb light with a particular ►polarization while transmitting light of a different polarization. Examples include tourmaline minerals and the commercially produced material Polaroid, which finds an everyday use in sunglasses: scattered and reflected light are partially polarized, so glare can be reduced by admitting only vertically polarized light through the lenses.

dichromate(VI) ion (dichromate ion) $Cr_2O_7{}^{2-}$ A very important reagent in organic chemistry, widely used as the potassium salt in acidic solution for oxidizing compounds such as alcohols or aldehydes. During the reaction, the orange dichromate(VI) ion changes to the green chromium(III) ion.

dicotyledon A flowering plant in which the embryo in the seed possesses two seed leaves (►cotyledons). **Dicotyledonous** plants generally have broad leaves with networks of veins, and floral parts arranged in multiples of two, four or five. Buttercups and sunflowers are examples. In some classifications they are now commonly regarded as a class (**Magnoliidae**) within the phylum ►Magnoliopsida. Compare ►monocotyledon.

dideoxy sequencing (Sanger's method) A technique for determining the sequence of bases in ►DNA. DNA is a polymer built from ►nucleoside triphosphates in which the sugar is ►deoxyribose: deoxyribose lacks the bottom right-hand oxygen atom in ►ribose (compare the diagrams for both entries). Dideoxyribose lacks the bottom *left*-hand oxygen atom as well. Because this oxygen atom is used to make the polymer chain, incorporating a dideoxy sugar stops the chain growing.

Sequencing is peformed by replication of the DNA *in vitro*. The reaction mixture contains single-stranded template DNA, a short stretch of synthetic single-stranded DNA (a ►**primer** that is ►complementary to the region of DNA where sequencing is to start), ►DNA polymerase, the four deoxynucleoside triphosphates (dNTPs) and a small proportion of the four corresponding, fluorescently labelled, dideoxy forms (ddNTPs). In the reaction, the primer ►base-pairs with the corresponding region of the template strand and the polymerase starts to extend the primer. It does this by incorporating the appropriate dNTPs to recreate double-stranded, base-paired DNA. However, the polymerase does not distinguish between dNTPs and ddNTPs. When a ddNTP happens to be incorporated, the new DNA strand can be extended no further. Thus a mixture of products is formed, each ending in a dideoxynucleotide. Since the new strands of these products are all of different lengths, they may be separated by gel ►electrophoresis. Each of the four ddNTPs carries a different fluorescent label, so the terminating dideoxynucleotide of a particular product is identified by the fluorescence it emits when stimulated. The sequence of terminating dideoxynucleotides is thus established, which is complementary to the sequence of the template DNA.

dieldrin The common name for a powerful contact insecticide (see the diagram). It caused the deaths, by ►bioaccumulation, of many birds of prey in the 1960s, which led to its eventual ban.

dieldrin

dielectric A polarizable medium, usually an insulator. When an electric field is applied to a dielectric, a ►dipole is induced in the atoms or molecules of the medium. This produces a ►polarization in the medium parallel to the applied field, which opposes it and reduces the electric field within the dielectric. ►►dielectric constant; electric displacement.

dielectric constant Symbol ε; unit F m^{-1}. A misleading but very common alternative name for the ►permittivity of a ►dielectric. Since the permittivity often depends on frequency for alternating fields, the term **dielectric function** is sometimes also used.

dielectric heating Heating of a ►dielectric by the application of a high-frequency electric field. Energy is dissipated in the dielectric by a process equivalent to ►hysteresis in magnetic materials: the ►polarization of the dielectric depends not only on the applied electric field strength, but also on whether it is increasing or decreasing.

dielectric strength The maximum strength of ►electric field that can be applied to an insulator without **dielectric breakdown** occurring. If the dielectric strength is exceeded, the electrons acquire enough energy to make transitions between ►energy bands, and become conduction electrons.

Diels–Alder reaction The reaction, a 1,4-cycloaddition, that occurs between a conjugated ►diene and a compound known as a **dienophile**. An example is shown in the diagram. It is one of the most important reactions in organic chemistry and the most important for the synthesis of cyclic compounds. The product has a specific stereochemistry, which is well understood in terms of the ►Woodward–Hoffmann rules. Named after Otto Paul Hermann Diels (1876–1954) and Kurt Alder (1902–58).

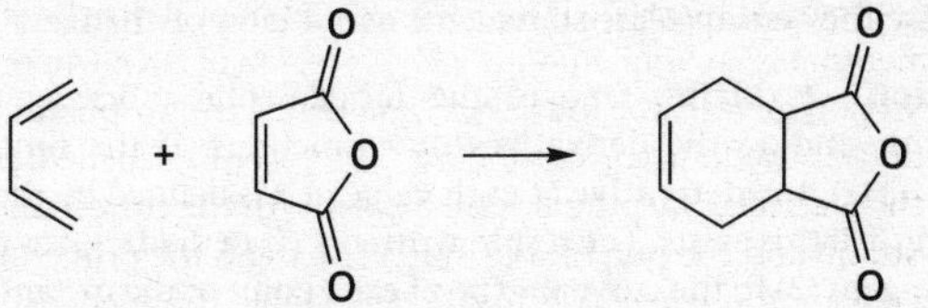

Diels–Alder reaction

Diesel cycle A four-stroke cycle that differs from the ►Otto cycle in three significant ways. First, the compression stroke causes typically twice the compression as happens in the Otto cycle. Second, in the power stroke the fuel ignites spontaneously without the need for a spark to initiate the reaction. And third, in the ideal Diesel cycle combustion occurs at constant pressure, in contrast to the ideal Otto cycle, in which combustion occurs at constant volume (►indicator diagram). Named after Rudolf Christian Karl Diesel (1858–1913).

diesel fuel The fuel for diesel engines which is produced from the fractional distillation of petroleum as the light gas oil fraction. It is a mixture of alkanes, typically containing about 16 carbon atoms in the hydrocarbon chain. ►►cetane number.

dietary fibre ►fibre (3).

diethylene glycol $HOCH_2CH_2OCH_2CH_2OH$ A colourless liquid used as a solvent, for cellulose nitrate, for example.

diethyl ether ►ethoxyethane.

difference The result of subtracting one quantity from another.

difference equation Alternative name for a ►recurrence relation.

differential For a function of one variable with the ►graph (2) $y = f(x)$, the quantity $dy = (df/dx)dx$ where dx is an increment of x. For functions of several variables, such as one with the graph $z = f(x, y)$, the total differential is d$z = (\partial f/\partial x)\,dx + (\partial f/\partial y)\,dy$. ➤differentiation.

differential amplifier An ➤amplifier whose output depends on the ➤potential difference across its input terminals. Most practical ➤operational amplifiers are differential amplifiers.

differential calculus ➤calculus.

differential coefficient ➤differentiation.

differential equation Any equation in which an unknown function appears as a derivative or partial derivative. The highest-order derivative that appears is called the ➤order (3) of the differential equation; thus the equations governing simple harmonic motion d^2x/d$t^2 = -\omega^2 x$ and Laplace's equation $\nabla^2\phi = 0$ are both second order. Solving a differential equation is equivalent to integrating it, so the solution involves arbitrary constants which may be found if ➤initial conditions or ➤boundary conditions are specified. Differential equations are an essential tool in science because they encapsulate statements about rates of change.

differentiation **1** (Math.) One of the fundamental processes of ➤calculus concerned with finding the derivatives of ➤functions. If the function has the ➤graph (2) $y = f(x)$, the **derivative** at each value of x is defined by the limit $\lim_{h\to 0}$ $[f(x+h) - f(x)]/h$, if this exists. The result, written $f'(x)$ or dy/dx, gives the ➤gradient (2) of the ➤tangent (2) to the curve $y = f(x)$ at each point or the instantaneous ➤rate of change of y with respect to x. The definition, together with shortcuts such as the ➤chain rule and the ➤product rule, may be used to calculate readily the derivative (or **differential coefficient** or **derived function**) of any ➤elementary function: for example,

$$\frac{d}{dx} x^n = n x^{n-1}, \quad \frac{d}{dx} \sin x = \cos x, \quad \frac{d}{dx} \ln x = 1/x.$$

Differentiating the first derivative gives the second derivative, $f''(x)$ or d^2y/dx^2, and higher-order derivatives; ➤partial derivatives for functions of more than one variable, and derivatives of ➤complex functions, may also be defined in a similar way. ➤Appendix table 3.
　2 (Biol.) The process by which actively dividing and growing cells assume morphological and physiological specializations. The classic example is that of a single fertilized egg, which divides to form cells from which all the tissues and organs of the body are derived.

diffraction The spreading out of a wave when it reaches an aperture or edge. Given the shape of the aperture (characteristic width d), and the wavelength λ and direction of the incident wave, it is possible to calculate the intensity of light at any point (at distance L) beyond the aperture. This can be approximated by ➤Fresnel diffraction when $L \gg d$, and further simplified to ➤Fraunhofer diffraction when $\lambda L \gg d^2$. ➤electron diffraction; X-ray diffraction.

diffraction grating A periodic array of apertures, usually consisting of narrow slits (width d), placed in the path of a beam of electromagnetic radiation to disperse it into a spectrum: to separate the radiation according to the wavelength λ. For a

➤monochromatic source at ➤normal incidence, the diffraction pattern of the grating is an array of sharp lines, at angles of diffraction θ_n given by $n\lambda = d \sin \theta_n$. If the source consists of several wavelengths, the diffraction grating will separate it by diffracting the different wavelengths through different angles. Gratings with many narrow slits are advantageous (➤resolution).

diffuse nebula ➤nebula.

diffusion The process by which a gas fills all the space available to it. Diffusion also takes place in liquids but, when not specified otherwise, 'diffusion' means gaseous diffusion. It occurs because it is entropically favourable for particles to spread out (➤entropy). An ideal gas diffuses, even though no energy change results (by the second law of ➤thermodynamics). Diffusion occurs most rapidly for lighter molecules (➤Graham's law), and this has been put to use in separating isotopes. For example, the ➤fissile nuclide ^{235}U can be separated from the more common ^{238}U by gaseous diffusion of the volatile compound uranium(VI) fluoride, UF_6; the small difference in mass means that the processing plant has to be large. ➤➤Fick's laws of diffusion.

diffusion coefficient ➤Fick's laws of diffusion.

diffusion equation A partial differential equation, involving a second-order space derivative and a first-order time derivative, of the form

$$D\nabla^2 \psi = \frac{\partial \psi}{\partial t},$$

where ∇^2 is the Laplacian (➤Laplace's equation). In one dimension this becomes

$$D\frac{\partial^2 \psi}{\partial x^2} = \frac{\partial \psi}{\partial t}.$$

Physically, with ψ identified as the concentration of, for example, gas molecules, the equation governs the change of this concentration (as a function of spatial coordinates) with time; D is known as the **diffusion coefficient**. ➤➤Biot–Fourier equation.

diffusion pump A type of ➤vacuum pump that uses a jet of very hot oil to adsorb molecules of air and remove them. It must have a **backing pump** (usually a rotary pump) to keep the high-pressure outlet at less than about 10 Pa. Pressures of 10^{-8} to 10^{-7} Pa are routinely attainable on the low-pressure side.

digestion The breakdown of large polymeric molecules into their monomeric constituents, usually by ➤hydrolysis. In particular the term is applied to the breakdown by enzymes of complex food molecules, such as proteins, so that they may be absorbed through the gut lining of an animal.

digit Any of the numbers $0, 1, \ldots, b-1$ which occur in a place-value number system with base b.

digital Describing a system, particularly an electronic device, that uses a discrete quantity to represent information, typically in a ➤binary representation.

digital computer A ➤computer that uses discrete pulses of electricity at well-defined voltages to store and process information. Internally, digital computers operate in base 2 (➤binary), and hence require two voltages, to represent the digits 0 and 1. Contrast ➤analogue computer. Except in a few specialized applications, modern computers are exclusively digital.

digital display A display that uses seven panels, illuminated in various combinations, to form the digits 0 to 9. Digital displays are often ➤LCDs, such as on a digital watch. One advantage of displaying speed in large digits is that it is easier to glance at the speedometer and read the speed quickly and accurately.

digital photography Photography in which the image is stored in digital form on an electronic medium, for example a still image on an SD (secure digital) card or a video stream on a hard disc. Digital still cameras have rendered traditional film cameras obsolescent: Nikon for example has committed fully to digital for the future.

digital recording Any method of recording information using discrete digits. For example, compact discs are digital since they store the magnitude of a sound as a number in binary form. This has theoretical advantages for the accuracy of sound reproduction, since the stored information can be recalled exactly, without suffering from degradation due to age or use.

digital-to-analogue converter (DAC) An electrical component designed to convert a binary number in a ➤register into a voltage for use in an analogue circuit. DACs are commonly used in circumstances where a computer is required to control an analogue circuit or system as in the output from a CD player.

diglyme $CH_3OCH_2CH_2OCH_2CH_2OCH_3$ The common name for a colourless liquid used as a high-temperature solvent. Being an ether, it is very unreactive.

dihaploid A condition in which the number of ➤chromosomes of plant cells in culture derived from ➤haploid tissues (such as microspores in anther culture) has been artificially doubled by the use of mutagens such as colchicine. Such a doubling is useful since mature plants derived from haploid culture are sterile. A doubling of the chromosome number renders them fertile and capable of subsequent propagation by sexual means.

dihedral angle The acute angle between two intersecting planes (see the diagram opposite). In aerodynamics, the planes are the left and right sides of the ➤aerofoil, and a wing with a positive dihedral angle therefore appears to kink at its midpoint, with the wing tips further off the ground than the midpoint. Having a dihedral wing increases the stability of the aeroplane against sideways motion.

dihydric alcohol ➤diol.

dihydroxybenzene (benzenediol) $C_6H_4(OH)_2$ There are three isomers of this molecule (see the diagram opposite), all with different traditional names. The ortho (1,2) isomer is **catechol**, which forms colourless crystals used in solution as a photographic developer. The meta (1,3) isomer is **resorcinol**, which forms colourless needles used in the production of diazo dyes and plasticizers. The para (1,4) isomer is **hydroquinone**, which forms colourless prisms also used as a photographic developer.

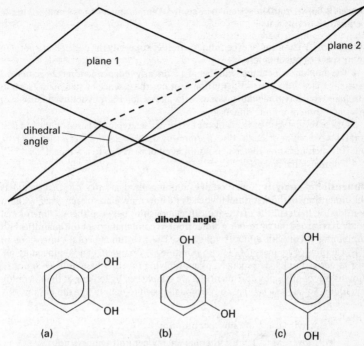

dihedral angle

(a)   (b)   (c)

dihydroxybenzene The three isomers: (a) catechol (systematic name benzene-1,2-diol), (b) resorcinol (benzene-1,3-diol) and (c) hydroquinone (benzene-1,4-diol).

2,3-dihydroxybutanedioic acid The IUPAC recommended name for ➤tartaric acid. Its sodium and potassium salt is ➤Rochelle salt.

3,4-dihydroxyphenylalanine ➤dopa.

dilation 1 A general word for an increase in a dimension.
2 ➤time dilation.

dilatometer A device, consisting of a bulb and graduated capillary, for measuring a change of volume of a fluid.

dilute solution A solution in which the ➤concentration is low. It often occurs in the study of ionic solutions in the context of an 'infinitely dilute solution', which is the hypothetical situation in which the added ions can be treated as completely independent. Normally they cause deviation from ideal behaviour because of interaction between the ions (➤Debye–Hückel theory).

dilution refrigerator An apparatus used in experimental low-temperature physics. It uses a mixture of ^{3}He and ^{4}He, which exists as two separate ➤phases at temperatures below about 1 K. The passage of ^{3}He from one phase to another is analogous to the evaporation of a liquid, and the ➤latent heat required for this process is drawn from the system to be cooled. Temperatures of a few millikelvin can be attained in this way.

dimension 1 The number of coordinates needed to specify the position of a point in some space of possibilities.

2 The maximum ➤cardinal number of a ➤linearly independent set of vectors in a ➤vector space. This abstract definition captures the essence of statements such as 'ordinary space is three-dimensions' and allows for the possibility of spaces of arbitrary finite or infinite dimensions.

3 One of several ways of assessing the extent of a set, typically based on the smallest total area or volume of a collection of arbitrarily small circles or spheres that cover the set. It is such measures that assign nonintegral dimensions to a ➤fractal.

4 ➤dimensional analysis.

dimensional analysis A method of examining physical situations that considers the dimensions of the quantities involved and how they relate to each other, without considering in detail the mathematical relationships between them. Dimensional analysis is carried out by forming dimensionless combinations of the quantities and combining them with arbitrary functions. The five fundamental dimensions in classical physics are mass (M), length (L), time (T), charge (Q) and temperature (θ). Just as derived units are produced by multiplying together base units (➤SI units), dimensions of other physical quantities are established by looking at their defining equations. Square brackets are conventionally used to denote 'the dimensions of'. For example:

$$[\text{velocity}] = \text{L T}^{-1} (\text{distance per unit time})$$

$$[\text{entropy}] = \text{M L}^2 \text{ T}^{-2} \ \theta^{-1} (\text{thermal energy per unit temperature}).$$

dimer A combination of two identical ➤monomers. This can occur for a variety of reasons. First, it can be caused by ➤hydrogen bonding, as in the dimer of ethanoic acid shown in the diagram. Second, it can be due to ➤coordinate bonding, as in the dimer of

dimer of ethanoic acid.

aluminium chloride. Third, it can be due to covalent bonding, as in ➤dinitrogen tetroxide.

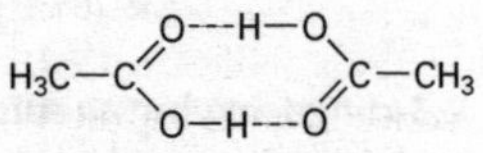

dimerization The process of forming a ➤dimer.

dimethylbenzene ➤xylene.

dimethylformamide (DMF) $HCON(CH_3)_2$ A colourless liquid which is an excellent solvent for a wide range of both organic and inorganic compounds including industrial materials. It is a ➤dipolar aprotic solvent.

dimethylglyoxime $(CH_3CNOH)_2$ A solid which forms colourless needles, used especially for the identification of nickel, which forms a dark-red crystalline complex.

dimethylsulfoxide (DMSO) $(CH_3)_2SO$ A colourless liquid extensively used as a solvent, especially in organic chemistry. It is a ➤dipolar aprotic solvent.

dimorphic 1 (Biol.) Describing an organism that exists in two clearly defined forms or **morphs**. Sexual dimorphism in humans and many other animal species is an example.

2 (Chem.) Having two crystalline forms.

_m_-dinitrobenzene (1,3-dinitrobenzene) $C_6H_4(NO_2)_2$ A solid which forms colourless crystals, made by nitration of nitrobenzene, the existing nitro group directing the second one preferentially to the ➤meta position.

dinitrogen oxide (nitrous oxide, laughing gas) NNO or N_2O A gas which has the physiological effect of relaxing the user. It was used in dentistry, and still is used combined with oxygen (➤Entonox) to relieve the pain of childbirth.

dinitrogen tetroxide (dinitrogen tetraoxide) N_2O_4 The dimer of nitrogen dioxide (see the diagram); it is a nearly colourless liquid which contrasts with the brown colour of the gaseous monomer.

dinitrogen tetroxide

2,4-dinitrophenylhydrazine (DNP) A reagent used frequently for rapid identification of a ➤carbonyl compound. Aldehydes and ketones form yellow-orange solid derivatives, called 2,4-dinitrophenylhydrazones (see the diagram), when reacted with DNP. The reaction occurs by a ➤condensation reaction.

dinosaur A major and diverse group of extinct vertebrates descended from ancestral reptiles, comprising two main orders, **Ornithischia** and **Saurischia**, differentiated by the arrangement of bones in the pelvic girdle. The dinosaurs dominated the fauna of the Earth from the early Jurassic until their sudden demise at the end of the Cretaceous, and

2,4-dinitrophenylhydrazine
General form of DNP derivative.

included the largest land animals that have ever lived, such as the herbivorous *Apatosaurus* and the carnivorous *Tyrannosaurus rex*. However, the group also contained many species which were no larger than modern chickens. Modern birds are descended from dinosaurs. ➤➤iridium.

diode An electrical component, typically based on a ➤p–n junction, with a current–voltage relationship that depends on the direction of current flow. A simple **semiconductor diode** exhibits the dependence

$$I = I_0 \left[\exp(eV/kT) - 1\right],$$

where e is the carrier (electron or hole) charge, k is the Boltzmann constant and T is the thermodynamic temperature. Thus for small positive V, the resistance is kT/eI_0 (the **forward slope resistance**), but for larger positive V (**forward bias**) the current increases rapidly (exponentially) with applied voltage, while for negative V (**reverse bias**) the current quickly reaches its saturation value, I_0. For reverse bias greater than the breakdown voltage, the relationship breaks down and the diode starts to conduct with a very low resistance. Diodes are used as ➤rectifiers. ➤➤Gunn diode; light-emitting diode; Zener diode.

dioecious Describing plants in which the male and female sex organs are found on separate individuals forming male and female plants. An example is the yew. Compare ➤monoecious.

diol (glycol, dihydric alcohol) A compound having two hydroxyl groups. The most famous example is ➤ethane-1,2-diol, 'antifreeze'.

dioptre Symbol D, equivalent to m^{-1}. A unit used to express the power of a ➤lens: the power in dioptres is the reciprocal of the ➤focal length in metres. Converging lenses have positive values, whereas diverging lenses have negative values.

dioxane A colourless flammable liquid, with the structure shown in the diagram, used as a solvent, for cellulose ethanoate for example.

dioxane

dioxin 2,3,7,8-tetrachlorodibenzo-*p*-dioxin (TCDD). A ➤teratogenic compound formed as an impurity during the synthesis of ➤2,4,5-T.

dioxygen (oxygen) O_2 While it is usually considered pedantic to use this technically correct name for the common form of the element oxygen, it is necessary when wishing to distinguish between this molecule and the other allotrope, trioxygen (ozone, O_3).

dioxygenyl ion O_2^+ The ion formed by removing a single electron from molecular oxygen. The very high energy required to do this is within 1% of that required to remove an electron from a xenon atom. This was important historically as Neil Bartlett, having made dioxygenyl hexafluoroplatinate ($O_2^+PtF_6^-$) by accident in 1962, set about making the xenon version intentionally, so starting the chemistry of the noble gases.

dip, magnetic ➤magnetic inclination.

dipeptide ➤peptide.

diphenylmethanone (benzophenone) $C_6H_5COC_6H_5$ Colourless prisms, made by a ➤Friedel–Crafts reaction of benzoyl chloride on benzene with an aluminium chloride catalyst. It is used in perfumery.

diphtheria An acute and highly infectious disease caused by the bacterium *Corynebacterium diphtheriae*, which infects the mucous membranes of the respiratory tract producing a powerful toxin, particularly in children. Inflammation of the membranes can cause asphyxiation and the toxin affects the heart and nervous system. Childhood vaccination stimulates the body to produce anti-toxin, providing immunity.

diploid A condition in which the ➤chromosomes in the nucleus of a cell exist as pairs. Each member of the pair carries information about the same genetic characters as the other. Such a pair of chromosomes is referred to as a ➤homologous pair. One of each pair is inherited from each parent. The **diploid number** of chromosomes in a cell is expressed as $2n$. In humans the diploid number is 46, and all body cells contain 46 chromosomes, with the exception of red blood cells, which have no nucleus. Such a diploid number is constant within a species. When ➤gametes are formed the diploid number is reduced by half to the ➤haploid number (n) by ➤meiosis. When gametes fuse at fertilization the diploid condition is restored.

diplotene ➤meiosis.

dipolar aprotic solvent (dipolar nonprotolytic solvent) A solvent capable of dissolving ionic substances because of the presence of a dipole, which allows the formation of an **ion-dipole** force. However, unlike the more common polar solvent water, dipolar aprotic solvents do not ionize to form hydrogen ions, which confers advantage in some circumstances. ➤Nucleophilic substitution by the S_N2 mechanism with a charged nucleophile is often faster in dipolar aprotic solvents.

dipole (electric dipole) Literally, 'two poles'. In electrostatics, the theoretical limit of a distribution of two charges Q and $-Q$, separated by a displacement $\boldsymbol{a}$ as $\boldsymbol{a}$ tends to 0, with $Q\boldsymbol{a}=\boldsymbol{p}$, the **dipole moment**, held constant. The➤electric potential at displacement $\boldsymbol{r}$ from a dipole is

$$\phi(\boldsymbol{r}) = \frac{\boldsymbol{p} \cdot \hat{\boldsymbol{r}}}{4\pi\varepsilon_0 r^2}$$

where $\hat{\boldsymbol{r}}$ is a unit vector in the direction of $\boldsymbol{r}$. Contrast this with the electrostatic potential of a point charge, which has just r in the denominator. When $\boldsymbol{a}$ is small but nonzero the dipole field is a good approximation for $r \gg a$. When a dipole is placed in an electric field, there is no overall force on the dipole, but there will be a ➤torque if the dipole is not aligned with the field. All heteronuclear ➤diatomic molecules, such as HCl, have a permanent electric dipole moment, whereas homonuclear ones, such as H_2 and Cl_2, do not. A molecule with a dipole moment will tend to orient itself when placed in an electric field. The presence of the electric field can also alter the magnitude of the dipole as the electron density of the molecule is reshaped. The actual dipole moment is given by a power series expansion in which the first term is the permanent dipole, and the second is equal to the ➤polarizability of the molecule multiplied by the electric field strength. ➤➤ magnetic dipole; multipole; spherical harmonic.

dipole–dipole force One of the two main types of ➤intermolecular force, existing between molecules with permanent electric dipole moments. The other main type, the ➤dispersion force, is usually stronger. ➤➤van der Waals forces.

dipole moment Symbol p; unit C m. ➤dipole.

Dirac delta function The function, denoted by $\delta(x)$, that is zero except at $x = 0$ and satisfies the integral

$$\int_{-\infty}^{\infty} \delta(x)\,\mathrm{d}x = 1.$$

This definition needs careful application to keep it rigorous. It is useful for solving partial differential equations such as arise in quantum mechanics. Named after Paul Adrien Maurice Dirac (1902–84).

Dirac equation An equation proposed by Dirac to govern the wavefunction of a fermion. The ➤Klein–Gordon equation was found to be unsatisfactory as a modification of the Schrödinger equation for spin-$\frac{1}{2}$ particles to take account of special relativity. The Dirac equation introduces a four-component spinor (still represented by ψ below) in place of the single wavefunction, and features only a first derivative with respect to space:

$$\left(\sum_{\mu} \mathrm{i}\gamma^{\mu}\frac{\partial\psi}{\partial x_{\mu}}\right) - \frac{mc}{\hbar}\psi = 0.$$

Here the coordinate x_{μ} is one of the four spacetime coordinates, γ^{μ} is one of the ➤Dirac matrices, m is the particle's mass, c is the speed of light and $\hbar = h/2\pi$, where h is the Planck constant.

Dirac matrices The four 4×4 matrices γ^{μ} appearing in the ➤Dirac equation. They are most easily represented as

$$\gamma^0 = \begin{bmatrix} I & 0 \\ 0 & -I \end{bmatrix} \quad \text{and} \quad \gamma^i = \begin{bmatrix} 0 & \sigma_i \\ -\sigma_i & 0 \end{bmatrix} \quad \text{for } i = 1\,2\,3,$$

where every entry is itself a 2×2 matrix and σ_i are the three ➤ Pauli matrices.

diradical An entity containing two unpaired electrons. The most famous example is ➤oxygen, O_2.

direct Describing the motion of a planet about the Sun in the same direction as the Earth. Compare ➤retrograde.

direct current (d.c.) In an electric circuit, a current that does not reverse direction periodically.

direct dye ➤dye.

direct-gap semiconductor A semiconductor in which the lowest energy transitions available are➤direct transitions.

direct hydration The manufacturing process for ethanol, in which the molecule is made by reaction between ethene and water vapour at about 70 atm and 300 °C over a catalyst of phosphoric acid supported on silica:

$$C_2H_4(g) + H_2O(g) \rightarrow CH_3CH_2OH(g).$$

direction cosines (direction ratios) The components of a ➤unit vector; they are so called because they give the cosines of the angles made by the vector with the coordinate axes.

directly proportional ➤variation.

directrix ➤conic.

direct transition A transition between electronic energy levels in a solid where the ➤crystal momentum of the initial and final states is the same. The transition therefore does not require the participation of a ➤phonon. Compare ➤indirect transition.

disaccharide A ➤carbohydrate molecule consisting of two ➤monosaccharides, joined as in a ➤glycoside. Common table sugar, ➤sucrose, is a disaccharide composed of the monosaccharides glucose and fructose.

discharge, electrical The liberation of energy from a distribution of ➤electric charge when the charge flows. The source of the discharge can range from a cumulonimbus cloud, from which the discharge manifests itself as lightning, to a simple battery, in which the discharge is a current doing useful work.

discharge tube ➤gas-discharge tube.

discontinuity A value of x at which a function $f(x)$ is not ➤continuous, that is the ➤limit $\lim_{x \to a} f(x)$ depends on whether a is approached from above or below. On a graph, it is a vertical break in a plot of $f(x)$ against x.

discontinuous variation Variation of a genetically determined feature in a population that can be ascribed to two or more distinct forms. For example, some varieties of pea have plants which are either tall (about 1 m) or short (about 0.5 m), with no intermediates. Human blood groups are another example. Compare ➤continuous variation.

discrete quantity A quantity that takes one of a finite or ➤countable set of values. For example, the number of planets in the Solar System is a discrete quantity.

discrete random variable A random variable that takes a finite or ➤countable set of values.

discriminator 1 A circuit that rejects signals (usually pulses) that fall outside a certain amplitude or frequency range.
2 A circuit that converts a frequency-modulated or phase-modulated signal into an amplitude-modulated one (➤modulation; phase modulation).

disinfectant A substance that kills or restricts the growth of certain bacteria. Disinfectants commonly do not kill all bacteria.

disintegration The emission of particles from an atomic ➤nucleus, either spontaneously in radioactive decay (➤radioactivity) or after a collision.

disintegration constant ➤exponential decay.

disjoint sets Two or more sets that have no elements in common.

dislocation A line defect in a crystal in which one part of the crystal is sheared, or 'slips', relative to another. In an **edge dislocation** (see the diagram) the dislocation travels parallel to the slip, in a **screw dislocation** it travels perpendicular to the slip.

disodium tetraborate ➤borax.

dispersed phase ➤colloid.

disperse dye ➤dye.

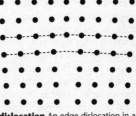

dislocation An edge dislocation in a crystal lattice.

dispersion In its most general sense, the change of the speed of a ➤wave motion in a medium with frequency. More specifically, dispersion refers to the separation of waves by frequency as a result of this property. In particular, the dispersion of light is the decomposition of a spectrum into its constituent frequency components by virtue of the difference in speed in a particular medium. ➤dispersion relation.

dispersion force (dispersion interaction, London force) The most important of the intermolecular forces. The origin of the force is quite complex. Any molecule when in the neighbourhood of another can have a temporary ➤dipole induced by the neighbour. These temporary dipoles can then 'get into step' and cause an attraction between the molecules. A dipole can even be induced in an individual atom; it is dispersion forces that hold the atoms of the noble gases together in their condensed phases. The magnitude of the dispersion force increases with increasing ➤polarizability of the species, which roughly increases with the number of electrons. This explains why the boiling points of the noble gases and the alkanes (see the diagram) increase steadily with the size of the atom or molecule.

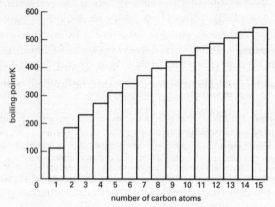

dispersion force The boiling points of the straight-chain alkanes.

dispersion medium ➤colloid.

dispersion relation The relationship between the ➤angular frequency ω of a wave and the magnitude k of its ➤wave-vector. It can be expressed in terms of the wave's speed, as $c = \omega/k$. For example, for water waves in deep water the speed can be shown to be $c(\omega) = g/\omega$, where g is the acceleration of free fall, and so $\omega(k) = \sqrt{gk}$. The speed of a wave-packet is given by $d\omega/dk$, which for water waves becomes $\frac{1}{2}\sqrt{g/k}$, and the rate at which the packet spreads ('disperses') is related to the second derivative of ω with respect to k. If the wave's speed is independent of angular frequency in a medium, then the medium is known as **nondispersive**. For light, the dispersion relation is often written in terms of the ➤ refractive index as a function of the wavelength, $n(\lambda)$. Typically (in **normal dispersion**), n decreases as λ increases (equivalent to c decreasing as k increases), but it can increase with increasing λ (**anomalous dispersion**).

dispersive power A measure of ➤dispersion of light in a medium. Given two wavelengths λ_1 and λ_2, the dispersive power, confusingly labelled ω by convention, is defined as

$$\omega = \frac{n(\lambda_1) - n(\lambda_2)}{\frac{1}{2}(n(\lambda_1) + n(\lambda_2)) - 1},$$

where $n(\lambda)$ is the ➤refractive index as a function of wavelength.

displacement A ➤vector that describes the position of one point with respect to a second point (often a reference point or ➤origin). It is usually thought of as a distance and a direction. When dealing with curved ➤spacetime in ➤general relativity, it is possible to invent a consistent idea of displacement without introducing the concept of distance (by defining a ➤metric).

displacement, electric ➤electric displacement.

displacement current The term $\partial D/\partial t$ in the fourth of ➤Maxwell's equations. Adding the displacement current to the current density j ensures that the ➤continuity equation for the conservation of charge is satisfied.

disproportionation Any reaction in which the ➤oxidation number of an element goes both up *and* down. Thus a disproportionation is a self-redox reaction. A simple example is the decomposition of hydrogen peroxide:

$$2H_2O_2(l) \rightarrow 2H_2O(l) + O_2(g).$$

The oxidation number of oxygen in peroxide is -1 which becomes -2 in water and 0 in the element oxygen.

dissipation The transfer of energy from a source to a medium in an irreversible way. The source can be something as obvious as energy in a battery (dissipation occurs in a resistance across it), or it can be the kinetic energy of a body in motion (dissipation occurs because of resistive forces).

dissociation A process in which a bond is broken (➤heterolytic fission; homolytic fission). When used in connection with electrolytic solutions, dissociation is used to mean the partial formation of ions from a ➤weak electrolyte. The '**acid dissociation constant**' is more usually termed ➤'acid ionization constant'. ➤Strong electrolytes are effectively completely ionized.

dissolution The process in which a solute dissolves in a solvent to form a ➤solution.

distance, measure of The concept of distance is almost instinctive, since we inhabit a space which has an apparently ➤Euclidean geometry and a working definition of distance will usually suffice (such as 'the number of metre rules that can be placed in line and end to end between two points'). In general, distance is more subtle than this, as ➤length contraction in ➤special relativity demonstrates, and it is necessary to extend the idea into more than three dimensions. Defining distance, by introducing a ➤metric to our concept of ➤spacetime, is a non-trivial statement about the nature of spacetime itself. ➤➤geodesic; relativity.

distance ratio ➤velocity ratio.

distillate The liquid produced by distillation.

distillation The process by which a volatile liquid is separated from a less volatile substance. For example, ethanol may be distilled from its solution with water. Simple distillation is successful if the solute has a low volatility. If the boiling points of the two substances are close, the technique of ►fractional distillation may be required.

distilled water Water that has been purified by distillation and hence contains far fewer dissolved ions.

distribution ►partition.

distributive That property of two binary operations, such as multiplication and addition, that enables brackets to be expanded in the familiar way: $a \times (b + c) = a \times b + a \times c$. In other words, it does not matter whether b and c are first added together and then multiplied by a or vice versa.

disulfide bond The covalent bond between two sulfurs in the **sulfydryl groups** of paired ►cysteine amino acids in a ►protein. The —S—S— ' bridge' is formed when the sulfydryl groups are oxidized. The oxidized form of cysteine is called ►cystine. These disulfide bonds are important in determining the overall three-dimensional shape and stability of many proteins, particularly ►immunoglobulins and ►enzymes.

diuretic Any substance, such as caffeine in coffee, that causes less water to be reabsorbed into the blood by the kidneys, with a subsequent increase in the volume of urine.

diurnal 1 By day as opposed to night, as applied to animals (opposite of nocturnal) and in phrases such as 'diurnal heating' (caused by the Sun).
   2 Daily, in the sense of once every 24 hours; in phrases such as 'diurnal variation' (of temperature) and 'diurnal motion' (the apparent motion of the stars).

diurnal rhythm A pattern of behaviour or physiological activity that exhibits regular periodicity over a period of 24 hours. Examples are the opening of flowers of many species during the day and their closing at night, and the daily feeding behaviour of bats. Such rhythms are controlled partly by environmental stimuli and partly by internal ►biological clocks.

divalent Having a valency of two. 'Divalent iron', for example, means iron with oxidation number +2.

divergence 1 The divergence of a ►vector field $v = v_1\mathbf{i} + v_2\mathbf{j} + v_3\mathbf{k}$ is the ►scalar field $\partial v_1/\partial x + \partial v_2/\partial y + \partial v_3/\partial z$, written div v or $\nabla \cdot v$.
   2 That property of a sequence or series that does not converge to a limit.

divergence theorem The identity $\int\int\int_R \text{div } F \, dV = \int\int_{\partial R} F \cdot n \, dS$, which relates the triple integral of the ►divergence of a ►vector field F over a three-dimensional region R to the surface integral of its ►normal (2) component $F \cdot n$ over the boundary ∂R of the region. The theorem is also known as **Gauss's theorem**, after Carl Friedrich Gauss (1777–1855). ►►Maxwell's equations.

divergent evolution The process by which distinct forms of an organism are derived from a common ancestor. Ultimately, divergent evolution may result in the formation of new species. ►►evolution.

diverging lens A lens designed to make a parallel beam of light diverge. Such lenses are used in spectacles to offset the effects of ➤myopia.

division The inverse operation to multiplication: finding that number (the **quotient**) which, when multiplied by one given number (the **divisor**), yields a second given number (the **dividend**). With dividend a and divisor b the quotient is denoted a/b.

divisor ➤division.

dizygotic twins ➤fraternal twins.

D layer The lowest layer of the ➤ionosphere, between altitudes of 60 and 90 km. It reflects low-frequency radio waves.

D lines The very prominent closely spaced pair of yellow lines (near 589 nm) that dominate the sodium emission spectrum, and are familiar in the glow of streetlights with sodium-vapour lamps. The transition responsible is from the 3p orbital in the excited atom to the ground state 3s, the line being split because of interaction between the orbital angular momentum (in the 3p orbital) and the spin angular momentum of the electron.

DMF ➤dimethylformamide.

DMSO ➤dimethylsulfoxide.

DNA Abbr. for deoxyribonucleic acid. A ➤polynucleotide molecule which comprises the genetic material of organisms, consisting of the sugar ➤deoxyribose, phosphate groups and nitrogenous bases. The molecule has a **sugar–phosphate backbone**, consisting of phosphate groups bonded to the third and fifth carbon atoms of neighbouring deoxyribose molecules, to each of which one of four different nitrogenous bases is attached. It is the specific sequence of these bases that encodes the information required to build and maintain the cells, and hence the bodies of all organisms. DNA is usually double-stranded, with each strand coiled alongside the other in a characteristic **double helix**. The two strands run in opposite directions, 3'–5' on one strand corresponding to 5'–3' on the other. The structural basis for the helix is ➤hydrogen bonding between specific pairs of bases in each strand. The nitrogenous bases are the ➤purines adenine (A) and guanine (G), and the ➤pyrimidines thymine (T) and cytosine (C). In double-stranded DNA, A pairs with T and G pairs with C (➤➤base pairing). The structure of DNA was elucidated in 1953 by James Watson, Francis Crick, Maurice Wilkins and Rosalind Franklin using ➤X-ray crystallography.

Most DNA in a cell is housed within the nucleus in the ➤chromosomes, but ➤chloroplasts and ➤mitochondria also have their own DNA. There are estimated to be six billion base pairs in the DNA in each ➤diploid human cell, with a combined length of about two metres. The information contained in the sequence of bases in DNA is used by the cell, via an RNA copy, to make proteins during ➤protein synthesis. A specific section of DNA that has a discrete function is called a ➤gene. The manipulation of DNA in the laboratory, particularly the ability to combine DNA from different sources and to introduce such DNA into cells, is the basis of ➤genetic engineering. Compare➤➤RNA.

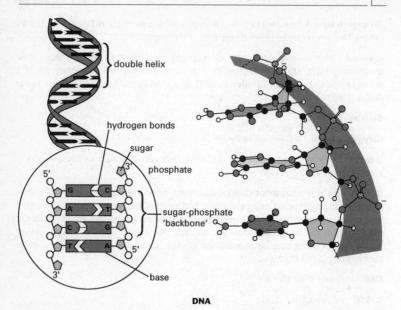

DNA

Complementary DNA (cDNA) is a DNA strand formed by copying an ➤RNA strand used as a template. The process requires the use of the enzyme ➤reverse transcriptase. The cDNA strand so formed can then be made into double-stranded DNA by the action of the enzyme ➤DNA polymerase, and cloned in appropriate host cells. A mixed population of cDNAs cloned in this way constitutes a gene library (➤gene cloning). cDNA libraries are useful because they enable the genes that are actually being expressed via their RNA in a particular cell type to be identified. This facility has many applications in cancer biology and other developmental studies.

Recombinant DNA is a DNA molecule containing segments from different sources. DNA from different organisms or from chemical synthesis can be cut with appropriate ➤restriction enzymes which form ➤sticky ends. If DNA from different sources is cut with the same restriction enzyme and mixed, the sticky ends join together by complementary ➤base pairing and can be further stabilized by the enzyme ➤DNA ligase. By such means, genes from different organisms can be combined in the same DNA molecule. ➤Genetic engineering makes extensive use of recombinant DNA.

DNAase An enzyme that breaks DNA into smaller fragments by the ➤hydrolysis of bonds between ➤nucleotides. ➤➤restriction enzyme.

DNA blotting (Southern blotting) A process that enables DNA fragments to be transferred from a slab of gel used in ➤electrophoresis to an inert membrane, usually of nitrocellulose, without disturbing the positions of the fragments. The term 'blotting' describes the overlaying of the membrane on the gel and the application of a pad to ensure even contact. On treatment with an appropriate solution, DNA moves from the gel on to the membrane where it is physically trapped. Once on the membrane it can be further manipulated to produce, for example, a ➤genetic fingerprint.

DNA fingerprint ➤genetic fingerprinting.

DNA insert A segment of ➤DNA that is joined to other DNA to form recombinant DNA. Usually the term refers to DNA inserted into a circular ➤plasmid which can be cloned and used as a vector (➤cloning vector).

DNA library A collection of various individually cloned DNA fragments from a specific organism which represents the entire genome of the target organism. **cDNA libraries** are formed from copies of mRNA and represent the coding sequences of the target organism; they are commonly used for studying eukaryotic genomes. Such libraries are useful since they represent the genes that are being expressed by cells at a given stage in their activity and confirm ➤intron–exon boundaries. Libraries representing relatively small genomes may be cloned in plasmid vectors, but most libraries from organisms with larger genomes are cloned using ➤lambda (λ) phage or ➤yeast artificial chromosomes.

DNA ligase An enzyme that repairs 'nicks' in the sugar–phosphate backbone of one strand of ➤DNA by forming covalent phosphodiester bonds. DNA ligase is used in particular to stabilize the bonds between the ➤sticky ends of fragments of DNA used in recombinant DNA technology.

DNA methylation The addition of methyl groups to ➤DNA by **methylating enzymes**; this blocks the ability of some ➤restriction enzymes to cut DNA if the methylation occurs in the restriction site. Bacteria protect their own DNA in this way.

DNA polymerase An ➤enzyme that elongates ➤DNA by incorporating individual ➤nucleotides in a precise sequence determined by a complementary template strand of DNA or ➤RNA. It can also degrade DNA into individual nucleotides. There are several different forms of the enzyme in ➤eukaryotic cells, differing in both their structure and the details of the reactions they catalyse. DNA polymerase from the ➤bacterium *Thermus aquaticus* (Taq polymerase) is stable at high temperatures and is used to amplify fragments of DNA in the ➤polymerase chain reaction.

DNA primer A short sequence of ➤nucleotides that initiates the copying of a fragment of ➤DNA, particularly in applications of the ➤polymerase chain reaction.

DNA probe (gene probe) A segment of single-stranded ➤DNA used to locate or identify specific fragments of DNA that have complementary sequences. DNA probes are identifiable by being radioactive or fluorescent, or by the production of a recognizable colour after suitable treatment. DNA probes are used to confirm the transfer of recombinant DNA into viruses growing in bacteria on agar plates in ➤genetic engineering, and to produce ➤genetic fingerprints after gel ➤electrophoresis of DNA fragments. They are used to identify specific sequences of nitrogenous bases within a ➤genome.

DNA profile ➤genetic fingerprinting.

DNA replication The synthesis *in vivo* of two identical double-stranded helical ➤DNA molecules from one original double strand. The original double strand separates into two single strands, each of which acts as a template for a new double strand under the influence of the enzyme ➤DNA polymerase. Since each of the original strands codes for one of the new strands, the process is said to be 'semi-conservative'. The basic principles of the process have been established by Arthur Kornberg, but the details are complex and involve a number of molecular mechanisms which are still being elucidated.

DNA sequencing The elucidation of the sequence of the four bases adenine, thymine, guanine and cytosine in a fragment of ➤DNA. The most widely used technique is ➤dideoxy sequencing. The sequence of bases in a nucleic acid determines its information content and can be used to compare DNA from different sources. Such information can be used to infer evolutionary relationships between organisms and to investigate gene function. The ➤human genome project has determined the entire base sequence of human DNA.

DNP ➤2,4-dinitrophenylhydrazine.

dodecahedron A solid that has twelve faces. A **regular dodecahedron** has faces that are regular pentagons and is one of the five ➤Platonic solids.

dolomite (magnesian limestone) A mineral consisting of magnesium and calcium carbonates, $MgCO_3 \cdot CaCO_3$, named after the mountain range in Northern Italy. Dolomite is a major ore of magnesium and, after heating to a high temperature (calcining), is a very common lining of furnaces for producing steel.

domain 1 (Phys.) A bounded volume of a ➤ferromagnetic material with uniform ➤magnetization. Neighbouring domains have different magnetization directions (see the diagram opposite). Although uniform magnetization is energetically favourable at short range, completely uniform magnetization has a very high total energy. The domain structure is energetically more costly at domain boundaries, but the absence of long-range magnetization compensates for this. An external magnetic field can cause the magnetization of all domains to align, and some remanent magnetization may remain when the field is removed (➤➤hysteresis).
   2 (Math.) ➤function.

dominant 1 Referring to the ➤allele that is expressed in the ➤phenotype of a ➤heterozygous individual. A sexually reproducing organism will inherit one allele for each character from each parent. If two different forms are present in the one organism, the allele that is actually expressed is dominant. For example, in pea plants the gene for height of plant exists in two allelic forms: tall and dwarf. Since tall is dominant to dwarf, a pea plant that contains both forms will be tall and outwardly indistinguishable from a plant that contains only the tall form. A dwarf plant will contain only the dwarf form of the gene, which is termed **recessive**. Sometimes, when two different alleles are present in the same organism they are expressed to produce an intermediate character. Examples of this codominance (**incomplete dominance**) include the human ➤blood groups, and the 'yellow green' character in plants such as tobacco and brassicas. ➤➤genotype; phenotype.
   2 The most abundant or conspicuous species in a community of organisms in a particular habitat. Thus in a particular woodland there may be a number of different tree species present, but if oak were significantly more common than other species it would be termed dominant.

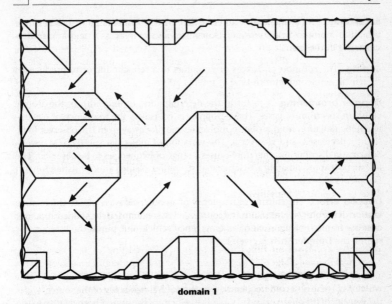

domain 1

3 Animals that live in social groups usually establish hierarchies in which individuals have different degrees of priority for access to food and breeding opportunities. Such a hierarchy is termed a **dominance hierarchy**. In groups of red deer and elephant seal, for example, there is usually one dominant male whose position is recognized and accepted by the rest of the group. In these species, and others like them, it is usually only the dominant male that will mate with females in the breeding season. In timber-wolf packs there is a dominant pair of male and female, who are the only pair to breed in the pack. Such dominance hierarchies reduce the overall level of competition and aggression within the social unit, although the seniority of the dominant animals will be challenged from time to time.

donor In an ➤extrinsic semiconductor, an impurity that adds an extra electron to the available carriers (compare ➤acceptor (2)). An ➤n-type semiconductor has an excess of donors (typically Group 15 elements, such as arsenic, introduced into silicon and germanium).

donor atom In a complex formed between a Lewis acid and a ➤Lewis base, the atom in the Lewis base that has provided the electron pair. For the nickel–ammonia complex, $[Ni(NH_3)_6]^{2+}(aq)$, nitrogen is the donor atom.

dopa (3,4-dihydroxyphenylalanine) An amino acid, normally isolated from certain plants. This natural product (L-dopa, see the diagram) is exclusively ➤laevorotatory, and is currently the main treatment for the debilitating illness Parkinson's disease. A derivative of the amino acid ➤tyrosine, dopa is found particularly in cells of the ➤adrenal glands, where it is a precursor of ➤dopamine, ➤noradrenaline and ➤adrenaline.

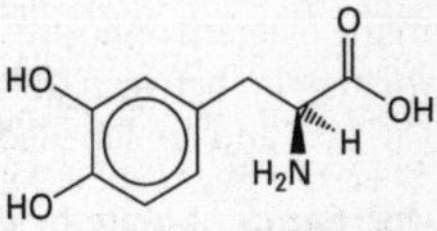

L-dopa

dopamine A ➤catecholamine intermediate in the synthesis of ➤noradrenaline from the ➤amino acid tyrosine. Dopamine also acts as a ➤neurotransmitter substance in the brain.

doping The addition of donors or acceptors to a semiconductor to produce an ➤extrinsic semiconductor (n-type or p-type).

Doppler broadening A factor in the determination of linewidth in the atomic emission spectrum of gases. The ➤Doppler effect causes the frequency of electromagnetic radiation from atoms or molecules moving away from the observer to be slightly decreased, and vice versa. The variation of molecular speeds (➤Maxwell–Boltzmann distribution) therefore causes a range of frequencies to be observed. The lineshape is approximately a ➤Gaussian distribution. Named after Christian Johann Doppler (1803–53).

Doppler effect The shifting of frequency of an observed wave due to the relative motion of the observer, medium and source. A classic example is the sudden apparent decrease in pitch of the siren of an emergency vehicle as it passes the listener. For sound, the **Doppler shift** is given by

$$\frac{f_{obs}}{f_0} = \frac{c - v_{obs}}{c - v_{source}},$$

where f_{obs} is the observed frequency, f_0 is the natural frequency of the source, v_{obs} is the velocity of the observer and v_{source} is the velocity of the source, both relative to the medium (in this instance c is the speed of sound). In the transmission and reception of electromagnetic radiation there is no medium, so only the relative velocity v of the observer to the source can feature in the equation:

$$\frac{f_{obs}}{f_0} = \frac{\sqrt{c - v}}{\sqrt{c + v}}.$$

The Doppler effect is a key tool in determining the relative speeds (and thus distances) of astronomical objects (➤redshift).

Doppler radar A form of ➤radar that employs the ➤Doppler effect to determine the relative speed of the object observed.

Doppler shift ➤Doppler effect.

d orbitals A set of ➤atomic orbitals, and therefore a set of solutions of the ➤Schrödinger equation. The d orbitals are filled across the d block of the ➤periodic table, and many d-block elements are of considerable technological importance. This set has an ➤orbital angular momentum quantum number, l, of 2. The number of orbitals in the set is always given by $2l + 1$, so is 5 in this case. Their shapes are shown in the diagram opposite. They all have the same energy in an isolated atom. Three of them point between the axes and are labelled d_{xy}, d_{yz} and d_{zx}. The other two lie along the axes and are labelled $d_{x^2-y^2}$ and d_{z^2}. ➤degenerate states.

dormancy A state or period in the life of an organism during which the rate of physiological processes is reduced to a minimum. Thus many plants have seeds which may remain dormant for months or even years before ➤germination, and animals become dormant during ➤hibernation.

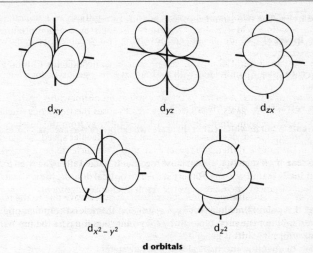

d_{xy} d_{yz} d_{zx}

$d_{x^2-y^2}$ d_{z^2}

d orbitals

dorsal Describing the upper surface, or a structure near the upper surface, of an organism, particularly an animal. In quadrupeds the dorsal surface is closest to the spine, and in bipedal forms such as birds and humans it is the backward-facing (posterior) surface. Compare ➤ventral.

dose A measure of the ➤ionizing radiation absorbed by a medium, particularly the human body. **Absorbed dose** (symbol D; unit gray) is the absorbed energy per unit mass of absorbing medium. **Exposure dose** (symbol X; unit C kg^{-1}, previously roentgen) is the total charge produced in dry air per unit mass of air. **Equivalent dose** (symbol H; unit sievert) is the absorbed dose multiplied by a factor representing the biological effect of the radiation received.

dosemeter (dosimeter) A device for measuring the ➤dose of radiation received, usually by a human subject. **Film badges** are the most common form: a film that is sensitive to the radiation is worn in the form of a film badge, and the blackening of the film over a period of time is measured.

dot-and-cross diagram ➤Lewis structure.

dot product ➤scalar product.

double-angle formulae Formulae such as sin $2A$ = 2 sin A cos A, cos $2A$ = $\cos^2 A - \sin^2 A$ and tan $2A$ = 2tan $A/(1 - \tan^2 A)$ that relate the trigonometric functions of twice a given angle to trigonometric functions of the angle.

double bond A bond in which two pairs of electrons hold two atoms together. The simplest examples occur in the oxygen molecule, O=O, and the ethene molecule, $H_2C{=}CH_2$. The C=C double bond is stronger than a C—C single bond, but not usually twice as strong (➤electrophilic addition). The presence of a double bond raises the possibility of ➤*cis–trans* isomerism.

double decomposition Now more commonly referred to as ➤ionic association.

double helix The two-stranded coiled structure of ➤DNA.

double integral A multiple integral involving two variables, typically written $\int\int f(x, y)\,\mathrm{d}x\,\mathrm{d}y$.

double recessive An individual with a ➤homozygous ➤genotype for a ➤recessive character.

double refraction ➤birefringence.

double salt A salt in which either there are two cations or two anions. One example is ➤Mohr's salt.

double star Two stars that appear close together in the sky. They may be a single, gravitationally bound system (➤binary star) or an **optical double**, the two stars lying at different distances, their closeness the result of chance alignment.

doublet **1** A pair of lines in a spectrum of any sort that have a common origin but have been split in frequency, typically by spin-orbit coupling. The sodium ➤D lines are an example.
   2 A pair of quantum-mechanical ➤degenerate states.

down quark One of the six ➤quarks of the ➤standard model; with its partner the ➤up quark, the down quark is a constituent of both the proton and the neutron.

Downs cell An electrolytic cell used for the production of metallic sodium. As sodium is so reactive, it is not possible to produce the metal by reduction of the aqueous solution. Hence the Downs cell uses a molten electrolyte of sodium chloride (plus a little calcium chloride to lower the melting point). Sodium is produced at the cathode and chlorine at the anode. The products need to be kept apart as they react so vigorously together. Named after (James) Cloyd Downs (1885–1957).

Down's syndrome A clinical condition of humans caused by an extra copy of ➤chromosome 21 (**trisomy 21**) named after the physician (John) Langdon Haydon Down (1828–96), who made the first detailed study. The condition is characterized by mental retardation, a broad face with slant eyes, weakened muscles and limbs with short digits. It was one of the first congenital conditions that could be diagnosed by ➤amniocentesis. Modern diagnoses of such conditions are by ➤genetic fingerprint techniques.

Dow process A method of extracting magnesium which relies on the fact that magnesium hydroxide is less soluble than calcium hydroxide; the addition of calcium hydroxide to an aqueous solution containing magnesium ions causes magnesium hydroxide to precipitate. The hydroxide can then be reacted with acid to give a solution containing magnesium ions. Named after Herbert Henry Dow (1866–1930).

drag In aerodynamics, the force on a body moving through a fluid that acts in the opposite direction to the motion. Typically the drag on a body is proportional to the square of its speed through the fluid. If the body is an ➤aerofoil a second term, called **induced drag**, proportional to the square of its lift, also contributes, due to the formation of ➤vortices at the ends of the aerofoil. Compare ➤lift.

drain The terminal of a field-effect transistor at which the conventional current enters (and thus the electrons leave, or drain away). It can be thought of as the positive terminal.

drift chamber A detector of elementary particles that uses the time taken for electrons, produced by the particle by ionization, to drift to a wire in a uniform electric field to deduce the particle's trajectory.

dropping mercury cathode ➤polarography.

Drosophila The common **fruit fly**, *Drosophila melanogaster*. This insect is extensively used in genetic investigations to elucidate patterns of inheritance and the functions of ➤genes. Their rapid life cycle, large numbers of offspring, small number of chromosomes (eight) and the ease with which they can be maintained in the laboratory make them ideal for genetic analysis. Work with mutants of characters such as eye colour and wing shape established the basic principles by which the positions of genes on ➤chromosomes can be mapped. The salivary glands of *Drosophila* larvae have so-called giant chromosomes which are also used to study chromosome structure and function.

Drude model A model of electrons in a metal proposed by Paul Karl Ludwig Drude (1863–1906) in 1900. It treats the loosely bound conduction electrons classically, as if they composed an ➤ideal gas, free to move within the crystal but colliding with each other and the ion cores, which are assumed to be immobile.

dry adiabatic lapse rate (DALR) As a volume of ➤unsaturated air rises in the atmosphere without exchanging heat with its surroundings, the resultant change in pressure causes the air to cool; the decrease in temperature per unit height is the dry adiabatic lapse rate. The DALR at sea level and moderate temperatures is typically 10 K per 1000 m.

dry cell The most common type of battery, the details of which are shown in the diagram. It is called 'dry' as no liquid is formed inside.

dry ice The name for the solid phase of carbon dioxide, so called because the solid sublimes to gaseous carbon dioxide, therefore not becoming wet in the process.

Ds Symbol for the element ➤darmstadtium.

dubnium Symbol Db. The element with atomic number 105.

ductile Describing the property, possessed by a typical metal, of being able to be drawn out into a wire.

ductless gland ➤endocrine gland.

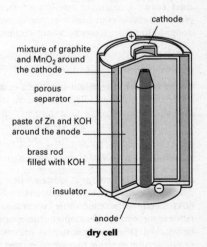

cathode

mixture of graphite and MnO_2 around the cathode

porous separator

paste of Zn and KOH around the anode

brass rod filled with KOH

insulator

anode

dry cell

Dulong and Petit's law A classical expression for the heat capacity of a crystal due to its ▶lattice vibrations, originally found by Pierre Louis Dulong (1785–1838) and Alexis Thérèse Petit (1791–1820), which takes no account of the quantum-mechanical nature of the ▶normal modes (▶phonons). The result is delightfully simple: the molar heat capacity is $3R$, i.e. a contribution of $3k$ per ion (k is the ▶Boltzmann constant, R the gas constant), independent of the nature of the crystal. It is improved upon for low temperatures by the ▶Debye model.

Dumas's method An obsolete method of ▶gravimetric analysis for measuring the molar mass of a gas. A ▶mass spectrometer is now used in preference. Named after Jean-Baptiste André Dumas (1800–84).

dummy variable A variable on which the actual value of a given expression does not in fact depend. For example, x is a dummy variable in $\int_0^b ax^2 \, dx = \frac{1}{3}ab^3$.

duodenum The anterior (nearest the stomach) region of the small intestine of mammals. Food enters the duodenum from the stomach via the **pyloric sphincter**. Ducts from the liver (bile duct) and the pancreas (pancreatic duct) deliver bile and pancreatic juice into the duodenum. Its surface area is greatly increased by ▶villi, and the walls contain numerous secretory **Brunner's glands**. The duodenum is the site of digestion of fats, carbohydrates and proteins, and also of some absorption. ▶▶gut.

duplex A communications link between two computers that is capable of handling the transmission of information in both directions. A **half-duplex** link can support only one direction at a time, but simultaneous transmission in both directions is possible with a **full-duplex** link. Compare ▶simplex.

Duralumin An important aluminium alloy with 4% by mass of copper, and some manganese and magnesium.

dust core A magnetic core (for a ▶transformer or similar device) made of a powdered material of high ▶magnetic permeability, like ▶ferrite. A dust core is designed to reduce losses by ▶eddy currents.

dutch metal An alloy of copper and zinc which comes in very thin sheets. It can spectacularly inflame in chlorine gas at room temperature.

dwarf (dwarf star, main-sequence star) A star on the main sequence of the ▶Hertzsprung–Russell diagram. Dwarfs are 'normal' stars, and are so called to distinguish them from ▶giant stars; 90% of the stars in the Milky Way ▶galaxy are dwarfs. The coolest of them are sometimes called **red dwarfs**. The Sun is a typical dwarf star. ▶▶brown dwarf; white dwarf.

Dy Symbol for the element ▶dysprosium.

dye An intensely coloured substance that can be firmly attached to a fabric; good dyes are 'fast' to water, light and detergents. Some **direct dyes**, such as diazo dyes, can bind directly to the fabric; other dyes require the use of **mordants** which activate the fabric in some way, for example by incorporating a basic group when an acid dye is to be attached. **Disperse dyes**, such as anthraquinones, are insoluble dyes applied as a fine dispersion in water; they are often used for cellulose ethanoate and nylon fibres. **Vat dyes**, such as indigo, are insoluble dyes used for dyeing cotton.

dynamic equilibrium A ▸steady-state situation in which a reverse process is occurring at such a rate that it exactly balances the corresponding forward process. All chemical equilibria are dynamic in the sense that, once equilibrium is reached, individual molecules continue to react even though there is no net change in the overall composition. ▸▸equilibrium constant.

dynamic friction ▸Friction that acts between a body and a surface when the body is moving. Dynamic friction originates from the need to deform microscopic irregularities in the sliding surfaces as they move past each other. Compare ▸static friction.

dynamic pressure The kinetic energy per unit volume for a fluid that flows steadily. For an incompressible fluid, it is the total pressure that would be exerted on a body if that body were to bring the fluid to rest, less the ▸static pressure. ▸▸Bernoulli's theorem.

dynamics The study of bodies in motion and the forces acting on them (compare ▸statics; kinematics).

dynamite The archetypal explosive, made by absorbing ▸nitroglycerine on ▸kieselguhr. This was much safer to handle than neat nitroglycerine. It was discovered by Alfred Nobel in 1867 and envisaged by him as a boon to help mining companies. His revulsion at its application to war was the reason he added a peace prize to his scientific ▸Nobel prizes.

dynamo A device for generating electricity from simple rotational motion of a coil in a magnetic field. By using multiple coils and carbon brushes that remain fixed as the coils rotate (connecting alternately to one brush then the other), it is possible to generate ▸direct current (with a small alternating 'ripple voltage' superimposed).

dyne An outdated (c.g.s.) unit of force equal to 10^5 N (which is 1 g cm s^{-2}).

dysprosium Symbol Dy. The element with atomic number 66 and relative atomic mass 162.5. It is a typical lanthanide element and as such has a common ▸oxidation number of +3, as in the yellow Dy^{3+} ion.

dystrophic A term used to describe lakes and rivers which, although relatively low in mineral nutrients, receive quantities of organic matter derived from, for example, peat deposits. Peaty materials in solution tend to give dystrophic waters a distinct brown colouration. Compare ▸eutrophic; mesotrophic; oligotrophic.

E

e A fundamental mathematical constant equal to the limit $\lim_{n \to \infty} (1 + 1/n)^n$ and to the infinite sum $\sum_{n=0}^{\infty} 1/n!$ It is a ➤transcendental number with decimal expansion beginning 2.718 281 828 459. . . It is the ➤base (3) of natural ➤logarithms and the value of the ➤exponential function, e^x, at $x = 1$.

e⁻ Symbol for the ➤electron (occasionally the negative sign is omitted).

e⁺ Symbol for the ➤positron.

e Symbol for elementary charge, one of the ➤fundamental constants, which has the approximate value 1.602×10^{-19} C. The ➤electronic charge is $-e$.

E Symbol for the prefix ➤exa−.

E 1 Symbol for ➤energy.
   2 Symbol for ➤electrode potential.
   3 Symbol for ➤Young's modulus.

E Symbol for ➤electric field strength.

E_a Symbol for ➤activation energy.

ear The vertebrate sense organ responsible for the detection of sound (see the diagram opposite). In mammals the ear is divided into three regions: the outer, middle and inner ear. The **outer ear** comprises the **pinna** (external lobes) and **auditory canal** (external meatus), and is responsible for focusing sound waves on the eardrum (**tympanum**), which then vibrates. The vibrations are carried across the **middle ear** by the **ear ossicles**, the smallest bones in the body. The three ossicles in order are the **malleus** (hammer), **incus** (anvil) and **stapes** (stirrup). Vibrations from the stapes transmitted to the **oval window** cause pressure waves to be initiated in the fluid (**perilymph**) that fills the **inner ear**. These pressure waves are converted into vibrations in the fluid (**endolymph**) of the **cochlea**, in which the **organ of Corti** converts them into nerve impulses. From here the nerve impulses are relayed to the brain by the **auditory nerve**. Excess vibrations in the perilymph are damped by the **round window**.

The inner ear also contains the **semicircular canals**, which detect changes in position and assist in the maintenance of balance. These consist of three endolymph-filled canals arranged at right angles to one another and joined at the sac-like **utriculus**. This is a continuation of the **sacculus**, which connects to the base of the cochlea. The canals have swellings (**ampullae**) which contain hair cells capped by a cone of jelly (**cupula**). Movement of fluid in the canals deflects these cells, which generate nerve impulses which are transmitted to the brain. Impulses from the three canals on each side are interpreted to provide information on the orientation of the body. The feeling of giddiness is caused by the fluid continuing to move and deflect the cupulas even though the body has stopped moving. This is possibly also the cause of motion- or sea-sickness. The **Eustachian tube** connects the middle ear with the pharynx (the back of the throat); its function is to equalize pressure on either side of the eardrum. The basic principle of the ear is the same in all vertebrates, but reptiles, amphibians, birds and some aquatic mammals lack an outer ear; snakes, for example, have no eardrum.

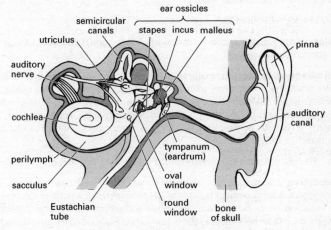

ear Simplified diagram of the human ear.

earth The ➤electric potential of the surface of the Earth. An electrical device is **earthed** if it is connected by a low-resistance path to the Earth. It is often convenient to use earth as the arbitrary zero for potential.

Earth The third closest planet to the Sun, with an orbital period of 1 year, and a rotational period of 1 day. Its average distance from the Sun is the ➤astronomical unit. The Earth has a radius of 6378 km at its equator, and 6357 km at its poles, resulting in a slight ➤equatorial bulge. Its mass is 5.98×10^{24} kg. ➤Radioactive dating indicates an age of about 4.6×10^9 years. The planet consists of an iron **inner core** at a temperature of about 7000 K, solidified under a pressure of about 4×10^{11} Pa; this is enclosed in a liquid iron **outer core** which extends to a radius of about 3500 km. Most of the mass of the Earth is in the rocky **mantle** outside the core. The thickness of the outer, solid **crust** is from 10 km (beneath the oceans) to 50 km (beneath the continents). These shells form the **lithosphere**; outside it is the ➤hydrosphere. The Earth's one natural satellite is the ➤Moon. ➤➤atmosphere; geomagnetism; geophysics; plate tectonics; Appendix table 4.

earthquake A disturbance of the Earth's surface caused by a ➤seismic wave. ➤➤Richter scale.

earthshine A faint illumination of the part of the visible face of the Moon not directly lit by the Sun. It is caused by sunlight reflected by the Earth. The phenomenon is descriptively called 'the old moon in the new moon's arms'.

ebullioscopy The name given to the obsolete process whereby the elevation of boiling point (➤colligative properties) is used to determine ➤molar mass. A ➤mass spectrometer is now used for the purpose.

eccentricity A constant that determines both the type and shape of a ➤conic. The eccentricities of the elliptical orbits of the planets are shown in Appendix 4.

ECF Abbr. for ➤extracellular fluid.

ECG Abbr. for ➤electrocardiograph.

echelon A very high resolution ➤diffraction grating made from a stack of identical glass plates, offset like steps.

echinoderm A member of the major animal ➤phylum **Echinodermata**, containing radially symmetrical organisms with spiny skins, having a pressurized fluid canal system operating hydraulic 'tube feet'. The phylum contains four classes which include the sea urchins, the starfish, the brittle stars and the sea cucumbers. The symmetry is typically pentaradiate, but there are many species of starfish that have more than five arms.

echo The reflected wave that results from a sound wave incident on a boundary.

echolocation A technique for finding the distance, and sometimes direction, of a remote object. A measurement is made of the time taken for a pulse of electromagnetic radiation or sound to reach the object and return from it after reflection. Examples are ➤radar and ➤sonar, and echolocation is also used by animals such as bats, dolphins and some cave-nesting birds.

eclipse The passage of one astronomical body through the shadow of another (see the diagram). An eclipse of the Moon (**lunar eclipse**) occurs when the Moon passes through the shadow cast by the Earth in the Sun's light. The opportunity for a lunar

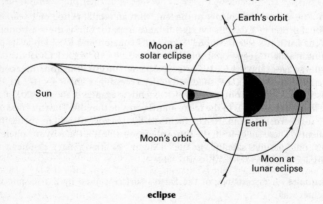

eclipse

eclipse occurs about twice each year. The Moon takes up to 3.5 hours to pass right through the Earth's shadow. A lunar eclipse is visible from wherever the Moon is above the horizon. A **solar eclipse** occurs when the shadow of the Moon is cast across a part of the Earth's surface. Since the ➤umbra of the Moon's shadow is about 300 km across on average at the Earth's surface, the region from which a **total solar eclipse** is visible is small; from any given spot it is visible on average once every 300 years. A partial solar eclipse, in which the Sun is not entirely hidden by the Moon, can be viewed over a much larger area by observers in the ➤penumbra. An **annular eclipse** is a solar eclipse in which a ring of the Sun's photosphere remains visible when the Earth, Moon, and Sun are aligned, and the Moon is near ➤apogee. ➤➤occultation.

ecliptic The plane of the orbit of the Earth about the Sun. (See the diagram at ➤celestial sphere.)

E. coli The bacterium *Escherichia coli*. It is a common gut bacterium, some types of which cause forms of food poisoning. A large number of physiologically and genetically defined strains are known, and it is commonly used in laboratory studies. Its historical significance is that work on the nutritional requirements of a range of ➤mutants in the 1950s led to the establishment of the relationship between genes and specific cellular functions. *E. coli* is now routinely used in ➤genetic engineering. The cells have a single circular chromosome consisting of a double strand of DNA containing approximately 4.5 million base pairs (➤base pairing). This has been extensively investigated and mapped, and the exact positions of over a thousand genes have been located.

ecological niche ➤niche.

ecology Broadly, the study of organisms in relation to their ➤environment. Ecology encompasses a wide range of scientific disciplines since the functioning of an organism in its environment is determined by its structure and physiology, and also by its behaviour. These functions in turn are determined by genetics, and as such reflect the organism's evolutionary history.

ecosystem A fundamental concept in ecology relating to a community of organisms which interact with each other and with their physical and chemical ➤environment to produce a characteristic pattern of energy flow through the living system. The pattern of energy flow is determined by the arrangement and composition of various trophic levels in ➤food chains. At the first trophic level are the ➤producers (usually green plants), which convert energy from sunlight into ➤biomass and stored energy-rich food. The second trophic level (primary ➤consumers) are the ➤herbivores, which feed on the producers. The third and subsequent trophic levels contain the carnivorous secondary and subsequent consumers. Different ecosystems are characterized by different numbers of, and different compositions of, trophic levels. Different patterns of energy flow also depend on species composition. In an oak wood, for example, the bulk of primary production is stored as the bulk of trees and a subsequently large 'standing crop'. By contrast, marine ecosystems are characterized by small standing crops, and the bulk of productivity by single-celled ➤diatoms is continuously consumed by planktonic consumers, of which there may be many levels leading to the 'top carnivores' such as sharks. The producers and consumers are joined by a third important group, the

➤decomposers, responsible for the breakdown of dead remains and excretory products, thus bringing about the recycling of essential elements. Ecosystems are dynamic and rely on an equilibrium between producers and consumers. Stable (balanced) ecosystems such as forests are characterized by a balance between the energy fixed by the green plants and that used in the maintenance of the system. ➤➤carbon cycle; nitrogen cycle.

ectoderm The outermost of three ➤germ layers of an animal ➤embryo. The ectoderm gives rise to the skin and its derivatives such as hooves, claws and nails, the nervous system, and the ➤adrenal medulla. The ectoderm is also the outer of the two germ layers of the body wall of Hydra.

Ectoprocta An animal ➤phylum including the 'sea mats', a marine group consisting of colonies of polyps commonly found on seaweeds in the intertidal zone. Formerly known as **Bryozoa**.

ectothermy ➤poikilothermy.

eddy current A current produced by electromagnetic ➤induction. An alternating magnetic field passing through a material gives rise to loops within which the magnetic flux is varying; thus there is an ➤electromotive force induced around these loops, and if the material is a conductor then a current will flow. Eddy currents are responsible for the loss of energy in ➤transformers as heat. ➤➤induction heating.

edge A line along which two faces of a ➤polyhedron meet.

editor A computer program which facilitates the editing of data. A common example is a text editor, with which text can be entered, altered, and moved around or deleted in blocks. The text is usually saved in plain ➤ASCII format. ➤➤word processor.

Edman degradation A technique for analysing a polypeptide by sequentially removing one residue at a time from the amino end of the peptide by reaction with phenyl isothiocyanate, C_6H_5NCS. Named after Pehr Edman (1916–77).

EDTA The tetrasodium salt of ethylenediaminete-traacetic acid (see the diagram) is a very useful ligand that can envelop many metal ions, forming a ➤chelate. EDTA is used in solution for the extraction of DNA from plant and animal tissue. Its function is to bind to metal cofactors that might be needed by enzymes, thus reducing enzymic degradation of the extracted DNA.

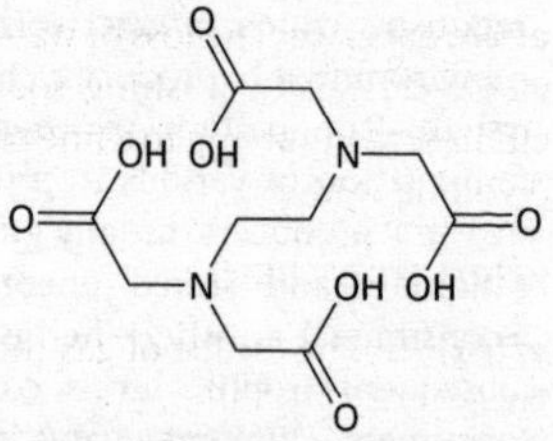

EDTA The ligand is the anion formed from the acid shown here by the loss of four hydrogen atoms.

EEG Abbr. for ➤electroencephalograph.

effective nuclear charge The net charge experienced by an electron in an atom. It is the difference between the actual nuclear charge (the number of protons in the nucleus) of the atom and the **shielding constant**, which represents the shielding effect of the other electrons in the atom. The shielding constant is hard to calculate, so simplifications are used to give approximate values. The effective nuclear charge controls the size of the atom, its ionization energy and many other physical properties. In general, it increases towards the right across a period and increases down a group.

effective potential A ➤potential (2) constructed to take account of all forces acting on a body (including ➤fictitious forces).

effective resistance The resistance of an electrical component to an alternating current. It takes account of all the mechanisms by which energy can be dissipated in the component, including therefore loss due to ➤eddy currents and ➤hysteresis.

effective temperature Symbol T_e. The temperature calculated for a star as if it were a black body (➤black-body radiation) of the same size, radiating with the same energy and over the same wavelength ranges. As the spectra of stars and black bodies are very similar, effective temperature is a good approximation to a star's actual surface temperature. For the Sun, T_e is about 5800 K. **Radiation temperature** is the effective temperature calculated for a particular wavelength range.

effector A structure in the body that, when suitably stimulated by a nerve impulse or a hormone, responds by producing a detectable event. Effectors include muscles and ➤endocrine glands.

efferent Describing a structure, such as a blood vessel, leading away from an organ. **Efferent nerves** carry information away from the brain to, for example, the muscles.

effervescence Bubbling of a liquid caused by release of a gas, not by boiling. 'Liver salts' are effervescent because carbon dioxide is evolved when the hydrogencarbonate and citric acid which compose it are added to water.

efficiency Symbol η. The ratio of the useful ➤work done by a machine or engine to its total energy input. An efficiency of less than 1, as is seen in all practical machines and engines, is a result of some energy being dissipated as unusable ➤heat or work. The distinction between useful and unusable heat or work is somewhat arbitrary: an incandescent light bulb is an inefficient source of light, but, particularly if placed in a sealed box, it makes a very efficient heater.

efflorescence The loss of water from a crystal over a period of time. The large clear glassy crystals of washing soda, sodium carbonate decahydrate, $Na_2CO_3 \cdot 10H_2O$, effloresce, crumbling to a fine powder of the monohydrate. A similar effect can occur on stone walls.

effort ➤machine.

effusion The escape of gas through a hole. The rate of effusion obeys ➤Graham's law.

egg The fertilized or unfertilized ➤ovum, and associated structures such as the shell, after it is released from the body. The term is also used synonymously with 'ovum'.

egg membrane One of a series of structures (depending on the group) around the egg of an animal in addition to the ➤plasma membrane. Egg membranes have a range of functions, including controlling the entry of sperm and protection.

EHF Abbr. for extremely high frequency (➤frequency band).

eigenfunction An ➤eigenvector for an ➤operator acting on a ➤vector space whose elements are functions. For example, $e^{i\omega x}$ is an eigenfunction of the differential operator d^2/dx^2 since $d^2(e^{i\omega x})/dx^2 = -\omega^2 e^{i\omega x}$.

eigenvalue ➤eigenvector.

eigenvector A nonzero vector v associated with a matrix or linear operator A for which $Av = \lambda v$ for some scalar λ; λ is called an **eigenvalue** of A. Eigenvectors play a key role in the mathematical foundations of quantum mechanics in terms of operators on ➤Hilbert space.

eighteen-electron rule A common electronic structure for compounds of elements in the first row of the ➤d block is eighteen electrons outside a noble gas core. An example is nickel carbonyl, $Ni(CO)_4$.

Einstein coefficients Two quantities used in the theory of the ➤laser, representing the rate at which an electron makes a transition due to **spontaneous emission** or ➤stimulated emission. Named after Albert Einstein (1879–1955).

Einstein field equation The pivotal equation of ➤general relativity: $G = 8\pi T$, where G is the ➤Einstein tensor and T is the ➤stress–energy tensor. It is the mathematical expression of the assertion that matter controls the curvature of space. In the limit of low speeds and stresses, all but one element of the tensor equation are insignificant, and it reduces to the ➤Poisson equation $\nabla^2 \phi = 4\pi G \rho$, where ϕ is the scalar gravitational potential and ρ is the density. The solution to the Poisson equation is Newton's well-known ➤inverse square law for the ➤gravitational field. ➤➤gravity.

einsteinium Symbol Es. The element, named after A. Einstein, with atomic number 99 and most stable isotope 252, which is a member of the actinide elements. The most common ➤oxidation number is +3, as in the chloride $EsCl_3$ and the oxide Es_2O_3. No important uses are known for the element or its compounds.

Einstein mass–energy relation The relation loosely described by the equation $E = mc^2$. As an equation it is more famous than useful. The c^2 is no more than a scaling factor, and the essence of the relationship, pivotal in ➤relativity, is that energy and mass are of the same physical dimension (just as space and time are of the same dimension in relativity). The ➤rest mass of a system is equivalent to its energy in its rest frame (in which its centre of mass is stationary). Thus a particle at rest still has an energy associated with it (➤energy–momentum). This relationship made Einstein believe that it would be possible to convert mass into energy, and his 1939 letter advising President Roosevelt that Germany had the capability to make an exceptionally powerful bomb, the atom bomb, led to the Manhattan Project. Although it is named after Einstein, the concept of relativistic mass was introduced one year earlier by Hendrik Antoon Lorentz (1853–1928).

Einstein shift A shift in the frequency of radiation from a massive body to a lower than natural frequency. Essentially, some of an escaping photon's kinetic energy is transformed into gravitational potential energy, and, having a lower energy, it emits lower-frequency light.

Einstein tensor Symbol G. A ➤tensor that describes the curvature of ➤spacetime in ➤general relativity. ➤➤Einstein field equation.

ejecta The debris thrown up after an impact on a planet or satellite. It forms a ring (known as the **ejecta blanket**) around the resulting crater, and sometimes also radial rays.

eka- A prefix applied to the undiscovered elements that Mendeleyev predicted for his ➤periodic table. As *eka* means 'first' in Sanskrit, *eka*-silicon is the unknown element first below silicon, named germanium when it was discovered.

elastance Unit F^{-1}. The reciprocal of ➤capacitance.

elastic collision A collision in which the total ➤kinetic energy of the colliding particles is conserved. Compare ➤inelastic collision.

elastic cross-section The ➤scattering cross-section that is due to ➤elastic scattering.

elasticity The tendency of a body to resume its original shape and size after it has been deformed. ➤➤elastic modulus; Hooke's law; compare ➤plastic deformation.

elastic limit The highest ➤stress that it is possible to apply to a material without deforming it permanently. ➤➤yield point.

elastic modulus The ratio of the ➤stress applied to a material to the ➤strain induced in it. In an ➤anisotropic material, stress and strain are both second-rank ➤tensors, and the elastic modulus is therefore in general a fourth-rank tensor. Even in an isotropic material, more than one component may be required. ➤➤bulk modulus; Poisson's ratio; shear modulus; Young's modulus.

elastic scattering The ➤scattering of particles by a potential such that their ➤kinetic energy is conserved.

elastomer A material that tends to be elastic even for fairly large deformations. Apart from natural rubber, important elastomers include ➤neoprene rubber, the copolymer ➤SBR, and the ➤silicone elastomers.

E layer ➤Heaviside–Kennelly layer.

electret A piece of permanently polarized dielectric, the electrostatic analogue of a permanent ➤magnet. Electrets are commonly used in microphones.

electric arc A luminous discharge between two electrodes with a high potential difference between them. The arc is sustained by the ionization of vaporized material from the electrodes.

electric-arc furnace An apparatus used to manufacture high-quality alloy steels in which the energy is provided by an electric arc. It has now been superseded by the ➤basic oxygen process.

electric charge Symbol q or Q; unit ➤coulomb, C. A property of a body that causes attractive and repulsive forces between it and other bodies with charge, according to ➤Coulomb's law. It may be thought of as the source of an ➤electric field. Charge can be positive or negative: bodies with the same sign of charge repel each other, those with opposite signs attract. Charge exists as integral multiples of the ➤electronic charge. ➤➤charge density.

electric constant An old name for the ➤permittivity of free space.

electric current Symbol I; unit ➤ampere, A. As a phenomenon, the movement of ➤electric charge, or, as a physical quantity, the *rate* of movement of electric charge. A current flows when an ➤electromotive force is applied to an electrical ➤circuit (2). Although current is normally considered as the flow of electrons in a ➤conductor, a current also exists when electrons move in free space or when ions flow in ➤electrolysis. The direction in which current is conventionally regarded as flowing is in fact opposite to the actual flow of electrons. ➤➤Joule heating; semiconductor.

electric dipole moment ➤dipole.

electric displacement (electric flux density) Symbol D. A quantity introduced in classical ➤electromagnetism that describes the ➤electric field due to external or free charges, as opposed to the ➤polarization charge induced by the presence of the field itself. Electric displacement is defined as

$$D = \varepsilon_0 E + P = \varepsilon E,$$

where E is the electric field strength, P is the polarization, ε_0 is the ➤permittivity of free space, and ε is the ➤permittivity. Outside a ➤dielectric, the displacement is just a recasting of the electric field in different units, corresponding to multiplication by the fundamental constant ε_0. In a dielectric, given that $P = \chi\varepsilon_0 E$ (where χ is the susceptibility of the dielectric), we can define permittivity ε such that

$$\varepsilon = \varepsilon_0(1 + \chi).$$

electric energy The energy stored in an ➤electric field, equal to $\frac{1}{2}\varepsilon_0 E^2$ per unit volume where E is the magnitude of the electric field E. Electric energy is stored when a ➤capacitor is charged. In a ➤dielectric, the local density of electric energy is given by $\frac{1}{2} D \cdot E$, where D is the electric displacement.

electric field One of the two ➤fields associated with electromagnetic interaction between objects. Because the electrostatic force on a body is, according to ➤Coulomb's law, proportional to the charge on the body, it is convenient to define a force per unit charge that would act on any body brought into the same position. This is a ➤vector field known as the **electric field strength**, E (unit V m^{-1}). Thus, for a charge q at the origin,

$$E(r) = \frac{q}{4\pi\varepsilon_0 r^2}\hat{r},$$

where $\hat{r}$ is a unit vector in the direction of r, r is the magnitude of r and ε_0 is the ➤permittivity of free space. The idea of electric field strength is so close to the heart of electromagnetism that it is usual to consider an electric field as an abstract entity rather than just the value of a physical quantity. It is helpful to think of charges as the source of electric fields which subsequently act on other charges. The electric field due to a distribution of charge with ➤charge density $\rho(r)$ can be found by finding an appropriate solution to the first of ➤Maxwell's equations.

electric flux The ➤flux of ➤electric displacement at a surface. From the first of ➤Maxwell's equations (in integral form), it is clear that electric flux has the dimensions of ➤electric charge.

electric flux density An alternative name (more consistent with the names for the corresponding magnetic quantities) for ➤electric displacement.

electricity The general name given to phenomena associated with electric currents, electric charges and electric fields.

electricity, frictional (triboelectricity) Electricity produced when charge is transferred from one surface to another by rubbing. The frictional forces cause ➤ionization of atoms on the surface.

electric-light bulb A device for producing light by radiation from a very hot filament (typically of tungsten), heated by an electric current, in a glass bulb containing an unreactive gas (e.g. argon) to prevent oxidation.

electric motor A device for turning electrical ➤potential energy into mechanical ➤kinetic energy. It uses a *magnetic* field to produce a force. ➤➤universal motor.

electric organ A modified group of muscle cells in certain types of fish, such as electric eels, that produce electric currents. Different types of electric organ produce different voltages. Low-voltage pulses are used for navigating and prey recognition in water where visibility is reduced. High-voltage emissions are used to stun prey and for defence.

electric polarization Symbol P; unit $C\,m^{-2}$. The electric ➤dipole per unit volume in a ➤dielectric. ➤➤electric displacement.

electric potential Symbol ϕ or V; unit ➤volt, V. A ➤scalar field whose ➤gradient (3) describes a static ➤electric field: $E = -\nabla\phi$. The difference between the potential at two points is therefore the energy required per unit charge to move a charged body from one point to the other. Note that in the presence of changing magnetic fields the electric field is not a ➤conservative field, and a simple electric potential cannot be defined. ➤➤potential difference.

electric power Symbol P. The rate at which an electric circuit does ➤work. In any component, the power is equal to the product of the potential difference across the component and the current flowing through it. In an a.c. circuit, where both these quantities are time-dependent, the instantaneous power will also be time-dependent, so it must be averaged over one cycle to obtain a sensible value (➤loss factor).

electric spark A discharge between two conductors with a high potential difference between them, caused by a ➤breakdown of an ➤insulator (typically air).

electric susceptibility Symbol χ. In a ➤dielectric, the ➤electric polarization induced per unit applied ➤electric field, divided also by the ➤permittivity of free space, ε_0, to make the quantity dimensionless.

electrocardiograph (ECG) A trace made on a chart recorder of the electrical activity of the heart. Electrodes are fastened over the heart and to the arms and legs, and collect signals which can be amplified and viewed on an ➤oscilloscope and printed as a trace on a chart. The pattern of the trace is useful in diagnosing irregularities of heart function and disease.

electrochemical cell ➤cell (2).

electrochemical equivalent ➤Faraday's laws.

electrochemical series (electromotive series) A listing of the elements in order of their ease of discharge by electrolysis, the most difficult to discharge being placed at the top of the series. It parallels the order of the ➤standard electrode potentials of the metals, with the occasional exception, such as sodium and calcium being swapped, because the electrochemical series is a *kinetic* quantity whereas the standard electrode potential is a thermodynamic quantity. The common metals appear in the order sodium, calcium, magnesium, aluminium, zinc, iron, lead, copper (mnemonic 'schools can make all zombies into living creatures'). The series is useful for predicting reactivity of an element with water and acids. A metal higher in the series will displace one lower in the series; the displacement of copper by zinc is used in the ➤Daniell cell to generate electricity.

The electrochemical series is valid only for aqueous solutions, and should not be applied outside this realm. (It is often misused to predict the stability of solid carbonates.)

electrochemistry The study of the interconversion of electrical energy and chemical energy in an electrochemical cell. Electrical energy may be used to cause chemical change in an electrolytic cell (➤Faraday's laws). Conversely, chemical change may be used to produce electricity in a ➤battery or galvanic cell, such as the ➤Daniell cell.

electrode A ➤conductor that is used to carry ➤electric current to a medium of lower conductivity. Electrodes are inserted into an electrolyte for electrolysis and attached to a semiconductor in an electronic device. ➤➤anode; cathode; half-cell.

electrode potential Often used loosely to mean ➤standard electrode potential. ➤➤Nernst equation.

electrodynamics The study of electromagnetic phenomena associated with moving charges. Compare ➤electrostatics.

electrodynamometer An instrument, similar to an ammeter, that uses the interaction between a fixed and a moving coil to measure the square of an electric current (usually with the aim of measuring power).

electroencephalograph (electroencephalogram, EEG) A trace on an oscilloscope or chart recording the electrical activity of the cortex of the ➤brain. The activity is detected by amplification of signals picked up by electrodes fixed to the scalp. Interpretation of the pattern of the trace assists in determining brain function, and is also used to diagnose malfunction in, for example, stroke patients.

electrokinetic potential (zeta potential) In an ➤ionic solution, the electrostatic potential at the interface between the ionic layer adsorbed on a charged surface and the bulk ➤electrolyte.

electroluminescence ➤Fluorescence in which the excitation is caused by bombardment by electrons.

electrolysis The passage of electric current through an **electrolyte** (a molten chemical or an aqueous solution of a chemical), causing chemical change at the electrodes. A simple **electrolytic cell** is shown in the diagram. The two electrodes are connected to a battery and they dip into the electrolyte. The electrolyte allows the current to flow and is usually either a solution containing ions or a fused ionic salt, as in either case the ions are free to move when the current flows. (A solid ionic salt also contains ions but they are locked into position in the crystal lattice.) The ➤anode is the site of oxidation; the ➤cathode is the site of reduction. For the quantitative laws, see ➤Faraday's laws.

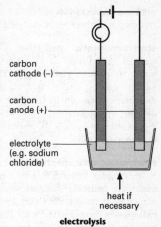

carbon cathode (–)

carbon anode (+)

electrolyte (e.g. sodium chloride)

heat if necessary

electrolysis

Electrolysis is used for extracting the most reactive metals, as they cannot be obtained in any other way. Its introduction in the first decade of the 19th century allowed the discovery of many previously unextractable elements, such as ➤fluorine; Humphry Davy discovered more than anyone else (seven), all by electrolysis. Sodium is made in the ➤Downs cell and aluminium in the ➤Hall–Héroult process. Electrolysis is also used in the ➤chlor–alkali industry to manufacture chlorine and sodium hydroxide.

electrolyte ➤electrolysis. Further subdivision of electrolytes into **strong electrolytes** and **weak electrolytes** is analogous to the subdivision of acids (➤strong acid; weak acid).

electrolytic capacitor A capacitor in which the ➤dielectric is a layer of metal oxide deposited on the anode by ➤electrolysis. The cathode may be the electrolyte itself. Electrolytic capacitors provide a high capacitance in a limited volume, but they suffer from high leakage currents.

electrolytic conductivity ➤conductivity, electrical.

electromagnet A magnet produced by passing a current through a wire wound around an iron core. The ferromagnetic iron causes a large field in the gap between the poles. Some of the highest magnetic fields obtainable in the laboratory are produced by superconducting electromagnets (➤superconductivity).

electromagnetic induction ➤induction, electromagnetic.

electromagnetic interaction The interaction (one of the four ➤fundamental interactions) between particles with ➤electric charge. ➤➤electroweak theory.

electromagnetic pump A pump for fluids that conduct electricity. A current is passed through the fluid across the pipe through which the fluid is flowing, and a magnetic field is applied perpendicular to the current, also across the pipe. The resulting ➤Lorentz force therefore acts along the tube and creates the pressure required to pump the liquid. Such pumps are used for circulating the liquid sodium used as a coolant in some nuclear reactors. It is troublefree as there are no moving parts.

electromagnetic radiation Radiation emitted in the form of an ➤electromagnetic wave. It includes visible light. The ➤wave–particle duality implies that electromagnetic radiation can equally be treated as made up of discrete fundamental particles called ➤photons. ➤➤black-body radiation; electromagnetic spectrum.

electromagnetic spectrum The name given to the range of frequencies over which electromagnetic radiation can propagate (see the diagram). It ranges from very low frequency radio waves (from about 3×10^3 Hz) to gamma rays (up to about 3×10^{21} Hz). ➤➤frequency band.

electromagnetic units (e.m.u.) An old system of units for electromagnetic quantities, based in part on the ➤c.g.s. system. In this system the permeability of free space μ_0 is defined as 1, rather than having the value of $4\pi \times 10^{-7}$ H m^{-1}, as in SI units.

electromagnetic wave A ➤wave that consists of an ➤electric field in conjunction with a ➤magnetic field oscillating with the same frequency. The simplest solution to ➤Maxwell's equations in this case is a ➤plane wave with both fields perpendicular to the direction of propagation and perpendicular to each other. Such a wave travels at a constant speed in a vacuum (c, the ➤speed of light in a vacuum), and is the only commonly observed wave that is able to propagate without a ➤medium. The magnitudes of the electric (E) and magnetic (H) fields are related thus:

$$E^2/H^2 = Z_0^2 = \mu_0/\varepsilon_0.$$

frequency/Hz		wavelength	
10^{20}		3pm	
	γ-rays		
10^{19}			
	X-rays		violet
10^{18}			420
10^{17}	3nm		blue
	far ultraviolet		470
10^{16}			
	ultraviolet		green
10^{15}	300nm		530
		visible	
10^{14}			yellow
	infrared		580
10^{13}	0.03mm		orange
	far infrared		620
10^{12}			
			red
10^{11}	3mm		700
	microwave		
10^{10}	3cm		
10^{9}	30cm		
10^{8}	3m		
10^{7}	radio		
10^{6}	300m		
10^{5}			

electromagnetic spectrum

The permittivity (ε_0) and permeability (μ_0) of free space are replaced by the appropriate values if the wave travels in a medium; Z_0 is known as the impedance of free space (about 377 Ω).

electromagnetism The area of physics that deals with ➤electric fields and ➤magnetic fields. It encompasses, among other topics, ➤electrostatics and ➤electrodynamics. ➤➤Maxwell's equations.

electrometer A device for measuring voltage; it has a very high ➤impedance, and thus draws almost no current from the source.

electromotive force (e.m.f.) Unit volt, V. In a simple electric circuit, the total ➤potential difference across the voltage sources. E.m.f. can be considered to be the driving force for the current in a circuit, and can be defined around any arbitrary closed loop L (not just loops of conductors forming circuits) as a ➤line integral:

$$\int_L \boldsymbol{E} \cdot \mathrm{d}\boldsymbol{l}.$$

electromotive series ➤electrochemical series.

electron Symbol e or e^-. An ➤elementary particle which is a constituent of all atoms, occupying an ➤atomic orbital around the nucleus. It was discovered by J. J. Thomson in 1897. The electron is a ➤lepton and therefore a ➤fermion, with spin quantum number $\frac{1}{2}$. It has a mass of 9.109×10^{-31} kg, more conveniently written as a ➤rest mass of 0.5110 MeV/c^2 for the purpose of high-energy physics. It carries a charge of -1.602×10^{-19} C. A vast amount of ➤solid state physics concerns the behaviour of electrons. Electrons are responsible for carrying electric current (and heat) in metals; since the charge is negative the electrons flow in the opposite direction to the conventional electric current. The behaviour of electrons in the presence of more than one nucleus is the essence of all chemistry (➤chemical bonding). Electron transfer from one substance to another constitutes a ➤redox reaction. Movement of ➤electron pairs constitutes a reaction between Lewis acids and bases (➤Lewis theory). ➤➤free electron; Appendix table 2.

electron affinity A less precise but common name for ➤electron-gain enthalpy. A problem with the term is the uncertainty over its sign: a positive sign is often used if heat is released. The term 'enthalpy' is unambiguous: heat release corresponds to a negative enthalpy change. ➤➤free electron; positron.

electron capture A nuclear process, governed by the ➤weak interaction, in which an electron combines with a proton to produce a neutron and a neutrino. Sometimes the term is used loosely of any process in which an electron is trapped.

electron-deficient compound A compound in which there are apparently not enough electrons for all the atoms on the basis of conventional two-centre, two-electron bonding, because there are delocalized multi-centre bonds. ➤➤delocalization.

electron density The ➤probability distribution for the location of a single electron, or the ➤superposition of probability distributions for multi-electron systems. It arises because the probabilistic nature of quantum mechanics makes it impossible to determine the position of an electron with certainty. It is used as the ➤charge density for many applications in atomic physics and chemistry (➤Hartree theory). It is usually measured by ➤X-ray diffraction on solids.

electron diffraction The ➤diffraction of electrons by matter. The wave–particle duality of electrons means that, just as photons exhibit diffraction in an electromagnetic wave, the wavefunction of electrons is diffracted at an aperture or edge that blocks the propagation of the wave. Historically, observation of electron diffraction gave the experimental proof of ►de Broglie's equation. Currently, the technique is often used to identify the molecular structure of gases; for example electron diffraction shows that the two BCl_2 ends of diboron tetrachloride, B_2Cl_4, are at right angles in the gas phase, whereas they are coplanar in the solid state. ►Low-energy electron diffraction is used to study surfaces.

electronegativity An approximate measure of the extent to which an atom attracts a shared ►electron pair. There is no rigorous definition and several scales exist, of which the most common are the original **Pauling scale** (see the table) and the **Mulliken scale**. Values on the Mulliken scale are proportional to the difference between the ionization enthalpy and the ►electron-gain enthalpy; the former has a more complicated definition, but gives similar values. The two electrons of a covalent bond between two dissimilar elements will be unequally shared, the more **electronegative** atom having a partial negative charge. The hydrogen chloride molecule, for example, is polarized thus: $H^{\delta+}$—$Cl^{\delta-}$. The magnitude of the dipole moment in ►debyes is approximately equal to the difference in electronegativity.

electronegativity Values for common elements on the Pauling scale.

H						
2.20						
Li	Be	B	C	N	O	F
0.98	1.57	2.04	2.55	3.04	3.44	3.98
Na	Mg	Al	Si	P	S	Cl
0.93	1.31	1.61	1.90	2.19	2.58	3.16
K	Ca					Br
0.82	1.00					2.96
Rb	Sr					I
0.82	0.95					2.66

electron–electron interaction The interaction between electrons; in a multi-electron system it is a very difficult area of quantum mechanics. A term must be added to the ►Hamiltonian in the ►Schrödinger equation to reflect this, but the extra potential is due to the electron density and is itself dependent on the solution to the equation, which therefore becomes ►nonlinear. The ►exchange interaction is a further complication. The electron–electron interaction is sometimes treated as a ►perturbation, though in many cases it is by no means small.

electron-gain enthalpy (electron affinity) The ►standard enthalpy change accompanying the attachment of an electron to an isolated atom (or ion) of the element E in the gas phase:

$$E(g) + e^-(g) \rightarrow E^-(g).$$

Electron-gain enthalpy is important in the ►Born–Haber cycle.

electron gas A ►degenerate gas of electrons. ►►Fermi gas.

electron gun A device for producing a stream of electrons. Most electron guns generate ➤free electrons by ➤thermal emission from a hot filament and accelerate them through a high potential difference.

electron–hole pair An electron together with the associated ➤hole in a solid state system. Electron–hole pairs occur after an ➤excitation causes the electron to move to a higher energy level.

electronic charge The charge carried by a single electron, -1.602×10^{-19} C.

electronic configuration ➤electronic structure.

electronics The study, construction and application of devices whose principle of operation is based on the motion of electrons, typically in ➤semiconductors. At a small scale, the subject is known as ➤microelectronics, which deals with the silicon devices that lie at the heart of computers. At a larger scale, entire communications networks are considered.

electronic structure (electronic configuration) A specification of the ➤atomic orbitals occupied by electrons. For example, in sodium the lowest-energy structure is written $1s^2 2s^2 2p^6 3s^1$. The order of filling the orbitals follows the ➤aufbau principle. The superscripts indicate the number of electrons in each ➤subshell, denoted by the shell number and the letter that indicates the ➤orbital angular momentum quantum number. (In a simplified notation, subshell labels are lost and the structure is written as 2.8.1, which conveniently lays stress on the number of electrons outside a **closed shell**.) Even more detail can be specified if the subshell is only partially filled. The structure of oxygen is $1s^2 2s^2 2p^4$; the four electrons in the 2p subshell are in fact arranged with two in one orbital and one in each of the other two orbitals, following ➤Hund's rules.

electronic transition A transition in which an electron moves from one orbital to another. The absorption of a photon can trigger an electronic transition to a higher energy level. The energy of the photon (hf) and the two energy levels in the atom or molecule (difference ΔE) are related by the **Bohr frequency condition**:

$$\Delta E = hf.$$

electron lens A system of electric or magnetic fields (or both) designed to focus an electron beam, analogously to the way a conventional lens focuses light.

electron microscope An instrument that creates a high-magnification image of an object by employing a beam of electrons (rather than light as in a conventional optical microscope). Because of ➤wave–particle duality, the electrons reflect, refract and diffract at the sample like waves, with a wavelength inversely proportional to their momentum. Because the electrons are charged and can be accelerated to high momenta using electric fields, it is possible to produce beams of electrons with much shorter wavelengths than the easily available sources of light, enabling smaller details to be resolved. Electrons with an energy of 50 keV have a wavelength of approximately 0.05 nm. ➤➤scanning electron microscope; transmission electron microscope.

electron multiplier A device for detecting and counting electrons. A single primary electron is accelerated towards an anode, where the impact of the electron causes the emission of several ►secondary electrons. These are in turn accelerated towards an anode at a higher positive potential, and the bombardment further multiplies the number of electrons produced. The process is repeated until there is a measurable pulse of current which can be counted in a conventional circuit.

electron optics The study of the refraction of electrons in electromagnetic fields. ►Wave–particle duality means that the electrons can be treated as waves and, as with light in ►geometric optics, it is possible to construct rays to represent their path. The rays refract when the electrons change speed where the electric field changes, as light does when it passes into a medium with different refractive index. It is thus possible to construct lenses to focus the electrons and perform many of the other functions that would normally be associated with optical systems.

electron pair An exceptionally important idea in chemistry, that electrons are almost always paired. Why this is the case is not obvious (►pairing of electrons). The ►Lewis theory of acids and bases attempts to explain reactions in terms of the motion of electron pairs.

electron paramagnetic resonance ►electron spin resonance.

electron-probe microanalysis A method of performing a chemical analysis on a surface by measuring the ►X-ray spectrum produced by electron bombardment. A focused electron beam (diameter typically 1 μm) is used to enable parts of the surface to be analysed selectively.

electron spectroscopy A name given to a range of techniques whose common feature is the detection of the distribution of the energy of emitted electrons. ►Wave–particle duality makes 'spectroscopy' an appropriate term. Versions of the technique include ►photoelectron spectroscopy and Auger electron spectroscopy (►Auger effect).

electron spin resonance (ESR, electron paramagnetic resonance, EPR) A version of ►spectroscopy used to investigate the structure of ►radicals, species with unpaired electrons, by inducing transitions between states of different ►spin. It is the equivalent for electrons of ►nuclear magnetic resonance. As in NMR, a photon is absorbed because of the transition of electron spin from one of its two possible states in a magnetic field to the other. Because the electron has a smaller mass than a proton, the absorption is at higher frequencies and ESR occurs in the microwave region of the electromagnetic spectrum. It is used in archaeology to measure the age of a specimen by studying electrons displaced by natural radiation and then trapped in the structure.

electron-transport chain A linked sequence of reversible ►redox reactions, also called the **respiratory chain**, that is responsible for the energy conversions occurring during ►aerobic respiration. During respiration a series of reactions in the ►Krebs cycle produces reduced ►coenzymes. The electrons involved in this reduction are passed along the chain in a characteristic 'zigzag' pattern, following the location of components on the inner membrane of the ►mitochondrion, and are finally taken up by oxygen, in association with hydrogen ions, to form water. Some of the redox reactions are coupled to the formation of the energy-transfer compound ►ATP. A similar chain of electron-transfer compounds exists in the ►chloroplast, which is responsible for the harnessing of energy during ➤photosynthesis. ➤➤cytochrome; FAD.

electronvolt (electron volt) Symbol eV. A unit of energy equal to the energy gained when an electron is accelerated through a potential difference of 1 volt, hence 1 eV $= 1.602 \times 10^{-19}$ J. It is a useful unit in atomic physics, for example to measure ionization energies (that for hydrogen is 13.6 eV).

electro-optical effect ➤Kerr effect. ➤➤Faraday (electro-optical) effect.

electrophile An electron-pair acceptor. The terminology is used almost exclusively in organic chemistry but an electrophile is nearly indistinguishable from a ➤Lewis acid. Important examples of electrophiles are positive ions such as H^+ or CH_3CO^+ (which is the active electrophile in the ➤Friedel–Crafts reaction using CH_3COCl).

electrophilic addition The common mechanism for reaction of the double bond in an ➤alkene. ➤Electrophiles attack the high electron density of the double bond after which in a second (fast) step a nucleophile reacts to give the addition product. For example, when HBr(aq) adds to ethene, the electrophile H^+ reacts with C_2H_4 to form $CH_3CH_2^+$, which then adds a bromide ion to give CH_3CH_2Br.

electrophilic substitution The common mechanism for reaction of aromatic hydrocarbons such as benzene. ➤Electrophiles, such as CH_3CO^+ in the ➤Friedel–Crafts reaction, attack the high electron density of the ring after which in a second (fast) step the intermediate carbocation deprotonates to reform the very stable aromatic ring. Thus benzene, C_6H_6, reacts with CH_3CO^+ to form $C_6H_5COCH_3$, after deprotonation.

electrophoresis A laboratory technique used to separate mixtures of molecules, such as proteins and nucleic acids in a suspension, by their charge-to-mass ratio. The mixture is added to an inert medium such as an agarose or acrylamide gel in an appropriate ➤buffer solution, and is subjected to an electric field. The charged molecules then move through the gel towards the appropriate electrode. **Gel electrophoresis** of fragments of DNA is routinely used to produce ➤genetic fingerprints and to carry out ➤DNA sequencing.

electroplating The coating of one metal onto another by making the object to be plated the cathode in an electrolytic cell and depositing the metal by reduction of its ions. For example, it is a common way of plating chromium onto steel.

electroporation A technique used in molecular biology and ➤genetic engineering to encourage the uptake of DNA in recipient cells. The cells are immersed in a solution of DNA, and a rapid pulse of high voltage is applied. This is thought to open up channels in the ➤plasma membrane through which DNA can pass.

electropositive This adjective is used in two ways. 'Electropositive metal' is most often used to mean that the metal is easily ionized, as with the elements of Group 1. An alternative usage, which should be discouraged, treats electropositive as the opposite of electronegative (➤electronegativity).

electroscope An instrument for detecting electric charge. An example is the ➤gold-leaf electroscope.

electrostatic field An ➤electric field that does not vary with time.

electrostatic force The force that exists between static charges. ➤Coulomb's law.

electrostatic generator A machine for producing potential differences by separating positive and negative electric charge.

electrostatic induction The production of electric charge on the surface of a conductor in the presence of an electric field. A conductor in equilibrium (passing no current) has no internal electric field, so charge accumulates on its surface to cancel exactly the externally applied field. ➤➤Faraday cage.

electrostatics The physics of electric fields in the absence of currents or moving charges. Compare ➤electrodynamics.

electrostatic units (e.s.u.) An obsolete system of units for electricity and magnetism, based on the ➤c.g.s. system. In this system the ➤permittivity of free space is defined to be 1, rather than having the value 8.854×10^{-12} F m^{-1}, as in SI units.

electrostriction An expansion or contraction of a ➤dielectric in the presence of an ➤electric field.

electrovalent bonding An obsolete name for ➤ionic bonding.

electroweak theory (Glashow–Weinberg–Salam model) A unified theory for the electromagnetic and weak interactions (➤fundamental interactions). It is a ➤gauge theory, mediated by the photon, ➤W boson and ➤Z boson.

element **1** (Chem.) A substance that cannot be broken down into two or more substances by any means. All elements contain atoms which have just one atomic number, although the masses can vary (➤isotope). The elements are listed in the ➤periodic table in order of their atomic numbers. ➤➤Appendix table 5.
2 (Math.) A member of a ➤set. The notation $a \in A$ means that a is an element of the set A.
3 (Math.) An individual entry of a ➤matrix.
4 (Math.) The expression following an integral sign. Thus in $\int y \, dx$, the expression $y \, dx$ is an element of area. It is often helpful to think of the area represented by the integral as a 'sum' of the elements.

elementary function Loosely speaking, any ➤function that can be obtained by pressing one or more buttons in sequence on a standard scientific calculator. Thus the term includes ➤polynomials, ➤rational functions, ➤exponential and ➤logarithmic functions, and ➤trigonometric and ➤hyperbolic functions (and their inverses), as well as combinations such as $x^3 \sin(2e^x + 3)$. ➤Derivatives of elementary functions are also elementary, but the same is not necessarily true of integrals; the integral $\int \exp(-x^2) \, dx$ and any ➤Bessel function are non-elementary (➤special function).

elementary particle (fundamental particle) A particle, constituting matter, that cannot be broken down into a smaller particle. Just which particles are considered to be elementary depends on how far science has progressed. Thus the Greeks conjectured that atoms were elementary particles; in the 19th century chemists regarded atoms as elementary; and physicists in the first half of the 20th century would have labelled protons and neutrons as elementary. Now, the

➤standard model breaks these and other ➤hadrons into ➤quarks; the ➤leptons are considered elementary, as are the ➤gauge bosons (see the table below).

elevation The height above mean sea level of a point on the Earth's surface.

elevation, angle of The angle made with the horizontal plane by a line from a given point to an object or point. If the object is below the horizon, the term **angle of depression** is used.

Particle	Symbol	Mass MeV/c^2	Lifetime s	Discovered	Quark Content
Gauge bosons					
Photon	γ	0	Stable	1923	
W^+/W^-	W^+/W^-	80400	10^{-25}	1983	
Z^0	Z^0	91200	10^{-25}	1983	
Gluon	g	0	Stable	1979	
Leptons					
Electron	e^-	0.511	Stable	1897	
Electron neutrino	ν_e	0	Stable	1956	
Muon	μ^-	105.7	2×10^{-6}	1937	
Muon neutrino	ν_μ	0	Stable	1962	
Tauon	τ^-	1784	3×10^{-13}	1975	
Tauon neutrino	ν_τ	0	Stable	(postulated)	
Mesons					
Pion	π^0	135	8×10^{-17}	1949	u$\bar{u}$
	π^+	140	3×10^{-8}	1947	u$\bar{d}$
Kaon	K^+	494	10^{-8}	1947	u$\bar{s}$
	K^0	498	10^{-9}	1947	d$\bar{s}$
D	D^0	1865	10^{-12}	1976	c$\bar{u}$
	D^+	1869	10^{-12}	1976	c$\bar{d}$
J	J	3097	10^{-20}	1974	c$\bar{c}$
B	B^+	5278	10^{-12}	1986	u$\bar{b}$
	B^0	5279	10^{-12}	1986	d$\bar{b}$
Upsilon	Y	9460	10^{-20}	1977	b$\bar{b}$
Baryons					
Proton	p	938.3	Stable	by 1920	uud
Neutron	n	939.6	896	1932	udd
Lambda	Λ	1116	3×10^{-10}	1951	uds
Sigma	Σ^+	1189	8×10^{-10}	1953	uus
	Σ^0	1192	7×10^{-20}	1956	uds
	Σ^-	1197	10^{-10}	1953	dds
Xi	Ξ^0	1315	3×10^{-10}	1959	uss
	Ξ^-	1321	2×10^{-10}	1952	dss
Omega-minus	Ω^-	1672	10^{-10}	1964	sss
Charmed lambda	Λ_c	2285	2×10^{-13}	1975	udc

The quarks that explain the nature of these particles are: up (u), down (d), strange (s), charm (c), bottom (b) and top (t); a baryon is made up of three quarks; a meson is made up of a quark and an antiquark (indicated by an overbar).

elevation of boiling point ➤colligative properties.

elimination reaction A reaction in which a small part of a molecule is removed to leave a simpler molecule, as for example in the elimination of HBr from $(CH_3)_3CBr$ to form $(CH_3)_2C{=}CH_2$.

ELISA Acronym for ➤enzyme-linked immunosorbent assay.

ellipse A ➤conic with eccentricity $e < 1$. An ellipse may also be defined as the locus of a point that moves so that the sum of its distances from two fixed points (➤focus (2)) is constant. It is an oval-shaped curve with two perpendicular axes of symmetry, with respect to which its Cartesian equation is $x^2/a^2 + y^2/b^2 = 1$ where $2a$ and $2b$ are the lengths of the major and minor axes. It encloses an area πab, but there is no simple formula for its perimeter. The orbits of the planets are ellipses.

ellipsoid An egg-shaped three-dimensional surface which has elliptical cross-sections parallel to the three coordinate planes. For a suitable choice of axes, its equation is $x^2/a^2 + y^2/b^2 + z^2/c^2 = 1$. A ➤spheroid is the special case where at least two of a, b and c are equal.

elliptical galaxy A ➤galaxy that appears to be ellipsoidal or spheroidal, with no trace of spiral structure.

elliptically polarized The most general state of ➤polarization of light. ➤➤circularly polarized; plane-polarized.

elliptic integral A class of integrals involving square roots of cubic and quartic polynomials which cannot be evaluated in terms of ➤elementary functions. Examples of problems which give rise to elliptic integrals are: finding the arc length of a sine curve; finding the perimeter of an ellipse; and finding the exact period of a simple pendulum.

El Niño (El Niño Southern Oscillation, ENSO) A perturbation in the normal ocean current system operating in the Pacific caused by atmospheric pressure oscillations over northern Australia and Tahiti and associated changes in wind patterns. The net effect is that the cold, nutrient-rich water of the Peruvian current is diverted by warm, nutrient-poor water flowing eastwards in the equatorial zone. El Niño ('the child') events tend to happen around Christmas and occur approximately every 7 or 8 years, although recent ENSO events have been more frequent. Associated ecological effects include the depression of fishing industries on the west coast of South America and often severe impacts on the ecosystems of the Galapagos Islands. El Niño has also been implicated in global climate events such as the collapse of the monsoon system, unseasonal flooding and droughts in many parts of the world.

email (e-mail) Abbr. for electronic mail. A message system that uses computers and computer networks as the intermediary. A textual message (often, in modern systems, with associated data files, moving visual images or sound recordings) can be sent to a particular email address using a computer network or a ➤modem.

emanation The obsolete name for the element ➤radium.

Embden–Meyerhof pathway ➤glycolysis. Named after Gustav Georg Embden (1874–1933) and Otto Fritz Meyerhof (1884–1951).

embolism An obstruction of a blood vessel caused by an **embolus** (blockage). An embolus may be a blood clot, a clump of bacteria, a fat droplet, an air bubble or other foreign body that enters the body as a result of trauma. The effect is to restrict blood flow to the area served by the vessel, which may result in an **infarct** or region of dead tissue. Compare ➤thrombosis.

embryo The earliest stage of development of a plant or animal. In humans the term is used for the first two months of existence in the ➤uterus; from then on the term ➤foetus is used. The embryo results from sequential divisions of cells derived from the ➤zygote. In animals the pattern of development of the embryo differs from one taxonomic group to another, depending mainly on the amount of yolk in the egg. In the ➤chordates a distinct sequence of events results in the proliferation of three ➤germ layers (the **endoderm**, **mesoderm** and **ectoderm**) from which all body structures are subsequently derived. In plants of the ➤Magnoliopsida and ➤Pinopsida the early stages of development result in an embryo in the ➤seed.

embryology The study of embryos, in particular with respect to their development into adult forms.

embryo-sac (embryosac) A large oval cell in the ovule of plants of the ➤Magnoliopsida in which ➤meiosis typically gives rise to eight ➤haploid nuclei. One of these nuclei is the functional female ➤gamete. The embryo-sac is functionally homologous to the gametophyte generation of spore-bearing plants such as ferns.

e.m.f. Abbr. for ➤electromotive force.

emission Generally, the outflow of a stream of particles (including electromagnetic radiation, as photons) from a system. ➤➤photoemission; stimulated emission.

emission nebula ➤nebula.

emission spectroscopy A form of ➤spectroscopy that measures the distribution in frequency (or energy) of emission, usually of electromagnetic radiation. Its most important variant is atomic spectroscopy (➤sodium spectrum; hydrogen atom emission spectrum).

emissivity Symbol ε. The ratio of the power per unit area of electromagnetic radiation emitted by a body to the power per unit area emitted by a perfect black body (➤black-body radiation) at the same temperature. ➤➤Stefan–Boltzmann constant.

emittance An obsolete name for ➤exitance.

empirical formula ➤formula.

empirical temperature Symbol θ. An arbitrary scale of relative temperature. If the empirical temperature of system A is greater than that of system B, then, in the absence of external influences, heat will flow from A to B when the two are in thermal contact.

empty set Symbol ∅. The set that contains no elements.

e.m.u. Abbr. for ➤electromagnetic units.

emulsifying agent ➤emulsion.

emulsion A ➤colloid in which the dispersed phase and the dispersion medium are both liquids. In most common emulsions, water or an aqueous solution is one of the liquids and oil is the other. To stabilize the emulsion, a third substance, an **emulsifying agent**, is often needed. For an oil-in-water emulsion, soaps can perform this function.

enamel The outer surface of a ➤tooth, forming the exposed part of the crown. Enamel is the hardest substance in the body, and is composed largely of phosphate and carbonate salts of calcium.

enantiomer ➤optical activity.

enantiotropy A form of ➤allotropy in which at least two phases can be stable at 1 bar pressure and different temperatures. Examples include sulfur and tin. Compare ➤monotropy.

endergonic process A process during which the ➤Gibbs energy of a system increases. It is the Gibbs energy counterpart of an ➤endothermic process. An endergonic process must be driven by a linked ➤exergonic process.

endocrine gland (ductless gland) A specialized gland in an animal's body that produces a ➤hormone. An endocrine gland has no duct (compare ➤exocrine gland), and the hormone is secreted directly into the bloodstream, in which it is transported round the body. Endocrine glands control many bodily mechanisms for development and ➤homeostasis such as growth and sexual development, ➤metabolic rate, blood sugar concentration and the regulation of the ➤menstrual cycle. The ➤pituitary gland, an endocrine gland situated at the base of the brain, controls the activities of many of the other hormone-producing glands. Most control and regulation by endocrine glands is by a process of negative ➤feedback in which the output of hormone from a particular gland is inhibited by an increase in concentration of another hormone produced by a gland that it stimulates. For example, the pituitary gland produces thyroid stimulating hormone (TSH), which induces thyroxine production by the thyroid gland. The output of TSH from the pituitary gland is itself inhibited by increased concentrations of thyroxine. By such means the concentrations of hormones in the bloodstream are finely regulated.

endocrinology The study of ➤endocrine glands and of the ➤hormones they secrete.

endocytosis The process whereby the ➤plasma membrane of a cell invaginates and pinches off to form a small, intracellular, membrane-bound vesicle. Endocytosis is used by some single-celled organisms to obtain food in solid or liquid form from the surrounding environment.

endoderm ➤embryo.

endoergic process A nuclear process that results in a net inflow of energy.

endogenous Describing a substance that is produced by and in an organism. Compare ➤exogenous.

endogenous rhythm ➤biological clock.

endonuclease ➤restriction enzyme.

endoplasmic reticulum A system of interconnected networks of membrane-bound tubules, vesicles and sacs (cisternae) found in the ➤cytoplasm of a cell. There are two main types of endoplasmic reticulum: rough (granular) and smooth (agranular). The **rough endoplasmic reticulum** contains ➤ribosomes and is concerned with ➤protein synthesis. The **smooth endoplasmic reticulum** is involved in a number of cellular synthetic processes, for example of ➤steroids. The endoplasmic reticulum also acts as an internal transport system by which substances can be rapidly transferred from one part of a cell to another.

endorphin (enkephalin) One of a class of ➤peptide ➤neurotransmitter substances, secreted by the ➤pituitary gland, associated with pain receptors in the body. The action of endorphins is similar to that of ➤morphine derivatives.

endoskeleton Any structure within the body of an animal that provides support and a framework for the attachment of muscles. In the vertebrates a combination of bone and cartilage comprises the skeleton; in the ➤echinoderms and ➤annelids the endoskeleton is hydraulic. Compare ➤exoskeleton.

endosperm A tissue surrounding and nourishing the ➤embryo in the ➤Magnoliopsida. It is derived from the fusion of two ➤haploid nuclei in the ➤embryo-sac and a pollen grain nucleus, thus forming a triploid ($3n$) structure that contains three sets of chromosomes. In nonendospermic seeds the endosperm is completely absorbed by the developing embryo and stored as food reserves in the ➤cotyledons, as in peas and beans. In endospermic seeds the endosperm is partly retained until after germination. Endospermic seeds include cereals such as wheat and maize, castor and coconut, which has an unusual liquid endosperm, the 'milk'.

endospore A (usually) resistant spore formed within the cell of a bacterium, allowing it to survive in unfavourable conditions.

endothelium A single layer of flattened epithelial cells lining the internal surface of blood and lymph vessels. Endothelial cells have a role in the regulation of blood pressure, and individuals who experience high blood pressure or are susceptible to coronary heart disease often have defects in the functioning of the endothelium.

endothermic A reaction or process during which heat is absorbed from the surroundings; thus the ➤enthalpy change in the system is positive. Simple examples include the dissolution of solids such as potassium nitrate in water and the decomposition of compounds such as carbonates. ➤Phase transitions, such as melting and boiling, are other examples, as is the most important, ➤photosynthesis. Compare the more common ➤exothermic reactions.

endothermy ➤homoiothermy.

end-point The point during a ➤titration when the ➤indicator changes most rapidly, often from one colour to another. If the indicator has been chosen carefully, the end-point will be very close to the ➤equivalence point.

-ene The suffix that indicates a double bond, as in the simplest ➤alkene, ethene $CH_2=CH_2$.

energy Symbol E; unit ➤joule, J. A physical quantity that measures the capacity of a system for doing ➤work. In ➤classical physics a system can have ➤kinetic energy and ➤potential energy. In a thermodynamic system it is not convenient to distinguish the two, and we can identify only an ➤internal energy. In quantum mechanics energy plays an even more fundamental role: the energy of a quantum-mechanical stationary state is an ➤eigenvalue of the ➤Hamiltonian. The law of ➤conservation of energy often makes examining the energy of a mechanical system a simpler way of deducing its physical properties than solving its equation(s) of motion.

energy band A range of allowed energies of single-electron states in a solid, particularly a metal. Energy bands are an exclusively quantum-mechanical phenomenon. The origin of energy bands can be considered from two opposing standpoints.

The **tight-binding approximation** starts from the observation that single atoms in isolation have discrete, well-separated energy levels. If two identical atoms are brought together, the interaction between them gives rise to two levels with slightly different energies for each single electron level of the isolated atom. As the number of atoms in the system increases, the energy levels spread still further until, for the large number of atoms close together in the solid state, the energy levels have spread into a band that has such a small gap between individual levels that it is effectively a continuous range of states (see diagram (a) below).

The **nearly-free-electron approximation** starts from the picture of a solid as free electrons contained in a box, and examines the effect of the periodic potential of the ions on the electron wavefunctions. Those wavefunctions with wavelength close to the interatomic spacing will be most affected, and an ➤energy gap opens up in the allowed energy states (see diagram (b) opposite).

Either picture is an approximation, and sophisticated techniques are required to calculate the ➤band structure of a solid. In general, though, the more tightly bound an electron in an isolated atom, the narrower the energy band when the atom combines with others as a solid.

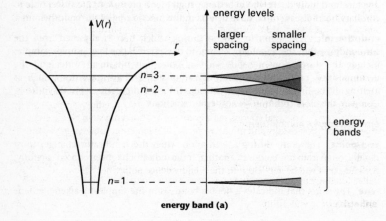

energy band (a)

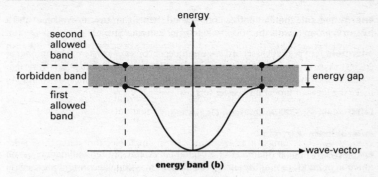

energy band (b)

energy density The energy per unit volume in a region of space. It is a useful concept in electromagnetism where energy density depends on the electric and magnetic field strengths. In relativity, energy density also includes mass–energy (➤Einstein mass–energy relation) and this makes up one component of the ➤stress–energy tensor.

energy flux The ➤flux of energy through a surface. ➤➤Poynting vector.

energy gap (band gap) The range of energy between the ➤energy bands of a solid, particularly between the valence band and the conduction band in a semiconductor. It is the size of the energy gap that determines whether a material is an insulator or a semiconductor: the smaller the gap, the more states of the conduction band will be occupied by electrons at room temperature. The semiconductors silicon and gallium arsenide both have small energy gaps of the order of 1 eV.

energy level One of the allowed energies of a quantum-mechanical system, typically an atom or a molecule. Unlike systems in ➤classical physics, only certain values of energy are permitted for systems in quantum mechanics; these are the energies of quantum-mechanical stationary states (➤quantum state). The energy is the ➤eigenvalue of the ➤Hamiltonian. In a hydrogen atom, the energy levels E_n are given by the expression E_1/n^2, where E_1 is the energy of the ground state (with absolute value about -13.6 eV) and n is the ➤principal quantum number.

energy–momentum Symbol p. A ➤four-vector (E, $p_x c$, $p_y c$, $p_z c$) of great importance in relativistic mechanics, comprising the energy of a particle as the time-like component and its momentum (multiplied by c, unless using ➤natural units) as the space-like components. The ➤norm of the vector is the ➤rest mass of the system (multiplied by c^2 unless using natural units), $m_0 c^2$, leading to the equation $E^2 - p^2 c^2 = m_0^2 c^4$. The total energy–momentum of an isolated system is conserved (in the same way that energy and momentum are individually conserved in Newtonian mechanics).

engine A device for transforming energy (usually chemical) into kinetic energy. ➤heat engine; internal combustion engine; steam engine.

enkephalin ➤endorphin.

enol A molecule that has both a double bond (hence the 'en-', as in alkene) and a hydroxyl group (hence the '-ol'). ≫keto–enol tautomerism.

ensemble A hypothetical set of a large number of copies of a system. The number of copies in a particular ≫microstate is proportional to the ≫probability that the system is in that microstate. The concept of an ensemble is used in ≫statistical mechanics to make the ideas of probability easier to grasp.

ENSO Abbr. for El Niño Southern Oscillation.

enterokinase ≫trypsin.

enthalpy Symbol H (identical to a capital Greek eta). A thermodynamic ≫state function intended to simplify energy calculations in systems at constant pressure (p) but variable volume (V). The enthalpy of a system is defined as

$$H = U + pV,$$

where U is the ≫internal energy of the system. Whereas U is normally considered as a function of ≫entropy (S) and V, H is a function of S and p. Enthalpy has the advantage of being independent of the way in which the system reached that state, making it useful for thermodynamic accounting. The name enthalpy is derived from the Greek for 'inner warmth'. A process in which the **enthalpy change** ΔH is negative is known as ≫exothermic (releasing heat), while if ΔH is positive the process is ≫endothermic (absorbing heat). ≫Hess's law; standard enthalpy change.

entomology The scientific study of ≫insects. Since insects are the most numerous organisms, entomology is a very large field and is commonly subdivided into more specific disciplines, for example **lepidoptery** – the study of butterflies and moths; **coleoptery** – the study of beetles.

Entonox The tradename for a mixture of ≫dinitrogen oxide and oxygen used as an ≫analgesic, particularly in childbirth during the first stages of labour.

entrainment The movement of particles dragged along by a fluid.

entropy Symbol S. An exceptionally important thermodynamic ≫state function, also used as a concept in information theory and probability theory.

In ≫thermodynamics, the entropy change dS in a system in thermal contact with its surroundings at temperature T is defined (for a ≫reversible process) as dq/T, where dq is the heat transferred to the system. The quantity is significant because the second law of ≫thermodynamics can be expressed in terms of it: the total entropy of an isolated system can never decrease. This places significant constraints on the processes that nature allows to happen. Thus, the temperature of a plate of hot food left in a cool room will always decrease, never increase.

In ≫statistical mechanics, which concerns the microscopic world, entropy is used as a measure of microscopic disorder: the entropy S of a system is loosely defined as $S = k \ln W$, where W is the number of ways in which the total energy can be distributed and k is the Boltzmann constant.

The two seemingly distinct definitions of entropy are brought together empirically by the observation that processes involving an increase in thermodynamic entropy in isolated systems also increase their disorder, spreading out their energy over a larger volume or number of particles. ➤Diffusion is a process governed by the increase in entropy caused as the particles spread out.

envelope For a family of curves $f(x, y, c) = 0$, a curve every point of which shares a tangent with a curve of the family. The equation of the envelope is obtained by eliminating c from the simultaneous equations $f(x, y, c) = 0$ and $\partial f(x, y, c)/\partial c = 0$.

environment An all-encompassing term describing the conditions in which an organism lives or a process operates. The environment of a particular organism is composed of nonliving (abiotic) factors such as climate and mineral composition of the soil, and living (biotic) factors such as predators and other organisms with which it interacts. The term is also applied to the internal conditions in an organism.

enzyme A ➤protein, produced by a cell, that acts as a biological ➤catalyst by accelerating specific biochemical reactions without itself being altered. Enzymes are highly specific, and a particular enzyme will catalyse only a single reaction. Their specificity stems from their mode of action, in which the reactant (the **substrate**) fits into a specific site (the **active site**) in the enzyme molecule. The relationship between the substrate and the active site is sometimes likened to a key fitting into a lock (the **lock-and-key model**). However, enzymes are not rigid, and the relationship is better conceptualized as a hand fitting into a glove. The **enzyme–substrate complex** is a short-lived intermediate stage that greatly enhances the rate of reaction, typically by a factor of many thousand, by lowering the ➤activation energy of the reaction. Enzymes that consist of a single ➤polypeptide chain are called **monomeric**; those formed from more than one polypeptide subunit (**protomers**) are called **oligomeric**.

Many enzymes also require a variety of nonprotein ➤cofactors for their activity. Enzymes function within narrow limits of temperature and ➤pH. Most function best at around 37 °C, and are inactivated on prolonged exposure to temperatures above 45 °C or by inappropriate pH.

The enzyme pepsin, which hydrolyses protein chains in the mammalian stomach, thus initiating digestion, works at an optimum pH of about 2; more commonly, most function best in the range pH 6–8. Enzyme action is further influenced by various ➤inhibitors which compete with the natural substrate for the active site, or by metal ions such as copper or lead, which can cause irreversible damage to active sites and are thus toxic. Control of enzyme activity in cells is essential for the normal growth, development and functioning of organisms, since ultimately all reactions in cells are mediated by enzymes. Since enzymes are proteins, they are the direct products of ➤genes, and their synthesis and structure is determined by ➤DNA. Their action is also regulated by a variety of negative ➤feedback mechanisms governed by the relative concentrations of substrates and products. In some ➤metabolic pathways the end product of the pathway acts as an inhibitor of the first enzyme in the sequence, thus exercising a fine-control mechanism.

Enzymes are used in a wide range of industrial processes in which their high specificity and low-temperature action make them economically desirable. Enzymes used in industry are often immobilized in small beads of a suitable inert polymer and can be reused several times. High-fructose corn syrup, used as a sweetener in the manufacture of soft drinks, is produced from glucose syrup by passing it over a bed of immobilized glucose isomerase, an enzyme that converts glucose into fructose.

The classification of enzymes and examples of their reactions are given in the table.

Class	Reactions catalysed and example
Oxidoreductases	Redox reactions; e.g. succinate dehydrogenase $^-O_2CCH_2CH_2CO_2^-$ becomes *trans*-$^-O_2CCH=CHCO_2^-$
Transferases	Transfer of functional groups; e.g. aminotransferase $RCH(NH_3^+)CO_2^- + R'COCO_2^-$ becomes $RCOCO_2^- + R'CH(NH_3^+)CO_2^-$
Hydrolases	Hydrolysis reactions; e.g. peptidase (amylase; nuclease; lipase; protease) $RCONHR' + H_2O$ becomes $RCO_2^- + R'NH_3^+$
Lyases	Addition to a double bond or the reverse reaction, i.e. formation of a double bond by cleavage; e.g. decarboxylase $RCH(NH_3^+)CO_2^-$ becomes $RCH_2NH_2 + CO_2$
Isomerases	Interconversion of isomers; e.g. triosephosphate isomerase $OCHCH(OH)CH_2OPO_3^{2-}$ becomes $HOCH_2COCH_2OPO_3^{2-}$
Ligases	Condensation reactions coupled with ATP hydrolysis; e.g. DNA ligase

enzyme-linked immunosorbent assay (ELISA) A diagnostic test that uses the high specificity of ➤enzymes and ➤antibodies to detect the presence of specific substances in a sample, usually by the production of a distinctive colour change. ELISA tests are increasingly being used in a wide range of applications such as pregnancy tests, the detection of ➤HIV antibodies in blood and the early detection of diseases such as potato blight in plants.

Eocene See table at ➤era.

eon ➤era.

EOR (XOR) Abbr. for exclusive OR: ➤truth table.

ephemeral An organism that has a short-lived growing season or adult phase. Examples are desert-living plants and some garden weeds which complete their life cycle rapidly when conditions are favourable, but otherwise remain in a state of ➤dormancy, usually as seeds. Typical animal examples are insects such as mayflies (order Ephemeroptera), which in the adult phase of their life cycle lack feeding mouthparts and live for only a few hours, during which time the females must mate and lay eggs.

ephemeris time A system of time for astronomical observations that moves forward at a uniform rate, as opposed to a system based on the motions of planets and stars. The ➤second was defined in terms of ephemeris time until the introduction of ➤International Atomic Time.

epicentre The point on the surface of the Earth that is directly above the point of origin of a ➤seismic wave.

epicotyl The part of the embryo of a germinating seed that is above the point of insertion of the ►cotyledons. The epicotyl comprises the plumule, which gives rise to the shoot system.

epicyclic gears A system of gears in which the axle of only one cog is fixed in space; the other cogs revolve around the fixed one.

epicycloid The path traced out by a point on a circle (radius r) rolling around the outside of a fixed circle (radius R). Special names are given to specific choices of the ratio R/r: the **cardioid** has ratio 1, and the **nephroid** has ratio 2. Epicycloids occur in the Ptolemaic system of astronomy, which was supplanted by the ►heliocentric system.

epidermis The external cellular covering layer of a multicellular organism. In plants it is one cell thick and may have a noncellular waxy cuticle covering it. The epidermis of invertebrate animals is frequently one cell thick, and may secrete cuticle – a noncellular protective covering layer. The epidermis of vertebrates is the multilayered external zone of the ►skin.

epigamic Describing an animal characteristic that serves to attract a mate.

epimer For a molecule containing more than one ►asymmetric carbon atom, each of the isomers differing only in the orientation around one of the carbon atoms.

epinephrine An alternative name for ►adrenaline which is more common in the USA.

epiphytic ►root (2).

episome A segment of DNA that can exist more or less independently in a cell. It may behave autonomously and replicate independently of the ►chromosomes, or it may become incorporated into them and become functionally indistinguishable from the rest of the ►genome. Episomic elements in bacteria are known as ►plasmids.

epitaxy The building of a covering of one substance by layer-by-layer deposition on the surface of a single crystal, usually of a different substance (known as the substrate (2)), in such a way that the deposited layer takes its orientation and structure from the substrate. Epitaxy is used to build semiconductor structures. ►►molecular beam epitaxy.

epithelium The tissue covering all body surfaces and the lining of internal organs and cavities. There are several types of epithelium, each with particular functions. The simplest is **squamous epithelium** lining the inner surface of the lungs; it is one cell thick and facilitates gas exchange. The epidermal layer of the skin and the lining of the mouth is made of **stratified epithelium**, which resists abrasion. The epithelial lining of the gut is highly glandular, and secretes the digestive juices as well as facilitating absorption of food. In the tubules of the kidney, **cuboidal epithelium** has a variety of physiological and regulatory functions, including the provision of a pathway for the reabsorption of water into the blood.

epoch ►era.

epoxide (oxirane) An organic molecule with an oxygen atom present in a ring, most commonly a three-membered ring, as in ➤epoxyethane. **Epoxidation** is often performed by reacting an alkene with a reagent such as perbenzoic acid, the ring being formed where the C=C double bond was in the alkene.

epoxyethane (ethylene oxide) The epoxide with two carbon atoms (see the diagram); it is a colourless flammable gas. Industrially it is formed by oxidation of ethene with oxygen, using a silver catalyst at 250 °C and 20 atm. Its main use is in the preparation of ➤ethane-1,2-diol (common antifreeze) by acidic hydrolysis of epoxyethane.

H_2C——CH_2
epoxyethane

epoxy resin One of an important class of polymers made by the condensation polymerization of epoxides such as 1-chloro-2,3-epoxypropane (see the diagram) with alcohols that contain several hydroxyl groups. The polymers are ➤thermosetting, tough and chemically resistant. Their many uses include adhesives, protective coverings and composites.

H_2C——CH——CH_2Cl
epoxy resin
1-chloro-2,3-epoxypropane

EPR Abbr. for electron paramagnetic resonance (➤electron spin resonance).

EPROM Acronym for erasable programmable read-only memory. A type of semiconductor device that can have its contents programmed electrically, and deleted using ultraviolet light. Once programmed, the chip acts as a normal ➤ROM. The process of deletion and reprogramming can be carried out many times; with a PROM the programming is irreversible.

Epsom salts The traditional name for magnesium sulfate heptahydrate, $MgSO_4 \cdot 7H_2O$. The name commemorates the town in Surrey. It exists as colourless crystals, used to produce the effect of snow in films.

Epstein–Barr virus ➤Burkitt's lymphoma; herpes virus. Named after (Michael) Anthony Epstein (b. 1921) and Yvonne M. Barr (b. 1932).

equation A statement that asserts the equality of two mathematical expressions. Unless they are identities, equations are only true for certain values of the variables involved; finding them is called **solving the equation**. Equations that occur often are given special names reflecting either their nature (as in 'quadratic equation') or their discoverer (as in the ➤Schrödinger equation).

equation, chemical A symbolic representation describing the changes that take place during a chemical reaction. Conventionally the substances that react are placed on the left-hand side of the equation and the products are placed on the right. Hence the ammonia synthesis is written as

$$N_2 + 3H_2 \rightarrow 2NH_3,$$

and can be interpreted to mean that one nitrogen molecule reacts with three hydrogen molecules to form two ammonia molecules. However, all reactions actually reach a position of ➤equilibrium. Although for most reactions the equilibrium is well over to the right-hand side, the ammonia synthesis, for example, produces only around 15% ammonia under normal manufacturing conditions. Hence the simple distinction between reactants and products becomes blurred. An equation can be used to calculate quantitatively what mass of one reactant will react with a given mass of the other (➤mole calculations).

equation(s) of motion **1** An equation or set of equations that describe the variation of the coordinates of a system with time. The equations are usually differential equations. A typical equation of motion is an application of Newton's second law (➤Newton's laws of motion), relating the second time derivative of the position of a body to the forces acting on the body. For example, for a particle (position x, mass m) on a spring (with spring constant k), the equation of motion is

$$\frac{d^2x}{dt^2} = -\frac{k}{m}x.$$

In more complex systems, all the coordinates and their time derivatives may be interrelated. ➤Hamilton's equations are equations of motion, as are ➤Euler's equations. Even the ➤Schrödinger equation is an equation of motion as it relates the time derivative of a state to a function of the state itself.

2 ➤motion under uniform acceleration.

equation of state An equation interrelating the variables of a thermodynamic system, typically the pressure, volume, temperature and amount of substance of a gas. The simplest is the ➤ideal gas equation; another is the ➤van der Waals equation.

equation of time The difference between the ➤apparent solar time and ➤mean solar time that results from the eccentricity of the Earth's orbit and the fact that the Sun follows the ➤ecliptic, not the celestial equator. The equation of time is a physical quantity, not an equation in the mathematical sense. Its value varies from $+16$ to -14 min over the course of the year.

equator The ➤great circle on the surface of the Earth whose plane is perpendicular to the Earth's rotational axis. It is the line of zero latitude. ➤➤celestial equator.

equatorial At or in the region of the Earth's ➤equator.

equatorial bulge The excess radius of the Earth at the equator (about 21 km) over that at the poles. It results from forces generated by the rotation of the Earth. The equatorial bulge means that the gravitational field around the Earth does not conform to a perfect ➤inverse square law, so satellite orbits are not perfect ellipses.

equilateral triangle A triangle with all three sides of equal length, and hence all interior angles equal to 60°.

equilibrium (plural **equilibria**) A state of a system that has no natural tendency to change without outside influence. A **stable equilibrium** exists where a small displacement of a system causes it to return to the equilibrium position. A system in **unstable equilibrium** tends to move further away from equilibrium when displaced. A system in **neutral equilibrium** does not change after a small displacement. A simple mechanical system is in equilibrium when no ➤resultant forces cause accelerations. ➤➤dynamic equilibrium; thermal equilibrium.

equilibrium, chemical That composition of a chemical reaction where there is no further net change in the reaction mixture. Individual molecules continue to react, but for every forward reaction there is a compensating backward one; the equilibrium is dynamic.

The **equilibrium constant** for the reaction

$$aA + bB \rightleftharpoons cC + dD$$

is defined as

$$K = [C]^c[D]^d / [A]^a[B]^b,$$

where [X] represents the concentration (in more precise work, the ➤activity should be used) of X at equilibrium. The equilibrium constant is a constant for a given reaction at a given temperature and any composition will evolve until the concentrations satisfy the equilibrium constant (➤➤K).

The equilibrium composition corresponds to the lowest ➤Gibbs energy for the reaction. The equilibrium constant K is related to the ➤standard Gibbs energy change by the equation

$$\Delta G^{\ominus} = -RT \ln K,$$

where R is the gas constant and T the thermodynamic temperature.

The only way an equilibrium constant can be altered is by changing the temperature. When the temperature is increased, reactions shift in the endothermic direction (➤Le Chatelier's principle).

equinox Either of the two points on the ➤celestial sphere where the ➤ecliptic intersects the ➤celestial equator. At the equinoxes, the Sun crosses the celestial equator in its apparent motion around the ecliptic. It is then overhead to an observer on the Earth's equator, and day (sunrise to sunset) and night are of about the same length at all points on the Earth. The **vernal equinox** is where the Sun crosses the equator from south to north, on or around 21 March, and the **autumnal equinox** is where it crosses the equator from north to south, on or around 23 September. These two dates are also known as the equinoxes.

equipartition of energy The energy in a system as a function of temperature is equally divided according to the number of ➤degrees of freedom of the system, as a consequence of the ➤Boltzmann distribution. The **classical equipartition theorem** states that, for each degree of freedom that contributes a term of the form $\alpha(dq/dt)^2$ (where q is a coordinate) to the total energy, a contribution of $\frac{1}{2}kT$ is made to the thermal energy of the system (where k is the ➤Boltzmann constant and T is the ➤thermodynamic temperature), independent of any of the constants of proportionality, α. For example, a perfect monatomic gas of N particles has $3N$ degrees of freedom (three position coordinates for each atom) and therefore has energy $\frac{3}{2}NkT$. The result is independent of the mass of the atoms. The prediction is fairly accurate for simple real gases, but the theorem has to be modified substantially to account for a quantum system, as in its classical form it assumes that there is a continuous range of energy available.

equipotential A surface joining points of equal potential. For a ➤scalar potential, the field derived from the potential is perpendicular to the equipotentials.

equivalence point The composition during a ➤titration where there is just enough of one reagent to react with the other exactly. They are then present in a molar ratio identical to that in the chemical equation. Ideally this corresponds closely to the ➤end-point, which is where the ➤indicator changes colour.

equivalence principle The fundamental principle of ➤general relativity on which all its results depend: the assertion that the laws of physics are the same in any nonrotating, freely falling ➤inertial frame. Thus within a closed laboratory there is no experiment that can distinguish whether the laboratory is on a spaceship far removed from all gravitational fields, or is falling freely in a lift-shaft in a uniform gravitational field. ➤gravitational mass.

equivalent ➤Faraday's laws.

equivalent circuit A combination of voltage or current sources and resistors that has the same behaviour as a more complicated electrical component. The diagram shows the equivalent circuit of a ➤field-effect transistor.

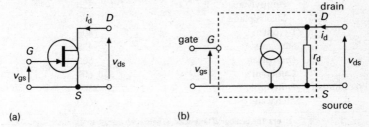

(a)   (b)

equivalent circuit (a) Field-effect transistor and (b) its equivalent circuit. The two intersecting circles represent a voltage source.

equivalent proportions, law of ➤reciprocal proportions, law of.

equivalent weight The mass of an element that can combine with or displace one gram of hydrogen. The term is obsolete, having been replaced by the concept of ➤molar mass.

Er Symbol for the element ➤erbium.

era (geological era) One of the major periods of geological time into which the Earth's history is divided. The four recognized eras (**Precambrian, Palaeozoic, Mesozoic** and **Cenozoic**) are further subdivided into **periods** (see the table overleaf). The periods are defined by the type of rock and the presence of characteristic fossils which relate to the global climatic conditions at the time. The two most recent periods, the Tertiary and the Quaternary, are further divided into **epochs**.

erbium Symbol Er. The element with atomic number 68 and relative atomic mass 167.3, which is a typical lanthanide. Hence it shows one dominant ➤oxidation number, +3, as in the pink solid oxide Er_2O_3 which is used in coloured glass.

erecting system An optical system, usually a prism, used in an instrument to invert an image that would otherwise appear to be the wrong way up.

Era	Period	Epoch	Approximate time since start in millions of years before the present
Cenozoic	Quaternary	Holocene (recent)	0.01
		Pleistocene	1.6
	Tertiary	Pliocene	5
		Miocene	23
		Oligocene	36
		Eocene	57
		Palaeocene	66
Mesozoic	Cretaceous		144
	Jurassic		208
	Triassic		245
Palaeozoic	Permian		286
	Carboniferous		
	Pennsylvanian		320
	Mississipian		360
	Devonian		408
	Silurian		438
	Ordovician		505
	Cambrian		570–600
Precambrian	Proterozoic		2500
	Archaean		4600

era The geological time-scale. (There is no consistent system of epochs for periods before the Tertiary.)

erg The c.g.s. unit of work, equal to one ➤dyne centimetre. One erg is equivalent to 10^{-7} J.

ergocalciferol ➤vitamin D.

ergodic system A system that eventually returns to the same state it started in, even if the period is very long. The supposed ergodic nature of thermodynamic systems (e.g. the trajectory in $6N$-dimensional ➤phase space of the coordinates of N particles of a gas) is often used in statistical mechanics as a justification for replacing ➤expectation values with averages over time.

ergonomics The study of the interaction between humans and their environment, particularly in the workplace. It includes the optimization of the arrangement of controls on, for example, cars and aircraft, and the design of seating.

ergosterol ➤vitamin D.

eriochrome black T A useful ➤complexometric indicator which forms a red complex with calcium or magnesium ions. When these complex ions are titrated against ➤EDTA, complexation with EDTA eventually liberates the free dye, which is blue in mildly alkaline solution; this colour change indicates the end-point of the titration. Such a titration can be used to analyse hard water quantitatively.

error of measurement A deviation of an experimentally measured quantity from the real value of the physical quantity being studied. Errors fall into two broad categories: **systematic errors**, which remain the same if an experiment is repeated, and **random errors**, which differ between successive repetitions. The only way of avoiding systematic errors is by careful design of an experiment. Random errors can be reduced by repeating a measurement a large number of times and formally analysed using ➤statistics. Informally, the random error in a measurement can be estimated by taking the ➤standard deviation of a large number of such measurements. For example, $a = 0.20 \pm 0.03$ m indicates an estimated random error of 0.03 m in the value of a. Errors can be combined to give estimates of errors in derived quantities. For example, if a quantity a is independently measured n times with estimated random error Δa in each measurement, the error in the mean value of a is $\Delta a / \sqrt{n}$.

erythro- A prefix indicating an isomer that resembles ➤erythrose in the position of the groups on adjacent asymmetric carbon atoms.

erythroblast A cell, originating in the red bone marrow, that gives rise to a red blood cell.

erythrocyte A red blood cell. Erythrocytes are the most common type of cell in the blood, with approximately 5 million per cubic millimetre. Erythrocytes are biconcave discs and carry the iron-containing protein ➤haemoglobin, which carries oxygen in the blood. They are unique among mammalian cells in not containing a ➤nucleus (2). Erythrocytes are produced in the red bone marrow of the ribs, vertebrae and hip bones. They live for about 120 days, and are then recycled by the ➤liver and the ➤spleen. Some of the breakdown products are excreted by the liver in the bile as bile pigments and give the characteristic colour to faeces. The body produces more erythrocytes in response to low partial pressures of oxygen, for example at high altitudes. Altitude acclimatization programmes depend partly on this, as does the (short-lived) effectiveness of high-altitude training for athletes. Compare ➤leucocyte.

erythrose $CHOCH(OH)CH(OH)CH_2OH$ A tetrose ➤monosaccharide in which the two hydroxyl groups are on the same side (see the diagram). Compare ➤threose.

CHO
H——OH
H——OH
CH_2OH

D-erythrose

Es Symbol for the element ➤einsteinium.

Esaki diode ➤tunnel diode. Named after Leo Esaki (b. 1925).

ESCA Abbr. for electron spectroscopy for chemical analysis. It is a version of ➤photoelectron spectroscopy.

escape velocity (escape speed) Symbol v_e. For a particular massive body (especially a star, planet or satellite), the speed an object needs to acquire to have enough energy so that its speed never falls to zero, no matter how far away from the body it moves. The object can therefore escape from the gravitational field of the body. The kinetic energy of the object has to be greater than the potential energy difference between its position a distance r from the body (of mass M) and infinity:

$$v_e = \sqrt{\frac{2GM}{r}},$$

where G is the gravitational constant. The escape velocity at the surface of the Earth is about 11 km s^{-1}.

Escherichia coli ➤*E. coli.*

ESR (esr) 1 Abbr. for ➤electron spin resonance.
 2 Abbr. for erythrocyte sedimentation rate.

essential (indispensable) amino acids ➤Amino acids that cannot be synthesized in sufficient quantity by an organism itself and therefore must be obtained from the diet. The amino acids recognized as being indispensable for adult humans include arginine, isoleucine, leucine, lysine, methionine, phenylalanine, threonine, tryptophan and valine. In addition, growing children (and possibly adults) require histidine. ➤➤Appendix table 7.

essential element ➤trace element.

essential fatty acids Certain carboxylic acids that cannot be synthesized by animals and are therefore required in the diet. For example, a lack of linoleic acid results in stunted growth and hair loss in many animals unless it is supplied in the diet.

essential oil In plants, a scented, terpene-based oil, secreted from glandular cells and possessing a characteristic aroma. The familiar aroma of mint and citrus fruit are examples.

ester An organic compound formed by the combination of an acid and an alcohol. The most important examples are **carboxylic esters** formed from carboxylic acids, typically in an ➤esterification reaction. (Inorganic esters also exist, one example of which is ethyl hydrogen sulfate, $CH_3CH_2OSO_2OH$. The most important phosphate esters are the ➤nucleotides.) A higher yield can be obtained by using the corresponding acyl chloride (➤acyl halide) in place of the acid; for example, ethanoyl chloride and ethanol give a high yield of ethyl ethanoate. Typically, esters are volatile liquids with a characteristic smell, most commonly resembling glue or fruits. Pear-drop essence, 3-methylbutyl ethanoate, is particularly distinctive. Esters are commonly used as solvents, flavouring agents and perfumes. Their most important reaction is hydrolysis back to the acid and alcohol from which they were made.

esterification The process in which an alcohol and a carboxylic acid are converted into an ester and water, often with acid catalysis. As an example, ethanol and ethanoic acid form ethyl ethanoate and water:

$$CH_3CH_2OH + CH_3COOH \rightleftharpoons CH_3COOCH_2CH_3 + H_2O.$$

As is common, this equilibrium mixture is only just in favour of the right-hand side ($K = 4$ at 100 °C; ➤equilibrium, chemical).

estrogen US spelling of ➤oestrogen.

e.s.u. Abbr. for ➤electrostatic unit.

Et A common abbreviation for an ➤ethyl group.

etalon An ➤interferometer consisting of two parallel layers of glass with a gap between them. The successive partial reflection of light between the layers results in constructive interference only if the distance between reflections is a multiple of the wavelength. Since this distance is determined by the angle of incidence of the ray to the layers, the etalon transmits different wavelengths at different angles, and thus creates a spectrum.

ethanal (acetaldehyde) CH_3CHO The ➤aldehyde with two carbon atoms, which is a colourless liquid with a characteristic smell of apples. It is manufactured by the ➤Wacker process. It can be made in the laboratory by careful oxidation of ethanol, CH_3CH_2OH, using acidified potassium manganate(VII). The care required involves removing the ethanal as it is formed (by distillation) as ethanal can be further oxidized to ethanoic acid, CH_3COOH. Ethanal is slowly formed when ethanol (alcohol) is oxidized in air and it contributes significantly to hangovers, as it is toxic. Addition of dilute acid causes polymerization to ➤metaldehyde.

ethanamide (acetamide) CH_3CONH_2 The amide with two carbon atoms, which forms colourless crystals which smell of mice. Its relatively high melting point is due to hydrogen bonding between the CO group and the NH group.

ethane C_2H_6 The alkane with two carbon atoms, which is a colourless odourless gas. It is found in the gas fraction produced by fractional distillation of petroleum. It is also a minor constituent in 'wet' natural gas. It shows the typical properties of ➤alkanes, combustion being the most important.

ethanedioic acid (oxalic acid) $(COOH)_2$ The simplest dicarboxylic acid, which exists as toxic white crystals. It can be dehydrated to a mixture of carbon monoxide and carbon dioxide.

ethane-1,2-diol (ethylene glycol) $HOCH_2CH_2OH$ A ➤diol with two carbon atoms, which is a colourless liquid. It is manufactured by the acidic hydrolysis of ➤epoxyethane. Its dominant use is as antifreeze. ➤diethylene glycol.

ethanoate (acetate) A salt or ester of ethanoic acid. Examples include sodium ethanoate, $Na^+CH_3CO_2^-$, and ➤ethyl ethanoate.

ethanoic acid (acetic acid) CH_3COOH The carboxylic acid with two carbon atoms, which is a colourless liquid with a characteristic pungent smell strongly reminiscent of vinegar. Its melting point is very close to room temperature. The frozen substance looks like a glass, hence the description **glacial ethanoic acid**. It dissolves readily in water: vinegar is a 5% solution, often coloured by caramel. It shows the typical properties of a ➤weak acid, for example neutralizing sodium hydroxide to form sodium ethanoate, $Na^+CH_3CO_2^-$.

ethanoic anhydride (acetic anhydride) $(CH_3CO)_2O$ A colourless liquid very useful as an ➤ethanoylating agent. It is used in the synthesis of aspirin (see the diagram overleaf) as it has the advantage of having a more controllable rate than ➤ethanoyl chloride.

$$(CH_3CO)_2O \ + \ HO \overset{COOH}{\underset{\text{2-hydroxybenzoic}}{\bigcirc}} \longrightarrow CH_3COO \overset{COOH}{\underset{}{\bigcirc}} \ + \ CH_3COOH$$

ethanoic
anhydride

2-hydroxybenzoic
acid

2-ethanoyloxybenzoic acid
(acetylsalicylic acid, aspirin)

ethanoic
acid

ethanoic anhydride Its role in the synthesis of aspirin.

ethanol (ethyl alcohol, alcohol) CH_3CH_2OH The alcohol with two carbon atoms, which is the archetypal alcohol; in common usage the word 'alcohol' means this specific chemical. It is a colourless liquid with a relatively high boiling point of $78\,°C$ (the molecule methoxymethane, CH_3OCH_3, which has the same mass boils at $-23\,°C$, because it lacks the hydrogen bonding present in ethanol). Ethanol is manufactured by ➤direct hydration of ethene or by ➤fermentation if petroleum sources are scarce, as in Brazil. Because ethanol forms an azeotropic mixture with water, the mixture cannot be concentrated above 96% (the ➤azeotrope) by fractional distillation.

Ethanol is oxidized to ethanal and then ethanoic acid, by acidified potassium manganate(VII) for example. Wine goes off by a similar process, turning vinegary. Ethanol can react with carboxylic acids in an ➤esterification reaction, often catalysed by concentrated sulfuric acid. Sulfuric acid and ethanol react in a variety of ways, depending on their relative amounts and the temperature (see the diagram), giving ethyl hydrogen sulfate, ethoxyethane and ethene, for example.

Ethanol's most common use is in alcoholic drinks. Physiologically it acts as a depressant, despite its apparent stimulant effect caused by removal of inhibitions. Another major use is as a solvent, the second most important after water. Ethanol is useful because it has a dipole, can form hydrogen bonds and has moderately large ➤dispersion forces. ➤Tincture of iodine, for example, is an alcoholic solution. Yet another use is as a fuel, **gasohol**, especially in countries poor in oil.

$$H_2C{=}CH_2$$

↕ excess acid (170°C)

$$CH_3CH_2OH \underset{\text{excess water}}{\overset{\text{concentrated } H_2SO_4}{\rightleftharpoons}} CH_3CH_2OSO_2OH$$

↕ excess alcohol (140°C)

$$(CH_3CH_2)_2O$$

ethanol Its reaction with sulfuric acid.

ethanoylating agent (acetylating agent) A compound, such as ethanoyl chloride or ethanoic anhydride, capable of replacing a hydrogen atom with an **ethanoyl group**, $—COCH_3$. For example, ethanol, CH_3CH_2OH, is ethanoylated to form ethyl ethanoate, $CH_3CH_2OCOCH_3$. The hydrogen atoms that can be replaced are typically those attached to oxygen atoms in alcohols or phenols or to nitrogen atoms in amines. The last reaction, as in the simplest case of CH_3NH_2 being converted to $CH_3NHCOCH_3$, is useful for the protection of amine groups during amino acid synthesis (►protecting group).

ethanoylation The replacement of a hydrogen atom with an **ethanoyl group**, —$COCH_3$, using an ►ethanoylating agent.

ethanoyl chloride (acetyl chloride) CH_3COCl The acyl chloride with two carbon atoms, which is a colourless liquid which fumes because of its rapid hydrolysis in moist air. The vapour is ►lachrymatory as a similar hydrolysis reaction occurs in the water round the eye, forming ethanoic acid. Ethanoyl chloride is a very important ►ethanoylating agent.

ethene (ethylene) C_2H_4 The simplest alkene, having two carbon atoms, which is a colourless flammable gas. It is made industrially by the ►cracking of a fraction, typically ►naphtha, from the fractional distillation of petroleum. It is much used for ripening fruit and manufacturing other chemicals. For example, ►direct hydration of ethene gives ethanol, whereas oxidation gives epoxyethane and thence ethane-1,2-diol (common antifreeze). Polymerization gives ►polyethylene.

ethene

ether 1 (Chem.) One of the class of organic compounds of general formula ROR', where R and R' are alkyl groups. A common method of preparation is ►Williamson's ether synthesis. The typical compound is ►ethoxyethane, $CH_3CH_2OCH_2CH_3$, and the term 'ether' is commonly used to refer to this specific ether. These molecules react much less readily than the isomeric alcohols. For example, methoxymethane is isomeric with ethanol but fails to be reduced by sodium or oxidized by acidified potassium dichromate(VI) or to react with ethanoic acid. This general lack of reactivity makes ethers valuable as solvents. One danger, however, is that they are both volatile and flammable.

2 (also spelt **aether**) The hypothetical medium for electromagnetic waves in 19th-century theories of electromagnetism, supposed to pervade all of space. Although without physical substance (because such waves travel in a vacuum), the ether was believed to be at rest in a particular ►inertial frame, the preferred frame of 'absolute space'. ►Special relativity, supported by experimental verification of the invariant speed of light in a vacuum (by, for example, the ►Michelson–Morley experiment) put paid to the existence of an ether, and with it any concept of absolute space.

Ethernet A computer network standard in which data is carried at rates of up to 100 megabits (100×2^{20} binary digits) per second.

ethology The study of animal behaviour. A major principle of ethology is that animal behaviour has been subject to the process of ►evolution by ►natural selection. The behaviour of animals has evolved in such a way that it promotes the chances of the survival of their offspring. For example, the adults of many ground-nesting birds will put themselves at risk by distracting a predator from eggs or chicks. ►►altruism.

ethoxyethane (diethyl ether, ether) $CH_3CH_2OCH_2CH_3$ The most important example of the class of ethers, which is a colourless liquid with a characteristic odour. It exhibits the common unreactivity of ethers. It is one of the most widely used solvents, dissolving organic solutes well because of its high dispersion forces. It used to be used as an anaesthetic, but has been supplanted by ➤halothane.

ethyl acetate ➤ethyl ethanoate.

ethyl acetoacetate ➤ethyl 3-oxobutanoate.

ethyl alcohol ➤ethanol.

ethylamine (aminoethane) $CH_3CH_2NH_2$ The amine with two carbon atoms, which is a colourless liquid with a characteristic fishy smell. As with all amines, its most important property is that it is a ➤Lewis base, neutralizing acids and forming complexes with copper(II) ions, for example.

ethylbenzene $C_6H_5CH_2CH_3$ A substituted aromatic compound, which is a colourless liquid. It is made by the ➤Friedel–Crafts reaction. Most of its properties are similar to those of benzene, and therefore electrophilic substitution is its dominant reaction. It differs from benzene in two ways. First, the side chain can be oxidized with acidified potassium manganate(VII) to benzoic acid, C_6H_5COOH. Second, on chlorination in ultraviolet light, the two hydrogens next to the ring can be chlorinated preferentially (forming $C_6H_5CCl_2CH_3$) because the intermediate ➤radical can be delocalized with the ring.

ethyl chloride ➤chloroethane.

ethylene A very common alternative name for ➤ethene.

ethylenediaminetetraacetic acid ➤EDTA.

ethylene glycol ➤ethane-1,2-diol.

ethyl ethanoate (ethyl acetate) $CH_3COOCH_2CH_3$ A very important ester, which is a colourless liquid with a pleasant odour reminiscent of glue. It is made by the esterification reaction of ethanoic acid and ethanol. It is widely used as a solvent, especially for cellulose varnishes and adhesives. Its pleasant smell disguises the fact that long-term exposure is harmful.

ethyl group The alkyl group with two carbon atoms, $-CH_2CH_3$ or $-C_2H_5$.

ethyl 3-oxobutanoate (ethyl acetoacetate, acetoacetic ester) A very important intermediate (see the diagram) in organic synthesis, which is a colourless liquid.

$$CH_3\overset{O}{\overset{\|}{C}}-CH_2-\overset{O}{\overset{\|}{C}}OCH_2CH_3$$

ethyl 3-oxobutanoate

ethyne (acetylene) C_2H_2 The simplest alkyne, which is a colourless flammable gas. The flame produced by combusting ethyne in oxygen (the 'oxyacetylene' flame) is the hottest of any easily transported source. Ethyne can be converted into a range of useful materials (such as chloroethene), usually by ➤electrophilic addition. The terminal hydrogen is easily replaced by a metal to form **ethynides** (commonly called **acetylides**), some of which, silver acetylide for example, explode on being touched.

etiolation A physiological response in plants which are grown in the absence of light. Etiolated plants are yellowed, lacking chlorophyll, and are elongated and spindly. Light is essential for chlorophyll synthesis and also inhibits the synthesis of ➤gibberellin, which causes cell elongation in shoots.

Eu Symbol for the element ➤europium.

Eubacteria A major and diverse category of ➤bacteria, including such significant genera as *Escherichia* (➤*E. coli*), and *Rhizobium*.

eucaryotic ➤eukaryotic.

Euclidean geometry The familiar geometry of points, lines and planes studied at school. Historically it was based on postulates and axioms set out in Euclid's *Elements*, the most influential textbook ever written. Euclidean geometry gives an accurate description of the world of everyday life but generalizations such as ➤Riemannian geometry are needed for the curved spacetime of ➤general relativity. Named after Euclid (*c*. 300–260 BC).

eudiometer A graduated glass tube used to measure gas volumes.

eukaryotic (eucaryotic) Describing a cell (a **eukaryote**) that contains a ➤nucleus (2) bounded by a double ➤plasma membrane and other membrane-bound ➤organelles such as ➤chloroplasts and ➤mitochondria. ➤Prokaryotic cells lack these features. The distinction between these two cell types represents a fundamental division in the levels of organization of living organisms. Eukaryotic organisms comprise the multicellular kingdoms of plants, animals and fungi, and the unicellular kingdom of protoctistans, which includes yeasts.

Euler angles The three angles that characterize the orientation of one set of ➤Cartesian coordinate axes to another. Any orientation of axes can be described by a rotation about each of the axes in turn (in a specified order), with the three angles of rotation being the Euler angles. One application is to the study of the rotation of ➤rigid bodies, where one set of axes defines an inertial frame and the second is fixed in the rotating body. Named after Leonhard Euler (1707–83).

Euler's formula ➤polyhedron.

europium Symbol Eu. The element with atomic number 63 and relative atomic mass 152.0, which is a typical ➤lanthanide. As expected, its most common ➤oxidation number is +3 as in the oxide Eu_2O_3, a pink solid. The aqueous ion is also pink-coloured and europium compounds are used as the red phosphors in colour TV sets. It is unusual in having another common oxidation number of +2, because of the stability of the half-filled f subshell in Eu^{2+}.

Eustachian tube ➤ear. Named after Bartolommeo Eustachio (1520–74).

eutectic When two liquids are mixed, the mixture typically freezes at a lower temperature than either pure substance. That composition which has the lowest melting point is called the eutectic from the Greek for 'easily melted'. Eutectics have some useful applications. For example, the eutectic mixture of tin and lead is used as solder. The eutectic mixture of salt and ice was used to define zero on the ➤Fahrenheit scale.

Eutheria ➤placenta.

eutrophic Describing lake or river water that has become enriched by (in particular) nitrates and phosphates. To some extent, **eutrophication** is a natural process occurring as minerals dissolve into water courses as they age. However, the term has become associated with a form of pollution in which nutrients are leached into lakes and rivers from agricultural and industrial sources stimulating the growth of certain types of ➤algae and ➤cyanobacteria, which in turn causes excess ➤biological oxygen demand, threatening the survival of some fish species. Some of these algae and cyanobacteria produce extracellular toxins, and under extreme conditions this has led to the precautionary closure of reservoirs to water sports and other public access. Compare ➤dystrophic; mesotrophic; oligotrophic.

eV Symbol for the unit ➤electronvolt.

evanescent wave A wave that is decaying in amplitude, having passed into a medium that attenuates it.

evaporation The process in which a liquid turns to a gas below its boiling point. Evaporation continues until the ➤saturated vapour pressure of the liquid is reached. Evaporation can be used to separate a solute from a volatile solvent. Crystals are often produced by evaporating water to form a hot concentrated solution; when the solution cools, crystals form.

even–even nucleus A nucleus with an even number of protons and an even number of neutrons.

even function A function f for which $f(-x) = f(x)$ for all values of x. Its graph is symmetrical about the y axis.

even number An ➤integer that on division by 2 gives another integer.

even–odd nucleus A nucleus with an even number of protons and an odd number of neutrons.

event In relativity, a point in ➤spacetime, usually a point at which an occurrence of some sort takes place. For example, the decay of a neutron is an event, though observers in different frames of reference may (according to ➤special relativity) disagree on its time or location. ➤➤world line.

event horizon The boundary of a ➤black hole with the rest of the Universe, within which events are unable to have a causal effect on events outside the black hole. The event horizon is where the ➤escape velocity becomes equal to the speed of light.

evergreen Describing plants that retain their leaves throughout the winter in temperate regions. Most conifers are evergreen (exceptions include the larch), but many flowering plants such as holly also fall into this category. Compare ➤deciduous.

evolution The process by which living organisms have developed from earlier ancestral forms. Evidence for evolution comes indirectly from studies of fossil records, comparative anatomy, biochemistry and physiology, and directly by various experiments observing the behaviour and survival of organisms in their natural environments. Various theories of evolution have been proposed, the most widely accepted mechanism being natural selection (➤Darwinism). Darwin's explanation of evolution remains the central theme of current evolutionary thinking. It has now been assimilated with modern genetics to form what is referred to as **neo-Darwinism**, in which evolution may be defined as the change in allele frequency in a population over time. **Speciation**, the establishment of new species, arises when such frequency changes result in the genetic difference between populations being sufficiently significant to prevent interbreeding between them. Two main additional lines of thinking have recently been applied to the evolutionary debate. The first of these seeks to explain the major breaks in the fossil record, such as that at the end of the Cretaceous period which signalled the end of the dinosaurs, by suggesting that long periods of stability have been interrupted by relatively short bursts of rapid evolutionary change. Such change is a response to rapid environmental change (such as a drying of world climate) and is referred to as **punctuated equilibrium.** The second reinforces the basic idea of natural selection, and sees the evolutionary process as being driven by the accumulation of random mutations which produce small but significant changes in an organism's ≻phenotype; this has been tested by mathematical and computer models. Evolution is a major unifying concept in virtually all areas of biology. ≫ethology.

exa- Symbol E. The SI prefix for 10^{18} base units. It is an uncommon prefix because the numbers involved are so large: the age of the Universe is of the order of an exasecond.

EXAFS Abbr. for extended X-ray absorption fine structure (≻X-ray absorption spectroscopy).

excess electron An electron in a ≻semiconductor that has come from ionization of an impurity rather than the pure semiconductor.

exchange interaction A purely quantum-mechanical effect that causes an effective force between ≻identical particles, such as electrons, depending on their ≻spin. Two electrons in the same atom must have a wavefunction that is antisymmetric overall under the ≻exchange operator. If their spins are parallel, the spatial part of the wavefunction is antisymmetric under the exchange operator. If their spins are antiparallel, the spatial part is symmetric and this state has a higher energy than the antisymmetric spatial wavefunction associated with parallel spin.

exchange operator A quantum-mechanical operator that exchanges identical particles. It has only two possible ≻eigenvalues, +1 and −1, called **even** and **odd**.

excimer An excited combination of two atoms. Excimers are used in some of the most powerful laser systems currently available, such as the KrF excimer laser, which radiates at 249 nm.

excitation A change of state of a quantum system from a lower energy state (usually the ≻ground state) to a higher energy state. The change is caused by an externally imposed perturbation, such as an oscillating electromagnetic field (perhaps in the form of light).

excited state A state, usually electronic, that is of higher energy than the ≻ground state.

exciton A bound electron state in a ≻semiconductor based around a ≻defect in the crystal structure.

exclusion principle ➤Pauli principle.

excretion The expulsion of the waste materials of metabolic processes from the body of an organism. Waste materials in animals include nitrogenous compounds such as ammonia, urea and uric acid from ➤protein metabolism, and carbon dioxide from respiration. Overall, green plants excrete oxygen as a waste product of photosynthesis, yet reabsorb a proportion of this in their respiration. The voiding of faeces from the gut is strictly not excretion, since faeces are largely indigestible material which has not been generated by metabolic activities in cells. However, the breakdown products of red blood cell metabolism are excreted in bile, which drains into the intestine and is carried out with the faeces.

exergonic process A process during which the ➤Gibbs energy of a system decreases. It is the Gibbs energy counterpart of an ➤exothermic process. Compare ➤endergonic process.

exitance (emittance) Symbol M. The flux (either ➤radiant flux or ➤luminous flux) emitted per unit area of a surface.

exocrine gland Any gland in an animal body that delivers its secretion to its place of function via a duct. Exocrine glands include most of those that produce digestive juices, and the sebaceous and sweat glands in the skin. Compare ➤endocrine gland.

exocytosis The process by which the membrane of an intracellular vesicle fuses with the ➤plasma membrane, thereby removing its contents from the cell into the extracellular fluid. It is by this means, for example, that secretory cells such as those producing digestive juices in the mammalian gut deliver their products to where they are used.

exoergic process A nuclear process that results in a net outflow of energy.

exogamy ➤outbreeding.

exogenous Describing a substance that is applied to an organism or culture of cells from external sources. Compare ➤endogenous.

exon A ➤DNA sequence that encodes the information to direct ➤protein synthesis. In the late 1970s it became apparent that in a ➤eukaryotic cell the DNA does not consist of single contiguous lengths of the molecule that code for proteins, as is the case in bacteria. In eukaryotes the coding regions are discontinuous and interrupted by sections of noncoding DNA called ➤introns.

exonuclease An ➤enzyme that hydrolyses DNA by removing individual nucleotides from the ends of DNA strands. Compare ➤endonuclease; ➤➤restriction enzyme.

exoskeleton Any outer layer of an organism that provides support and a framework for the attachment of muscles. The most familiar examples are in the ➤arthropods: the exoskeletons of, for example, insects are composed of ➤chitin, and those of crustaceans such as crayfish are typically impregnated with calcium salts to produce a 'shell'. Compare ➤endoskeleton.

exosphere The outer layer of the Earth's atmosphere, from about 400 km altitude upwards.

exothermic Describing a process during which heat is given out to the surroundings. This is in contrast to ➤endothermic processes, during which heat is taken in. Exothermic processes are much more common than endothermic ones as energy is more widely dispersed as a result. Examples of exothermic processes are neutralization reactions and, most importantly, combustion reactions. The economic importance of combustion reactions cannot be overstressed as they power our transport, heat our homes and supply most of our electricity. ➤thermodynamics.

expanded polystyrene The polymer ➤polystyrene manufactured as a foam. It is expanded by impregnating it with pentane and then heating it in steam.

expansion of the Universe The observation that galaxies beyond the ➤local group are receding from us. This is deduced from the ➤redshift in their spectra, which is interpreted as being caused by the ➤Doppler effect. ➤Hubble constant.

expansivity (coefficient of expansion) The ratio of the change of size of an object to its original size per unit temperature rise. The object's 'size' can be its length (**linear expansivity**), its area (**superficial expansivity**) or its volume (**volume expansivity**).

expectation (mean) For a ➤discrete random variable X, the sum $E(X) = \Sigma_x x P(X = x)$, where $P(X = x)$ is the probability that X has the value x and the sum is over all such values. For a continuous random variable, with ➤probability density function f, it is given by the integral $\int x f(x)\, dx$.

expert system A computer system that can provide expertise in some specialized field of knowledge. The knowledge of human specialists is stored in the computer system, including imprecise and probabilistic reasoning: this is the knowledge base of the expert system. It also has an inference engine, which allows it to follow lines of reasoning and request information in order to deduce the likely situation from the information presented to it. An explanation program is usually included, so that the user can ask for justification for the system's questions and reasoning, and a language processor allows the user to converse in natural language. One expert system has been designed to help with the determination of chemical structures, and another can give diagnostic advice on human diseases.

explant A piece of tissue removed from an organism to initiate a culture or to transplant as a ➤graft.

explicit equation An equation of a curve of the form $y = f(x)$, where the dependent variable y occurs only once. Compare ➤implicit equation.

explosion A very rapid release of energy and/or matter. A ➤chain reaction such as that between hydrogen and oxygen illustrates both modes of explosion. First, a chain-branching step produces two reactive ➤radicals from one, which then each produce two more, and so on; a rapid escalation in the number of chain carriers causes an explosively quick reaction. Second, a thermal explosion can result if so much energy is released in a small volume that the molecules cannot spread the energy fast enough. Compare ➤implosion.

exponent Another name for ➤index.

exponential curve A curve with equation $y = Ae^{kx}$ for constants A and k. Also known as a growth curve or decay curve according to whether $k > 0$ or $k < 0$ respectively. **Exponential growth** applies to a pattern of cell or population growth in which the rate of increase is proportional to the number of individuals and their reproductive rate. Theoretically all populations are capable of exponential growth and an exponential phase occurs in the early stages of the growth of, for example, cells in tissue culture or bacteria growing on a nutrient medium. In practice the exponential phase is eventually moderated by natural control factors such as space, the availability of food and the accumulation of waste in the local environment. This levelling-off results in a characteristic S-shaped **sigmoid growth curve**.

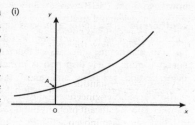

(i)

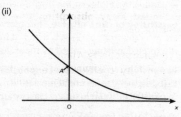

(ii)

exponential curve The graph $y = Ae^{kx}$ with (i) $k > 0$ (exponential growth) and (ii) $k < 0$ (exponential decay).

exponential decay A decay process of a quantity N, with respect to time t, of the form $\mathrm{d}N/\mathrm{d}t = -\lambda N$, where λ is known as the **decay constant** or **disintegration constant**. The solution to this first-order differential equation is $N = N_0\, e^{-\lambda t}$, where N_0 is the value of N at time $t = 0$. In the context of radioactivity, the ➤half-life of the parent nuclide is related to the decay constant by $\lambda = \ln 2/t_{\frac{1}{2}}$.

exponential function The real or complex function denoted by e or exp. It is defined by the **exponential series**

$$\sum_{n=0}^{\infty} x^n/n!$$

and is equal to its own derivative. Because $y = Ae^{kx}$ satisfies the differential equation $\mathrm{d}y/\mathrm{d}x = ky$, exponential functions occur in many growth and decay situations. **Euler's formula** $e^{ix} = \cos x + i \sin x$ links the exponential and trigonometric functions. The inverse of e is the natural logarithm function, ln. Since $a^x = e^{x\ln a}$, a general exponential quantity a^x can always be expressed in terms of e^x.

exposure meter (light meter) A device for measuring the intensity of light for the purpose of deciding what combinations of aperture and shutter speed will give a correctly exposed photograph.

expression Anything expressed in algebraic form, as in an equation or formula.

expression vector A ➤plasmid or other recombinant DNA ➤vector (4) that permits a gene to be expressed when inserted into a host cell. Many important therapeutic proteins have been produced by the use of expression vectors. For example, human growth hormone (HGH) was formerly available only from the pituitary glands of cadavers, and demand greatly exceeded supply. Once the gene for HGH and its cDNA had been cloned, it was possible to insert it into a plasmid containing a ➤promoter (2) which would allow the inserted gene to be expressed in the bacterium ➤*E. coli.* Such *E. coli* cells can then replicate and can be cultured in large vessels, yielding cells which are capable of synthesizing HGH, now being produced relatively inexpensively in quantities to satisfy demand. Such techniques were early targets of the ➤genetic engineering industry. A wide variety of other therapeutic and commercially significant proteins can now be manufactured in a similar way. These include tissue plasminogen activator for heart attack treatment, erythropoietin for anaemia and ➤insulin for diabetes.

extended X-ray absorption fine structure (EXAFS) ➤X-ray absorption spectroscopy.

extensive property A property of a thermodynamic system that scales as the size of the system. Thus volume is an extensive property, while pressure is not. If the system is doubled, the volume is doubled while the pressure remains the same (pressure is an ➤intensive property).

extensometer A device for measuring the increase in length (the **extension**) of a body when it is stretched.

extinction The termination of the existence of a species or population. Extinction is an inevitable function of the evolutionary process in which new species arise and replace previously existing ones. Whilst species extinction is an inexorable process, there have been five 'mass extinction' events in the geological record characterized by relatively rapid extinction of animal groups at higher taxonomic levels, for example families. Such extinction events are thought to be associated with rapid environmental change such as shifts in the world's climate, sea level change, periods of intense volcanic activity or impact by meteors. The most famous of these occurred at the **Cretaceous–Tertiary (K/T)** boundary which brought about the demise of the dinosaurs. There appears to be no evidence for associated mass extinctions of plants. Currently the rate of extinction due to human-induced environmental change is approaching that seen in the 'big five', leading to the notion that we are now approaching the 'sixth extinction'.

extinction coefficient ➤Beer's law.

extracellular fluid (ECF) The fluid that bathes the internal cells of a multicellular organism. It consists of water with a wide variety of dissolved substances including oxygen and carbon dioxide, various mineral ions, foods and vitamins, hormones and wastes. An important characteristic of the ECF in mammals is that its composition, temperature and pH remain remarkably constant, despite fluctuating external conditions, due to ➤homeostasis.

extraction ➤solvent extraction.

extraordinary ray A ray vector in an ➤anisotropic optical medium that is not parallel to the corresponding ➤wave-vector. ➤birefringence; compare ➤ordinary ray.

extrapolation The prediction of the value of a function at a point outside the range of some known values of the function. It is a technique which must be used with considerable care when applied to real data since it can lead to unwarranted predictions.

extrinsic semiconductor A ➤semiconductor that has been doped (➤doping). A large proportion of the carriers in an extrinsic semiconductor come from impurities. Compare ➤intrinsic semiconductor.

eye An animal sense organ responsible for the reception of light. Eyes vary considerably in structure and complexity. Many single-celled organisms such as *Euglena* have a light-sensitive 'eye-spot'. More advanced eyes exist in the ➤arthropods, which have simple ocelli and multi-faceted compound eyes made up of individual units called **ommatidia**. The most sophisticated eyes are found in the cephalopod ➤molluscs and ➤vertebrates. In these groups, eyes are paired and often provide binocular vision facilitating the judgement of distance, which is particularly important for predators. Vertebrate eyes are more or less spherical, with the shape maintained by a tough outer wall (the **sclera**), under which is the **choroid**, a blood-rich nutritive layer, and inner cavities filled with liquids: the **aqueous humour** between the cornea and the lens, and the jelly-like **vitreous humour** between the lens and the retina (see the diagram). Light rays pass into the eye through the **cornea** and **pupil**, and are refracted to bring them into focus. The cornea is protected and maintained by an outer membrane, the **conjunctiva**, and is responsible for about 70% of the refraction required for focus. Fine focusing is

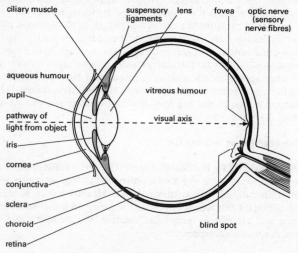

eye Vertical section through the human eye.

achieved by the **lens**, the focal length of which can be adjusted by the action of the **ciliary muscles**, which adjust the tension in the **suspensory ligaments**. Light rays are brought to focus on the **retina**, which contains light-sensitive cells of two types. **Rods** are sensitive to low light intensities but do not respond to colour; **cones** are receptive to higher light intensities and are concentrated in the **fovea** ('yellow spot'). Different cones are sensitive to different wavelengths of light and allow for ➤colour vision. Rods and cones are sensitized by light to generate impulses which pass along the **optic nerve** to the brain, where they are interpreted as images in the visual cortex. On the retina is a **blind spot**, where blood vessels and nerve fibres emerge to form the optic nerve. The **lachrymal gland** is a gland, associated with the eye, which produces tears.

eyepiece In an optical instrument, the lens or combination of lenses closest to the observer. Its arrangement is usually with the image from the objective in its focal plane so the observed virtual image appears at infinity.

F

f Symbol for the prefix ➤femto-.

ƒ Symbol for ➤frequency.

F 1 Symbol for the element ➤fluorine.
 2 Symbol for the unit ➤farad.

°F Symbol for degree Fahrenheit (➤Fahrenheit scale).

F Symbol for the ➤Faraday constant.

F Symbol for ➤force.

F1 (F2) generation The first (second) filial generation: the offspring in the first (second) generation of a ➤cross between selected parents in a breeding investigation.

face One of the flat surfaces of a ➤polyhedron.

face-centred cubic (f.c.c., cubic close packing) One of the two close-packed structures (➤close packing). It is adopted by many elements, such as aluminium, copper, silver and gold.

facsimile ➤fax.

factor An ➤integer or ➤polynomial that divides without remainder into another integer or polynomial: thus 7 is a factor of 21 because $21 = 3 \times 7$ and $x - 1$ is a factor of $x^3 - 7x + 6$ because $x^3 - 7x + 6 = (x-1)(x^2 + x - 6)$. Finding all the factors of a polynomial depends on what coefficients are allowed; thus $x^4 + 1$ has no factors with ➤rational coefficients but factorizes as $(x^2 - \sqrt{2}x + 1)(x^2 + \sqrt{2}x + 1)$ with ➤real coefficients.

factor VIII One of a number of blood clotting factors that function to seal wounds when blood vessels are damaged. Factor VIII specifically initiates the sequence of reactions by which prothrombin is converted to thrombin, causing a blood clot to form. Individuals who are unable to make sufficient factor VIII suffer from the genetically determined condition ➤haemophilia.

factor formulae Formulae such as:

$$\sin A + \sin B = 2 \sin \tfrac{1}{2}(A + B) \cos \tfrac{1}{2}(A - B)$$

and

$$\cos A - \cos B = -2 \sin \tfrac{1}{2}(A + B) \sin \tfrac{1}{2}(A - B),$$

which relate sums and differences of ➤trigonometric functions to products of such functions.

factorial If n is a whole number, then $n!$ (read as 'n factorial') is the product $1 \times 2 \times \ldots \times n$; for example $4! = 24$. By convention, $0!$ is taken to be 1.

faculae Bright areas of the Sun's photosphere, often associated with sunspots.

FAD Abbr. for flavin adenine dinucleotide. A ➤coenzyme forming the ➤prosthetic group of enzymes involved in ➤redox reactions in cells such as occur during ➤respiration. It acts as an acceptor for hydrogen (FAD is the oxidized form, and $FADH_2$ the reduced form) which it can then pass on to a series of reactions in which ➤ATP is generated. ➤➤electron-transport chain.

Fahrenheit scale A scale of temperature that takes as its zero the lowest temperature Gabriel Daniel Fahrenheit (1686–1736) could reach, the freezing point of a ➤eutectic mixture of salt and water. On this scale the melting and boiling points of pure water are $32\,°F$ and $212\,°F$ respectively.

fail-safe Designed in such a way that, in the event of a component failing, no danger results from the failure.

Fajans' rules Rules that allow qualitative estimates of the extent of covalent bonding present in formally ionic compounds. For the cation, covalent character increases with increasing ➤charge density. For the anion, covalent character increases with increased ➤polarizability, which increases with the number of electrons. So for an ➤isoelectronic series such as NaF, MgO, AlN, SiC, the covalent character increases to the right; by AlN the ionic model provides a very poor description of the bonding. Named after Kasimir Fajans (1887–1975).

Fallopian tube ➤oviduct. Named after Gabriele Fallopio (1523–62).

fall-out The radioactive debris from a nuclear explosion. Among the most dangerous fall-out radionuclides are strontium-90 and ➤iodine-131.

family A unit of classification (➤taxonomy) of organisms that contains one or more closely related genera. In turn, families are grouped together in orders.

farad Symbol F. The SI unit of ➤capacitance. A 1 farad capacitor stores a charge of 1 coulomb when the potential difference is 1 volt. For practical purposes it is an inconveniently large unit, and capacitance is often measured in microfarads, or even picofarads. Named after Michael Faraday (1791–1867).

Faraday cage A closed conducting surface, for example a wire mesh, surrounding electrical equipment (or possibly people) which ensures that no electric field exists inside it due to the influence of sources outside it. The conducting enclosure is at the same ➤potential all over its surface, so the electric field inside it is zero.

Faraday constant The charge carried per mole of electrons: $F = eL = 9.649 \times 10^4$ C mol^{-1}. ➤Faraday's laws; ➤➤Appendix table 2.

Faraday (electro-optical) effect The rotation of the plane of polarization of an electromagnetic wave travelling through a material in the presence of a magnetic field. Such an effect is observed with visible light in certain flint glasses and in quartz, as well as in ferrites with microwaves. ►Kerr effect.

Faraday's law (of induction) The ►electromotive force induced in a circuit is proportional to the rate of change of ►magnetic flux through the circuit. This is a direct consequence of the third of ►Maxwell's equations.

Faraday's laws (of electrolysis) Faraday's first law is that the mass of a substance liberated at an electrode is proportional to the charge passed. The constant of proportionality Z in the equation linking the mass m liberated during electrolysis to the charge passed (equal to the product of the current I and the time t), is called the **electrochemical equivalent**: $m = ZIt$.

Faraday's second law is that, to liberate one mole of an element, a whole number of moles of electrons must be passed (►Faraday constant).

far-field diffraction ►Fraunhofer diffraction.

fast breeder reactor ►fast reactor.

fast fission Nuclear fission induced by neutrons with an energy exceeding the threshold for ^{238}U (about 1.5 MeV). ►fast reactor.

fast Fourier transform (FFT) A technique used, for example, in spectroscopy, for the rapid computation of a ►Fourier transform. The number of calculations required by the algorithm for a sample of N points is proportional, not to N^2, as would be expected by a simple application of the formula for the Fourier transform, but to $N \ln N$.

fast neutron A neutron with a kinetic energy exceeding about 0.1 MeV.

fast reactor (fast breeder reactor) A nuclear reactor without a ►moderator, in which ►nuclear fission is induced by a ►fast neutron rather than a thermal neutron. The fuel is about 20% plutonium (^{239}Pu), and more plutonium is generated in the reactor (hence '**breeder**'). A fast reactor can use a much greater proportion (typically 75%) of the energy in its fuel than can a thermal reactor (1%).

fat ►lipid.

fatigue (of metals) ►metal fatigue.

fatty acid An aliphatic ►carboxylic acid with a hydrocarbon chain of varying length joined to a terminal carboxyl (—COOH) group. Fatty acids are usually unbranched and those with even numbers of carbon atoms between 14 and 22 carbons long react with glycerol to form a ►lipid. The **saturated fatty acids** have no double bonds and include stearic acid and palmitic acid. The **unsaturated fatty acids** have one double bond and include oleic acid while **polyunsaturated fatty acids** include linoleic acid and linolenic acid. The oxidation of fatty acids, by a process known as **beta-oxidation** that removes successive pairs of carbon atoms from the carboxylic acid chain to release energy for the cell, releases approximately twice as much energy per unit mass as the oxidation of carbohydrate.

fauna The collective term for animal life existing in a particular habitat, region or time. Compare ►flora.

fax (facsimile) A method of transmitting a ➤bitmap, usually over telephone lines. A fax machine consists of a scanner to turn documents on paper into bitmap images, a ➤modem to transmit the encoded data, and a printer to create a hard copy of any bitmaps received.

f-block elements The elements in which the highest-energy orbital being filled is an f orbital. The main subdivision is into the ➤lanthanides and the ➤actinides.

f.c.c. Abbr. for ➤face-centred cubic.

Fe Symbol for the element ➤iron (from the Latin *ferrum*).

feedback A basic method of control, employed in many areas, particularly electronics, in which a function of the output of the device being controlled is used as an input to the controlling system. An amplifier using **positive feedback** adds a portion of the output to the input stage, creating the possibility of instability. An amplifier using **negative feedback** subtracts a portion of the output from the input stage to make the system ➤stable. Negative feedback also arises in populations of organisms: more predators will reduce the number of prey, but this limits the number of predators that can be supported. In many biological systems the production of a ➤hormone often stimulates a process that in turn inhibits the production of the hormone itself, thus finely regulating the amount of hormone present. The control of blood sugar concentration is a good example (➤diabetes). The biosynthesis of the amino acid isoleucine in bacteria, for example, demonstrates **feedback inhibition**, since when the concentration of isoleucine reaches a critical level it binds to the regulatory site on the enzyme involved, thereby halting the enzyme's function.

Fehling's solution A solution used to distinguish between aldehydes and ketones and to identify reducing sugars. It is made by adding copper(II) sulfate to an alkaline solution containing ➤Rochelle salt. The blue complexed copper(II) ion formed can be reduced to a brick-red precipitate of copper(I) oxide, Cu_2O, by aldehydes and reducing sugars. Named after Hermann Christian von Fehling (1812–85). ➤➤Benedict's solution.

feldspars Complex aluminosilicates, which are the most common constituents of igneous rocks. The alkali feldspars such as orthoclase, $KAlSi_3O_8$, have potassium as the dominant counter-ion whereas in plagioclase feldspars such as anorthite, $CaAl_2Si_2O_8$, calcium and/or sodium dominate.

FEM Abbr. for field-emission microscope (➤field emission).

femto- The SI prefix meaning 10^{-15} base units; for example, 1 fs is equivalent to 10^{-15} s. Protons and neutrons are approximately 1 fm in diameter.

Fenton's reagent A mixture of hydrogen peroxide and iron(II) sulfate used in organic synthesis, for example to introduce hydroxyl groups into aromatic rings. Named after Henry John Horstman Fenton (1854–1929).

Fermat's last theorem The assertion that the equation $x^n + y^n = z^n$ has no solutions with n, x, y, z whole numbers and $n > 2$. It was conjectured by Pierre de Fermat (1601–65) and was one of the most famous unproved statements in mathematics until its proof by Andrew Wiles in 1994.

Fermat's principle of least time The path taken by light passing through an optical system minimizes the time of travel of the beam. It is by virtue of Fermat's principle that geometrical optics is based on the paths of rays; without it, a full analysis of the wave optics would be required for every case.

fermentation A series of enzyme-catalysed reactions occurring under ►anaerobic conditions in certain cells (particularly yeasts) in which organic compounds such as ►glucose are converted into simpler substances, such as ►carbon dioxide and ►ethanol, with the release of energy. Fermentation is involved in bread-making where the carbon dioxide produced by the yeast causes dough to rise. **Malolactic fermentation**, the conversion of malic acid to lactic acid, is also relevant to winemaking. ►►glycolysis.

fermenter (bioreactor) A stainless steel or glass container in which large numbers of ►microorganisms can be grown under carefully controlled conditions in artificial culture. Fermenters are widely used in ►biotechnology to produce large colonies of organisms yielding substances of commercial interest, such as antibiotics.

Fermi–Dirac statistics The name given to the way in which ►fermions fill energy levels at nonzero temperature, T. The probability that a state with energy E_i is occupied by a fermion is given by

$$p_i = [\exp(E_i - \mu)/kT + 1]^{-1},$$

where μ is the chemical potential of the system (which can be determined by ►normalizing the probability distribution for all states) and k is the ►Boltzmann constant. Named after Enrico Fermi (1901–54) and Paul Adrien Maurice Dirac (1902–84).

Fermi energy Symbol E_F. In a solid state system (particularly a metal), the energy of the highest occupied ►energy level, the **Fermi level**, at a temperature of absolute zero. If the energies of the states of the system are plotted against the three-dimensional ►crystal momentum corresponding to each state, the contour corresponding to the Fermi energy is known as the **Fermi surface**. For an alkali metal (such as sodium) with a simple ►band structure, the electrons behave almost as ►free electrons, and the Fermi surface is spherical (a **Fermi sphere**). The Fermi energy of sodium is approximately 5 eV (with respect to the bottom of the conduction band). At nonzero temperatures, ►Fermi–Dirac statistics ensure that some states above the Fermi level will be occupied, and some below the Fermi level will be empty.

Fermi gas The name given to a hypothetical system of free ►fermions in three dimensions. The sea of electrons in the conduction band of an alkali metal such as sodium is, to a good approximation, a Fermi gas.

fermion In quantum mechanics, a particle of half-integral ►spin. A system of two identical fermions changes the sign of its ►wavefunction under the operation of exchange of particles (►identical particles). As a consequence, two fermions cannot occupy the same quantum state, a feature known as the ►Pauli principle. Electrons, protons and neutrons are all fermions, with spin $\frac{1}{2}$. Named after E. Fermi. Compare ►boson.

Fermi surface ►Fermi energy.

fermium Symbol Fm. The element with atomic number 100 and most stable isotope 257, which is one of the transuranium elements. It is produced by neutron irradiation of californium and has no significant uses. Named after E. Fermi.

ferrate The oxoanion FeO_4^{2-}, in which iron has oxidation number +6. The ferrate ion is purple in aqueous solution and is stable only in strongly alkaline conditions.

ferric Traditional name identifying the higher common oxidation number of iron, now designated by iron(III).

ferric alum $K_2SO_4 \cdot Fe_2(SO_4)_3 \cdot 24H_2O$ An ➤alum containing Fe^{3+} ions in place of Al^{3+}.

ferricyanide ion Traditional name for the ➤hexacyanoferrate(III) ion.

ferrimagnetism A type of magnetic ordering, similar in nature to ➤antiferromagnetism, in which magnetic moments on adjacent sites are aligned antiparallel, but have different magnitudes, thus giving an overall magnetic moment. ➤Ferrites (from which the name of the effect is derived) exhibit ferrimagnetism. Compare ➤ferromagnetism.

ferrite A mixed oxide formed by fusion of, for example, calcium carbonate with iron(III) oxide to give $CaFe_2O_4$. They are used to construct radio aerials and were used for core memory in some early computers.

ferro- Prefix indicating an alloy (or compound) of iron or steel. Common examples of alloys are ferrovanadium, ferrochromium and ferromanganese.

ferrocene The archetypal ➤sandwich compound in which an iron ion is sandwiched between two cyclopentadienyl ligands, as shown in the diagram.

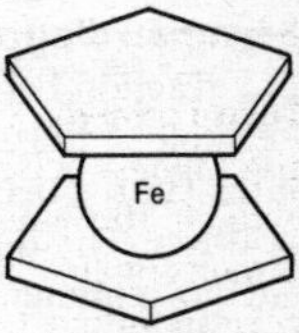

ferrocyanide ion Traditional name for the ➤hexacyanoferrate(II) ion.

ferroelectric A material (such as $BaTiO_3$) that exhibits behaviour analogous to ➤ferromagnetism, but with electric dipole moments instead of magnetic dipole moments.

ferrocene

ferromagnetism The phenomenon exhibited by some materials, for example iron (after which it is named), which gives the material a large magnetic susceptibility, as well as allowing it to be permanently magnetized through the mechanism of ➤hysteresis. A **ferromagnetic** state is composed of ➤domains, which act like tiny permanent magnets that can be individually aligned. In the presence of an external magnetic field the individual magnets all align parallel to the field, giving a high value of the ➤magnetic flux density within the material, and thus a high ➤magnetic permeability. Above a critical temperature, the **Curie temperature**, the spontaneous alignment of magnetic moments is no longer favourable, and the ferromagnetic material instead exhibits ➤paramagnetism. ➤➤antiferromagnetism; Curie–Weiss law; ferrimagnetism.

ferrous Traditional name identifying the lower common oxidation number of iron, now designated by iron(II).

ferroxyl indicator An indicator used to identify corrosion by rusting. It contains hexacyanoferrate(III) ions used to identify iron(II) ions (hence *ferr*-), a blue colour indicating the region around the ➤anode, and phenolphthalein used to identify hydr*oxyl* ions, a pink colour indicating the region around the ➤cathode.

fertile Describing a nuclide that can be transformed into a ➤fissile nuclide in a series of nuclear reactions. For example ^{238}U is fertile, but not itself fissile; in a nuclear reactor it can capture a neutron to become ^{239}U, which undergoes ➤beta decay over a period of days via ^{239}Np to ^{239}Pu, which is fissile.

fertilization (syngamy) The process in the life cycle of a sexually reproducing organism in which two ➤gametes fuse to form a ➤zygote.

fertilizer A substance added to the soil to increase the yield of crops. The three essential elements needed for plant growth are nitrogen, phosphorus and potassium, provided by 'NPK' fertilizers. The most important sources of each element are ammonium nitrate, ammonium sulfate and urea (for nitrogen), superphosphate of lime (for phosphorus) and potassium chloride (for potassium). Organic fertilizers include animal dung. While the benefits of fertilizers are enormous, enabling significant improvement in the overall standard of living for whole countries, care must be exercised in their application (➤eutrophic).

FET Abbr. for ➤field-effect transistor.

fetus The US spelling of ➤foetus. (It is gaining currency in the UK.)

Feynman diagram A schematic representation of a reaction between elementary particles. For example, the diagram here shows the interaction between an electron and a positron. The diagrams were introduced as an aid to enumerating terms in the detailed calculations of interactions. Named after Richard Phillips Feynman (1918–88).

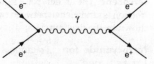

Feynman diagram for one mechanism of electron–positron scattering.

FFT Abbr. for ➤fast Fourier transform.

Fibonacci number A member of the sequence 0, 1, 1, 2, 3, 5, 8, . . . in which every term is the sum of the previous two. The ratio of successive terms tends to the **golden ratio**, $\frac{1}{2}(1 + \sqrt{5})$. Named after Leonardo Fibonacci (*c.* 1170–*c.* 1250).

fibre 1 A material that exists in long thin strands. A biological example is muscle fibre. Important organic chemical examples are natural polymers such as cotton, and artificial polymers such as ➤polyesters, ➤nylon and ➤Kevlar. Inorganic chemical examples are glass fibre and ➤asbestos.

 2 A lignified cell with support and protection functions in the stems of flowering plants and conifers.

 3 Material of plant origin, comprising mostly ➤cellulose, which cannot be digested by humans. **Dietary fibre (roughage)** has an important function in stimulating the mechanical action of the gut.

fibre, optical A long filament of glass (typically 10 μm in diameter) used primarily for digital telecommunications. An optical fibre may be thought of as a 'light pipe', carrying whatever signal is introduced at one end to the other end, regardless of the curves in the fibre. The signal is carried by light pulses which are transmitted with very little loss along the fibre. The ➤bandwidth of optical fibre cabling can be very much greater than the equivalent size of electrical conductor. Optical fibres are used in medical instruments inserted into the body to obtain a view of internal structures.

fibrin An insoluble fibrous protein that is precipitated at the site of wounds to form ➤blood clots. Clots reduce blood loss and seal off the blood vessel to prevent ➤infection.

fibrinogen The soluble precursor form of ➤fibrin, into which it is converted by thrombin.

fibula ➤tibia.

Fick's laws of diffusion Laws governing ➤diffusion in gases and solutions. Fick's first law states that the ➤flux of matter J_z in the direction z is proportional to the concentration gradient dc/dz in that direction, the coefficient of proportionality being the **diffusion coefficient** D:

$$J_z = -Ddc/dz.$$

Fick's second law is now called the ➤diffusion equation. Named after Adolph Eugen Fick (1829–1901).

fictitious force A force introduced into a frame of reference to compensate for the acceleration of the frame relative to an ➤inertial frame. ➤Newton's laws of motion hold only in an inertial frame, but in any case it is often more convenient to consider motion in an accelerating frame of reference (such as a frame rotating with the Earth) by including fictitious forces such as ➤centrifugal force and ➤Coriolis force. In an inertial frame S, Newton's second law of motion can be written as $F = m(a + A)$, where a is the acceleration of the body in the accelerating frame S´, and A is the acceleration of S´ with respect to the inertial frame S. The equation can be rearranged to give $F - mA = ma$, where $-mA$ now is a fictitious force that acts on every body within the frame. These two equations are entirely equivalent algebraically, but the concept of adding forces together to find a resultant acceleration is more natural than splitting the resultant acceleration into two parts (a and A). Note that the choice is one frame or the other, and that one may not include, for example, both a fictitious centrifugal force and a centripetal acceleration. In particular, if the rotation is simple and there is no acceleration in the frame S´ ($a = 0$), there is little point in choosing that frame; it has therefore become standard practice to solve all such problems in an inertial frame (using, for example, a centripetal acceleration rather than a centrifugal force). However, fictitious forces do not have a corresponding equal and opposite force as required by Newton's third law.

field A physical quantity whose value is a function of spatial position. The quantities that are most frequently referred to as fields – ➤electric field, ➤magnetic field and ➤gravitational field – are ➤**vector fields**, where the quantity at a specified point is a vector. A vector field may be represented as continuous ➤lines of flux, provided it has zero ➤divergence; this is always true for magnetic fields (see the diagram overleaf),

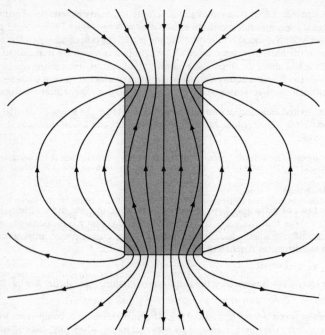

field The magnetic field of a bar magnet. (By convention, the arrows on magnetic field lines indicate the direction in which a north pole would move at particular points in the field. This is why, in this diagram, the field lines apparently reverse direction at the edges of the magnet.)

and is true for electric fields in regions where no charge is present. A more general way of picturing a field is by drawing an arrow to represent its magnitude and direction at a number of sample points. ➤scalar field; vector analysis.

field-effect transistor (FET) A class of ➤transistors in which current flows from a **source** to a **drain** via a channel whose resistance can be controlled by applying a voltage to a ➤gate. In a junction gate FET, known as a **JUGFET** or **JFET** (diagram (a) opposite), this is achieved by surrounding the n-type channel with p-type material (or vice versa) connected to the gate; the width of the ➤depletion layer controls the resistance of the channel. In an insulated gate FET, known as an **IGFET** (diagram (b)), when a gate voltage above a threshold level is applied, the channel is formed in p-type material (with low doping concentration) between the highly-doped n-type source and drain regions. An IGFET is typically a metal oxide semiconductor device, and hence is also called a **MOSFET**.

field emission The emission of electrons from a solid in the presence of a large electric field at the surface. A **field-emission microscope** (FEM) utilizes this phenomenon to map the ➤work function of areas on the surface of a solid, using a fine metal tip at high negative potential.

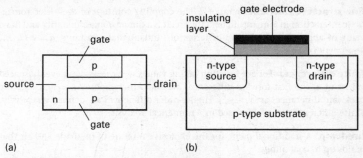

field-effect transistor (a) Junction gate (JUGFET) and (b) insulated gate (IGFET) field-effect transistors.

field ionization The ionization of gas molecules at a surface in the presence of a large electric field. A **field-ion microscope** (FIM) uses a fine metal tip held at high positive potential just above a surface to induce field ionization in low-pressure helium in the chamber of the microscope. This allows individual atoms of the surface to be detected as regions of increased field ionization.

file In computing, a block of data recorded on a **backing store**, such as a floppy disk or a hard disk. The file may be a program, or data for use by a program, such as text created on a text editor or word processor, or a spreadsheet, etc.

film badge ➤dosemeter.

filter 1 An electrical device designed to block signals in one or more chosen frequency bands, while transmitting signals in other bands. The simplest passive filters work on the principle that the ➤impedance of a capacitor is inversely proportional to frequency, while the impedance of an inductor is directly proportional to it. A radio receiver employs a filter for tuning.
2 A device, typically a sheet of glass or plastic, inserted into an optical system to change the relative intensities of different wavelengths of the light. It might, for example, transmit only red light.
3 ➤filtration.

filtrate The clear solution that has passed through after ➤filtration. The solid left behind is called the **residue**.

filtration The process by which a solid is separated out of a suspension, typically by passing it through a filter funnel with filter paper in it. The paper retains the solid while the solution passes through. This may be done under suction using a ➤Buchner funnel. Industrial filtration using large sand beds is an early stage in the recycling of water. ➤gel filtration.

FIM Abbr. for field-ion microscope (➤field ionization).

fine structure Structure observed in spectral lines in absorption or emission spectroscopy caused by the excitation of rotational and vibrational states. Compare ➤hyperfine structure.

fine-structure constant Symbol α. The coupling constant of the ➤electromagnetic interaction in ➤quantum field theory. It is a dimensionless quantity and has a value of approximately $1/137$. In terms of fundamental constants, $\alpha = e^2/2\varepsilon_0 hc$. ➤➤Appendix table 2.

finite differences For a ➤function taking values $y_0, y_1, \ldots, y_n$ at evenly spaced values of x, the first forward differences are the ➤increments $\Delta y_k = y_{k+1} - y_k$; the backward differences are $y_k - y_{k-1}$. Higher-order differences are defined recursively. Finite differences are widely used in ➤numerical analysis.

firedamp A traditional name for the explosive mixture of methane and air that builds up in coal mines.

fire extinguisher Any substance capable of putting out fires. The nature of the fire is very important. For example, although water is normally a very good extinguisher it increases the danger in the case of electrical fires. One way in which extinguishers work is by depriving the fire of oxygen: carbon dioxide is denser than air and does not support combustion, thus smothering the fire. They can also reduce the number of reactive intermediates in the flame, as is the case for ➤BCF.

Fischer projection A projection convention for depicting molecules whereby the bonds drawn horizontally are, in three dimensions, coming out of the page; the bonds drawn vertically lie in the plane of or behind the page (see the diagram). Molecules are intrinsically three-dimensional (as are molecular models) and hence their representation on a flat sheet of paper is not trivial. Named after Emil Hermann Fischer (1852–1919). ➤➤Newman projection.

Fischer projection (left) compared with a three-dimensional representation of the 2,3-dibromobutane molecule.

Fischer–Tropsch synthesis The industrial manufacture of methanol from ➤synthesis gas using a catalyst of dicobalt octacarbonyl, $Co_2(CO)_8$. This synthesis is likely to assume greater significance as oil reserves deplete. It was developed in the 1920s in Germany by Franz Josef Emil Fischer (1877–1947) and Hans Tropsch (1889–1935).

fish Aquatic vertebrates with typically scaly bodies and fins for propulsion and stability. There are two major groups of fish: the sharks and rays with cartilaginous skeletons (Chondrichthyes), and the bony fish (Osteichthyes), which include trout, cod and plaice.

fissile Describing a nuclide capable of undergoing ➤nuclear fission by interacting with a low-energy neutron. Some actinide nuclides with odd nucleon number tend to be fissile (^{233}U, ^{235}U, ^{239}Pu, ^{241}Pu). Compare ➤fissionable.

fission 1 A form of asexual reproduction, particularly in single-celled organisms, involving the division of the cell into two or more identical new cells called daughter cells.
 2 ➤nuclear fission.

fissionable Describing a nuclide capable of undergoing ➤nuclear fission by interacting with a neutron having enough energy to overcome the energy barrier for the fission of the compound nucleus. Some actinide nuclides with even nucleon number tend to be fissionable (^{232}Th, ^{238}U, ^{240}Pu, ^{242}Pu). Compare ➤fissile.

fitness The success of an organism in surviving to produce offspring. The fitness of organisms depends on their particular characteristics and the demands made on them by the environment. ➤evolution.

fixation ➤nitrogen cycle.

fixed stars Stars that appear to be at rest with respect to the ➤celestial sphere (those having no detectable ➤proper motion). The concept of fixed stars is confined almost entirely to their providing an ➤inertial frame against which other motion can be assessed, in circumstances where a frame moving with the Earth or the Sun is inappropriate. ➤➤Mach's principle.

flag 1 (Phys.) A single line in an electronic circuit which can be either on (high) or off (low).
  2 (Comput.) In ➤low-level languages, the abstraction of the physical flag.

flagellum (plural **flagella)** ➤undulipodium.

flame test A method for identifying certain elements, especially the ➤s-block elements, in which a sample of the solid being investigated is placed on a carefully cleaned wire and heated in a flame. The elements emit characteristic colours: lithium (crimson), sodium (yellow), potassium (lilac), rubidium (red), caesium (blue), calcium (brick-red), strontium (crimson) and barium (yellow-green). Many of these colours are familiar from fireworks, which rely on the same spectral transitions.

flammable ➤inflammable.

flare, solar A sudden burst of energy from the Sun's corona. Flares are believed to be triggered by discontinuities in the magnetic field associated with sunspots. Electrons are ejected upwards and downwards. The outbound electrons produce radio frequency radiation as they pass through the coronal gases, while the inbound ones heat the chromosphere below, causing X-ray emission and the optically visible flare.

flash photolysis An exceptionally important experimental method in chemical kinetics: it involves creating a species very rapidly using a **photoflash** and then observing its subsequent decay using a **specflash**. Most modern versions use lasers, the two flashes being created by splitting the laser beam and sending one of the split beams a short distance, the delay being typically nanoseconds.

flash point The lowest temperature above which a liquid will catch fire, measured under carefully controlled conditions.

flat As a concept for a two-dimensional surface, 'flat' has an obvious, everyday meaning, enshrined in the axioms of ➤Euclidean geometry (that parallel lines do not meet, etc.). However, in the context of four-dimensional spacetime, the idea of flatness has a similar, if generalized, meaning: that the ➤metric is the same as the simple metric of ➤Minkowski spacetime, and that the ideas of ➤special relativity are applicable. Compare ➤general relativity.

flavin Any derivative of vitamin B$_2$ (**riboflavin**) that functions as a ➤coenzyme in the ➤electron-transport chain. ➤FAD; flavoprotein; and see the table at ➤vitamin.

flavin adenine dinucleotide ➤FAD.

flavonoids A large group of phenolic compounds, including ➤anthocyanins, formed in plants. They form a wide spectrum of pigments ranging from colourless (white) through to reds and purples in petals, fruits and leaves. Flavonoids have a range of biological properties including antimicrobial, insecticidal and antioxidant, the latter making them a focus of interest in the treatment of heart disease.

flavoprotein One of a group of conjugated ➤proteins in which a ➤flavin is bound as a ➤prosthetic group. In particular, flavoproteins act as dehydrogenase enzymes in the ➤electron-transport chain.

flavour The distinguishing feature of the six different ➤quarks of the ➤standard model. The flavours are **up, down, strange, charm, bottom** and **top**.

F layer ➤Appleton layer.

flint An opaque form of ➤quartz, which is used in ceramics, glasses and road construction.

flip-flop (bistable) Describing an electronic circuit which may assume one of two possible states.

floating Of an electrical device or part of a circuit, not connected to ➤earth or any other source of ➤potential.

flocculation The process by which particles in a ➤colloid aggregate into clumps.

flop Abbr. of floating point operation. This corresponds to the processing by a computer of one arithmetic operation on real numbers (stored internally in a format known as '**floating point**'), as opposed to on integers. This leads to measures of processor speed in terms of **flops** (floating point operations per second). Supercomputers are typically capable of many gigaflops.

floppy disk ➤magnetic storage.

flora The collective term for plant life existing in a particular habitat, region or time. Compare ➤fauna.

flotation A common method of concentrating ores by grinding with a frothing agent, floating on water and agitating with compressed air.

flotation, principle of A special case of ➤Archimedes' principle: the mass of water displaced by a floating body is equal to the mass of the body itself.

flow chart A pictorial representation of the logical design of a computer program. Different-shaped boxes are used for different sorts of operations, such as decisions, processing operations and complex functions (see the diagram). For the details of complex programs, flow charts become quite cumbersome.

flower The organ, found exclusively in the ►Magnoliopsida, that contains the structures associated with sexual reproduction (see the diagram). Flowers typically consist of a number of concentric whorls comprising the **calyx** (sepals), the **corolla** (petals), the **androecium** (stamens) and the **gynoecium** (carpels). In some forms the calyx and corolla are indistinguishable from each other, and together form the **perianth**. Flowers are variously adapted to promote pollination (►pollen), either by insects or other animals, or by wind.

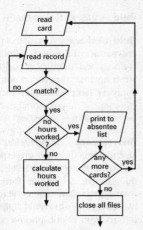

flow chart

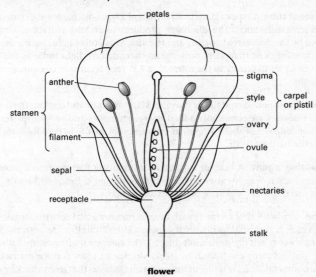

flower

flowering plant ►Magnoliopsida.

flowers of sulfur A fine yellow powder formed by condensing sulfur vapour.

fl oz Abbr. for the obsolete unit **fluid ounce**, still used in cookery: 1 fl oz = 28.41 cm^3.

fluctuations Changes, usually assumed to be random, in the value of a quantity with time.

fluid A state of matter that is able to flow – a liquid or gas.

fluid mechanics A branch of physics and engineering dealing with the ➤statics and ➤dynamics of a ➤fluid. A fluid in motion can be characterized by its velocity field (velocity as a function of position) but because of the complex nature of the forces affecting fluids (in general, forces of both compression and ➤viscosity) the result of applying basic principles such as Newton's second law is a set of non-linear equations. Computational methods therefore play a large part in fluid dynamics. ➤➤Navier–Stokes equation.

fluorescence The emission of light by a substance which is itself exposed to electromagnetic radiation, especially light. The incident light excites electrons to higher electronic energy levels, and the subsequent return to the ground state causes fluorescence. A common **fluorescent** substance is **fluorescein**, a dye showing an intense green fluorescence. Fluorescent 'whitening' agents added to detergents work by converting ultraviolet light incident on them into visible light. Fluorescence differs from ➤phosphorescence in that fluorescence ceases within about a microsecond of the stimulus ceasing, whereas phosphorescence can persist for minutes.

fluorescent tube A type of lamp that consists of a ➤gas-discharge tube coated with a fluorescent substance. The discharge produces ultraviolet radiation, which is absorbed by the fluorescent coating and re-radiated as visible light. Such a device is much more efficient than (and outlasts) an incandescent light bulb, as the latter radiates much of its energy in the infrared (a 7 W tube is considered equivalent to a 40 W bulb).

fluoride A compound containing fluorine. Many are ionic and contain the fluoride ion, F^-. Fluorides are commonly added to water supplies and/or toothpastes (at a recommended level of about 1 ppm) as conversion of apatite in dental enamel into fluorapatite hardens teeth.

fluorinating agent A reagent capable of introducing fluorine into a molecule. Examples include fluorine itself, hydrogen fluoride and chlorine trifluoride, ClF_3. The last can fluorinate asbestos, platinum and xenon.

fluorine Symbol F. The element with atomic number 9 and relative atomic mass 19.00, which is the ➤halogen with the lowest atomic number. As with the other halogens it exists as a diatomic molecule, F_2. The element itself, a nearly colourless gas first isolated using electrolysis by Henri Moissan in 1886, is the most violently reactive nonmetal, often reacting in the cold with elements that resist attack by other halogens or even oxygen. Thus it explodes at $-200\,°C$ in the dark with hydrogen, whereas chlorine requires activation either by heating or exposure to light before it will explode with hydrogen. There are two reasons for this extreme reactivity. First, the fluorine–fluorine bond is unusually weak, a behaviour caused by repulsion between the three lone pairs on each atom. Second, fluorine forms very strong single covalent bonds with many atoms and produces the largest lattice energy of any singly charged ion: both of these features are due to the small size of the fluorine atom and fluoride ion.

As fluorine is the most ➤electronegative element of all, in *all* its compounds it has oxidation number −1. This means that in the compound with oxygen, OF_2, oxygen is forced to show the exceptionally unusual oxidation number of +2. Fluorine has the ability to cause many elements to show unusually high oxidation numbers; xenon hexafluoride, XeF_6, iodine heptafluoride, IF_7, and silver(II) fluoride, AgF_2, do not have chlorine analogues. Important compounds include ➤fluorite, CaF_2, uranium(VI) fluoride, UF_6, the polymers ➤Kel-F and ➤PTFE, the newly introduced ➤hydro-fluorocarbons (which are replacing ➤chlorofluorocarbons), and the anaesthetic ➤halothane.

fluorite structure The structure shown in the diagram, which is one of the most important for compounds of stoichiometry AB_2, named after **fluorite**, CaF_2, the main mineral source of fluorine. It consists of a ➤face-centred cubic array of calcium ions together with fluoride ions in all the tetrahedral holes. The ➤antifluorite structure, with the cation and anion positions swapped, is also common, as in the alkali metal oxides such as Na_2O.

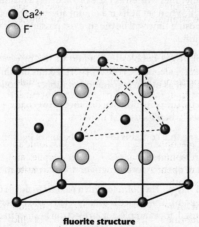

● Ca^{2+}
○ F^-

fluorite structure

fluorodinitrobenzene ➤Sanger's reagent.

fluorosulfonic acid $HOSO_2F$ A major constituent of ➤superacid mixtures.

fluorspar Alternative name for fluorite, CaF_2.

Fluothane A tradename for ➤halothane.

flux In general, the amount of something flowing through a specified surface per unit time. Thus **mass flux** is the mass of fluid passing through, say, a pipe per unit time. It can be evaluated as a surface integral,

$$\int_s \rho v \cdot dS,$$

where v is the velocity field of the fluid, S is a cross-sectional surface of the pipe and ρ is the density of the fluid. This is abstracted to the idea of electric and magnetic flux (ϕ), defined as the surface integrals of D (➤electric displacement) and B (➤magnetic flux density) respectively.

flux density, magnetic ➤magnetic flux density.

fluxional molecules Molecules that are easily rearranged from one form to another. For example, the fluorine atoms in the molecule PF_5 have been shown by ➤nuclear magnetic resonance to shift between the vertical and horizontal positions in the ➤trigonal bipyramidal structure (➤valence-shell electron-pair repulsion theory).

fluxmeter A device for measuring changes in magnetic ➤flux.

Fm Symbol for the element ➤fermium.

FM Abbr. for frequency ➤modulation.

f-number The ratio of the ➤focal length of a lens to its ➤aperture. A camera lens designated f 2.8 (sometimes written as $f/2.8$ or $f{:}2.8$) has a maximum aperture $1/2.8$ times its focal length. When set at its maximum aperture the lens will let the most light in, and the exposure time will be the shortest possible; the ➤depth of field will then be at a minimum.

foam A dispersion of gas bubbles in a liquid in the presence of surface-active stabilizers such as lauryl alcohol, $CH_3(CH_2)_{11}OH$. Foams are important in shaving and fire-fighting, as well as in the preparation of rubber. ➤➤colloid.

focal length Of a lens, the perpendicular distance between the ➤focal plane and the plane of the lens itself when focused at infinity.

focal plane The plane that contains the foci of beams of light that are incident on a lens at different angles.

focal ratio The ratio of the ➤focal length of a lens to its ➤aperture.

focus (plural foci) 1 (Phys.) The point to which parallel incident rays are refracted by a device (see the diagram). The device is usually a lens, and the rays are usually light, but the term is also used, for example, for an electron beam refracted by electric and magnetic fields in an ➤electron microscope.
  2 (Math.) ➤conic.

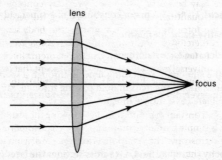

focus 1

foetus (fetus) The embryo of a mammal that has reached a stage of development in the ►uterus in which most of the adult features are recognizable. Specifically in humans it refers to the stage of development after the appearance of bone cells, a process occurring 7 to 8 weeks after ►fertilization.

fog A dispersion of a liquid in a gas; in particular, water droplets with radii of around 1–10 μm in air, reducing visibility to 1 km or less. Fogs occur naturally, but when pollutants are also present ►smog can result.

folic acid A water-soluble ►vitamin of the B group which acts as a ►coenzyme in various processes involving the metabolism of ►purines and ►pyrimidines. It is found particularly in green leafy vegetables. A deficiency of folic acid in the diet results in a form of anaemia and folic acid is often given during pregnancy to reduce the risk of birth defects such as spina bifida.

follicle stimulating hormone (FSH) A ►glycoprotein ►hormone secreted from the anterior lobe of the ►pituitary gland, which stimulates the development of egg-containing follicles in the ►ovary (2) and the development of sperm in the ►testes. ►►menstrual cycle.

food chain A sequence of organisms linked by their interdependence for food, in which energy and mineral elements are transferred from one organism to the next. Food chains exist in all communities of organisms, and are crosslinked in complex **food webs** because many organisms eat more than one food. At the beginning of virtually all food chains are green plants, which convert sunlight into chemical energy stored during ►photosynthesis. These plants are then eaten by herbivorous animals (**primary consumers**), which in turn are preyed upon by carnivores (**secondary consumers**). Aquatic food chains are particularly complex and may involve a large number of such so-called **trophic levels**. ►►ecosystem.

fool's gold ►pyrites.

foot An Imperial unit of length: 1 ft = 12 in. (►inch).

forbidden transition A transition between two ►quantum states that is not permitted by the ►selection rules for the system.

force Symbol F; unit ►newton, N. An influence that causes a body to change its momentum. Force may be a straightforward, everyday concept, but it is almost impossible to define quantitatively without trivializing the second of ►Newton's laws of motion: the force on a body is equal to the body's rate of change of momentum, and is therefore a vector quantity. It is possible to regard force as the result of the action of a field, either electromagnetic (►Lorentz force) or gravitational (►Newton's law of universal gravitation). Force is a concept that loses its usefulness in the worlds of quantum mechanics and general relativity, but it remains indispensable in the everyday macroscopic world of engineering and mechanics.

force constant The constant of proportionality k linking the restoring force f to the displacement x in ►simple harmonic motion: $f = -kx$. One important application is in ►vibrational spectroscopy. The vibrational frequency ω of a molecule is related to the force constant by the equation $\omega = \sqrt{(k/\mu)}$, where μ is the ►reduced mass of the molecule; the higher the force constant, the stronger the chemical bond.

forced oscillation A steady state oscillation, sustained by a **driving force** appropriate to the system in question. A mass on a spring can be forced to oscillate by a mechanical driving force, an ➤LC circuit by a signal generator producing a sinusoidal voltage output, and a diatomic molecule by an electromagnetic field (as in ➤infrared spectroscopy). The peak of the response amplitude as the frequency is varied is called the ➤resonance. ➤➤Q factor.

force ratio ➤mechanical advantage.

forebrain ➤brain.

forensic science The application of scientific techniques to legal investigations. Traditional forensic science includes, for example, fingerprint analysis and the identification of fragments of clothing fibres, paint and other materials left at the scene of a crime. More advanced techniques such as ➤genetic fingerprinting have greatly increased the ability to identify individuals from small tissue or blood samples.

formaldehyde Traditional name for ➤methanal.

formalin A solution of methanal (formaldehyde) in water. This was used to preserve biological specimens.

formic acid Traditional name for ➤methanoic acid, the name deriving from the Latin *formica*, ant, as ants were an early natural source of the acid.

formula **1** (Chem.) The formula of a compound expresses the number of atoms of each element present. The simplest, the **empirical formula** (EF), simply gives the ratio of atoms. The **molecular formula** (MF) gives the actual number present. For many compounds, especially organic ones, further information is needed about how the atoms are bonded together and this is given by the **structural formula** (SF). For butane, the EF is C_2H_5, the MF is C_4H_{10}, and the SF is $CH_3CH_2CH_2CH_3$. Its ➤isomer methylpropane $(CH_3)_3CH$ has the same EF and MF but a different SF. In simple inorganic compounds, round brackets are used if a polyatomic ion such as sulfate or nitrate is present more than once, for example $Al_2(SO_4)_3$ or $Pb(NO_3)_2$. Complex ions are indicated by the use of square brackets as in $[Cu(NH_3)_4(H_2O)_2]^{2+}$.
   2 (Math.) A formal mathematical expression of some general rule or law. Examples are: $A = \pi r^2$ for the area A of a circle with radius r; the trigonometrical identity $\sin 2x = 2 \sin x \cos x$; the nth term of the sequence 1, 2, 4, 8, ... expressed as 2^{n-1}.

formylation The introduction of a —CHO group into a molecule.

Fortran Acronym for formula translation. A ➤high-level computer language, based on algebraic representations and especially suited to scientific, mathematical and engineering applications. Efficient Fortran ➤compilers have been available, leading to fast programs.

forward bias ➤diode.

fossil The remains of an organism that lived in the past, preserved in rock, amber, ice or peat deposits. Generally only the hard parts of organisms become fossilized, structures such as bones and wood undergoing mineralization. The fossil record has provided much of the evidence for ➤evolution.

fossil fuel A fuel, such as petroleum, coal, peat or natural gas, that was originally formed by the decomposition of organisms. The supply of fossil fuels is finite; compare ➤renewable energy sources.

Foucault pendulum A simple pendulum designed to demonstrate that the Earth rotates. The pendulum swings in a plane fixed in an ➤inertial frame, but because the Earth rotates in that frame the plane of the pendulum appears to rotate with respect to the laboratory frame of reference. Named after (Jean Bernard) Léon Foucault (1819–68).

four-colour theorem The theorem that any geographical map may be unambiguously coloured using no more than four different colours. It was first proved in 1976, but this proof was controversial in that it made such heavy use of computers that the details could not be fully checked by human beings.

Fourier analysis The application of the Fourier series and the Fourier transform, for example to the ➤differential equations which arise in the analysis of wave motion. Named after (Jean-Baptiste) Joseph Fourier (1768–1830).

Fourier series The expression of an arbitrary ➤periodic function as a series of sine and cosine 'harmonics'. On the interval $-\pi$ to π, the Fourier series of $f(x)$ is

$$\tfrac{1}{2}a_0 + \sum_{n=1}^{\infty} a_n \cos nx + \sum_{n=1}^{\infty} b_n \sin nx,$$

where the **Fourier coefficients** are given by

$$a_n = \frac{1}{\pi} \int_{-\pi}^{\pi} f(t) \cos nt \, \mathrm{d}t, \ b_n = \frac{1}{\pi} \int_{-\pi}^{\pi} f(t) \sin nt \, \mathrm{d}t.$$

For a wide class of functions, the Fourier series converges to the function, which makes it a powerful tool for investigating any oscillating system. Intervals other than $-\pi$ to π may be handled in a similar way.

Fourier transform The ➤integral transform taking a ➤function f to a function F, given by

$$F(y) = \int_{-\infty}^{\infty} f(x) \, \mathrm{e}^{-\mathrm{i}xy} \, \mathrm{d}x.$$

The fact that f can be recovered from F by the inversion formula

$$f(x) = \frac{1}{2\pi} \int_{-\infty}^{\infty} F(y) \, \mathrm{e}^{\mathrm{i}xy} \, \mathrm{d}y$$

and the use of standard tables or ➤residue calculations make it a useful tool in solving differential equations.

four-vector A vector quantity with three space-like components and a time-like component (which is multiplied by c, the speed of light in a vacuum, to give it the same dimension). For example, the four-vector displacement of an event at (x, y, z) at time t is written (ct, x, y, z). The interval between the events, the magnitude of the vector, is $\sqrt{x^2 + y^2 + z^2 - c^2 t^2}$ (note the minus sign). A four-vector transforms from one ➤inertial frame to another according to the ➤Lorentz transformation. The magnitude of a four-vector is obtained using the ➤metric for four-dimensional spacetime. It is the fact that a four-vector has a time-like component that leads to time sometimes being referred to as the **fourth dimension**. ➤➤special relativity.

fovea ➤eye.

Fr Symbol for the element ➤francium.

fractal An object with fractional dimension. This is a surprising concept until one appreciates the subtlety of the definition of dimension. Many formal definitions are possible, concerned, for example, with the space-filling properties of an object. Fractals are infinitely detailed, and display self-similarity (i.e. they have a similar, though usually not identical, appearance at any magnification), as shown by the well-known ➤Mandelbrot set. Fractals are often associated with ➤chaos theory.

fraction A number, or algebraic expression, which is the ➤quotient of two integers or algebraic expressions. A fraction is written a/b where a is called the **numerator** and b the **denominator**.

fractional crystallization A separation technique that utilizes the small difference in solubility between two solids. If a solution of the two is cooled, crystallization will concentrate one of them preferentially. The most famous application of the technique was Marie Curie's separation of radium from barium by 4000 fractional crystallizations of the sulfates.

fractional distillation (fractionation) A separation technique that utilizes the small difference in boiling points between two volatile liquids. Commercially, a tall **fractionating column** with **bubble caps** is used (see the diagram opposite), where the ascending hot vapour can exchange energy with the descending liquid mixture. As long as the column is tall enough, separation can in principle be achieved as long as an ➤azeotrope is not formed. Two important examples are the fractional distillation of liquid air to produce liquid oxygen and liquid nitrogen, and the fractional distillation of petroleum, the first step in oil refining.

fragmentation pattern In a ➤mass spectrometer, the pattern of fragments produced when the ions formed at the start break apart as they fly through the apparatus. These fragments provide a 'fingerprint' for each molecule, and computerized matching to known patterns can rapidly identify even complex molecules.

frame of reference A set of axes and coordinates for ➤events as constructed by one particular observer. A different observer may choose a different set of axes or a different origin, or may be moving with respect to the first observer. ➤➤inertial frame.

francium Symbol Fr. The element with atomic number 87 and most stable isotope 223, which is the alkali metal with the largest atomic number. Its chemistry is dominated by its radioactivity. The expected single oxidation number of +1 has been confirmed, although no common compounds exist in nature.

Franck–Condon principle In spectroscopy, the principle that the time for an electron to make a transition to a higher energy state is much faster than the period of

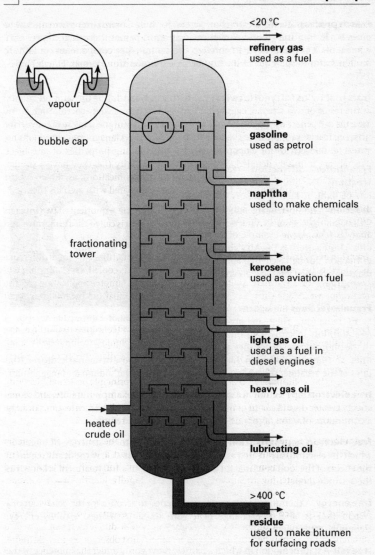

fractional distillation The products of the fractional distillation of petroleum and their uses.

vibrations of molecules. As a result, it can be assumed that the nuclei are in the same positions during the transition. Named after James Franck (1882–1964) and Edward Uhler Condon (1902–74).

Frasch process The main extraction process for sulfur, which uses three concentric pipes bored into the ground. Superheated steam is passed through the outer one, while a blast of compressed air through the centre pipe forces a foam containing molten sulfur to spew out of the middle pipe. Named after Herman Frasch (1851–1914).

fraternal twins (dizygotic twins) Twins that develop during the same pregnancy as the result of two separate eggs being fertilized by two separate sperm. Such twins are thus not genetically identical and are not necessarily of the same sex. They share no more features in common than would be expected between normal brothers and sisters of different ages. Compare ➤identical twins.

Fraunhofer diffraction (far-field diffraction) The limiting case of the ➤diffraction of light at large distances from a small diffracting aperture. The intensity of the diffracted light may be found by ➤Fourier analysis. The classic example of a Fraunhofer diffraction pattern is the **Airy ring** pattern for diffraction through a circular aperture (see the diagram). Named after Joseph von Fraunhofer (1787–1826).

Fraunhofer lines Lines in the spectrum of the Sun's radiation. They are absorption lines (➤absorption spectroscopy) due to elements in the photosphere, hence they appear as dark

Fraunhofer diffraction Airy rings produced by diffraction of light passing through a small circular aperture.

lines. The most prominent in the visible spectrum are those of neutral hydrogen (the lines of the ➤Balmer series), ionized calcium, and neutral sodium and magnesium.

free electron An electron in a quantum state that is not a ➤bound state, and so has energy greater than the limit of the potential energy at infinity. Free electrons occupy a continuum of states, as opposed to the discrete bound states.

free-electron approximation A model of the electronic structure of metals in which the electrons are treated as being in a flat potential well, without any regard to the effects of the local potentials of the ion cores. The classical treatment is known as the ➤Drude model.

free energy A thermodynamic function designed to account for the behaviour of a system that is able to exchange heat with its surroundings. ➤Gibbs energy; Helmholtz free energy.

free fall A state of motion in which a particle moves only under the influence of the local gravitational field. ➤➤acceleration of free fall.

free radical Alternative name for ➤radical, the word 'free' being redundant.

free space A volume that has no particles within it (including particles that are sometimes considered as waves, such as ➤photons). It is thus free of all electromagnetic fields and interactions.

freeze drying A technique for preserving material of plant or animal origin by deep freezing and then drying under reduced pressure so that the frozen water molecules undergo ➤sublimation. The technique preserves a high degree of the original structure and is used extensively in the dehydrated food industry, as well as for preserving museum specimens.

freezing ➤change of state.

freezing point ➤melting point.

freezing point depression ➤colligative properties.

Frenkel defect A ➤point defect in a crystal where a particle occupies an ➤interstitial site, leaving a ➤vacancy. Named after Yakov Ilyich Frenkel (1894–1952).

Freon Tradename for a class of ➤chlorofluorocarbons.

frequency Symbol f or ν. For a periodic phenomenon, the number of times the same event occurs per unit time, that is the reciprocal of the period of the phenomenon. Typically the phenomenon is an oscillation or wave. Its SI unit is the hertz (Hz), where 1 Hz is equivalent to 1 s^{-1}. The term is sometimes loosely used to refer to ➤angular frequency (for which the use of the hertz should be avoided) $\omega = 2\pi f$.

frequency band A range of frequencies of ➤radio frequency waves. The different bands and their uses are given in the table.

Band	Abbr.	Frequencies	Wavelengths	Typical uses
Very low	VLF	3–30 kHz	10–100 km	Long-range navigation
Low	LF	30–300 kHz	1–10 km	Commercial radio, short-range navigation
Medium	MF	300 kHz–3 MHz	100 m–1 km	Commercial radio, short-range navigation
High	HF	3–30 MHz	10–100 m	Long-range mobile communications
Very high	VHF	30–300 MHz	1–10 m	Commercial radio, short-range mobile communications and navigation
Ultra high	UHF	300 MHz–3 GHz	100 mm–1 m	Short-range mobile communications, commercial TV, radar
Super high	SHF	3–30 GHz	10–100 mm	Radar
Extremely high	EHF	30–300 GHz	1–10 mm	Radar and satellite communications

frequency distribution A tabulation of data giving the number (**frequency**) of data items for each value or in each of a number of class intervals. In the latter case, the mid-class value is often taken as representing that class in calculations.

frequency modulation (FM) ➤modulation.

frequency response (transfer function) The (complex) amplitude of response of a system per unit amplitude of stimulus at a single frequency as a function of that frequency.

Fresnel diffraction ➤Diffraction when the distance from the diffracting aperture is too short for the condition for ➤Fraunhofer diffraction to hold, but still much greater than the size of the aperture itself. The expression for the amplitude of Fresnel diffraction takes into account the quadratic term in the variation of phase across the aperture. This makes an exact solution impossible except for special cases where the appropriate integral can be evaluated explicitly. Named after Augustin Jean Fresnel (1788–1827).

Fresnel lens A lens made from a large number of small sections in the plane of the lens. Each section has the same curvature that the convex surface of a full lens would have, but it is a fraction of the thickness. It was originally designed to reduce the weight of glass required for a lighthouse lens.

friction A force opposing the motion of one surface over another. The frictional force is parallel to the sliding surfaces and proportional in magnitude to the force of interaction perpendicular to the surfaces (the constant of proportionality being the ➤coefficient of friction). ➤➤dynamic friction; rolling friction; static friction.

Friedel–Crafts reaction An exceptionally widely applicable and useful reaction in aromatic organic chemistry in which a benzene ring has either an acyl group or an alkyl group added. The reagents used are the acyl chloride (or alkyl chloride) and a ➤Lewis acid catalyst such as aluminium chloride. The common reagent mixture of ethanoyl chloride, CH_3COCl, plus $AlCl_3$ introduces an ethanoyl group into the ring: benzene becomes phenylethanone, $C_6H_5COCH_3$. Named after Charles Friedel (1832–99) and James Mason Crafts (1839–1917).

fringes Bands of alternating high and low intensity of light resulting from ➤diffraction or ➤interference.

frontier orbitals Although all the orbitals of two reactants that possess the correct symmetry can and do overlap to some extent, the overlap is dominated by the two orbitals closest in energy. These frontier orbitals are the highest occupied molecular orbital (**HOMO**) of the ➤nucleophile and the lowest unoccupied molecular orbital (**LUMO**) of the ➤electrophile. The HOMO–LUMO interaction dominates the bonding in the product.

frost Frozen dew. Compare ➤hoarfrost.

fructose $C_6H_{12}O_6$ A very important ➤monosaccharide, which is a ketohexose, that is, contains six carbon atoms and a ketone group in the open-chain form. In natural products it exists in the ➤furanose form (➤sucrose) while it crystallizes in the ➤pyranose form. Fructose is the sugar commonly found in ripe fruits and it is an intermediate in many biochemical pathways involving the metabolism of glucose.

fruit fly ➤*Drosophila*.

frustum The portion of a solid figure such as a ➤cone that is cut off by two parallel planes. 'Frustum' is often misspelt 'frustrum'.

FSD Abbr. for ➤full-scale deflection.

FSH Abbr. for ➤follicle stimulating hormone.

ft Symbol for the unit ➤foot.

FTP Abbr. for file transfer protocol. The ➤Internet's protocol for transferring files. ➤➤HTTP; www.

fuel cell A cell in which the reactants are provided from external reservoirs rather than being installed once and for all when the cell is made. An example is the Bacon cell, which uses hydrogen and oxygen as the reactants with carbon electrodes and a platinum catalyst.

fuel element One of the blocks (usually rods) of fissile material that, with the ➤moderator (if present), form the core of a ➤nuclear reactor.

fugacity The effective pressure of a gas, taking into account deviations from the criteria for an ➤ideal gas. Many of the equations that apply to ideal gases can be modified to apply to real gases simply by replacing the pressure by the fugacity.

fulcrum ➤lever.

fullerite ➤buckminsterfullerene.

full-scale deflection (FSD) The maximum value that can be recorded by a particular measuring instrument.

full-wave rectifier A ➤rectifier circuit that reverses the sign of the negative part of the cycle of an alternating input voltage. Compare ➤half-wave rectifier.

fulminate of mercury ➤mercury(II) fulminate.

fumaric acid $HOOCCH=CHCOOH$ The traditional name for *trans*-butenedioic acid, which exists as colourless needles. The *cis* form of the acid is traditionally called ➤maleic acid.

function In general, a rule for assigning to the ➤elements (2) of one ➤set X (the **domain**) elements of another set Y (the **codomain**). A function is indicated by notation such as $f: X \rightarrow Y$ or $y = f(x)$ and the nature of the sets involved may be indicated by qualifying phrases such as 'a function of a real variable'. The **range** of a function f is the ➤set of **values**, $f(x)$, for x in the domain of f.

functional group The group of atoms responsible for the characteristic reactions of organic compounds; different functional groups distinguish the different classes of compound. For example, all alcohols contain the —OH functional group. **Functional group isomerism** is a form of isomerism in which the functional group varies, as for methoxymethane, CH_3OCH_3, and ethanol, CH_3CH_2OH.

fundamental The lowest frequency of a complex waveform, for example a musical note, that typically consists of a superposition of several frequencies. If a note is pure, not a chord, the other frequencies (➤harmonics or ➤overtones) are multiples of the fundamental.

fundamental constant A value that appears as a constant of proportionality in a fundamental physical law, independent of location in the Universe. There are three broad categories of fundamental constant. The seven most important are the speed of light (c), the Planck constant (h), the Boltzmann constant (k), the permittivity of free space (ε_0), the mass of the electron (m_e), the elementary charge (e) and the gravitational constant (G). Of these seven, four could be considered as defining equivalences between fundamental physical quantities as new laws are discovered. Taking energy as the basis, for example, c relates mass to energy in ➤ special relativity, h relates time to energy in ➤quantum mechanics, k relates temperature to energy in ➤statistical mechanics and ε_0 relates charge to energy in ➤electrostatics. The other three are either properties of ➤elementary particles (m_e and e) or coupling constants of the ➤fundamental interactions (G). The values of these and other fundamental constants are given in Appendix table 2.

fundamental interactions The four ways in which, it is currently believed, ➤elementary particles interact with one another at the deepest level: ➤gravity, ➤electromagnetism, the ➤weak interaction and the ➤strong interaction. Various ➤unified field theories have attempted to relate them to one another. The ➤electroweak theory suggests that at a deep level electromagnetism and the weak interaction are related. The strong interaction is described in a similar way in the ➤standard model, allowing further unification. Gravitation is more recalcitrant, as it appears significantly different to the others; a 'theory of everything' is the postulated unification of all four interactions. ➤➤grand unified theory.

fundamental particle ➤elementary particle.

fundamental theorem of algebra The statement that every ➤polynomial equation of the nth ➤degree (2) with ➤complex number ➤coefficients has at most n distinct ➤roots all of which are complex numbers. Thus although the ➤real numbers have to be extended to the complex numbers in order that equations such as $x^2 + 1 = 0$ may be solved, no further extension of the complex numbers is needed to solve all other polynomial equations.

fundamental theorem of arithmetic The statement that every whole number can be expressed uniquely as a product of ➤prime numbers. For example, $4312 = 2^3 \times 7^2 \times 11$.

fundamental theorem of calculus The theorem that, with appropriate hypotheses about a ➤function F and its ➤derivative f,

$$F(b) - F(a) = \int_a^b f(x)\,dx.$$

It provides the theoretical justification for the statement that ➤'integration is the reverse of ➤differentiation'.

fundamental unit ➤base unit.

fungus (plural fungi) A member of the major ➤kingdom Fungi (or Mycota) containing ➤eukaryotic organisms that lack ➤chlorophyll and reproduce by ➤spores. Fungi obtain their nutrition as either ➤saprotrophs or parasites (➤symbiosis) since they are unable to synthesize their own food. The characteristic structural unit of a fungus is a filamentary structure called a **hypha**. The cells of the hypha are bounded by a wall which generally lacks cellulose, and may contain derivatives of ➤chitin. Fungi are important in the process of decay and in the recycling of elements in nature, but some cause economically significant disease in crops. Potato blight, which caused the famines in Ireland in the middle of the 19th century, is caused by the fungus *Phytophthora infestans*. Fungal infections of animals include ringworm and athlete's foot.

funicle ➤ovule.

furan A colourless liquid with the structure shown in the diagram; it is of less importance than its hydrogenated derivative tetrahydrofuran.

furan

furanose

furanose A structure present in many saccharides, such as sucrose, comprising a ring with five members, four carbons and one oxygen atom (see the diagram). Compare ➤pyranose; ➤➤ribose.

furfural The aldehyde derivative of ➤furan, which is a colourless liquid used as a solvent for extracting mineral oils and crude rosin.

fuse, electrical A device designed to protect an electrical circuit from a high current. It consists of a metal filament or strip that melts if it passes more than a designated current and thus breaks the circuit. Compare ➤circuit-breaker.

fused When referring to electrolytes, the term means 'molten'.

fusel oils A variable mixture mainly of alcohols containing typically from three to six carbon atoms. Although more toxic than ethanol, they tend to induce vomiting rapidly and so cause fewer fatalities. They do, however, contribute to hangovers.

fusion 1 The act of melting. The ➤enthalpy change associated with the transition is called the **enthalpy of fusion**. ➤➤fused.
   2 ➤nuclear fusion.

fuzzy logic A branch of logic which, instead of only allowing the two extreme logical states of true and false, has a continuum of possible states between these extremes. Fuzzy logic has applications in ➤expert systems and ➤artificial intelligence.

G

g **1** Symbol for the unit ➤gram.
  2 (Chem.) Label used in chemical equations to indicate the gaseous state.

g Symbol for the ➤acceleration of free fall. Its value at the Earth's surface is approximately 9.8 m s^{-2}.

γ **1** The ratio of the ➤specific heat capacity of a substance at constant pressure to its specific heat capacity at constant volume; it is a dimensionless quantity. Its value for a simple gas depends on the number of ➤degrees of freedom, F, of the gas particles as $\gamma = 1 + 2/F$. It appears in the expression for the ➤adiabatic expansion of a gas: pV^{γ} = constant.
  2 The dimensionless quantity $1/\sqrt{(1 - v^2/c^2)}$ that appears in the ➤Lorentz transformation.

G **1** Symbol for the prefix ➤giga-.
  2 Symbol for the unit ➤gauss.
  3 Symbol for the purine base ➤guanine.

G **1** Symbol for ➤Gibbs energy.
  2 Symbol for the gravitational constant (➤Newton's law of (universal) gravitation).

Ga Symbol for the element ➤gallium.

GABA (gamma-aminobutyric acid) A ➤neurotransmitter substance active in inhibitory ➤synapses in the human brain. Many benzodiazapine tranquillizers such as Valium produce their effect via GABA receptors.

Gabriel synthesis The formation of a primary ➤amine from a halogenoalkane by reaction with potassium phthalimide (see the diagram) and subsequent hydrolysis. It is better than the conventional synthesis using ammonia, which gives secondary and tertiary amines as well. Named after Siegmund Gabriel (1851–1924).

Gabriel synthesis
Potassium phthalimide.

gadolinium Symbol Gd. The element with atomic number 64 and relative atomic mass 157.3; it is a ➤lanthanide and is named after an early investigator of the lanthanide elements, Johan Gadolin (1760–1852). Gadolinium, as expected for a lanthanide, shows compounds almost exclusively with ➤oxidation number +3, such as $GdCl_3$ (which forms colourless crystals). Gd^{3+} has the largest paramagnetism of all ions, as it has seven unpaired electrons.

gain The ratio of the output signal strength of an amplifier to the input signal strength. The gain of an imperfect amplifier varies with the frequency of the signal. In a more general context, gain can be applied to any system whose output is a multiple of its input. Gain can be measured in ➤decibels (dB).

galactose A ➤monosaccharide that is part of the disaccharide sugar ➤lactose, found in milk. It is also present in agar and seaweed.

galaxy A vast assemblage of ➤stars, interstellar matter (➤interstellar medium) and ➤dark matter. The galaxy in which the Sun is located is named 'the Galaxy', with a capital 'G'. It is a large ➤spiral galaxy over 100 000 light years in diameter, containing about 10^{11} stars. From our vantage point nearly 30 000 light years from the Galactic centre, the structure of the Galactic disc is visible as the **Milky Way** (a name sometimes given to the Galaxy itself). The ➤Magellanic Clouds are companion galaxies of our own, and all belong to the ➤local group of galaxies. Apart from spirals,

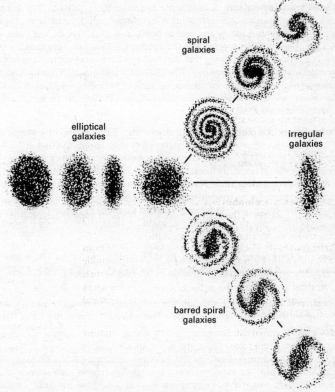

galaxy The conventional classification of galaxies in the so-called tuning fork diagram. (It does not imply an evolutionary sequence.)

the other main types are ➤elliptical galaxies and irregular galaxies. In the traditional classification of galaxies, illustrated in the **tuning fork diagram** shown overleaf, ellipticals are ranked by their apparent degree of ellipticity, and spirals by how tightly their spiral arms are wound. ➤➤Seyfert galaxy.

galena PbS An ore, lead(II) sulfide, which is black with a metallic lustre.

Galilean satellites Traditional name for the four largest satellites of Jupiter: Io, Europa, Ganymede and Callisto. They were discovered by Galileo Galilei (1564–1642).

Galilean spacetime The ➤spacetime of ➤Newtonian mechanics. It has three space dimensions and one time dimension, considered entirely separate. It is a satisfactory model of the world in which we live, provided movement is restricted to speeds much less than the ➤speed of light. Compare ➤Minkowski spacetime.

Galilean telescope ➤telescope.

Galilean transformation A coordinate transformation between two ➤inertial frames (in ➤Galilean spacetime) moving with constant velocity with respect to each other. If an observer A places an event at point (x, y, z) at time t, then observer B, moving at speed v with respect to A in the direction of the x axis, measures the event at $(x - vt, y, z)$ at time t (assuming that the observers have synchronized the origin of their coordinate systems at time $t = 0$). The Galilean transformation is simply a formal expression of the naïve rule for relative velocities (➤velocity, relative). It does not preserve the speed of anything, including light, between all frames. Compare ➤Lorentz transformation.

gallium Symbol Ga. The element with atomic number 31 and relative atomic mass 69.72, which is in Group 13, immediately below aluminium. Melting at 30 °C yet boiling only at 2400 °C, it has the longest liquid range of any element. Like aluminium, gallium is mainly found with oxidation number +3, as in $GaCl_3$, which, like $AlCl_3$, can dimerize. There is an unimportant and very unstable +1 oxidation number, as expected from the ➤inert pair effect. The most important compounds of gallium are the compounds between it and two elements in Group 15, phosphorus and especially arsenic. **Gallium arsenide**, GaAs, is a very important ➤semiconductor, rivalling silicon in usefulness.

gallon An Imperial unit of volume. Its value is different in the UK, where it is obsolescent, and the USA: 6 US gallons ≈ 5 UK gallons. 1 UK gallon = 4.546 dm^3; 1 US gallon = 3.785 dm^3.

galvanic cell A cell capable of producing electric current from chemical reactions occurring within it. An example is the ➤Daniell cell. Named after Luigi Galvani (1737–98).

galvanize To treat iron or steel with zinc, which is a more reactive metal and so donates electrons preferentially on demand, thus protecting the iron or steel from rusting.

galvanometer An instrument for measuring small values of electric current. ➤➤ballistic galvanometer.

gamete A ➤haploid germ cell that is specialized for ➤fertilization. Eggs and sperm are gametes and are structurally differentiated; in humans, each contains 23 ➤chromosomes.

gametophyte The ➤haploid, ➤gamete-forming phase in the life cycle of many plants including all land plants, the mosses, ferns and many algae. The gametophyte of mosses is the leafy green plant, and in ferns it is an independent scale-like structure, separate from the adult ➤sporophyte. In the ➤Magnoliopsida, the female gametophyte is reduced to a few cells in the ➤embryo-sac in the ovule, and the male gametophyte is represented by the pollen grain. ➤➤alternation of generations.

gamma-aminobutyric acid ➤GABA.

gamma decay The mode of radioactive decay (➤radioactivity) in which ➤gamma rays are emitted. Neither the atomic number nor the nucleon number of the nuclide changes in this form of decay; the nuclide merely releases energy as a photon in a transition from an excited state to, usually, its ground state.

gamma function The ➤function defined by

$$\Gamma(x) = \int_0^\infty t^{x-1} e^{-t} \mathrm{d}t$$

for real numbers $x > 1$; the definition may be extended to ➤complex numbers. Since $\Gamma(n + 1) = n!$ for whole numbers n, the gamma function may be regarded as a generalized ➤factorial function.

gamma globulin ➤globulin.

gamma-iron The allotrope of iron that is stable between 910 and 1400 °C. ➤➤iron–carbon phase diagram.

gamma ray Very high frequency electromagnetic radiation emitted when an excited nucleus undergoes ➤gamma decay. Gamma rays are very damaging to human tissue, and exposure to them needs strict monitoring. Gamma rays can be detected by means of a ➤Geiger counter.

ganged circuits Two or more circuits whose controls are mechanically linked together for simultaneous operation.

ganglion (plural ganglia) A small mass of nerve tissue consisting of the cell bodies of a number of nerve cells. In many groups of animals other than ➤chordates, such as molluscs, the central nervous system consists of a series of ganglia joined by nerve cords. Ganglia also occur in the nervous systems of chordates.

gap ➤energy gap.

garnet A group of silicates containing two metal ions, one with ➤oxidation number +2 and one with oxidation number +3, used as gemstones. The neodymium(III) ion in the compound yttrium aluminium garnet (YAG) forms a useful laser system.

gas The state of matter that lacks structure, the particles behaving essentially independently of one another, and expanding to fill any available volume. The ➤ideal gas model explains many of the properties of gases quantitatively. Below its critical temperature, a gas can be liquefied by increasing the pressure or decreasing the temperature (➤critical point). The word comes from the Greek for 'chaos'.

gas chromatography (GC) A form of ➤chromatography in which the mobile phase is gaseous. One version is **gas–liquid chromatography** (GLC), while another uses the output of a gas chromatograph as input to a mass spectrometer (GCMS), which powerful combination is used to analyse complex mixtures.

gas constant (universal gas constant) Symbol R. The constant of proportionality in the ➤ideal gas equation; its value is $8.315 \, \mathrm{J \, K^{-1} \, mol^{-1}}$. It is directly related to the ➤Boltzmann constant k by the formula $R = Lk$, where L is the ➤Avogadro constant. ➤➤Appendix table 2.

gas-cooled reactor A nuclear reactor cooled by gas. Early versions used ➤Magnox fuel elements.

gas-discharge tube A tube containing a gas that, under the influence of a high electric field, conducts electricity. The gas becomes conducting because it undergoes ➤ionization, which requires an external trigger (such as a source of ultraviolet radiation) but is in many cases self-sustaining. The study of gas-discharge tubes played a central role in the early progress of atomic physics, in particular in the discovery of the electron.

gas exchange (gaseous exchange) A process that enables organisms to exchange gases produced by ➤metabolism with essential gases in their environment. For example, ➤respiration produces carbon dioxide, which would become toxic if allowed to accumulate; oxygen from the environment is required for respiration. Any surface that facilitates the removal of carbon dioxide and the uptake of oxygen is a **gas exchange surface.** In very small aquatic organisms the surface of the body itself suffices, but larger organisms (which have a lower surface-to-volume ratio) usually possess specialized internal organs. Gas exchange organs such as ➤gills and ➤lungs characteristically have large surface areas in relation to the volume of the organism. They also have moist surfaces since gases can diffuse into cells only in solution, and, where appropriate, they have associated with them well-developed ➤vascular systems. In plants the surfaces of stems, roots and, in particular, leaves act as gas exchange organs for both respiration and ➤photosynthesis. ➤➤stoma.

gasification of fuels A set of processes that convert solid or liquid fuels into gaseous ones for ease of transmission. An example is the ➤Lurgi process.

gas laws Laws that describe quantitatively how the pressure, volume, temperature and amount (in moles) of a gas depend on one another. The two most important are ➤Boyle's law, which states how the volume of a fixed mass of gas depends on the pressure at a fixed temperature, and ➤Charles's law, which states how the volume of a fixed mass of gas depends on the temperature at a fixed pressure. ➤➤ideal gas equation.

gas oil A common name for one of the heavier fractions of petroleum, often being subdivided into **light gas oil** (containing about 16–20 carbon atoms) and **heavy gas oil** (containing about 20–25 carbon atoms). Its main use is as an industrial heating fuel.

gasoline (petrol) The technical name of the fraction of ➤petroleum containing around five to eight carbon atoms (➤naphtha). The shortened form of the name of the fraction, 'gas', is used in the USA in place of 'petrol'. The exact composition of a gasoline and its percentage of the whole sample depends on the geographical location of the source of the petroleum.

gastrin A ➤hormone, secreted by cells in the lining of the stomach and small intestine, which stimulates the release of gastric juice, containing hydrochloric acid and the ➤protease enzyme pepsin. Oversecretion (particularly in response to alcohol) and high levels of protein may lead to the formation of gastric ulcers.

gas turbine A form of ➤turbine in which the energy of hot, expanded gas is used to drive a turbine wheel. Gas turbines are used in power plants and to power ships and jet aircraft.

gate 1 One of the three electrodes of a ➤field-effect transistor. The gate is the connection that is used to control the behaviour of the transistor, either by varying the gate voltage or the current through it.
   2 ➤logic gate.

gauge An arbitrary quantity in any sort of ➤potential (2) that can be chosen at will or by convention because it has no effect on any observable physical quantity. For example, a constant value ϕ_0 can be added to an electrostatic potential $\phi(r)$ at every point without changing the value of its gradient, the electric field $E(r)$. Similarly, the gradient of any scalar function $a(r)$ can be added to the magnetic vector potential $A(r)$ without changing the value of curl $A(r)$, the magnetic flux density $B(r)$. ➤➤gauge theory.

gauge boson A particle corresponding to the field that causes interactions between other subatomic particles. The gauge boson for the electromagnetic field is the ➤photon; for the ➤strong interaction it is the ➤gluon; for the ➤weak interaction the ➤W boson and ➤Z boson; for gravity the ➤graviton is postulated, but has never been observed.

gauge theory One of several theories of ➤fundamental interactions. The ➤invariance of a physical system to the choice of ➤gauge for any of its potentials has profound consequences in ➤quantum mechanics. In the same way that the invariance of a system to translations and rotations in free space leads to the conservation of momentum and angular momentum respectively, the gauge invariance of the potential and the wavefunction lead to the fundamental conservation laws of quantum mechanics.

gauss Symbol G. The c.g.s. unit of magnetic field strength. It is related to the SI unit, the tesla, by 10 000 G = 1 T. It is often a more convenient unit than the tesla: for example, the Earth's magnetic field is around half a gauss. Named after Carl Friedrich Gauss (1777–1855).

Gaussian distribution Alternative name (often used by physicists) for the ➤normal distribution.

Gauss's law The total flux of ➤electric displacement at a closed surface is equal to the total charge enclosed by the surface. This results from applying the ➤divergence theorem to the first of ➤Maxwell's equations:

$$\int_S \boldsymbol{D} \cdot d\boldsymbol{S} = \int_V (\nabla \cdot \boldsymbol{D}) dV = \int_V \rho_{\text{free}} dV = Q,$$

where ρ is the charge density and therefore Q is the total charge enclosed within the surface S. The name is also used for the corresponding result for gravity: the total flux of ➤gravitational field at a closed surface is $-4\pi G$ times the total mass enclosed by the surface, where G is the ➤gravitational constant.

Gay-Lussac's law For a reaction in the gas phase, the volumes of gases that react (and the volumes of the products, if gaseous) are in a simple whole-number ratio. Although introduced by Joseph-Louis Gay-Lussac (1778–1850) before modern atomic theory was introduced, the explanation is now obvious. Species combine in whole-number ratios of their amounts of substance in moles and, as the ➤molar volume of all gases is constant, the volumes of gases that react must also be in a simple whole-number ratio.

GB ➤nerve gas.

GC Abbr. for ➤gas chromatography.

GCMS ➤gas chromatography.

Gd Symbol for the element ➤gadolinium.

Ge Symbol for the element ➤germanium.

Geiger counter (Geiger–Müller tube) An instrument for counting photons of ionizing radiation (see the diagram). It consists of a tube containing a gas (typically argon) at low pressure; a potential difference is maintained between a central rod (the anode) and the casing (the cathode). The electric field in the tube is just below the

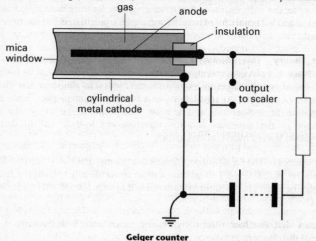

Geiger counter

threshold required for ionization, so even a low-energy electron can ionize an atom of the gas. When incident radiation ionizes a gas atom, the free electron is accelerated towards the anode, causing more ionization on its way, which in turn causes further ionization. Thus a pulse of current flows, which is amplified and counted. Named after Hans Wilhelm Geiger (1882–1945) and Walther Müller (1905–79).

Geiger–Marsden experiment A classic experiment, suggested by Ernest Rutherford, to observe the effect of firing alpha particles at thin sheets of gold foil; Rutherford was forced to introduce the concept of the atomic ➤nucleus to explain the results. The vast majority of the alpha particles passed through the foil (and therefore through the gold atoms), but about 1 in 8000 were deflected back. Named after H. W. Geiger and Ernest Marsden (1889–1970).

gel ➤colloid.

gelatin A water-soluble ➤protein derived from animal ➤collagen which, on heating in water and cooling, sets to a gel. It is widely used in the food industry, particularly in 'jellies'; as a medium for the preparation of drugs in the pharmaceutical industry; in adhesives and photographic emulsions; and in the preparation of media for growing bacteria in the laboratory.

gel electrophoresis ➤electrophoresis.

gel filtration A type of ➤chromatography in which the mixture of compounds to be separated is applied to a gel, usually packed in a glass column, and washed through with an appropriate solvent. The gel, which may be of cellulose or agarose origin, separates the ➤components (2) according to their different molecular sizes.

gelignite A form of dynamite used as an explosive.

gem- **(geminal-)** The prefix applied to compounds in which two identical substituents are attached to the same carbon atom. The *gem*-diols, for example, are less stable than their ➤*vic* isomers.

gene The fundamental unit of inheritance and function in a cell. The term 'gene' is used differently in different contexts. In classical genetics involving plant and animal breeding the gene is regarded as a unit of heredity: in peas, for example, 'tall' and 'dwarf' characters are controlled by different forms of the same gene called ➤alleles (➤➤Mendel's laws). A gene can also be considered as a discrete unit of a chromosome that determines the production of a specific ➤protein. The gene is now defined as a specific section of ➤DNA that codes for a recognizable cellular product, either ➤RNA or ➤polypeptide.

gene cloning A technique that enables multiple copies to be made of specific ➤DNA fragments and provides the ability to recover substantial amounts of single DNA fragments from extremely complex mixtures. The target DNA to be cloned is first cut, using ➤restriction enzymes. The resulting fragments are then added to a solution containing DNA from a suitable ➤vector (4) that has been cut with the same restriction enzyme. Commonly used vectors are ➤phages or ➤plasmids. The target DNA fragments recombine with the plasmid fragments, which can then be inserted into cells such as a bacterium or yeast. These multiply, producing copies of the inserted DNA. Since each plasmid can combine with only one of the many fragments of the target DNA, this can give rise to multiple individual clones of specific DNA fragments. DNA fragments cloned from a single source are collectively referred to as a ➤gene library.

gene expression The use of information encoded in the ➤DNA of ➤genes to produce functional molecules, either ➤RNA or proteins. For proteins, expression consists of the processes of ➤transcription, RNA processing (➤splicing) and ➤translation. Expression may be regulated at any of these levels, but is most commonly controlled at the initiation of transcription, depending on cell type, developmental stage, cell cycle stage or environmental cues.

gene library A collection of ➤DNA inserts cloned into a suitable ➤vector (4). A **cDNA library** contains inserts derived from complementary ➤DNA, while a **genomic library** contains inserts that are fragments of genomic DNA (➤genome). Gene libraries are extensively used in the cloning of genes, and in genome sequencing projects.

gene pool The sum total of all ➤genes in an interbreeding population at any given time. Different genes occur with varying frequencies depending on the selection pressure that the population may be experiencing. The isolation of specific gene pools with different gene frequencies is a major factor in the process of ➤evolution.

gene probe ➤DNA probe.

generalized coordinates A set of coordinates used for a dynamic system which are not necessarily the three spatial coordinates of each particle in the system. For example, a system of two masses joined by a rigid rod might most conveniently be described using the three coordinates of the centre of mass and two angles to describe the direction of the rod. For studying a coupled oscillating system, the amplitudes of each ➤normal mode provide a very useful set of generalized coordinates. ➤➤degrees of freedom; equations of motion; Lagrangian.

general (theory of) relativity The extension of ➤special relativity to regions of spacetime that are no longer ➤flat. The curvature of spacetime is caused by mass and manifests itself as ➤gravity. The mathematical theory revolves around the ➤Einstein field equation. ➤➤equivalence principle; geodesic.

general solution A solution to a ➤differential equation containing arbitrary constants equal in number to the ➤order (3) of the equation.

general term A formula that generates the terms of a ➤sequence when successive whole numbers are substituted into it.

generating function The generating function of a sequence of numbers or functions a_n is the associated power series $\sum_{n=0}^{\infty} a_n t^n$. If it can be expressed in closed form, properties of the sequence can often be easily deduced from the generating function.

generation time In ➤nuclear fission reactions, the mean time between the production of a neutron by fission and its participation in the fission of a different nucleus (creating a new **generation** of neutrons).

generator A device for transforming mechanical kinetic energy into electrical potential energy. ➤dynamo.

gene therapy A field of ➤genetic engineering in which diseases caused by genetic defects are identified and corrective treatment given. A functional gene can be inserted into an embryo or into mature cells where a defect has been identified. For example, treatment is being developed for sufferers of cystic fibrosis in which a DNA preparation is inhaled and becomes incorporated into the cells lining the respiratory tract, restoring some degree of normal function to affected cells.

genetically modified organism (GMO) An organism that has novel genes added (using ➤genetic engineering). **Golden rice** has added genes that produce high levels of ➤β-carotene (which is converted in the body into Vitamin A): this produces the yellow colour. Golden rice could potentially help the approximately 100 million people suffering from Vitamin A Deficiency. Genetic modification is however currently *highly* controversial, so much so that the term **Frankenstein food** has been coined for foodstuffs that have been genetically modified.

genetic code Information which is the basis of all ➤gene functions encoded as a linear sequence of the four ➤nitrogenous bases that form part of the structure of ➤DNA. After ➤transcription, when DNA is copied into ➤RNA, the information is encoded as sequences of three bases called **triplets** or **codons**. Each codon codes for a specific ➤amino acid to be incorporated into a specific position in the ➤polypeptide product of the gene. Since there are four nucleotides in RNA (and DNA), there are 64 possible codons. Polypeptide products of genes contain 20 amino acids, so some amino acids are coded for by more than one codon (see the table overleaf). The ➤termination codons halt the translation of a piece of RNA and signal the end of a gene.

genetic drift Fluctuations in gene frequencies in a population with respect to time due to random processes rather than ➤selection pressures. Such processes are dependent on the probabilities of mating between individuals with particular ➤genotypes. Drift is particularly evident in small, genetically isolated populations (such as on islands) in which the average genotype might diverge significantly from that of a larger parent population from which it may have been derived. Genetic drift was first described and quantified by mathematical models by Sewall Wright (1889–1988).

genetic engineering The technology that enables DNA fragments from different sources to be combined to make recombinant ➤DNA and insertion into cells, thus altering the function of the recipient ➤transgenic cells. Applications of genetic engineering include the production of human hormones in bacterial cells or in other mammalian systems such as in sheep's milk; the insertion of genes into crop plants to modify some property of the crop (e.g. to increase the shelf life of tomatoes), to make them disease-resistant, or to produce novel products such as plastics; and ➤gene therapy to alleviate the symptoms of genetically caused disease. Some aspects of genetic engineering have raised ethical and moral issues such as the possibility of breeding pigs, engineered to match human immune systems, as a source of organs for transplant; and the manipulation of crop plants to be resistant to certain chemical herbicides.

	AGA					
	AGG					
GCA	CGA					
GCG	CGG					
GCU	CGU	AAU	GAU	UGU	CAA	GAA
GCC	CGC	AAC	GAC	UGC	CAG	GAG

| Amino acid | Ala | Arg | Asn | Asp | Cys | Gln | Glu |

			UUA			
			UUG			
GGA			CUA			
GGG		AUA	CUG		(Start)	
GGU	CAU	AUU	CUU	AAA		UUU
GGC	CAC	AUC	CUC	AAG	AUG	UUC

| Amino acid | Gly | His | Ile | Leu | Lys | Met | Phe |

	AGU					
	AGC					
CCA	UCA	ACA			GUA	
CCG	UCG	ACG			GUG	UAA
CCU	UCU	ACU		UAU	GUU	UAG
CCC	UCC	ACC	UGG	UAC	GUC	UGA

| Amino acid | Pro | Ser | Thr | Trp | Tyr | Val | (Stop) |

Termination

genetic code The 64 possible 3-base triplets, called codons, derived from combinations of the four bases present in RNA and the amino acids encoded by them. The code is degenerate in that some amino acids are coded for by more than one codon. The codons UAA, UAG and UGA are termed 'stop' or termination codons, and they signal the end of an open reading frame. The codon AUG, coding for methionine, is the initiation codon, starting an open reading frame.

genetic fingerprinting (DNA fingerprinting) A technique which enables genetic relationships between close relatives or the identity of an individual to be established. The term 'fingerprint' refers to the original method that produced from gel ►electrophoresis a series of dark bands on an X-ray film by using radioactively labelled DNA (autoradiography). The modern variant of this method, called a **DNA profile**, consists of the determination of the lengths of four or five selected chromosomal fragments, called **Variable Number Tandem Repeats** (or **VNTR**); chemiluminescent probes have largely replaced radioactive ones. DNA profiling, in association with PCR (►polymerase chain reaction) can be used to identify individuals from very small samples of DNA left in fingerprints, or in blood or other body fluid samples. It thus has a wide range of applications in forensic science.

genetic imprinting The phenomenon whereby the behaviour of an ➤allele in a ➤diploid organism depends on the parent from which the allele was inherited. Thus an imprinted gene might be expressed from the allele inherited from the father, but not from the allele inherited from the mother. The imprinted state is maintained by the pattern of ➤DNA methylation of the gene. Imprinted genes are generally found on ➤autosomes. Compare ➤sex-linkage.

genetic map (linkage map) An ordering of a set of ➤genes, with the distance between two adjacent genes based on the frequency of meiotic recombination (➤meiosis) between them. A genetic map is an abstract concept, but the order of genes corresponds to their order on chromosomes. The distances, however, are not directly related to physical position on chromosomes, since recombination frequency varies along the length of a chromosome. A genetic map may consist of several **linkage groups**; genes in one linkage group segregate randomly with respect to genes in other linkage groups. Thus there is at least one linkage group per chromosome. Compare ➤physical map.

genetics The study of heredity and the ➤gene. Modern genetics includes the field of molecular genetics, which encompasses the study and manipulation of ➤DNA.

genome The total genetic complement of a cell or organism. **Genomic DNA** is the sum total of all of an individual's DNA present in one cell. The total human genome consists of an estimated 3 billion ➤base pairs (6 billion in double-stranded DNA). ➤➤human genome project; megabase.

genotype A statement of the genetic constitution of an organism that describes in shorthand form the ➤alleles present. Thus a tall pea plant may have the genotype TT, which indicates that only T (tall) alleles are present, or Tt, which indicates that half the alleles present are T and half are t (dwarf). ➤➤heterozygous; ➤➤homozygous; compare ➤phenotype.

genus (plural **genera)** A unit of classification (➤taxonomy) which contains one or many ➤species. Genera are grouped together in a ➤family.

geocentric system An incorrect theory of the structure of the Solar System which held that the Earth is at the centre. It has been decisively replaced by the ➤heliocentric system.

geodesic The path of shortest length between two points in a ➤Riemannian geometry; especially used to mean the shortest interval between two events in ➤spacetime. A geodesic is the four-dimensional equivalent of a straight line; however, the **interval**, the four-dimensional spacetime equivalent of length in three dimensions, is defined by the ➤metric which at any given point depends on the curvature of spacetime. Thus geodesics do not necessarily appear to be straight lines in our more familiar three-dimensional space. Light rays travel along geodesics, and it was the observation of the apparent bending of light that was one of the early experimental confirmations of ➤general relativity.

geological era ➤era.

geology The study of the origin, distribution and form of the rocks of the Earth's crust. Geology is further subdivided into specialities such as palaeontology, the study of fossils.

geomagnetism The phenomenon of the ➤Earth's magnetic field. The origin of the field is the slight difference in the rotation speeds of the inner and outer cores, plus convection currents in the molten outer core. This produces an approximate ➤magnetic dipole with its axis at about 17° to the rotational axis of the Earth. The alignment of magnetic compasses with the magnetic field is an important aid in the navigation of ships and aircraft, except close to the poles. The magnitude of the magnetic field is typically about 50 μT, with its direction determined by the ➤magnetic declination and ➤magnetic inclination at any point on the Earth's surface. Studies of magnetic minerals in rocks (magnetism in rocks indicates the alignment of the Earth's magnetic field when they were formed) show that **polarity reversals** have occurred in the past, in which the field has flipped through 180°.

geometric isomerism ➤cis–trans isomerism.

geometric mean The geometric mean of the positive numbers $a_1, a_2, \ldots, a_n$ is the quantity $(a_1 a_2 \ldots a_n)^{1/n}$.

geometric optics The study of light in the form of ➤rays (compare wave ➤optics). Geometric optics treats reflection and refraction of light, but not diffraction or interference. It is the usual model used in the design of optical instruments containing lenses. ➤➤Fermat's principle of least time.

geometric progression A ➤sequence in which the ➤ratio between successive terms is constant, e.g. the number of radioactive atoms halves during each ➤half-life. Compare ➤arithmetic progression.

geometry ➤Euclidean geometry; non-Euclidean geometry; projective geometry; Riemannian geometry.

geophysics The physics of the Earth, including its formation, its dynamics, and the structure of its surface and atmosphere.

geostationary orbit A near-circular satellite orbit, with a period of exactly one day, in the plane of the Earth's equator. The satellite is overhead at the same point on the surface of the Earth at all times. From the third of ➤Kepler's laws it is clear that this orbit is only possible at one particular radius, 42 200 km. Compare ➤geosynchronous orbit.

geostrophic flow Flow of a rotating fluid where the dominant force is the ➤Coriolis force. In meteorology, a **geostrophic wind** is one blowing parallel to the ➤isobars, which results from a balance between pressure-gradient and Coriolis forces at the surface of the Earth.

geosynchronous orbit A near-circular satellite orbit with a period of exactly one day. The satellite remains above the same ➤meridian at all times. Compare ➤geostationary orbit.

geothermal energy Energy derived from the transfer of heat from the core to the surface of the Earth.

geraniol A terpene alcohol responsible for the smell of spearmint.

germane GeH_4 The germanium analogue of methane, CH_4, which is also a colourless gas.

germanium Symbol Ge. The element with atomic number 32 and relative atomic mass 72.61, which lies immediately below silicon in Group 14. Its chemistry is dominated by compounds with ➤oxidation number +4, as for silicon. Like silicon, it forms a colourless volatile liquid chloride, $GeCl_4$, which is rapidly hydrolysed, and a white solid oxide, GeO_2, which is amphoteric rather than acidic, following the usual trend down a group. Germanates exist that resemble silicates, although their complexity is not so profound. The most useful substance is the element itself which is, again like silicon, a semiconductor with widespread uses in the electronics industry.

germ cell A cell that gives rise to a ➤gamete.

germination The first phase of development of a ➤seed or spore when growth resumes after a dormant period. The process is initiated by the uptake of water through the seed or spore coat. This mobilizes food reserves and enzymes, allowing ➤metabolism and subsequent growth.

germ layers The layers of cells in an animal ➤embryo which are destined to give rise to particular tissues and organs in the adult.

germ theory of disease A unifying concept that all diseases are caused by microorganisms (traditionally referred to as 'germs'). The theory was expounded and developed by the French chemist Louis Pasteur (1822–95).

gestation The period of time between the implantation of a fertilized egg into the ➤uterus wall of a mammal and the birth of the young. In humans the gestation period is 40 weeks (approximately nine months).

getter One material used to remove another from a container by chemical bonding. One example is the use of a titanium sublimation pump to achieve an ultra-high vacuum: titanium, the getter, is evaporated by heating on to the walls of the chamber where it bonds with any gas molecules that adsorb. Thermionic valves incorporate lithium or magnesium to combine with any remaining nitrogen or oxygen.

GeV Symbol for gigaelectronvolt (10^9 ➤electronvolts); the spelt-out form is rarely written or spoken. Its value is about 1.6×10^{-10} J.

GH Abbr. for ➤growth hormone.

ghosts In a spectrum, peaks that are artefacts of the spectrometer itself, such as lines due to imperfections in a grating.

giant (giant star) A star that lies above the main sequence on the ➤Hertzsprung–Russell diagram. Giants are so called to distinguish them from ➤dwarf stars. They have a lower surface temperature and higher luminosity than dwarfs, and are therefore much larger. Giants evolve from dwarf stars when their hydrogen content is exhausted and nuclear fusion of heavier elements begins. ➤➤red giant; supergiant.

gibberellin (gibberellic acid, GA) A class of plant growth substances originally isolated from the fungus *Gibberella fujikuroi*, which infects and causes deformities in rice. There is a range of gibberellins, of which the most studied is GA_3. Gibberellins have a range of dramatic effects on plant growth, particularly in the elongation of cells in growing shoots. When seedlings of genetically dwarf varieties of plants are treated with GA they grow tall and are physically indistinguishable from tall varieties. GA is also involved in the breaking of dormancy in many seeds and buds. At the molecular level GA is implicated in the promotion of genes coding for starch-digesting enzymes (amylases) in seeds.

Gibbs energy (Gibbs free energy, Gibbs function) Symbol G. An important quantity in thermodynamics, defined by the equation

$$G = H - TS,$$

where H is the ➤enthalpy, T the ➤thermodynamic temperature and S the ➤entropy. Under conditions of constant pressure and temperature, the direction of a reaction's progress is in the direction of lower Gibbs energy. At constant temperature, the Gibbs energy change ΔG is related to the enthalpy change ΔH and the entropy change ΔS by

$$\Delta G = \Delta H - T\Delta S.$$

Strictly, the Gibbs energy change is measured at a specific composition and is really a differential quantity: equilibrium occurs when the Gibbs energy is a minimum (see the diagram). The ➤standard Gibbs energy change, $\Delta G^{\ominus}$, is important in chemical equilibrium as there is a direct link between it and the ➤equilibrium constant, K, for a reaction:

$$\Delta G^{\ominus} = -RT \ln K.$$

Named after (Josiah) Willard Gibbs (1839–1903).

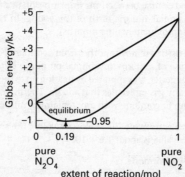

Gibbs energy The straight line has a slope equal to the standard Gibbs energy change for the decomposition of pure N_2O_4 into pure NO_2. The Gibbs energy change at any intermediate extent of reaction is affected by the entropy of mixing of reactants and products, in this case the dimer and monomer, respectively. This is a positive entropy change, and corresponds to a negative Gibbs energy change. The minimum Gibbs energy corresponds to chemical equilibrium.

Gibbs–Helmholtz equation A thermodynamic relation from which the ➤Gibbs energy change may be calculated as a function of temperature:

$$\frac{d}{dT}\left(\frac{\Delta G}{T}\right) = -\frac{\Delta H}{T^2}.$$

Named after J. W. Gibbs and Hermann Ludwig Ferdinand von Helmholtz (1821–94).

Gibbs isotherm ➤van't Hoff isotherm.

gibbsite A native form of aluminium hydroxide, $Al(OH)_3$.

GIF Abbr. for graphics interchange format. A standard format for a compressed graphics file that supports 256 colours, which is used extensively on the ➤Internet in Web pages (➤www).

giga- A prefix in the SI system meaning 10^9 base units, for example 1 GHz is equivalent to 10^9 Hz.

gill 1 A gas-exchange organ in many aquatic animals. Gills serve to increase the surface area available for gas exchange by diffusion. The annelids, molluscs and arthropods have a variety of body extensions and projections which are referred to as gills. Gills are most advanced in fish, where they are supported by skeletal structures in the ➤pharynx; complex circulatory systems promote efficient gas exchange as water is passed over the gills, through the mouth and out through the gill slits. The possession of gill slits at an early stage of embryonic development is a characteristic feature of a ➤chordate.
  2 A spore-bearing structure in the Basidiomycota, a phylum of fungi which includes the familiar mushrooms and toadstools. Gills form the characteristic radial pattern on the underside.

gizzard The anterior part of the gut of many animals, particularly those that lack teeth, in which food is ground up. The gizzard of crustaceans such as crabs has tooth-like internal projections, whereas that of birds contains grit ingested with or in addition to the food.

glacial Glass-like, as in 'glacial ethanoic acid', whose melting point is sufficiently close to room temperature that during cold winter nights it solidifies to a glassy solid.

gland 1 A specialized multicellular structure that produces a substance needed for the normal functioning of an animal. ➤endocrine gland; exocrine gland.
  2 A spheroidal structure on or below the surface of a plant organ, or borne on a hair, that produces a variety of substances. The hairs on stinging nettles and on many insectivorous plants are examples of glands, as are the nectaries which produce sugary solutions to attract animal pollinators.

Glashow–Weinberg–Salam model ➤electroweak theory. Named after Sheldon Lee Glashow (b. 1932), Steven Weinberg (b. 1933) and Abdus Salam (1926–96).

glass A hard but brittle amorphous solid phase of a substance, often described loosely as a ➤supercooled liquid. The everyday meaning of the word refers to the translucent glass formed when molten silica or silicates are cooled. A glass has no fixed melting point, softening over a range of temperatures; this behaviour allows the material to be 'blown' into different shapes. The most common type of glass is made by fusing sand, essentially silica, SiO_2, with calcium and sodium carbonates. ➤Pyrex also contains boron oxide, B_2O_3, and is a **borosilicate glass**. Specialized glasses incorporate small quantities of other metallic compounds in the melt either to colour the glass or to impart specific properties. Extra-low dispersion (ED) glass is used in high-quality optical instruments. **Laminated glass** consists of two pieces of glass sandwiched round a plastic filling, and does not splinter upon impact, a vital property for windscreens.

glass electrode An electrode made out of a highly specialized glass that is porous to hydrogen ions (➤pH meter).

glass fibre A material made from thin fibres of glass (less than 1 μm in diameter) bonded together with resin. It is used where strong but light materials are an advantage, for example in some cars, boats and aircraft. ➤➤fibre, optical.

Glauber's salt Sodium sulfate decahydrate, $Na_2SO_4 \cdot 10H_2O$. Named after Johann Rudolf Glauber (1604–70).

glaucoma A clinical condition of the ➤eye in which the fluid (aqueous humour) between the cornea and the lens exerts a pressure on the lens causing sight defects and other symptoms such as headaches. It can be treated either by drugs in the early stages or by surgery.

GLC Abbr. for gas–liquid chromatography (➤gas chromatography).

Gln Abbr. for ➤glutamine.

global positioning system (GPS) A navigation system that allows extremely accurate measurement of position by comparing the time of travel of a radio signal from a number of satellites (with known orbits). Depending on the number of signals received from different satellites, the mean error can be as low as a few metres.

global warming A theory of climatology (now officially endorsed by the premier scientific society in each of the G8 countries) which suggests that the mean temperature of the Earth's atmosphere is rising under the influence of the enhanced ➤greenhouse effect caused by the emission of greenhouse gases. When Tom Avery recreated in 2005 Robert Peary's first successful attempt in 1909 to reach the North Pole, he could only hope to do the *outward* journey (both expeditions took 37 days). The return journey was impossible as too much open water would have been present: the average ice depth has fallen *by a third* in less than a century.

globular cluster ➤cluster.

globulin A water-soluble, spheroidal class of ➤proteins. Globulins constitute the majority of molecular species of biologically active proteins in living systems such as ➤enzymes and ➤antibodies, and are of four main types (termed alpha 1, alpha 2, beta and gamma). **Gamma globulins** are a group of important antibodies produced in the liver, spleen and bone marrow, and are used, for example, in inoculations for protection against hepatitis.

glomerulus (plural **glomeruli**) ➤kidney.

glottis ➤larynx.

glove box A sealed box with gloves incorporated into one side, enabling dangerous or sensitive materials to be handled without risk of contamination.

Glu Abbr. for ►glutamic acid.

glucagon A ►hormone secreted by the pancreas involved in the regulation of ►glucose concentration in the blood. ►diabetes.

glucose A common white crystalline ►monosaccharide hexose sugar (see the diagram), widely distributed in nature. In many organisms it forms the starting point for the process of ►respiration and, as such, is a common energy source in living systems. It is the '**blood sugar**' in mammalian blood. Glucose is optically active and its ability to rotate polarized light gives it its archaic name of 'dextrose'. Glucose is also a subunit of many disaccharides such as ►sucrose and polysaccharides such as ►cellulose.

D-glucose The hydroxyl at carbon atom 1 is in the ►beta position.

glucosinolate One of a number of about 70 derivatives of ►glucose containing nitrogen and sulfur, particularly found in plants of the brassica family (cabbages, mustards and rape), producing their characteristic flavours. They have been implicated in the defence of these plants against fungal and insect attack; artificial stimulation of glucosinolate production by treatment of plants with substances such as aspirin is proving a useful method of disease control.

gluon The ►gauge boson of the ►strong interaction. Protons and neutrons, for example, interact by gluon exchange.

glutamic acid (Glu) A common ►amino acid. One derivative of this dibasic acid, ►monosodium glutamate, is a widely used flavour enhancer in the food industry. ►►Appendix table 7.

glutamine (Gln) A common ►amino acid, the acid amide of glutamic acid. ►►Appendix table 7.

gluten A complex protein of variable composition occurring in wheat grains. The exact gluten content and composition of the grain determines the character of the flour made from it, and whether it is best suited for making bread, pasta or biscuits. The lining of the gut is sensitive to gluten in people with **coeliac disease**, and their diets must be gluten-free.

Glx Abbr. for either glutamine or glutamic acid.

Gly Abbr. for ►glycine.

glyceraldehyde A molecule closely related to ➤glycerol, and formed from it by mild oxidation. The identification of the ➤optical activity of glyceraldehyde was a crucial step to working out the structures of all carbohydrates, which can be related back to that of glyceraldehyde (see the diagram). Glyceraldehyde is a key intermediate in many biochemical pathways.

$$CHO$$
$$H \longrightarrow OH$$
$$CH_2OH$$

D-glyceraldehyde

glyceride An ester of glycerol. One, two or three acid groups can be attached. Lipids are triglycerides (now called ➤triacylglycerols).

glycerol (glycerine, propane-1,2,3-triol) A trihydric alcohol which is a fundamental component of ➤lipids found in living systems (see the diagram). Commonly three ➤fatty acids are bound to a glycerol molecule to form ➤triacylglycerols, which are important as structural components of membranes, as sources of energy, and as metabolic intermediates in many biochemical pathways.

$$CH_2OH$$
$$CHOH$$
$$CH_2OH$$

glycerol

glyceryl trinitrate ➤nitroglycerine.

glycine (Gly) NH_2CH_2COOH The simplest ➤amino acid. It can be made from bromoethanoic acid using the ➤Gabriel synthesis. The colourless crystals dissolve in water and, unusually for an amino acid, the solution is not optically active, as two hydrogens are bonded to the central carbon atom. ➤➤Appendix table 7.

glycocalyx ➤plasma membrane.

glycogen A ➤polysaccharide found in animals, where it is used as a storage form of ➤glucose. Glycogen is closely related to plant ➤starch but is more branched and generally larger.

glycol Traditional name for a ➤diol.

glycolipid One of a number of various types of ➤lipid in which a fatty acid chain is linked to a sugar, usually glucose or galactose. Glycolipids form an important layer on the outer surface of ➤ plasma membranes.

glycolysis (Embden–Meyerhof pathway) A series of reactions in which ➤glucose is converted into two molecules of ➤pyruvic acid which are then converted into either lactate or into ethanol and carbon dioxide (during ➤fermentation). The process is ➤anaerobic and represents the first phase in the breakdown of glucose during ➤respiration. A number of genetically inherited errors of metabolism are caused by defects in ➤genes controlling the synthesis of ➤enzymes used in the pathway, as is the case in haemolytic anaemia.

glycoprotein A molecule consisting of a protein unit complexed with a carbohydrate. Glycoproteins form important structural and functional components of ➤plasma membranes. The addition of the carbohydrate to the protein is initiated in the ➤endoplasmic reticulum and is completed in the ➤Golgi body.

glycoside A molecule consisting of a sugar unit such as glucose joined with another molecule such that the aldehyde group in the sugar is replaced by a different group, such as ➤acetal (see ➤sucrose for a diagram). The individual sugar molecules in a disaccharide, such as sucrose, or a ➤polysaccharide are joined by **glycosidic bonds**. Glycosides are very varied in structure and have a wide range of functions. The plant pigment ➤anthocyanins are glycosides, as are several substances with medical significance, for example digitalin used as a heart stimulant.

gnotobiotic Referring to the germ-free conditions in which experimental animals may be kept, particularly those which may be artificially infected with disease.

Gödel's theorems Two seminal results in mathematical logic. **Gödel's incompleteness theorem** concerns a noncontradictory set of ➤axioms sufficiently rich to capture the arithmetic properties of whole numbers; it asserts that there are true statements about whole numbers that cannot be deduced from the axioms and so are undecidable within that set of axioms. The other theorem states that it is impossible to prove, within the system, that the set of axioms is noncontradictory. Named after Kurt Gödel (1906–78).

goitre ➤thyroid gland.

gold Symbol Au (from the Latin *aurum*). The element with atomic number 79 and relative atomic mass 197.0, which is in Group 11, below copper and silver. In common with its ➤congeners it is a metal with a high conductivity and a low reactivity, hence its use for high-quality electrical connectors. Its low reactivity and attractive lustre account for its use in jewellery. The cyanide process, which relies on the formation of the linear complex $[Au(CN)_2]^-$, can extract gold at lower concentration than was possible by simple 'panning'. Gold is the most malleable element, which is why gold foil was chosen for the ➤Geiger–Marsden experiment.

Its chemistry is dominated by two ➤oxidation numbers. It shows oxidation number +1, as in the chloride AuCl, a yellow crystalline solid. Whereas the higher common oxidation number for copper is +2, that for gold is +3, as in the chloride $AuCl_3$, a red solid. Both oxidation numbers can also be found in complexes, usually linear for gold(I), as in the cyanide complex above, or square planar for gold(III), such as the tetrachloroaurate(III) ion $AuCl_4^-$.

golden ratio (golden number) ➤Fibonacci number.

gold-leaf electroscope An apparatus used in electrostatics to show the presence of electric charge. It consists of a foil strip (traditionally of gold leaf) fixed at its top edge to an insulated metal rod, enclosed in a draughtproof glass-sided case. When a charged object is brought near the top of the rod, the bottom edge of the strip moves away from the rod; the cause is mutual electrostatic repulsion, as both rod and foil acquire the same sign of charge.

Goldschmidt process The reduction of metal oxides with aluminium (➤thermit reaction). Named after Hans Goldschmidt (1861–1923).

Golgi body (Golgi apparatus) An ➤organelle of ➤eukaryotic cells consisting of a series of stacked pouches (**cisternae**) complexed with networks of tubules. It appears that components of membranes are synthesized and transported in the Golgi body. Varying numbers are found in cells, from one or two to over a hundred. The Golgi body is also the centre for ➤glycoprotein metabolism, and is involved in ➤mitosis and ➤meiosis. In plant cells Golgi bodies play a major role in the synthesis of cellulose derivatives used in the structure of cell walls. Named after Camillo Golgi (1843–1926).

gonad An animal organ that produces ➤gametes. ➤Testes produce sperm, and ➤ovaries produce eggs. The **gonadotrophins** are ➤hormones produced by the anterior ➤pituitary gland, and control the functioning of the gonads as well as aspects of the function and development of the whole animal (e.g. ➤follicle stimulating hormone; luteinizing hormone).

Gondwana (Gondwanaland) One of the two major continental land masses postulated to exist in Mesozoic times (the other being ➤Laurasia), consisting of the present-day South America, Africa, India, Australasia and Antarctica. Gondwana fragmented into the present-day continents by the movements of massive plates (➤continental drift). The presence of Gondwana and its subsequent division has been a major factor influencing the distribution of groups of animals such as marsupials and the flightless birds of New Zealand.

goniometer An instrument for measuring bearings or angles used, for example, in X-ray diffraction.

Gouy balance A balance for measuring the magnetization of a paramagnetic substance. The substance's attraction into a magnetic field is balanced by the gravitational attraction of a counterweight. Named after Louis-Georges Gouy (1854–1926).

governor The general name for a device that limits the speed of a machine to a particular value or range of values. For example, a constant-speed propeller fitted to an aircraft piston engine uses a governor: if the speed of the engine increases, a system of weights causes the pitch of the propeller to become coarser, increasing the load and thus reducing the speed. This is a good example of negative ➤feedback.

GPS Abbr. for ➤global positioning system.

grad A differential operator; the name is an abbreviation for the ➤gradient (3) of a scalar field.

grade (grad) A measure of angle such that there are 100 grades in a right angle.

gradient 1 A measure of the slope of a straight line given by the constant value of 'change in y value/change in x value' for any two points on the line.

2 The gradient of a curve at a point is the gradient of its tangent at that point, equal to the ➤derivative, dy/dx.

3 The gradient of a ➤scalar field ϕ is the ➤vector field

$$\text{grad } \phi = \nabla\phi = \frac{\partial\phi}{\partial x}\mathbf{i} + \frac{\partial\phi}{\partial y}\mathbf{j} + \frac{\partial\phi}{\partial z}\mathbf{k}.$$

graduated An instrument is graduated if there are marks on it to indicate a scale of measurements. Examples are a measuring cylinder and a burette; a pipette, on the other hand, is not usually graduated.

graft (transplant) A piece of tissue that is isolated and joined to another tissue, either in the same or a different organism, so that subsequent growth results in fusion of the tissues. Grafts are used extensively in horticulture as an artificial means of asexual propagation of disease-resistant root stock, particularly in roses and fruit trees. Grafting is also used surgically in the treatment of burns, where skin from one part of the body is grafted to repair damage to another part. The transplanting of tissue from one individual to another (e.g. in bone marrow or organ transplants) is called an **allograft**.

Graham's law The law describing the rate of diffusion or effusion of an ideal gas: the rate is inversely proportional to the square root of the density of the gas (or, equivalently, its molar mass, as the molar volume of a gas is constant). Named after Thomas Graham (1805–69).

grain ➤polycrystalline.

gram (gramme) Symbol g. A metric unit of mass. The SI base unit is 1 kg (equivalent to 1000 g); the base unit in the older c.g.s. system is the gram itself.

gram-atom An obsolete term for ➤atomic weight, expressed in grams. Similar terms were **gram-equivalent** for ➤equivalent weight, and **gram-molecule** for ➤molecular weight. All these terms have been superseded by ➤molar mass.

gram molecular volume An obsolete term now replaced by the ➤molar volume. Similarly, **gram molecular weight** has given way to ➤molar mass.

Gram's stain A technique used in the classification of ➤bacteria that relies on the differential ability of bacterial cells to take up and retain certain stains. **Gram-positive** bacteria appear violet. **Gram-negative** bacteria appear red or pink. The abilities to retain dyes differ because Gram-positive bacteria have a thicker wall outside the cytoplasmic membrane, which traps more of the strain. There are also physiological differences between the two types of cell, which may otherwise be morphologically similar. Gram-positive bacteria are more sensitive to penicillin and some form resistant spores; Gram-negative bacteria are more readily attacked by antibodies and do not form spores. Named after Hans Christian Joachim Gram (1853–1938).

grand unified theory (GUT) A much-sought-after theory relating three of the four ➤fundamental interactions; it must therefore unify the ➤electroweak theory and the ➤strong interaction.

granite An important igneous rock used for building houses. A particular danger associated with granite is its capacity to retain radon, causing a build-up in natural environmental radioactivity.

granum (plural **grana**) ➤chloroplast.

graph 1 A diagram to show the relationship between the values of ➤variables by plotting points with respect to the scales on coordinate axes (➤coordinate system) representing the variables.
  2 A diagram to illustrate a ➤function f by means of a ➤curve joining the points (x, y) for which $y = f(x)$. Functions of two variables may similarly be represented by curved surfaces.
  3 ➤graph theory.

graphic formula A full structural ➤formula.

graphite (plumbago, black lead) The allotrope (➤allotropy) of carbon which is most thermodynamically stable at room temperature and pressure. Its structure consists of layers of hexagonal sheets of carbon atoms (see the diagram). ➤Delocalization within each layer causes graphite to conduct electricity, unlike carbon's other main allotrope, ➤diamond. The conductivity is five thousand times better along the layers than perpendicular to them. Together with its slight chemical reactivity, this makes graphite a common choice for the electrode material in

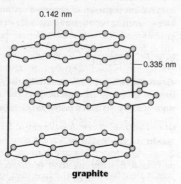
graphite

electrolysis. It does, however, combust, so in the ➤Hall–Héroult process, for example, the graphite electrodes need to be replaced regularly. Other uses include the 'lead' in pencils, where layers can be rubbed off onto the paper (which explains its peculiar ancient names of black lead and, via Latin, plumbago).

graph theory That branch of mathematics concerned with the properties of graphs, that is finite ➤sets of points (**vertices** or **nodes**) joined by lines (**edges** or **arcs**).

graticule A grid of fine lines set at the focal point of an optical instrument, usually a microscope, used to aid the measurement of lengths in the image.

grating An optical aperture consisting of an array of many fine lines. ➤diffraction grating.

gravimetric analysis A method of quantitative analysis that involves the precipitation of a highly insoluble compound followed by weighing of the dried precipitate formed.

gravitation ➤gravity.

gravitational collapse ➤supernova.

gravitational constant ➤Newton's law of (universal) gravitation; Appendix table 2.

gravitational field Symbol g. The gravitational force acting per unit mass. The gravitational field may vary with position, and so is a ➤vector field. Because it also stems from an ➤inverse square law, its properties are similar to those of the ➤electric field, which plays the corresponding role in electrostatics. The gravitational field therefore obeys an equation analogous to the first of ➤Maxwell's equations:

$$\nabla \cdot g(\mathbf{r}) = -4\pi G\rho(\mathbf{r}),$$

where $\rho(\mathbf{r})$ is the mass density at position $\mathbf{r}$. In particular, ➤Gauss's law also applies to the gravitational field.

gravitational interaction The weakest of the four ➤fundamental interactions, corresponding to the force of ➤gravity. The ratio of the repulsive ➤electrostatic force and the attractive gravitational force between two protons is about $10^{36} : 1$. Gravitation is therefore observed only where charges are extremely well balanced, which tends to be when at least one of the bodies is extremely massive, such as a star or planet.

gravitational mass A measure of the mass of a body that is proportional to the gravitational force between it and other bodies. The distinction between gravitational mass and ➤inertial mass is usually unnecessary because for any body they are both known, empirically, to be equal to a very high precision. This equality is suggested by ➤Mach's principle and follows from the ➤equivalence principle.

gravitational wave A wave emitted from an accelerating mass (analogous to the way an electromagnetic wave is emitted from an accelerating charge). In terms of general relativity, the wave is an oscillation of the ➤metric in time and space. The particle corresponding to it is the graviton, as the photon corresponds to the electromagnetic wave. No gravitational waves have yet been detected, since the amplitude of such waves is very small compared to other sources of vibration.

gravitino ➤supergravity.

graviton The hypothetical ➤quantum of the gravitational field. It is postulated to have zero mass and a spin of two.

gravity (gravitation) The phenomenon that causes an attractive force between any two bodies that have ➤mass. The theory of gravitation in nonrelativistic physics is based around ➤Newton's law of gravitation. ➤General relativity treats gravitation in circumstances in which the approximations of nonrelativistic physics break down around objects of high density, such as ➤neutron stars or ➤black holes.

gray Symbol Gy. The SI unit of radiation dose equivalent to 100 rad. Named after Louis Harold Gray (1905–65).

great circle A circle on the surface of a sphere whose plane passes through the centre of the sphere. The shortest distance between two points on a sphere is a segment of a great circle: thus great-circle routes over the Earth's surface are used for ships and aircraft covering large distances. If a grid of latitude and longitude is superimposed on the sphere, as is conventional for the surface of the Earth, the great-circle route will pass closer to the nearer pole than a track of constant bearing (so an aircraft flying from London to Los Angeles will first fly north-west over Scotland, even though Los Angeles is much further south than London).

greenhouse effect The effect that causes a greenhouse to have a higher temperature on a sunny day than the air outside. Heat is transmitted into the greenhouse in the form of high-frequency, short-wavelength radiation from the Sun. The interior of the greenhouse warms as it absorbs the radiation, and re-emits it as longer-wavelength, lower-frequency infrared radiation, but this cannot be radiated from the greenhouse at the same rate because the panes of glass block the infrared radiation and trap the heat inside. The name 'greenhouse effect' is therefore given to any process where re-radiation of incoming heat is restricted in a similar way. The atmospheres of planets such as Venus and the Earth cause a greenhouse effect because they contain **greenhouse gases** (such as water, carbon dioxide and methane) which absorb in the infrared. Note that the greenhouse effect keeps the mean temperature of the Earth some 35 K higher than it would be without the atmosphere, and is thus essential to the continued existence of most life forms. A higher level of greenhouse gases, caused for example by more carbon dioxide present due to the combustion of fossil fuels, leads to an **enhanced greenhouse effect** and causes ➤global warming. The level of carbon dioxide has risen by about a quarter over the last thirty years and this represents a significant rise, since it exceeds the natural variation between the seasons. The enhanced level of carbon dioxide is experimentally established; the long-term climatic effect is currently uncertain.

Green's function A type of function introduced by George Green (1793–1841) to facilitate the solution of differential equations such as Laplace's equation. It typically represents the response of a system to an input that has the form of a ➤Dirac delta function.

green vitriol A green salt derived from 'oil of vitriol' (sulfuric acid), hydrated iron(II) sulfate, $FeSO_4 \cdot 7H_2O$.

Gregorian calendar The calendar in common use in the Western world. Each year has 365 days, but to compensate for the extra fraction of a day making up a year, every fourth year, known as a **leap year**, has 366 days. This keeps the calendar closely in step with the ➤tropical year, but for fine adjustment every hundredth year that is not a multiple of four hundred is *not* a leap year. This makes the mean calendar year equal to 365.2425 days, which is within one part per million of the tropical year. Named after Pope Gregory XIII (1502–85).

grey matter ➤brain; spinal cord.

grid An electrode other than the anode and cathode in a ➤thermionic valve, usually introduced into the valve for control purposes.

Grignard reaction One of the five most useful synthetic reactions in organic chemistry, capable of forming several classes of molecules (see the table). The **Grignard reagents** are aryl or alkyl magnesium halides, such as ethylmagnesium bromide, CH_3CH_2MgBr, made *in situ* from bromoethane and magnesium in an ether solvent. What makes them so useful is that the carbon–magnesium bond is polarized such that the carbon is slightly *negatively* charged, unlike the situation in halogenoalkanes, alcohols, aldehydes, ketones, etc. As a result, the Grignard reagent can act as a ➤nucleophile. Named after (François Auguste) Victor Grignard (1871–1935).

Grignard reaction Products from the reaction of a Grignard reagent RMgBr with carbonyl compounds.

Reactant	Hydrolysis product	Yield
Methanal (formaldehyde), HCHO	Primary alcohol, RCH_2OH	Good
Other aldehydes, R′CHO	Secondary alcohol, RR′CHOH	Good
Ketones, R′R″CO	Tertiary alcohol, RR′R″COH	Good to poor
Carbon dioxide, CO_2	Carboxylic acid, RCOOH	Good

ground In an electrical circuit, an alternative (chiefly US) name for ➤earth.

ground state The lowest energy state of a quantum-mechanical system. Unlike the lowest energy possible in a classical system, the ground state in quantum mechanics has a non-zero kinetic energy (the ➤zero-point energy). Compare ➤excited state.

ground wave An electromagnetic wave (usually in the ➤high-frequency band or lower) that travels close to the ground, diffracting around the curvature of the surface of the Earth. For ➤VHF and above, the diffraction is negligible and the path followed by the wave is almost a straight line; thus the absence of a ground wave makes VHF communications possible at short-range only. Compare ➤ionospheric wave.

group 1 (Chem.) A vertical column in the ➤periodic table containing elements that resemble each other. The elements of Group 1, the alkali metals, are very alike, being reactive metals which form alkaline hydroxides and ionically bonded compounds exclusively with oxidation number +1. Group 18 contains all the noble gases. For the ➤p-block nonmetals, there are more significant variations within the group, but there remain close similarities. For example, in Group 17, the ➤halogens, the most common oxidation number is -1 in every case and the sodium halides are white crystalline solids which appear almost identical.

The numbering of groups was changed in 1986. Before then the alkali metals were numbered as Group I and the halogens as Group VII. Confusion arose because historically a letter was occasionally added for the p-block elements; UK authors used Group VIIB, and US authors used Group VIIA. To avoid this confusion and to make more explicit use of the knowledge about electronic structure arising from the Schrödinger equation, the groups were renumbered from 1 to 18. ➤➤Appendix table 5.

2 (Math.) A mathematical object consisting of a ➤set of ➤elements equipped with an ➤associative ➤binary operation which is not necessarily ➤commutative but for which there is an ➤identity (2) element and ➤inverses for every element. ➤➤group theory.

group representation A ➤homomorphism from a given ➤group (2) to a group of invertible matrices. Because matrices are well-understood objects, representation theory is a powerful tool for investigating the structure of abstract groups.

group speed The speed at which a ➤wave packet travels in a particular medium. The group speed is given by $d\omega/dk$, where ω is the angular frequency and k is the magnitude of the wave-vector. The corresponding vector quantity is **group velocity**. If the medium exhibits ➤dispersion, the group speed differs from the ➤phase speed ω/k.

group theory The investigation of groups, typically through the study of their **subgroups** and ➤group representations. Groups may be identified in many contexts. In particular, they give a powerful method for analysing symmetries and so find extensive applications in crystallography and quantum theory. One striking triumph for group theory was the prediction of the existence of the ➤omega-minus particle before it was found in 1964.

growth A wide-ranging concept in biology. The term can refer to an increase in size, biomass, or number of cells present in an individual or in a colony; or when applied to populations, an increase in the numbers of individual organisms. The term should therefore be used with qualification.

growth hormone (GH, somatotrophin) A ➤polypeptide ➤hormone, secreted from the anterior lobe of the ➤pituitary gland, that helps to control ➤collagen metabolism in bones, particularly affecting their length. Overproduction of GH in early life results in gigantism, while underproduction results in dwarfism. GH for clinical treatment of dwarfism is now made by ➤genetic engineering techniques. **Bovine somatotrophin** (BST) may be added to some animal foodstuffs to improve milk and meat yield in cattle.

guanine Symbol G. A ➤purine base found in ➤DNA and ➤RNA.

guano The accumulated droppings of seabirds. In some areas, notably on the coast of Peru and Chile, the accumulations are sufficiently deep to make for viable commercial exploitation as a source of agricultural phosphate fertilizer.

guanosine A ➤nucleoside consisting of the base ➤guanine covalently bonded to the sugar ➤ribose (or to ➤deoxyribose in **deoxyguanosine**).

guard cell ➤stoma.

Guldberg and Waage's law ➤mass action, law of. Named after Cato Maximilian Guldberg (1836–1902) and Peter Waage (1833–1900).

gum One of a number of carbohydrate derivatives in plants that form sticky solutions in water. Gums are typically produced in response to wounding, and their natural function is probably temporary blockage of such damage. Plant gums are widely used in a variety of adhesives, and also as binding agents for tablets in the pharmaceutical industry.

guncotton An explosive made by nitrating cellulose.

Gunn diode A ➤microwave oscillator based on the negative differential resistance property of gallium arsenide in the ➤Gunn effect. Named after John Battiscombe Gunn (b. 1928). Compare ➤tunnel diode.

Gunn effect The scattering of electrons in the ➤conduction band in gallium arsenide above an energy of about 0.36 eV into a subsidiary minimum of energy in the ➤band structure. The electrons in the subsidiary minimum have a much higher effective mass than when in the bottom of the conduction band, so above a threshold of about 300 kV m^{-1} their ➤mobility decreases markedly with increasing electric field. At the macroscopic level this looks like a negative resistance, and is the basis for the ➤Gunn diode.

gunpowder A traditional explosive made from potassium nitrate, charcoal and sulfur. Although relatively insensitive to shock, unlike nitroglycerine, it ignites readily. It was invented by the Chinese and revolutionized warfare.

gut A hollow tube in animals, leading from the mouth to the anus, derived from endoderm and mesoderm in the ➤embryo, that functions in the digestion and absorption of food. The gut is organized into zones, each with a characteristic structure and function (see the diagram opposite). In humans the **mouth** leads via the **oesophagus** to the **stomach**. In the mouth, food is physically broken down by the action of the teeth, and secretions containing ➤amylase from the salivary glands commence the digestion of carbohydrates. Food is propelled along the gut by ➤peristaltic waves of contraction of the smooth ➤muscle in the gut lining. In the

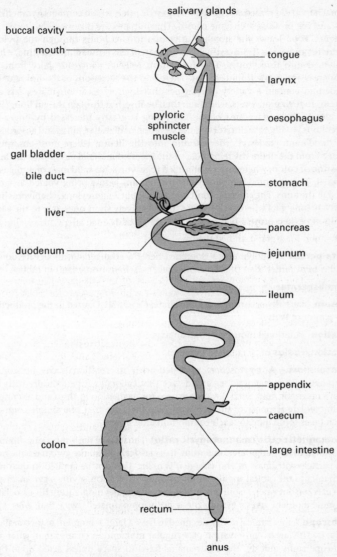

gut Schematic diagram of the human gut.

stomach food is churned and mixed with **gastric juice**, which contains hydrochloric acid and the protease ➤enzyme pepsin. This initiates the digestion (hydrolysis) of proteins. Food leaves the stomach after one to two hours through the **pyloric sphincter** and enters the **intestine**. The first part of this is the **small intestine**, which is divided into three zones. The **duodenum** receives pancreatic juice from the ➤pancreas and ➤bile from the liver. These and the secretions of the lining of the duodenum contain a variety of enzymes that hydrolyse carbohydrates, fats and proteins to their monomers, which can then be absorbed through the gut lining into the blood. The surface area of the gut lining is greatly increased by finger-like projections, ➤villi, to facilitate both the secretion of digestive juices and absorption. The duodenum grades via the **jejunum** into the **ileum**, where most absorption occurs. From the ileum the remainder of the ingested material passes into the **large intestine** or **colon**, where water is absorbed to leave a semi-solid mass of undigested material. This is stored temporarily in the **rectum** before being voided as faeces through the **anus**. The gut in an adult human is about 7 metres long. Herbivores have relatively long guts whereas carnivores have relatively short ones, due to the easier digestion of their protein-rich food intake. ➤➤gizzard.

GUT Abbr. for ➤grand unified theory.

gutta-percha The polymer of ➤isoprene where the arrangement about each double bond is *trans* rather than the *cis* orientation (➤*cis–trans* isomerism) in rubber.

Gymnospermae ➤Pinopsida.

gypsum Naturally occurring calcium sulfate, $CaSO_4 \cdot 2H_2O$, used in the production of ➤plaster of Paris.

gyration Rotational motion around an axis.

gyration, radius of ➤radius of gyration.

gyrocompass A ➤gyroscope, mounted with its rotational axis horizontal, combined with a magnetic sensing device that calibrates it periodically with the Earth's magnetic field. Such devices are used for navigation in ships and aircraft, as the immediate response of the gyroscope is superior to that of a simple compass, which takes some time to reach equilibrium after a turn.

gyromagnetic ratio (magnetogyric ratio) The ratio of the ➤magnetic moment of a system to its ➤angular momentum. In a classical system the gyromagnetic ratio of a particle with mass m and charge q is $q/2m$. This can be applied in quantum mechanics to the orbital angular momentum of an electron, giving a gyromagnetic ratio of $e/2m$; but when ➤spin is considered as an angular momentum the associated magnetic moment gives a gyromagnetic ratio approximately twice that, e/m.

gyroscope A mechanical device designed to have a large ➤moment of inertia about one axis. The law of conservation of ➤angular momentum ensures that, when the gyroscope spins quickly, the axis remains fixed in space unless acted upon by a ➤couple with a component perpendicular to the axis. Gyroscopes are used where it is important to retain a sense of direction, hence they are used for instrumentation and navigation in both aircraft and ships. ➤➤nutation; precession.

h Symbol for the ➤Planck constant; $\hbar$ denotes the quantity $h/2\pi$.

H Symbol for the element ➤hydrogen.

H⁺ Symbol for the ➤hydrogen ion. Equally it may be interpreted as the symbol for the proton, since only a proton remains when a hydrogen atom has lost one electron to become the ion (➤protonation). $H^+(aq)$ is the characteristic particle that all acids provide. ➤➤oxonium ion.

H Symbol for ➤enthalpy.

H Symbol for ➤magnetic field strength.

H5N1 A strain of avian influenza ('bird flu') which is currently considered as posing a high risk of causing a major pandemic, as a similar strain was the cause of the 1918 Spanish flu pandemic. The notation refers to the versions of the two main ➤antigens: H stands for **haemagglutinin** and N for **neuraminidase**, both of which are transmembrane ➤glycoproteins of the viral envelope.

Haber–Bosch process (Haber process) The manufacturing process for ➤ammonia, NH_3, which is synthesized from nitrogen and hydrogen. Its main feature is the use of high pressure, as the equilibrium produces only two moles of ammonia gas for every three moles of hydrogen gas and one mole of nitrogen gas. According to ➤Le Chatelier's principle, increasing pressure forces the equilibrium towards product. The conditions normally chosen are around 250 atm and 450 °C with a catalyst of iron. Named after Fritz Haber (1868–1934) and Carl Bosch (1874–1940), who won Nobel prizes independently; Haber devised the small-scale synthesis and Bosch made the high-pressure industrial process work.

habit (crystal habit) The external shape of a crystal.

habitat The locality where an organism normally lives, such as a pond or a wood. Compare ➤ecosystem; environment; niche.

Hadfield steel An alloy steel that has the convenient property that its strength increases as it is worked, a feature put to good use in excavators and safes. Named after Robert Abbott Hadfield (1858–1940).

Hadley cell A circulation pattern (in a vertical plane) of air in the Earth's atmosphere. Hot air rises at the equator and moves away about 30° to the north and south, where it descends and returns at surface level towards the equator, causing the trade winds in the equatorial belt. Named after George Hadley (1685–1768).

hadron A subatomic particle that experiences the ➤strong interaction. A hadron is either a ➤baryon or a ➤meson. The ➤standard model distinguishes them from ➤leptons and ➤gauge bosons in that hadrons are composed of ➤quarks. Baryons have half-integral spin, whereas mesons have integral spin.

haem (US **heme**) An iron-containing ➤porphyrin group forming an integral part of various pigments such as ➤haemoglobin and ➤cytochromes. The haem acts as the ➤prosthetic group, making it possible to carry oxygen or take part in redox reactions in cells.

haematite A common brown-coloured ore of iron, essentially iron(III) oxide, Fe_2O_3.

haemoglobin (Hb) The oxygen-carrying protein in the blood of vertebrates and some other organisms. Haemoglobin occupies some 90% of the volume of mammalian ➤erythrocytes. Its main adult form consists of four ➤polypeptide sub-units (two alpha chains and two beta chains). In addition each chain has a nonprotein iron-containing ➤haem prosthetic group. It is the interaction of oxygen with the iron of the ➤haem group that facilitates the uptake and transport of this gas in the blood. Humans have different types of haemoglobin, notably a foetal and an adult form. These have different affinities for oxygen, facilitating gas exchange between maternal and foetal circulations across the ➤placenta. Haemoglobin also transports carbon dioxide and has the ability to bind hydrogen ions, a property which is essential for the maintenance of blood pH. It thus plays an important role in ➤homeostasis. The genetically inherited disease ➤**sickle cell anaemia** is caused by the substitution of the amino acid valine in position 6 of the beta chains, replacing glutamine in this position in normal haemoglobin.

haemophilia A clinical condition in which the blood fails to clot after injury due to a deficiency of ➤factor VIII, one of a number of proteins involved in the clotting reaction. The condition is determined by a ➤recessive gene and is 'sex-linked', only males being affected. This is because the gene responsible is located on the X ➤chromosome. In males, who have an X and a Y chromosome, there is no dominant masking gene on the Y chromosome. Haemophilia is a rare condition, but it was formerly common in the royal families of Europe into which the offspring of Queen Victoria married. Victoria was a ➤carrier of the gene, and the condition can be traced through several generations of her descendants.

hafnium Symbol Hf. The element with atomic number 72 and relative atomic mass 178.5, which is in the third row of the d block of the periodic table. It occurs mainly with oxidation number +4, as does its ➤congener titanium.

hahnium The obsolete name of the element 105, now called dubnium. Named after Otto Hahn (1879–1968). ➤postactinide elements.

hair **1** A keratinized thread-like structure consisting of dead epidermal cells arising from hair follicles in mammalian skin, and having a variety of functions. Hair is variously distributed over the bodies of mammals, including humans, and helps to conserve body temperature by trapping an insulating layer of air. The hair of polar bears is hollow to enhance this property. The whiskers of cats are an example of the sensory function of hairs.
2 Any of a number of cellular structures arising from the epidermis of plants. For example, root hairs are extensions of epidermal cells that increase the functional

surface area of roots; and a variety of hairs (**trichomes**) are found on stems and leaves, where they may serve to reduce water loss or possess ➤glands (2) associated with defence, as in the hairs of stinging nettles.

half-cell An electrolytic cell consists of two half-cells, in one of which there is a relative tendency for oxidation to occur and in the other reduction. In the simplest case of a metal/metal ion half-cell, a metal electrode dips into a solution of its ions.

half-equation (half-reaction) By combining a reduction half-equation with an oxidation half-equation, it is possible to build up the full equation for ➤redox reactions by simply multiplying by suitable factors to balance the numbers of electrons in the two half-equations. An example of a reduction half-equation is that involving manganate(VII) ion:

$$MnO_4{}^-(aq) + 8H^+(aq) + 5e^- \rightarrow Mn^{2+}(aq) + 4H_2O(l).$$

Note how the number of electrons corresponds to the difference in the oxidation numbers of the element being affected (+7 and +2 in the example above).

half-life Symbol $t_{\frac{1}{2}}$. For a radioactive decay process, the (constant) time taken for half the radioactive atoms of the parent nuclide to decay into the ➤daughter nuclide. Radioactive decay processes are the most important examples of first-order reactions (➤order (4)), for which the concentration varies exponentially. The half-life of ^{14}C is 5730 years, which makes it important in ➤radiocarbon dating. The term can also be used for other orders of reaction, for which it is defined as the time taken for the concentration of a substance to fall to half its original value.

half-reaction ➤half-equation.

half-thickness The thickness of attenuating or absorbing material required to halve the intensity of a beam or wave. ➤➤Beer's law; Lambert's laws.

half-wave plate A plate of ➤anisotropic crystal, such as mica, whose ➤birefringence introduces a ➤phase difference of 180° between light polarized parallel to each of its axes. ➤➤quarter-wave plate.

half-wave potential ➤polarography.

half-wave rectifier A ➤rectifier circuit that outputs a voltage for only half of each full cycle of an alternating input. For the other half of the cycle the output is zero. Compare ➤full-wave rectifier.

halide Symbol for the halide ion: Hal^- or X^-. A compound with one of the halogens having oxidation number −1, common examples being fluorides, chlorides, bromides and iodides.

halite structure An alternative name for the ➤rock-salt structure. Halite is an alternative name for sodium chloride, NaCl.

Hall effect The production of an ➤electric field across a material through which an electric current is flowing and on which a ➤magnetic field is acting. The electric field, E_x, is perpendicular to both the current, j_z, and the

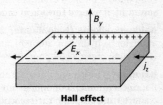

Hall effect

magnetic field, B_y (see the diagram overleaf). The force applied to the charge carriers by the electric field exactly balances the ▸Lorentz force from the magnetic field. The strength of the electric field is given by

$$E_x = R_H B_y j_z,$$

where R_H is known as the **Hall coefficient**. In a classical model of the Hall effect, R_H is simply $1/nq$, where n is the number of charge carriers per unit volume and q is their charge. The Hall coefficient changes sign with the sign of the charge carrier, and therefore provides an important means of investigating the electronic structure of the solid state. In particular, the positive Hall coefficients exhibited by metals such as magnesium and aluminium are a clear indication that a naïve picture of a sea of conduction electrons is inappropriate because the majority carriers are clearly positively charged (and are, in fact, ▸holes). Named after Edwin Herbert Hall (1855–1938).

Halley's Comet The best known ▸comet, with an orbital period of 76 years, which last reappeared in 1986. Whereas most comets are named after t eir discoverer, this one is named after the astronomer who first calculated its orb t, Edmond Halley (1656–1742). Its orbital motion is ▸retrograde.

Hall–Héroult process The manufacturing process for aluminium. The metal cannot be produced by carbon reduction of its oxide, which is too stable thermodynamically, so electrolysis is used. The ore used is purified alumina, Al_2O_3, which is dissolved in molten cryolite, sodium hexafluoroaluminate, Na_3AlF_6, as the melting point of pure alumina is extremely high. Even so, the temperature of the melt is around $950\,°C$. Aluminium is produced at the cathode, from where it is siphoned off. The use of cryolite proved the decisive technological breakthrough and was suggested independently by Charles Martin Hall (1863–1914) and Paul Louis Toussaint Héroult (1863–1914) in the same year, when both were twenty-three years old.

Hall probe An instrument, based on the ▸Hall effect, for measuring ▸magnetic flux density. It measures the electric field produced per unit current flowing (in a ▸semiconductor channel), which is proportional to the magnetic flux density. Named after E. H. Hall.

haloalkane ▸halogenoalkane.

haloform A generic name for the compounds $CHHal_3$, the most common example being the famous anaesthetic chloroform. ▸▸iodoform test.

halogenation A process involving the addition of a halogen or a hydrogen halide molecule or the substitution of a halogen atom for a hydrogen atom in an organic compound. Common examples include the chlorination of an alkane and aromatic substitution using a Lewis acid catalyst.

halogenoalkane (haloalkane) A compound formed by substituting a halogen atom for a hydrogen atom in an alkane. For some examples, see ▸chlorine-containing organic compounds. ▸▸BCF; CFC; iodoform.

halogens The generic name for the elements of Group 17 of the ▸periodic table: fluorine, chlorine, bromine, iodine and astatine. They are named from their most characteristic property: their reaction with metals to form salts (Greek *halx*), which are called 'halides'. Fluorine is the strongest oxidant among the elements; the other halogens too act as oxidants usually, but they become weaker oxidants down the group. The most important halogen in industry is chlorine; its manufacture is described under ➤chlor–alkali industry.

halon ➤CFC.

halothane (Fluothane) The most frequently used general anaesthetic, 2-bromo-2-chloro-1,1,1-trifluoroethane, $CF_3CHBrCl$. It is remarkably safe, although long-term exposure of operating-theatre workers leads to liver damage.

Hamiltonian Symbol H. In quantum mechanics, the ➤operator (2) for energy: ➤Schrödinger equation. The Hamiltonian contains information about the dynamics of a system.

In classical mechanics, the Hamiltonian is related to the ➤Lagrangian. Applying ➤Hamilton's principle to the Lagrangian leads to ➤Hamilton's equations. Named after William Rowan Hamilton (1805–65).

Hamilton's equations The equations of motion of a system in classical mechanics written in terms of its ➤Hamiltonian, **generalized coordinates** q_i and **generalized momenta** p_i. For each generalized coordinate,

$$\dot{p}_i = \frac{\partial H}{\partial q_i} \text{ and } \dot{q}_i = \frac{\partial H}{\partial p_i}.$$

where $\dot{p}_i = \mathrm{d}p_i/\mathrm{d}t$ and $\dot{q}_i = \mathrm{d}q_i/\mathrm{d}t$.

These are first-order differential equations which describe the path of any system in ➤phase space.

Hamilton's principle The motion of any system between two specified times (t_1 and t_2) follows the path in ➤phase space that makes the ➤action of the system a stationary value where the action integral I is defined as

$$I = \int_{t_1}^{t_2} L \, \mathrm{d}t,$$

where L is the ➤Lagrangian of the system. It is also sometimes called the **principle of least action**.

haploid The condition in a cell in which the ➤diploid number of chromosomes is reduced by half. This occurs in higher organisms when ➤gametes are formed by ➤meiosis. Haploid cells thus contain one of each ➤homologous pair of chromosomes contained in the diploid cell. When gametes combine at ➤fertilization, the diploid number is restored.

hapto The number of atoms in a Lewis base actually forming a ➤coordinate bond with the Lewis acid. For example, the cyclopentadienyl ring in ➤ferrocene is penta-hapto, that is, all five carbon atoms are bonded. Compare ➤1,2-diaminoethane, which has only two atoms coordinated.

haptotropism ➤thigmotropism.

hard and soft acids and bases An empirical subdivision of ➤Lewis acids and bases such that hard acids tend to react best with hard bases, and soft acids with soft bases.

The origin of the interaction for hard species depends primarily on their charges whereas the polarizability of species is relevant to their softness.

hard disk ➤magnetic storage.

hardness ➤Mohs scale.

hardness of water ➤hard water.

hard radiation An imprecise name given to ➤ionizing radiation of high frequency which tends to be more penetrating than radiation of lower frequency. Compare ➤soft radiation.

hardware Any physical component of a computer system. Keyboards, silicon chips, integrated circuits, printers and monitors are all examples of hardware. Compare ➤software.

hard water Water that contains certain dissolved ions, most importantly calcium and magnesium ions, which cause it to lather poorly with soap, forming a scum. The hardness may be divided into two classes, temporary and permanent. **Temporary hardness** can be removed by boiling, which causes decomposition of hydrogencarbonate (bicarbonate) salts, as in the following equation, depositing a precipitate of the carbonate (causing scale):

$$Ca(HCO_3)_2(aq) \rightarrow CaCO_3(s) + H_2O(l) + CO_2(g).$$

Permanent hardness cannot be removed by boiling; instead an ➤ion-exchange resin must be used, which can also remove temporary hardness.

harmonic 1 Of an oscillating system, having a period of oscillation independent of the amplitude of oscillation (➤harmonic oscillator).
2 In a ➤waveform, a sinusoidal oscillation that is an integer multiple of the ➤fundamental. The first harmonic is the fundamental itself. A ➤superposition of sound waves with frequencies related in this way tends to sound pleasant.

harmonic function Any solution of ➤Laplace's equation.

harmonic mean The harmonic mean H of the positive numbers $a_1, \ldots, a_n$ is defined by

$$H = \frac{n}{1/a_1 + \cdots + 1/a_n}.$$

Harmonic means occur in some rate problems: the overall average speed of two laps at 110 mph and 140 mph is 123.2 mph, the harmonic mean of 110 and 140.

harmonic motion ➤harmonic oscillator; simple harmonic motion.

harmonic oscillator An oscillator that has a constant period. An oscillating system is harmonic if the restoring force is proportional to displacement: $f = -kx$.
Using ➤Newton's second law of motion,

$$f = ma = m\frac{d^2x}{dt^2}.$$

One solution to the equation

$$\frac{d^2x}{dt^2} = -\left(\frac{k}{m}\right)x$$

is $x = x_0 \sin \omega t$ with the ➤angular frequency $\omega = \sqrt{k/m}$.

hartree A unit of energy useful in atomic physics: 1 hartree = 27.2 eV. The binding energy of the electron in the ground state of the hydrogen atom is exactly $\frac{1}{2}$ hartree. Named after Douglas Rayner Hartree (1897–1958). ➤➤Appendix table 2.

Hartree–Fock theory (self-consistent field theory) An extension of ➤Hartree theory; it uses a multi-electron wavefunction and therefore takes account of the ➤exchange interaction, ignored in simple ➤Hartree theory. Named after D. R. Hartree and Vladimir Alexandrovich Fock (1898–1974).

Hartree theory An iterative method for finding the ➤ground state of a multi-electron system in ➤quantum mechanics. The electrons are considered to occupy a set of one-electron levels, represented by simple wavefunctions $\psi_i(r)$. Then, with a term in the Hamiltonian that represents the interaction between the electrons as simply the potential from the charge density of all the other electrons, $\sum_{i \neq 1} |\psi_i(r)|^2$, the Schrödinger equation is solved for $\psi_1(r)$. This procedure is repeated until the wavefunctions are consistent with the charge density that generates them.

Harvard classification A system, devised by Edward C. Pickering at Harvard University in 1891, for classifying stars by their ➤atomic emission spectra. It ranks stars by their ➤effective temperature (using Wien's law ➤black-body radiation), from O (emitting predominantly blue He lines at greater than 30 000 K) through B, A, F, G, K, to M (emitting red lines at about 3000 K). Classes R, S and N were subsequently added after M. The Harvard system was refined in 1943 into the Morgan–Keenan classification, which is the system currently in use.

hassium Symbol Hs. The element with atomic number 108.

hausmannite A brown ore of manganese, Mn_3O_4.

Hawking radiation Radiative emission predicted by Stephen William Hawking (b. 1942) to come from black holes as one of a particle/antiparticle pair falls into the black hole while the other escapes.

hazchem symbols Symbols warning the public about hazardous chemicals (see the diagram overleaf).

Hb Abbr. for ➤haemoglobin.

h.c.p. Abbr. for hexagonal close-packed or hexagonal ➤close packing.

HDPE Abbr. for high-density ➤polyethylene.

He Symbol for the element ➤helium.

health physics The application of physics to health and safety. The effect of ➤ionizing radiation is a particular concern.

heart A muscular pump that assists in blood circulation in animal phyla possessing a vascular system. At its simplest, a heart may be a modified blood vessel with muscular walls (as in an ➤ annelid). Vertebrate hearts have three basic components: **cardiac**

explosive   flammable   toxic   corrosive

radioactive   oxidizing   harmful   irritant

hazchem symbols

muscle (organized into discrete chambers), **valves** and a **pacemaker**. There are four chambers in the mammalian heart (see the diagram opposite). The upper **atria** have relatively thin walls, whereas those of the lower **ventricles** are much thicker. Blood is moved through the various chambers as they alternatively contract and relax. Blood enters the right atrium from the **venae cavae** and passes through the **tricuspid valve** to the right ventricle. From here it is pumped via the **pulmonary artery** to the lungs. Oxygenated blood returns from the lungs via the **pulmonary vein** to the left atrium, from where it passes through the **bicuspid valve** into the left ventricle. The left ventricle pumps blood via the **aorta** and arteries to the organs of the body. Deoxygenated blood returns from the organs via the veins which drain into the venae cavae. Blood thus passes through the heart twice on every circuit in a characteristic 'double circulation'. Fish have only a single circulation, the blood passing from heart to gills, to body organs, and back to the heart again. The advantage of the double circulation is that blood can be at higher pressure, thus supplying the extra energy demands of ➤homoiothermy (warm-bloodedness) in mammals.

Cardiac muscle has a unique cellular conformation in which the muscle fibres are cross-linked to form a net-like structure. Cardiac muscle has an automatic ability to contract without external stimulus. However, the pattern and rate of contraction are controlled by the pacemaker. This is a group of specialized cardiac muscle cells, forming the sino-atrial node in the wall of the right atrium, which respond to impulses from the vagus and accelerator nerves; these respectively slow or accelerate the rate of heartbeat. Contraction of a heart chamber is called **systole**, and relaxation **diastole**. The sino-atrial node communicates with the rest of the heart via the **Purkinje fibres** to control cardiac output. The valves are flaps of connective tissue, often supported by tendons ('heart strings') which promote the one-way flow of blood through the heart.

heat Symbol Q; unit J. Energy transferred to a system that is not identifiable as ➤work at the macroscopic level. The heat transferred to a system is the difference between the increase in the ➤internal energy of the system and the work done on the

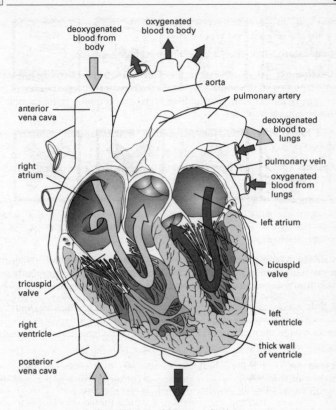

heart Diagram of the mammalian heart.

system. Like force, heat is a concept that is familiar in everyday life, but is impossible to define without resorting to what is, at first sight, a circular definition. ➤thermodynamics.

heat capacity Symbol C_V or C_p; unit J K^{-1}. The ➤heat transferred to a system per unit rise in temperature, defined in terms of the conditions imposed on the system at the time. For an ideal gas of volume V, at pressure p and at temperature T, the heat capacities at constant volume and constant pressure are

$$C_V = \left[\frac{\partial U}{\partial T}\right]_V \text{ and } C_p = \left[\frac{\partial H}{\partial T}\right]_p,$$

where U is its ➤internal energy and H its ➤enthalpy. ➤specific heat capacity.

heat engine A device for turning ➤heat into useful ➤work. A simple consequence of the second law of ➤thermodynamics is that heat cannot be turned into work without some side-effect, in this case a change in the temperature difference between two thermal **reservoirs**. ➤➤Carnot cycle.

heat equation ➤Biot–Fourier equation; ➤➤diffusion equation.

heat exchanger A device for making use of unwanted heat which is the by-product of another process. Its most usual form is a thermal contact between the exhaust gases of an engine or furnace and a cold fluid; the gases heat the fluid as in, for example, the heating systems of most cars.

heat of reaction (combustion, formation, neutralization, solution, etc.) ➤standard enthalpy change, etc.

heat pump A device that compresses a fluid to raise its temperature, and then uses it to heat a cooler system.

heat radiation The process by which heat is emitted as ➤electromagnetic radiation, or the radiation itself. Radiation is the only method of heat transfer possible through a vacuum. ➤➤black-body radiation; Planck's law; Stefan–Boltzmann constant.

heat shield A layer of low thermal conductivity and high reflectance designed to resist the transfer of heat. A heat shield of ceramic tiles is used to protect space shuttles from being burnt up as they re-enter the Earth's atmosphere.

heat sink An object with a large ➤heat capacity that receives heat by conduction from an electrical component and dissipates the heat to its surroundings, chiefly by convection. Most microprocessors in personal computers have heat sinks to prevent them from overheating.

heat transfer The transfer of heat from a hotter object to a cooler one. Heat can be transferred in three ways: by ➤conduction, ➤convection or ➤heat radiation. A change in ➤entropy is associated with any heat transfer.

Heaviside–Kennelly layer (E layer) A layer of the ➤ionosphere, between altitudes of about 90 and 150 km. It reflects medium-frequency radio waves. Named after Oliver Heaviside (1850–1925) and Arthur Edwin Kennelly (1861–1939).

heavy hydrogen Alternative name for ➤deuterium.

heavy metal A nonrigorous term to signify a metal of relatively high relative atomic mass. Typical examples include mercury, thallium and lead; they tend to be human poisons.

heavy water D_2O Water in which the normal hydrogen isotope has been substituted by the heavier ➤deuterium isotope. Heavy water is an important ➤moderator in nuclear reactors.

hectare Symbol ha. A unit of area common in agriculture: 1 ha = 10 000 m^2.

Heisenberg uncertainty principle (indeterminacy principle) If the position of an electron (for example) is known exactly, then its momentum would be completely unknown and hence its position at the next instant would be incalculable. The principle thus puts bounds on the limit of our knowledge, which came as a profound shock to the scientists in the 1920s when it was proposed by Werner Karl Heisenberg (1901–76). More precisely stated,

$$\Delta x \Delta p_x \geqslant \frac{1}{2}\hbar,$$

where Δx is the uncertainty in the position in the x direction, Δp_x is the uncertainty in the momentum in the same direction, $\hbar = h/2\pi$ with h the ➤Planck constant. Other pairs of observables are linked by similar uncertainty relations.

helicity A quantum-mechanical ➤operator (2) (with an associated ➤quantum number) defined by $p \cdot J$, where p is momentum and J ➤angular momentum. The helicity of a particle determines whether it is 'left- or right-handed'. Most particles can be either, but neutrinos are exclusively left-handed (helicity $-\frac{1}{2}$).

heliocentric system A theory of the structure of the Solar System which places the Sun (Greek *hēlios*) at the centre. Nicholas Copernicus revived the concept of a heliocentric solar system, originally due to the Greeks, in his book *De revolutionibus orbium coelestium*, published in 1543 (the year he died).

helium Symbol He. The element with atomic number 2 and relative atomic mass 4.003, which is the ➤noble gas with the smallest atomic number. Its name commemorates the fact that it was first identified from its emission spectrum from the Sun (Greek *hēlios*). It is completely unreactive and this, together with its low density, is the major reason for its success as a lifting gas for airships or birthday balloons. Helium's other uses include helium-neon lasers, which are among the most powerful currently available. Helium is a colourless gas which has a very low boiling point of just 4.2 K, as the forces between the atoms are only very weak dispersion forces. Liquid helium is a most unusual substance, as it undergoes a phase transition below 2.18 K to a ➤superfluid phase, capable of climbing up the side of a containing vessel. It cannot be solidified below 25 atm pressure either, another unique property among the elements.

helix The space ➤curve that has the shape of a spring and parametric equations $x = r \cos t$, $y = r \sin t$, $z = ct$ where r and c are constants.

Helmholtz coils A coaxial pair of cylindrical current-carrying coils, separated by exactly the radius of the coils. This produces an almost uniform magnetic field. Named after Hermann Ludwig Ferdinand von Helmholtz (1821–94).

Helmholtz free energy Symbol F. A thermodynamic ➤state function of a system defined by $F = U - TS$, where U is the system's internal energy, T its thermodynamic temperature and S its entropy. When all the ➤extensive properties of a system (except entropy) are held constant, and the system is held at constant temperature (by putting it in thermal contact with a heat reservoir), F cannot increase. Compare ➤Gibbs energy; ➤➤free energy.

helper T cell (T helper cell) A lymphocyte cell of the ➤immune system. They assist the ➤plasma cells produced by the B lymphocytes to produce antibodies. ➤➤T cell.

heme US spelling of ➤haem.

hemiacetal A molecule in which an —OH group and an —OR group are joined to a single carbon atom which has one hydrogen and one alkyl group also attached (see the diagram), a half-way house to an acetal. It is derived from an aldehyde by addition of an alcohol across the carbonyl group. Hemiacetal structures feature in several carbohydrates: ►ribose.

hemiacetal A cyclic example.

hemihydrate A compound containing only one molecule of water of crystallization per two formula units, the most important example being plaster of Paris, $2CaSO_4 \cdot H_2O$.

hemiketal A hemiketal is to a ketone as a ►hemiacetal is to an aldehyde.

hemisphere One of the two portions of a sphere cut off by a plane through its centre.

Henderson–Hasselbalch equation An equation relating the ►pH of a ►buffer solution to the pK_a (►acid ionization constant) of the acid and the concentrations of the un-ionized acid A and its salt S:

$$pH = pK_a + \log_{10}([S]/[A]).$$

A key point is that when equal concentrations of the acid and salt are present, the pH of the solution equals the pK_a of the parent acid. Named after Lawrence Joseph Henderson (1878–1942) and Karl Albert Hasselbalch (1874–1962).

henry Symbol H. The SI derived unit of ►inductance. One henry is equal to 1 volt second per ampere. Named after Joseph Henry (1797–1878).

Henry's law The mass of gas dissolved in a solvent is directly proportional to the pressure of the gas, the constant of proportionality being called the **Henry constant**. Named after William Henry (1775–1836).

heparin An amino-sugar derivative produced by mast cells and acting as an anticoagulant in the blood.

heptane C_7H_{16} The straight-chain alkane with seven carbon atoms, which is a colourless liquid. It is a typical alkane, commercially obtained in the ►gasoline fraction.

herbaceous Describing plants that lack permanent woody tissues.

herbivore An animal that feeds exclusively on plants. Herbivores are the primary consumers in ►food chains. Compare ►carnivore; omnivore.

hereditary Describing information or traits that may be passed through reproductive processes from one generation to the next. Such information might be contained within the nuclear ►genome, but might also be carried by extranuclear ►genes contained in the ►cytoplasm or ►organelles of eggs. The term is also applied to materials that might pass across the placenta from mother to foetus, for example infective agents or toxins such as drugs.

hermaphrodite An organism that possesses both male and female sets of functional sex organs. The state is commonly found in plants, where, for example, flowers have organs which produce both male and female gametes. It is also found in certain groups of animals such as tapeworms and snails, but it is rare in vertebrates.

hermetic seal A seal that prevents air from penetrating.

Hermite polynomials The sequence of polynomials

$$H_n(x) = (-1)^n \exp(x^2) \frac{\mathrm{d}^n}{\mathrm{d}x^n} \exp(-x^2)$$

that satisfies the differential equation $\mathrm{d}^2 y/\mathrm{d}x^2 - 2x \ \mathrm{d}y/\mathrm{d}x + 2ny = 0$. Hermite polynomials are important in quantum mechanics where they occur in the solutions to the ➤Schrödinger equation for a ➤harmonic oscillator. Named after Charles Hermite (1822–1901).

Hermitian matrix A square ➤matrix with complex number entries, every element of which is equal to the complex conjugate of the corresponding element of its ➤transpose.

heroin ➤morphine.

Heron's formula The formula $\sqrt{s(s-a)(s-b)(s-c)}$ for the area of a triangle with sides of length a, b, c and semiperimeter $s = \frac{1}{2}(a + b + c)$. Named after Heron of Alexandria (*fl.* 1st century AD).

herpes virus One of a group of DNA-containing viruses which includes herpes simplex, which causes cold sores, and the ➤Epstein–Barr virus, which causes glandular fever and is implicated in some cancers.

hertz Symbol Hz. The SI derived unit of ➤frequency, named after Heinrich Rudolf Hertz (1857–94), who first detected electromagnetic waves. One hertz is equal to one cycle per second. The hertz is unusual in that it is a simple reciprocal of the second, and in most similar cases the fundamental unit is kept in that form (thus the SI unit for wavenumber is m^{-1}). However, the hertz carries with it the idea of whole cycles, which contrasts with the use of s^{-1} as the unit of ➤angular frequency (in effect, radians per second).

Hertzsprung–Russell diagram Abbr. HR diagram. A plot of the absolute visual ➤magnitude (2) of stars against their ➤spectral type (see the diagram) and thus, equivalently, of their luminosity against their surface temperature. It is the most important graph in astrophysics. From its position on the HR diagram, the size of a star can be deduced. Most stars fall on the **main sequence**; these are nearly all ➤dwarf stars (among them the Sun). Above the main sequence lie the ➤giant and ➤supergiant stars, and below it are the

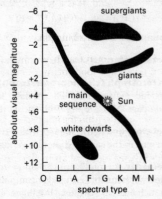

Hertzsprung–Russell diagram

➤white dwarfs. Named after Ejnar Hertzsprung (1873–1967) and Henry Norris Russell (1877–1957).

Hess's law (law of constant heat summation) The ➤standard enthalpy change is independent of the route taken from the reactants to the products. This follows from the first law of ➤thermodynamics and, with tables of standard enthalpy changes of formation, makes it possible to calculate the standard enthalpy changes to be expected in a huge range of possible reactions. It underpins the success of the ➤Born–Haber cycle for understanding ionic bonding. Named after Germain Henri Hess (1802–50).

heterochromatin ➤chromatin.

heterocyclic aromatic compound An ➤aromatic compound that contains a benzene ring with one or more carbon atoms replaced by another element, frequently nitrogen. A simple example is ➤pyridine, and more important ones include the four nitrogenous bases in DNA: ➤adenine; cytosine; guanine; thymine.

heterodyne receiver A receiver of radio waves that uses the ➤beats between the ➤carrier-wave frequency and a locally generated signal to produce its output. ➤➤superheterodyne receiver.

heterogeneous catalyst A ➤catalyst that is in a different phase from the reactants. Important examples include solid metal surfaces used for catalysing gaseous reactions, notably iron in the ➤Haber–Bosch process.

heterojunction A junction between two types of ➤semiconductor of different polarities. ➤p–n junction.

heterolysis ➤heterolytic fission.

heterolytic fission (heterolysis) Breaking of a bond such that the two shared electrons are both taken by one of the atoms. A simple example is the first step in unimolecular nucleophilic substitution, for example

$$(CH_3)_3CBr \rightarrow (CH_3)_3C^+ + Br^-.$$

Compare ➤homolytic fission.

heteropolar An obsolete synonym for ionic: ➤ionic bonding.

heterotroph (chemoorganotroph) An organism that feeds on pre-synthesized organic matter. Thus all animals, fungi and most bacteria are heterotrophic. Compare ➤autotroph; ➤➤chemosynthesis.

heterozygous The state in which two different ➤alleles for a particular gene are present in the same ➤diploid ➤genome. Compare ➤homozygous.

heuristic An intelligent trial-and-error approach, as opposed to a rigid algorithmic method. Heuristics are used in computer programs which can learn from experience, in such fields as ➤artificial intelligence, and their nonrigorous nature is also suited to ➤expert systems.

hexaammine A complex containing six ammonia molecules coordinated to the central metal ion, as in hexaamminenickel(II) ion, $[Ni(NH_3)_6]^{2+}$. This complex is octahedral: ➤valence-shell electron-pair repulsion theory.

hexaaqua (hexaaquo) A complex containing six water molecules coordinated to the central metal ion. Examples abound, especially with the d-block elements, for example $[Ni(H_2O)_6]^{2+}$. This complex is octahedral: ➤valence-shell electron-pair repulsion theory.

hexachlorocyclohexane ➤benzene hexachloride.

hexachlorophane (hexachlorophene) $(C_6H_2Cl_3O)_2CH_2$ White crystals, used as a disinfectant.

hexacyanoferrate(II) ion $[Fe(CN)_6]^{4-}$ An aqueous solution of hexacyanoferrate(II) ions is used to test for iron(III) ions, a characteristic ➤Prussian blue precipitate confirming the presence of iron(III) ions. A more common use of sodium hexacyanoferrate(II) is as the anticaking agent in dishwasher salt.

hexacyanoferrate(III) ion $[Fe(CN)_6]^{3-}$ An aqueous solution of hexacyano-ferrate(III) ions is used to test for iron(II) ions, a characteristic ➤Prussian blue precipitate confirming the presence of iron(II) ions.

hexadecane ➤cetane number.

hexadecimal (Hex) The ➤base 16 number system in which the digits are denoted 0, 1,…, 9, A, B, C, D, E, F, so that, for example, 1DC represents $1 \times 16^2 + 13 \times 16 + 12 = 476$. Importantly for applications in computing, conversion between hexadecimal and ➤binary is straightforwardly done in blocks of four digits; for example, to convert 1DC to binary, simply convert each digit: 1 to 1, D to 1101, C to 1100 to give 111011100.

hexafluoroaluminate ion AlF_6^{3-} ➤cryolite. Its structure is octahedral, as predicted by the ➤valence-shell electron-pair repulsion theory.

hexagon A ➤polygon that has six sides. A honeycomb is **hexagonal**.

hexagonal close packing (hexagonal close-packed, h.c.p.) ➤close packing.

hexahelicene A molecule containing six fused benzene rings. It is interesting because the repulsion between the rings forces the molecule into a helical shape. As the helix can be either right- or left-handed, ➤optical activity results, a variation on the conventional asymmetric carbon atom.

hexahydrate A crystal with six molecules of water of crystallization, for example $MgCl_2\cdot6H_2O$.

hexane C_6H_{14} The straight-chain alkane with six carbon atoms, which is a colourless liquid.

hexanedioic acid (adipic acid) $HOOC(CH_2)_4COOH$ The straight-chain dicarboxylic acid with six carbon atoms, which exists as colourless needles. The diacyl chloride formed from it is used to make nylon-6,6.

hexose ➤aldohexose; ketohexose.

Hf Symbol for the element ➤hafnium.

HF Abbr. for ➤high frequency.

H **field** ➤magnetic field strength.

Hg Symbol for the element ➤mercury, from its Latin name *hydrargyrum*.

hibernation A physiological state in homoiothermic (warm-blooded) animals in which the metabolic rate is drastically lowered, along with body temperature, to reduce energy demand during winter conditions in temperate regions. True hibernation occurs in, for example, hedgehogs and bats in the UK. Compare ➤aestivation.

Higgs boson (Higgs particle) The massive ➤boson that is postulated to explain the nonzero masses of the ➤W boson and ➤Z boson that mediate the ➤weak interaction according to the ➤standard model, in the same way that the photon mediates the electromagnetic interaction. Higgs proposed a theory in which particles acquire mass by moving through space and interacting with a background field, the so-called **Higgs field**. Discovery of the Higgs boson, the particle associated with this field, would corroborate the theory. Named after Peter Ware Higgs (b. 1929).

high-energy physics ➤particle physics.

highest common factor The largest ➤natural number that is a factor of two given integers.

high frequency (HF) The radio frequency range between 3 and 30 MHz. ➤➤frequency band.

high-level language Any computer language whose syntax is not dependent on the hardware on which the language is run, and which therefore has the advantage that a program written in it is theoretically portable between different computer systems. High-level languages cannot be executed directly, requiring a ➤compiler or ➤interpreter, but they are designed to be easier to program in than ➤low-level languages, and they are often designed for specific purposes. ➤➤BASIC; C; Fortran; LISP; Pascal.

high-performance liquid chromatography ➤chromatography.

high-spin complex A complex in which the number of unpaired spins is a maximum value for the number of electrons. Fluoride ion is one of the most likely ligands to cause high-spin complexes. Compare ➤low-spin complex. ➤➤spectrochemical series.

high tension (HT) Alternative phrase for high voltage.

Hilbert space A real or complex ➤vector space with a ➤scalar product that is complete in the sense that any sequence of vectors that is getting closer and closer together in fact converges to a vector in the space. Infinite-dimensional Hilbert spaces and ➤operators on them are the bedrock of the modern mathematical formulation of ➤quantum mechanics. Named after David Hilbert (1862–1943).

hindbrain ➤brain.

hindrance ➤steric hindrance.

His Abbr. for ➤histidine.

histamine A substance produced by ➤mast cells of the immune system, in response to a local infection, which dilates blood capillaries so that plasma and white blood cells can pass into the affected area. This produces a local swelling or inflammation. Insect bites and various ➤allergies often stimulate the release of histamine.

histidine (His) An important ➤amino acid, often responsible for the complicated ➤tertiary structure of proteins. For example, the histidine near the active site of ➤haemoglobin is crucial to its biochemical behaviour. ➤➤Appendix table 7.

histocompatibility The degree of tolerance of an organism to the implantation of foreign tissue, for example in an organ transplant. The recognition of foreign tissue depends on the ability of cells of the immune system to identify histocompatibility ➤antigens on the surface of transplanted cells. ➤➤human leucocyte antigen system; major histocompatibility complex.

histogram A graphical representation of a ➤frequency distribution by means of abutting rectangles (see the diagram). The rectangles have widths equal to those of each category or **class interval** and heights proportional to the frequency in each interval, so that the area of each rectangle represents its class frequency.

density /g cm^{-3}	0–5	5–10	10–15	15–25
frequency	21	35	10	6
frequency density	4.2	7.0	2.0	0.6

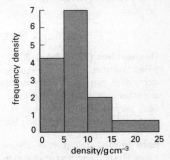

histology The study of the tissues of organisms.

histone One of a class of basic proteins that combine with ➤DNA to form ➤chromatin. Histones are involved in the mechanisms that control ➤transcription.

histogram representing the density of elements with atomic number less than 85 which are solids at 298 K.

HIV (human immunodeficiency virus) A ➤virus that infects ➤T cells of the immune system via the T4 ➤antigen site, rendering the body's ability to ward off infection less effective. Such infection usually leads to the development of ➤AIDS. HIV can be transmitted from person to person by contact with infected blood, semen or vaginal fluid. It can also pass across the ➤placenta from mother to foetus. If ➤antibodies to HIV are detected in the blood, the person is said to be **HIV positive**. HIV is a ➤retrovirus, with RNA as its genome. Recent evidence suggests that the source of the pandemic may have been a crossing between species of a similar virus among chimpanzees living in south-east Cameroon.

HLA system ➤human leucocyte antigen system.

Ho Symbol for the element ➤holmium.

hoarfrost   Ice produced by direct ➤sublimation of water vapour from the atmosphere, at temperatures below the ➤triple point of water. Ordinary frost is simply frozen dew.

Hoffman voltameter   An electrolytic apparatus, typically using dilute sulfuric acid as electrolyte, which effectively causes the electrolysis of water, thus producing hydrogen and oxygen, in a two-to-one ratio by volume, at the two electrodes. Named after Friedrich Hoffman (1660–1742).

Hofmann degradation   A reaction that reduces the length of a carbon chain by one carbon atom, via the Hofmann rearrangement. Named after August Wilhelm von Hofmann (1818–92).

Hofmann rearrangement   A molecular rearrangement that produces an amine from an amide using bromine in sodium hydroxide. The sodium hydroxide first extracts a proton from the amide, bromine substitutes where the hydrogen was, and then another proton is extracted, resulting in the species which then rearranges as shown in the diagram opposite. The rearrangement is used to make 2-aminobenzoic acid, $H_2NC_6H_4COOH$, an intermediate used to make dyes.

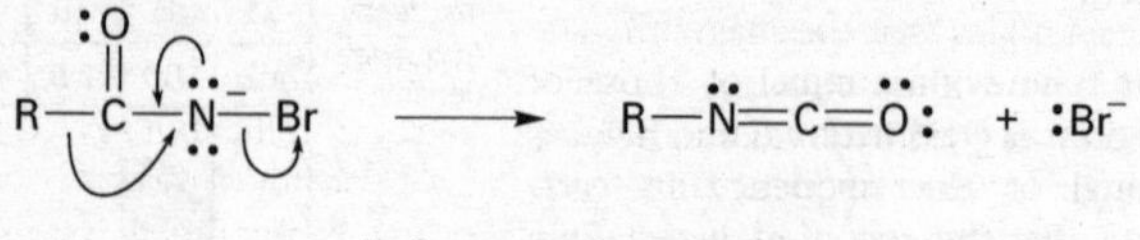

Hofmann rearrangement

Hogness box (TATA box)   A sequence of ➤nucleotides (TATAAAA) lying some 25 bases 'upstream' from a ➤eukaryotic gene which acts as a ➤promoter of gene expression, essentially as a 'switch' to control gene activity. It is suspected that there are other controlling sequences. In ➤prokaryotic genes the **Pribnow box** (TATAAT box) plays a similar role, but is located closer to the gene sequence, about 10 nucleotides away. Named after David Swenson Hogness (b. 1925).

hole   In solid state physics, an unoccupied state in a predominantly full ➤energy band. The behaviour of electrons in nearly full bands makes it possible to model the system as a small number of positively charged particles filling states at the bottom of a nearly empty band. Hole states are most important in ➤semiconductors. At nonzero temperatures, ➤Fermi–Dirac statistics imply that the number of occupied states in the conduction band equals the number of empty states in the valence band, giving rise to equal numbers of positive and negative carriers. By means of ➤doping, holes can be made the majority carriers of charge in the material, creating ➤p-type semiconductors.

holmium   Symbol Ho. The element with atomic number 67 and relative atomic mass 164.9, which is one of the lanthanides. It shares their common features, notably the dominance of oxidation number +3 in its compounds, such as Ho_2O_3, a grey insoluble powder, and the yellow aqueous ion Ho^{3+}.

Holocene   See table at ➤era.

holoenzyme   The catalytically active form of an enzyme comprising the polypeptide enzyme molecule together with any cofactors.

hologram   A recorded ➤interference pattern produced by the ➤superposition of a wave from an object and a reference wave. If this pattern is illuminated with a coherent source of light similar to the reference wave, one of the orders of ➤diffraction from the pattern will recreate the original object as a virtual image, which is effectively a three-dimensional representation. ➤➤holography.

holography   A method of recording and reproducing three-dimensional images of objects by creating a hologram. Sophisticated modern recording techniques allow holograms to be viewed with natural light rather than a special monochromatic source, and they have become commonplace on, for example, credit cards, as they are difficult to forge.

holophytic   Describing an ➤autotroph that feeds like plants, that is by photosynthesis. Compare ➤holozoic.

holozoic   Describing a ➤heterotroph that feeds like animals, that is by ingesting food into a gut or other specialized chamber where digestion occurs. Compare ➤holophytic.

homeobox   One of a number of ➤DNA sequences of about 180 base pairs in the coding regions of ➤homeotic genes, first identified in fruit flies but also conserved in a wide range of other species. These sequences code for ➤transcription factors controlling the expression of genes involved in embryonic development, for example determining that a fruit fly appendage should be either a leg or an antenna. Genes containing the homeobox have also been shown to be important in mammalian development.

homeomorphism   A continuous function between two ➤topological spaces that has a continuous inverse. Topologically, homeomorphic spaces are indistinguishable, which gives rise to the definition of a topologist as someone who cannot tell the difference between a doughnut and a teacup!

homeostasis   The state arising from the maintenance of a controlled environment within cells. All cells maintain internal conditions which differ to some extent from those of their environment. However, the more advanced the organism, the greater the sophistication of this ability. Mammals and birds can maintain a relatively constant internal temperature within fixed limits, usually in excess of that of the environment. The internal control of ionic concentration, regulation of pH and the removal of wastes from the body during excretion are all necessary for homeostasis. Virtually all aspects of growth and development are under homeostatic control. In turn, virtually all homeostatic mechanisms are regulated by hormones produced by ➤endocrine glands.

homeothermy   A variant spelling of ➤homoiothermy.

homeotic genes   ➤Genes, first identified in fruit flies, that are responsible for the correct development of specific body parts, for example determining that a leg forms in an appropriate place. Homeotic genes are thought to have a highly conserved function in the embryonic development of a wide range of animal species, including humans.

hominid A member of the Hominidae, a family of primates including humans and their recent ancestors placed in the genus ➤*Homo*. The family also includes chimpanzees, gorillas and orang-utans.

Homo The genus of hominids that includes modern ➤humans (*Homo sapiens*) and ancestral forms known from the fossil record, such as *Homo habilis* and *Homo erectus*. There is still much debate about the relationships between the various fossils placed in the genus, but they are believed to represent a line distinct from the Australopithecines (genus *Australopithecus*), a group of ape-like primates that coexisted.

HOMO Abbr. for highest occupied molecular orbital (➤frontier orbitals).

homogeneous Literally 'the same all through'. For example, a ➤phase is homogeneous if its composition is the same at all places.

homogeneous catalyst A ➤catalyst that is in the same phase as the reactants. Probably the most common examples are acid- and alkali-catalysed processes, such as hydrolysis. A novel homogeneous hydrogenation catalyst is ➤Wilkinson's catalyst; to be contrasted with the more common heterogeneous nickel catalyst.

homoiothermy (endothermy, warm-bloodedness) The ability of an animal to maintain a constant internal body temperature despite changes in the temperature of its external environment. Mammals and birds are **homoiotherms** and maintain temperature by conserving heat generated as a result of metabolic activity, particularly in organs such as muscles and the liver. Homoiothermy is a good example of ➤homeostasis.

homologous Describing structures in organisms that have a common ancestral origin. Thus the legs of quadruped land animals and the flippers of seals are homologous since the internal arrangements of the bones of each are derived from the same general pattern.

homologous pair A pair of ➤chromosomes that contain the same linear sequence of ➤genes. The chromosomes in the cells of a ➤diploid organism exist as homologous pairs, one member of each pair being inherited from each parent. Although they carry the same sequence of genes, they may not carry identical information since the genes on each chromosome may exist as different ➤alleles. The term is also applied to DNA segments that contain the same linear sequences of base pairs (➤base pairing).

homologous series A series of compounds related to each other by the addition or removal of a CH_2 group. Thus propane, $CH_3CH_2CH_3$, and butane, $CH_3CH_2CH_2CH_3$, are **homologues** in the alkane series.

homologue ➤homologous series.

homolytic fission (homolysis) Breaking of a bond such that the two atoms involved each retain one of the shared pair of electrons, thus giving rise to two

➤radicals. A good example is in the photochemical chlorination of alkanes, where the initiation step is homolytic fission of the chlorine molecule, caused by light.

homomorphism A function between two mathematical structures that preserves the binary operations defining the structures. For example ➤lg is a homomorphism from the ➤group of positive real numbers under multiplication to the group of real numbers under addition because lg xy = lg x + lg y.

homonuclear diatomic molecule A molecule containing two identical atoms, for example Cl_2, O_2 or H_2. Such a molecule cannot possess a permanent electric dipole moment and so the forces between homonuclear diatomic molecules are entirely due to ➤dispersion forces.

homopolar Obsolete synonym for covalent (➤covalent bonding).

homozygous The state in which the ➤alleles for a particular gene carried on each ➤chromosome of a ➤homologous pair in a ➤diploid ➤genome are the same. Compare ➤heterozygous.

Hooke's law A simple law of ➤elasticity stating that the ➤strain in a material is proportional to the ➤stress causing it, provided the ➤limit of proportionality is not reached. Named after Robert Hooke (1635–1703).

horizon The line dividing the Earth and sky. It may be defined in different ways. The optical horizon is the circle on the surface of the Earth that is the limit of vision from a particular point because of the curvature of the Earth. For an observer at a height h, the straight-line distance to the horizon is approximately $\sqrt{2Rh}$ where R is the radius of the Earth (for $h \ll R$). However, the curvature of light by ➤refraction in the Earth's atmosphere increases this distance by about 10%. The astronomical horizon is the ➤great circle on the ➤celestial sphere that divides the stars that are observable from a particular point on the Earth from those that are not. It is the intersection of the celestial sphere with the plane perpendicular to a line joining the observer to the ➤zenith.

horizontal In a plane perpendicular to the local ➤gravitational field.

hormone A substance produced by an ➤endocrine gland that is transported in the bloodstream to a target organ or group of cells where it exerts its effect. Hormones are usually either proteins (such as ➤insulin) or steroid derivatives, and are effective at very low concentrations. They control aspects of growth and metabolism such as the development of musculature and the breaking of the voice during puberty in males, and the regulation of the menstrual cycle in females.

horsepower Symbol hp. A traditional unit of ➤power supposedly equal to the power available from a single horse driving a pump or mill. One horsepower is equal to 745.7 watts. It remains a popular unit for measuring the power of ➤internal combustion engines and aero engines.

hour Symbol h. A unit of time equal to 3600 ➤seconds. The unit dates from classical times and, as well as being in common use world-wide, is acknowledged as a supplementary unit in ➤SI units.

hp Symbol for ➤horsepower.

Hs Symbol for the element ➤hassium.

HT Abbr. for ➤high tension.

HTML Abbr. for hypertext markup language. The computer file format that tells a Web browser how to display Web pages (➤www). ⋙HTTP.

HTTP Abbr. for hypertext transfer protocol. The ➤Internet's protocol for transferring Web pages (➤www). ⋙FTP.

Hubble constant Symbol H_0. A measure of the rate of ➤expansion of the Universe. It is found by dividing the rate v at which galaxies recede from our own by their distance r; the relation $v = H_0 r$ is known as **Hubble's law**. Using a convenient set of units for astronomy, the value of H_0 is estimated to be between 50 and 100 kilometres per second per megaparsec. The reciprocal of H_0 is a time, the **Hubble time**, which allows a naïve estimate (between 9 and 18 billion years) of the age of the Universe (assuming its present rate of expansion is constant). Named after Edwin Powell Hubble (1889–1953). ⋙parsec; redshift.

Hückel theory A simple, approximate molecular orbital theory most applicable to aromatic compounds such as benzene. Neighbouring atoms are assumed to overlap and bond strongly and equally, whereas atoms that are not immediate neighbours are assumed not to interact. Named after Erich Armand Arthur Joseph Hückel (1896–1980).

hue ➤colour.

human (man) The modern species *Homo sapiens*, which includes all modern human groups. Modern humans evolved in Africa some 2 to 3 million years ago from a group of hominid apes. Studies of the evolution of mankind have been hampered by fragmentary fossil evidence. Modern humans are distinguished from ancestral lines of humans by the relatively larger cranium and by the shape of the skull and jaws.

human genome project An international initiative to elucidate the sequence of the estimated three billion base pairs of the haploid human ➤genome. The project has involved the mapping of DNA sequence information obtained from a number of different laboratories (notably the Whitehead Institute in Cambridge, Mass., and the Sanger Institute in Cambridge, UK) and the first draft of the full sequence was published in 2001. It is hoped that the project will lead to a greater understanding of genetic processes in humans, for example inherited metabolic diseases such as cystic fibrosis, and the functioning of the ➤immune system. At the same time the project is also seen as controversial since it raises issues of ownership of genetic information and knowledge. For example, an individual who has been screened for, and shown to possess, genes associated with a greater risk of genetic disease or cancer might be denied access to life insurance. ⋙DNA sequencing.

human immunodeficiency virus ➤HIV.

human leucocyte antigen system (HLA system) A group of four genes that code for antigens (**histocompatibility antigens**) present on the surface of plasma membranes that enable an organism to differentiate between its own tissues and invading or transplanted tissues.

Hume-Rothery rules Rules that enable the composition of alloys to be rationalized by focusing on the ratio of valence electrons to atoms. A common ratio is 3:2, with examples such as CuZn, AuCd, Cu_3Al and Cu_5Sn. The rules were first propounded in 1926 by William Hume-Rothery (1899–1968).

humerus The bone of the upper forearm, between the shoulder blade and the elbow. ➤pentadactyl limb.

humidity ➤absolute humidity; relative humidity.

humus Decaying organic matter derived from the remains of plants and animals. It forms an important structural and functional component of soil.

hundredweight Symbol cwt. A unit in the ➤avoirdupois system of weights and measures. 1 cwt = 112 lb (pounds) in the UK, though the more intuitive **short hundredweight** (100 lb) is often used in the US.

Hund's rules A set of guidelines for finding the lowest energy state of an atom. The rules are empirical, but have theoretical justification in quantum mechanics. **Hund's first rule** (also called the **law of maximum multiplicity**) asserts that the state with the highest spin has the lowest energy; where several states have the same spin, the state with the highest orbital angular momentum has the lowest energy. **Hund's second rule** (derived from considering the energy of ➤spin–orbit coupling) is that, in the lowest energy state of a shell that is less than half-full, the total ➤angular momentum quantum number J has its minimum value; in a shell that is more than half-full, J has its maximum value. Named after Friedrich Hermann Hund (1896–1997).

hunting The oscillation of a variable governed by a control system, caused by the controller overcorrecting for deviations from the desired value.

Huygens' construction (Huygens' principle) A geometrical method of solving the ➤wave equation. Each point on the wavefront W is considered to be the source of a spherical wave – a **Huygens wavelet** (see the diagram). A circle that is tangential to all the Huygens wavelets after a time t is the position W' of the wavefront after time t. Named after Christiaan Huygens (1629–95).

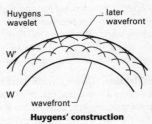

Huygens' construction

hybridization **1** The pairing of two complementary strands of single-stranded ➤DNA, giving rise to a hybrid molecule.

2 The interbreeding of two genetically distinct varieties of plant or breeds of animal to form offspring. Such hybrid offspring exhibit combinations of features of both parents. Hybrids between plants (F1 hybrids) frequently show hybrid vigour: they are more vigorous than their parents and are valued in horticulture. For example, hybrid wheats show increased vigour and higher yields than the original parents. Hybrids between animals are usually sterile. For example, a cross between a horse and a donkey produces a sterile mule.

3 (Chem.) An explanation common in organic chemistry whereby sets of atomic orbitals are combined to form hybrid orbitals, which can then overlap with other orbitals. For example, the 2s and three 2p orbitals of carbon in methane are combined into four sp^3 hybrids, which then overlap with the hydrogen 1s orbitals to form the observed shape. It is essential to note several points about this common but most unsatisfactory theory. It is incapable of prediction; the assignment of hybridization depends on knowing the structure of the molecule. If the structure *is* known then the ➤point group (2) of the molecule can be used to assign unambiguously the orbitals involved in the bonding. The methane molecule does not have four equal-energy bonds as predicted by the hybridization theory; rather it has one orbital more stable than the other three (➤delocalization).

hybridoma ➤antibody.

hydantoin The dione formed from ➤imidazole.

hydrate A compound containing water, be it a hydrated salt such as hydrated copper(II) sulfate, $CuSO_4 \cdot 5H_2O$, or an organic hydrate such as the hypnotic chloral hydrate, $CCl_3CH(OH)_2$.

hydrated ➤hydrate. ➤➤hexahydrate; pentahydrate.

hydration In organic chemistry, the process of addition of water to a molecule. For example, ethanol is manufactured by the ➤direct hydration of ethene:

$$C_2H_4 + H_2O \rightarrow CH_3CH_2OH.$$

In inorganic chemistry it is used to describe the process of accumulation of water molecules by ions to form a hydrated ion, such as $[Fe(H_2O)_6]^{2+}$. Compare ➤hydrolysis.

hydraulic press A machine that achieves a favourable ➤mechanical advantage by using different areas of piston at each end of a column of fluid. A force is applied to the small piston; the pressure is the same at the other end of the column, but acts on a much larger piston and thus applies a much larger force (the pressure multiplied by the area of the piston). The small piston must, of course, apply the force over a much greater distance if the force is to do any ➤work.

hydrazine H_2NNH_2 A hydride of nitrogen containing two nitrogen atoms, which is found as a colourless liquid. Its formula is related to ammonia in the same way as hydrogen peroxide is related to water. Hydrazine is an important reductant which was the fuel for the rocket engines in the V2 programme in Germany during the Second World War. ➤➤2,4-dinitrophenylhydrazine; hypergolic.

hydrazinium The ion formed from hydrazine or its derivatives, the simplest example being $H_2NNH_3^+$.

hydrazoic acid HN_3 The corresponding acid to ➤azides, which is found as a colourless liquid. Some heavy-metal salts, such as lead(II) azide, are used as detonators.

hydrazone The resulting condensation product from the reaction of hydrazine with aldehydes or ketones. Similar substituted hydrazones from ►2,4-dinitrophenyl-hydrazine are yellow-orange precipitates used to identify aldehydes and ketones.

hydride A compound containing hydrogen. Binary hydrides can be classified as follows: saline hydrides, covalent hydrides and interstitial hydrides. **Saline hydrides**, typically those of the s-block metals, are salt-like in that they are ionic solids containing the H^- ion, which hydrolyse in water to give hydrogen gas:

$$H^- + H_2O \rightarrow OH^- + H_2.$$

Covalent hydrides are covalently bonded instead and this class includes the most famous hydrides such as water, ammonia, methane and hydrogen chloride. **Interstitial hydrides** are named after the observed structure of the hydrides, where the hydrogen atoms occupy the gaps (interstices) in the metal's lattice. Interstitial hydrides are formed by some of the ►transition metals such as titanium, although many transition metals have no binary hydride. They are also commonly ►nonstoichiometric, in that titanium hydride has a formula TiH_x, where x is around 1.7. **Complex hydrides** also exist and are typified by the important reductants lithium tetrahydridoaluminate, $LiAlH_4$, and sodium tetrahydridoborate, $NaBH_4$.

hydriodic acid HI(aq) The acid made by dissolving hydrogen iodide in water. It is analogous to hydrochloric acid. It is a useful reductant in aqueous solution, a brown colour indicating that iodine has been produced by oxidation.

hydro- From the Greek *hudros*, a prefix meaning 'of water'. It is extremely common, reflecting the universal significance of water.

hydroboration A very important synthetic process in which a borane such as diborane, B_2H_6, in tetrahydrofuran is added to an alkene; the resulting species can be oxidized with hydrogen peroxide to produce an alcohol. For example, hex-1-ene forms hexan-1-ol; acid-catalysed addition of water follows ►Markovnikov's rule to give a different isomer, hexan-2-ol.

hydrobromic acid HBr(aq) The acid made by dissolving hydrogen bromide in water. It is analogous to hydrochloric acid.

hydrocarbon A compound containing carbon and hydrogen atoms only. The first subdivision of hydrocarbons is into ►aromatic hydrocarbons (such as ►benzene) and ►aliphatic hydrocarbons, which can then be subdivided into alkanes, alkenes, alkynes and alicyclic compounds.

The economic importance of the combustion of hydrocarbons can scarcely be overemphasized. Methane is the dominant component in natural gas. Petroleum is essentially simply a mixture of various hydrocarbons and fractions from its ►fractional distillation include petrol, diesel, kerosene and gas oil. The combustion of all of these fuels accounts for the vast majority of transportation energy costs, by aircraft, boat or car, as well as a significant proportion of heating costs both industrially and in the home.

hydrochloric acid HCl(aq) The acid made by dissolving hydrogen chloride in water. It is one of the three common mineral acids in the laboratory, coming in two strengths, dilute (typically 1 mol dm^{-3}) and concentrated. The concentrated acid is highly corrosive, but dilution lessens the danger and the dilute acid is a common reagent.

hydrocortisone A ►cortisone molecule with a hydroxyl group at carbon atom 11. It is used as an intravenous injection or applied topically to treat inflammation.

hydrocracking A variant of ►cracking in which hydrogen and a hydrogenation catalyst are present, ensuring that the products are alkanes, rather than the usual mixture of alkanes and alkenes. It is used to produce high-grade petrol.

hydrocyanic acid (prussic acid) HCN(aq) The acid formed when hydrogen cyanide dissolves in water. It is infamous as a poison.

hydrodealkylation An industrial process in which hydrogen is used to break a carbon–carbon bond in an aromatic hydrocarbon. This usually difficult process requires a catalyst such as platinum supported on alumina, Al_2O_3, at a temperature of about 600 °C. Under these conditions, methylbenzene reacts with hydrogen to give benzene and methane.

hydrodynamics A synonym for fluid dynamics (►fluid mechanics). The term was originally applied to the dynamics of water, but is now commonly used to refer to the behaviour of any moving fluid.

hydroelectric power Electric power generated by using the kinetic energy of flowing water. Hydroelectric power stations employ huge water ►turbines to drive electric ►generators. Most use a high-altitude reservoir from which the flow rate can be controlled and electricity generated at the rate required to meet fluctuating demand.

hydrofluoric acid HF(aq) The acid formed when hydrogen fluoride dissolves in water. It is a ►weak acid, unlike the corresponding hydrochloric, hydrobromic and hydriodic acids, partly due to the very strong bond between hydrogen and fluorine.

hydrofluorocarbon (HFC) A fluorinated hydrocarbon, which therefore contains carbon, hydrogen and fluorine only. HFCs are assuming increasing importance as they appear to be the best substitutes for the ►chlorofluorocarbons, which are criticized for depleting the ►ozone layer. One of the most useful examples is HFC 134a, CH_2FCF_3, which in addition to zero depletion has low toxicity and is nonflammable. It has recently been introduced in air conditioning systems, aerosols and for foam blowing.

hydrogen Symbol H. The first chemical element, with atomic number 1 and relative atomic mass 1.008, just above unity partly because in addition to normal hydrogen (or protium) there is a small percentage of ►deuterium, D, which has a mass of 2 units. There is a negligible percentage of the third isotope, the dangerously radioactive tritium, T. Hydrogen atoms were the early focus of theories of atomic structure and indeed the ►Bohr model worked well for hydrogen, despite failing significantly for others. Hydrogen is normally found as a colourless gas, in which hydrogen atoms combine together to give hydrogen molecules, H_2. Hydrogen is produced in the laboratory by adding a dilute acid to a reactive metal, typically zinc and dilute hydrochloric acid. The test for hydrogen is to put a lighted splint at the neck of a test-tube of the gas: a squeaky pop is a positive test, the pop caused by the rapid expansion of gas from the strongly exothermic combustion of the hydrogen. This susceptibility to violent combustion was a real worry for early airship designers who used the low density of hydrogen (it is the least dense gas) to gain buoyancy. After the tragic destruction of the *Hindenburg* by fire, this fundamental design problem necessitated the switch to the nonflammable helium.

Large quantities of hydrogen are used in the ➤Haber–Bosch process. The current industrial manufacture of hydrogen starts from methane and steam, using a variety of catalysts.

hydrogenation The process in which hydrogen gas is added to a molecule. The most usual examples come from organic chemistry, as in the conversion of an alkene into an alkane, using the ➤heterogeneous catalyst nickel (or palladium and platinum) or else the homogeneous ➤Wilkinson's catalyst. Hydrogenation is used to convert oils into fats for the production of margarine. ➤➤hydrocracking.

hydrogen atom emission spectrum The emission spectrum of atomic hydrogen (produced by passing a discharge through low-pressure hydrogen gas and observing the emitted light through a diffraction grating) consists of various series of lines which can be quantified using the ➤Rydberg–Ritz equation. Its spectrum was crucial to the elucidation of the structure of atoms around the beginning of the 20th century. ➤➤Balmer series; Lyman series.

hydrogen bomb ➤nuclear weapons.

hydrogen bonding A specific form of bonding in which a hydrogen atom bonded to one of the three most ➤electronegative small elements (fluorine, oxygen and nitrogen) is able to form a **hydrogen bond** with another one of these three elements.

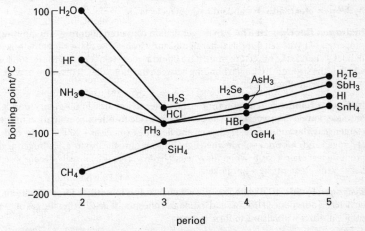

hydrogen bonding As the dispersion forces in them increase, the hydrides of the Group 14 elements (CH_4 to SnH_4) show increasing boiling points with increasing atomic number. Whereas this trend is generally reflected in the hydrides of elements in Groups 15 to 17 from period 3 onwards, the anomalously high boiling points of NH_3, HF and H_2O signify the hydrogen bonding in these molecules.

Hydrogen bonds are present, for example, in the structure of ➤ice. The origin of the bonding is often attributed (simplistically) to the significant dipole in, say, the O—H bond, causing a high partial positive charge on the hydrogen.

Hydrogen bonding has profound effects, causing, for example, the low density of ➤ice. In the structure of the DNA molecule specific ➤base pairing enables the molecule to be replicated and hence provide the basis for all biological reproduction. More prosaically, hydrogen bonding causes significant increases in the melting and boiling points relative to similar molecules (see the diagram overleaf); it also causes increased viscosity, for example of ➤glycerol.

hydrogen bromide HBr A colourless pungent gas which dissolves in water to form ➤hydrobromic acid. It cannot be produced from bromides and concentrated sulfuric acid because it is too easily oxidized to bromine, so concentrated phosphoric acid is used instead. Either the gas or the solution reacts with alkalis to form bromides.

hydrogencarbonate ion (bicarbonate ion) HCO_3^- or, more precisely, $HOCO_2^-$. An important ion in aqueous solution, causing temporary hardness of water (➤hard water). It is the ➤conjugate base of carbonic acid, H_2CO_3, and the ➤conjugate acid of carbonate ion CO_3^{2-}. One of the two most important ➤buffer solutions that act near neutral pH is H_2CO_3/HCO_3^-.

hydrogen chloride HCl A colourless pungent gas which dissolves in water to form ➤hydrochloric acid. It can be produced from a solid chloride and concentrated sulfuric acid. Either the gas or the solution reacts with alkalis to form chlorides.

hydrogen cyanide HCN(g) A colourless gas smelling of bitter almonds which dissolves in water to form ➤hydrocyanic acid. It is a notorious poison.

hydrogen electrode ➤standard hydrogen electrode.

hydrogen fluoride HF The ➤hydrogen halide containing fluorine. The physical properties of hydrogen fluoride differ significantly from those of the other hydrogen halides. Whereas the others are gases, it is a liquid at room temperature, the reason for which is the strong ➤hydrogen bonding between the molecules. It dissolves in water to form ➤hydrofluoric acid.

hydrogen halide A general term for the molecules hydrogen fluoride, chloride, bromide, iodide (and astatide). They can all be made by direct combination of the elements, with varying ease. Hydrogen and fluorine explode at $-200\,°C$ in the dark. Hydrogen and chlorine explode when light is shone onto the mixture. Hydrogen and bromine react fairly easily when lit, whereas hydrogen iodide is easily decomposed back to the elements by a hot poker.

hydrogen iodide HI A colourless gas which dissolves in water to form ➤hydriodic acid. It can be produced from a solid iodide and phosphoric acid. Either the gas or the solution reacts with alkalis to form iodides.

hydrogen ion H^+ The hydrogen ➤cation is essentially a bare proton, the species left when a hydrogen atom has lost its only electron. This gives it unique properties. For example, in water the attraction of H^+ to water is so strong that the species present is better termed H_3O^+, the ➤oxonium ion, and even this is further hydrated. The concentration of aqueous hydrogen ions is particularly important and is measured by the ►pH.

hydrogenolysis The breaking of a molecule by reaction with hydrogen gas. This is a much rarer process than the similar-sounding 'hydrolysis'.

hydrogen overvoltage (hydrogen overpotential) The excess voltage over and above that predicted using the ►standard electrode potential required to produce a reasonable rate of evolution of hydrogen gas during electrolysis. Electrolysis of an aqueous solution is likely to produce hydrogen gas, as the hydrogen ion is more easily reduced than the ions from most metals (►standard electrode potential). However, the need for two atoms to meet and bond to form H_2 provides a *kinetic* barrier to evolution of the gas.

hydrogen peroxide H_2O_2 A colourless viscous liquid most commonly seen as an aqueous solution. The common label '20-volume' indicates that the aqueous hydrogen peroxide can evolve 20 times its own volume of oxygen, when it decomposes catalytically (with manganese(IV) oxide, for example). It can act both as an oxidant, for example oxidizing acidic iodide ions to a brown solution of iodine, and as a reductant, for example reducing acidic manganate(VII) ions to manganese(II) ions. It is the parent acid of peroxides.

hydrogen spectrum ►hydrogen atom emission spectrum.

hydrogensulfate ion (bisulfate ion) HSO_4^- or, more precisely, $HOSO_3^-$ It is the ►conjugate base of sulfuric acid and the conjugate acid of sulfate ion.

hydrogen sulfide (sulfuretted hydrogen) H_2S A colourless poisonous gas with a characteristic smell of bad eggs. In the laboratory the test for hydrogen sulfide is that acidified potassium dichromate(VI) on filter paper held in the gas turns from orange to green, as hydrogen sulfide is a reductant. However, ►sulfur dioxide also gives this test; the best way of distinguishing between them is to test with filter paper soaked in aqueous lead(II) ions. Hydrogen sulfide gives a dark brown stain of insoluble lead(II) sulfide whereas sulfur dioxide produces no change.

hydrological cycle (water cycle) The circulation of water around the surface of the Earth by evaporation, condensation, precipitation and the flow of liquid water on and under the surface.

hydrology The study of water and water vapour in the context of the atmosphere, climate and geography of the Earth. About 97% of all the Earth's water is in the oceans.

hydrolysis The process in which a molecule is broken down by water. In inorganic chemistry this usually occurs with covalent halides, commonly chlorides. For example, phosphorus trichloride violently hydrolyses to form hydrogen chloride gas and phosphonic acid.

In organic chemistry, the process is frequently catalysed either by acid or base or both. ►Esters, for example, are best hydrolysed by base, as the more powerful ►nucleophile hydroxide ion attacks the ester.

hydrolytic process A process that involves ►hydrolysis.

hydrometer An instrument for measuring the ➤density of a liquid. The simplest hydrometers use the principle of ➤flotation.

hydronium ion Alternative name for the ➤oxonium ion.

hydrophilic (literally '**water-loving**') Describing a molecule or part of a molecule that has an affinity for water. ➤hydrophobic interaction.

hydrophilic colloid A colloid that resists coagulation on the addition of small quantities of electrolytes (compare ➤hydrophobic colloid).

hydrophobic (literally '**water-hating**') Describing a molecule or part of a molecule that repels water. ➤hydrophobic interaction.

hydrophobic colloid A colloid that coagulates on the addition of small quantities of electrolytes (compare ➤hydrophilic colloid).

hydrophobic interaction An interaction between a hydrophobic ('water-hating') part of a molecule and an aqueous environment. Such interactions are particularly significant in establishing the three-dimensional shape of enzymes in solution. Enzyme molecules commonly have the polypeptide chain folded to form a hydrophobic core comprising amino acids that have hydrophobic side chains (e.g. that of phenylalanine) with the hydrophilic ('water-loving') side chains (such as those of glutamic or aspartic acids) on the surface forming weak bonds with surrounding water molecules. Hydrophobic and hydrophilic interactions are also important in maintaining the integrity of biological ➤membranes.

hydroponics A technique for culturing plants without using soil. Plants are grown in suitable containers in a soil-less medium, and the roots are irrigated in a variety of ways by a solution containing formulations of nutrients designed specifically for the plant in question. The composition of the solution can be monitored and manipulated to suit particular stages of growth. Many commercial glasshouse crops such as tomatoes and cucumbers are cultivated in this way.

hydroquinone A colourless solid (see the diagram) used as a photographic developer. ➤dihydroxybenzene.

OH

hydrosphere The portion of the Earth's crust made up of water and its solutes (e.g. sodium and magnesium chlorides). It includes groundwater, as well as the oceans and icecaps.

OH

hydroquinone

hydrostatic equation The differential equation

$$\frac{\mathrm{d}p(z)}{\mathrm{d}z} = -\rho(z)g,$$

relating pressure p to height z in a compressible fluid, of density ρ, under the influence of a gravitational field of magnitude g. It also applies to the variation of pressure with altitude in the atmosphere.

hydrostatics The study of ➤fluids at rest.

hydrous Containing water. Compare ➤anhydrous.

hydroxide ion OH⁻ The ion central to alkaline behaviour in aqueous solution. Although the logical formula for this exceptionally important ion is HO⁻, giving the elements in order of increasing electronegativity, the usage of the alternative order OH⁻ is so ingrained in chemistry that the habit is hard to break. The most common soluble hydroxides are those of the ➤s block of the periodic table, such as sodium hydroxide, NaOH, and calcium hydroxide, $Ca(OH)_2$. Most other metal hydroxides are insoluble in water, such as copper(II) hydroxide.

hydroxonium ion ➤oxonium ion.

hydroxy- A prefix in organic chemistry signifying the presence of an —OH group, as in the 2,3-dihydroxybutanedioate ion (more commonly called the ➤tartrate ion).

hydroxyapatite An ➤apatite with hydroxide ion present, $Ca_5(OH)(PO_4)_3$.

hydroxybenzene A significantly less common alternative name for ➤phenol.

hydroxylamine NH_2OH A colourless solid, which explodes on heating, used to test for aldehydes and ketones with which it forms ➤oximes.

hydroxyl group The group —OH which occurs in several organic compounds, such as ➤ethanol and ➤phenol.

hygrometer An instrument for measuring the ➤humidity of air. ➤➤dew-point hygrometer.

hygroscopic Describing substances that absorb water from the atmosphere, but do not dissolve in the water (if they do, they are said to exhibit ➤deliquescence). Examples include magnesium perchlorate, $MgClO_4$.

hyperbola A ➤conic with eccentricity $e > 1$. It has two branches and two axes of symmetry with respect to which its Cartesian equation is $x^2/a^2 - y^2/b^2 = 1$; the lines $y = bx/a$ and $y = -bx/a$ are ➤asymptotes. A **rectangular hyperbola** has perpendicular asymptotes with respect to which its equation is $xy = c^2$, where c is a constant. The orbits of some comets are **hyperbolic**.

hyperbolic functions Six functions named by putting the word 'hyperbolic' in front of the six ➤trigonometric functions (as in hyperbolic sine), abbreviated to sinh, cosh, cosech, sech, tanh and coth, which are defined by $\sinh x = \frac{1}{2}(e^x - e^{-x})$, $\cosh x = \frac{1}{2}(e^x + e^{-x})$, $\operatorname{cosech} x = 1/\sinh x$, $\operatorname{sech} x = 1/\cosh x$, $\tanh x = \sinh x/\cosh x$, $\coth x = 1/\tanh x$. Hyperbolic functions are so called because $(a \cosh t, b \sinh t)$ lies on a hyperbola.

hyperboloid A surface in three dimensions which has a hyperbolic cross-section perpendicular to two coordinate planes. A cooling tower is shaped like a hyperboloid.

hypercharge Symbol Y. A ➤quantum number for subatomic particles defined as the sum of the ➤baryon number and the ➤strangeness.

hyperconjugation A theory to explain the stability of ➤carbocations considering the interaction of ➤sigma bonds with ➤pi bonds. The interaction stabilizes carbocations by ➤delocalization.

hyperfine structure Structure observed in spectral lines in absorption or emission spectroscopy caused by the excitation of nuclear states (different energy levels of the atomic nucleus). Compare ➤fine structure.

hypergolic Capable of spontaneous inflammation on contact; hypergolic propellants are useful for spacecraft propulsion. An example is the mixture of dinitrogen tetroxide, N_2O_4, and unsymmetrical dimethylhydrazine, $(CH_3)_2NNH_2$, which was used by the Apollo astronauts to leave the Moon's surface.

hypermetropia (long-sight) The inability of the eye to focus at short distances. It is commonly caused by a foreshortened eyeball, and is corrected by spectacles with a ➤convex lens. Compare ➤myopia.

hyperon Any ➤baryon (e.g. the ➤lambda particle), except the proton or neutron, that does not decay by the ➤strong interaction.

hypersonic Travelling at at least five times the speed of sound. Compare ➤supersonic.

hypersphere An extension to more than three dimensions of the idea of a ➤sphere. For example, the equation $x_1^2 + x_2^2 + x_3^2 + x_4^2 = R^2$ describes a hypersphere, radius R, in four-dimensional space with coordinate system (x_1, x_2, x_3, x_4).

hypertext Computer text within a Web page (➤www) which has links to other Web pages or sites, accessible by clicking the mouse on the link. The link is called a hyperlink. ➤➤HTML; HTTP.

hypha (plural **hyphae)** ➤mycelium.

hypnotic A substance capable of sending a person to sleep. Typical examples are the barbiturates.

hypo A traditional name for sodium thiosulfate, $Na_2S_2O_3$, used as a photographic fix.

hypo- In biology, a prefix meaning 'attached or inserted below' a structure, as in the hypocotyl, the part of a seedling below the ➤cotyledons, or referring to a lower physical state, as in **hypothermia**, a lowering of body temperature.

hypobromous acid (bromic(I) acid) HOBr The oxoacid of bromine in which bromine has oxidation number +1, hence its alternative name, which forms a colourless solution. The parent acid of **hypobromites**.

hypochlorous acid (chloric(I) acid) HOCl The oxoacid of chlorine in which chlorine has oxidation number +1, hence its alternative name, which forms a colourless solution. The parent acid of **hypochlorites**, which are used as bleaches.

hypogeal A pattern of germination in which the ➤cotyledons remain beneath the ground.

hypoiodous acid (iodic(I) acid) HOI The oxoacid of iodine in which iodine has oxidation number +1, hence its alternative name, which forms a colourless solution. The acid is less stable than hypochlorous acid, decomposing easily on heating to form iodic acid and hydriodic acid.

hypotenuse The longest side of a ➤right-angled triangle.

hypothalamus ➤brain.

hypothesis A provisional supposition, of questionable validity, that is used as a basis for further logical development. A hypothesis is tested by seeking experimental verification of predictions made using the hypothesis.

hypothesis test A statistical procedure designed to assess the degree of plausibility of a given statement or **null hypothesis**, concerning, for example, the value of the population mean, on the basis of the evidence provided by a **sample.** Thus a test conducted at the 5% **significance level** means that the null hypothesis will be accepted if the given sample value is one of those which would be expected from 95% of all random samples. It is important to realize that a hypothesis test cannot lead to a definite conclusion.

hypsometer An apparatus for calibrating a thermometer to the ➤steam point.

hysteresis The irreversibility of the ➤magnetization in ➤ferromagnetism, a result of the irreversibility of ➤domain boundary movements. When an external magnetic field is applied to a ferromagnetic material, the dependence of the magnetic flux density B on the applied field H is as shown in the diagram. The value of B reaches a limit (known as ➤saturation). If the applied field is reduced and subsequently reversed, B does not follow the original path back to the starting point. The value of B at R is known as the **residual flux density** (entirely a consequence of the magnetization of the material, M); this is the phenomenon of **remanence.** A subsequent reversal of applied field makes B follow the

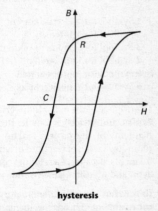

hysteresis

arrowed path; the value of H at C is the **coercivity** of the material. The entire loop is called the **hysteresis curve**; energy (proportional to the area enclosed by the loop) is expended in cycling through the hysteresis curve. Similar phenomena, in which the value of a dependent quantity varies according to whether the independent quantity is increasing or decreasing, are sometimes also termed hysteresis.

Hz Symbol for the unit ➤hertz.

H zone ➤muscle.

i The square root of minus one. Hence $i^2 = -1$. ➤complex number.

î A ➤unit vector in the direction of the positive x axis.

I Symbol for the element ➤iodine.

I **1** Symbol for ➤electric current.
   2 Symbol for ➤intensity.
   3 Symbol for ➤moment of inertia.
   4 Symbol for ionic strength (➤Debye–Hückel theory).
   5 Symbol for ➤isotopic spin.
   6 Symbol for ➤nuclear spin.

I band One of the major transverse bands making up the repeated striated pattern of cardiac and skeletal ➤muscle; the 'I' stands for 'isotropic'. I bands consist of thin filaments of the protein ➤actin. During muscle contraction a ➤nerve impulse promotes a series of reactions in which actin filaments link with the thicker ➤myosin filaments. Because of a resulting conformational change in the myosin filaments, the actin and myosin filaments slide past each other, causing the muscle to contract.

ibuprofen An anti-inflammatory drug used as a pain-killer in the treatment of headache, menstrual pain, rheumatoid arthritis and other conditions.

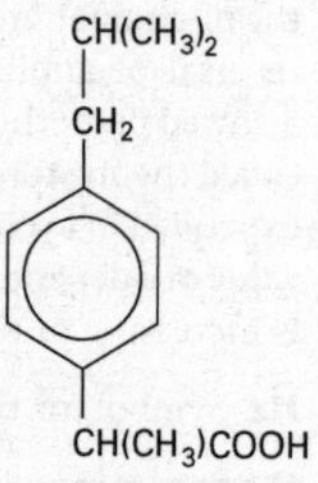

CH(CH₃)₂
|
CH₂

CH(CH₃)COOH
ibuprofen

-ic A suffix labelling acids and ions with an element's higher oxidation number (compare ➤-ous). It was once the common label, but has been superseded in many cases by nomenclature based on specifying oxidation numbers. For example ferric was the old name for iron(III). Examples still in common use include nitric acid and sulfuric acid.

IC Abbr. for ➤integrated circuit.

ice The solid phase of water. Water is the only substance that has different names for all three of its phases. Ice is almost unique in that its density is *less* than that of liquid water. Nevertheless, this is vital for life on Earth: as ice floats on water, ponds freeze downwards from the top and aquatic life can survive. Its low density is caused by strong ➤hydrogen bonding, which holds the oxygen atoms in an open structure (see the diagram opposite; hydrogen bonds are conventionally indicated by dashed

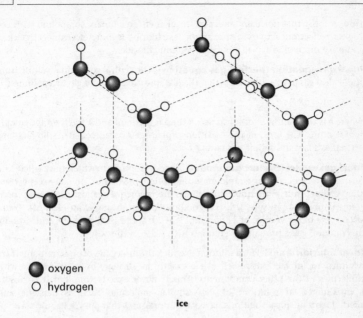

● oxygen
○ hydrogen

ice

lines). At very high pressures phase changes occur, to give other crystal structures (distinct from normal ice); these are labelled by roman numerals, as in ice II.

ice age A period when glacial ice covered significant areas of the Earth. For example, during the ➤Pleistocene ice age, around one-third of the Earth's surface was iced over.

Iceland spar Very pure crystalline calcite, $CaCO_3$. Its transparent crystals are unusual in showing ➤birefringence.

ice point The temperature at which ice and water are in equilibrium at standard atmospheric pressure (101 325 Pa). Its value, 273.15 K, is just 0.01 K below the ➤triple point of water.

icon In computing, a small graphical representation used in ➤WIMP environments. Icons depict objects or actions that will be executed if the user moves the pointer over the icon and clicks a button on the mouse.

icosahedral Having the shape of an icosahedron. The crystal structure of ➤boron involves icosahedral B_{12} units.

icosahedron A solid that has twenty faces. A **regular icosahedron** has faces that are ➤equilateral triangles and is one of the five ➤Platonic solids. The ➤buckminsterfullerene molecule has the shape of a **truncated icosahedron**.

ICSH Abbr. for interstitial cell stimulating hormone (➤luteinizing hormone).

-ide A suffix that normally indicates the element in a binary compound with the higher ➤electronegativity. For example, as chlorine is more electronegative than sodium, common salt, NaCl, is called sodium chloride.

ideal gas equation (perfect gas equation) The central equation resulting from the ➤ideal gas model. It predicts that the pressure p exerted by a gas in a volume V is

$$pV = nRT,$$

where n is the ➤amount of substance, R the gas constant and T the thermodynamic temperature. The form of the equation explains ➤Avogadro's law, ➤Boyle's law, ➤Charles's law and ➤Dalton's law.

ideal gas model (perfect gas model) A model for a gas in which the particles are assumed to be in constant, rapid, random motion and do not interact except when they collide; thus ➤intermolecular forces are ignored. For most gases near room temperature and pressure, this is a good model. For example, an ideal gas would occupy a ➤molar volume of 24 dm^3 mol^{-1} at 20 °C and 1 bar pressure, and virtually all common gases have molar volumes within 1% of this value.

ideal solution A model for a solution in which the molecules of one component are assumed to interact with each other exactly as strongly as they interact with molecules of the other components. This is never exactly true, but it is closely approximated in solutions of very similar molecules such as benzene and methylbenzene. Ideal solutions have vapour pressures that obey ➤Raoult's law.

identical particles A fundamental concept of ➤quantum mechanics: that the square ➤modulus of the ➤wavefunction of a system composed of two indistinguishable particles (such as two electrons) must be ➤invariant under a ➤symmetry transformation that exchanges the two particles. Thus the wavefunction itself must be **symmetric** (unchanged after the operation) or **antisymmetric** (multiplied by -1 after the operation). Identical particles with a symmetric wavefunction are called ➤bosons; identical particles with an antisymmetric wavefunction are called ➤fermions. Because they are indistinguishable, we cannot label electrons in a system as 'electron 1' and 'electron 2', we can only talk of particular quantum states being empty or occupied by *an* electron.

identical twins (monozygotic twins) The result of the fertilization of an egg by a sperm and the subsequent division of the ➤zygote into two cells each of which develops into two independent ➤embryos. Since each of the cells derived from the zygote has the same ➤genome, the resultant individuals will be identical to each other. Compare ➤fraternal twins.

identity **1** A statement, sometimes indicated by $\equiv$, that two mathematical expressions give the same result whatever values are substituted for their variables.

2 The unique element e in a ➤group such that $a * e = e * a = a$ for all a in the group, where $*$ is the group operation.

3 That square ➤matrix with elements 1 down the ➤leading diagonal and 0 everywhere else.

4 That ➤transformation which leaves the elements of a given ➤set unchanged.

IE Abbr. for ➤ionization energy.

iff A contraction of the phrase 'if and only if' to connect two statements each of which may be derived from the other. The two statements are thus logically equivalent.

Ig Abbr. for ➤immunoglobulin.

igneous rock One of the three main types of rock: igneous rocks form when molten ➤magma cools and crystallizes. **Extrusive igneous rock**, such as basalt, forms at the Earth's surface; rapid cooling results in small crystals. **Intrusive igneous rock**, such as ➤granite, solidifies beneath the Earth's surface; slow cooling results in large crystals. Compare ➤metamorphic rock; sedimentary rock.

ignition temperature The temperature at or above which a substance will combust.

IL-1, IL-2 Abbrs. for ➤interleukin 1, ➤interleukin 2.

Ile Abbr. for ➤isoleucine.

ileum ➤gut.

ilium ➤pelvic girdle.

illumination (illuminance) Symbol E; unit lux. The ➤luminous flux falling on a surface per unit area.

ilmenite Traditional name for a ➤mixed oxide of iron and titanium, whose formula is $FeTiO_3$. It is a common black ore from which both metals can be extracted.

image A construct of geometrical optics describing a point or set of points, other than the source of light, at which rays converge – a **real image** – or from which they appear to diverge – a **virtual image** (see the diagram).

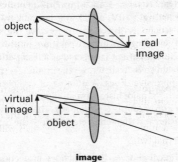

image

imaginary axis The y axis in the ➤Argand diagram.

imaginary number A ➤complex number with real part zero, which therefore lies on the ➤imaginary axis.

imaginary part The part y in the ➤complex number $z = x + iy$.

imago The adult stage of an insect that exhibits complete metamorphosis, such as a moth or butterfly.

imidazole $C_3H_4N_2$ A colourless solid five-membered heterocyclic aromatic compound (see the diagram). Derivatives include the amino acid ➤histidine.

imidazole

immersion objective A lens system in microscopy in which cedarwood oil fills the cavity between the lowest lens (the ➤objective) and the cover glass over the sample. The oil has a refractive index between the refractive indices of the two glasses, and this helps to reduce reflection losses and increase the ➤numerical aperture.

immiscible Describing a combination of liquids that do not mix with each other. A common example is oil and water. An agitated mixture of two immiscible liquids exerts a vapour pressure equal to the sum of the vapour pressures of each liquid.

immune response The reaction of the body's ➤immune system to foreign cells or substances that may be potentially dangerous or cause disease. Immune responses are the cause of many allergic reactions, and are also responsible for causing the rejection of grafted or transplanted organs and tissues in mammals.

immune system A sophisticated and diverse system found in vertebrates whose primary task is to defend them against foreign organisms, usually ➤viruses and ➤bacteria, by recognizing the difference between 'self' and 'non-self'. The differentiated cells that form the immune system are the ➤leucocytes and include ➤macrophages, which engulf and digest foreign matter by ➤phagocytosis, and ➤lymphocytes, which aid the recognition of invading organisms. Some lymphocytes (in particular the B cells, some of which mature into plasma cells) produce ➤antibodies which selectively bind to a particular arrangement of atoms on the surface of foreign organisms called an ➤antigen. The reaction between an antibody and an antigen is termed an ➤immune response (➤T cell). Immune deficiency viruses, including ➤HIV, attack the immune system and render the animal susceptible to secondary infections to produce the symptoms of ➤AIDS. Problems frequently arise during transplant operations in which tissue from a donor implanted into the recipient patient is rejected as a result of the immune system's ability to recognize foreign cells. In such cases the immune response is moderated by the use of immunosuppressive drugs. A number of components of the immune system are encoded by the ➤major histocompatibility complex (MHC). This is a large cluster of ➤genes on ➤chromosome 6 in humans. There are many variations of the genes found in this cluster, and their characterization is the basis of tissue typing for transplant operations.

immunization A clinical process that provides protection against disease on subsequent exposure to infection. There are two types of immunization, active and passive. **Active immunization**, pioneered by Edward Jenner 200 years ago, promotes the production of circulating ➤antibodies against foreign ➤antigens by injecting a small quantity of modified antigen into the bloodstream. Vaccines against viral diseases, for example, consist of injecting either inactivated ➤virus particles or specially attenuated strains of the virus. **Passive immunization** is the application of antibodies from an immune individual to the nonimmune patient. This treatment is used when an individual has been, or probably will be, exposed to an infectious disease and there is insufficient time for active immunization. Passive immunization is still used for tetanus, hepatitis A, rabies and measles.

immunoassay One of a number of techniques for estimating the amount of a ➤protein or other ➤antigen in a sample using the specificity of ➤antibodies. Usually the reaction is designed to produce a colour so that the colour intensity can be used to estimate the amount of antibody bound to the antigen and hence the amount of antigen present. Immunoassays have a wide range of applications in clinical and diagnostic testing. ➤➤enzyme-linked immunosorbent assay.

immunoglobulin (Ig) One of a very diverse group of ➤antibodies produced by ➤B cells. Each Ig molecule is composed of two identical light chains and two identical heavy chains, held together by ➤disulfide bonds. There are five classes of heavy chain, each one possessing different properties. These heavy chains determine the classes of Ig: IgA, IgD, IgE, IgG and IgM. IgE is responsible for allergic reactions such as asthma, hay fever and the skin condition urticaria.

impact parameter Symbol s. In classical ➤scattering theory, the distance of closest approach of two particles that would be observed if there were no interaction and the particles were to continue with constant velocity (see the diagram). In reality, if the interaction is repulsive the particles will never get this close; if attractive they will get closer than the impact parameter.

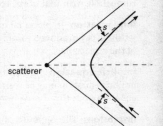

impact parameter

impedance Symbol Z; unit Ω. In general, the property, expressible as a complex number, of a simple ➤linear system that relates the ➤amplitude and ➤phase (2) of its response to the amplitude and phase of a stimulus. Specifically in electrical systems, it is the complex quantity Z that relates the applied voltage V and current I, both in complex form (➤complex number): $V = ZI$. A ➤capacitor has an impedance $1/i\omega C$, an ➤inductance $i\omega L$ and a ➤ resistance R, where ω is the angular frequency of the current or voltage. As with resistances in a ➤d.c.circuit, if two or more impedances are in ➤series (3), their impedances add; if in ➤parallel (4) their reciprocals add:

$$Z_{\text{series}} = Z_1 + Z_2 + \ldots, \quad Z_{\text{parallel}} = \cfrac{1}{\cfrac{1}{Z_1} + \cfrac{1}{Z_2} + \ldots}.$$

impedance matching The minimization of power loss caused by the internal ➤impedance of a power supply. When a power supply with an internal impedance is used, a load with low impedance will cause much of the power to be dissipated in the supply. However, a load with high impedance will cause a low current to flow, and again little power will be dissipated in the load. Maximum power is dissipated in the load when its impedance is equal to the ➤complex conjugate of the impedance of the supply.

Imperial units An older system of units, though still in common use, particularly in the US (➤US Customary Units). Distances are measured in inches (symbol in. or ″, defined as 0.0254 m), feet (symbol ft or ′, = 12 in.), yards (= 3 ft) or miles (= 1760 yd = 1609.344 m). Masses are measured in ounces (symbol oz, = 0.028350 kg), pounds (symbol lb, = 16 oz = 0.453592 kg) and tons (= 2240 lb = 1016.05 kg) (UK definitions). The disadvantage of Imperial units compared with ➤SI units is the unmemorable set of conversion factors, particularly with derived units. Their use seems harmless, but a mix of metric and Imperial units, as is used in aviation, can be awkward. The crew of a Boeing 767 learnt this in 1983, when the 22 300 kg of fuel which they thought had been loaded was in fact 22 300 lb, and the aeroplane lost all power. The captain glided to a successful landing.

impermeable Resistant to the passage of some or all fluids. For example, gas-permeable contact lenses are impermeable to water but permit the eye to receive oxygen from the air.

implant The introduction of foreign tissue or a foreign organ into a host. The host will not reject the implant if the donor's ➤major histocompatibility complex is compatible with that of the recipient.

implantation In placental mammals, including humans, the event in which the fertilized ovum adheres to and embeds itself in the uterine wall before the formation of the ➤placenta.

implicit equation An equation of a curve of the form $f(x, y) = 0$. Sometimes an implicit equation may be rearranged to give y as an explicit function of x (➤explicit equation), but this is not always possible.

implosion The opposite of an ➤explosion; in an implosion the volume of a substance is suddenly reduced. Potentially its most important application is in ➤nuclear fusion. A pellet of ➤deuterium and ➤tritium implodes under the impact of a laser beam, to generate energy; to achieve this much more energy has to be put in than is generated, but the hope is that the energy deficit will eventually be overcome.

imprinting 1 In ➤ethology, the instinctive response of the young of certain animal species which enables them to recognize a parent. It is particularly strong in geese and ducks: newly hatched chicks will adopt the first moving animal they see as a parent, even if that animal is a human. The phenomenon was studied in particular by Konrad Lorenz.
    2 ➤genetic imprinting.

improper integral (infinite integral) A ➤definite integral in which infinity intrudes, either as a ➤limit of integration or as a ➤singularity of the ➤integrand. If meaning can be given to the integral by taking a suitable limit, it is said to be 'convergent'; otherwise it is 'divergent'.

improper rotation A symmetry operation in which an object can be returned to its original shape by rotation about an axis followed by reflection in a perpendicular plane.

impulse 1 (Phys.) Symbol J. The change in linear momentum P of a system caused by a ➤force F. The impulse imparted to a system can be found by integrating the applied force with respect to time:

$$J = \Delta P = \int F(t)\, dt.$$

Thus a given impulse can be associated with a weak force acting over a long time, or a strong force acting over a short time; the latter is termed an **impulsive force**. ➤➤Newton's laws of motion.

2 (Biol.) ➤nerve impulse.

impurity scattering Scattering of electrons in a solid caused by the presence of impurities. At low temperatures, impurity scattering, which is largely independent of temperature, is the dominant factor in determining the resistivity of a conductor; a perfect metallic crystal at 0 K would have no resistance.

in. Symbol for the unit ➤inch.

In Symbol for the element ➤indium.

inbreeding A system of sexual reproduction in which male and female gametes from one individual fuse (self-fertilization) or when breeding occurs between individuals in a small population. Inbreeding populations are characterized by having a high proportion of genes existing in a ➤homozygous condition. Such an increase in the numbers of mutant or lethal ➤recessive genes can result in **inbreeding depression**, in which individuals are weakened or show lower biological ➤fitness. In human populations this can result in increased incidences of genetic disorders, and virtually all human societies have social or religious taboos to prevent inbreeding between closely related individuals.

incandescence The emission of light by virtue of the high temperature of the source. ➤black-body radiation.

incendiary device A device that can explode violently and set fire to other objects.

inch Symbol in. (with a full stop). An ➤Imperial unit of distance. It is now defined as *exactly* 2.54 cm. It will probably remain in everyday use for many years, particularly in the US.

incidence angle (angle of incidence) The angle between an incoming ray or trajectory and a ➤normal to the surface it strikes. ➤Snell's law.

inclination **1** The angle between a celestial body's ➤axis of rotation and a line perpendicular to the plane of its orbit. The Earth's axial inclination is about $23\frac{1}{2}°$.
2 ➤magnetic inclination.

incoherent wave Any wave, but most usually an electromagnetic wave, consisting of the sum of two or more contributions of random ➤phase (2). Compare ➤coherence.

incomplete dominance ➤dominant.

incompressible Referring to a fluid, of constant density. Incompressibility of a liquid is often a necessary simplifying assumption in ➤fluid mechanics. It can be expressed mathematically in terms of the ➤velocity as a nondivergent flow: div $v = \nabla \cdot v = 0$. A magnetic field can be pictured as an incompressible fluid, as the second of ➤Maxwell's equations is of this form.

increment A change in a quantity. If the quantity is y, then Δy denotes any increment in y; δy denotes a *small* increment in y. A negative increment is sometimes called a 'decrement'. Compare ➤differential.

incubation 1 The development of an ➤infection from the time at which an organism is exposed to the disease-causing agent until the appearance of the first symptoms.

2 The placing of a culture of, for example, bacteria into a suitable environment to promote growth.

3 The use by (particularly) birds of their body warmth to promote the development of eggs.

incus ➤ear.

indefinite integral An ➤integral without specified ➤limits of integration which, as a result, is only determined to within an arbitrary constant. Compare ➤definite integral. ➤➤Appendix table 3.

indene C_9H_8 A colourless liquid bicyclic aromatic hydrocarbon (see the diagram), used as a solvent. The carbon atom indicated can be deprotonated as the resulting anion is ➤delocalized, and so the hydrocarbon is acidic.

indene

independent Two events A and B are independent if $P(A \cap B) = P(A)P(B)$ or, equivalently, if $P(A|B) = P(A|B')$ (➤probability for notation). This refines the intuitive idea that events should be independent if the occurrence of one has no bearing on the occurrence of the other.

independent assortment ➤meiosis.

independent migration of ions, law of ➤Kohlrausch's law.

independent variable ➤variable.

indeterminacy principle ➤Heisenberg uncertainty principle.

indeterminate An expression that, for some values of a variable or variables, becomes meaningless. Meaning can sometimes be restored by taking a ➤limit, as in $\sin x/x$ tends to 1 as x tends to 0, but this is *not* always possible.

index (plural indices) A number indicating the ➤power (2) to which a given number or expression is raised. Indices are written as right superscripts and obey the rules $x^a x^b = x^{a+b}$, $x^a/x^b = x^{a-b}$, $(x^a)^b = x^{ab}$, which enable meaning to be assigned to all real and, indeed, complex indices. Particularly important cases include $x^0 = 1$, $x^{-n} = 1/x^n$ and $x^{1/n} = \sqrt[n]{x}$, the nth root of x.

indicator A chemical that changes colour to show when a process has occurred. The most common indicator is one which changes colour as the ➤pH of a solution is varied. Such an **acid–base indicator** has a pK_{in} value; if the pH is lower than pK_{in} by about one unit then the solution will appear 'acidic' and if the pH is higher by about one unit it will appear 'alkaline'. Litmus, a very common indicator, has a pK_{in} near 7. Phenolphthalein has a pK_{in} value of 9.3 whereas methyl orange has the value 3.7. The last two are included among the six components of **universal indicator**, which changes colour several times over the pH range 3 to 11: red at 3, green at 7 and violet at 11. Other types of indicator exist: the dye ➤eriochrome black T acts as an indicator for ➤complexometric titrations with ➤EDTA. ➤end-point.

indicator diagram A diagram, first used by James Watt to describe his steam engine, showing how the pressure and volume of a gas vary in an engine. Examples include those for the ➤Otto cycle and ➤Diesel cycle, compared in the diagram. The larger shaded area of the Diesel cycle indicates its intrinsic advantage in fuel economy.

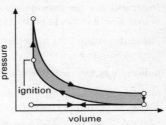

(a) Otto cycle

indicator species A species which is characteristic of particular environmental conditions and which can be used to recognize those conditions. For example, the sphagnum mosses are associated with low-pH conditions on wet moorlands, and fish such as trout are characteristic of well-oxygenated waters with little pollution.

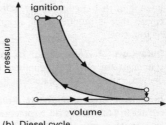

(b) Diesel cycle

indicator diagram for (a) the Otto cycle and (b) the Diesel cycle.

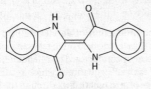

indigo

indigo A blue dye (see the above diagram), which gives its name to one of the seven colours in the visible spectrum, originally made by natural acidification of plants of the ➤genus *Indigofera*.

indirect hydration An older way of making ethanol from ethene, by absorption in concentrated sulfuric acid followed by the addition of water. The modern method is ➤direct hydration.

indirect transition A transition between electronic energy levels in a solid that involves the transfer of ➤crystal momentum from a ➤phonon to an electron. Compare ➤direct transition.

indistinguishable particles ➤identical particles.

indium Symbol In. The element with atomic number 49 and relative atomic mass 114.8, so called because of an indigo blue line in its atomic emission spectrum. It is in Group 13, the group that includes aluminium. The element is a very soft metal and is therefore used in metal-to-metal joints in ultra-high vacuum systems where any liquid would have too high a vapour pressure. Like aluminium, it has an oxidation number of +3 in most compounds. However, because of the ➤inert pair effect it also shows an oxidation number of +1; thus indium forms two chlorides, $InCl_3$ and $InCl$. Compounds with elements in Group 15, such as InSb, are used as semiconductors.

indole C_8H_7N A yellow solid bicyclic heterocyclic aromatic compound (see the diagram). Despite its presence in intestinal putrefaction, it is used in orange blossom perfume.

indole

induced current The current that results from an e.m.f. produced by electromagnetic ➤induction.

induced e.m.f. ➤induction, electromagnetic.

inducer A molecule involved in ➤operon systems in cells. Inducers bind with ➤repressor proteins, thus rendering them unable to act in preventing ➤transcription. Inducers are usually the natural ➤substrate for the particular enzyme, the synthesis of which is being controlled.

inductance Unit ➤henry, H. The magnetic flux through a circuit per unit current flowing in that circuit (strictly, **self-inductance**, symbol L) or in a neighbouring circuit that is producing the magnetic field (**mutual inductance**, symbol M). ➤➤inductor.

induction A mathematical technique for proving assertions. The truth of a statement for all whole numbers is deduced from its truth for $n = 1$ and a demonstration that the statement for $n + 1$ is true if the statement for n is true.

induction, electromagnetic Broadly, the production of an electric field by a change in a magnetic field. In practical terms the electric field often takes the form of an ➤electromotive force (e.m.f.) around a loop of conductor. The phenomenon of induction is summed up in the third of ➤Maxwell's equations, which relates the e.m.f. around such a loop to the rate of change of ➤magnetic flux through the loop. Induction has engineering applications in many fields (➤transformer and the entries below). ➤➤Faraday's law.

induction coil A pair of coils arranged like a ➤transformer with a small number of turns in the ➤primary coil and many more in the ➤secondary coil. When the relatively low e.m.f. in the primary is cut off and the current is suddenly interrupted, a very large e.m.f. is produced (by ➤induction) in the secondary. This technique is used to produce the high voltages required for sparks in a petrol engine from a low-voltage battery.

induction heating (radio-frequency heating) A form of ➤Joule heating in which the e.m.f. is produced by ➤induction, and the currents are therefore ➤eddy currents. It is particularly useful for melting metals because the currents are produced in the molten metal itself.

induction motor An alternating current electric motor in which the torque is produced by the interaction between the current in the ➤secondary coil and the magnetic field produced by the ➤primary coil. ➤➤rotor; stator.

inductive effect The polarization of one bond caused by the polarization in an adjacent bond. Relative to a hydrogen atom, some groups of atoms such as a methyl group are **electron-releasing**; such groups stabilize ➤carbocations and make S_N1

(➤nucleophilic substitution) and ➤electrophilic substitution reactions faster. Other groups such as a nitro group are **electron-withdrawing** and they slow down electrophilic substitution.

inductor A component of an electric circuit designed to have a known ➤inductance. The voltage developed across an inductor is given by $V = -L \, dI/dt$, where L is the self-inductance, I the current and t the time. Inductors are normally made from coils of conductor.

inelastic collision A collision in which the total kinetic energy of the colliding particles is not conserved. A good example is a collision between two cars. The total energy is conserved, according to the law of ➤conservation of energy; most of it is absorbed by distortion of the cars' bodyshells, while some energy is dissipated as sound.

inelastic cross-section The contribution to the ➤scattering cross-section of a scattering process in which some of the ➤kinetic energy of the scattered particles is lost.

inequality A statement that one quantity is greater than or equal to another. An important example is the ➤Cauchy–Schwarz inequality.

inert gas An obsolete name for a noble gas. The name had to be abandoned in 1962 when the first compounds of the noble gases were made.➤➤xenon.

inertia The tendency of a body to continue with constant (including zero) velocity unless acted upon by a force. It is not a physical quantity, such as energy or temperature, simply an abstract noun; its use as a synonym for either ➤inertial mass or ➤momentum is incorrect. ➤➤moment of inertia.

inertial frame A coordinate system that is not accelerating. In an inertial frame the laws of ➤special relativity hold, and, in the limit of low velocities, ➤Newton's laws of motion hold.

inertial guidance (inertial navigation) A system of determining position in space by sensing ➤acceleration, and integrating twice with respect to time by computer to find the distance travelled.

inertial mass The physical quantity ➤mass used in the laws of dynamics. It measures the resistance of a body to being accelerated by applied forces. Compare ➤gravitational mass.

inert pair effect An effect shown by the heavier elements in Groups 13, 14 and 15 of the periodic table where they have compounds with an oxidation number of $N-12$, where N is the group number. It thus appears that one pair of electrons is 'inert'. The true explanation is more complex and results from a subtle balance of increased ionization energy compared with increased lattice or bond energy. An example is lead(ii), an important oxidation number for lead.

infection The introduction into an organism of an agent of disease, such as a ➤bacterium or ➤virus.

inferior Describing a structure that lies below (or ➤posterior to) a reference point in an organism. For example, the inferior vena cava (posterior vena cava) is the main vein of the body below the ➤heart. Compare ➤superior.

infinite integral Another name for an ➤improper integral.

infinite series ➤series.

infinite set A ➤set that does not have a finite number of elements. ➤cardinal number; countable.

infinitesimal An ➤increment whose ➤limit is zero; also used informally for an 'infinitely small quantity'. In older books, ➤calculus is sometimes called 'infinitesimal calculus'.

infinity Symbol ∞. The mathematical idea pertaining to any variable, quantity or geometrical location that is greater than, or beyond, a given bound.

inflammable Describing a substance that can burst into flame. (The alternative term **flammable** is perhaps preferable.)

inflationary Universe A modification of the ➤big bang theory that suggests that the Universe experienced a very rapid increase in size at the age of about 10^{-35} s.

inflection (inflexion), point of A point on a curve at which the curve crosses its ➤tangent (2). Inflections occur at some or all points satisfying the equation $d^2y/dx^2 = 0$. They may or may not be ➤stationary points as well.

inflorescence Collectively, the flowers on a single plant and the pattern in which they are arranged.

information theory The formal study of information, encompassing, for example, the study of noise, theoretical transmission capabilities of communication links and compression of data. Information content is measured in terms of entropy, which is given an analogous definition to that of Boltzmann's microscopic definition of ➤entropy in thermodynamics.

infrared A region of the ➤electromagnetic spectrum where the frequency is below the red end of the visible spectrum, about 4×10^{14} Hz, corresponding to a wavelength of greater than about 700 nm. Conventionally, the infrared extends down to frequencies of about 3×10^{11} Hz (wavelength 1 mm), below which come ➤radio frequencies. **Infrared cameras** respond to thermal emissions from the subject, and hence can function in the dark.

infrared spectroscopy A form of ➤absorption spectroscopy that utilizes electromagnetic radiation in the ➤infrared region. Absorption is caused by excitation of the ➤normal modes of molecules (plus associated rotational transitions). It is extremely useful for identifying molecules, and is routinely used in analytical and organic chemical laboratories. The diagram opposite shows the **infrared spectrum** of ethyl ethanoate.

infrasound Sound with a frequency of less than 16 Hz, which is the normal lower limit of human hearing.

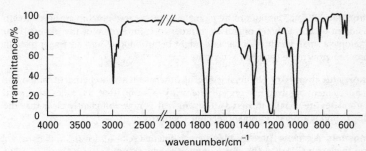

infrared spectroscopy The infrared spectrum of ethyl ethanoate.

inheritance The transmission of information in the form of ➤genes from parents to offspring. The only characteristics that can be passed on by inheritance are those encoded in the DNA present in ➤gametes; traits acquired during the lifetime of the parent, such as muscle development in athletes, cannot be passed on.

inhibition 1 (Biol.) ➤competitive inhibition; noncompetitive inhibition.
  2 (Chem.) ➤inhibitor.

inhibitor A substance that reduces the rate or effectiveness of a process. For example, leaded petrol contained an inhibitor, ➤tetraethyllead, that reduces ➤knocking.

inhomogeneous semiconductor A semiconductor in which the concentration of ➤donor and ➤acceptor (2) impurities varies with position. A simple example is a ➤p–n junction.

initial conditions ➤Boundary conditions for a ➤differential equation specifying values of the solution and possibly some of its derivatives (➤differential) at an initial time.

injection The introduction of a fluid into a part of the body (such as skin, subcutaneous tissue, muscle or blood) via a syringe. Injections are commonly used in inoculations or to administer drugs or anaesthetics. ➤➤inoculation.

injection moulding A process for shaping ➤thermoplastic polymers in which they are softened by heating and then injected into a cool mould.

inner ear ➤ear.

inner planets (terrestrial planets) A collective name for the four planets nearest the Sun, within the asteroid belt (Mercury, Venus, Earth and Mars). Compare ➤outer planets.

inner product ➤scalar product.

inner-sphere complex A complex in which the ➤ligand is directly bonded to the metal ion.

inoculation The introduction of living cells or an ➤antigen into the body in order to stimulate an ➤immune response to confer immunity against a specific disease.

inoculum A small amount of living material introduced onto an artificial growth medium or into a living organism in order to culture more of the material. For example a small quantity of bacteria might be introduced onto an ➤agar plate in order to grow more bacteria of the same type.

inorganic chemistry The chemistry of all other elements except that of carbon, the main element involved in ➤organic chemistry. (Carbon itself, its oxides, and metal carbonates are usually treated as inorganic.) It is now abundantly clear that the distinction is artificial (➤organometallic chemistry).

inositols A generic name for nine stereoisomeric cyclo-hexanehexols $C_6H_{12}O_6$, the most common isomer being shown in the diagram. They are essential dietary components for animals, being part of the vitamin B complex.

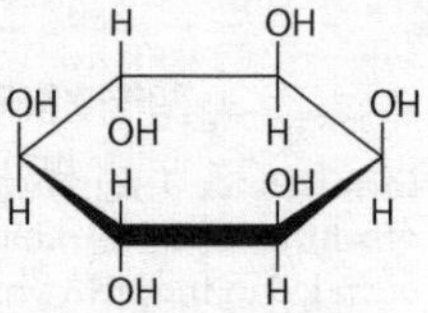

meso-**inositol**

input impedance The ➤impedance across the input of an electronic device, particularly an amplifier. An ideal amplifier has infinite input impedance, and therefore draws no current.

insect A member of the major ➤class Insecta of the ➤arthropods. Insects are characterized by a body divided into three distinct regions: the **head**, **thorax** and **abdomen**. The thorax bears three pairs of legs and, typically, two pairs of wings. Insects are the most numerous of all the animals, in terms of both numbers of species and numbers of individuals.

insecticide A compound or mixture that kills insects. The first effective one, ➤benzene hexachloride, was discovered by Michael Faraday in 1825. The World Health Organization estimated that 5 million lives were saved in the first eight years of the use of ➤DDT to kill mosquitoes.

insemination The introduction of semen into the uterus. **Artificial insemination** (AI) is the introduction of semen by means of a syringe or similar apparatus. AI is widely used in cattle breeding, and in some forms of human infertility treatment.

insertion reaction A reaction in which a reagent inserts itself into a bond. A common example is the reaction of ➤methylene, CH_2, with double bonds in alkenes to form cyclopropane derivatives.

insertion sequence One of a number of different kinds of ➤transposon.

insolation Heating of a surface, particularly the surface of the Earth, by heat radiation from the Sun. It is insolation on a typical summer's day that causes moist air to rise and form ➤clouds. ➤➤solar constant.

insoluble Describing a substance that dissolves to a negligible extent in a particular solvent (by default, water). Ionically bonded substances tend to be insoluble in organic solvents because the solvation energy due to dispersion forces cannot compensate for the loss of lattice energy.

instability ➤unstable.

instantaneous frequency The rate of change of the ➤phase (2) of an oscillation. For a simple sinusoidal disturbance, the instantaneous ➤frequency is simply the frequency of the oscillation, but for more complicated waveforms the rate of change of the phase can vary with time.

instinct A pattern or type of behaviour that is innate rather than learned. Instinctive behaviour often has survival value; an example is the 'freezing' of motion by well-camouflaged chicks of ground-nesting birds when threatened by predators. The migratory behaviour of many birds and other animals is another example.

insulation The use of an ➤insulator to prevent the flow of electric current (**electrical insulation**) or heat (**thermal insulation**), for example in a sheath around a current-carrying wire, or the jacket around a domestic hot-water tank. The term is also applied to the insulating material itself.

insulator A material that does not conduct electricity. Insulators have no sea of ➤conduction electrons. Compare ➤metal; semiconductor.

insulin A ➤polypeptide ➤hormone secreted by the beta cells of the ➤islets of Langerhans in the pancreas. It increases the rate of synthesis of ➤glycogen, ➤fatty acids and ➤proteins; it stimulates glycolysis and promotes the entry of glucose, other sugars and amino acids into fat and muscle cells. Insulin deficiency results in ➤*diabetes mellitus*, which affects several hundred million people world-wide. Insulin was first extracted from pancreas tissue by Frederick Banting in 1922. It was the first protein for which the ➤amino acid sequence was determined (by Frederick Sanger in 1955).

integer One of the numbers 0, ±1, ±2,… which is either a ➤natural number or the negative of a natural number.

integral **1** The adjective from ➤integer.
   2 ➤integration. ➤➤Appendix table 3.

integral calculus ➤calculus.

integrand ➤integration.

integrated circuit (IC) An electronic circuit on a semiconductor, usually of silicon but sometimes of gallium arsenide. Very large numbers of components can be created on a single chip, using VLSI (➤very large scale integration) techniques.

integrating circuit A simple ➤analogue circuit whose output V_{out} is the indefinite integral, with respect to time, of its input V_{in} (see the diagram).

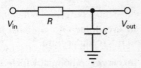

integrating circuit

integration Historically conceived as the inverse operation to ➤differentiation, when it was realized that finding the area under a curve required the reverse process to that of finding tangents. Since then, refinements such as **Riemann integration** and **Lebesgue integration** have given an independent meaning to integration so that the connection with differentiation becomes the ➤fundamental theorem of calculus. The **indefinite integral** of a function $f(x)$ is denoted $\int f(x)\,dx$; if this equals $F(x)$, then the derivative of the **integral** $F(x)$ is the **integrand** $f(x)$. The integral $F(x)$ is determined up to the addition of an arbitrary **constant of integration** because a constant differentiates to zero. The difference between the values of the integral at two values of x is a **definite integral** written $\int_a^b f(x)\, dx$ and equal to $F(b) - F(a)$, where a and b are the **limits of integration**. In principle, the hard problem of determining whether a function has an integral that is a familiar ►elementary function is fully solved by the Risch algorithm and versions of this are used in computer algebra software packages, but the theory is subtle; for example, $x \sin x$ and $\sqrt{\tan x}$ have elementary integrals but $x \tan x$ and $\sqrt{\sin x}$ do not. Some functions such as the ►gamma function are best defined as integrals and, for many applications, ►numerical integration or a ►Monte Carlo method will give an acceptably accurate answer to an intractable integral. Integration is used to find areas, volumes, arc lengths, expectations, etc., and to solve ►differential equations. ►►Appendix table 3.

integration by parts A technique for the integration of certain products of functions based on the identity

$$\int u \frac{dv}{dx}\, dx = uv - \int v \frac{du}{dx}\, dx.$$

integration by substitution ►change of variables.

integument ►ovule.

intensity Symbol I. A quantity that describes some factor, usually rate of flow of energy, transmitted by wave motion. Intensity is proportional to the square modulus of the wave's amplitude, or the sum of such quantities for a wave composed of more than one frequency component.

intensive property In thermodynamics, a property that is independent of the size of the system under consideration. Examples include temperature and pressure; volume is an ►extensive property.

interaction In general, a situation in which the behaviour of one ►system affects and is affected by another system. Elementary particles can interact in four ways: ►fundamental interactions.

intercalation compounds Compounds, particularly of graphite, where other particles are incorporated in the crystal structure with only the spacing between the layers changing significantly. The first example found (in 1926) was the bronze-coloured C_8K.

intercellular Of a substance or structure occurring between rather than in cells, as in **intercellular fluid**.

intercept A point at which a graph cuts an axis.

interface 1 (Phys.) The border between two phases. At the interface between a liquid and its vapour, ►surface tension holds the molecules into the liquid.

  2 (Comput.) A connecting part of a computer system. A **hardware interface** might enable the computer to be connected to a printer or a monitor, while a **user interface** such as a ►WIMP is a software interface which provides the 'connection' between the user and computer.

interference 1 The phenomenon caused by the addition of waves from two or more coherent sources (►coherence), where the ►phase difference varies with position or angle. With a phase difference of zero, the interference is **constructive** and the amplitudes simply add together arithmetically; with a phase difference of 180°, it is **destructive** and the resulting amplitude may be zero (see the diagram). ►►beats; fringes; Young's fringes.

2 Disturbance of a radio signal by either ►noise or ►static (2).

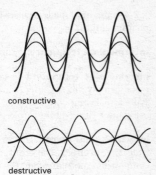

constructive

destructive

interference 1

interferometer A spectrometer that uses interference to determine the frequency spectrum of electromagnetic radiation.

interferon One of a group of small proteins synthesized by certain ►T cells of vertebrates, which confer resistance to many ►viruses. There are three types of interferon in humans, produced by cells of the ►immune system. The production of interferon by ►genetic engineering techniques has been exploited for its use in the treatment of some kinds of cancer.

interhalogen A compound formed between the halogens themselves. There are a range of these, from all the simple AB compounds such as ICl through several of the form ICl_3 to the largest, IF_7. The central atom in the last two cases is the less electronegative atom.

interleukin 1 (IL-1) The protein secreted by ►macrophages of the ►immune system after activation by ►antigens. IL-1 stimulates ►B cell activation.

interleukin 2 (IL-2) The protein secreted by activated ►helper T cells of the ►immune system. This protein stimulates these cells, causing them to proliferate.

intermediate 1 The particle that corresponds to the energy 'well' in reactions with ►mechanisms that involve two (or more) stages. It is important to stress the difference between an intermediate, which is (in principle at least) capable of being isolated, and the two ►transition states either side, which are not. An example is the **Wheland intermediate** (see the diagram) in aromatic ►electrophilic substitution. In the nitration of benzene, the reactants are C_6H_6 and NO_2^+, and the products are $C_6H_5NO_2$ and H^+. The Wheland intermediate has the structure shown in the diagram.

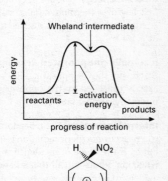

intermediate 1 The Wheland intermediate for the nitration of benzene.

2 A compound made *en route* in the production of another chemical (such as a pharmaceutical).

intermediate frequency ➤superheterodyne.

intermetallic compound A compound of two or more metals. Normally two metals simply mix together in arbitrary ratios to form ➤alloys. However, there are some actual compounds formed, such as CuZn. Their formulae can be rationalized using the ➤Hume-Rothery rules.

intermolecular forces The forces between molecules; their existence explains a wide range of effects. The stronger the forces, the higher the boiling point of a liquid composed of molecules, and the more serious the deviations a gas shows from the ➤ideal gas model. The main types of intermolecular force are ➤van der Waals forces, including ➤dipole–dipole forces and ➤dispersion forces, and ➤hydrogen bonding.

internal combustion engine An engine in which the combustion of fuel takes place within a cylinder, the hot gases forcing a piston downwards during the power stroke. Most internal combustion engines operate either on the ➤Otto cycle, and use petrol as fuel, or on the ➤Diesel cycle, and use diesel as fuel.

internal conversion 1 The partial conversion of radiant energy into heat. When a substance has gained energy by absorbing a photon, some of this energy may be reemitted in the form of another photon, as in ➤fluorescence. This second photon has a lower frequency because some of the energy is lost nonradiatively as the excited substance sheds energy to reach a more stable state. Much detail about such internal conversion remains obscure.
 2 The ejection of an electron (then termed a **conversion electron**) from an atom caused by the decay of the nucleus from an excited state.

internal energy Symbol U, or occasionally E. The energy stored in a system. The first law of ➤thermodynamics states that the change ΔU in a system's internal energy is equal to the heat q added to the system plus the work w done on the system: $\Delta U = q + w$.

internally compensated Referring to molecules with two identical ➤chiral centres that do not show ➤optical activity because the rotation by one end of the molecule is exactly cancelled by the opposite rotation by the other end. ➤*Meso*-isomers are internally compensated.

internal reflection The reflection of radiation at the interface between two media when electromagnetic radiation is incident from the medium with higher refractive index. If the angle of incidence is greater than the critical angle, ➤total internal reflection occurs and all the radiation is reflected.

internal resistance The resistance of a power source due to components within the source itself. It can be measured as the voltage across the output with a load of infinite resistance, divided by the current drawn by a load of zero resistance. ➤➤equivalent circuit.

International Atomic Time (TAI) An internationally agreed scale of time based on atomic clocks kept at Le Bureau Internationale de l'Heure in Paris. ➤atomic clock; second.

International Date Line The boundary between time zones that are 12 and 13 hours ahead of ➤Universal Time (e.g. Eastern Russia) and those 11 hours behind UT (e.g. Alaska); it passes close to the 180° ➤meridian.

international standard atmosphere (ISA) A model of the Earth's atmosphere in which the temperature at sea level is defined to be 15 °C, and the pressure 1013.25 mbar. The temperature decreases by 6.50 °C per km altitude up to 11 km (the ➤tropopause), above which it is assumed to be constant (−56.5 °C) up to 20 km, and thereafter to rise by 1 °C per km altitude up to 32 km. The ISA is the standard against which barometric altimeters are defined.

international system of units ➤SI units.

Internet A worldwide computer ➤network which provides information, bulletin boards, on-line discussion arenas and electronic mail for millions of users. The Internet can be accessed using even a very basic computer attached to a telephone line via a ➤modem. The Internet forms the basis of the 'information superhighway' and consists of many small networks interconnected by gateways. ➤HTML; Java; www.

interphase The state of a cell nucleus which is between cell divisions, particularly in rapidly dividing tissues such as root tips in plants or during embryonic growth. It is during the interphase that the nucleus is exerting control over the metabolic processes in the cell, and therefore the commonly applied term of '**resting nucleus**' for this phase is inaccurate and inappropriate. ➤meiosis; mitosis.

interpolation The prediction of the value of a function at a point within the range of some known values of the function. This is often done by fitting a curve to the data values: **linear interpolation** corresponds to fitting a straight line. ➤extrapolation.

interpreter A computer ➤program that translates another program written in a ➤high-level language into executable instructions. Crucially this is done *at the same time* as the program is being run (contrast ➤compiler). This makes an interpreted program slower than one that has been compiled, but it has the advantage that interpreters may be used to run programs directly.

interquartile range ➤percentile.

interrupted continuous wave A radio transmission that is interrupted periodically in such a way that the frequency of interruption is audible.

intersection That ➤set consisting of those elements that are common to a family of sets. The intersection of two sets A and B is denoted $A \cap B$.

interstellar medium Gas (predominantly hydrogen) and dust particles present in ➤interstellar space. It consists principally of warm (about 8000 K), very low density (about 10^5 particles m^{-3}), partially ionized gas, interspersed with cooler, denser clouds of neutral gas and dust (close to absolute zero, 10^{16} particles m^{-3}), and hot, low-density regions (10^8 K, 1 particle m^{-3}). Individual molecules have been detected including ether, $CH_3CH_2OCH_2CH_3$.

interstellar space The vast volumes of space between the ➤stars in a ➤galaxy.

interstitial ➤defect.

interstitial cell stimulating hormone (ICSH) ➤luteinizing hormone.

interstitial compound A compound in which the atoms of one small element fit into the gaps (**interstices**) in the crystal structure of another. Examples include the hydrides of the d-block elements, such as titanium hydride. They are commonly ➤nonstoichiometric.

intersystem crossing ➤phosphorescence.

interval **1** ➤geodesic.
   2 The ➤set of ➤real numbers between two given values. The notation (a, b) denotes the interval from a to b excluding a and b; $[a, b]$ denotes the interval from a to b including a and b.

intestine ➤gut.

intra- Prefix meaning 'within'. Hence, **intracellular fluid** is the fluid inside cells (compare ➤intercellular). **Intramolecular forces** are the forces (between atoms) within molecules as opposed to those between molecules, the ➤intermolecular forces; intramolecular forces are usually due to covalent bonding.

intrinsic semiconductor A pure ➤semiconductor that has not been doped, for example pure silicon. Compare ➤extrinsic semiconductor.

intron A transcribed (➤transcription) intervening sequence of DNA which is then excised from the functional RNA before the latter's translation.

inulin A ➤polysaccharide polymer of fructose, occurring as a storage compound in the roots of plants of the family Asteraceae, such as Jerusalem artichokes and dahlias.

in vacuo Latin phrase meaning 'in a vacuum'.

Invar Tradename for an alloy of steel with 36% by mass nickel. It has a particularly low ➤coefficient of expansion, and hence is used for parts of precision instruments to minimize the effect of changes in temperature.

invariant A property or coordinate of a system that is unchanged after a transformation. For example, the ➤rest mass of a system is invariant under a ➤Lorentz transformation, and ➤time is invariant under a ➤Galilean transformation (but not under a Lorentz transformation).

inverse element That element of a group, denoted by a^{-1}, for which $a^{-1} * a = a * a^{-1} = e$, where $*$ is the group operation and e is the ➤identity (2).

inverse function The function $f^{-1}: Y \to X$ such that, for $x \in X$ and $y \in Y$, and $f:X \to Y$, $f^{-1}(y) = x$ if $f(x) = y$. This definition only makes sense if f is a ➤one-to-one correspondence, which accounts for the care needed, for example, when defining ➤inverse trigonometric functions.

inverse hyperbolic functions Loosely speaking, the inverse functions of the six ►hyperbolic functions; but some care is needed to get uniquely defined functions. The definition of hyperbolic functions in terms of the ►exponential function means that the inverse hyperbolic functions can be expressed in terms of logarithms. An example of the notation used is $\sinh^{-1} x$ or arcsinh x.

inversely proportional ►variation.

inverse matrix That matrix associated with a ►nonsingular matrix such that its product with a given matrix is the ►identity matrix.

inverse spinel A crystal structure in which the ions are arranged as in a ►spinel, except that the B cations are now half in the tetrahedral holes and half in the octahedral holes in the oxide array. A famous example is ►magnetite, Fe_3O_4, in which Fe^{3+} ions are the B cations.

inverse square law A force that varies as the reciprocal of the square of the distance from a specified point obeys an inverse square law. ►Newton's law of gravitation and ►Coulomb's law are examples.

inverse trigonometric functions Loosely speaking, the inverse functions of the six ►trigonometric functions; but some care is needed to get uniquely defined functions. For example, the inverse sine function, written $\sin^{-1} x$ or arcsin x, only makes sense for values of x between -1 and $+1$ and is defined to be that value y between $-90°$ and $+90°$ for which $x = \sin y$.

inversion **1** (Math.) A ►transformation (2) of a system in which one or more ►coordinates is multiplied by -1. In **temporal inversion**, time is multiplied by -1. In **spatial inversion**, all three space coordinates are multiplied by -1. If an object is left unchanged by spatial inversion through a point (a **centre of symmetry**), the object has **inversion symmetry**.

2 (Geog.) An atmospheric condition in the troposphere in which temperature *increases* with altitude (compare ►international standard atmosphere). The trapping of cold air near the ground, below an **inversion layer**, causes ►fogs.

inversion layer **1** (Phys.) A layer in a ►semiconductor in which the majority carriers have the opposite charge to the normal carriers. In a ►p-type semiconductor, the positive charge carriers are usually much more numerous than the negative carriers. If, in a layer near a surface or interface, an electric field causes the negative carriers to be in the majority, the layer is an inversion layer. Similarly, a layer of ►n-type semiconductor in which positive carriers are in the majority is an inversion layer.

2 (Geog.) ►inversion (2).

inversion of configuration Inversion of configuration occurs when a ►dextro-rotatory stereoisomer is changed into the corresponding ►laevorotatory stereoisomer (or vice versa), causing the substance to rotate the plane of plane-polarized light in the opposite direction. The discovery of ►Walden inversion was helpful in identifying the common mechanism of ►nucleophilic substitution.

inversion temperature The temperature above which the ►Joule–Thomson effect causes an expanding gas to heat up rather than cool. Below the inversion temperature, expansion can be used to liquefy a gas.

invertase The enzyme in pancreatic juice and in yeast that catalyses the conversion of cane sugar into ➤invert sugar.

invertebrate General term of convenience given to an animal species that is not a member of the ➤chordate subphylum Vertebrata. In modern classification systems the term has no taxonomic status. ➤Appendix table 8.

inverter A simple ➤digital component whose output is the logical negation of its input.

invert sugar An equal mixture of glucose and fructose formed by hydrolysis of sucrose catalysed by ➤invertase. It is called *invert* sugar as it rotates the plane of plane-polarized light in the opposite direction to sucrose (➤optical activity).

in vitro Latin for 'in glass'. Describing a biological process that can be made to occur in an artificial environment, outside a living organism. For example, *in vitro* ➤fertilization is a technique in which ➤gametes can be mixed in an appropriate liquid medium in a shallow dish to achieve fusion of eggs and sperm. The fertilized egg so formed can then be ➤implanted into the uterus. If successful, this can enable couples who cannot achieve fertilization naturally to have a 'test tube baby'.

in vivo Latin for 'in life'. Describing a biological process conducted or occurring within a living organism.

involuntary muscle ➤muscle.

Io ➤Jupiter.

I/O Abbr. for input/output, describing the transfer of data to or from a computer's memory. A touch-sensitive screen is an example of an I/O peripheral, and a ➤terminal is an I/O system.

iodate ion (iodate(v) ion) IO_3^- An oxoanion containing iodine with oxidation number +5. The ion and its parent acid HIO_3 are strong oxidants.

iodic(ı) acid ➤hypoiodous acid.

iodide ion I^- The ➤anion formed by iodine. It is present in solid iodides, as well as in solution. The test for the aqueous ion is to add acidified silver nitrate and look for a yellow precipitate of silver iodide. In acidic solution iodide ion is used to test for strong oxidants, which produce a brown solution of iodine.

iodine Symbol I. The element with atomic number 53 and relative atomic mass 126.9, which is a member of the halogens, Group 17, and hence a nonmetal. It is a black solid at room temperature, of formula I_2, with a purple vapour, the colour of which is responsible for its name (Greek *iōdēs*). Its concentration in solution is measured by titration with standard sodium thiosulfate using the reaction

$$I_2(aq) + 2S_2O_3{}^{2-}(aq) \rightarrow 2I^-(aq) + S_4O_6{}^{2-}(aq).$$

Iodine is commonly found as the iodide ion, as in 'iodized salt'. In the human body, iodine is critically important for the function of the ➤thyroid gland. The effectiveness of antiradiation tablets depends on this – the most dangerous radioisotope released in a small nuclear accident is ^{131}I and uptake of this can be blocked by saturating the thyroid with iodine by ingesting KIO_3. There are many important compounds of iodine, examples of which include ➤iodine(v) oxide, I_2O_5, potassium iodide, KI, and iodine trichloride, ICl_3. Many organic compounds also contain iodine, despite its low abundance in nature. One example is ➤iodoform.

iodine number The result of a test for the degree of unsaturation in fats and oils which involves measuring the percentage by mass of iodine absorbed by a sample under specific conditions.

iodine(v) oxide I_2O_5 A white solid used to estimate carbon monoxide, which reacts with it to form iodine.

iodoform (triiodomethane) CHI_3 A yellow solid sparingly soluble in water with a characteristic 'hospital' smell. It is the iodine equivalent of chloroform. The **iodoform test** was important for the identification of the groupings CH_3CO- and $CH_3CH(OH)-$ in organic compounds as iodoform is formed when sodium hydroxide and iodine are added to the compound.

ion A species that carries an electric charge. There are two types of ion. Positive charge is carried by **positive ions** (also often called **cations**), such as Na^+ or Fe^{3+}, which have fewer electrons than their parent atoms. Negative charge is carried by **negative ions** (also often called **anions**), such as Cl^- or O^{2-}, which have more electrons than their parent atoms. Polyatomic ions also exist, such as ammonium NH_4^+, nitrate NO_3^- and sulfate SO_4^{2-}. Ions are responsible for a wide range of effects: for example, they cause lightning and the different taste of various bottled waters. ➤➤ionic bonding.

ion exchange A process in which ions are exchanged between one place and another. A common application is in water softening. A cation exchange resin, often a ➤zeolite, has sodium ions bound to a huge polyatomic anion. Hard water, which contains calcium and magnesium ions, is run through the resin. The calcium and magnesium ions leave the water and attach to the anion as they have a higher charge (Ca^{2+} or Mg^{2+}) than the sodium ions (Na^+), two of which go into the water to preserve charge balance. Anion exchange resins also exist.

ion gauge ➤ionization gauge.

ionic association (double decomposition) When two solutions containing ions are mixed, sometimes a precipitate is formed by association of two of the ions. Examples include the common tests for chloride ion and for metal ions in ➤qualitative analysis:

$$Ag^+(aq) + Cl^-(aq) \rightarrow AgCl(s)$$
$$Fe^{2+}(aq) + 2OH^-(aq) \rightarrow Fe(OH)_2(s).$$

ionic atmosphere Ions in solution attract ions of the opposite charge hence causing there to be an 'atmosphere' surrounding any one ion, which mainly contains the ions of opposite charge. This impedes the movement of the ions. The first successful attempt to take this into account in describing the properties of ionic solutions was the ➤Debye–Hückel theory.

ionic bonding One of the two important models of ➤chemical bonding. In its extreme form, an electron is *transferred* from one atom to another; the resulting **ions** have a strong electrostatic attraction to each other and hence stick together. For example, sodium and chlorine react together vigorously, to produce sodium chloride. Sodium has one electron in its outer shell, chlorine has seven, one short of the number in the nearest noble gas. The sodium atom can transfer its outer electron to the chlorine atom; the resulting Na^+ and Cl^- ions attract each other.

Compounds with ionic bonding have the following properties. They are crystalline solids because the ions need to pack in a regular fashion, positive ions next to negative ions and so on. The solids are, however, ➤brittle, as a small shear will bring ions of the same charge next to each other. They have high melting and boiling points, because the force between the ions is great. They are usually soluble in water, as the hydration energy it provides is sufficient to compensate for the lattice energy. Ionic compounds conduct electricity when in solution or molten, as the ions are then free to move. Ionic bonding is one model, to be contrasted with ➤covalent bonding where *sharing* of electrons occurs. Most compounds have bonding which falls between these two extremes, but this should not detract from the enormous success that these simple models have had since they were introduced in 1916.

ionic product for water (ionic product of water) Symbol K_w. The presence of hydrogen ions and hydroxide ions in water, albeit to only a very small extent, is recognized in the equilibrium constant $K_w = [H^+][OH^-]$. At 25 °C, the value of this constant is 1.008×10^{-14} mol^2 dm^{-6}. In neutral water the hydrogen ion and hydroxide ion concentrations are equal, with a value very close to 1×10^{-7} mol dm^{-3}. From the definition of ➤pH, this means neutral water has a pH of 7.0 at 25 °C.

ionic radius The radius of an ion, normally quoted for the solid state. Consistent values that are reasonably transferable from one compound to another are available. For example, the sodium ion Na^+ has a radius of 102 pm and the chloride ion Cl^- a radius of 180 pm. Positive ions are smaller than their parent atoms whereas negative ions are larger than their parent atoms. Ionic radii tend to increase on going down a group in the periodic table.

ionic strength ➤Debye–Hückel theory.

ionic theory The theory concerned with the behaviour of ions in solution, for example their electrical conductivity.

ionization The process in which an atom or molecule is turned into an ion, by the loss or gain of one or more electrons.

ionization chamber The part of an apparatus where ions are formed. In the mass spectrometer, an electron beam is used to ionize molecules. The rapidly moving electrons have enough energy to remove one electron from a molecule on impact.

ionization energy (IE) The energy involved in the loss of an electron from a particle. The most important is the **first ionization energy** of an atom, which is the minimum energy required to remove one electron from an isolated atom, E, of an element in the gas phase:

$$E(g) \rightarrow E^+(g) + e^-(g).$$

It is normally expressed as a value per mole; hydrogen atoms, for example, require 1310 kJ mol^{-1}.

ionization gauge (ion gauge) A device for measuring pressures of 10^{-3} Pa and lower. Electrons are accelerated through a positively charged grid towards a negatively charged plate. An electron that strikes a gas molecule may create a positively charged ion that is attracted to the plate. This ion current at the plate is measured and related to the pressure of the gas.

ionization potential An obsolete term, which was traditionally measured in electronvolts (eV), related to ➤ionization energy. As 1 eV corresponds to 96.5 kJ mol^{-1}, the ionization potential for hydrogen atoms is 13.6 eV.

ionizing radiation Radiation that can cause ➤ionization. Alpha particles (helium nuclei, ➤alpha decay), beta particles (electrons, ➤beta decay) and ➤gamma rays (high-energy photons) are all examples of ionizing radiation.

ion microprobe analysis A technique for investigating surfaces. On bombardment with a stream of electrons, ions are blasted off the surface and subsequently analysed by ➤mass spectrometry.

ionosphere The part of the Earth's ➤atmosphere above the stratosphere, from about 60 km altitude. The ionosphere contains free ions, atoms and electrons produced by the bombardment of the atmosphere by ➤ionizing radiation. ➤➤Appleton layer; D layer; Heaviside–Kennelly layer.

ionospheric wave (sky wave) A radio wave (of ➤high frequency (HF) and lower frequencies) reflected back to Earth by the ➤ionosphere. The charged particles in the ionosphere both cause ➤attenuation and reflect electromagnetic waves. HF communications rely on these reflections. ➤Medium-frequency and ➤low-frequency signals tend to be attenuated rather than reflected. Compare ➤ground wave.

ion pump A pump used to maintain a pressure of 10^{-3} Pa or lower. Atoms of gas colliding with plate electrodes at high potentials (several kilovolts) are ionized and remain on the plates.

Ir Symbol for the element ➤iridium.

IR Abbr. for ➤infrared.

iridium Symbol Ir. The element with atomic number 77 and relative atomic mass 192.2, which is a member of the ➤platinum metals. The element itself is a hard shiny silvery metal, used for filaments where high emission at low currents is required. The anomalously high concentration of iridium in a thin layer of rock deposited 66 million years ago is a clue to the fate of the ➤dinosaurs. It suggests that there was an impact by a large iridium-rich meteorite that threw dust into the atmosphere, obscuring the Sun for perhaps as long as two years. This would have so altered the climate that large animals like the dinosaurs would not have survived. The dust eventually settled out to produce the enriched iridium layer. The impact site has been tentatively identified as the 180 km diameter Chicxulub crater at the edge of the Gulf of Mexico.

iris A variable aperture in an optical system used to control the total intensity of incoming light. The coloured part of the human ➤eye is the (original) iris.

iron Symbol Fe from the Latin *ferrum*. The element with atomic number 26 and relative atomic mass 55.85, which is a first-row ➤transition metal. It is the most important metallic element, constituting nearly 90% of all the metal we use. (The Iron Age succeeded the Bronze Age around 1100 BC.) However, iron does have drawbacks, most notably that it rusts. The element itself, extracted in a ➤blast furnace, is familiar, but its alloy with a few per cent of carbon plus traces of other elements, ➤steel, is much more common. The ➤iron–carbon phase diagram is very complex and just some of the phases involved are austenite, pearlite and cementite. Being a transition metal, iron has a variety of oxidation numbers. The most common are +2 and+3, the compounds of which are usually coloured pale green and red-brown respectively, as in the chlorides $FeCl_2$ and $FeCl_3$. These oxidation numbers can be found in solution as the hydrated ions Fe^{2+}(aq) and Fe^{3+}(aq), as well as in solid compounds such as the oxides FeO, a black solid, and Fe_2O_3, a red-brown solid. (The compound ➤magnetite, Fe_3O_4, contains both oxidation numbers.) Complexes are common, too, as in the ➤hexacyanoferrate ions. The higher oxidation number is more common with strong oxidants, so FeF_3 is the common fluoride, whereas the weaker oxidant iodine only produces FeI_2. Iron's other oxidation numbers are less important, but examples do exist, such as iron pentacarbonyl, $Fe(CO)_5$, with oxidation number 0, and the violet ferrate(VI) ion, FeO_4^{2-}, with oxidation number +6.

iron–carbon phase diagram A ➤phase diagram for representing the composition of steels (see the diagram opposite). Pure iron can exist in three different phases, labelled α, γ, and δ in order of increasing temperature. The compound Fe_3C is called cementite; other important phases include austenite and pearlite.

iron-containing organic compounds The most important single compound of iron is ➤haemoglobin, the pigment in blood. Other more recent discoveries include ➤ferrocene, one of the first ➤sandwich compounds, in which an iron ion is sandwiched between two cyclopentadienyl ligands.

iron pyrites ➤pyrites.

ironstone An ore of iron: iron(II) carbonate, $FeCO_3$.

irradiance The generalization of the concept of ➤illumination: the ➤flux of incident electromagnetic radiation (not just light) per unit surface area.

irradiation Exposure to radiation, particularly ➤ionizing radiation. Irradiation of perishable foods is now used to increase their useful lifetime.

irrational number A real number, such as $\sqrt{2}$, e or π, that is not a ➤rational number.

irritability The response of organisms to ➤stimuli. Irritability is one of the defining characteristics of living organisms.

irrotational Describing a ➤vector field *v* for which, at all points over which it is defined, curl $v = \nabla \times v = 0$. This is a necessary and sufficient condition for *v* to be the gradient of a scalar potential. For example, in the absence of changing magnetic

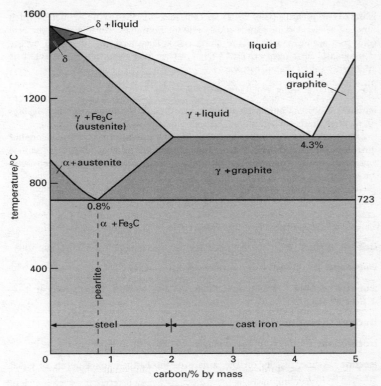

iron–carbon phase diagram for low carbon contents.

fields, the third of ➤Maxwell's equations becomes curl $E = \nabla \times E = 0$, which allows us to define an electrostatic potential ϕ such that $E = -\nabla\phi$.

ISA Abbr. for ➤international standard atmosphere.

isatin $C_8H_5O_2N$ An orange solid bicyclic aromatic compound, used in making dyes. It is a diketone derivative of ➤indole.

ischium ➤pelvic girdle.

ISDN Abbr. for integrated services digital network. A telecommunications network, usually of ➤fibre optic cables, with a ➤bandwidth broad enough to handle simultaneous transmission of voice, video and computer data. ISDN networks send all information digitally, unlike older telephone systems which used analogue transmission. They allow much higher speed access to the ➤Internet than a ➤modem provides.

isenthalpic Describing a process in which the ➤enthalpy remains constant.

isentropic Describing a process in which the ➤entropy remains constant.

islets of Langerhans (islet cells) Specialized cells located in the ➤pancreas which function as ➤endocrine glands and secrete the hormones ➤insulin (from the beta cells) and ➤glucagon (from the alpha cells). They were first described by the physiologist Paul Langerhans (1847–88). The term 'islet' refers to their characteristic appearance under the microscope.

iso- A prefix meaning 'equal', from the Greek *isos*.

isobar 1 (Geog.) A line joining points that are at equal atmospheric pressure. Isobars on weather charts are labelled by the appropriate pressure (e.g. 980 mbar).
  2 (Phys.) One of two or more nuclides that have the same number of ➤nucleons but differ in ➤atomic number. A ➤daughter nuclide resulting from ➤beta decay will be an isobar of the parent nuclide; for example, the three nuclei in the following process are isobars:

$$^{77}_{32}\text{Ge} \rightarrow\, ^{77}_{33}\text{As} + e^- + \bar{\nu}_e$$
$$\downarrow$$
$$^{77}_{34}\text{Se} + e^- + \bar{\nu}_e.$$

isobaric surface A surface in three dimensions joining points of equal pressure.

isobutane Traditional name for methylpropane, $(CH_3)_3CH$.

isobutyl alcohol Traditional name for the alcohol 2-methylpropan-1-ol, $(CH_3)_2CHCH_2OH$.

isochore ➤van't Hoff isochore.

isochronous A process that is coincident in time with another process.

isocline A line, usually on the surface of the Earth, joining points of equal ➤magnetic inclination.

isocyanate A compound containing the group —NCO. An important reaction that produces an isocyanate as an intermediate is the ➤Hofmann rearrangement.

isocyanide Traditional name for an ➤isonitrile.

isodynamic line A line, usually on the surface of the Earth, joining points with equal horizontal components of the Earth's ➤magnetic field.

isoelectric point (pI) The ➤pH at which the net charge on a molecule with groups capable of being protonated or deprotonated is zero. It is particularly significant in ➤amino acids and proteins; at its isoelectric point a protein does not migrate in an electric field during ➤electrophoresis. Most proteins have an isoelectric point around neutral but that for ➤pepsin is near 1 and those of protamines, proteins found in sperm, are around 12. The isoelectric point is a function of the relative numbers of positively and negatively charged groups and also of the ionic composition of the solution, but not of the concentration of the protein.

isoelectronic Describing species having an equal number of electrons. The structures of isoelectronic species are closely similar. For example, NH_4^+ is isoelectronic with CH_4, and is also tetrahedral.

isoenzyme (isozyme) An ➤enzyme that occurs in multiple forms in an organism at different stages in the organism's life cycle or in different tissues or organs. For example, lactate dehydrogenase (LDH) exists in a number of different forms, each of which depends on the arrangement of four **protomer** subunits. In humans different forms of LDH exist in skeletal muscle and heart tissue, and this difference is exploited diagnostically in heart attack patients, since heart LDH can be detected in blood serum after myocardial infarction (the death of an area of heart tissue).

isogonal line ➤magnetic meridian.

isolating mechanism One of a set of circumstances that separates members of a population and prevents them from interbreeding with the parent population. Isolating mechanisms include physical barriers such as geographical and temporal factors, as well as genetic barriers caused by chromosomal aberrations such as polyploidy (➤polyploid).

isolation The separation of part of an electrical system in such a way as to prevent current flowing from it to earth even under conditions of high potential difference.

isolation method A method first introduced by Wilhelm Ostwald in which the dependence of the rate of a reaction on the concentration of each reactant is investigated by adding an excess of the others. This isolates the effect of a particular reactant, as the concentrations of the others will change little.

isoleucine (Ile) A common ➤amino acid. ➤➤Appendix table 7.

isomer One of a series of compounds that have the same molecular formula, that is, they contain exactly the same number of atoms of every element. The type of isomer can then be identified more precisely. **Structural isomers** have the same molecular formula but different structural formulae. For example, butane, $CH_3CH_2CH_2CH_3$, and methylpropane, $(CH_3)_3CH$, are structural isomers, both having the molecular formula C_4H_{10} but different structures. **Stereoisomers** even have the same structural formula but have a different orientation in space. For example, isomers related by ➤*cis–trans* isomerism are stereoisomers as they differ only in the arrangement of the groups at each end of the double bond relative to each other. ➤optical activity.

isomer, nuclear One of two or more nuclei with the same atomic number and nucleon number, but different energy states.

isomerization The conversion of a molecule into one of its isomers. The most important example occurs when the retinal isomers related by ➤*cis–trans* isomerism are interconverted when a photon hits the retina of the eye, which ultimately triggers the mechanism of vision.

isometric process A process that takes place at constant volume.

isomorphic **1** (Math.) ➤isomorphism.
    2 (Chem.) A synonym for ➤isomorphous.

isomorphism A ➤homomorphism between two mathematical structures, such as groups, which is also a ➤one-to-one correspondence. Two **isomorphic** groups are usually considered to be the same.

isomorphous Having the same crystal structure. The ➤alums, formed by elements that exist as M^{3+} ions (such as aluminium, chromium and iron), are isomorphous. A crystal of one alum can often act as a seed crystal on which another alum can grow.

isomorphous replacement A method whereby the crystal structure of a huge molecule is investigated by replacing a small atom by a larger atom that is assumed not to alter the crystal structure. The large atom dominates the ➤X-ray diffraction pattern and hence enables the structure to be worked out in detail.

isonitrile (isocyanide) An isomer of a nitrile in which nitrogen is bonded to the alkyl group rather than carbon, that is RNC rather than RCN. There are also inorganic examples, such as AgNC rather than AgCN.

iso-octane An old-fashioned name for a particular isomer of octane, 2,2,4-trimethylpentane (see the diagram at ➤IUPAC). It is the standard whose ➤octane number is set arbitrarily to 100.

isoprene (2-methylbuta-1,3-diene) C_5H_8 A colourless liquid hydrocarbon (see the diagram). Note the two double bonds in the molecule. On polymerization, if the remaining double bonds (one per monomer) are all arranged in the *cis* orientation, the isoprene polymer is ➤rubber. If they are *trans*, the polymer is ➤gutta-percha. The **isoprene rule** suggests that the structure of essential oils such as ➤terpenes is based on simple multiples of the isoprene structure (although plants do not synthesize terpenes from isoprene).

isopropyl The nonsystematic but convenient name to describe the structure —$CH(CH_3)_2$. Isopropyl alcohol, which has the structure $(CH_3)_2CHOH$, is a solvent, used, for example, to clean record styli and CDs.

isosceles triangle A triangle with two sides of equal length and hence having two equal angles.

isosmotic ➤isotonic.

isospin ➤isotopic spin.

isotactic polymer When a polymer has a group attached to every second carbon atom, as in polystyrene, the groups can be arranged in a number of ways. In the isotactic arrangement, all the groups lie on the same side of the polymer chain. Compare ➤atactic polymer and ➤syndiotactic polymer (where there is a diagram).

isotherm 1 A line linking points of equal temperature on, for example, a weather chart or a phase diagram.
2 ➤van't Hoff isotherm.

isothermal Describing a process in which the temperature remains constant.

isotonic (isosmotic) Describing a medium that contains the same concentration of osmotically active molecules as, for example, a cell or tissue immersed in it. The immersed cells neither gain nor lose an appreciable amount of water by ➤osmosis. Isotonic solutions thus provide an osmotic buffer or support for cells. This is particularly important in the artificial culture of animal cells or manipulation of the ➤protoplasts of plant cells.

isotope One of two or more atomic species having the same ➤atomic number but different ➤nucleon numbers. Isotopes are therefore variants of the same element as the number of protons is the same; however, the number of neutrons, and hence the mass of the atom, varies. For example, chlorine has two naturally occurring isotopes: ^{35}Cl and ^{37}Cl. As the former is three times as abundant as the latter, and the ➤relative atomic mass is the weighted mean of the isotopes, the relative atomic mass for chlorine is about 35.5. Isotopes can be stable or radioactive (➤radioisotope). Isotopes can be separated electromagnetically (because the force deflecting the particles is the same but their mass is different), by laser excitation (as the laser frequency can be fine-tuned to excite only one isotope) or by gaseous diffusion (the heavier isotope diffusing more slowly). The latter method was formerly used to separate $^{238}UF_6$ from $^{235}UF_6$ to prepare enriched uranium for nuclear reactors.

isotope effect Any effect caused by the substitution of one ➤isotope for another. Normally this substitution has little effect on the chemistry of an element, but there are minor effects. The most important results from substituting ➤deuterium for hydrogen; the deuterated compound will react up to seven times slower, as a result of the lower ➤zero-point energy of the bond to the heavier isotope. Such a large **kinetic isotope effect** indicates clearly that the bond to hydrogen is broken in the ➤rate-limiting step of the reaction.

isotopic spin (isospin) Symbol I. A ➤quantum number that is conserved in the ➤strong interaction. Neutrons and protons both have a total isospin of $\frac{1}{2}$. A component of isospin I_3 is also defined, and is a quantum number conserved in both the strong and electromagnetic interactions; the neutron has $I_3 = -\frac{1}{2}$, the proton $I_3 = +\frac{1}{2}$. Isotopic spin is related to ➤spin only in that both quantities have similar mathematical properties.

isotropic Equal or identical in all directions, and thus ➤invariant under rotation. A liquid is an isotropic material, but a crystalline solid is not, as it has crystal axes. Free space is isotropic. Compare ➤anisotropic; ➤➤homogeneous.

isozyme ➤isoenzyme.

IT Abbr. for **information technology**, the branch of technology concerned with the use of computers and other telecommunications technology for the transmission, reception, storage and handling of information. **ICT** is an abbreviation for information and communication technology.

-ite A suffix labelling oxoanions with an element's lower oxidation number. It was once the common label, but has been superseded in many cases. Examples still in common use include ➤nitrite and ➤sulfite. Compare ➤-ate.

iteration A method for solving an equation in which approximations (**iterates**) to a solution are successively refined in a systematic way. The most famous example is the ➤Newton–Raphson method. Iterative methods are easy to implement on computers and can lead to complex phenomena such as the ➤Mandelbrot set.

IUPAC Acronym for International Union of Pure and Applied Chemistry. The nomenclature of organic chemistry is decided by IUPAC. The simple rules for IUPAC nomenclature are:

(i) find the longest continuous carbon chain (the spine);

(ii) name the substituents off the spine;

(iii) specify their location on the spine, keeping the numbers involved as small as possible at the first point of difference.

The diagram shows a molecule with a chain of five carbon atoms, so it is a substituted pentane. There are three methyl groups off the chain, making it a trimethylpentane. Two methyl groups are bonded to the second carbon atom and one to the fourth; so the complete name is 2,2,4-trimethylpentane.

$$H_3C-\underset{\underset{CH_3}{|}}{\overset{\overset{CH_3}{|}}{C}}-CH_2-\underset{\underset{CH_3}{|}}{CH}-CH_3$$

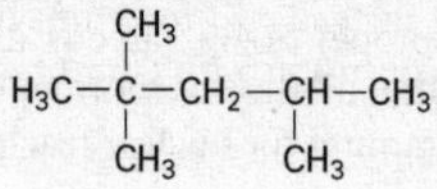

IUPAC The systematic name for iso-octane is 2,2,4-trimethylpentane.

J

j A symbol sometimes used instead of i for ➤imaginary numbers, especially by electrical engineers to avoid confusion with current.

j A ➤unit vector in the direction of the positive y axis.

j **1** Symbol for ➤angular momentum.
2 Symbol for ➤current density.

J Symbol for the unit ➤joule.

J Symbol for total ➤angular momentum quantum number.

Jacobian The Jacobian of n functions $f_1, ..., f_n$ is the ➤determinant $|\partial f_i / \partial x_j|$. The Jacobian arises when making a change of variables in a multiple integral. Named after Carl Gustav Jacob Jacobi (1804–51).

Jacob–Monod theory ➤operon. Named after François Jacob (b. 1920) and Jacques Lucien Monod (1910–76).

jade An ➤amphibole silicate containing calcium and magnesium.

Jahn–Teller effect The distortion of nonlinear molecules or complexes to avoid ➤degenerate electronic states. Some important examples are complexes of metal ions containing nine d electrons, such as Cu^{2+}. The copper(II) ion in an octahedral complex has four short bonds and two longer ones, rather than six bonds of the same length. Named after Hermann Arthur Jahn (1907–79) and Edward Teller (1908–2003).

jamming A deliberate attempt to create noise to hinder reception of a radio or radar signal by broadcasting a signal that interferes with it.

jasper A red form of impure quartz which is a common gemstone.

jaundice A clinical condition due to a disorder of the liver, particularly associated with the metabolism of ➤bile. The condition is indicated by characteristic yellowing of the skin.

Java A type of programming language developed by Sun Microsystems which can be embedded within Web pages (➤www). It allows the user to do more complex things than are possible with ➤HTML.

jejunum ➤gut.

JET Acronym for ➤Joint European Torus.

jet engine A type of aero engine that works by jet propulsion. Several types exist, such as the **turbojet**, which is based on a ➤turbine compressor, and **ramjet**.

jet propulsion A method of propulsion that employs the law of ➤conservation of momentum. A jet of fluid (typically water or air) is expelled from a vehicle in one direction, causing a change in the momentum of the vehicle equal and opposite to that of the jet.

jet stream A high-level thermal wind that flows parallel to the boundary between a warm airmass and a cold airmass. The wind speed in the core of the jet stream, which occurs just below the ➤tropopause on the warm side of the interface, can reach almost 400 km h^{-1}. The strong currents within the jet stream can cause clear-air turbulence, a hazard to passenger aircraft. The front between cold polar airmasses and warm tropical airmasses over the Atlantic leads to a jet stream at variable latitudes, and is largely responsible for the shorter journey times for west-bound transatlantic flights than east-bound.

***jj* coupling** ➤spin-orbit coupling.

Johnson noise A fluctuating voltage, caused by the thermal motion of conduction electrons, measured across a resistance. A simple classical model leads to the expression $\Delta V = \sqrt{4RkT}$, where ΔV is the root mean square voltage fluctuation, R the resistance of the conductor, k the ➤Boltzmann constant and T the thermodynamic temperature. Named after John Bertrand Johnson (1887–1970), who discovered the effect in 1928.

joint A point of conjunction of elements of animal skeletons. In the ➤endoskeletons of vertebrates, joints may be fused and immovable, such as those between the bones of the skull, or capable of flexing in one or more planes, like the elbow and hip joints. The joint consists of associated tissues, including cartilage and ligaments, with a lubricating ➤synovial fluid in movable joints. (In the ➤exoskeletons of arthropods, joints are of a different form: the limbs have peg-and-socket linkages to allow movement.) The flexing of skeletal elements at joints is by the action of ➤antagonistic muscles.

Joint European Torus (JET) An experimental nuclear reactor built to study the feasibility of power generation by controlled ➤nuclear fusion. The first step was taken in 1991, when ➤tritium was first added to the plasma and 40 kW of fusion power was obtained.

Jones reductor A glass tube filled with amalgamated zinc and used to reduce solutions of certain elements, such as titanium, vanadium, chromium, iron, copper and uranium, to their lower oxidation numbers before analysis by titration with an appropriate oxidant. Named after Clemens Jones, who was awarded a patent in 1892.

Josephson effect A phenomenon observed at very low temperatures in which a current flows between two superconductors through a thin insulating layer, an arrangement known as a **Josephson junction**. In the absence of a voltage between the superconductors, a direct current flows whose value is sensitive to any applied magnetic field; if a voltage is applied across the junction, an alternating current flows, with a well-determined frequency that depends on the voltage. The Josephson effect can be used to measure the Planck constant. Named after Brian David Josephson (b. 1940).

joule Symbol J. The SI unit of ➤energy. One joule is transformed when a force of one newton moves an object through one metre in the direction of the force. Named after James Prescott Joule (1818–89), who measured the ➤mechanical equivalent of heat.

Joule heating (Joulean heating) The heating of a resistive material (resistance R) through which a current I is passing. The energy transformed to heat per unit time is given by I^2R.

Joule–Thomson effect (Joule–Kelvin effect) When gases are expanded through a small hole or throttle, some of their internal energy is transformed into kinetic energy, causing a change in temperature. The expansion is ➤isenthalpic, and the effect arises because of the existence of ➤intermolecular forces. Above the ➤inversion temperature this causes a heating of the gas; below the inversion temperature it causes a cooling. Named after J. P. Joule and William Thomson, Baron Kelvin (1824–1907).

Jovian Relating to the planet ➤Jupiter.

JPEG Abbr. for Joint Photographic Expert Group. A computer graphics file format with variable compression of images with up to 16.7 million colours.

J/ψ particle A meson consisting of a charm ➤quark and its antiquark. It was the first particle discovered that contained one of the heavier quarks (charm, bottom and top). It was discovered at Brookhaven and Stanford independently in 1974; the two research groups chose the names J and ψ respectively (hence the dual name).

J segment One of a series of short ➤DNA sequences that code for part of the variable region of an ➤antibody. The 'J' indicates that it joins the ➤C segment and ➤V segment.

JUGFET Abbr. for junction gate ➤field-effect transistor.

jugular vein One of a pair of veins draining the brain and joining with the subclavian veins before discharging into the anterior vena cava (➤heart).

jumping gene ➤transposon.

junction diode ➤diode; p–n junction.

junction transistor Another name for a ➤bipolar junction transistor.

Jupiter The fifth planet from the Sun, and the largest. Jupiter has an equatorial radius of 71 492 km, more than 11 times that of the Earth, and a mass twice as great as that of all the other planets put together. The planet's orbital period is 11.86 years, and its average distance from the Sun is 5.2 AU. Its rotation period of 9 h 51 min is faster than that for any other planet. It has a magnetic field about 14 times as strong as the Earth's, and a gravitational field 2.4 times as strong. Jupiter's atmosphere is composed mainly of hydrogen. The **Great Red Spot** is a huge anticyclone in the atmosphere. Jupiter has 16 known satellites in all, four (Io, Europa, Ganymede and Callisto) with radii in excess of 1000 km. ➤➤Appendix table 4.

Jurassic See table at ➤era.

juvenile hormone (JH) One of a number of insect hormones produced by the **corpora allata** (glandular bodies on either side of the oesophagus), which control the sequence of larval or nymphal stages in the life cycle of an insect. The suppression of the production of JH causes the change to the adult form. The action of JH has been extensively studied in the blood-sucking bug *Rhodnius*.

K

k Prefix signifying a thousand base units, for example 1 kg is equivalent to 1000 g.

k **1** Symbol for the ➤Boltzmann constant (k_B is also used).
2 Symbol for ➤wavenumber.

k A ➤unit vector in the direction of the positive z axis.

k Symbol for ➤wave-vector.

K **1** Symbol for the element ➤potassium, from the Latin *kalium*.
2 Symbol for the unit ➤kelvin.
3 (Comput.) The memory of a computer is expressed in K (kilobytes) where 1 K is 1024 bytes.

K Symbol for various ➤equilibrium constants, usually modified with a subscript descriptor to indicate the type of equilibrium constant: K_a, ➤acid ionization constant; K_b, ➤base ionization constant; K_{in} (➤indicator); K_{sp}, ➤solubility product; K_{stab}, ➤stability constant of a complex; K_w, ➤ionic product for water.

kainite An ore of potassium and magnesium of approximate formula $MgSO_4 \cdot KCl \cdot 3H_2O$.

kaolin A native aluminosilicate named after the Chinese for 'high ridge', the original site where it was discovered, and used to make ceramics.

kaon ➤K meson.

karyotype A description of the number and form of the ➤chromosomes of an organism. A photograph or diagram of a preparation of the chromosomes viewed under the microscope, usually arranged in their ➤homologous pairs, is normally presented as a **karyogram**.

katharometer An instrument for measuring the ➤thermal conductivity of a gas. It is frequently used as the detector in gas chromatography.

K capture A process in which a nucleus absorbs one electron from the electron cloud surrounding it and emits a ➤neutrino, for example: $^7_4Be + e^- \rightarrow {}^7_3Li + \nu_e$. The process is so named because the electron usually comes from the innermost ➤shell, the K shell.

KE Abbr. for ➤kinetic energy.

keeper (armature) A ferromagnetic bar placed across the poles of a permanent magnet. A keeper completes the ➤magnetic circuit and helps to minimize ➤demagnetization.

Kekulé structure The original structure proposed by Friedrich August Kekulé (1829–96) for benzene (see the diagram). Although it is less often used nowadays, it is still common in certain circumstances, for example to indicate the motion of electron pairs during reaction mechanisms. ➤delocalization.

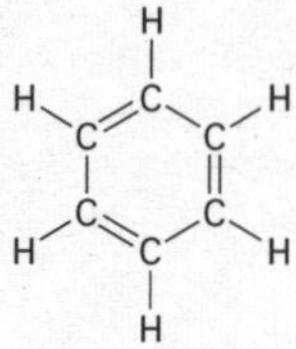

Kekulé structure for benzene.

Kel-F Tradename for a ➤thermoplastic polymeric material that resists the attack of many chemicals, notably fluorine.

kelvin Symbol K. The SI unit of ➤thermodynamic temperature. (Note that there is no degree sign; compare °C.) Named after William Thomson, Baron Kelvin (1824–1907).

Kelvin balance ➤current balance.

Kelvin scale The scale of thermodynamic temperature. The lower fixed point is ➤absolute zero at zero ➤kelvin; the higher fixed point is the ➤triple point of water at exactly 273.16 K. The melting point of ice is 273.15 K. Temperatures t on the Celsius scale can be converted to temperatures T on the Kelvin scale: $T/K = t/°C + 273.15$.

Kepler's laws Three laws of planetary motion introduced by the astronomer Johannes Kepler (1571–1630):

(i) Each planet moves in an orbit that is an ellipse with the Sun at one focus (➤conic). At the time, this was of particular significance because it superseded the concept of circular orbits, and gave the first real indication that the Sun controls the movement of the planets.

(ii) A line joining the Sun to a planet sweeps out area (as shaded in the diagram opposite) at a constant rate. This is a consequence of the law of ➤conservation of angular momentum.

(iii) The square of the period of the orbit is proportional to the cube of the length of the major axis of the ellipse. This resulted in the first accurate prediction of the relative distances of planets from the Sun (since their orbital periods were well known).

Kepler's laws were explained later in the 17th century by ➤Newton's law of gravitation.

keratin A tough fibrous ➤protein found particularly in the ➤epidermis of vertebrate skin and forming structures such as hair, claws, hooves and fingernails.

kernel 1 The set of elements of a ➤group (2) that are mapped by a ➤homomorphism to the ➤identity (2) element of another group. In particular, the kernel of a ➤matrix ➤transformation is the subspace of vectors that are mapped to the ➤zero vector.

2 A function k of two variables relating two functions by an ➤integral transform of the form $f(x) = \int k(x, t)g(t)\, dt$. The Fourier and Laplace transforms are of this type.

kerosene (kerosine) A fraction of petroleum which is burnt in turbine engines. It is a mixture of hydrocarbons with a range of roughly 11 to 14 carbon atoms. The old

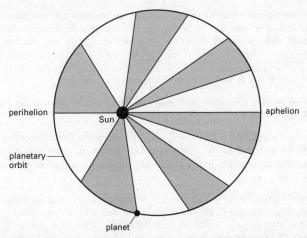

Kepler's laws Kepler's second law states that a planet orbiting the Sun sweeps out equal areas in equal time intervals: each sector marked in the diagram is of equal area, and each arc is traversed in the same time.

name for this fraction was 'paraffin', but that fell into disuse due to confusion with the homologous series of paraffins (now called alkanes). In the early decades of the 20th century, this fraction was much used both for heating and lighting (kerosene lamps).

Kerr cell A cell of liquid nitrobenzene between crossed ▸Polaroids across which an electric field can be applied. The electro-optical ▸Kerr effect allows light to pass only in the presence of the field, and the cell thus can act as a fast shutter with no moving parts. Named after John Kerr (1824–1907).

Kerr effect Two separate effects in optics. In the **electro-optical Kerr effect** (which is exploited in the ▸Kerr cell), an electric field applied to certain isotropic materials (such as glass or nitrobenzene) causes the material to have different ▸dielectric constants parallel to and perpendicular to the field, and to show ▸birefringence. It can therefore cause ▸plane-polarized light transmitted through the material to become ▸elliptically polarized. In the **magneto-optical Kerr effect**, the reflection of plane-polarized light from a magnetic surface causes a similar change in polarization.

ketene A colourless gas of formula $H_2C=C=O$ used in the manufacture of aspirin.

keto–enol tautomerism An important step in the reactions of ▸ketones, for example with iodine. In this often rate-limiting step, a ketone such as CH_3COCH_3, under acidic or alkaline catalysis, becomes an ▸enol, $H_2C=C(OH)CH_3$, which then reacts quickly with the iodine.

ketohexose A▸monosaccharide having a ketone structure at the carbonyl carbon and with six carbon atoms. An example is ▸fructose.

ketone An organic molecule that contains a carbonyl group, CO, attached to two alkyl groups. The simplest is propanone, CH_3COCH_3. In ketones apart from propanone and butanone, a number is needed to specify the position of the carbonyl group. Pentan-2-one is $CH_3COCH_2CH_2CH_3$ whereas pentan-3-one is $CH_3CH_2COCH_2CH_3$. The level of ketones in urine is an important test for diagnosing diabetes as well as complications during pregnancy.

ketose A ➤monosaccharide with a ketone group in the open-chain form. An example is ➤fructose.

Kevlar An ➤aramid fibre which is very strong for its weight. The benzene rings in the structure are arranged in the ➤para positions, conferring great tensile strength. Kevlar is used, for example, in Grand Prix cars, often as a ➤composite material with ➤carbon fibre.

keyboard In computing, an input device consisting of a set of keys used for entering text and using the machine. The majority of keyboards in English-speaking countries use the 'qwerty' layout (so called because of the order of letters at the top left of the keyboard). Many keyboards conform to a standard layout with 103 keys, where the additional keys are for functions concerned with use of the computer, or for fast numerical entry.

kg Symbol for the unit ➤kilogram.

kHz Symbol for the unit ➤kilohertz.

kidney The major organ of fluid regulation (➤osmoregulation) and nitrogenous excretion in animals. In vertebrates the functional units are tubular structures called nephrons (see the diagram opposite), of which there may be a million in each kidney. ➤Blood is filtered at high pressure so that some of the plasma passes into the nephrons via the **Bowman's capsules** from the **glomeruli** (clusters of capillaries). As the fluid is passed along the length of the nephron, useful substances such as water, glucose and some minerals are recovered. The nephron can selectively reabsorb both water and minerals and regulate hydrogen ion concentration, thus controlling blood ➤pH. The exact functioning of the kidney thus depends on the physiological conditions existing in the rest of the body at a particular time. The resultant fluid (**urine**) is drained from the nephrons into the **pelvis** of the kidney, and from there, via the **ureter**, to the **bladder**, where it is temporarily stored before passing to the exterior of the body via the **urethra**. The average human body produces about 1.5 litres of urine per day. The arrangement of the nephrons divides the kidney into two zones – the outer **cortex** and the inner **medulla**. The kidney is an important organ of ➤homeostasis, and its function is under the control of ➤hormones.

kieselguhr (diatomaceous earth, diatomite) A sedimentary mineral deposit formed from the skeletons of ➤diatoms. It has several industrial uses, for example in the manufacture of toothpaste and polishes, and in filters for purifying liquids.

killer T cell A specialized ➤lymphocyte cell of the ➤immune system that carries receptor molecules on its surface. These interact directly with other cells that may carry foreign ➤antigens, for example virus-infected cells that have viral particles on

(a)

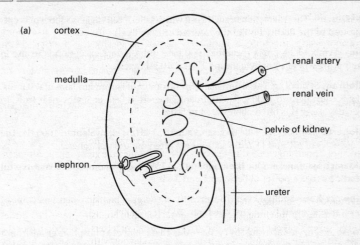

- cortex
- medulla
- nephron
- renal artery
- renal vein
- pelvis of kidney
- ureter

(b)

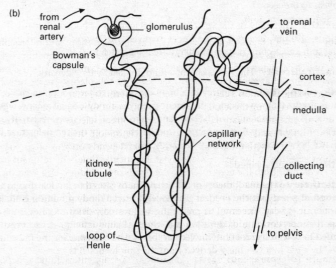

- from renal artery
- Bowman's capsule
- glomerulus
- to renal vein
- cortex
- medulla
- capillary network
- kidney tubule
- collecting duct
- loop of Henle
- to pelvis

kidney (a) Vertical section through the kidneys, and (b) detail of nephron.

their surface. The interaction of such a cell with a killer T cell triggers a series of events that lead to the destruction of the infected cell. ≫T cell.

kilo- Symbol k. A prefix meaning 1000 base units. A rare exception is its use in computing where a kilobyte means 2^{10} (i.e. 1024) bytes.

kilogram Symbol kg. The SI unit of mass, and one of the seven fundamental SI units: 1 kg is equivalent to 1000 g. The standard kilogram is kept at Sèvres, near Paris. In everyday usage it is often abbreviated to '**kilo**'.

kilohertz Symbol kHz. One thousand ≻hertz: 1 kHz is equivalent to 1000 Hz. The frequencies of MF and LF radio stations are often quoted in kilohertz.

kilowatt Symbol kW. One thousand watts: 1 kW is equivalent to 1000 W. A typical electric fire has a power of between 1 and 3 kW.

kilowatt-hour Symbol kWh. The quantity of energy expended when a machine of power one kilowatt is used for one hour: 1 kWh = 3600 kJ.

kinase An enzyme activated by cyclic ≻AMP that removes a phosphate group from a high-energy phosphate compound (such as ≻ATP) and transfers it to another substance, frequently activating the latter. For example, in the initial stage of ≻glycolysis a phosphate group is transferred to a ≻glucose molecule from ATP by a kinase enzyme and this phosphorylation provides the activation energy to enable glycolysis to proceed.

kinematics The study of the velocity and acceleration of bodies, particularly in collisions, without reference to the forces involved; the laws of ≻conservation of momentum and ≻conservation of energy are used instead.

kinematic viscosity The ≻viscosity of a fluid divided by its ≻density.

kinetic energy (KE) The energy of a body due solely to its motion in space. In classical physics (at low velocities) the kinetic energy of a body is equal to $\frac{1}{2}mv^2$, where m is the mass (≻inertial mass) of the body and v is its speed. In ≻special relativity the concept of total energy is more natural, and kinetic energy is then defined as the difference between the total energy of a body and its ≻rest energy.

kinetics ≻chemical kinetics.

kinetic theory of gases A theory of the behaviour of gases that models the gas as a collection of point particles (≻ideal gas model) in rapid random motion colliding elastically. It is very successful at predicting various properties of gases, such as transport properties, rate of diffusion and thermal conductivity. It can even be extended to take account of intermolecular forces. The equation for the pressure p exerted by a gas in a volume V derived from the simple model is

$$pV = \tfrac{1}{3}Nm\overline{c^2},$$

where N is the number of particles, each of mass m, and $\overline{c^2}$ indicates the mean of the squares of their speeds (the mean square speed). As the mean square speed is proportional to the ≻thermodynamic temperature (in kelvin), this model predicts the ≻ideal gas equation.

kinetochore Multiprotein complex assembled on the ➤centromere of each chromatid, by which chromosomes are attached to the microtubules of the spindle during ➤mitosis and ➤meiosis, and which facilitates the movement apart of the chromatids. The correct conformation of the kinetochore is an important factor in the correct distribution of chromosomes to daughter cells during such cell divisions. There are over 40 different proteins in the simplest kinetochore, including histones which bind with chromatid DNA.

kingdom The largest of the categories used in the classification of organisms (➤taxonomy). Currently all organisms are grouped into one of five kingdoms depending on their cellular organization and mode of nutrition: Animalia (animals), Plantae (plants), Fungi, Protoctista (single-celled ➤eukaryotic organisms) and Monera (➤prokaryotic organisms, including bacteria). The boundaries between some kingdoms, particularly the Protoctista and the Plantae, are not clear and the division is open to revision.

kinin ➤cytokinin.

Kipp's apparatus An apparatus for producing a controlled stream of gas, especially carbon dioxide. It has three bulbs on top of each other. The top and bottom ones are connected and contain acid. In the middle bulb is a solid reagent (such as calcium carbonate for the production of carbon dioxide). When a tap in the middle bulb is shut, the pressure of gas forces the acid out of the bulb, thus stopping the reaction. When open, the pressure is released and the acid in the top bulb falls to the bottom one under gravity, which then overflows into the middle one. Named after Petrus Jacobus Kipp (1808–64).

Kirchhoff's laws Two laws that assist in the evaluation of currents and voltages in circuits and networks (see the diagram).

(i) The current leaving a **node** is equal to the sum of the currents entering it (it is sensible to consider the current flowing in each **loop** of the circuit separately).

(ii) The total of the potential differences between nodes in a closed loop is equal to the total ➤electromotive force in that loop.

Named after Gustav Robert Kirchhoff (1824–87).

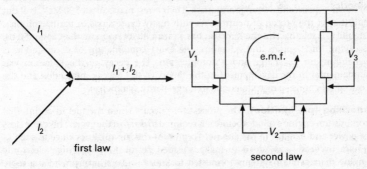

first law

second law

Kirchhoff's laws In the second law, e.m.f. = $V_1 + V_2 + V_3$.

Kjeldahl's method A method for measuring the proportion of nitrogen, particularly in an organic compound, by reacting the compound with concentrated sulfuric acid, which produces ammonium sulfate. To this mixture, excess alkali (NaOH) is added and the resulting ammonia distilled into standard acid. Named after Johan Gustav Christoffer Kjeldahl (1849–1900).

Klein bottle The closed one-sided surface formed by imagining the neck of a tapering bottle to pass through the side of the bottle and to be sealed to its base (see the diagram). It can also be formed by gluing two ➤Möbius strips together. Named after (Christian) Felix Klein (1849–1925).

Klein bottle

Klein–Gordon equation A ➤relativistic equivalent of the ➤Schrödinger equation of quantum mechanics. For a free particle it is:

$$\left[\frac{\partial^2}{\partial t^2} - \nabla^2 + m^2\right] \phi(x, t) = 0,$$

where ϕ is the ➤wavefunction and m is the mass of the particle. For plane-wave states the Klein–Gordon equation has negative-energy solutions and so does not lead to a tenable relativistic wave-mechanical theory for spin-$\frac{1}{2}$ particles like the electron (for which the ➤Dirac equation is essential). However, it does lead to an acceptable formulation of ➤quantum field theory for scalar ➤bosons. Named after Oskar Benjamin Klein (1894–1977) and Walter Gordon (1893–1940).

K lines Lines in an ➤X-ray spectrum resulting from an electronic transition which ends with the electron on the innermost shell of the atom, the ➤K shell.

klystron A generator or amplifier of ➤microwaves. A klystron employs high-frequency radio waves to modulate the velocity of an electron beam in a sealed tube, and then uses that ➤modulation to generate electromagnetic radiation in the microwave band.

K mesons (kaons) A family of stable ➤mesons (such as K^+ and K^0) made up of an up or down ➤quark and a strange antiquark. Their ➤rest mass, about 500 MeV, is about half that of the proton or neutron. As with many expressions in particle physics, 'stability' is relative and the K meson has a mean life of no more than about 50 ns, decaying into ➤pions and ➤leptons, the exact combination of decay products depending on its charge and ➤isotopic spin. The decay of the K meson was fundamental in the development of the theory of the ➤weak interaction and the concept of charge–parity violation (➤charge–parity symmetry).

knocking (pre-ignition) The process that occurs when the fuel in an internal combustion engine explodes before it is required to do so in the cycle. This causes loss of power and damage to the engine. To prevent this an **antiknock** agent, such as ➤tetraethyllead, was added to make ➤leaded petrol. More recently, electronic engine management systems connected to knock sensors have provided a more sophisticated method of dealing with this problem.

Knoevenagel reaction A method for increasing the chain length of an aldehyde by two carbon atoms. The aldehyde is warmed with malonic acid, $CH_2(COOH)_2$, in ▶pyridine. After hydrolysis and heating, an unsaturated acid is formed. Thus RCHO becomes $RCH=CHCOOH$. Named after Emil Knoevenagel (1865–1921).

knot A unit of speed, one ▶nautical mile per hour, used extensively in aviation and nautical applications. It is equivalent to a speed of 1.852 km h^{-1}.

knot theory A branch of ▶topology dealing with the properties of knots treated as simple closed curves embedded in three-dimensional space. In the 1980s, some very powerful methods for classifying knots were discovered with applications to the mechanisms of ▶DNA replication and to ▶quantum field theory.

Knudsen flow A model of the flow of gas in a container at a very low density. If the ▶mean free path of the gas is much greater than the dimensions of the container, the flow is governed by collisions between the molecules of gas and the container, not by intermolecular collisions. Thus the rate of flow does not depend on the ▶viscosity of the gas. Named after Martin Hans Christian Knudsen (1871–1949).

Kohlrausch's law (law of independent migration of ions) The molar ▶conductivity, at infinite dilution, of a compound is the sum of the molar conductivities of the individual ions present in the solution. Named after Friedrich Wilhelm Georg Kohlrausch (1840–1910).

Kolbe reaction The electrolysis of a salt of a carboxylic acid to produce alkanes; the yield is poor. For example, ethanoate ion produces small quantities of ethane as follows:

$$2CH_3CO_2^- \rightarrow C_2H_6 + 2CO_2 + 2e^-.$$

Named after Adolf Wilhelm Hermann Kolbe (1818–84).

Kr Symbol for the element ▶krypton.

Krebs cycle (citric acid cycle, tricarboxylic acid cycle, TCA cycle) A metabolic pathway for the complete ▶oxidation of ▶pyruvic acid comprising a series of reactions occurring during ▶aerobic respiration in the ▶mitochondria of cells. The cycle commences with the synthesis of citrate from ▶acetyl coenzyme A, derived from pyruvate (itself derived from ▶glucose), and oxaloacetate. A series of decarboxylations and molecular rearrangements regenerates oxaloacetate and reduces oxidized coenzymes. Subsequent reoxidation of these coenzymes by the ▶electron-transport chain on the membranes of the mitochondrion generates ▶ATP. The main stages of the cycle are shown in the diagram overleaf. The cycle also acts as a central pathway of metabolism, and intermediates can be diverted into, for example, fat and amino acid metabolism. In turn the cycle can be fed by the resultant products of the breakdown of these substances; the cycle does not operate in isolation but is a dynamic part of cell metabolism in all aerobically respiring organisms. The efficiency of the cycle can be estimated by calculating how many ATP molecules are formed in one turn of the cycle, and by comparing the energy represented by each molecule with the energy involved in the direct combustion of glucose. Theoretically 38 ATP molecules can be produced during one turn, but this is somewhat artificial

Krebs cycle

step	molecules of coenzyme per molecule of glucose	ATP equiv.
A	2(NADH + H⁺) (+2ATP)	6 + 2
B	2(NADH + H⁺)	6
C	2(NADH + H⁺)	6
D	2(NADH + H⁺)	6
E		2
F	2FADH₂	4
G	2(NADH + H⁺)	6
Total ATP yield		38

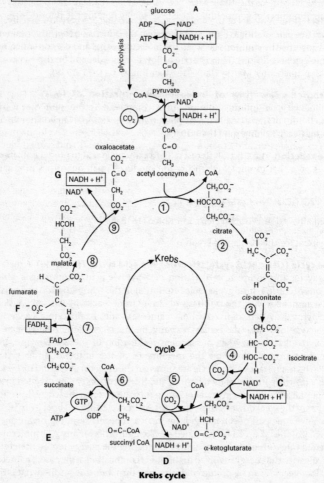

Krebs cycle

because of the dynamic nature of the operation of the cycle in relation to other metabolic pathways. Named after Hans Adolf Krebs (1900–81). ➤➤FAD; NAD⁺.

Kroll process The production of ➤titanium by reduction of titanium(IV) chloride with magnesium. Named after Wilhelm Justin Kroll (1889–1973), who invented the process in 1940.

Kronecker delta The function δ_{ij} taking the values 1 if $i = j$ and 0 if $i \neq j$. It is a convenient shorthand, for example when dealing with the properties of ➤orthonormal vectors. Named after Leopold Kronecker (1823–91).

krypton Symbol Kr. The element with atomic number 36 and relative atomic mass 83.8, which is one of the noble gases and as such is virtually inert. In recent years a very small number of compounds have been made, notably the fluoride KrF_2.

K shell An outdated term for the innermost ➤shell of an atom, the one whose ➤principal quantum number is 1. The terminology is still in common use in Auger electron spectroscopy (➤Auger effect).

***k*-space** A coordinate system based on ➤crystal momentum and ➤wave-vector. Since the crystal momentum of an electron in a conductor is quantized, it is often useful to refer to and plot its three components as if they were vectors on a grid of spatial coordinates, with the corresponding concepts of surface (➤Fermi energy) and volume. The idea of quantum states in *k*-space is essential to the development of the theory of conduction and other transport properties in metals and semiconductors. Values of *k* corresponding to ➤reciprocal lattice vectors represent quantum states. ➤➤reciprocal lattice.

Kundt's tube An apparatus for determining the speed of sound in a gas under different conditions (of pressure, temperature and humidity). It uses a resonance technique to find the wavelength of standing waves in a tube of variable length. Named after August Eduard Eberhard Adolph Kundt (1839–94).

kW Symbol for the unit ➤kilowatt.

kWh Symbol for the unit ➤kilowatt-hour.

L

l **1** Symbol for ➤litre, which is not often used as the cubic decimetre (dm³) is preferred in ➤SI units.

   2 (Chem.) Label used in chemical equations to indicate the liquid state.

L **1** Symbol for the ➤Avogadro constant. ➤➤Loschmidt constant.

  2 Symbol for ➤luminance.

La Symbol for the element ➤lanthanum.

labelling The use of a specific isotope, which can be either radioactive or nonradioactive, to mark an element in order to follow its course during a reaction. The progress of a nonradioactive label can be followed by ➤mass spectrometry. The progress of a radioactive label can be monitored by using, for example, a ➤Geiger counter. The radioisotope ^{14}C may be used to follow carbon assimilation in photosynthesis. Radioactive labelling with the isotope ^{131}I is used to investigate thyroid disorders. Labelling was used to establish that ➤DNA is the genetic material of ➤viruses. Viral ➤protein and DNA of a ➤bacteriophage that attacks *E. coli* were labelled with radiosulfur and radiophosphorus respectively. By following the distribution of radioactivity when *E. coli* cells were infected, it was possible to show that of the two components of the virus only the DNA entered the bacterial cells on which the virus was cultured. ➤➤marker (2).

labile Describing a ligand in a complex that is easily replaced by another ligand. Complexes of s- and f-block metals are usually labile. Lability can be explained by looking at the electronic structure of the central ion; d^3 and low-spin d^6 octahedral ➤complexes are generally **nonlabile**, as the lower-energy orbitals are respectively half-filled or filled.

labium (plural **labia)** A lip-shaped structure, in particular the lower lip of an insect's mouthparts, used in feeding. It is also used for the lower lip of flowers of the order Lamiales.

labrum (plural **labra)** The upper lip of an insect's mouthparts, used in feeding.

lachrymal gland ➤eye.

lachrymatory Describing a substance (such as ethanoyl chloride, CH_3COCl, or ➤CS gas) that causes the eye's lachrymal gland to produce tears.

lac **operon** An ➤operon mechanism operating in cells (particularly in ➤*E. coli*) to regulate the expression of sequences of ➤DNA associated with the synthesis of enzymes involved in the ➤metabolism of the sugar lactose.

lacquer A protective coating formed by evaporation of the solvent from a solution of a ➤resin.

lactam A cyclic amide, an example of which is ➤isatin. ➤Penicillin has a four-membered lactam ring.

lactate **1** (Chem.) The anion formed from lactic acid.
  2 (Biol.) To produce milk by ➤lactation.

lactation The production of milk in ➤mammals by the mammary glands. This usually occurs after the birth of young, and is in response to suckling on the teat (nipple). Lactation is controlled by the hormones ➤prolactin and ➤oxytocin.

lactic acid The traditional name for the compound 2-hydroxypropanoic acid, $CH_3CH(OH)COOH$, so named from the Latin for 'milk', since sour milk was the source from which it was first extracted. It is formed in the body during ➤anaerobic respiration; it is the molecule responsible for cramp in actively respiring muscle. The molecule exhibits ➤optical activity as the second carbon atom (emphasized above) is bonded to four different groups.

lactide A compound containing a six-membered ring formed by self-esterification of two molecules of a carboxylic acid that has a hydroxyl group on the carbon atom next to the carboxyl group. The structure is named after the archetypal compound formed from lactic acid.

lactone An internal ➤ester formed by dehydration of a carboxylic acid that has a hydroxyl group on the third or fourth carbon atom from the carboxyl group.

lactose A ➤disaccharide sugar present in milk. It is hydrolysed by the enzyme lactase (β-galactosidase) into the monosaccharides glucose and galactose. **Lactose intolerance** is a condition in which an individual is unable to synthesize lactase; when milk is ingested, abdominal cramps and diarrhoea result. Lactose can be fermented to lactic acid.

laevorotatory Symbol *l* or (–). A molecule with one ➤chiral centre has two enantiomers (➤optical activity): the laevorotatory enantiomer rotates the plane of plane-polarized light to the left (*laevus* is the Latin for 'left'). Compare ➤dextro-rotatory.

Lagrange (undetermined) multipliers A ➤calculus technique for finding stationary values of a function of several ➤variables subject to one or more constraints. For example, for $f(x, y)$ subject to the constraint $g(x, y) = 0$, a Lagrange multiplier λ is introduced and the stationary point(s) found from the simultaneous equations:

$$g = 0, \quad \frac{\partial f}{\partial x} = \lambda \frac{\partial g}{\partial x}, \quad \frac{\partial f}{\partial y} = \lambda \frac{\partial g}{\partial y}.$$

Named after Joseph Louis Lagrange (1736–1813).

Lagrange's theorem In a finite ➤group (2), the ➤order of every subgroup divides the order of the whole group.

Lagrangian Symbol *L*. A function of the ➤generalized coordinates of a mechanical system, used in ➤Hamilton's principle to derive equations of motion for the system. For a nonrelativistic system it is defined as the difference between the kinetic energy of the system and its potential energy: $L = T - V$.

Lamarckism The 19th-century rival of the theory of evolution by ➤natural selection which claimed that organisms are capable of passing on to their offspring characteristics they have acquired during their lifetime. For example, if a mouse's tail were cut off, it would be possible for its offspring to have shorter tails. Lamarckism now has no scientific credibility and is not supported by modern experimental genetics. Named after Jean-Baptiste Pierre Antoine de Lamarck (1744–1829).

lambda particle A ➤hyperon, with spin $\frac{1}{2}$, and a mass (1.116 GeV/c^2) about 10% more than that of a proton, but with ➤strangeness −1 and zero ➤isotopic spin. In the ➤standard model it is composed of one up ➤quark, one down quark and one strange quark.

lambda phage A ➤bacteriophage that infects ➤*E. coli*; the DNA of lambda phage has been fully sequenced and is routinely used in laboratory studies of DNA.

lambda point A point on a ➤phase diagram where the conjunction of phase boundaries resembles the Greek letter lambda (λ). One example occurs at the temperature at which the two liquid phases of ^{4}He coexist. ➤➤superfluid.

Lambert's laws A set of laws governing the intensity of radiation. The illumination of a surface lit by rays from a point source is inversely proportional to the square of the distance between the source and the surface, and proportional to the cosine of the angle the rays make to a ➤normal to the surface. The intensity of radiation passing through an absorbing medium (or, for visible light, the luminous intensity) decreases as e^{-ad}, where *d* is the distance travelled through the medium and *a* is the **linear absorption coefficient**. Named after Johann Heinrich Lambert (1728–77). ➤➤Beer's law.

Lamb shift A very small energy separation in the energy levels of the hydrogen atom. Neither the ➤Schrödinger equation nor the ➤Dirac equation could account for any energy gap; the quantitative explanation provided by Willis Eugene Lamb (1913–2008), in terms of a quantum interaction between the electron and electromagnetic radiation, was an important step in the development of ➤quantum electrodynamics.

lamella (plural **lamellae)** A biological structure in the form of multiple layers or folds with a large functional surface area. Examples are the folded membranes in ➤chloroplasts which capture light during ➤photosynthesis; the spore-bearing surfaces of the gills of mushrooms and other fungi; and various membranous structures in animals, particularly those on which the structure of bone is formed. The **middle lamella** is the zone between individual plant cells containing a matrix of pectic substances which glues cells together.

lamina (plural **laminae)** **1** (Phys.) A very thin, flat sheet or layer.

2 (Biol.) In plants, the flat expanded part of a ➤leaf that provides the site for ➤photosynthesis. The term is also applied to the expanded ➤thallus of red, green and brown seaweeds (such as the kelps).

laminar flow Steady flow of a fluid with a low ➤Reynolds number, where all particles move in parallel layers (laminae) along ➤streamlines. The ➤vorticity of laminar flow is always zero: there are no closed loops of motion. Compare ➤turbulent flow.

laminated iron Iron in the form of sheets interleaved with thin layers of insulator to prevent ➤eddy currents from forming. It is used in the core of ➤transformers.

Landau levels Energy levels of an electron in a magnetic field due to the ➤quantization of the energy associated with motion perpendicular to the field. Motion parallel to the field is unaffected since the ➤Lorentz force on electrons moving parallel to the field is zero. Named after Lev Davidovich Landau (1908–68).

Langmuir isotherm An equation that describes the adsorption of a gas on a surface, it being assumed that every surface site is equivalent. The fraction f of the surface covered is given by

$$f = \frac{Kp}{1 + Kp},$$

where K is a constant and p is the pressure of gas. The adsorption of carbon monoxide on charcoal follows the Langmuir isotherm closely. Named after Irving Langmuir (1881–1957).

lanthanide (lanthanoid, lanthanon, rare earth) Common symbol Ln. One of the fourteen elements following lanthanum in the ➤periodic table. They are very similar to each other; in every case their most common oxidation number is +3. Many were first discovered in two oxides found near Ytterby in Sweden. The atoms and ions gradually shrink as the atomic number rises since the added electrons do not shield the extra positive charge fully (➤effective nuclear charge). This **lanthanide contraction** is crucial to the separation of the Ln^{3+} ions by ion-exchange chromatography. ➤➤Appendix table 5.

lanthanide shift reagent A lanthanide ion, frequently europium(III), introduced to shift the ➤nuclear magnetic resonance spectrum, which it does by altering the local magnetic field experienced by the nucleus under examination.

lanthanoid Alternative name for ➤lanthanide.

lanthanon Alternative name for ➤lanthanide.

lanthanum Symbol La. The element with atomic number 57 and relative atomic mass 138.9, which is a grey, lustrous metal. Its chemistry is typical of the fourteen elements after it, the ➤lanthanides. Its dominant oxidation number is +3, as the balance of ionization energy and hydration energy favours this oxidation number. Its compounds such as the white solids La_2O_3 and $LaCl_3$ are predominantly ionic in character.

lapis lazuli (lazurite) A semiprecious stone, coloured blue or green. The colour is due to sulfur radical anions, such as S_3^- and S_2^-.

Laplace's equation The partial differential equation $\nabla^2\phi = 0$ where ∇^2 is the ➤Laplacian. It is satisfied by the gravitational and electrostatic potentials (away from mass or charge sources). With appropriate and physically natural boundary conditions, the solutions to Laplace's equation are unique. Named after Pierre Simon de Laplace (1749–1827).

Laplace transform The integral transform $\int_0^\infty e^{-xt} f(t)\, dt$. It is used to solve certain types of ➤differential equations by converting them into algebraic expressions for the Laplace transform of the unknown function, which can then be found from standard tables.

Laplacian (Laplace operator) The Laplacian of a real-valued function f of one variable is $\partial^2 f/\partial x^2$, of two variables is $\partial^2 f/\partial x^2 + \partial^2 f/\partial y^2$, and of three variables is $\partial^2 f/\partial x^2 + \partial^2 f/\partial y^2 + \partial^2 f/\partial z^2$. It is often abbreviated to $\nabla^2 f$, where the operator ∇^2 is called 'del-squared'.

lapse rate The rate at which temperature decreases with altitude in the atmosphere. ➤➤dry adiabatic lapse rate.

Larmor precession The ➤precession of a magnetic moment (such as an orbiting charge) in a magnetic field. For a charge q of mass m moving in a magnetic field with flux density B, the angular precession rate, the **Larmor frequency**, is given by:

$$\omega_L = qB/2m.$$

Named after Joseph Larmor (1857–1942).

larva (plural **larvae)** The juvenile form of many animal groups that develops after the hatching of eggs. Larvae are usually sexually immature and frequently differ markedly from the adult form. In sessile aquatic forms such as barnacles, the larva is free-living and plays an important role in dispersing the species over large distances. Other examples are the tadpoles of amphibia, the caterpillars of moths and butterflies, and the maggots of flies. The larva develops into the adult form by ➤metamorphosis. The larvae of some animal groups, such as echinoderms, have been extensively studied and provide an evolutionary link between ➤chordates and nonchordates.

larynx The enlarged anterior part of the ➤trachea housing a slit-like opening, the **glottis**, connecting the mouth cavity with the trachea. In most mammals the larynx houses two membranous folds, the ➤vocal cords, which vibrate as air passes over them to produce sound.

laser A device for producing an intense, highly coherent source of light; the name is an acronym for light amplification by stimulated emission of radiation. Electrons are excited into a higher energy level in sufficient numbers for a ➤population inversion to occur. Then, when stimulated to drop to the lower energy level by a passing photon, each excited atom emits its energy and produces a photon of very precise frequency. Lasers can be **pulsed**, if the energy is produced in a single rapid burst, or **continuous**. The four main characteristics of a laser are its ➤monochromatic nature,

➤coherence, high power and narrow beam. The monochromatic property is especially important for ➤isotope separation, as the laser can excite the vibrational modes of a compound of just one of the isotopes. **Tunable dye lasers** can have their monochromatic emission tuned across the visible spectrum. Similarly, lasers have greatly extended the usefulness of Raman spectroscopy (➤Raman effect). The fact that laser light is coherent makes ➤holography possible. Pulsed lasers can deliver their energy in a very short time. This has proved useful for extending the technique of ➤flash photolysis down to nanosecond and even picosecond timescales. Laser beams are used in surgery on the retina, for measuring the Moon's distance and for reading the pits on a compact disc.

laser printer A printer which can quickly and cheaply produce high-quality images from a computer, using similar principles to a xerographic photocopier. A low-powered laser is used to build up an electrostatic copy of the image on a photosensitive rotating drum, just as an image is built up on a television screen by a raster ➤scan. The charged areas on the drum then attract a fine powder (**toner**), which is transferred to the paper and heated to fix the image.

Lassaigne's test A method for detecting the presence of several elements in an organic compound. The compound is first fused with sodium and then elements such as chlorine are detected as their anions. Named after Jean Louis Lassaigne (1800–59).

latent heat The heat transferred to or from a substance's surroundings when the substance changes ➤phase. The heat transferred to a solid to melt it is the **latent heat of fusion**, and the heat transferred to a liquid to vaporize it is the **latent heat of vaporization**. The energy input is required to overcome the attractive forces between the particles.

lateral inversion ➤inversion.

latex A colloidal solution (➤colloid) of varied composition found in special ducts in the stems, roots and leaves of several plant families. It is the white juice exuding from cut dandelion stems and leaves, and is a characteristic feature of the Euphorbiaceae, the family that includes the commercially exploited rubber tree (*Hevea brasiliensis*). Rubber latex can be processed in a number of ways and is used in the production of a number of commodities including rubber gloves, condoms and aircraft tyres.

latitude One of the angular coordinates used to specify a point on the surface of the Earth. The latitude of a location is the angle between the equatorial plane and a line from the centre of the Earth to that location. Compare ➤longitude; ➤➤celestial sphere.

latitude, line of A circle on the surface of the Earth that joins points of equal ➤latitude. The equator is the line of zero latitude.

lattice The characteristic regular arrangement of particles in a crystalline solid. The lattice can be produced by repeated ➤translations (2) of the ➤unit cell. ➤➤Bravais lattice.

lattice energy A colloquial term for the more precise ➤lattice enthalpy.

lattice enthalpy The enthalpy change required to turn a crystal into a gas of ions; it is the ➤standard enthalpy change for processes such as

$$NaCl(s) \rightarrow Na^+(g) + Cl^-(g),$$

for which the lattice enthalpy is $+787 \text{ kJ mol}^{-1}$. When gas-phase ions collapse into a lattice, the enthalpy change, called the **lattice formation enthalpy**, has the same magnitude but the opposite sign. The lattice enthalpy depends on two main factors: the larger the charge on the ions, the larger the lattice enthalpy; and the smaller the separation between the ions, the larger the lattice enthalpy. ➤➤Madelung constant.

lattice plane A plane in a ➤Bravais lattice on which lattice points lie. Since the lattice is of infinite extent, there is an infinite number of lattice points in the plane. Each lattice plane corresponds to a ➤reciprocal lattice vector. ➤➤Miller indices.

lattice vector Any member of the set of vectors that form a ➤Bravais lattice. It is thus any integer combination of ➤primitive vectors.

lattice vibration A collective oscillation of the ions of a crystal lattice. A quantum-mechanical treatment of these ➤normal modes of the crystal views them as particles called ➤phonons.

laughing gas ➤dinitrogen oxide.

launch window ➤window.

Laurasia One of the two major continental land masses postulated to exist in Mesozoic times (the other being ➤Gondwana), consisting of the present-day North America, Europe and Asia. ➤➤continental drift.

lauroyl peroxide A compound which easily decomposes into two $CH_3(CH_2)_{10}CO^\bullet$ radicals, thus being able to initiate ➤radical polymerization reactions.

lava ➤Magma that originates from a volcano.

law See under most appropriate name: for example, for *law of mass action*, ➤mass action, law of.

lawrencium Symbol Lr. The element with atomic number 103 and most stable isotope 262, named after Ernest Orlando Lawrence (1901–58), the inventor of the ➤cyclotron. It is the last of the fourteen elements after actinium that constitute the ➤actinides. Its chemistry is restricted, because of its radioactivity.

lazurite ➤lapis lazuli.

lb Symbol for the unit ➤pound.

LCAO Abbr. for ➤linear combination of atomic orbitals.

LC circuit A ➤resonant circuit containing inductors and capacitors. Such arrangements are often used in electrical ➤filters. If resistors are included as well, the circuit is called an **LCR circuit**.

LCD Abbr. for ➤liquid crystal display.

LCM Abbr. for ➤least common multiple.

LD$_{50}$ ➤median lethal dose.

LDPE Abbr. for low-density ➤polyethylene.

leaching The process by which soluble components are removed from insoluble ones by the action of a solvent. An important example is the leaching of metal ions into groundwaters. For example, ➤acid rain can leach aluminium ions from soil; the combination of acidity and high aluminium level is very damaging to fish as aluminium compounds subsequently precipitate in their gills and interfere with gas exchange, hindering their breathing.

lead Symbol Pb (from the Latin *plumbum*). The chemical element with atomic number 82 and relative atomic mass 207.2, which is in Group 14 of the periodic table. As such many of its compounds have oxidation number +4, such as lead(IV) oxide, PbO$_2$. Because of the ➤inert pair effect, it has another oxidation number, +2, examples of which include lead(II) oxide, PbO, and lead(II) chromate, PbCrO$_4$, the yellow pigment (chrome yellow) used to mark roads. The element itself has been known for thousands of years as it is a soft unreactive metal which is easily reduced from its oxide ore. Its early uses were in making water courses (the word 'plumber' reflects this), roofs of churches and drinking vessels. However, lead is toxic and concern about this has led to the use of unleaded petrol.

There are three oxides of lead, the polymorphic yellow/red solid lead(II) oxide PbO, the black solid lead(IV) oxide PbO$_2$, and the famous red solid ➤red lead, Pb$_3$O$_4$, which is a mixed oxide containing both oxidation states of lead.

lead accumulator A ➤lead–acid battery.

lead–acid battery A battery in which lead plates are used for the electrodes. The cathode is plated with lead(IV) oxide, which acts as the oxidant. The electrolyte is sulfuric acid. The lead is oxidized to lead(II) sulfate and the lead(IV) oxide is reduced to the same compound. The chemical reactions can be reversed if an external electricity supply is used, and so this is a rechargeable battery and is used in cars. ➤cell (2).

lead-chamber process An obsolete method of making sulfuric acid, which has now been replaced by the ➤contact process.

leaded petrol Petrol that contains the additive ➤tetraethyllead.

leading diagonal The elements $a_{11}, ..., a_{nn}$ in an $n \times n$ ➤square matrix.

lead tetraethyl ➤tetraethyllead.

leaf A major organ of plants, providing the site for ➤photosynthesis. Leaves are very diverse in their shape and form and in their arrangement on the plant stem. The leaf comprises the ➤lamina (2), an expanded surface providing an enhanced area for light capture; the midrib and veins, which contain xylem and phloem of the ➤vascular system and assist in the support of the lamina; and a leaf base joining the stem, which may be extended into a petiole or leaf stalk. The increased surface area of leaves also allows for ➤transpiration. All leaves of the ➤Magnoliopsida have a bud (axillary bud) in the angle (axil) between the base and the stem, and are derived from meristematic (➤meristem) leaf buttress cells on either side of the stem apex in the apical bud. The 'leaves' of mosses are not ➤homologous with the leaves of higher plants. Leaves may be modified into a variety of specialized structures, such as the traps of insectivorous plants, or reduced to spines, as in cacti.

least action, principle of ➤Hamilton's principle.

least common multiple (lowest common multiple) The smallest whole number of which two given integers are factors.

least-squares method A procedure for fitting lines or curves of a specific type to a set of data points by finding the line or curve that minimizes the sum of the squares of the data points from it. The least-squares line through $(x_1, y_1),\ldots, (x_n, y_n)$ has the equation $y - \bar{y} = s_{xy}(x - \bar{x})/s_x^2$, where $\bar{x}$ and $\bar{y}$ are the means, s_{xy} is the ➤covariance and s_x^2 is the variance. The technique is strictly appropriate only when the random errors of measurement obey a ➤normal distribution.

least time ➤Fermat's principle of least time.

leaving group ➤nucleophilic substitution.

Lebesgue integral A general conception of ➤integration due to Henri Léon Lebesgue (1875–1941) in which a 'length' (**Lebesgue measure**) is assigned to more general ➤sets than ➤intervals (2) and then such measurable sets are used in place of the defining rectangles of the ➤Riemann integral. ➤Sequences and ➤series of Lebesgue-integrable functions have good convergence properties and the space L^2 of functions whose squares are Lebesgue-integrable is the ➤Hilbert space at the heart of the mathematical foundations of ➤quantum theory.

Leblanc process An obsolete method for making sodium carbonate. It has been replaced by the ➤Solvay process, which is significantly cheaper. Named after Nicolas Leblanc (1742–1806).

Le Chatelier's principle If a chemical equilibrium is subjected to a constraint, such as a change in the temperature, the equilibrium will tend to shift to minimize the effect of the constraint. For example, it will move in the endothermic direction when the temperature is increased, as energy is absorbed and the rise in temperature diminished. Named after Henri Louis Le Chatelier (1850–1936). ➤➤van't Hoff isochore.

lecithin (phosphatidylcholine) A ➤phospholipid found extensively in the ➤membranes of animal and plant cells. It also occurs as a surface-active agent in the lungs, helping to maintain open airways, and as a component of ➤bile, rendering ➤cholesterol more soluble.

Leclanché cell A cell in which the anode is amalgamated zinc and the cathode carbon. The electrolyte is ammonium chloride, with manganese(IV) oxide as depolarizer. The e.m.f. is about 1.5 V and a variation of this cell is the common ➤dry cell. Named after Georges Leclanché (1839–82).

LED Abbr. for ➤light-emitting diode.

LEED Abbr. for ➤low-energy electron diffraction.

Legendre polynomials The sequence of polynomials given by

$$P_n(x) = \frac{1}{2^n n!} \frac{d^n}{dx^n} (x^2 - 1)^n$$

that occur as solutions of **Legendre's equation**

$$(1 - x^2)\frac{d^2 y}{dx^2} - 2x\frac{dy}{dx} + n(n + 1)y = 0,$$

which arises when solving ➤Laplace's equation in situations that have rotational symmetry. Named after Adrien-Marie Legendre (1752–1833).

leguminous plant (legume) A plant of the pea family producing a characteristic fruit (pod) called a legume. Some members have symbiotic nitrogen-fixing ➤ bacteria associated with their roots and are important in maintaining soil fertility in crop-rotation schemes. Their seeds are rich in protein and some produce commercially important oils. Examples are peas, beans, clover, alfalfa and soybean.

lemma A subsidiary result proved as a stepping-stone in the proof of a ➤theorem.

lemniscate (lemniscate of Bernoulli) ➤Cassinian curve.

L enantiomer For an α-amino acid, the L enantiomer has the —COOH, —R and —NH$_2$ groups arranged anticlockwise, when viewed along the C—H bond. A superior terminology is explained under ➤ (R–S) system. One of the most fascinating features of nature is that the optically active amino acids in the human body are almost exclusively the L isomers. It has been suggested that the 'handedness' of the ➤weak interaction might account for this. Compare ➤D enantiomer.

length contraction The apparent reduction in length of fast-moving objects as predicted by ➤special relativity. An object of length L in its ➤rest frame has an apparent length of $L\sqrt{1 - v^2/c^2}$ in a frame moving at speed v with respect to its rest frame. ➤Lorentz–Fitzgerald contraction.

lens A device that uses ➤refraction to change the direction of ➤rays (usually of light) that pass through it. Lenses cause parallel beams to converge or diverge (see the diagram). An optical lens is constructed from a medium with a higher ➤refractive index than its surroundings; glass is most often used. The surfaces of a lens are usually either flat or sections of a sphere. ➤focal length; focus.

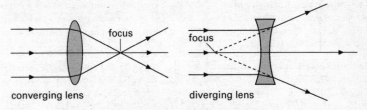

converging lens   diverging lens

lens (a) Converging and (b) diverging.

lenticel A raised pore on the surface of plant organs such as woody stems which allows ➤gas exchange to take place between the atmosphere and the internal tissues of the stem. The pore is filled with cork cells produced by the cork cambium and generally forms over the site of a ➤stoma. The minute brown flecks on the stems of elder and skins of apples are lenticels.

Lenz's law The principle that, when the ➤magnetic flux through a circuit changes, any current produced by electromagnetic ➤induction acts so as to oppose the change. This is a direct consequence of the minus sign in the third of ➤Maxwell's equations. Named after Heinrich Friedrich Emil Lenz (1804–65).

lepton A ➤fermion that is not affected by the strong nuclear force (➤strong interaction). The electron is a lepton, as are the ➤muon, the ➤tauon and the various ➤neutrinos. ➤➤elementary particle.

leptotene ➤meiosis.

LET Abbr. for ➤linear energy transfer.

Leu Abbr. for ➤leucine.

leucine (Leu) A common ➤amino acid. ➤➤Appendix table 7.

leucocyte (US **leukocyte)** A white blood cell. Leucocytes are nucleated, and lack ➤haemoglobin. They form part of the ➤immune system. Special cells in the bone marrow (**multipotent stem cells**) give rise to two functional groups of white cell: those that confer natural immunity, such as ➤macrophages, and those that give rise to ➤lymphocytes, which provide adaptive immunity. Compare ➤erythrocyte.

leucoplast One of a variety of ➤organelles in plant cells (typically not exposed to light) that lack pigment. Leucoplasts have a variety of metabolic and storage functions, including storage of oils, protein and carbohydrate.

leukaemia ➤Cancer of the blood cells. It is generally believed that tumours occur because the blood cells of a specific lineage fail to mature completely. Since blood cell precursors are capable of infinite division, a tumour results.

lever A mechanical device that allows a small force to overcome a larger one by increasing the distance over which the small force must work (see the diagram). The point about which the lever pivots is the **fulcrum**. The lever is the simplest possible ➤machine.

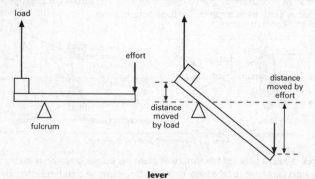

lever

Lewis acid An electron pair acceptor. The simplest example is the hydrogen ion H^+. Other less obvious examples are $AlCl_3$ and BCl_3, which are used in the ➤Friedel–Crafts reaction. Named after Gilbert Newton Lewis (1875–1946).

Lewis base An electron pair donor. Examples include the hydroxide ion OH^-, the halide ions and uncharged molecules with lone pairs, such as ammonia, NH_3. All form complexes with metals, such as $[Al(OH)_4]^-$, AlF_6^{3-} and $[Ni(NH_3)_6]^{2+}$. There is little difference between the concepts of Lewis base and ➤nucleophile.

Lewis structure A diagram showing the arrangement of electrons in a molecule. It usually only shows the outermost shell for each atom, as in the diagram. The electrons are not identified by dots or crosses as in the alternative **dot-and-cross diagrams**, as the latter give the false impression that it is possible to distinguish the origin of one electron from the other when paired.

$$\overset{\textstyle H}{\underset{}{\vdots O \vdots H}}$$

Lewis structure of water.

Lewis theory (of acids and bases) Gilbert Newton Lewis (1875–1946) introduced a very important definition of acids and bases in 1923: a Lewis acid is an electron pair acceptor and a Lewis base is an electron pair donor. It provided a different focus from the alternative definition (produced in the same year) in the ➤Brønsted–Lowry theory. The classification subsumes common acid–base reactions as the hydrogen ion H^+ is an electron pair acceptor. However, it also includes ➤complexes formed by metal ions with ligands. It is helpful to view the reactions of substances under the twin headings of ➤redox and Lewis acid–base. So, for example, ammonia behaves as a reductant and as a Lewis base. Being a Lewis base, it will react with H^+ to form the ammonium ion NH_4^+ and form complexes with metal ions, as in $[Zn(NH_3)_4]^{2+}$.

LF Abbr. for ➤low frequency.

lg Abbr. for ➤logarithm to the base 10, $\log_{10}$.

LH Abbr. for ➤luteinizing hormone.

Li Symbol for the element ➤lithium.

lichen One of a group of specialized organisms comprising a mutualistic ➤symbiosis between a ➤fungus and an ➤alga or ➤cyanobacterium. Lichens are highly successful organisms. They have colonized environments that are hostile to other forms of life, such as deserts and high mountain-tops. They are frequently found growing on old stone walls and gravestones, but are highly sensitive to pollution, the presence of certain lichens being used as ➤indicator species for sulfur dioxide levels in the atmosphere.

Liebig condenser (condenser) An apparatus in which cold water passes against gravity around the outside of a central pipe (see the diagram), causing heat to be exchanged

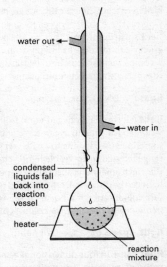

Liebig condenser When the Liebig condenser is arranged vertically, it can be used to ➤reflux two reactants together.

with the hot vapour inside, and thereby producing condensation of the vapour. Named after Justus von Liebig (1803–73).

life cycle A series of events in the existence of an organism from its formation as a result of ►fertilization or ►cell division to an equivalent stage in the next generation. Life cycles are diverse, but conform to basic patterns within each class of organism.

lifetime Symbol τ. One way of expressing the parameter of ►exponential decay. The lifetime is the reciprocal of the ►decay constant.

lift A force, perpendicular to the direction of motion, acting on an object moving through a fluid. It results from a difference in the net force from pressure on opposite sides of the object, which must either be asymmetric itself or orientated with its symmetry plane at an angle to the direction of motion. ►Aerofoils are designed to produce lift. Compare ►drag.

ligament Elastic connective tissue, chiefly derived from the protein ►collagen, that binds bones together at vertebrate joints.

ligand The electron pair donor in a ►complex. There is no necessity for the ligands to be identical; for example, the copper–ammine complex $[Cu(NH_3)_4(H_2O)_2]^{2+}$ has both ammonia and water as ligands.

ligand field theory A theory explaining the important characteristics of ►complexes, such as colour and paramagnetism, in terms of the splitting of energy levels caused by the ►ligands. In an octahedral complex, such as $[Ti(H_2O)_6]^{3+}$, the five d orbitals on the central metal ion, which were originally ►degenerate states, split in energy when the ligands are present, as two of the orbitals point at the ligands, whereas three point between the ligands (see the diagram at ►d orbitals). This **ligand field splitting** means that an electron excited from its ground state in the lower energy orbital absorbs energy from incident light (►d–d transition), causing the ►complementary colour to be transmitted. The titanium complex above absorbs in the green, and hence looks purple.

ligase ►DNA ligase; enzyme.

light Electromagnetic radiation from the region of the ►electromagnetic spectrum visible to the human eye (a wavelength of about 380 to 780 nm). The term 'light' is sometimes used to refer to a wider range of electromagnetic radiation, with the visible range referred to as 'visible light'. The speed of propagation is constant throughout the electromagnetic spectrum (►speed of light). Being electromagnetic radiation, light can be diffracted, reflected and refracted. It consists of a stream of photons, each carrying an energy E equal to hf, where h is the ►Planck constant and f is the frequency of the light. ►►colour; colour vision.

light, speed of ►speed of light.

light cone In four-dimensional ►spacetime, the surface that represents all possible paths of photons originating from a particular event. There can be no causal connection between the originating event and an event outside the light cone. ►►causality.

light-emitting diode (LED) A ➤p–n junction in which recombination of ➤electron–hole pairs causes light to be emitted. The intensity of the light depends on the number of excess carriers, which in turn depends on the ➤bias current.

lightning An electric ➤breakdown of the atmosphere that occurs between a cumulonimbus cloud and the Earth, or another cloud, or even a different part of the same cloud. The breakdown causes a visible flash together with a sound wave (**thunder**).

lightning conductor An earthed, pointed, metal structure on the top of a tall building or other elevated structure that is designed to protect against ➤lightning. An electric field in the atmosphere causes the conductor to accumulate induced charge at the point, which intensifies the field and causes the ➤breakdown to occur there rather than on other parts of the structure.

light reaction (light-dependent reaction) One of the two series of reactions (the other is the ➤dark reaction) that occur in ➤chloroplasts during photosynthesis. Energy from sunlight energizes an electron in chlorophyll, enabling it to travel along an oxidation chain. The electron transfer results in the pumping of protons across the internal thylakoid membranes of the chloroplast. The light reaction generates ➤ATP and the reduced ➤coenzyme NADPH, which pass to the dark reaction to enable the incorporation of carbon dioxide into ➤carbohydrate.

light year Symbol l.y. A convenient unit of distance (not time) in astronomy. One light year is the distance that light travels (in a vacuum) in one year; 1 l.y. = 9.46×10^{15} m. ➤➤parsec.

lignin A polymeric compound of carbohydrate derivatives forming the wood of trees. Lignin is deposited in the walls of a variety of plant cells, such as ➤xylem vessels and **sclerenchyma** – elongated tapered cells which provide rigidity, support and protection, particularly in stems and leaves.

lime Calcium oxide, CaO, is called quicklime, or sometimes simply lime. **Slaked lime** is calcium hydroxide, $Ca(OH)_2$.

limelight The bright light given off when quicklime, CaO, is strongly heated. This was the original source of illumination in theatres, hence the phrase 'being in the limelight'.

limestone A common mineral in sedimentary rocks. It is mainly composed of calcium carbonate, $CaCO_3$; indeed 'limestone' is often used to mean that substance. Limestone is one of the ingredients required in the ➤blast furnace for making iron.

limewater A clear solution of calcium hydroxide, $Ca(OH)_2(aq)$. It is alkaline and can therefore be used to test for the acidic gas carbon dioxide, which reacts to form an insoluble precipitate of calcium carbonate, causing the limewater to appear milky or cloudy:

$$Ca(OH)_2(aq) + CO_2(g) \rightarrow CaCO_3(s) + H_2O(l).$$

limit A number to which a given sequence converges (➤convergence). Formally, the sequence (a_n) converges to the limit l, written $\lim_{n\to\infty} a_n = l$ or $a_n \to l$ as $n \to \infty$, if for every $\varepsilon > 0$ there can be found a number N for which $|a_n - l| < \varepsilon$ for every $n \geqslant N$. This is the basic notion behind every limiting operation in mathematics. Thus an infinite series converges (to its sum) if the sequence formed by the partial sums of its first n terms has a limit; and a function $f(x)$ has the limit l as x tends to a if $f(x_n) \to l$ for all sequences with $x_n \to a$.

limit of integration One of the endpoints of an integral over which a definite integral (➤integration) is calculated.

limit of proportionality The maximum extension of a spring or material beyond which ➤Hooke's law does not apply. ➤➤elastic limit.

limnology The study of the biology, physics and chemistry of lakes and rivers.

linac Acronym for ➤linear accelerator.

Lindlar catalyst A catalyst, consisting of platinum supported on barium sulfate, used for hydrogenation of alkynes to *cis*-alkenes. Named after Herbert H. M. Lindlar (b. 1909).

line 'Line' usually implies a straight line, although sometimes it is used to denote a curve (as in ➤'line integral'). In two-dimensional Cartesian coordinates, the equation of a **straight line** is $y = mx + c$, where m is its gradient and c is the intercept on the y axis.

linear Literally 'like a straight line', 'linear' is also used to describe a dependence of one variable on another where a plot of one against the other would appear as a straight line. ➤linear relationship.

linear absorption coefficient ➤Lambert's laws.

linear accelerator (linac) A particle ➤accelerator that uses ➤radio-frequency electric fields to accelerate charged particles (typically electrons, protons or ions) in a straight line. The electrodes are arranged in such a way that the charged particles move with the peak electric field as it moves in a travelling wave down the linac. Magnetic fields are used to focus the particle beam.

linear combination An expression of the form $c_1 v_1 + \cdots + c_n v_n$, where the c_i are ➤scalars and v_i are ➤vectors (2). ➤➤superposition.

linear combination of atomic orbitals (LCAO) An approximate method for calculating a molecular orbital, using a superposition of orbitals on individual atoms as a starting point. It can be improved by allowing for interaction between the original atomic orbitals.

linear differential equation A ➤differential equation of the form $a_n y^{(n)} + a_{n-1} y^{(n-1)} + \ldots + a_1 y^{(1)} + a_0 y = f$, where the $y^{(i)}$ denote successive ➤derivatives of y and the a_i are functions of x alone. The general solution of the ➤homogeneous equation in which $f(x) = 0$ is called the **complementary function**; the general solution of the original equation is then the sum of the complementary function and any ➤ particular integral.

linear energy transfer (LET) The average energy transferred to a medium per unit distance travelled through the medium by a charged particle.

linear equations A set of simultaneous equations that may be written in the form $Ax = b$ for a (known) column vector b and a matrix A and an unknown column vector x. For three unknowns, each equation is of the form $ax + by + cz = d$, so solving the equations corresponds geometrically to finding the points common to the planes represented by them.

linearly independent set A set of vectors with the property that the only way a linear combination of vectors from the set can be the zero vector is for all the scalars involved to be zero. Otherwise the set is **linearly dependent**, with the result that at least one vector in the set can be expressed as a linear combination of the others.

linear motor An ➤induction motor with a linear (as opposed to the more conventional cylindrical) ➤stator and ➤ rotor.

linear programming The branch of mathematics concerned with methods of finding the maximum and minimum values of linear expressions of several variables, subject to constraints on the variables expressed as inequalities. Such problems occur frequently in connection with the optimal allocation of resources in operational research and economics.

linear relationship A relationship between two variables whose graph is a straight line. The special case in which the line goes through the origin occurs when one variable is directly ➤proportional to the other.

linear transformation A ➤transformation (2) that maps each position vector x to a vector Ax for some fixed matrix A.

line defect ➤defect.

line integral An integral evaluated over a curve in the plane or space, typically of the form $\int_A^B v \cdot dr = \int_A^B v_1 dx + v_2 dy + v_3 dz$ for a ➤vector field $v = v_1 i + v_2 j + v_3 k$ and a curve with endpoints A and B.

line-of-sight velocity ➤radial velocity.

lineshape The characteristic shape of a peak in a ➤line spectrum. Various mechanisms, for example ➤Doppler broadening, cause lines to have a nonzero width; some energy loss mechanisms can cause lines to have an asymmetric shape.

lines of flux A representation of a ➤field as lines in the direction of the field. The number of lines per unit area (width for a two-dimensional representation) is proportional to the strength of the field. This representation can be used only for fields with zero ➤divergence.

lines of force Lines of flux of an electric field, representing the direction of the force between charges. It is strictly inappropriate to use the term for a magnetic field, because the ➤Lorentz force acts perpendicular to the magnetic flux density.

line spectrum A spectrum that appears as a series of sharp peaks (**spectral lines**) on a low-intensity background. Atomic emission spectra, for example, are line spectra, as the lines result from transitions between well-defined energy levels. The lines can be used to identify elements, as in an emission spectrophotometer. The fact that lines are seen provided early experimental support for ➤quantum mechanics.

linewidth The width (either in frequency or energy) of a peak in a ➤line spectrum.

linkage map ➤genetic map.

Linnaean system A system of biological nomenclature devised by the Swedish naturalist Carl von Linné (1707–78), whose name in Latin form was Carolus Linnaeus. The basic principle of the Linnaean system is still in use today as the **binomial system**, in which individual organisms are defined by a scientific, Latinized name consisting of the name of the ➤genus and the name of the ➤species, as in, for example, *Homo sapiens*. ➤➤taxonomy.

linoleic acid A polyunsaturated ➤fatty acid with two double bonds, *cis,cis*-octadeca-9,12-dienoic acid, common in the human diet and essential for the biosynthesis of the prostaglandins.

linolenic acid A polyunsaturated ➤fatty acid with three double bonds, *cis,cis,cis*-octadeca-9,12,15-trienoic acid.

linotype An alloy of about 80% lead with antimony used to set type in the hot metal printing process.

linseed oil The oil obtained from the dried seeds of flax. It consists mainly of glycerides of ➤linolenic acid and is used routinely in making lacquers and varnishes.

Liouville's theorem A theorem of classical mechanics that has very important consequences in ➤statistical mechanics: for any set of points in ➤phase space bounded by a closed surface, the ➤trajectories of the points are such that the volume of phase space occupied by the set is ➤invariant over time. Thus elements of phase space behave like an ➤incompressible fluid. Named after Joseph Liouville (1809–82).

lipase An ➤enzyme that hydrolyses fats and oils (➤lipids). In the human ➤gut, lipase enzymes are found in the pancreatic juice and the secretions from the wall of the small intestine.

lipid A generic term for a water-insoluble biomolecule that can be extracted using nonpolar solvents. Examples include fats and oils (which are ➤triacylglycerols containing fatty acids), ➤waxes, ➤steroids, ➤glycolipids, fat-soluble ➤vitamins, ➤carotenes and ➤chlorophylls. Fats are important in animals as an energy source and reserve, and also under the skin for insulation, as in the blubber of whales. Lipids are one of the four most important classes of molecules in the human body.

lipoprotein A conjugated molecule consisting of a ➤protein complexed with a ➤lipid. They exist in a wide range of form and density. Some have a transport function within cells and others are implicated in the ➤metabolism of ➤cholesterol, where the relationship between different forms known as low density (**LDL**), high density (**HDL**), and very high density (**VHDL**) is a determinant of heart disease. Regular exercise raises HDL and counteracts the effects of high levels of cholesterol.

liposome An artificially constructed microscopic spherical ➤vesicle made in the laboratory by adding an aqueous solution to a ➤phospholipid gel. The wall of the vesicle is selectively permeable and behaves in some respects like a living cell membrane. Liposomes are used to deliver drugs that have high toxicity to specific targets in tissues and cells.

liquefied petroleum gas ➤LPG.

liquid One of the three principal states of matter, intermediate both in structure and molecular motion between a solid and a gas. When a solid is heated, at first the particles simply vibrate more energetically at their lattice positions. Changes occur at the melting point of the solid as the particles move away from their lattice positions; at the boiling point of the liquid, more extreme changes occur as the particles become essentially independent of each other. A liquid takes the shape of its containing vessel and a ➤meniscus forms (➤➤surface tension). The liquid can be poured out of the vessel although the ease of pouring varies with the ➤viscosity of the liquid. All liquids can be solidified by cooling at 1 atm pressure, except for liquid helium, which has some unusual properties including ➤superfluid behaviour.

liquid chromatography ➤chromatography.

liquid crystal A state of matter characterized by a fluidity similar to that of liquids, but with some degree of order, albeit less than in a crystal. The compounds that form liquid crystals have long rod-like molecules or form flat platelets. Three different types exist: **smectic, nematic** and **cholesteric** (chiral nematic), which have different degrees of order (see the diagram). Cholesteric liquid crystals find vital applications in early detection of breast tumours, which show up as hot spots and hence different colours of the liquid crystal, and in measuring the temperature of children too young for mercury thermometers to be used.

(a)   (b)

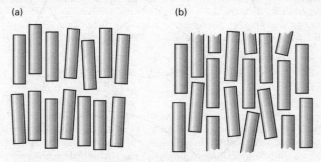

liquid crystal (a) Smectic and (b) nematic.

liquid crystal display (LCD) A display based on ➤liquid crystals that is commonly used in watches, cameras, calculators, portable computers and video recorders together with flatpanel TVs. The application of an electric field can change the alignment of the liquid crystal, which in turn changes the polarization state of light that the crystal absorbs, or its reflectivity. Thus application of a voltage can cause a thin film of liquid crystal to change from clear to opaque.

liquid drop model A model of the atomic nucleus as a droplet of liquid, giving an empirical expression for its ➤binding energy (2) in terms of its mass number A, neutron number N and atomic number Z. The binding energy is the sum of a number of terms including a surface area term (proportional to $A^{2/3}$) and a Coulomb repulsion (proportional to $Z^2 A^{-1/3}$).

liquidus In a solid–liquid ➤phase diagram, a line indicating the commencement of solidification or the completion of melting of a particular phase. Compare ➤solidus.

LISP Acronym for list processing. A computer language much used in artificial intelligence research, since it operates mainly on symbols rather than mathematical expressions. The data structures used are lists, which can consist of other lists and symbols, and hence there is a strong element of recursion.

Lissajous figures Parametric plots of the form

$$x = \sin m\omega t, \quad y = A \sin(n\omega t + \phi),$$

where m and n are integers. The figures may take up many different shapes. For $n = m$ the figures are ellipses of eccentricity determined by A and ϕ, reaching the limit of a straight line for $\phi = 0$ or π. Patterns like those shown in the diagram can be observed on an oscilloscope by connecting variable frequency supplies to the X and Y inputs. Named after Jules Antoine Lissajous (1822–80).

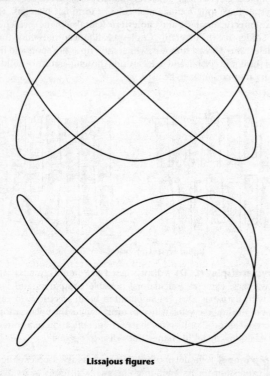

Lissajous figures

Listeria A genus of rod-shaped, motile ➤bacteria. It includes *Listeria monocytogenes*, whose name is derived from the appearance of large numbers of monocytes in the blood of infected organisms. One species causes the disease **listeriosis**, the clinical symptoms of which include a type of meningitis. Infections in the uterus can cause birth defects. The bacterium is highly resistant to both physical and chemical treatment. Potential sources of listeriosis include contaminated faeces and some foodstuffs such as soft cheese. Named after Joseph Lister (1827–1912).

lithal A shorthand for ➤lithium tetrahydridoaluminate.

litharge A yellow ore of lead: lead(II) oxide, PbO.

lithium Symbol Li. The element with atomic number 3 and relative atomic mass 6.941, which is in Group 1, the alkali metals. The element itself is a shiny metal. Its standard electrode potential is the most negative of all metals, implying that thermodynamically it is the strongest reductant. However, its kinetic reactivity is lower than that of the other alkali metals and the metal reacts only slowly with cold water, reducing it to hydrogen gas. This anomaly can be traced to the small size of the lithium cation, which is also responsible for other differences between lithium and the other alkali metals; for example, lithium carbonate is more easily decomposed, because of the high lattice energy of its oxide, Li_2O. Lithium carbonate finds application in the **lithium therapy** for manic depression, which seems to be associated with the regulation of the sodium/potassium ion balance.

lithium alkyls Strong reductants formed when alkyl groups are bonded to lithium atoms, such as *tert*-butyllithium, $(CH_3)_3CLi$.

lithium tetrahydridoaluminate (lithium aluminium hydride, lithal) $LiAlH_4$ A powerful reductant encountered very widely in organic chemistry, used in ether solution followed by addition of water. For example, lithal can reduce aldehydes and ketones to the corresponding alcohols. It can also reduce nitriles RCN to primary amines RCH_2NH_2, and reduce esters RCOOR′ to two alcohols, RCH_2OH and R′OH.

lithography A technique in which a pattern is transferred to the surface of a ➤substrate (2) (typically a semiconductor for use as an electronic component) by exposure to radiation in the form of light (**photolithography**), X-rays or an electron beam.

lithosphere ➤Earth.

litmus A very common acid–base ➤indicator, made from lichens. Its pK_{in} is very close to pH 7. Acids turn moist litmus paper red whereas alkalis turn it blue.

litre Symbol l or L. A metric unit of volume equal to the volume occupied by 1 kg of water at its maximum density. Standards for the kilogram and metre were established before high-precision measurements were possible, and the consequent discrepancy between the intended and actual size of the kilogram makes a litre equal to 1.000 028 dm^3.

live A conductor or circuit that is not at ➤earth potential. ➤➤neutral (2).

liver A large, lobed organ in animals with a wide range of metabolic, secretory and storage functions. Livers exist in a wide range of animal groups but they are not all ➤homologous. In vertebrates the liver arises as an outgrowth of the gut. In humans it is the largest internal organ, lying in the abdominal cavity just below the diaphragm. The functions of the liver include the production of ➤bile, the storage of ➤glycogen, the ►metabolism of amino acids, the storage of fat-soluble vitamins such as A and D, the detoxification of poisons such as alcohol, and the production of urea. The high level of biochemical activity in the liver produces heat, and the liver is important in temperature regulation in warm-blooded animals.

lm Symbol for ►lumen.

ln Abbr. for ►logarithm to the base e, $\log_e$.

load 1 The part of an electric circuit in which power is dissipated by design, usually as the output of a power supply or electronic device. For example, a loudspeaker is a load when it is driven by an amplifier.
   2 ►machine.

lobe The range of angles between the ►nulls or minima of a radiation pattern.

local group A cluster of approximately 30 galaxies, including the **Milky Way**. A large proportion of the mass of the local group is contained in just two of the galaxies, the Milky Way and the Andromeda Galaxy (M31).

localized Describing an electron whose wavefunction is concentrated on just one or two atomic sites. Compare delocalized (►delocalization).

local oscillator The internal oscillator of a ►heterodyne or ►superheterodyne radio receiver. The output of the local oscillator is combined with the received signal to produce a ►beat frequency signal.

locant A letter (e.g. p-) or number (e.g. 2-) indicating the position of a substituent in a chain. Examples include ►PABA and propan-2-ol.

lock-and-key model ►enzyme.

locomotion Movement of an organism from one location to another. It may be effected, for example, by ►undulipodia; more complex mechanisms of locomotion involve ►muscle contractions and specialized structures such as limbs.

locus 1 (Math.) A curve, surface or region that arises as the path of a point moving subject to certain constraints.
   2 (Biol.) The position of a particular gene on a chromosome.

lodestone The ancient name for the oxide of iron of formula Fe_3O_4. As the slightly more modern name magnetite implies, it is magnetic; lodestone was the material from which ships' compasses were originally made.

log Abbr. for ►logarithm. The base should normally be specified.

logarithm For a fixed number a (the ►base (3)) the logarithm of x to the base a, written as $\log_a x$, is that exponent y for which $x = a^y$. Every number is thus expressed as an exponent of the fixed number a so, from the rules of ►indices, the rules of logarithms follow:

$$\log_a(xy) = \log_a x + \log_a y,$$
$$\log_a(x/y) = \log_a x - \log_a y,$$
$$\log_a x^n = n \log_a x.$$

Two important bases are e for the ➤natural logarithm and 10 for the common logarithm. Historically, the development of logarithms in the 17th century greatly facilitated calculations, for they reduced the labour of multiplying and dividing numbers to that of adding or subtracting their logarithms. This role has been taken over by the scientific calculator, but logarithms are still important in calculus and in the analysis of data. For example, a relationship of the form $y = ab^x$ yields a straight line graph if $\log y$ is plotted against x. Logarithms occur in some scientific definitions, for example that of ➤pH. ➤➤Arrhenius equation; Nernst equation.

logarithmic series The ➤series $\sum_{n=1}^{\infty} (-1)^{n-1} x^n/n$, which converges to $\ln(1 + x)$ for values of x between -1 and 1 (➤convergence).

logic The theory behind the use of ➤ logic gates and other circuits to evaluate Boolean expressions and perform mathematical calculations on a digital computer (➤➤Boolean algebra). The functions of logic usually operate on binary values, since this is the easiest way to implement them as logic gates. The NAND logic function is false (0) if and only if all of the inputs are true. NAND gates are particularly significant since they can be used to implement any other logic function, and hence a computer can be built from them alone.

logic gate (gate, logical element, switching gate) A component that carries out a logic function such as NAND. Thus logic gates generally have at least two inputs and produce one output which depends on the states of the inputs. The logic states (i.e. true or false) of the inputs and output are specified in binary by using different voltages. Logic gates form the basis of digital computers. Currently they are implemented electronically, but optical gates have been built, giving hope for extremely fast computers that would use light beams instead of electrical currents. ➤➤truth table.

LOGO A computer programming language, derived from ➤LISP, used for teaching the fundamentals of programming to children.

log-off (logout) Ending a session at a computer terminal.

log-on (login) Starting a session at a computer terminal, typically by entering a password to identify the user to the host computer in order to allow access to system resources and privileged files.

London force ➤dispersion force. Named after Fritz Wolfgang London (1900–54).

lone pair (nonbonding pair) A pair of electrons not shared with another atom. Examples can be seen from the ➤Lewis structure for water (see diagram at Lewis structure). For water two lone pairs exist, whereas for ammonia only one exists. The possession of a lone pair makes a molecule both a ➤Lewis base, and hence a ➤ligand, and a ➤nucleophile. ➤➤valence-shell electron-pair repulsion theory.

long-day plant ➤photoperiodism.

longitude The ➤azimuth used to specify a point on the surface of the Earth. The longitude of a location is the angle between the plane of the ➤prime meridian and the plane of a ➤great circle through both the location and the Earth's axis. It is usually measured as an angle east or west of the prime meridian. Compare ➤latitude; ➤➤celestial sphere.

longitude, line of A line joining points on the surface of the Earth of equal ➤longitude; that is, half of a great circle with the axis of the Earth as a diameter. ➤➤meridian.

longitudinal Along the length or in the direction of travel of an object.

longitudinal wave A wave whose displacement is in the direction of motion of the wave. A sound wave, for example, is a longitudinal wave. Compare ➤transverse wave.

long sight ➤hypermetropia.

long terminal repeat (LTR) A double-stranded ➤DNA sequence, generally several hundred base pairs long (➤base pairing). LTRs are repeated at the ends of ➤retrovirus and ➤retrotransposon DNA.

Lorentz–Fitzgerald contraction An empirical expression for the observed contraction in length of an object moving through the ➤ether (2). The hypothesis behind this expression was superseded by ➤special relativity, which included a prediction of ➤length contraction equal to the Lorentz–Fitzgerald contraction. Named after Hendrik Antoon Lorentz (1853–1928) and George Francis Fitzgerald (1851–1901).

Lorentz force The force F that is exerted on a charged particle in an electric and/or magnetic field:

$$F = q(E + v \times B),$$

where q is the charge on the particle, E is the electric field strength, v is the particle's velocity and B is the magnetic flux density. If a magnetic field is absent, the Lorentz force reduces to the Coulomb force (➤Coulomb's law).

Lorentzian function A function $f(\omega) = a^2/(a^2 + \omega^2)$, where a is a width parameter. The Lorentzian (see the diagram opposite) occurs frequently in expressions for lineshapes in spectroscopy by virtue of being the ➤Fourier transform of an exponential decay.

Lorentz transformation The transformation that, according to ➤special relativity, transforms from the ➤inertial frame S to frame S′ moving at constant velocity v with respect to S. If v is taken to be in the positive x direction, then the transformation for an event (x, y, z, t) is given by

$$x' = \gamma(x - vt), \quad y' = y, \quad z' = z \quad t' = \gamma(t - vx/c^2),$$
$$\gamma = 1/\sqrt{1 - v^2/c^2}.$$

This differs fundamentally from the ➤ Galilean transformation which it replaces, in that it dispels the notion of absolute time: time becomes a coordinate which is manipulated in such a transformation just like distance. Direct consequences of the Lorentz transformation include ➤length contraction and ➤time dilation.

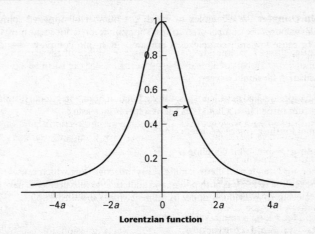

Lorentzian function

Loschmidt constant (Loschmidt number) The number of particles in 1 cm^3 of an ideal gas (➤ideal gas model) at ➤STP. It is a similar concept to the ➤Avogadro constant, which explains the origin of its symbol L. Named after Johann Joseph Loschmidt (1821–95).

loss factor (power factor) The ratio of the mean ➤power dissipated in a load to the ➤apparent power. The loss factor is given by $\cos\phi$, where ϕ is the ➤phase difference between the current through the load and the voltage across it.

loudspeaker A device that produces a sound wave from an electrical signal. Typically a small coil is located at the centre of a diaphragm in a magnetic field. An alternating current through the speaker causes the coil, and thus the diaphragm, to oscillate, producing a sound wave.

low-energy electron diffraction (LEED) A technique used to determine the structure of a surface at the atomic scale by ➤diffraction of a plane wave of electrons. The diffraction pattern can be related to the geometry of the surface. Qualitatively, it is useful for finding the symmetry axes of a surface, and quantitatively it can yield information on bond distances and a detailed geometric reconstruction of the surface. A typical beam energy of 100 eV gives a wavelength of the electrons of about 0.1 nm.

lowering of vapour pressure ➤colligative properties.

lowest common multiple Alternative name for ➤least common multiple.

low frequency (LF) The radio frequency range from 30 to 300 kHz. ➤frequency band.

low-level language Any computer language at a level corresponding to the computer's hardware. Low-level languages are thus processor-specific. Typically there will be one instruction for each ➤machine code equivalent, but often extra instructions are included to ease the task of the programmer.

low-spin complex A ➤complex in which the number of unpaired spins is a minimum value for the number of electrons. The cyanide ion is one of the most likely ligands to cause low-spin complexes. Compare ➤high-spin complex. ➤➤spectrochemical series.

LOX Acronym for liquid oxygen.

LPG Abbr. for liquefied petroleum gas, the gas fraction from ➤fractional distillation of petroleum turned into a liquid. Its use as a fuel is increasing.

Lr Symbol for the element ➤lawrencium.

LS **coupling** ➤spin–orbit coupling.

LSD Abbr. for lysergic acid diethylamide. A psychoactive drug which causes severe hallucinogenic activity. It is both potent and addictive; serious accidents have been known to occur to individuals under its influence (on a so-called 'trip').

LTR Abbr. for ➤long terminal repeat.

Lu Symbol for the element ➤lutetium.

Lucite Tradename for a polymer made from methyl methacrylate (➤polyalkene). It is extensively used where transparency coupled to chemical resistance is important, such as for aircraft windscreens.

lumen **1** (Phys.) Symbol lm. The SI derived unit of ➤luminous flux. It is the luminous flux emitted by a uniform source of one candela into unit solid angle. Thus the total luminous flux emitted from such a source is 4π lm.
  2 (Biol.) The internal cavity of a biological structure such as a blood vessel, the gut of an animal, or the xylem vessels of plants.

luminance Symbol L; unit cd m^{-2}. The ➤luminous intensity of light emitted from a source per unit surface area. The luminance of a source may therefore vary across its surface.

luminescence The emission of electromagnetic radiation from a material caused by transitions between energy levels in the material. Luminescence can arise from chemical reactions (**chemiluminescence**), life processes (➤bioluminescence), or the crushing of crystals (**triboluminescence**). Compare ➤incandescence. ➤➤fluorescence; phosphorescence.

luminosity A measure of the total intensity of light emitted by a star. ➤➤Hertzsprung–Russell diagram; magnitude.

luminous flux Symbol ϕ_v; unit ➤lumen, lm. The total power of light emitted or received by a body or surface. It is restricted to visible wavelengths (as indicated by the subscript) and adjusted to take account of the sensitivity of the human eye.

luminous intensity Symbol I_v; unit ➤candela, cd. The power of light emitted from a source (➤luminous flux) per unit solid angle.

LUMO Abbr. for lowest unoccupied molecular orbital (➤frontier orbitals).

lunar eclipse ➤eclipse.

lunar month ➤month.

lune **1** The crescent-shaped region inside one but outside the other of two overlapping circles.

2 One of two portions of the surface of a sphere that lies between two ➤great circles.

lung A sac-like organ of ➤gas exchange, particularly in air-breathing vertebrates, formed in the ➤embryo as a branch of the ➤pharynx. Lungs are not muscular, and are ventilated by muscles of the ➤diaphragm (2) and chest wall causing volume and pressure changes in the enclosed air. Air passes in and out of the lungs through the ➤trachea and **bronchi**. These branch repeatedly, forming **bronchioles**, and eventually end in small sacs called **alveoli**. Alveoli have very thin, moist linings and are richly supplied with blood vessels, facilitating gas exchange with the blood. In birds the lungs are reduced, and air sacs, of different origin and structure, are more prominent. The development of the blood system supplying the lungs is paralleled by changes in the structure of the heart, and reflects the evolution of the vertebrates to a terrestrial existence and the trend towards ➤homoiothermy. Lungs always occur in pairs, except in terrestrial ➤molluscs, which have a single lung. Spiders possess a series of ➤gill-like structures in paired cavities called 'lung-books'.

Lurgi process The complete gasification of coal using a stream of oxygen and superheated steam at high pressure. Named after the Lurgi Company in Germany who devised the process in the 1930s.

luteinizing hormone (LH, interstitial cell stimulating hormone, ICSH) A mammalian ➤hormone, produced by the anterior ➤pituitary gland, that stimulates the events in the ➤menstrual cycle leading to ➤ovulation in the female ovary and the production of ➤androgens in the male testis. Production of LH is in turn controlled by the hypothalamus (➤brain).

lutetium Symbol Lu. The element with atomic number 71 and relative atomic mass 175.0, which is a lanthanide. Its common oxidation number therefore is +3, as in the oxide Lu_2O_3, a white powder, or the chloride, $LuCl_3$, white crystals.

lux Symbol lx. The SI derived unit of ➤illumination, equal to one lumen per square metre.

Lw Obsolete symbol for the element ➤lawrencium. Lr is now accepted internationally.

lx Symbol for ➤lux.

l.y. Symbol for ➤light year.

Lyman series A series of lines in the ➤hydrogen atom emission spectrum, produced when the electron falls to the orbital with ➤principal quantum number 1. The ➤convergence limit of the Lyman series corresponds to the ionization energy of the hydrogen atom, enabling this to be measured spectroscopically. Named after Theodore Lyman (1874–1954). ➤➤Balmer series; Rydberg–Ritz equation.

lymph A clear fluid in the vessels of the ➤lymphatic system which is rich in proteins and ➤leucocytes. Lymph does not contain ➤erythrocytes or ➤platelets.

lymphatic system A complex of vessels and ducts carrying ➤lymph, forming a secondary circulatory system to the ➤blood in vertebrates. The system consists of a network of fine lymph capillaries draining into larger ducts, and it functions in the circulation and distribution of ➤leucocytes in the body. In some vertebrates (not birds or mammals) the system includes a pulsating muscular vessel referred to as a 'heart'; otherwise circulation of lymph occurs by squeezing of the vessels during body movement aided by non-return valves. Lymph is formed by tissue fluid (➤extracellular fluid) being squeezed under pressure into the lymph capillaries in the major organs. Lymph is returned to the blood via a duct into the subclavian vein near the right side of the heart. The lymphatic system also has a number of nodes or **lymph glands**, situated particularly in the armpit and groin regions; they contain lymphocytes which act as part of the body's ➤immune system. Cancer of the lymph nodes is called **Hodgkin's lymphoma**.

lymphocyte One of a number of types of ➤leucocyte found in the blood, ➤lymphatic system and ➤thymus gland. Lymphocytes are important component cells of the ➤immune system. ➤➤ B cell; T cell.

lymphokine A soluble chemical released from ➤lymphocytes after activation by ➤antigens. Those produced by ➤T cells are termed ➤interleukins.

lymphoma ➤cancer.

lyophilic Describing a ➤colloid in which the dispersed phase is solvated by the dispersion medium. Compare ➤lyophobic.

lyophilization (quick-freeze drying) A method used by microbiologists to keep certain microorganisms alive for long periods of time. A stable preparation of microbes is created which are first rapidly frozen and then subjected to a high vacuum to draw off their moisture.

lyophobic Describing a ➤colloid in which the dispersed phase is unsolvated by the dispersion medium, and which can therefore be coagulated easily. Compare ➤lyophilic.

Lys Abbr. for ➤lysine.

lysergic acid diethylamide ➤LSD.

lysine (Lys) A common ➤amino acid. ➤➤Appendix table 7.

-lysis A suffix from the Greek for 'breaking': by electrical means in **electrolysis**, for example, by thermal means in **thermolysis**, or by light in **photolysis**.

lysis The rupture of a cell by **osmotic shock** (the bursting of a cell as a result of an excessive ingress of water by ➤osmosis), or in response to infection by a virus or by antibiotic action on the cell wall. ➤➤bacteriophage.

lysogenic Describing a cell of a ➤bacterium that has been infected by a ➤bacteriophage and that is capable of producing new infectious bacteriophage particles without undergoing ➤lysis. Lysogenic bacteria are immune to infection by the same or similar viruses. Subsequent lysis of such cells may be induced by exposure to ultraviolet light or chemical agents. Bacteriophages that produce such infections are termed 'temperate', and include lambda phage.

lysosome One of a variety of membrane-bound ➤organelles in ➤eukaryotic cells containing various hydrolytic enzymes. They function by bringing about the degradation of substances taken into the cell for digestion (in the cells lining parts of the gut, for example) and are also used to remove substances in the cell that have become inappropriate for its current physiological or developmental state. They are formed by budding of the membrane system of the ➤Golgi body.

lysozyme A globular protein which catalyses the hydrolysis of specific bonds in the cell walls of certain bacteria, causing their ➤lysis. It thus has an antiseptic effect and is found in tears, saliva and other body fluids of mammals. It was the first protein whose complete three-dimensional structure was determined (as of 2008, over 50 000 structures are known).

M

m   **1** Symbol for the prefix ➤milli-.
    2 Symbol for ➤metre.

m-   Symbol for the ➤meta isomer.

m   **1** Symbol for ➤mass.
    2 Symbol for ➤magnetic moment.

M   **1** Symbol for the prefix ➤mega-.
    2 (Chem.) Symbol for the obsolete term ➤molarity.

M   **1** Symbol for ➤molar mass.
    2 Symbol for ➤Mach.
    3 Symbol for ➤magnetization.

M_m Alternative symbol for ➤molar mass.

M_r Symbol for ➤relative molecular mass.

Ma   Alternative symbol for ➤Mach.

μ Symbol for the prefix ➤micro-.

machine   A device designed to allow a small force (the **effort**) to move a larger force (the **load**). ➤efficiency; lever; mechanical advantage; velocity ratio.

machine code   The coding system which represents the repertoire of instructions that can be performed by a microprocessor.

Mach number   Symbol *Ma* or *M*. The speed of a body moving through a fluid expressed as a fraction of the speed of sound in the fluid. In dry air at $0\,^\circ$C, Mach 1 = $331.4\ \mathrm{m\ s^{-1}}$ or $1193\ \mathrm{km\ h^{-1}}$. Jet passenger aircraft cruise at about Mach 0.8; Mach 1 is about $1065\ \mathrm{km\ h^{-1}}$ at $-55\,^\circ$C, a temperature typical of the lower stratosphere in which they fly. The vehicle *Thrust SSC* reached Mach 1 on land in 1997. Named after Ernst Mach (1838–1916).

Mach's principle   The theory that all motion is in some way relative motion, and that if all forces are taken into account then any frame of reference can be treated as an ➤inertial frame. For example, Mach's principle asserts that there is no such thing as a nonrotating frame; also, the laws of physics in a rotating frame are the same as in an inertial frame, and the ➤centrifugal force experienced in the rotating frame comes from interactions within that frame. An alternative statement of the principle is that the ➤inertial mass of any particle is determined by the gravitational attraction of all the other mass in the Universe. ➤gravitational mass.

Maclaurin expansion The representation of a function as a ➤power (2) series $\sum_{n=0}^{\infty} f^{(n)}(0)x^n/n!$ where $f^{(n)}(0)$ represents the value of the nth derivative of $f(x)$ at $x = 0$. It is the ➤Taylor expansion of $f(x)$ about $x = 0$. Named after Colin Maclaurin (1698–1746).

McLeod gauge A vacuum pressure gauge suitable for measuring pressures down to about 10^{-3} Pa. It works by compressing the volume of the gas by a known high factor and measuring the higher pressure; the original pressure of the gas is then found from ➤Boyle's law. McLeod gauges are often used to calibrate direct-reading gauges. Named after Herbert McLeod (1841–1923).

macro- Prefix from the Greek *makros*, meaning 'large'.

macromolecule A large molecule typically containing more than about 1000 atoms. ➤Polymers are the commonest examples of macromolecules.

macronutrient Any substance required for the normal growth of organisms (particularly plants) in quantities larger than for ➤trace elements (typically >5 ppm). Examples are nitrogen (in the form of nitrate or ammonium), phosphorus (in the form of phosphate), sulfur (in the form of sulfate), calcium, potassium and iron.

macrophage One of a number of types of ➤phagocyte of the ➤immune system that can ingest bacteria and cell fragments. They are typically found in connective and blood-forming tissues, and circulate in the blood.

macroscopic On a scale comparable with everyday objects and events. Thus a 1 kg mass or a 1 m rule is a macroscopic object, whereas an atom or an electron is not. Temperature is a macroscopic quantity (and is meaningless for a very small system), whereas ➤spin angular momentum is not. Macroscopic properties can usually be well approximated by a classical treatment of a system, whereas quantum mechanics is often necessary to predict ➤microscopic properties.

macrostate The state of a thermodynamic system as defined by its ➤macroscopic variables. For example, the macrostate of a gas might be described by its volume and temperature. Each macrostate can correspond to large numbers of ➤microstates.

Madelung constant The constant of proportionality in the equation for the ➤lattice enthalpy relating the total energy of a crystal to the energy of attraction of just one ion of each element in it. It is an infinite sum over all the ion–ion interactions. For the ➤rock-salt (NaCl) structure, the Madelung constant is 1.748. Named after Erwin Madelung (1881–1972), who first calculated it in 1918.

Magellanic Clouds Two small irregular ➤galaxies that are gravitational companions of our Galaxy. They are visible as faint patches of light in the sky of the southern hemisphere, and were first reported to Europeans by the crews commanded by the explorer Ferdinand Magellan (1480–1521).

magic numbers The numbers of protons or neutrons that tend to be present in a stable nucleus. Empirically, the numbers are found to be 2, 8, 20, 28, 50 and 82. For example, tin, with 50 protons, has ten stable isotopes. The magic numbers are the numbers of protons or neutrons required to fill ➤nuclear energy levels that have a significant energy gap above them. Thus they can be compared to the numbers of electrons found in the atoms of a ➤noble gas (2, 10, 18, 36,…). Lead has a magic number of protons, which is one reason why it is the stable end-product of ➤radioactive decay.

maglev Acronym for ➤magnetic levitation.

magma The molten material making up the liquid outer core of the Earth. If magma penetrates to the surface of the Earth, it can create a volcano, from which ➤lava flows.

magnesia An old name for magnesium oxide, MgO. ➤milk of magnesia.

magnesium Symbol Mg. The element with atomic number 12 and relative atomic mass 24.31, which is in Group 2 of the periodic table. It is a shiny metal, which gradually dulls in air due to the formation of an oxide layer. It is a strong reductant, although its kinetic reactivity is lower than that of calcium. It reacts only slowly with cold water, but naturally reacts faster in steam. It can react directly with nitrogen, when heated strongly, as the resulting magnesium nitride has a large ➤lattice energy due to the high charge on both the magnesium and nitride ions, and their relatively small sizes.

The ions of both magnesium and calcium are responsible for ➤hard water. Many of magnesium's compounds contain the magnesium ion, Mg^{2+}. Unless the anion imparts colour, magnesium compounds are white solids, such as the chloride $MgCl_2$ and the oxide MgO.

The small size of the atom and ion explains most of magnesium's differences from calcium. Its carbonate, $MgCO_3$, decomposes more readily than $CaCO_3$ because the lattice energy of the resulting oxide, MgO, is greater. Magnesium ion's small size contributes to its strong bonding in one of life's most important molecules, ➤chlorophyll.

Magnesium is the constructional metal with the lowest density and therefore is used where low mass is desirable, as in car wheels. However, a danger is that magnesium burns fiercely at high temperatures, so magnesium wheels can catch fire in a severe accident. This high reactivity with oxygen is used in flares.

magnet, permanent A piece of metal that has a ➤magnetization permanently present and therefore is surrounded by a ➤magnetic field. It is a ➤ferromagnet that has been left with a permanent magnetization by ➤hysteresis. Almost all permanent magnets are made of ➤iron and its alloys (such as ➤steel); the rest are made from cobalt or nickel alloys. Compare ➤electromagnet; temporary magnetism.

magnetic bottle An arrangement of ➤magnetic fields used to confine ➤plasma.

magnetic circuit A closed loop of ➤magnetic field. Because magnetic fields have zero ➤divergence (➤Maxwell's equations), they have no sources or sinks and must form complete circuits.

magnetic constant A former name for the ➤permeability of free space.

magnetic declination (magnetic variation) At a particular location, the angle between the horizontal component of the Earth's magnetic field (➤geomagnetism) and the geographical north pole. It is expressed as **degrees east or west,** depending on whether ➤magnetic north is east or west of true north. Compare ➤magnetic inclination.

magnetic dip ➤magnetic inclination.

magnetic dipole The limit of a current loop around a surface S carrying a current I as S becomes vanishingly small and the **magnetic dipole moment** $m = IS$ is held constant. This phenomenon is the magnetic equivalent of the electric ➤dipole; no corresponding ➤magnetic monopole has been identified. When placed in a magnetic field, the torque on the dipole aligns it along magnetic field lines. ➤magneton.

magnetic domain ➤domain.

magnetic field A ➤field that is concentrated around electric currents, ➤magnetic dipoles and permanent ➤magnets and determines their effect on, in particular, moving charges (see diagram at ➤field). Two different physical quantities, both vectors, are associated with a magnetic field: the ➤magnetic flux density B is the more fundamental of the two, but for historical reasons the ➤magnetic field strength H takes the name that seems more closely related to the phenomenon itself. In the absence of magnetization the two are related by $B = \mu_0 H$, where μ_0 is the ➤permeability of free space. ➤Lorentz force.

magnetic field strength (H field) Symbol H. A quantity related to the ➤magnetic flux density B by the formula $H = B/\mu_0 - M$, where μ_0 is the ➤permeability of free space and M is the ➤magnetization. It can be calculated from the currents producing the field by the ➤Biot–Savart law or ➤Ampère's theorem. ➤magnetic field.

magnetic flux Symbol ϕ; unit Wb, ➤weber. A measure of the total ➤magnetic flux density B passing through a surface S. The magnetic flux is given by

$$\phi = \int_s \mathbf{B} \cdot \mathrm{d}\mathbf{S},$$

or, for a uniform field, the component of B perpendicular to the surface multiplied by the area of the surface. Magnetic flux is particularly relevant in electromagnetic ➤induction, and appears in ➤Faraday's law (of induction).

magnetic flux density (B field) Symbol B; unit T, ➤tesla. A vector ➤field that acts on moving charges producing a force perpendicular to both the field and the direction of motion of the charge (➤Lorentz force). Compare ➤magnetic field strength; ➤magnetic field.

magnetic force The part of the ➤Lorentz force arising from the magnetic field, or the force of attraction or repulsion between ➤magnetic poles.

magnetic inclination (magnetic dip) The angle between the Earth's local magnetic field and the horizontal. It varies from zero at the magnetic equator to 90° at the magnetic poles. It causes a freely suspended magnet to lie out of the horizontal plane, which is inconvenient in a magnetic ➤compass. ➤➤geomagnetism.

magnetic induction An obsolete name for ➤magnetic flux density.

magnetic intensity An obsolete name for ➤magnetic field strength.

magnetic levitation (maglev) The use of a magnetic force to act against gravity and lift a heavy object. The principle is used in **maglev trains** to overcome friction by raising the train clear of the track. ➤Meissner effect.

magnetic lines of flux A representation of the ➤magnetic flux density as a set of continuous ➤lines of flux.

magnetic meridian (isogonal line) A line, usually on the Earth's surface, of constant ➤magnetic declination, analogous to a ➤meridian. The analogy is not exact: magnetic meridians are by no means straight.

magnetic mirror A high magnetic field designed to reverse the direction of charged particles entering a particular region. Such devices are used in plasma physics.

magnetic moment (magnetic dipole moment) Symbol m; unit A m^2. ➤magnetic dipole.

magnetic monopole The magnetic equivalent of a point ➤electric charge. In classical electromagnetism, ➤Maxwell's equations imply that magnetic monopoles cannot exist. However, some theories in ➤particle physics predict the existence of magnetic monopoles.

magnetic north North as indicated by a compass. A compass needle points towards the magnetic north pole, which does not quite coincide with the geographical north pole (➤magnetic declination).

magnetic permeability Symbol μ. The ratio of the ➤magnetic flux density B to the ➤magnetic field strength H in a material. The ratio of μ to μ_0, the ➤permeability of free space, is known as the **relative permeability**, μ_r. Care should be taken as the subscript r is often dropped in practice. The relative permeability is related to the ➤magnetic susceptibility χ by $\mu_r = 1 + \chi$. ➤Ferromagnetic materials tend to have very high relative permeabilities (of the order of 10^4), whereas for ➤paramagnetic and ➤diamagnetic materials μ_r is just greater than and just less than one, respectively.

magnetic pole 1 (Phys.) One end of a permanent magnet from which ➤magnetic lines of flux appear to emanate. Poles are designated **north** (N), or **south** (S), according to whether the lines originate from or end at the pole. **Like** poles (N/N or S/S) repel each other; **unlike** poles (N/S) attract each other. A magnetic pole is a source of ➤magnetic field strength H, whereas the absence of ➤magnetic monopoles means that the ➤magnetic flux density B must always form closed loops (see diagram at ➤field).
  2 (Geophys.) One of the two points on the surface of the Earth towards which all magnetic compasses align themselves. The pole located in northern Canada is in fact a south pole, and thus it attracts the north pole of a magnetic compass.

magnetic potential One of two quantities used in calculating ➤magnetic flux density B. The **magnetic scalar potential** ϕ_m (unit A) is completely analogous to the electrostatic potential: $B/\mu = H = \text{grad } \phi_m$. The **magnetic vector potential** A (unit T m) is defined by $B = \text{curl } A$. Like all potentials it has an arbitrary ➤gauge, and the conventional choice for A is to make div $A = 0$ everywhere.

magnetic quantum number Symbol m_l. A quantum number associated with atomic structure that takes all integer values from l to $-l$, where l is the ➤orbital angular momentum quantum number. For example, for d orbitals $l = 2$, and so m_l can take the values 2, 1, 0, -1, -2: a total of five orbitals. Hence there are five d orbitals, which are degenerate (➤degenerate states) in the absence of a magnetic field.

magnetic resistance ➤reluctance.

magnetic resonance imaging (MRI) A diagnostic technique based on the principles of ➤nuclear magnetic resonance which produces the clearest pictures of the human brain. It is also much safer than using X-rays.

magnetic storage The storage of data by a computer system on magnetic surfaces. Several different types of magnetic storage are common, including **hard disks** (with capacities measured in gigabytes), **floppy disks** (typically storing megabytes of data) and magnetic tape (again with capacities measured in gigabytes). The principle is broadly the same: an inert ➤substrate (2) (often a polyester) is coated with a layer of magnetic film (often magnetite; ➤lodestone) in an inert binder. Data are stored in binary form, with each binary digit being represented by the polarity of the magnetization of a particular area of the coating.

magnetic susceptibility Symbol χ. The ratio of the magnetization M of a magnetic material to the applied magnetic field strength H. Susceptibility is dimensionless; it is positive for ➤ferromagnetic and ➤paramagnetic materials, negative for ➤diamagnetic materials. ➤magnetic permeability.

magnetic variation ➤magnetic declination.

magnetism The generic name for all phenomena associated with magnetic fields. ➤➤antiferromagnetism; diamagnetism; ferrimagnetism; ferromagnetism; magnet, permanent; paramagnetism.

magnetism, terrestrial ➤geomagnetism.

magnetite ➤lodestone.

magnetization Symbol M; unit A m^{-1}. The ➤magnetic dipole per unit volume within a magnetic material. ➤magnetic field; magnetic flux density.

magneto A mechanical device that produces a spark by electromagnetic ➤induction, usually in an ➤internal combustion engine. Magnetos have the advantage that they do not require a source of electrical power. They are used in aircraft piston engines to ensure that an electrical failure does not cause the engine to stop.

magnetogyric ratio ➤gyromagnetic ratio.

magnetohydrodynamics (MHD) The study of the action of a ➤magnetic field on a conducting fluid or ➤plasma.

magnetometer A device for measuring a ➤magnetic field, for example a ➤Hall probe.

magnetomotive force (m.m.f.) Symbol F_m; unit A. The magnetic equivalent of ➤electromotive force. It is defined for a loop L as:

$$F_m = \oint_L \boldsymbol{H} \cdot d\boldsymbol{l},$$

where $\boldsymbol{H}$ is the magnetic field strength and $d\boldsymbol{l}$ is an element of the loop. The fourth of ➤Maxwell's equations implies that this is equal to the total electric current passing through (not around) the loop. The magnetic equivalent of ➤Ohm's law is $F_m = R\phi$, where R is the total ➤reluctance of a circuit and ϕ the magnetic flux around it.

magneton Symbol μ. A unit of ➤magnetic dipole moment useful in atomic physics. The **Bohr magneton**, given by $\mu_B = e\hbar/2m_e$, is the magnetic dipole moment of the electron, where e is its charge, $\hbar = h/2\pi$ where h is the Planck constant, and m_e is the mass of the electron (➤Appendix table 2). The **nuclear magneton**, given by $\mu_N = e\hbar/2m_p$, is the magnetic dipole moment of the proton, where m_p is the mass of the proton.

magneto-optical effect Any optical effect that depends on the application of a magnetic field. Examples are the magneto-optical ➤Kerr effect and the ➤Faraday effect.

magnetosphere The region around the Earth within which the trajectories of charged particles in the ➤solar wind are determined predominantly by the Earth's magnetic field. Trapped solar wind particles cause ➤aurorae. The magnetosphere is asymmetric (see the diagram opposite): where the solar wind is first disturbed there is a shock wave called the **bow shock**, while on the 'leeward' side the solar wind draws out the magnetosphere into a long **magnetotail**. The magnetosphere includes the ➤Van Allen Belts.

magnetostriction A change of shape or size of a ➤ferromagnetic material induced by a magnetic field. It is caused by ➤domain boundary movements associated with the magnetization. Magnetostriction is a major cause of vibration in transformers, but is also used to produce ➤ultrasound in a **magnetostriction oscillator** by deliberately causing a ferromagnet to oscillate at a chosen frequency by applying a magnetic field to it.

magnetron A ➤microwave ➤oscillator based on ➤thermionic emission. Magnetrons are used as the source for signals in radar.

magnification A measure of the enlargement of an object produced by imaging. **Linear magnification** is the ratio of the size (as a linear dimension) of the image to the size of the object.

magnifying glass A convex ➤lens designed to produce an image with high magnification so that an object can be viewed in more detail.

magnitude 1 (Math.) The length of a vector. In particular, the vector $x\boldsymbol{i} + y\boldsymbol{j} + z\boldsymbol{k}$ has magnitude $\sqrt{x^2 + y^2 + z^2}$.

2 (Astron.) A measure of the brightness of a star or other celestial object. Brighter stars have lower magnitude values. The modern magnitude scale is a rigorous quantification of the Greek astronomer Hipparchus' classification in which the

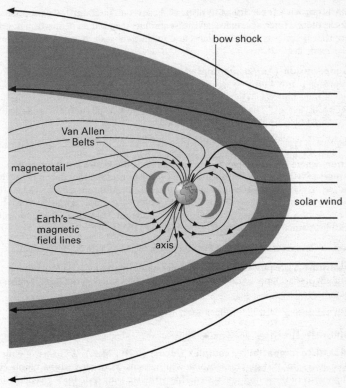

magnetosphere The main features of the Earth's magnetosphere.

brightest visible stars were of first magnitude, and the faintest visible were sixth magnitude. A magnitude difference of 5 now corresponds to a brightness difference of 100, and the scale is extended in both directions to accommodate brighter and fainter objects. **Apparent** magnitude (symbol m) indicates an object's brightness as seen from the Earth. On this scale the Sun has an apparent magnitude of −26.7, the full Moon −12.7, Venus as high as −4.7 and the brightest star, Sirius, −1.46. Apparent magnitudes are now determined using ➤photometers; the most commonly used of several variants is the **V magnitude** (m_V), measured across a yellow-green bandwidth centred on 545 nm. **Absolute magnitude** (M) takes into account an object's distance from the Earth, and is thus a measure of its intrinsic brightness (for a star, its ➤luminosity). It is normally reckoned as the V magnitude the object would have if it were a distance of 10 ➤parsecs from the Earth:

$$M_V = m_V - 5\log_{10}(d/10),$$

where d is the object's actual distance in parsecs. For the Sun, much closer than 10 parsecs, $M_V = +4.8$, while the star Betelgeuse appears ($m_V = +0.5$) much dimmer than it really is ($M_V = -5.1$) because of its distance, believed to be about 130 parsecs. Some celestial objects (such as comets) have their absolute magnitudes defined differently. Note that despite its name, the absolute magnitude scale is still a relative one. ➤spectral class.

Magnoliopsida (Angiospermophyta) A phylum within the plant kingdom comprising plants with a ➤vascular system that bear ➤flowers during at least some stage of their life cycle. The phylum is further subdivided into ➤dicotyledons and ➤monocotyledons. The Magnoliopsida is the largest phylum of land plants, represented by over 250 000 species worldwide.

Magnox A type of magnesium alloy used in some nuclear fission reactors (called, therefore, **Magnox reactors**) to encase ➤fuel elements.

Magnus effect The effect on a spinning cylinder or sphere that moves through a fluid, in which a force acts perpendicular to the direction of motion and to the direction of spin. This is used to advantage in sports, for example tennis, in which the trajectory of a ball hit with topspin is very different from that of one hit with backspin. Named after Heinrich Gustav Magnus (1802–70).

main A conductor used for the distribution of electrical power. ➤ring main.

mainframe A large and expensive computer system, typically capable of supporting hundreds of simultaneous on-line users, and usually connected to a large number of ➤peripherals.

main sequence ➤Hertzsprung–Russell diagram. ➤dwarf.

major axis The longer axis of an ➤ellipse.

major histocompatibility complex (locus) (MHC, MHL) A cluster of closely linked genes (located on chromosome 6 in humans) associated with a number of components of the ➤immune system. In particular, they code for ➤glycoproteins associated with the recognition of antigens by ➤lymphocytes, and include the ➤human leucocyte antigen system. Other vertebrate species have similar gene clusters.

majority carrier Whichever type of ➤carrier (2), either positive ➤holes or negative electrons, outnumbers the other type in a ➤semiconductor. Thus holes are the majority carrier in ➤p-type semiconductors, whereas electrons are the majority carrier in ➤n-type semiconductors. Compare ➤minority carriers.

malachite A blue ore of copper, basic copper(II) carbonate of approximate formula $Cu(OH)_2·CuCO_3$.

malaria A debilitating and widespread (particularly in the tropics) disease caused by a ➤protoctistan parasite of the genus *Plasmodium*. Malaria is estimated to be responsible for over a million deaths a year. The most commonly fatal form of malaria is caused by the species *Plasmodium falciparum*. The life cycle is complex, involving phases in the blood and in liver cells, and also in mosquitoes of the genus *Anopheles* which act as ➤vectors (3) responsible for the spread of the disease. Infectious phases of the parasite are injected into the body when mosquitoes feed on blood. Efforts to control malaria have been directed mainly at eradicating mosquitoes by insecticide sprays such as ➤DDT, or preventing bites by the use of fine-mesh protective nets. Drug treatment of the disease in the body traditionally used compounds based on ➤quinine, but today synthetic derivatives of chloroquine are used.

maleic acid An old name for *cis*-butenedioic acid, which exists as a colourless solid. As the carboxylic acid groups are next to each other in the ➤*cis* form, it can dehydrate much more easily on heating than the *trans* form (➤fumaric acid), producing maleic anhydride.

maleic anhydride The anhydride from maleic acid (see the diagram), a colourless solid. It is widely used as a dienophile in the ➤Diels–Alder reaction.

malleable Describing a material that can be beaten into a different shape. Metals are malleable, but ionic crystals are not. Two very malleable metals are gold and indium. Compare ➤brittle.

maleic anhydride

Malpighian layer ➤skin. Named after Marcello Malpighi (1628–94).

Malus's law The intensity of light transmitted through two ➤polarizers is proportional to $\cos^2 \theta$, where θ is the angle between the axes of the polarizers. Named after Étienne Louis Malus (1775–1812).

mammal A member of the major vertebrate class Mammalia distinguished by ➤homoiothermy, the presence of **mammary glands** to suckle the young, the presence of body hair (generally) and specialized arrangements of the bones of the skull. There are three groups: the **monotremes**, such as the duck-billed platypus, which lay eggs; the **marsupials**, such as the kangaroo, whose young are born at a very early stage of development, and mature in an external pouch; and the **placentates**, including horses, dogs and humans, whose young develop internally in the uterus, nourished by the ➤placenta.

mammary glands ➤mammal.

man ➤human.

Mandelbrot set A particularly famous ➤fractal, discovered by Benoit B. Mandelbrot (b. 1924), variously described as the most complex and most beautiful object in mathematics. For each point c on the complex plane, a simple iteration is carried out: $z_0 = 0$, $z_{n+1} = z_n^2 + c$. If the values of z_n remain bounded (i.e. do not tend to infinity), then the point at c is coloured black; otherwise it is coloured white (see the diagram overleaf). The role of computers in the discovery of the Mandelbrot set is typical of their essential part in research into fractals and chaotic systems.

manganate(VI) ion (manganate ion) The ion with formula MnO_4^{2-}; hence manganese has oxidation number +6. Compare ➤manganate(VII) ion.

manganate(VII) ion (permanganate ion) The ion with formula MnO_4^-; hence manganese has oxidation number +7. It is coloured deep purple by a ➤charge transfer mechanism. It is a strong oxidant in acidic solution; when reduced it usually forms a nearly colourless solution of manganese(II); the retention of a purple colour in the solution provides an easy way of detecting the end-point in a redox titration. Its

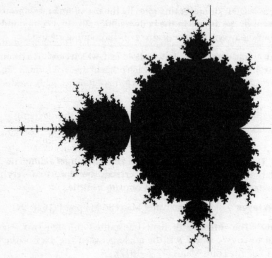

Mandelbrot set

oxidizing ability is also useful in organic chemistry, for example to oxidize alcohols. Compare ➤manganate(VI) ion.

manganese Symbol Mn. The element with atomic number 25 and relative atomic mass 54.94 which is in the first row of the transition metals and shows their typical properties. First, it has variable oxidation numbers: common ones are +2 in the Mn^{2+}(aq) ion, +4 in the solid ➤manganese(IV) oxide, and +7 in the ➤manganate(VII) ion. This leads to its use in ➤catalysis; MnO_2 is a catalyst for the decomposition of hydrogen peroxide. Colour is common in its compounds, although the pink Mn^{2+} ion is particularly pale because the ion is high-spin d^5 and so all the five orbitals have one electron each, making the ➤d–d transition spin-forbidden. In contrast, the ion MnO_4^- is deep purple, but this is due to ➤charge transfer rather than a d–d transition. Complexes are numerous and include the aqueous ion, written more correctly as $[Mn(H_2O)_6]^{2+}$, which is strongly ➤paramagnetic because of the five unpaired electrons. Major potential sources of manganese together with other metals are the **manganese nodules** present on the seabed. Currently their extraction is uneconomical, but perhaps in the future they will become a common source of manganese.

manganese(IV) oxide (manganese dioxide) MnO_2 A brown-black solid which is a useful oxidant, as in the oxidation of hydrochloric acid to ➤chlorine, and a catalyst, as in the decomposition of hydrogen peroxide. When it is formed by reduction of manganate(VII) ions it appears as a brown suspension.

manometer A device for measuring pressure in terms of the difference in height of linked columns of fluid. A typical manometer is a U-tube filled with mercury or water, exposed to the pressure to be measured at one end and evacuated at the other. The pressure is equal to ρgh, where ρ is the density of the fluid, g is the ➤acceleration of free fall and h is the height difference between the columns. Such a method of measurement gave rise to units such as the ➤torr.

mantle The layer between the Earth's core and its crust. The mantle contains about two-thirds of the Earth's mass. It consists of heavy iron- and magnesium-bearing silicates, and is the source of ➤magma.

mapping Another name for a ➤function or a ➤transformation (2).

map projection A method of representing the spherical surface of the Earth on a flat map. There are a large number of different ways of doing this, but the useful ones fall into three classes. For a **cylindrical projection** the surface of the Earth is projected on to the surface of a cylinder whose section is a ➤great circle, usually the ➤equator; lines of ➤latitude appear as straight lines in this case. For an **azimuthal projection** the surface is projected on to a tangent plane at some point. For a **conical projection**, the surface is projected onto a cone, usually with its vertex on the axis of the Earth beyond a pole. A projection is **conformal** only if the scale is equal in all directions at every point, with the result that angles and bearings can be measured reliably from the map. ➤➤Mercator projection.

marble A crystalline form of calcium carbonate, $CaCO_3$.

mare (plural **maria**) ➤Moon.

Mariotte's law The name by which ➤Boyle's law is generally known in continental Europe. Named after Edmé Mariotte (*c.* 1620–84).

marker 1 A molecule of known size that is used to calibrate the characteristics of gels used in gel ➤electrophoresis. It is particularly applied to ➤DNA samples applied to a gel as standards to find the size of unknown fragments derived, for example, from digestion with a ➤restriction enzyme. Commercial mixtures of DNA fragments of varying lengths are available as 'ladders', so-called because of the typical regular pattern they produce on a gel. ➤Restriction fragment length polymorphisms (RFLPs) are also widely used as genetic markers.

2 A ➤gene used to identify a particular fragment of DNA during, for example, ➤transgenic work in ➤genetic engineering applications. The marker gene confers a ready method of identification, such as the production of a colour on suitable treatment, or resistance to an ➤antibiotic so that cells containing the fragment will grow on a medium into which the appropriate antibiotic is incorporated. Cells that do not contain the fragment are not able to grow, providing a useful screening process to identify cells into which the fragment of DNA in question has been inserted. ➤➤labelling.

Markov chain A sequence of random events from a fixed collection of events in which each event is only influenced by its immediate predecessor. A Markov chain is fully described by its **transition matrix**, with elements p_{ij} giving the probability that event j had event i as predecessor. Named after Andrei Andreiëvich Markov (1856–1922).

Markovnikov's rule The dominant product when an unsymmetrical molecule adds to an unsymmetrical alkene will be the one where the hydrogen has added to the end of the bond with more hydrogens already. For example, HBr can add to

$CH_3CH{=}CH_2$ to form two possible products ($CH_3CH_2CH_2Br$ or $CH_3CHBrCH_3$). In this case, $CH_3CHBrCH_3$ will predominate. The explanation of the rule depends on recognizing that the two possible ►carbocations formed on protonation of the alkene have different stabilities due to the electron-releasing nature of alkyl groups. Named after Vladimir Vasilyevich Markovnikov (1838–1904).

Mars The fourth planet from the Sun. Mars has a radius of 3397 km (about half the Earth's radius) and orbits at 1.52 AU with an orbital period of 1.88 years. Its rotational period (24 h 37 min) and axial inclination (25°) are both remarkably close to those of the Earth. The eccentricity of its orbit is 0.09, much greater than the Earth's, and the planet is much closer to the Sun in its southern hemisphere's summer; this results in a significant difference in climate between the two hemispheres. Its atmosphere, with a surface pressure of only 600 Pa, is 95% carbon dioxide. The surface temperature varies between about -130 and about 20 °C. Carbon dioxide sublimes at -78 °C, and there are clouds composed of solid carbon dioxide. At some time in its past, water abounded on Mars: its surface features include ancient, dry river beds and flood plains. The largest known volcano in the solar system, the 600 km diameter Olympus Mons, is on Mars. It has two small satellites. ►►Appendix table 4.

Marsh's test The first successful test for detecting arsenic depending on conversion of the arsenic compound into the volatile hydride AsH_3, which can be decomposed to a brown stain on heating. This test made arsenic less popular as a poison! Named after James Marsh (1794–1846), who discovered it in 1836.

marsupial ►mammal.

martensite A phase in the complicated ►iron–carbon phase diagram, consisting of a solid solution of carbon in iron with a tetragonal structure.

mascon A concentration of mass below the ►Moon's surface, first detected by spacecraft in lunar orbit whose motion was more irregular than expected for a uniform gravitational attraction. Mascons occur beneath lunar maria, and probably originated when ancient impacts removed lighter surface rock, and denser rock from the mantle bulged upwards.

maser Acronym for microwave amplification by stimulated emission of radiation. The maser is the equivalent of the ►laser for microwave frequencies.

mass Symbol m; unit kg, ►kilogram. A fundamental characteristic of a body, determined by the amount of matter it contains. It is usually defined as ►inertial mass but measured as ►gravitational mass. ►►conservation of mass, law of; Einstein mass–energy relation; momentum; rest mass.

mass action, law of An obsolete law suggesting that the rate of a reaction is proportional to the concentrations (originally called 'active masses') of the reactants. As several reactions are known where the ►order of the reaction is zero with respect to one reactant, this law must clearly be an oversimplification.

mass decrement The difference between the ►rest mass of a decaying particle and the sum of the rest masses of its decay products. It is thus equivalent (by the ►Einstein mass–energy relation) to the transfer of energy from mass energy to kinetic energy in such a process. Compare ►mass defect.

mass defect The difference between the sum of the ➤rest masses of the nucleons making up a stable nucleus and the rest mass of the nucleus as a whole. It is thus equivalent (by the ➤Einstein mass–energy relation) to the binding energy of the nucleus. Compare ➤mass decrement.

mass–energy relation ➤Einstein mass–energy relation.

massless particle A subatomic particle with no mass, for example the photon. A massless particle must travel at the speed of light, and therefore it has no zero-momentum states. Strictly, it is impossible to define a ➤spin for such a particle, only a ➤helicity, which combines spin and momentum.

mass number ➤nucleon number.

mass spectrograph An obsolete name for ➤mass spectrometer.

mass spectrometer An instrument for measuring the mass of a species, from single atoms to complex molecules. The simplest form of mass spectrometer consists of an ionization chamber where an electron is removed from the species and its positive ion is produced (see the diagram). The positive ions can then be accelerated in an electric field to a nearly uniform velocity before being deflected in a magnetic field. The extent of deflection of each ion depends on its mass. The detector is at a fixed position, so ions are detected by changing the magnetic field. The ➤fragmentation pattern, caused by the distinctive combination of fragments that break off a molecule in the apparatus, gives insight into the structure of the molecule. (➤Chromatography for a description of GCMS, gas chromatography–mass spectrometry.)

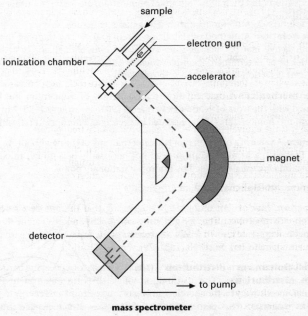

mass spectrometer

mass spectrum The output from a ➤mass spectrometer.

mast cell A large ➤amoeboid ➤leucocyte cell found in vertebrate connective tissues. Mast cells function as part of the ➤immune system; specifically, they are involved in allergic reactions and immunity to parasitic diseases.

matched load A load for an electrical power source with an ➤impedance equal to the ➤complex conjugate of the supply's internal impedance. ➤➤impedance matching; matched termination.

matched termination A load that terminates a ➤transmission line with the same impedance as the characteristic ➤impedance of the line itself. This ensures that there is no reflected wave. ➤➤impedance matching; matched load.

materials science The study of the structure, properties and function of materials. It includes ➤metallurgy, and also the science of nonmetals such as ➤polymers.

mathematics The interrelated study by logical means of number, pattern, shape and space. **Pure mathematics** deals with their generalizations and abstractions, and **applied mathematics** with applications to the physical and human worlds.

matrix A rectangular array of numbers enclosed in brackets. An $r \times s$ matrix has r (horizontal) rows and s (vertical) columns; an entry or **element** in the ith row and jth column is denoted a_{ij}. Matrices of the same shape may be added together by adding corresponding elements together. An $r \times s$ matrix A multiplies an $s \times t$ matrix B to give an $r \times t$ matrix AB with entries $(AB)_{ij} = \sum_k a_{ik} b_{kj}$. Matrices occur in connection with linear equations, but their most important applications are for representing ➤linear transformations.

matrix isolation A method of stabilizing reactive species by trapping them in a solid unreactive matrix, such as solid argon, for long enough for their spectral properties to be determined. Examples include the isolation of radicals and their subsequent detection by ➤electron spin resonance.

matrix mechanics A mathematical representation of ➤quantum mechanics, associated with the work of Werner Heisenberg (1901–76) around 1925–7 in which quantum-mechanical states are represented by vectors and the operators by matrices.

maximum A value of a function that is greater than any of the neighbouring values. Notice that such a **local maximum** need not necessarily be a **global maximum**. Local maxima are usually located by finding ➤stationary points.

maximum multiplicity, law of ➤Hund's rules.

maxwell Symbol Mx. The ➤c.g.s. unit of ➤magnetic flux. 1 Mx = 10^{-8} Wb (➤weber). Named after James Clerk Maxwell (1831–79).

Maxwell–Boltzmann distribution (Maxwellian distribution) The distribution of molecular speeds (or kinetic energies) in a gas (see the diagram). The distribution is not

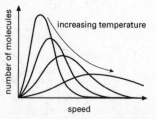

Maxwell–Boltzmann distribution

symmetrical: the fall is less steep than the rise to the maximum. At higher temperatures the most probable speed is higher, as is the mean speed. Significantly, there is a large increase in the probability of very high speeds, which explains why increasing temperature increases the rate of many reactions dramatically, since a greater proportion of molecules have the necessary ►activation energy. Named after J. C. Maxwell and Ludwig Edward Boltzmann (1844–1906).

Maxwell's demon An imaginary creature in a thought-experiment proposed by J. C. Maxwell. He imagined a miniature being who could see the motion of air molecules in a closed vessel divided into two portions by a partition with a small hole in it. The creature controls the opening and closing of the hole, allowing faster molecules into one side of the vessel while letting slower molecules through to the other side. Hence the creature is able, without any expenditure of energy, to raise the temperature of one side of the vessel while lowering the temperature of the other side, thereby lowering the total entropy of the system and apparently contradicting the second law of ►thermodynamics. Resolving this contradiction has provoked much debate; modern resolutions rely on considering the increasing entropy of the demon's mind as it makes its decision.

Maxwell's equations The four fundamental equations governing the classical theory of ►electromagnetism, relating the strengths of the electric field (E) and ►magnetic field (H), the ►electric displacement (D) and the ►magnetic flux density (B) to the ►charge density (ρ) and ►current density (j) that create the fields. Any presentation of the equations requires ►vector analysis, with the equations expressed in either differential or integral form. Both forms are given below:

$$\text{Maxwell I}: \quad \nabla \cdot D = \rho_{\text{free}}, \quad \int_S D \cdot dS = \int_V \rho_{\text{free}} dV.$$

The subscript 'free' is used in the first equation to denote that the charge density does not include charges produced by the polarization of dielectrics (because these charges are accounted for in the permittivity ε, which relates D to E). If all charges are considered, the equation becomes:

$$\text{Maxwell Ia}: \quad \nabla \cdot E = \frac{\rho}{\varepsilon_0}, \quad \int_S E \cdot dS = \int_V \frac{\rho}{\varepsilon_0} dV.$$

►Coulomb's law is simply the solution to this equation for a point charge, when ρ is a ►Dirac delta function in three dimensions. Physicists tend to call this physical law (in integral form) **Gauss's law**.

$$\text{Maxwell II}: \quad \nabla \cdot B = 0, \quad \int_S B \cdot dS = 0.$$

The second equation expresses the continuity of magnetic ►lines of flux, and the absence of ►magnetic monopoles.

$$\text{Maxwell III}: \quad \nabla \times E = -\frac{\partial B}{\partial t}, \quad \int_L E \cdot dl = -\int_S \frac{\partial B}{\partial t} \cdot dS.$$

In integral form this third equation is ►Faraday's law (of induction), and relates the ►e.m.f. around a closed loop to the rate of change of magnetic flux through the loop.

Maxwell IV : $\quad \nabla \times \boldsymbol{H} = \boldsymbol{j}_{\text{free}} + \dfrac{\partial \boldsymbol{D}}{\partial t}, \quad \displaystyle\int_L \boldsymbol{H} \cdot \mathrm{d}\boldsymbol{l} = \int_S \left(\boldsymbol{j}_{\text{free}} + \dfrac{\partial \boldsymbol{D}}{\partial t} \right) \cdot \mathrm{d}\boldsymbol{S}.$

The fourth equation describes the relationship between a current and the magnetic field that it generates, with the extra term $\partial \boldsymbol{D}/\partial t$, the ➤displacement current, required to conserve charge. The integral form is equivalent to ➤Ampère's theorem, and, just as Maxwell I can be solved for a point charge to obtain Coulomb's law, Maxwell IV can be solved for a current element to obtain the ➤Biot–Savart law.

Maxwell used these equations to predict the existence of electromagnetic waves travelling at the speed of light. One remarkable property of Maxwell's equations is that they are perfectly consistent with ➤special relativity, which was introduced by Einstein many years later.

Mb 1. (Comput.) Symbol for the unit ➤megabyte.
  2 (Biol.) Symbol for the unit ➤megabase.

mbar Symbol for the unit ➤millibar.

MBE Abbr. for ➤molecular beam epitaxy.

Md Symbol for the element ➤mendelevium.

Me Symbol for the ➤methyl group.

mean 1 Short for the ➤arithmetic mean.
  2 The mean of a ➤frequency distribution is $\sum f_i x_i / \sum f_i$, where f_i is the class frequency and x_i is the mid-class value.
  3 Another name for the ➤expectation of a random variable.

mean free path The average distance that a molecule travels in the gas phase (in the ➤mean free time) between collisions. For example, for carbon dioxide at room temperature and pressure the mean free path is about 55 nm. The term is also used for the average distance travelled by a particle from its creation to when it decays.

mean free time The mean time between collisions for a particle in a gas. ➤➤mean free path.

mean life The mean time for which a particle remains in one state before decay (for ➤subatomic particles) or recombination (for ➤electron–hole pairs). It is the reciprocal of the ➤decay constant. For standard ➤exponential decay, the mean life is the ➤half-life divided by ln 2.

mean solar time (local mean solar time) The time of day reckoned by the position of the Sun in the sky (➤apparent solar time), adjusted to smooth out the effects of eccentricity of the Earth's orbit and so make the scale of time uniform with respect to a standard clock. ➤➤equation of time.

mechanical advantage (force ratio) The ratio of the force applied by a machine to the force applied to it. ➤➤efficiency; velocity ratio.

mechanical equivalent of heat The ➤work needed to increase the temperature of a body by unit temperature. Its first accurate determination, by Joule in 1843, paved the way for an understanding of the underlying connection between work and heat, and laid the foundations of ➤thermodynamics. Its value is 4.2 joules per calorie.

mechanics The branch of physics concerned with the analysis of the behaviour of objects under the action of forces. The subject was first quantified as **Newtonian mechanics**, which dealt with the rotational and translational motion of ➤rigid bodies under ➤Newton's laws of motion. Since then, these methods have been extended and applied in other areas such as ➤fluid mechanics, ➤quantum mechanics and ➤statistical mechanics.

mechanism A description of a reaction that designates exactly how each step of the reaction occurs. Knowing the mechanism, it is possible to predict how small alterations to a molecule will affect its subsequent reaction. We know a great deal about many organic mechanisms, as these reactions tend to be rather slow. (We know much less about inorganic mechanisms, as they are significantly faster.) In the reaction between iodine and propanone, the ➤order of reaction with respect to iodine is zero, so iodine cannot be involved in the ➤rate-limiting step. The reaction is first order in the catalyst, H^+ ions. The mechanism involves a ➤keto–enol tautomerism, catalysed by acid, followed by a faster reaction with iodine. ➤➤nucleophilic substitution.

median 1 A line from a ➤vertex of a triangle to the midpoint of the opposite side. The three medians meet at the centroid of the triangle (➤centre of mass).
   2 The value below which half of a ➤frequency distribution or ➤probability distribution lies.

median lethal dose Symbol LD_{50}. The dose of ➤ionizing radiation or toxic material that kills 50% of a specified population of an organism (usually within a specified time) while leaving 50% alive. The ➤median (2) is taken because such distributions can be very asymmetric (if, for example, a subset of the population is resistant in some way), and the ➤mean (2) could give an unrepresentatively high value.

medical physics The application of physics to medicine, encompassing techniques such as ➤radiotherapy. Compare ➤health physics.

medicine A complex and wide-ranging branch of biological science concerned with the diagnosis and recognition of the clinical symptoms of disease or other body malfunction, and with the treatments required to restore normal function. Branches of medicine include anatomy, surgery, biochemistry, physiology, genetics and immunology, as well as areas concerned with specific body functions such as gynaecology.

medium The matter through which a wave propagates. For example, for radio transmission from a broadcast station to a receiver in the open, the medium is air.

medium frequency (MF) The radio frequency range between 300 kHz and 3 MHz. ➤➤frequency band.

medulla oblongata ➤brain.

mega- Symbol M. The SI prefix for a million units base; 1 MHz, for example, is equivalent to 10^6 Hz.

megabase Symbol Mb. A sequence of 10^6 bases in a DNA molecule. The human ➤genome contains an estimated 3000 Mb (thus 6 billion in double-stranded human DNA); that of baker's yeast (*Saccharomyces cerevisiae*) 12 Mb.

megabyte Symbol Mb. A term in computing, equal to 1024 kilobytes or 2^{20} bytes. The capacity of ➤RAM in personal computers is often of the order of megabytes.

megahertz Symbol MHz. A unit of frequency equal to a million ➤hertz: 1 MHz is equivalent to 10^6 Hz.

meiosis The process of cell division that results in the formation of ➤haploid cells from ➤diploid cells to produce, for example, ➤gametes and ➤spores (see the diagram opposite). Meiosis also produces new combinations of genes in the nuclei of the resultant cells. Unlike ➤mitosis, meiosis consists of two sequential divisions and produces four daughter cells. The first part of the first division (**prophase I**) is the most significant in establishing the principles of meiosis, and consists of a number of recognizable stages. However, these stages are continuous and grade into each other rather than being discrete steps. During **leptotene** the ➤chromosomes become more dense and take up stain more readily, and each migrates to pair with its homologue (➤homologous pair) at **zygotene** to form a **bivalent**. Each chromosome now replicates along its length to form two ➤chromatids attached at the centromere, and at **pachytene** the chromatids become intimately entwined with each other and may break at exactly homologous points to exchange fragments of chromatid with each other. This is referred to as 'crossing over', and the points where it occurs are the **chiasmata**. Each bivalent usually has at least one chiasma, and longer chromosomes may have two or three. Since crossing over involves the exchange of genetic information encoded in the DNA, this produces new combinations of genes. At **diplotene** the chromatids and chiasmata are clearly visible, and at **diakinesis** the four chromatids comprising each bivalent start to move apart.

At **metaphase I** the **spindle** has been formed as a result of ➤microtubule activity, and the bivalents become attached to its equator. During **anaphase I** the centromeres of each of the homologues move apart and migrate to the poles of the spindle, each carrying two chromatids. At this stage the centromeres are undivided. In **telophase I** the chromosomes group together such that each daughter nucleus contains the haploid number of centromeres, each with two chromatids attached. The cell now divides to produce two daughter cells; they may proceed directly to the second division, or may exist in a more or less prolonged **interphase**.

At **prophase II** the chromosomes reappear (if the cell entered an interphase), and at **metaphase II** another spindle forms and the centromeres become attached to its equator. At **anaphase II** the centromeres divide and migrate to the poles separating the individual chromatids. The various possible combinations of chromatids arriving at the poles in this way occur at random and this, coupled with crossing over, is the basis of the ➤variation (2) seen in the offspring of sexually reproducing organisms. At **telophase II** the chromatids (now chromosomes) become less dense, new nuclear membranes form and the cells divide to produce two daughter cells each, giving four cells in all.

The behaviour of chromosomes during the two divisions is entirely independent. There is an equal probability that either member of a homologous pair in the first

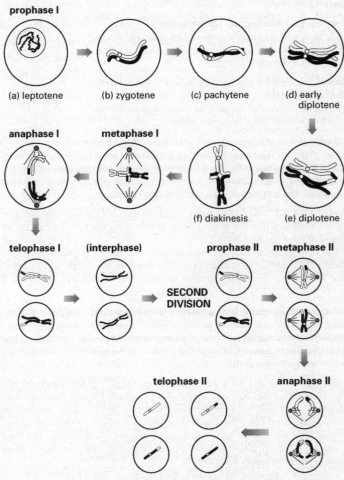

FIRST DIVISION

prophase I

(a) leptotene   (b) zygotene   (c) pachytene   (d) early diplotene

anaphase I   **metaphase I**

(f) diakinesis   (e) diplotene

telophase I   **(interphase)**   **SECOND DIVISION**   **prophase II**   **metaphase II**

telophase II   **anaphase II**

meiosis in a diploid cell with a pair of chromosomes

division, or either chromatid attached to any centromere in the second divisions, will migrate into either of the resultant cells. This is the basis of **independent assortment**, which is a fundamental principle in increasing genetic diversity (➤Mendel's laws).

Meissner effect The perfect ➤diamagnetism exhibited by a superconductor (➤superconductivity). No ➤magnetic flux can enter or leave a superconductor's surface. Named after (Fritz) Walther Meissner (1882–1974).

meitnerium Symbol Mt. The element with atomic number 109. Named after Lise Meitner (1878–1968).

melamine (2,4,6-triamino-1,3,5-triazine) Colourless crystals with the structure shown in the diagram, which react with methanal to form a thermosetting resin which is very resistant to heat and light.

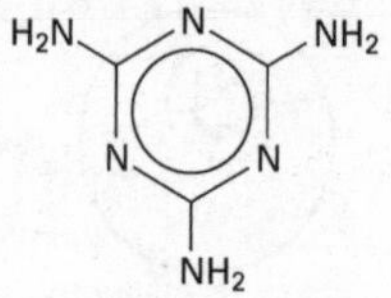

melamine

melanin A dark brown pigment of skin and hair in animals, particularly vertebrates, derived from the amino acid tyrosine. The absence of melanin in individuals unable to metabolize tyrosine is the cause of ➤albinism.

melatonin A ➤hormone produced by the ➤pineal gland of vertebrates. Melatonin controls rhythmic phenomena such as breeding cycles, and its production is regulated by daylight hours and light intensity. Recently, melatonin deficiency has been implicated in seasonal affective disorder (SAD), in which patients experience lethargy and depression in seasons of shorter day-length.

melting 1 (Phys.) The process during which a substance changes ➤phase from a solid to a liquid. While the phase change is occurring the temperature remains constant, at the ➤melting point.

   2 (Biol.) The separation, by heating, of the two strands of a double-stranded DNA molecule. The temperature at which this occurs (**melting temperature**, T_m) depends on the degree of ➤hydrogen bonding between the strands and hence on the proportion of A–T and G–C base pairs (➤base pairing).

melting point (melting temperature, freezing point) The temperature at which a liquid is in ➤dynamic equilibrium with its solid. The melting point at an atmospheric pressure of 1 atm is called the **normal melting point**. The melting point is a characteristic of a pure substance; the presence of impurities lowers the melting point. ➤➤colligative properties.

membrane 1 One of a variety of structures enclosing and compartmentalizing ➤cells. Membranes are extended sheets of ➤phospholipids and ➤proteins (**unit membranes**), the detailed arrangement of which depends on the location and function of the membrane. The interior membranes of cells and ➤organelles serve as templates for a wide range of processes such as ➤protein synthesis, the ➤light reaction of photosynthesis and ➤oxidative phosphorylation. The ➤plasma membrane is the boundary layer of cells.

   2 A structure separating two phases. An example is the selectively permeable membrane vital in ➤osmosis.

membrane cell ➤chlor–alkali industry.

memory ➤RAM; ROM; EPROM.

memory cell Special ➤B cells of the ➤immune system that retain the ability to recognize an ➤antigen over a period of time which may vary from months to years. When subsequently exposed to the same antigen they quickly produce ➤antibodies which prevent the antigen from producing a pathological effect. This is the principle behind the establishment of immunity following vaccination (➤vaccine).

mendelevium Symbol Md. The element with atomic number 101 and most stable isotope 258, named after the discoverer of the periodic table, Dmitri Ivanovich Mendeleyev (1834–1907). It is a ➤transuranium element and all isotopes are radioactive. Both Md^{2+} and Md^{3+} can be formed in aqueous solution.

Mendel's laws A set of principles in ➤genetics, established by Gregor Johann Mendel (1822–84). Mendel observed patterns of inheritance in pea plants. He found that characteristics of organisms are inherited as discrete units, rather than by a blending of characters in offspring, as had been previously accepted. For example, pea plants were observed to be either tall or dwarf, and tall plants crossed with dwarf plants produced offspring which were tall, rather than an intermediate 'blend' of medium height. The significance of Mendel's work was not fully realized until the 1890s; his 'laws' provide the foundation of plant and animal breeding today. In modern form, the two key laws are these:

Mendel's first law (law of segregation): Each of the two ➤alleles that an individual possesses at any one ➤locus will pass into separate gametes during ➤meiosis.

Mendel's second law (law of independent assortment): In the gametes, all possible combinations of alleles possessed by an organism are equally probable.

meninges ➤cerebrospinal fluid.

meningitis Inflammation of the membranes (**meninges**) that line the brain and spinal cord. Meningitis may be caused by trauma, or by bacterial or viral infection. Viral meningitis is a relatively benign condition but the bacterial, or meningococcal, infection caused by *Neisseria meningitidis* is a serious disease, particularly in children and young adults.

meniscus The form assumed by the upper surface of a liquid where it meets a solid vertical surface; for aqueous solutions in contact with glass, the water surface curves slightly upwards at the surface of the glass. For titrations involving aqueous solutions, the volume should be read from the bottom of the meniscus. Mercury in contact with glass has a meniscus that curves downwards rather than upwards because the liquid-liquid attractive forces are larger than the liquid-solid attractive forces.

menstrual cycle The approximately four-weekly ➤oestrous cycle in human females associated with the release of an egg from the ➤ovary (ovulation) and the preparation of the wall of the ➤uterus in readiness for a fertilized egg. The cycle is under the control of ➤hormones involving secretions from the ➤pituitary gland and the ovaries themselves. The various interactions between the hormones provide a good example of ➤feedback control. If no fertilized egg is present, the lining of the uterus breaks down and passes out through the ➤vagina over a period of five or six days (a '**period**') during menstruation.

mensuration The measurement and calculation of lengths, angles, areas and volumes of geometric figures.

menthol An optically active terpene alcohol (see the diagram). It is most famous as a nasal decongestant; the vapour is very good at clearing nasal passages.

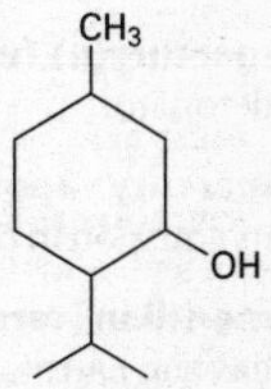

CH₃

OH

menthol

mercaptans The analogues of ➤alcohols, but with sulfur in place of oxygen. Their formulae are therefore RSH. The odour of the ethyl example, ethyl mercaptan CH_3CH_2SH, is said to resemble a combination of 'rotting cabbage, garlic, onions and sewer gas'.

Mercator projection A conformal (➤conformal transformation) cylindrical ➤map projection of the Earth's surface. It is characterized by the equations $x = \lambda$ and $y = \ln \tan (\frac{1}{4}\pi + \frac{1}{2}\phi)$ where λ is the longitude and ϕ is the latitude of the point to be represented by (x, y). The scale becomes progressively larger towards the poles. It was frequently used for short-range navigation as the grid of ➤meridians and parallels of latitude are perpendicular, making measurement of angles particularly easy. It is also easily adapted to take the eccentricity (➤conic) of the Earth into account (➤equatorial bulge). Named after Gerardus Mercator (1512–94), who first used it in 1569.

mercury Symbol Hg. The element with atomic number 80 and relative atomic mass 200.6, which is in Group 12. The origin of its symbol Hg is its Latin name *hydrargyrum*, 'liquid silver'; mercury is unique in being the only liquid metal at room temperature, which explains many of its uses, such as in thermometers.

The effect of mercury, a heavy metal poison, on humans is well documented. The phrase 'as mad as a hatter' originated because hatters used a mercury compound in their work. Despite this danger, dental fillings still often use ➤amalgams containing about 70% silver.

Many compounds of mercury are known, with two main oxidation numbers. First, there are compounds with oxidation number +2 (as for its congeners zinc and cadmium), as in $HgCl_2$, a corrosive sublimate, or the yellow-orange powder HgO. A characteristic second oxidation number is +1, as in the chloride Hg_2Cl_2, often called calomel. Note that the formula is not HgCl because it has been shown, first by Raman spectroscopy (➤Raman effect), that the cation is Hg_2^{2+}.

Mercury The closest planet to the Sun, at a distance of 0.39 AU, with a radius of 2440 km. Though it was originally believed that Mercury always kept the same face to the Sun (as the Moon does to the Earth), it was discovered in 1965 by radar observations that it has a rotational period of 58.6 days, exactly two-thirds of its 88-day ➤sidereal year. Its rotational axis is almost perpendicular to its orbital plane. Its surface is very much like that of the Moon, and it has an extremely thin atmosphere (10^{-7} Pa) of hydrogen and helium. The surface temperature variation is the greatest for any planet: from $-180\,°C$ (night) to $430\,°C$ (day). ➤➤Appendix table 4.

mercury cell An obsolete technology which was used to manufacture chlorine. ➤chlor–alkali industry.

mercury(II) fulminate Grey crystals of formula $Hg(CNO)_2 \cdot \frac{1}{2}H_2O$ used as a classic detonator.

mercury vapour lamp A lamp that produces light, with a significant proportion of its emission in the ultraviolet, by passing an electric current through mercury vapour.

meridian (terrestrial meridian) Any ➤great circle on the surface of the Earth that has the Earth's axis as a diameter. It is therefore the combination of two lines of constant ➤longitude.

meridian, celestial The ➤great circle passing through the ➤celestial poles and an observer's ➤zenith and ➤nadir. It cuts the horizon at points due north and south of the observer.

meristem An actively proliferating tissue in plants that gives rise to new cells which will ultimately differentiate and specialize. Meristems include the vascular and cork ➤cambia and the actively dividing cells of root tips and buds.

Merrifield synthesis The production of proteins by a solid-phase synthesis. The growing polypeptide chain is chemically attached by its C-terminal amino acid to polystyrene beads. As each new unit is attached, the reagents and by-products are simply washed away, leaving behind the growing polypeptide; this virtually eliminates laborious purification steps. The technique is easily automated. For developing the synthesis, (Robert) Bruce Merrifield (1921–2006) was awarded the 1984 Nobel prize for Chemistry.

Mersenne prime A ➤prime number of the form $2^n - 1$ where n, necessarily, is also a prime number. By 2008, 46 Mersenne primes had been discovered (8 in the 21st century), including the largest known prime which has $n = 43\,112\,609$. Named after Marin Mersenne (1588–1648).

mesencephalon ➤brain.

mesoderm ➤embryo.

***meso* isomer** When a molecule has two identical ➤chiral centres, three isomers exist: one that rotates the plane of plane-polarized light to the left, one that rotates it to the right, and the *meso* isomer (from the Greek for 'middle') that does not have any effect on the light because the rotation by one end is exactly cancelled by the opposite rotation by the other end. The isomer is said to be **'internally compensated'**. An example is *meso*-tartaric acid.

mesomeric effect An obsolete term for stabilization caused by ➤delocalization.

meson One of a class of ➤subatomic particle that consists of a ➤quark and an ➤antiquark. An example is the ➤pion. The name was intended to suggest a mass between that of the electron and the proton. The ➤muon was originally called the 'mu-meson', though it is now known to be a ➤lepton, not a meson at all. The mesons and the ➤baryons together form the ➤hadrons; mesons have integral spin, whereas baryons have half-integral spin.

mesophyll The cells of the central part of a leaf, between the upper and lower epidermal layers. The mesophyll is usually present as two layers, an upper **palisade mesophyll**, where most ➤photosynthesis occurs, and a lower **spongy mesophyll**, which has extensive air spaces promoting ➤gas exchange.

mesosphere The layer of the Earth's ➤atmosphere at an altitude of about 50 to 85 km, lying between the ➤stratosphere and the ➤thermosphere. The temperature drops from about $0\,°C$ at its base to about $-100\,°C$ at its top.

mesotrophic A term used to describe lakes and rivers of medium nutrient content. The category is defined by the assemblage of phytoplankton (➤plankton) species present. Compare ➤dystrophic; eutrophic; oligotrophic.

Mesozoic See table at ➤era.

messenger RNA (mRNA) Single-stranded ➤RNA which serves as the template for ➤protein synthesis. DNA is copied into RNA by ➤transcription; the resulting RNA molecules are processed (➤splicing) to give mRNA, which is decoded by ➤ribosomes and translated into ➤protein (➤translation). ➤➤genetic code.

Met Abbr. for ➤methionine.

meta Symbol *m*-. A locant indicating the relative positions of two substituents on a benzene ring. A simple example is meta-dinitrobenzene (see the diagram). The groups are separated by one carbon atom, as opposed to being neighbours in the ➤ortho isomer, or opposite in the ➤para isomer.

meta-dinitrobenzene

metabolic rate The rate at which chemical reactions occur in the cells of an organism. Metabolic rates vary considerably with activity and type of organism. Generally, the more active an organism, the higher its metabolic rate. **Endotherms** (such as birds and ➤mammals) have high energy costs and therefore rapid metabolic rates. **Ectotherms** (such as reptiles, fish and amphibians, as well as all nonvertebrates) of comparable body size have much lower metabolic rates because of their lower energy costs. Metabolic rate is usually measured as oxygen consumption per unit mass of body tissue in a given time. ➤➤homoiothermy; poikilothermy.

metabolism The totality of all the chemical reactions occurring in a cell, organ or organism. All living processes are based on complex series of chemical reactions under the control of ➤enzymes, and may be grouped under two main headings: those that build up complex molecules from simpler ones (➤anabolism), and those that degrade complex molecules into simpler ones (➤catabolism).

metabolite Any chemical that takes part in any chemical reaction associated with living processes (➤metabolism). In particular the term refers to specific substances which are required as basic raw materials for vital processes, such as ➤glucose in ➤respiration, or those compounds which occupy key positions in pathways such as the ➤Krebs cycle.

metal A substance held together by ➤metallic bonding and having the following characteristics, explained by the 'sea' of electrons: they are electrical and thermal conductors, malleable, ductile and shiny. The electrical conductivity arises because the sea of electrons can easily be made to move under the influence of an electric field, causing a current to flow.

The vast majority of the elements are metallic in their uncombined state. Except for the Group 1 elements, the main structures adopted by metals are cubic ➤close packing and hexagonal close packing. The Group 1 metals are less closely packed, crystallizing as ➤body-centred cubic.

metaldehyde A mixture of polymers, mainly the tetramer, of ➤ethanal (acetaldehyde). It has two main uses, as a fuel and as a slug poison.

metal fatigue A long-term decrease in the strength of a metal or alloy caused by repeated elastic stresses.

metallic bonding Bonding that occurs when atoms have sufficiently low ionization energies to allow one or more of their outermost electrons to be shared with their neighbouring atoms, the resulting 'sea' of electrons holding the remaining positive ions together. This is an extreme example of ➤delocalization.

metallocene A compound in which metal ions are sandwiched between two or more cyclopentadienyl rings. An example is ➤ferrocene.

metalloid An element with properties somewhere between those of a metal and a nonmetal. There is no unique way of distinguishing a metalloid from a true nonmetal but the most common is that metalloids are usually ➤semiconductors rather than insulators. Also, their oxides are usually amphoteric. The six elements most often classified as metalloids, namely silicon, germanium, arsenic, antimony, selenium and tellurium, occur in a diagonal pattern in the periodic table. Elements to the bottom left of the metalloids will be metals; those to the top right will be nonmetals.

metallurgy The study of metals and alloys. It includes the extraction of metals from their ores, and the modification of alloys to produce materials with superior properties. Metallurgy is sometimes divided into ferrous and nonferrous metallurgy, showing the dominant importance of iron and its alloys.

metamorphic rock One of the three main types of rock: metamorphic rock has been changed by a number of processes such as heating or high pressure. Metamorphic rocks are significant because they constitute a large part of the continental crust. Examples of metamorphic rock are marble and slate. Compare ➤igneous rock; sedimentary rock.

metamorphosis One or more changes in form during the life cycle of an organism, such as an amphibian or insect, in which the juvenile stages differ from the adult. Thus the frog's life history includes a transition from egg to tadpole to adult. The term **complete metamorphosis** is applied to insects such as butterflies in which the caterpillar stage is distinct from the adult. **Incomplete metamorphosis** describes the life histories of insects such as locusts in which the young go through a series of larval stages, each of which bears similarities to the adult.

metaphase ➤meiosis; mitosis.

metaphosphoric acid One of the acids formed by phosphorus with oxidation number +5, with formula HPO_3. ➤ phosphoric acid.

metastable An ➤equilibrium is metastable if there is another more stable equilibrium position that can be easily reached. An example is a coin standing on its edge, as a more stable equilibrium position is available with it lying flat, showing 'heads' or 'tails'.

metathesis An older name for a reaction in which two substances 'swap partners', so that AB + CD becomes AD + CB. For example, two soluble reactants can produce a precipitate on mixing. The preferred term for this particular process is ➤ionic association.

Metazoa An archaic term for Animalia, the kingdom of ➤animals.

meteor The tube of ionization briefly visible in the Earth's atmosphere, caused by the burning up of a ➤meteoroid. ➤➤meteor shower.

meteorite A rocky meteoroid that is sufficiently large not to burn up completely in the Earth's atmosphere and so reaches the Earth's surface.

meteoroid A small rock or dust particle, originating respectively from an asteroid or comet, on an orbit that intersects the Earth's. When a meteoroid enters the Earth's atmosphere it can become a ➤meteor or ➤meteorite.

meteorology The study of weather, including forecasting. It is, in essence, the physics of the ➤atmosphere, particularly with regard to the distribution of pressures and temperatures within it. Modern meteorology uses very powerful computers to develop forecasts from weather reports, but it suffers from the chaotic nature of the atmosphere: the development of weather systems is very sensitive to the initial conditions (➤chaos theory).

meteor shower The appearance on the same dates of the year of ➤meteors, seeming to radiate from the same point in the sky. They are observed when the Earth passes close to the orbit of a ➤comet, and occur as the Earth encounters dust particles shed by the comet that have spread around its orbit.

meter **1** An instrument for measuring some physical quantity. For example, a ➤voltmeter measures potential difference and an ➤ammeter electric current.
2 The US spelling of ➤metre.

methanal (formaldehyde) The molecule HCHO, the simplest ➤aldehyde. It is normally found in the lab as a colourless aqueous solution called ➤formalin. It is produced by partial oxidation of methanol. Its major use is in making polymeric materials, especially ➤phenol–formaldehyde resins.

methane CH_4 The simplest ➤alkane, which is a colourless gas. It has the four hydrogen atoms arranged at the corners of a tetrahedron (see the diagram). It is the major constituent of natural gas; indeed 'dry' natural gas contains at least 95% methane. Being an alkane, methane is generally unreactive, apart from burning well, a fact put to good use in central heating systems, water heaters and cookers.

methane

methanoic acid (formic acid) HCOOH The simplest carboxylic acid, which is a colourless liquid. It was originally isolated from ants, hence the name 'formic'. It is also found in stinging nettles, giving them their sting.

methanol (methyl alcohol) CH_3OH The simplest alcohol, which is a colourless liquid. It is chemically very similar to ethanol, and is also used as a fuel, particularly in Indy car racing in the US. To avoid government tax, it is often used instead of ethanol in solvents, such as ➤methylated spirit. Methanol has the serious effect of causing permanent damage to vision.

methionine (Met) A common ➤amino acid. ➤➤Appendix table 7.

methyl alcohol Traditional name for ➤methanol.

methylamine CH_3NH_2 The simplest amine, which is a colourless liquid with a smell like ammonia's. Also like ammonia it is alkaline, turning moist litmus blue, so that test is not definitive for ammonia. Methylamine can be distinguished from ammonia by reaction with nitrous acid, which produces nitrogen gas with the amine.

methylated spirit ('meths', US denatured alcohol) ➤Ethanol to which additives, especially ➤methanol and a dye, have been added to make it unpalatable. The addition of methanol makes it very dangerous to consume, as it can cause blindness and even death. It is used as a fuel and a solvent.

methylation ➤DNA methylation.

methylbenzene (toluene) $C_6H_5CH_3$ A very important aromatic molecule, derived from benzene by substituting a methyl group for a hydrogen atom (see the diagram), which exists as a colourless liquid. It is widely used as a solvent, especially for organic compounds, and as a raw material for making TNT (➤2,4,6-trinitromethylbenzene).

methylbenzene

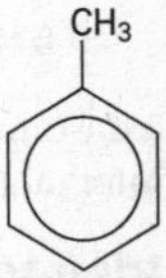

methylene The group $=CH_2$.

methylene blue A blue dye, which has green crystals, used as a redox indicator and to stain biological specimens.

methylene dichloride The old-fashioned name for ➤dichloromethane, CH_2Cl_2.

methyl group The group $—CH_3$, which is the simplest ➤alkyl group. Examples of common chemicals containing the group are methylbenzene, $C_6H_5CH_3$, and methylpropane, $(CH_3)_3CH$.

methyl 2-methylpropenoate (methyl methacrylate) A colourless liquid composed of the molecules shown in the diagram. Its transparent polymer Perspex (➤poly(methyl methacrylate)) is an ➤addition polymer.

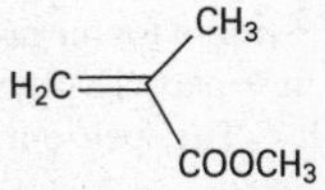

$$H_2C \diagup \diagdown \substack{CH_3 \\ COOCH_3}$$
methyl 2-methylpropenoate

methyl orange An ➤acid–base indicator whose pK_{in} is near 4, making it a useful indicator for titrations between a strong acid and a weak alkali. It is yellow if the pH is above 4 and red at a pH below 4. The colour change is not very noticeable and other ingredients are sometimes added to make **screened methyl orange**, which undergoes a more obvious colour change.

methylpropane (2-methylpropane, *iso*-butane) $(CH_3)_3CH$ The isomer of ➤butane with three carbon atoms in a row and a methyl group off the central carbon atom. Like butane, it is a colourless gas.

methyl *tert*-butyl ether ➤MTBE.

metre (US meter) Symbol m. The SI unit of length. The metre dates from the French Revolution, when the distance along the meridian through Paris from the equator to the north pole was defined to be 10 million metres. In 1984 the metre was redefined as the distance travelled by light in a vacuum in 1/299 792 458 s.

metric Symbol g. A ➤tensor that generalizes the concept of the ➤distance between two points. In ➤general relativity, where curvature of spacetime is caused by the presence of mass, it is the metric that expresses this curvature. In any coordinate system with an arbitrary number of dimensions, the distance (ds) between two points separated by an infinitesimal increment in each coordinate dx_m is given by $\mathrm{d}s^2 = \Sigma g_{mn}\,\mathrm{d}x_m\,\mathrm{d}x_n$. In three-dimensional Cartesian coordinates, the metric is the unit matrix (resulting in ➤Pythagoras' theorem). In four-dimensional spacetime the metric is

$$\begin{bmatrix} -1 & 0 & 0 & 0 \\ 0 & 1 & 0 & 0 \\ 0 & 0 & 1 & 0 \\ 0 & 0 & 0 & 1 \end{bmatrix}$$

with the time-like component occupying the first slot, giving rise to the definition of an **interval** (ds) in spacetime of $\mathrm{d}s^2 = -c^2\mathrm{d}t^2 + \mathrm{d}x^2 + \mathrm{d}y^2 + \mathrm{d}z^2$.

metric space A particular type of ➤topological space consisting of a ➤set equipped with a metric $d(x, y)$ which assigns a distance between pairs of points x and y that is 0 only if $x = y$; it is symmetric in that $d(x, y) = d(y, x)$; and satisfies the triangle inequality $d(x, y) \leq d(x, z) + d(z, y)$.

metric system A system of units based on the ➤metre as the unit of length. Several such systems were evolved, in particular the c.g.s. system and the m.k.s. system. The internationally agreed metric system in current use is called ➤SI units.

metric ton ➤tonne.

metrology The study of weights and measures.

MeV Symbol for mega- (10^6) ➤electronvolt, equal to about 1.6×10^{13} J. It is a useful unit in ➤particle physics, as the ➤rest energy of an electron is approximately 0.5 MeV. The spelt-out form is rarely written or spoken.

MF Abbr. for ➤medium frequency.

Mg Symbol for the element ➤magnesium.

MHC Abbr. for ➤major histocompatibility complex.

MHD Abbr. for ➤magnetohydrodynamics.

MHL Abbr. for ➤major histocompatibility locus (➤major histocompatibility complex).

mho An unofficial unit of conductance (identical to the ➤siemens). Conductance is the inverse of resistance; 'mho' is 'ohm' spelt backwards.

MHz Symbol for the unit ➤megahertz, commonly used for measuring radio frequencies.

mica A type of layered aluminosilicate. An important example of a mica is ➤muscovite. Micas flake easily, because of their layered structure.

micelle A subcellular particle made up of an aggregate of molecules in a ➤colloid. Typically the particle is spherical, with a ➤hydrophilic exterior and a ➤hydrophobic core. The structure is significant since the **micellar theory** proposes that the origin of the first cells in which key biological processes were isolated was as a simple membrane-bound droplet.

Michaelis–Menten equation An equation representing a general theory of ➤enzyme action and kinetics proposed by Leonor Michaelis (1875–1949) and Maud Leonora Menten (1879–1960) in 1913:

$$1/V = 1/V_{max} + (K_M/V_{max})/[S],$$

where V_{max} is the maximum value of the rate V and [S] is the substrate concentration. The **Michaelis constant** K_M represents the substrate concentration required for the reaction to attain half its maximum rate. The lower the constant, the higher the affinity of the enzyme for its substrate.

Michael reaction A very important reaction in organic chemistry for extending the carbon chain. A common reagent is ➤ethyl 3-oxobutanoate. The species to which it adds is of the form RCH=CHX where X is an acyl group. Named after Arthur Michael (1853–1942).

Michelson–Morley experiment An experiment designed to measure the speed at which the Earth moves through the ➤ether (2). The principle was based on the analogy that a swimmer swimming at a constant speed c in a river moving at speed v takes longer to swim from A to B (see the diagram) and back, than from C to D and back cross-stream. Thus a beam of light that travels parallel to the ether, reflects off a mirror, and returns antiparallel to the ether has a phase delay introduced relative to one moving perpendicular to the ether for both legs. This should be detectable as a shift in the ➤interference pattern between the two beams. However, no such shift could be found. This led to the inescapable conclusion either that the Earth was at rest with respect to the ether (which was unacceptable, given that the Earth moved with respect to all other heavenly bodies), or that the ether did not exist and that the speed of light in a vacuum was constant in any ➤inertial frame. This was used by Einstein as a basis for ➤special relativity. It was the most significant null result in the history of science. Named after Albert Abraham Michelson (1852–1931) and Edward Williams Morley (1838–1923).

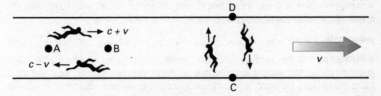

Michelson–Morley experiment

micro- Symbol μ. An SI prefix meaning one-millionth of a base unit; for example, 1 μg is equivalent to 10^{-6} g.

microbe ➤microorganism.

microbiology The branch of biology concerned with the study of ➤microorganisms.

microcomputer A computer based upon one or more ➤microprocessors, with memory and ➤peripherals. Microcomputers have become dramatically more sophisticated in recent years; many now possess megabytes of ➤RAM and have ➤multitasking ➤WIMP-based ➤operating systems. A typical microcomputer has a processing speed of tens of ➤MIPS per MHz, so is generally less powerful and physically smaller than a ➤minicomputer. The term is becoming obsolescent.

microcrystalline ➤crystalline.

microelectronics The field concerned with the fabrication of integrated circuits on semiconductor chips. VLSI (➤very large scale integration) allows in excess of a million transistors on a single silicon chip. A continuing goal in microelectronics is to fit more transistors closer together, so reducing capacitive effects.

micrometer A device designed to measure lengths to a very high precision.

micrometre (US **micrometer)** Symbol μm. A unit of distance equivalent to one-millionth of a metre. This is comparable to the wavelength of visible light, so to view objects at this scale is to push the best optical microscopes to their limit. It is now possible with laser cutting devices to manufacture to this scale, as in precision castings for turntables.

micron An obsolescent name for the ➤micrometre (10^{-6} m); the latter is preferred in SI units.

micronutrient ➤trace element.

microorganism (microbe) A generalized term for any unicellular organism, but particularly ➤bacteria. The term is often used incorrectly to include ➤viruses.

microphone A device for converting sound into an electrical signal. Typically the sound vibrates a diaphragm whose physical movement is detected by electromagnetic ➤induction (**moving-coil microphone**), by a change in the resistance of a material (**carbon microphone**) or by the generation of an ➤e.m.f. in a material showing the ➤piezoelectric effect (**crystal microphone**).

microprocessor A single integrated circuit, usually of silicon, which contains the ➤central processing unit that forms the basis of a ➤microcomputer.

micropyle ➤ovule.

microscope A device that allows a small object to be viewed under high magnification. An **optical microscope** achieves this by passing light reflected from the object through a combination of magnifying lenses. An ➤electron microscope produces images by using electrons and electron lenses. ➤➤atomic force microscope; proton microscope; scanning tunnelling microscope; ultramicroscope.

microscopic On a very small scale, typically requiring the use of a microscope. Compare ➤macroscopic.

microsecond Symbol μs. A unit of time equivalent to one-millionth of a second (10^{-6} s). Reactions on the microsecond time-scale can be investigated using laser ►flash photolysis.

microstate A concept in statistical mechanics: for a quantum-mechanical system, its ►quantum state; for a classical system, its position in ►phase space.

microtome A mechanical device for accurately cutting thin sections of material to be viewed under the microscope.

microtubule A hollow cylinder made from protein, about 25 nm in diameter, forming an essential part of the structure of the ►cytoplasm of cells. They are associated in particular with structures in cells that bring about movement such as cilia and flagella (►undulipodium). During ►meiosis and ►mitosis, microtubules form part of the spindle associated with the orientation and separation of ►chromosomes during cell division.

microwave A region of the ►electromagnetic spectrum lower in frequency than the infrared. The typical wavelength range is 1 mm to 30 cm. Microwave ovens operate on the principle that the absorbed microwave energy causes water molecules to rotate more violently, so heating the food.

microwave background ►cosmic background radiation.

microwave spectroscopy A form of ►absorption spectroscopy using ►microwave radiation. Absorption of photons occurs because of rotational transitions within molecules. Pure rotational spectra in the gas phase have been used to measure bond lengths in molecules such as HCl.

midbrain ►brain.

middle ear ►ear.

middle lamella The amorphous layer between adjacent cells of plants that binds the layers together. The main constituent is calcium pectate, a gum.

migration The seasonal movement of, usually, whole populations of organisms in response to environmental stimuli such as temperature and daylight hours. The annual movements of whales off the west coast of North America from Mexico to the Arctic and back, and those of swallows from central Africa to Europe to breed, are typical examples.

mile, nautical ►nautical mile.

mile, statute An Imperial unit of distance equal to 1760 ►yards. It is thought to be derived from the Roman unit *Milia passuum*, 1000 double paces of a foot-soldier. 1 mile = 1609.344 m.

milk A nutritious aqueous fluid produced by the mammary glands in order to feed young ►mammals. Milk is of varied composition, but typically contains proteins, fats and the sugar lactose in addition to minerals such as calcium and iron, vitamins and various antibodies.

milk of magnesia A white suspension of magnesium hydroxide, $Mg(OH)_2$, used to neutralize acid in the stomach, being a mild alkali.

Milky Way ➤galaxy.

Miller indices Symbol (*hkl*). A method of describing a particular plane through points in a ➤lattice. The Miller indices of a plane are the components of the shortest ➤reciprocal lattice vector that is perpendicular to the plane. To take a trivial example, the reciprocal lattice for a simple cubic lattice is itself simple cubic, so

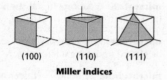

(100)   (110)   (111)
Miller indices

the planes are labelled by simple cubic lattice vectors: (100), (110) and (111) are three examples (see the diagram). Named after William Hallowes Miller (1801–80).

milli- Symbol m. An SI prefix meaning one-thousandth of a base unit; for example, 1 mm is equivalent to 10^{-3} m.

millibar Symbol mbar. A unit of pressure equal to one-thousandth of a bar. Atmospheric pressure is quoted in millibars in weather reports.

millilitre Symbol ml. This very common notation is rather inelegant as it means one-thousandth of a thousand cubic centimetres; the symbol cm^3 is preferable (➤litre).

millimetre (US millimeter) Symbol mm. A unit of distance equal to one-thousandth of a metre: 1 mm is equivalent to 10^{-3} m.

MIME Acronym for multipurpose Internet mail extension. The MIME type enables a Web browser (➤www) or an ➤email viewer to specify which application should run when a file is opened over the Internet.

mineral The source of an element as found in nature. Minerals are usually inorganic such as mineral green, copper(II) carbonate $CuCO_3$, or mineral butter, antimony(I) chloride $SbCl$, although a few organic examples exist.

mineral acid One of the common inorganic acids, typically hydrochloric, nitric or sulfuric acid.

minicomputer A computer intermediate in processing power between a ➤microcomputer and a ➤mainframe. However, these distinctions become increasingly blurred as computing power increases. In particular, the differences from microcomputers are ill-defined, since minicomputers are also based on microprocessors. In general, minicomputers can support more than one user simultaneously and are more expensive and slightly larger than desktop microcomputers. The term is becoming obsolescent.

minimum A value of a function that is less than any of the neighbouring values. Notice that such a **local minimum** need not necessarily be a **global minimum**. Local minima are usually located by finding ➤stationary points.

Minkowski spacetime The four-dimensional ➤spacetime that forms the basis of special relativity, in the absence of gravitational fields. It has three space dimensions and one time dimension, thus an ➤event occurring at point (x, y, z) at time t is given the coordinate (ct, x, y, z), where c is the speed of light (➤➤four-vector). Named after Hermann Minkowski (1864–1909).

minor axis The shorter axis of an ➤ellipse.

minority carrier Whichever type of ➤carrier (2), either positive ➤holes or negative electrons, is in the minority in a ➤semiconductor. Compare ➤majority carrier.

minor planet ➤asteroid.

minute 1 Symbol min. A unit of time equivalent to 60 ➤seconds.
2 Symbol ´. A unit of angle equivalent to 1/60 of a ➤degree.

Miocene See table at ➤era.

MIPS Abbr. for million instructions per second. A unit of computing power, used to measure the speed at which machine code instructions are processed.

mirror A reflective surface designed to form an optical ➤image of some sort. A plane mirror produces a virtual image behind the mirror. A **concave spherical mirror** produces a real image in front of the mirror, while a **convex spherical mirror** produces a virtual image behind it; in both cases the mirror suffers from ➤aberration and the image is imperfect. A **paraboloidal mirror** has no such limitation, and focuses a parallel beam at the focus of the paraboloid.

mirror image The result of a geometrical transformation of an object that includes an ➤inversion. Literally, the ➤image of an object reflected in a mirror is inverted in this way. The mirror image of a right-handed object is a left-handed object and vice versa. ➤chiral.

mirror plane The plane that remains invariant in a ➤reflection (2). It would be the plane of the mirror if the transformed object were the virtual ➤image of the original in a mirror.

miscibility The propensity of two liquids to mix and form a single liquid ➤phase. Because entropy increases on mixing, miscibility is the expected behaviour unless the intermolecular forces in the two liquids are very different, in which case the attraction to their own kind may cause phase separation. Two liquids which are **miscible** in all proportions are ethanol and water; on the other hand, oil and water are almost **immiscible**.

mist A colloidal suspension of a liquid in a gas (➤colloid). In particular, water droplets in the atmosphere are formed by the condensation of atmospheric water vapour. The droplets nucleate on dust (➤nucleation), which is normally present in sufficient quantity to allow mist to form whenever the temperature falls to the ➤dew point. Mist that forms away from the surface of the Earth is usually called **cloud**. Mist that reduces visibility to less than 1 km is termed ➤fog.

mitochondrion (plural **mitochondria)** The 'powerhouse' of a cell; a membrane-bound ➤organelle (see the diagram overleaf) in ➤eukaryotic cells which is the site of ➤aerobic respiration and the source of most ➤ATP by ➤oxidative phosphorylation.

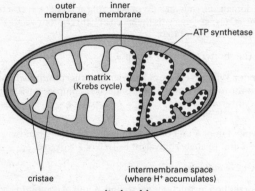

mitochondrion

Numbers in cells vary from one to several hundred and, generally, the more active the cell, the more mitochondria are present. They are cylindrical or spheroidal bodies 0.5–5 μm in diameter and up to 10 μm in length, bounded by a double membrane. The inner membrane has its surface area increased by inwardly protruding folds called **cristae**, which are the site of the ➤electron-transport chain. The enclosed fluid-filled space in the middle, the **matrix**, contains the enzymes and other components of the ➤Krebs cycle. Mitochondria contain their own ➤DNA, which codes for a number of mitochondrial proteins and has a number of applications in genetic engineering.

mitosis The process of cell division involved in growth that gives rise to an increase in the number of body cells in multicellular organs (see the diagram opposite). In single-celled organisms and in plants that can reproduce vegetatively, it is also important as a process which allows reproduction. Mitosis produces two cells which are genetically identical to, and contain the same number of ➤chromosomes in the ➤nucleus (2) as, the parent cell. The process comprises a nuclear division, usually followed by a division of the cytoplasm. The nuclear division proceeds by the formation of a **spindle** derived from ➤microtubules, and the replication of the chromosomes along their length to form two ➤chromatids, attached to each other at the **centromere**. The chromatids contract and become more dense, so that with suitable staining they are visible under the microscope. This constitutes **prophase**. During **metaphase** the nuclear membrane disintegrates, and the centromeres become attached to the spindle at the metaphase plate. At **anaphase** the centromeres divide and migrate to the poles of the spindle, separating each pair of chromatids which now become the new (daughter) chromosomes. In **telophase** a new nuclear membrane forms around each group of daughter chromosomes, which now become less dense and are no longer visible as discrete structures under the microscope. (These stages grade into each other, and are not discrete steps.) This completes mitosis, and two daughter cells are now normally produced by cytoplasmic division (cytokinesis) and enter interphase. The new cells may continue to divide or become specialized (differentiated). Mitosis normally takes one to three hours to complete. Compare ➤meiosis.

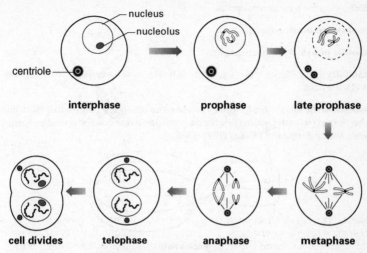

mitosis in a diploid cell with a pair of chromosomes.

Mitscherlich's law (of isomorphism) An approximate law that suggests that crystals composed of the same number of similar elements tend to show ➤isomorphism. Named after Eilhardt Mitscherlich (1794–1863).

mixed oxide A structure in which an array of oxide ions has two or more cations present. Examples include the ➤ferrites used in indoor aerials.

mixture A system consisting of more than one pure substance. An example is air, which consists of nitrogen, oxygen, argon, carbon dioxide, etc. Mixtures should be contrasted with ➤compounds, such as carbon dioxide. Mixtures do not have a fixed composition; the exact percentage of carbon dioxide in air changes by about 10% from summer to winter and in addition shows a rise over the last few decades (➤greenhouse effect). Liquid mixtures include ➤petroleum. Mixtures can usually be easily separated, as in the fractional distillation of liquid air or petroleum.

m.k.s. system A metric system of units based on the metre, kilogram and second. It was the precursor to ➤SI units.

ml Symbol for the unit ➤millilitre.

mm Symbol for the unit ➤millimetre.

m.m.f. Abbr. for ➤magnetomotive force.

mmHg An obsolete unit of pressure corresponding to the weight per unit area of a column of mercury (Hg) one millimetre (mm) high, now called the ➤torr. Normal atmospheric pressure is 760 mmHg.

Mn Symbol for the element ➤manganese.

Mo Symbol for the element ➤molybdenum.

MO Abbr. for ➤molecular orbital.

mobile element ➤transposon.

mobile phase ➤chromatography.

mobility In a conductor, the ratio of the drift velocity of a ➤carrier (2) to the applied ➤electric field.

Möbius strip The surface formed by taking a strip of paper, giving it a half-twist, and then joining its ends together (see the diagram). It has one side and one edge. Named after August Ferdinand Möbius (1790–1868).

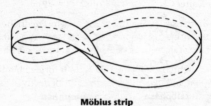

Möbius strip

mode 1 (Math.) The value(s) or class interval(s) containing the largest frequency in a ➤frequency distribution: there may be more than one mode.
　2 ➤normal mode.

modem Acronym for 'modulator–demodulator'. A device that allows a computer to transmit and receive data over a conventional telephone line. It works by converting digital information from the computer to audio signals. The receiving modem converts the audio signals back to digital form. This form of transmission is restrictively slow because of the severely limited ➤bandwidth of current telephone systems, which were not designed for the transmission of data. ➤fibre, optical; ISDN.

moderator A substance used in a ➤nuclear reactor to slow the neutrons down sufficiently for them to be captured by the nuclei. Common examples are boron, graphite and heavy water (D_2O). Boron solution was tipped over the remains of the Chernobyl reactor, the scene of a major nuclear accident in 1986, in an attempt to contain the damage.

modular arithmetic Addition and multiplication **modulo** n of two of the numbers $0, 1, \ldots, n - 1$ is defined by adding or multiplying them in the usual way, but only retaining the ➤remainder on division by n. For example, 5 multiplied by 6 is 2 modulo 7, because 30 leaves remainder 2 on division by 7.

modulation The coding of an electrical signal $f(t)$ within a **carrier** $c(t)$, which is usually a ➤wave (e.g. $c(t) = \sin \omega t$), to enable it to be transmitted in combination, $g(t)$. In **amplitude modulation** (AM) the carrier is multiplied by the signal: $g(t) = f(t) \sin \omega t$. In **frequency modulation** (FM) the signal acts on the frequency of the carrier: $g(t) = \sin[(\omega + f(t))t]$. In **phase modulation** the signal acts on the phase of the carrier: $g(t) = \sin(\omega t + f(t))$. When the combination reaches its destination (typically after transmission as a ➤radio wave) it is subjected to ➤ demodulation to extract the signal.

modulus **1** (Math.) The quantity $\sqrt{x^2 + y^2}$ associated with the complex number $x + iy$ which represents its distance from the origin on the ➤Argand diagram.
2 (Phys.) ➤elastic modulus.

Moho (Mohorovičić discontinuity) The boundary between the Earth's crust and ➤mantle. Named after Andrija Mohorovičić (1857–1936).

Mohr's salt A hydrated double salt, ammonium iron(II) sulfate, $(NH_4)_2SO_4$· $FeSO_4$·$6H_2O$, which is useful as it resists aerial oxidation in the solid state, unlike most iron(II) salts. Named after Carl Friedrich Mohr (1806–79).

Mohs scale A qualitative scale of hardness invented by Friedrich Mohs (1773–1839) in which a substance with a higher number on the scale can scratch all substances with a smaller number. Talc is given the value 1 on the scale, and diamond 10.

moiré pattern The light and dark banded pattern formed by superimposing two sets of alternating light and dark lines that are not perfectly aligned. A fold in net curtains made of moiré silk gives this effect.

mol Symbol for the unit ➤mole.

molality The amount of substance of a solute present per unit volume (usually one litre) of solvent; the volume measured is that of the solvent, *not* the solution. Molality is most useful in ➤colligative properties where the magnitude of the effects depends linearly on the molality.

molar conductivity ➤conductivity, electrical.

molarity Symbol M or M. An obsolete term that is synonymous with the molar ➤concentration in mol dm^{-3}.

molar mass Symbol M or M_m. The mass per mole of a substance. Because of the definition of the mole, the molar mass has the same numerical value as the substance's ➤relative formula mass, but has the units of g mol^{-1}. Carbon dioxide has an M_r of $12.01 + (2 \times 16.00) = 44.01$, so its molar mass, $M(CO_2)$, is 44.01 g mol^{-1}.

molar volume Symbol V_m. The volume occupied per mole of a substance. An ➤ideal gas would occupy a molar volume of about 24 dm^3 mol^{-1} at 20 °C and 1 bar. Almost all gases have molar volumes very close to this value.

mole Symbol mol. The SI unit of amount of substance. One mole is that amount of any substance that contains a number of specified species equal to the ➤Avogadro constant. The statement '1 mol of oxygen' is imprecise, since it has not been specified whether it is a mole of O atoms or O_2 molecules; 1 mol CO_2 means 1 mol of carbon dioxide molecules. The amount of substance can be calculated by dividing the mass of the substance by its molar mass or, for a gas, by dividing the volume of the substance by the molar volume.

molecular beam A beam of molecules at very low pressure in the gas phase. Since the molecules undergo very few collisions with other similar molecules, measurements can be made on essentially isolated molecules. Study of colliding molecular beams can provide very detailed information about reactions, as each colliding molecule can be prepared in a particular rotational state, for example.

molecular beam epitaxy (MBE) The slow deposition of thin layers (typically a few atoms thick) of an **evaporant** with great precision (to within a single layer of atoms) on a ➤**substrate** (2) in ultra-high vacuum. Typically, both the deposited layer and the substrate are semiconductors. The technique is used extensively in ➤microelectronics.

molecular biology The major branch of modern biology associated with the study of macromolecules such as ➤nucleic acids. It includes significant areas such as ➤genetic engineering and aspects of the study of, for example, ➤cancer and inherited diseases. It also has applications in ➤forensic science and in deducing relationships between organisms (➤taxonomy).

molecular formula The formula that specifies the number of atoms of each element present in a molecule. For example, ethene has an ➤empirical formula of CH_2 but a molecular formula of C_2H_4.

molecularity An important piece of information about a step in a reaction, giving the number of species (*not* necessarily molecules) present in the step. The rate-limiting step of an S_N2 reaction (➤nucleophilic substitution), for example, is bimolecular (signified by the 2), so two species react in that step.

molecular map ➤physical map.

molecular orbital (MO) Just as atomic orbitals are the solutions of the ➤Schrödinger equation for atoms, molecular orbitals are the corresponding solutions for molecules. The electrons in atoms occupy atomic orbitals that are commonly represented as regions around the nucleus where there is a high probability of finding the electrons. When atomic orbitals approach closely together, they **overlap** and form molecular orbitals. Two atomic orbitals can overlap to give two molecular orbitals. One of these molecular orbitals lies at a lower energy than the overlapping atomic orbitals and is called a **bonding molecular orbital**. The other lies at a higher energy and is called an **antibonding molecular orbital**. Each molecular orbital can hold one or two electrons, in accordance with the ➤Pauli principle. Molecular orbital theory can be used to explain the detailed electronic structure of ➤oxygen. ➤➤linear combination of atomic orbitals.

molecular sieve A sieve that operates at the molecular level, separating larger molecules from smaller ones. These are often made from ➤zeolites.

molecular weight Traditional name for ➤relative molecular mass.

molecule Two or more atoms chemically combined. Multitudes of important substances contain molecules, from the simple O_2 molecule through the H_2O molecule to the DNA molecule.

mole fraction A quantity x_i expressing the composition of a mixture in terms of the amounts of substance. It is given by $x_i = n_i/n$, where n_i is the amount of substance of component i and n is the total amount of substance. The mole fraction is thus dimensionless. ➤➤Dalton's law.

mollusc (US **mollusk)** A member of the animal ►phylum Mollusca, comprising organisms without segmented bodies and typically possessing a muscular foot. Examples include the gastropod snails and slugs and the lamellibranchs such as mussels and clams. The most advanced molluscs are the cephalopods, which include squid, cuttlefish and octopus.

molybdenum Symbol Mo. The element with atomic number 42 and relative atomic mass 95.94, which is in the second row of the ►transition metals and possesses the usual transition metal property of variable oxidation number. Examples of its compounds include molybdenum disulfide, MoS_2, black crystals which act as a viscosity enhancer, and the molybdates with oxidation number +6, for example the ion heptamolybdate $Mo_7O_{24}{}^{6-}$, which have a rich structural chemistry.

moment 1 (Phys.) ►moment of force.
2 (Phys.) In general, for an object composed of smaller parts, the sum of the values of some physical quantity associated with each part multiplied by the distance of the part from some specified axis or plane. Thus if the quantity is force acting on the part, the moment is the total ►moment of force; if the quantity is electrical charge, the moment is the ►dipole moment.
3 (Math.) ►Expectations of polynomial functions of a random variable X, of the form $E(X - a)^n$, where the constant a is normally zero or the mean of X. The sequence of all moments uniquely determines X.

moment, magnetic ►magnetic dipole.

moment of force A measure of the ability of a force to rotate an object on which it acts. The moment of a force can be defined with respect to any chosen axis (though a pivot or ►centre of mass is often most practical), and is the force multiplied by the perpendicular distance to the axis. There is a rotational analogue of ►Newton's second law of motion which states that the total moment of force or **torque** G_{TOT} applied to a body is equal to the rate of change of ►angular momentum L: $G_{TOT} = dL/dt$. ►►couple; moment of inertia.

moment of inertia For a ►rigid body rotating about an axis, the sum of the mass m_i of each part of the body multiplied by the square of the perpendicular distance r_i to the axis: $I = \sum_i m_i r_i^2$. It relates the ►angular momentum L of a body to its ►angular velocity ω as $L = I\omega$, and the torque G (►moment of force) to its ►angular acceleration α as $G = I\alpha$. For general three-dimensional rotations the situation is very much more complicated, and the equations cannot simply be treated as three components of three arbitrary perpendicular axes. The moment of inertia becomes a second-rank **inertia tensor** I, defined as $I_{xx} = \sum_i m_i(r_i^2 - x_i^2)$ and $I_{xy} = \sum_i -m_i x_i y_i$, where part i is at (x_i, y_i, z_i), a distance r_i from the origin, relating the angular momentum L to the angular velocity ω (which may not necessarily be in the same direction) as $L = I\omega$. The tensor is symmetric, so it can be diagonalized along three specific perpendicular directions known as the ►principal axes of the body.

momentum Symbol p. The product of the ►velocity v of a body and its ►inertial mass m. Its rate of change appears in the second of ►Newton's laws of motion and effectively defines the ►force acting on the body. In relativistic mechanics the quantity becomes a ►four-vector as the ►energy–momentum of the body.

momentum, angular ➤angular momentum.

momentum, conservation of ➤conservation of momentum, law of.

monatomic Consisting of only one atom. The ➤noble gases are monatomic.

Mond process The manufacturing process for nickel in which impure nickel is combined with carbon monoxide at $60\,°C$ to make the compound tetracarbonyl-nickel, $Ni(CO)_4$, which is subsequently decomposed to nickel at above $200\,°C$. Named after Ludwig Mond (1839–1909).

Monel A nickel alloy containing 68% nickel and 32% copper (by mass). It is useful for transporting fluorine, as it resists attack by this highly reactive element.

Monera The ➤kingdom containing all ➤prokaryotic organisms. It includes all ➤bacteria. The Monera are now considered to comprise two major groups: the ➤Archaea and the ➤Eubacteria.

monobasic acid An acid, such as nitric acid HNO_3, having only one hydrogen ion capable of reacting with a base or alkali. Compare ➤dibasic acid.

monochromatic Of one frequency (or, equivalently, of one wavelength). Visible light that is monochromatic has a single pure colour, as in a ➤laser. A perfect monochromatic source exhibits temporal ➤coherence.

monoclinic Describing a crystal system in which only two of the three angles between the unit vectors are right angles. Thus its appearance is of rectangular layers, offset as they are stacked. An example of a monoclinic crystal is that of sulfur, above the transition temperature of $96\,°C$. ➤Bravais lattice.

monoclonal antibody ➤antibody.

monocotyledon A plant of the class Liliidae of the ➤Magnoliopsida that has one seed leaf (cotyledon) in its seeds. **Monocotyledonous** plants typically have strap-like parallel-veined leaves, and floral parts arranged in multiples of three or six. Examples are cereals, fodder grasses, lilies and orchids. Compare ➤dicotyledon.

monocyte The largest of the ➤leucocytes in vertebrate blood. Monocytes have a large kidney-shaped nucleus and may give rise to ➤macrophages and produce growth factors such as ➤interleukin 1.

monodentate ligand A 'one-toothed' ligand, one that donates only a single electron pair to the central metal ion. Examples are water and ammonia. Compare ➤polydentate ligand.

monoecious Describing plants that have both male and female organs on the same individual. Compare ➤dioecious; ➤➤hermaphrodite.

monohybrid A cross or the offspring of a cross between two parents in which a character determined by a single pair of ➤alleles is transmitted by ➤inheritance.

monolayer A single layer of atoms on a surface, usually created by ➤epitaxy. Because monolayers are two-dimensional systems, they have some unusual properties. Monolayers of conductors, for example, may have different energy levels from their bulk equivalents.

monomer The simple molecule from which a polymer can be built by ➤polymerization. The monomer of polythene is ethene; the monomer of ➤PTFE is tetrafluoroethene.

monomeric ➤enzyme.

monophyletic Having a single common ancestry.

monosaccharide A sugar containing only one ring in its ring form. The most common formula is $C_6H_{12}O_6$, as for ➤glucose and ➤fructose. The table overleaf shows the C_3–C_6 ➤aldoses and the C_6 ➤ketoses.

monosodium glutamate The amino acid glutamic acid is dibasic and so can form two salts; the monosodium salt has had only one hydrogen replaced. It is perhaps the most common flavour enhancer, present in many snack and convenience foods, such as crisps.

monotonic The property of a function f over a specified interval $a < x < b$ for which either $df(x)/dx < 0$ for all $a < x < b$, or $df(x)/dx > 0$ for all $a < x < b$. A monotonic function has no ➤stationary points within the interval.

monotreme ➤mammal.

monotropy A form of ➤allotropy where only one phase is thermodynamically stable at room pressure throughout the temperature range. Phosphorus and carbon are **monotropic**, as red phosphorus and graphite are the only stable phases at all temperatures at 1 bar. Compare ➤enantiotropy.

monoxide A binary compound containing one oxygen atom, such as carbon monoxide, CO, or nitrogen monoxide, NO.

monozygotic twins ➤identical twins.

Monte Carlo method A probabilistic method for numerically evaluating an integral of one or more independent variables. Instead of evaluating the quantity to be integrated at an equally spaced grid of values, the quantity is evaluated at a set of *random* values and averaged over those values. It is particularly useful for problems with complicated boundaries where one need only determine if the random point chosen lies within the boundary or not (and is included in or excluded from the sum on that basis).

month A length of time approximately equal to the orbital period of the ➤Moon. Different months are defined according to different reference points. A **sidereal month** (about 27.32 days) is the length of an orbit relative to the ➤fixed stars. A **lunar month** is the time between two identical ➤phases of the Moon (about 29.53 days). A **calendar month** is a period of from 28 to 31 days in the ➤Gregorian calendar.

Moon The only natural satellite of the Earth. It has a radius of 1738 km and orbits at a distance of between 363 000 km (at ➤perigee) and 406 000 km (at ➤apogee). It keeps the same face turned towards the Earth as its orbital and rotation periods are the same. The Moon may have a small core of solid iron, surrounded by a mantle; the crust has an average thickness of 70 km. Viewed from the Earth, the surface of the Moon consists of dark areas known as **maria** (Latin for 'seas', actually solidified lava

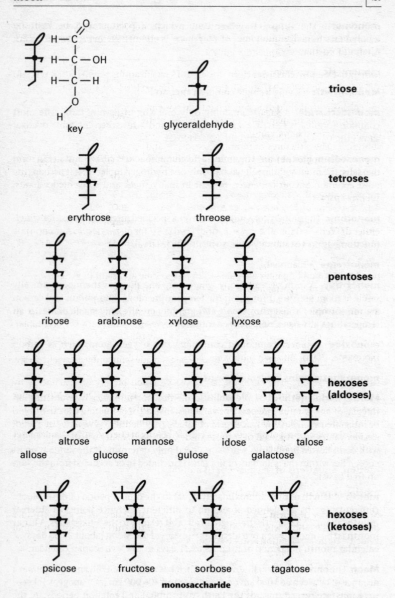

triose

glyceraldehyde

key

tetroses

erythrose   threose

pentoses

ribose   arabinose   xylose   lyxose

hexoses
(aldoses)

allose   altrose   mannose   gulose   idose   galactose   talose

glucose

hexoses
(ketoses)

psicose   fructose   sorbose   tagatose

monosaccharide

plains) and the lighter **lunar highlands**, covered with impact craters. Neil Armstrong was the first human to walk on the Moon, on 20 July 1969 during the successful Apollo 11 mission. ➤Appendix table 4.

Morgan–Keenan classification ➤Harvard classification. Named after William Wilson Morgan (1906–94) and Philip Childs Keenan (1908–2000).

morphine An alkaloid drug which has been both an enormous boon to mankind, in the relief of pain during severe trauma, and an abused, addictive substance. The structures of morphine, codeine and **heroin (diamorphine)** are very closely related (see the diagram).

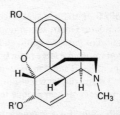

morphogenesis The origin of form in organisms. The development of a few cells in an embryo into a fully functioning eye is an example.

morphine The structure is that of morphine when R=R′=H, codeine when R=CH₃ and R′=H, and heroin when R=R′=CH₃CO.

morphology The study of shape and form in organisms.

mortar 1 A building material made from slaked lime and sand in water.
  2 The bowl in a mortar and pestle used to grind up solids.

Moseley's law The square root of the frequency of characteristic lines in an element's ➤X-ray spectrum is proportional to the atomic number (➤periodic table). Named after Henry Gwyn Jeffreys Moseley (1887–1915).

MOSFET A ➤field-effect transistor based on metal oxide semiconductor technology.

Mössbauer effect The emission of gamma radiation of very precise energy by atoms held rigidly in a crystal lattice. This effect is exploited in **Mössbauer spectroscopy**, in which these very sharp emission lines are used to elucidate the structure of nuclear energy levels (by looking at the splitting of lines) and the chemical environment of particular species (by looking at energy shifts). Named after Rudolf Ludwig Mössbauer (b. 1929).

motion, equations of ➤equations of motion.

motion, laws of ➤Newton's laws of motion.

motion under uniform acceleration The five equations describing motion under uniform acceleration, where s is the displacement, u the initial velocity, v the final velocity, a the acceleration and t the time, are:

$$v = u + at, \quad v^2 = u^2 + 2as,$$
$$s = \tfrac{1}{2}(u+v)t, \quad s = ut + \tfrac{1}{2}at^2, \quad s = vt - \tfrac{1}{2}at^2.$$

motor A device for producing mechanical (usually kinetic) energy from a power source.

motor neurone ➤neurone.

moving-coil ammeter An ►ammeter that uses the interaction of a moving coil of wire carrying the current with a fixed magnetic field to produce a deflection. Compare ►moving-magnet ammeter.

moving-magnet ammeter (moving-iron ammeter) An ►ammeter that uses the interaction of a moving magnet and a fixed coil of wire carrying the current to produce a deflection. Compare ►moving-coil ammeter.

MPEG Abbr. for Moving Picture Experts Group. A working group of the International Standards Organization responsible for development of standards for the representation of ►digital audio and video. The standards contain compression algorithms that result in a large reduction in the size of files used to store audio and video data. The MPEG-1 standard is used in compact discs and MP3 audio files, MPEG-2 is used in DVDs and digital television, while MPEG-4 is used in the delivery of video over the ►Internet and on mobile telephones.

MRI Abbr. for ►magnetic resonance imaging.

mRNA Abbr. for ►messenger RNA.

Mt Symbol for the element ►meitnerium.

MTBE Abbr. for ►methyl *tert*-butyl ether. It is increasingly used as an additive to petrol, which can raise the ►octane number significantly.

mucus An aqueous solution containing the protein **mucin** and various salts. It is secreted by goblet cells of **mucous membranes** which line a variety of body surfaces, including the walls of the gut, the vagina and the respiratory system, and the external surfaces of, for example, amphibians and ►molluscs such as slugs. The functions of mucus include lubrication, prevention of desiccation and protection.

Mulliken scale A scale for measuring ►electronegativities. Named after Robert Sanderson Mulliken (1896–1986).

multicellular Describing a tissue or organism composed of many cells.

multicentre bonds It is becoming clear that most bonds within molecules are in fact multicentre, that is they bond more than two atoms together. The first unambiguous case was in the molecule diborane, as the central hydrogen atoms are bonded to both boron atoms with just two electrons bonding the three atoms together (►three-centre bond). The early chemists thought that the critical feature of bonding was the involvement of two atoms, missing the more fundamental significance of the electron pair. Other examples of multicentre bonds include the six-centre lowest-energy molecular orbital in benzene. ►►delocalization.

multimeter An instrument for measuring electric ►currents and ►voltages, and sometimes also ►resistances. Modern multimeters are almost exclusively **digital multimeters** (DMMs).

multiple An integer a is a multiple of an integer b if b divides a with no remainder.

multiple bonding A situation in which at least two atoms are joined together by more than one pair of electrons. Double bonds exist in oxygen O=O and ethene CH_2=CH_2. A triple bond exists in nitrogen N≡N.

453   **muscle**

multiple integral An integral that involves more than one variable, such as a double or triple integral. They are evaluated by successively integrating with respect to each variable, treating the others as constants.

multiple proportions, law of When two elements combine together to form more than one compound, the masses of one element that combine with a fixed mass of the other are in a simple whole-number ratio: for example, 3:5 for PCl_3 and PCl_5. This was one of the great early laws of chemistry which were eventually explained by the ➤atomic theory.

multiple star ➤binary star.

multiplet 1 A group of lines in a ➤spectrum that are derived from the same basic transition, but with different shifts in frequency or energy causing individual lines to appear separate. Chemical shifts give rise to multiplets, and ➤spin–orbit coupling in incomplete shells (e.g. the 4f subshell of some ➤lanthanides) often creates a spectrum with many components.
    2 A group of subatomic particles in the ➤standard model that share the same ➤spin, ➤parity and ➤baryon number, but have different values of ➤hypercharge and ➤isotopic spin. They are formed from similar combinations of ➤quarks.

multiplication A fundamental operation in arithmetic in which two numbers a and b are combined to give a third, their **product**, written as $a \times b$ or $a \cdot b$ or ab. Initially defined by repeated addition of whole numbers, so that 2×3 means '2 lots of 3', multiplication is successively extended to fractions and to all real and complex numbers. Algebraic expressions are multiplied in accordance with the ➤distributive law.

multiplicity The number of items in a ➤multiplet.

multipole The generalization of a dipole, quadrupole, etc. Multipoles are of significance when a three-dimensional angular dependence is broken up into ➤spherical harmonics.

multitasking In computing, the ability of an ➤operating system apparently to handle several jobs at once. Frequently this is done by dividing the processor's time between the jobs, giving each a small **time slice**, so that while only one operation is ever being carried out at once, it seems to the user that the jobs are being executed simultaneously. Contrast ➤parallel processing, when several operations are physically carried out at the same time.

muon Symbol μ. The ➤lepton that, in the ➤standard model, corresponds to the ➤strange quark and ➤charm quark. It is essentially a more massive electron (charge -1, spin $\frac{1}{2}$, mass 105.7 MeV/c^2). Its ➤lifetime is measured in microseconds. It was discovered in 1937 in cosmic rays by Carl David Anderson (1905–91), and was initially wrongly classified as a ➤meson because of its mass.

murexide An indicator used especially for complexometric ➤titrations.

muscle One of a number of types of contractile tissue in animals that bring about movement (see the diagram overleaf). Muscles consist of muscle fibres which make up muscle tissue. The fibres are usually all oriented in one direction so that, when sufficient of them contract, the tissue shortens, so generating a force. Muscle fibres contain two special filamentous proteins, ➤actin and ➤myosin. When a nerve impulse reaches a muscle, calcium ions are released, which brings about a conformational change in the myosin filaments, causing the actin and myosin filaments to slide past each other (see diagram (a)). A number of other protein factors and energy from ➤ATP are also involved. Vertebrate muscle is of three types. **Striated muscle** (voluntary or skeletal muscle) is attached to the skeleton and brings about movement of bones, under conscious control. The term 'striated' derives from the appearance of a number of transverse bands (A and I bands, and H zones) and **Z lines**, which can be seen under the optical microscope and denote the origin and areas of overlap between the actin and myosin filaments (see diagram (b)). **Cardiac muscle** is found exclusively in the heart and has a number of special properties. **Smooth muscle** (or **involuntary muscle**) is under the control of the autonomic ➤nervous system and is found in the gut, arteries and the walls of other tubular structures in the body.

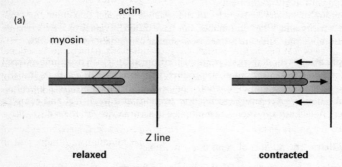

(a)

actin

myosin

Z line

relaxed   **contracted**

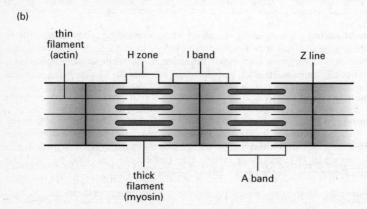

(b)

thin filament (actin)

H zone   I band   Z line

thick filament (myosin)

A band

muscle (a) Muscle contraction, showing the 'sliding filament mechanism'; change in the conformation of the protein myosin causes the actin and myosin filaments to slide past each other, shortening the muscle fibre. (b) Structure of striated muscle.

muscovite (white mica) An aluminosilicate containing potassium with the formula $KAl_2(OH)_2(Si_3AlO_{10})$.

mustard gas A war gas which causes severe burns on contact. Its formula is $ClCH_2CH_2SCH_2CH_2Cl$. It was first used in the First World War and its use was subsequently banned by the Geneva Protocol.

mutagen A chemical or other agent, such as X-rays, that increases the rate of ➤mutation by causing changes in the DNA.

mutant Any organism that possesses a ➤mutation.

mutation A change in the genetic material of a ➤cell or ➤virus. Mutations may be either major changes such as the loss or rearrangement of a large section of a ➤chromosome, or relatively minor changes involving the substitution or deletion of a single base in a section of ➤DNA, which may produce a change in the amino acid sequence of a protein. Both types may result in changes to the functioning of the cell. Mutations occur during cell division and arise either as a malfunction of the mechanisms of ➤meiosis and ➤mitosis, or as a copying error during the replication of DNA. If they occur in ➤gamete-forming cells, the changes can be inherited and result in new features in the offspring. Most mutations reduce the functional efficiency of cells. For example, in humans a substitution of a single base in the DNA that codes for haemoglobin produces a single change in the amino acid sequence of this molecule; this results in ➤sickle cell anaemia, in which the haemoglobin is less efficient at carrying oxygen. Mutations that add or delete single bases or pairs of bases interrupt the ➤reading frame and are termed **frameshift mutations**. Mutations may occasionally produce a favourable change in an organism and render it more likely to survive in its environment. It is in this respect that mutations are important in ➤evolution. ➤➤Darwinism; genetic code.

mutual inductance Symbol M_{12} (or similar); unit H, ➤henry. The ➤e.m.f. E_1 induced in circuit 1 by a change in the magnetic flux through that circuit when the current I_2 in circuit 2 is changed is given by:

$$E_1 = M_{12} \frac{dI_2}{dt}.$$

➤➤inductance; induction.

mutual induction The ➤induction of an ➤e.m.f. in a circuit by a change in the ➤magnetic flux through that circuit when the current in a different circuit is changed. ➤➤mutual inductance.

mutualism ➤symbiosis.

mutually exclusive Two events A and B are mutually exclusive if the ➤probability that both A and B occur is zero, from which it follows that $P(A \cup B) = P(A) + P(B)$.

mycelium The collective term for the vegetative body of a ➤fungus composed of hyphae. These are thread-like structures, consisting of a hyphal wall containing membrane-bound cytoplasm. They may be divided by cross-walls into discrete sections, depending on the type of fungus.

mycology The study of ➤fungi.

Mycota An alternative name for the ➤kingdom Fungi.

myelin sheath The fatty cylinder around the fibres of some nerve cells secreted by Schwann cells (➤neurone).

myeloid tissue The site of blood formation (haemopoiesis) in the red bone marrow of vertebrates. Myeloid tissue is the site of origin of several cell lines important in the ➤immune system.

myeloma A malignant ➤cancer of the ➤myeloid tissue causing a form of anaemia, usually in adults and the elderly.

myoglobin An iron-containing protein found in vertebrate muscle that can bind oxygen to provide a reserve in addition to the oxygen bound by ➤haemoglobin in actively respiring muscle. It is present to a greater or lesser extent in all vertebrates, including humans, but is particularly found in the muscles of diving mammals where it provides a supply of oxygen during extended periods of submersion.

myopia (short sight) The inability of the eye to focus on objects at a distance, often because the distance from the lens to the retina exceeds the ➤focal length of the lens in its most relaxed state. It is usually caused by an elongated eyeball, which causes the image to be brought to a focus in front of the retina. It is corrected by spectacles with a ➤concave lens. Compare ➤hypermetropia.

myosin One of the two major proteins found in ➤muscle; the other is ➤actin. It is a change in the conformation of myosin and its interaction with actin that is the basis of muscle contraction.

myriapod A member of the ➤arthropod class formerly called **Myriapoda**; this class is now regarded as two separate classes, Chilopoda and Diplopoda. The Chilopoda (centipedes) contains carnivorous animals with one pair of legs per segment. The Diplopoda (millipedes) contains primarily herbivorous animals with two pairs of legs per segment.

myxovirus One of a group of ➤RNA-containing ➤viruses that cause diseases in vertebrates, including humans. Measles, mumps and fowl pest are myxoviruses.

n Symbol for the prefix ➤nano-.

N **1** Symbol for the element ➤nitrogen.
   2 Symbol for the unit ➤newton.

ℕ Symbol for the set of ➤natural numbers.

Na Symbol for the element ➤sodium, derived from the Latin *natrium*.

NA Abbr. for ➤numerical aperture.

nabla ➤del.

NAD⁺ (nicotinamide adenine dinucleotide, coenzyme I) A ➤coenzyme which participates in many ➤redox reactions in biological systems. It is particularly important in certain reactions of the ➤Krebs cycle that involve ➤dehydrogenation. The nicotinamide ring of NAD⁺ accepts a hydrogen ion plus two electrons (effectively a hydride ion, H⁻) forming NADH. The ring is no longer stable, and therefore the added hydride ion is easily transferred to other molecules. Subsequent reoxidation of NADH by the ➤electron-transport chain is linked to the formation of ➤ATP. See the diagram overleaf.

NADH ➤NAD⁺.

nadir The point on the ➤celestial sphere directly opposite the ➤zenith.

NADP⁺ (nicotinamide adenine dinucleotide phosphate, coenzyme II) A derivative of ➤NAD⁺ which participates in ➤redox reactions during the ➤light reaction of photosynthesis. High-energy reactions cause the ➤photolysis of water, in which the hydrogen reduces NADP⁺ to NADPH and generates the oxygen released during photosynthesis. The reduced NADPH is used in the conversion of carbon dioxide to carbohydrate during the ➤dark reaction of photosynthesis.

NADPH ➤NADP⁺.

nano- Symbol n. The SI prefix meaning one-billionth of a base unit. For example, 1 ns is equivalent to 10^{-9} s. Its name comes from the Latin for 'dwarf'.

nanometre (US **nanometer)** Symbol nm. A unit of distance equivalent to 10^{-9} m. It is a useful unit for atomic dimensions as they are usually of the order of 0.1 nm, and also for optical wavelengths.

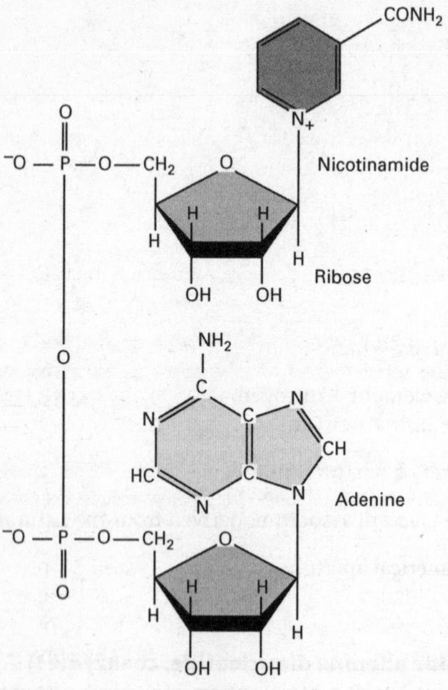

$CONH_2$

$^-O - P - O - CH_2$ Nicotinamide

Ribose
OH   OH

NH_2

Adenine

$^-O - P - O - CH_2$

OH   OH

NAD^+

nanosecond Symbol ns. A unit of time equivalent to 10^{-9} s. Molecular vibrations occur in a few nanoseconds, as do some of the fastest chemical reactions. Processes on this time-scale can be measured using laser ➤flash photolysis.

nanotechnology Technology involving objects with dimensions of the order of magnitude of a few atoms, thus typically around a distance of 1 nanometre. An important example is a **carbon nanotube**, which can be imagined as constructed by splitting a buckyball (➤buckminsterfullerene) into two and then adding a cylindrical section to join the two hemispheres together. Nanotubes are being investigated as possible vehicles for targeted delivery of drugs.

naphtha A fraction of petroleum obtained by ➤fractional distillation. Different oil companies use different names for the fractions with five to ten carbon atoms; the range from five to eight is often termed 'gasoline' and that from nine to ten 'naphtha'. Naphtha contains mainly alkanes, both straight-chain and branched. It is currently the favourite feedstock for further refining by ➤cracking.

naphthalene The simplest polynuclear aromatic hydrocarbon, containing two benzene rings fused together (see the diagram), which exists as a white crystalline solid with a penetrating smell, used in mothballs. There are two series of substitution positions for one group (➤naphthol).

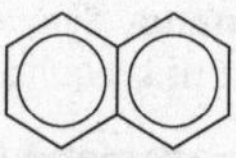

naphthalene

naphthol A derivative of naphthalene with a hydroxyl group either at the top of the molecule (1 or α) or at the side (2 or β) (see the diagram). The 2-isomer is more significant, being used as a test for aromatic amines via coupling reactions (➤diazo coupling). Both isomers are crystalline solids.

1-naphthol   **2-naphthol**

Napierian logarithm Alternative name for ➤natural logarithm, named after John Napier (1550–1617).

narcotic drug A drug that when administered to the body in moderate doses relieves pain, dulls the senses and induces sleep. In excessive doses it may cause stupor or convulsions. Typical narcotics include opium and its derivatives, including ➤morphine and heroin.

nascent hydrogen A fictitious concept; hydrogen at its 'moment of birth' was thought to be especially reactive. More likely, the reactions thought to be due to nascent hydrogen arose from hydrogen atoms or excited hydrogen molecules.

nastic movements (nasties) Movements of parts of plants in response to environmental stimuli that do not involve growth. The so-called 'sleep movement' of flowers (**nyctinastic movement**) involving the closing of petals at night is an example. The periodicity of nastic movements is usually controlled by an internal ➤biological clock.

native Describing an element found in nature uncombined with any other element. Gold is a common example, as are the ➤platinum metals.

natural frequency Symbol f_0 or ω_0 (natural angular frequency). The frequency at which a system would undergo oscillations in the absence of any driving force. The natural frequency is usually close to, but not exactly equal to, the resonant frequency (➤resonance). ➤➤simple harmonic motion.

natural gas A mixture containing dominantly methane, CH_4, together with some of the higher gaseous alkanes. During the next half century natural gas can be expected to provide a major share of the UK's energy needs, both for cooking and for heating.

natural logarithm The ➤logarithm to the ➤base e, written $\log_e$ or ln. It is the ➤inverse function of the ➤exponential function. ➤➤Arrhenius equation.

natural number Symbol $\mathbb{N}$. One of the whole numbers 1, 2, 3, ... Some authors also count zero as a natural number.

natural selection A crucial part of Darwin's theory of evolution (➤Darwinism), still used by biologists to explain evolutionary processes. It holds that individuals in a large population have different chances of survival determined by their genetic make-up: those with characteristics that weaken their capacity to survive (such as the inability to escape predators or obtain nutrients) will have less chance of reproducing and passing on their characteristics to succeeding generations. Evolution by natural selection can be observed directly and has been tested experimentally. It remains the central feature of modern evolutionary theory.

natural units A system of units, particularly useful in high-energy physics, in which certain fundamental constants are set to unity and physical quantities are measured with a reduced number of ➤dimensions (4). The constants usually chosen are $\hbar$ (the rationalized ➤Planck constant) and c (the speed of light). This allows a single unit to be used for the basic physical quantities of mass, length and time. For high-energy physics this is often the megaelectronvolt (MeV). Approximate conversion factors are as follows:

Mass : $m/\mathrm{kg} = 1.783 \times 10^{-30} \ m/\mathrm{MeV}$,

Length : $l/\mathrm{m} = 1.973 \times 10^{-13} \ l/\mathrm{MeV}^{-1}$,

Time : $t/\mathrm{s} = 6.582 \times 10^{-22} \ t/\mathrm{MeV}^{-1}$.

nautical mile A unit of distance intended to be equal to one ➤minute (2) of arc of latitude at the Earth's surface. As the Earth is not a perfect sphere, the distance varies with latitude. British sailors and aviators have used 6080 feet (1 minute at about 48°N) as a nautical mile, though the international standard is now 6076.115 feet (1 minute at 45°N). The nautical mile is still used at sea, and is the International Civil Aviation Organization's standard unit of distance for navigation.

Navier–Stokes equation The fundamental equation governing the flow of a viscous fluid, in the absence of external forces acting on the fluid:

$$\frac{\partial \boldsymbol{v}}{\partial t} + (\boldsymbol{v} \cdot \nabla)\boldsymbol{v} = -\frac{1}{\rho}\nabla p + \frac{\eta}{\rho}\nabla^2 \boldsymbol{v},$$

where $\boldsymbol{v}$ is the velocity of the fluid (as a function of position), p the pressure, ρ the density and η the ➤viscosity. Because the equation is nonlinear, it is notoriously difficult to solve. Named after Claude-Louis-Marie Henri Navier (1785–1836) and George Gabriel Stokes (1819–1903).

Nb Symbol for the element ➤niobium.

Nd Symbol for the element ➤neodymium.

Ne Symbol for the element ➤neon.

nearly-free-electron approximation ➤energy band.

nebula (plural **nebulae)** A cloud of gas in the interstellar medium. There are two main types of **diffuse nebula: emission nebulae,** which are self-luminous, and **reflection nebulae,** which shine by reflected light from nearby stars. Emission and reflection nebulae are together classed as **bright nebulae; dark nebulae** (such as the Coalsack) are not illuminated and are visible as darker areas against background stars. Other types include **planetary nebulae,** which are shells of gas shed by old stars.

necessary condition A condition that is necessary for the truth of a statement in the sense that it follows from the truth of that statement. Thus, for a quadrilateral to be a square, it is necessary for its sides to be equal in length.

nectar A sugar-rich secretion produced by the nectaries of insect-pollinated flowers. Nectar is an inducement to insects to visit the flower thereby promoting pollination.

Néel temperature Symbol T_N. The critical temperature below which a material that shows ➤antiferromagnetism can be magnetically ordered. For example, the Néel temperature of chromium is 311 K. Named after Louis Eugène Félix Néel (1904–2000).

negative **1** In most contexts, negative means less than zero or measured in the opposite sense to a given ➤positive sense; but it is also used to mean minus, as in 'negative two' for −2.

2 ➤electric charge.

negative feedback ➤feedback.

nekton The collective term for all the animals inhabiting open water habitats in oceans, lakes and rivers. Nekton include fish, and ➤pelagic mammals such as whales.

nematic liquid crystal ➤liquid crystal.

nematode A member of the major animal ➤phylum Nematoda comprising the unsegmented **roundworms**. The phylum includes a number of significant parasites such as *Ascaris*, as well as the ubiquitous soil-dwelling nematodes.

neodymium Symbol Nd. The element with atomic number 60 and relative atomic mass 144.2, which is one of the ➤lanthanides, and as such its chemistry is dominated by oxidation number +3. Its most important use is in laser systems using neodymium in yttrium aluminium garnet (YAG).

Neolithic (New Stone Age) An age in the history of humankind starting in approximately 8000 BC and continuing until about 2500 BC, when the transition to the Bronze Age began in some areas. During the Neolithic, people first developed agriculture and made sophisticated stone tools.

neon Symbol Ne. The element with atomic number 10 and relative atomic mass 20.18, which is one of the noble gases and as such is found in trace quantities in air. It is totally unreactive, as expected from its very high first ionization energy. Neon vapour is used in the classic 'neon lights', the different colours being due either to additives of other noble gases or to coloured glass.

neoplasm ➤cancer.

neoprene rubber A synthetic rubber made by polymerization of $CH_2{=}CClCH{=}CH_2$, which differs from ➤isoprene, the monomer of natural rubber, by having a chlorine atom in place of a methyl group.

neoteny The early onset of sexual maturity while other features of the animal remain embryonic. The term is particularly applied to the larvae of the salamander species *Ambystoma mexicanum*, the axolotl, which are capable of sexual reproduction before ➤metamorphosis. The term is often taken to include the retention of juvenile stages of animals that have high levels of parental care, such as humans.

nephroid ➤epicycloid.

nephron ➤kidney.

Neptune The eighth planet from the Sun, at an average distance of 30.1 AU. Like ➤Uranus, it probably consists of a solid core of rock with a radius of about 7000 km, surrounded by a mantle of liquid water, methane and ammonia. The radius is about 24 764 km. Its rotational period is 16 h 7 min; its orbital period is 164.8 years. Its atmosphere consists of about 85% hydrogen and 15% helium, with a trace of methane that gives it its blue colour; there are clouds of methane crystals at higher levels. Unlike Uranus, Neptune displayed some detail to the Voyager 2 probe which visited it in 1989, for example the **Great Dark Spot**, a huge anticyclone. Neptune has eight satellites, the largest of which, **Triton**, is probably a captured body. ➤Appendix table 4.

neptunium Symbol Np. The element with atomic number 93 and most stable isotope 237, which is the first of the ➤transuranium elements, discovered in the debris of the first nuclear explosion and since made in nuclear reactors. Uranium was named after Uranus, so the next planet from the Sun was the source of neptunium's name. Its chemistry is dominated by its radioactivity. The most interesting of its many oxidation numbers is the strongly oxidizing +7 state.

Nernst effect One of a number of thermomagnetic effects in conductors, in which a potential difference is observed when a temperature gradient is created in a conductor in a transverse magnetic field. Named after (Hermann) Walther Nernst (1864–1941).

Nernst equation An equation describing how the electrode potential E in a solution differs from its standard value $E^\ominus$ because of the concentration being different from unity:

$$E = E^\ominus - \frac{RT}{zF} \ln Q$$

Here Q is the ➤reaction quotient, R the gas constant, T the thermodynamic temperature, z the number of electrons transferred and F the Faraday constant. Since this is a logarithmic dependence on concentration, changes in concentration usually have little effect on E.

nerve cord The major tract of tissue, consisting of ➤neurones, that forms part of the central ➤nervous system. In vertebrates the nerve cord is ➤dorsal and forms the spinal cord, but in ➤arthropods and ➤annelids the nerve cord is ➤ventral.

nerve gas A gas that affects the central ➤nervous system, typically by blocking the function of the enzyme acetylcholinesterase. Two notorious nerve gases are **GB** or **sarin**, $(CH_3)_2CHO(CH_3)FP=O$, an odourless and colourless volatile gas which can kill within minutes of exposure; and **VX**, which is similarly effective but persists for longer.

nerve impulse A series of events propagated along the length of a ➤neurone that carries information in the form of an electrochemical signal. The impulse is generated after suitable stimulation of the neurone has produced a 'generator potential' above the threshold level, which causes a rapid influx of sodium ions. This influx brings about a reversal of the electric potential difference across the cell membrane, which in its resting state is maintained at about −70 mV. The ionic imbalance is propagated along the neurone in an 'all or nothing' fashion: the impulse is of the same intensity and charge differential irrespective of the intensity of stimulus. After the impulse has passed, the membrane is reset to its original potential difference by the active pumping of both sodium and potassium ions. This takes a short while and during this period the cell cannot be stimulated again. When an impulse reaches the end of a neurone, it may be transmitted across a narrow gap (a

➤synapse) to one or more adjoining neurones by means of the secretion of a ➤neurotransmitter.

nervous system A system of cells and tissues in multicellular organisms that enables appropriate **responses** to be made to a wide range of changes in the organism's environment. In ➤vertebrates, information is received via receptor cells and organs (such as touch receptors in the skin or light-sensitive cells in the eye) and relayed to the **central nervous system** (CNS). The CNS (consisting of the ➤brain and spinal cord) processes this information by, for example, referring to stored information in memory, and sends out ➤nerve impulses to ➤muscles and glands which bring about an appropriate response. The basic functional unit of the nervous system is the ➤neurone. Although the basic principles of nervous systems are essentially the same in all animals, the level of complexity of the system is a function of the degree of structural complexity in each animal type. In vertebrates the nervous system is organized in two divisions: the CNS and **peripheral nervous system** associated with conscious responses, and the **autonomic nervous system** controlling the smooth muscle (of the gut for example) and the activity of glands, providing the link between the nervous system and the ➤endocrine glands. The autonomic nervous system is further subdivided into the **sympathetic** and **parasympathetic** nervous systems. These have different origins and organization of their neurones, and generally have an antagonistic action to each other: the sympathetic system stimulating and the parasympathetic system inhibiting organ functions such as heart rate.

Nessler's reagent A reagent for detecting ammonia gas. It is made from a solution of mercury(II) iodide in potassium iodide. A positive test is the formation of a yellow or brown precipitate of $Hg_2NI\cdot H_2O$. Named after Julius Nessler (1827–1905).

network A connected set of computers. The extent of networks can vary enormously, ranging from a **local area network** (LAN), which might be confined to a single building, to a **long-haul network**, which could encompass the world. Networks are important because they allow high-speed communication between one computer and another, and they enable expensive resources such as printers to be shared by many users. Without networks, the usefulness of computers would be greatly diminished; networks are responsible for the realization of the concept of a 'global village'.

neural network (neural net) In computing, a processing structure inspired by biological neural networks in the brain. Neural networks are a growing area in artificial intelligence research since they are successful at the same sort of pattern recognition processes as the brain, such as those involved in seeing and hearing. The networks are not explicitly programmed. Instead, they are trained. Presented with examples of input, a suitable learning algorithm adjusts the weightings applied to the signals between the simulated neurones, until the output gradually approaches the correct form. Once trained, the net has built up its own internal representation of its area of knowledge and it can recognize unknown or incomplete examples of input.

neurohormone A ➤hormone produced by a nerve cell rather than an ➤endocrine gland. Examples include ➤noradrenaline and ➤vasopressin.

neurone (neuron) The basic cellular unit of the ➤nervous system, capable of carrying ➤nerve impulses. Neurones are of different types depending on their position in the nervous system. Most neurones have projections from the cell body, **dendrites**, which join to form the **dendron** on the input side of the cell and an **axon** on the output side. The dendron and axon may be surrounded by a fatty (**myelin**) sheath, secreted by **Schwann cells**. This may have gaps (**nodes of Ranvier**) along its length. **Sensory neurones** typically have a long dendron, with their cell body located close to the spinal cord; **motor neurones** have short dendrons and long axons.

neurotransmitter A chemical released into the space (➤synapse) between the end of a ➤neurone and one or more adjacent neurones when a ➤nerve impulse arrives. Stimulation of membrane receptors by the neurotransmitter results in a nerve impulse being generated in the adjacent neurone(s). Whether or not sufficient neurotransmitter generates successive nerve impulses depends on a number of factors and is the basis of coordination and control in the ➤nervous system. Neurotransmitters include ➤noradrenaline, ➤catecholamine, ➤dopamine and ➤acetylcholine. Many nerve poisons and some anaesthetics work by blocking the secretion of neurotransmitters into synapses or by inhibiting their resynthesis after transmission.

neutral **1** (Phys.) Having neither positive nor negative ➤electric charge.
2 (Phys.) One of the three lines of a mains power supply; the neutral line is designed to be close to ➤earth potential but is not connected to earth.
3 (Chem.) A solution that has a ➤pH of 7.

neutralization A reaction in which acids react with bases. The enthalpy of neutralization is defined per mole of water formed in the reaction; for strong acids reacting with strong bases the value is very close to 57 kJ mol^{-1}. The resultant salt solution may not, however, be 'neutral', that is at pH 7.

neutrino A class of elementary particles with zero charge and almost zero rest mass. There is positive experimental evidence for an **electron neutrino** associated with the electron and a **muon neutrino** associated with the muon. Theory suggests that there should be a third, the **tauon neutrino**, associated with the tauon. Neutrinos are ➤leptons. All are very elusive, travelling at the speed of light and passing through matter with virtually no interaction.

neutron A subatomic particle with no charge, a mass of 939.6 MeV/c^2 and a spin of $\frac{1}{2}$. Like the proton it is a ➤nucleon. It feels the ➤strong interaction, which binds it with the protons and other neutrons in the nucleus. The two particles are also close in mass, the ratio to the electron's mass being about 1839 for the neutron compared with 1836 for the proton. ➤Quark theory suggests that they are closely related, as the quark constitution of the proton is uud whereas that of the neutron is udd. A free neutron decays into a proton, an electron and an electron antineutrino with a half-life of about 10 min. A similar process occurring in the atomic nucleus gives rise to ➤beta decay. The discovery of the uncharged neutron by James Chadwick in 1932 finally explained the masses of atoms.

neutron diffraction As subatomic particles such as the neutron exhibit ➤wave–particle duality, they can be diffracted like light waves. Neutrons do not interact with the electrons in a crystal, so, in contrast with ➤electron diffraction, the diffraction pattern depends only on the positions of the nuclei and on their ➤spins. Neutrons from a ➤thermal reactor have a wavelength of about 0.1 nm, convenient for such diffraction experiments. Neutron diffraction is particularly useful for identifying the positions of hydrogen atoms in solids as they are very difficult to locate with the more common ➤X-ray diffraction technique.

neutron number The number of neutrons in an atomic nucleus. It is therefore the difference between the ➤nucleon number and the ➤atomic number of a ➤nuclide.

neutron star A highly compact object (typically 30 km diameter) consisting of densely packed neutrons held apart by degeneracy pressure (➤degenerate gas). Such objects are formed by the gravitational collapse of the core of a massive star that becomes a ➤supernova. A maximum theoretical mass exists, about 2.5 solar masses, above which the degeneracy pressure is not enough to prevent further collapse to a ➤black hole. ➤Pulsars are a form of observable neutron star.

Newlands' law of octaves John Alexander Reina Newlands (1837–98), one of the people who caught a glimpse of periodicity before Mendeleyev produced a definitive ➤periodic table, noticed that elements with similar properties reappeared every eighth element (because the ➤noble gases had not been discovered), as in musical scales. A good example is the first three alkali metals: lithium, sodium and potassium. However, the pattern was quickly broken, the next alkali metal being much further down the table.

Newman projection A projection showing the orientation of groups around a carbon–carbon bond (see the diagram). The view looks along the C—C bond. Compare ➤Fischer projection. Named after Melvin Spencer Newman (1908–93).

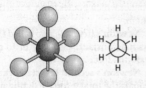

Newman projection The bonds shown joined together are those on the nearer carbon atom. The bonds to the other (hidden) carbon atom are shown touching the circle.

newton Symbol N. The SI derived unit of force: 1 N is the force required to give a mass of 1 kg an acceleration of 1 m s^{-2}. Named after Isaac Newton (1642–1727).

Newtonian fluid A fluid in which the ➤viscosity is constant at constant temperature and pressure. ➤Non-Newtonian fluids include ➤thixotropic fluids, such as nondrip paint, which become less viscous the faster they move.

Newtonian mechanics A formulation of mechanics based on ➤Newton's laws of motion. It is usually referred to in this way to distinguish it from ➤relativistic or ➤quantum mechanics.

Newtonian telescope ➤telescope.

Newton–Raphson method A method for solving the equation $f(x) = 0$ by iteration, in which successive approximations to the root are given by $x_{n+1} = x_n - f(x_n)/f'(x_n)$ with a suitable starting value x_0. It was first used by I. Newton and popularized in a textbook written by Joseph Raphson (1648–1715).

Newton's experimental law If two colliding masses are moving in a straight line then their respective velocities u_1, u_2 before impact and v_1, v_2 after impact are related by $v_2 - v_1 = e(u_1 - u_2)$, where e is a constant, the **coefficient of restitution.** The value of e is 1 for an ➤elastic collision and less than 1 for an ➤inelastic collision.

Newton's law of cooling An empirical law stating that the rate at which a body at temperature T loses heat to its surroundings at temperature T_s is proportional to $T - T_s$. For a body with a constant ➤heat capacity, this leads to an ➤exponential decay of temperature from T to T_s. The law is a good approximation for small temperature differences, particularly where conduction is the sole means of heat transfer.

Newton's law of (universal) gravitation The ➤inverse square law, discovered by Newton, relating the magnitude of the gravitational force F between two bodies of mass m_1 and mass m_2 to the inverse square of the distance r between them:

$$F = Gm_1m_2/r^2,$$

where G is a ➤fundamental constant, the **(universal) gravitational constant**, the value of which is 6.673×10^{-11} N m^2 kg^{-2}. For centuries, Newton's law of gravitation was almost unrivalled as a fundamental principle of physics. It explained ➤Kepler's laws quantitatively for the first time. It remains the appropriate expression for describing gravitation at low velocities and low densities, though modern theories of ➤gravity have progressed beyond Newton's (➤general relativity).

Newton's laws of motion The fundamental laws of mechanics that describe the way in which bodies in an ➤inertial frame move in response to the forces acting on them.

Newton's first law, also called **Galileo's law,** states that a body continues to move with a constant ➤velocity or to remain at rest unless acted on by an unbalanced external force.

Newton's second law states that the rate of change of ➤momentum $\boldsymbol{p}$ of a body equals the total ➤force $\boldsymbol{F}$ acting upon it:

$$\boldsymbol{F} = \mathrm{d}\boldsymbol{p}/\mathrm{d}t.$$

If, as is normally the case, the mass of the body is constant, $\boldsymbol{F} = \mathrm{d}(m\boldsymbol{v})/\mathrm{d}t$ reduces to $\boldsymbol{F} = m\mathrm{d}\boldsymbol{v}/\mathrm{d}t$ or

$$\boldsymbol{F} = m\boldsymbol{a},$$

where $\boldsymbol{a}$ is the ➤acceleration of the body. Note that the force and acceleration are vectors. The first law is the null case of the second law (if $\boldsymbol{F} = \boldsymbol{0}$ then $\boldsymbol{a} = \boldsymbol{0}$).

Newton's third law states that if a body A exerts a force $\boldsymbol{F}$ on body B, then body B exerts a force $-\boldsymbol{F}$ on body A.

Newton's rings The ➤interference pattern formed when a curved glass surface and a plane reflecting surface are very close to each other (see the diagram). Partial ➤internal reflection occurs at the curved surface, and the light that is transmitted is totally reflected at the plane surface. The air film between the two has a

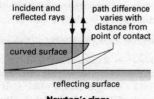

Newton's rings

thickness that depends on the position at which the reflection occurs, and thus the path difference depends on angle, creating the rings by constructive and destructive interference.

Ni Symbol for the element ➤nickel.

niacin The commercial name of nicotinamide, one of the B group ➤vitamins. Niacin is important as a component of NAD^+, and is the anti-pellagra vitamin. **Pellagra** is a ➤deficiency disease that induces a wide range of symptoms including weakness and lassitude, insomnia, weight loss, gastro-intestinal disorders and skin and mouth lesions. In later stages of the condition there is dysfunction of the ➤nervous system. The necessity for the vitamin was established in a series of classic experiments started in 1915 by Joseph Goldberger.

niche A fundamental concept in ecology which encapsulates the way of life of an individual species; a niche is simply defined as where an organism lives, but the term is extended to include the behaviour of the organism and its physiological functioning within an ➤ecosystem. Thus two species of bird such as blackbirds and thrushes may appear superficially to share a common niche since both species are found in gardens. However, since they rely on different food sources they are occupying different niches. Similarly, migratory birds such as swallows occupy very extensive niches. The **competitive exclusion principle** implies that two different species cannot exist stably in the same niche. Differential success of organisms in competition for the same niche is an important concept in ➤evolution.

Nichrome Tradename for an alloy of nickel and chromium used where it is necessary to resist high temperatures. Flame-test wires are often made of Nichrome.

nickel Symbol Ni. The element with atomic number 28 and relative atomic mass 58.69, which is in the first row of the ➤transition metals and shares their usual properties. Its range of oxidation numbers is unusually small however, being dominated by +2, as in the oxide NiO, a green solid, although higher states do exist, such as +3 in the complex K_3NiF_6, which forms violet crystals. In aqueous solution, $Ni^{2+}(aq)$ is a beautiful green colour. When ammine ligands are present the colour becomes blue-violet as $[Ni(NH_3)_6]^{2+}$ eventually forms. The most characteristic test for nickel is to add **dimethylglyoxime** to form a red precipitate. Note that the structure of the precipitate is square planar, which is a common stereochemistry for d^8 complexes. The element itself is made in the ➤Mond process and is used extensively for catalysing hydrogenation reactions and for coinage.

nickel arsenide structure (NiAs structure) An important crystal structure, in which nickel atoms are in the octahedral holes in a hexagonal close-packed array of arsenic atoms. Contrast the filling of the octahedral holes in a *cubic* close-packed array that describes the ➤rock-salt structure of NaCl.

nickel–cadmium cell (nickel–cadmium battery) An important battery which has become very common, to power cameras for example. It is rechargeable from the mains using a battery charger. The chemistry is currently not fully understood but a simplified equation for the overall process is

$$2NiOOH + Cd + 2H_2O \rightleftharpoons 2Ni(OH)_2 + Cd(OH)_2.$$

Discharging causes the products to dominate, whereas charging causes the reactants to dominate.

Nicol prism A ➤polarizer constructed from two crystals of ➤calcite which form a prism exhibiting ➤birefringence. Named after William Nicol (1768–1851).

nicotinamide adenine dinucleotide ➤NAD⁺.

nicotinamide adenine dinucleotide phosphate ➤NADP⁺.

nicotine An alkaloid (see the diagram) present in tobacco leaves. Nicotine in cigarette smoke acts as a stimulant and in the short term may relieve stress and anxiety; however, it is also addictive, which explains why people who smoke have great trouble stopping smoking. Smoking increases the risk of developing cancer.

nicotine

ninhydrin A reagent (see the diagram), often used in the past as a spray to develop chromatograms of peptides, as it gives a deep blue colour with amino acids. Health fears have arisen about its use, which has thus been reduced.

niobium Symbol Nb. The element with atomic number 41 and relative atomic mass 92.91, which is in the second row of the ➤transition metals. It was known as **columbium** until the 1940s. Its new name comes from Niobe, the daughter of Tantalus, to show its close relation to its congener ➤tantalum. Its most common oxidation number is +5, as in NbF_5 and Nb_2O_5, both white solids. The niobates are ➤mixed oxides. Examples include lithium niobate $LiNbO_3$; oscillators made from lithium niobate crystals control shutter speeds in many cameras.

ninhydrin

nitrate ion (nitrate(v) ion) The oxoanion of formula NO_3^-. As nitrogen has oxidation number +5, its alternative but less common name is nitrate(v). The anion does not produce any visible absorption, so only the cation can impart colour to nitrates, as in the blue solid copper(II) nitrate $Cu(NO_3)_2$. All common nitrates are soluble in water.

In aqueous solution the nitrate ion is difficult to identify, as there are no precipitation tests. One common test relies on the oxidizing properties of nitrate ion in acidic solution and involves adding freshly prepared iron(II) sulfate followed by concentrated sulfuric acid. This produces a **brown ring**, which gives its name to the test, because the NO produced on reduction of the nitrate forms a complex with iron(II) ions. Nitrates are vital fertilizers, but several dangers accompany overuse (➤eutrophic).

Thermal ➤decomposition of a nitrate usually produces an oxide, nitrogen dioxide and oxygen. The nitrates of sodium and potassium, however, initially decompose thermally into nitrites and oxygen.

nitrating mixture A mixture of concentrated nitric and sulfuric acids that is used to introduce the nitro group, $-NO_2$, into a molecule. A classic example is the conversion of benzene, C_6H_6, into nitrobenzene, $C_6H_5NO_2$. The ➤electrophile in this reaction has been shown to be the NO_2^+ ion.

nitration A reaction in aromatic organic chemistry in which a nitro group is introduced into a ring often by reaction with a nitrating mixture. Sometimes the nitro compound itself is not wanted but is reduced with tin and concentrated hydrochloric acid to the corresponding amine, such as phenylamine.

nitrazepam $C_{15}H_{11}N_3O_3$ The molecule most used to send people to sleep, as it is the active ingredient in Mogadon.

nitre ➤saltpetre.

nitric acid (nitric(v) acid) HNO_3 or $HONO_2$ The parent acid of nitrates, in which nitrogen has oxidation number +5, hence the alternative, less used, name. The azeotropic mixture of 65% nitric acid should be a colourless liquid, but it often looks yellow because decomposition produces some nitrogen dioxide. It is produced industrially on a massive scale via the ➤Ostwald process.

Nitric acid is a particularly important substance for three reasons. First, it is a monobasic strong acid, producing nitrates on neutralization, as in the reaction

$$NaOH + HNO_3 \rightarrow NaNO_3 + H_2O.$$

Second, it is an oxidant. Its strength as an oxidant increases with its concentration. Typical of its oxidizing reactions is the reaction of copper to form copper(II) nitrate. Copper does not produce hydrogen with acids and it is the possibility of reducing nitric acid to nitrogen oxide or dioxide that enables this reaction to take place. Third, when combined with concentrated sulfuric acid, it constitutes a ➤nitrating mixture, used to make nitroglycerine and TNT for example.

nitric oxide Alternative name for ➤nitrogen oxide.

nitride A binary compound with nitrogen as the more electronegative element. Two common examples are lithium nitride, Li_3N, and magnesium nitride, Mg_3N_2. In both the **nitride ion** N^{3-} is present. On addition to water this ion hydrolyses to produce ammonia gas. Transition metal nitrides, on the other hand, are nonstoichiometric interstitial compounds. Perhaps the most important nitride is the covalent molecule **boron nitride**, BN, one of the polymorphs of which, called borazon, is extremely hard, second only to diamond on the ➤Mohs scale.

nitrifying bacteria ➤nitrogen cycle.

nitrile An organic compound, RCN, containing the cyanide group. Nitriles are useful intermediates in organic synthesis as they can be hydrolysed in acidic solution to give carboxylic acids, RCOOH.

nitrite ion (nitrate(III) ion) Salts of nitrous acid contain the nitrite, NO_2^-, ion. As with the parent acid, nitrites can act as oxidants, being reduced to ammonia, or reductants, being oxidized to ➤nitrate ion.

nitro- A prefix indicating the presence of the ➤nitro group.

nitrobenzene $C_6H_5NO_2$ This yellow oil is made by reacting benzene with a ➤nitrating mixture. It undergoes substitution reactions similar to benzene's, albeit more slowly, giving on nitration *m*-dinitrobenzene (see the diagram at ➤meta).

nitrogen Symbol N. The element with atomic number 7 and relative atomic mass 14.01, which is at the top of Group 15 of the ➤periodic table. Nitrogen gas is covalently bonded, N_2, and constitutes about 78% of the air, from which it can be extracted by cooling followed by fractional distillation of the liquid air. Liquid nitrogen is the most useful refrigerant, as it is safe and cheap and has a low boiling point of $-196\,°C$. Industrial production of nitrogen is second only to that of sulfuric acid. As importantly, four of nitrogen's compounds are also among the top 15 chemicals manufactured in industry: ammonia, urea, nitric acid and ammonium nitrate. The use of ➤nitrogenous fertilizers revolutionized agriculture in the 20th century.

nitrogenase The bacterial enzyme catalysing the fixation of nitrogen (➤nitrogen cycle). This enzyme catalyses the reduction of atmospheric nitrogen to ammonia.

nitrogen cycle A series of chemical processes, mostly occurring in organisms, in which nitrogen atoms are circulated in nature (see the diagram opposite). Although nitrogen is an important component of many molecules of biological significance (such as proteins, nucleic acids and chlorophyll), it can be used directly by only a few specialized organisms capable of nitrogen fixation. **Nitrogen fixation** is the direct incorporation of gaseous nitrogen into organic molecules, such as amino acids, using **nitrogenase** enzymes, and is restricted to a few kinds of nitrogen-fixing bacteria, such as *Rhizobium*, and ➤cyanobacteria. In turn, these organisms release nitrate, which can be taken up by the roots of green plants and incorporated into proteins and other nitrogen-containing compounds. Plants are eaten by animals, which convert plant protein into animal protein. Nitrogenous excretion by animals and the decay of dead organisms releases nitrogen compounds back into the soil. These are degraded by the bacteria that cause decay, and are converted via ammonia and nitrite to nitrate by **nitrifying bacteria**. In poor, waterlogged soils, **denitrifying bacteria** may convert nitrate back into gaseous nitrogen. The biological nitrogen cycle is supplemented by the production of nitrogen oxides during thunderstorms and by the industrial synthesis of nitrogenous fertilizers. ➤leguminous plants.

nitrogen dioxide NO_2 A poisonous brown gas which is produced in many ways: in the environment as a result of reaction between nitrogen and oxygen in an internal combustion engine, in industry during the manufacture of nitric acid in the ➤Ostwald process, and in the laboratory when copper reacts with concentrated nitric acid or nitrates are thermally decomposed. It tends to dimerize and an equilibrium with the lighter-coloured dimer, dinitrogen tetroxide, is set up:

$$2NO_2(g) \rightleftharpoons N_2O_4(g).$$

nitrogen fixation ➤nitrogen cycle.

nitrogenous base A general term applied to the ➤purine and ➤pyrimidine bases present in ➤nucleic acids.

nitrogenous fertilizer A fertilizer capable of providing the soil with nitrogen in combined form. The most common are urea, ammonium nitrate and ammonium sulfate.

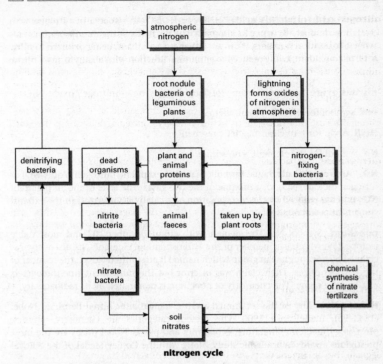

nitrogen cycle

nitrogen oxide (nitrogen monoxide, nitric oxide) NO A colourless gas which is produced when nitrogen and oxygen react in the region of a spark, as for example in lightning strikes or in an internal combustion engine. NO reacts very fast with atmospheric oxygen to form NO_2, as can be seen when NO is made from copper and dilute nitric acid in a test tube; there is a colourless layer of only a few centimetres before a brown gas, NO_2, is seen.

Given that NO is toxic, it is a surprise that this molecule functions as a ➤neurotransmitter and as a signal molecule between cells regulating blood vessel dilation (NO is the active metabolite produced from ➤nitroglycerine ingested to alleviate the pain caused by *angina pectoris*).

nitroglycerine The pale yellow oily liquid formed by the reaction between glycerine (➤glycerol) and a ➤nitrating mixture. It is an explosive, as on thermal ➤decomposition not only is much energy released but there is a massive increase in volume due to release of 29 gas molecules from every four nitroglycerine molecules:

$$4C_3H_5(NO_3)_3(l) \rightarrow 6N_2(g) + 12CO_2(g) + 10H_2O(g) + O_2(g).$$

nitro group —NO_2 It is common in ➤aromatic chemistry, in nitrobenzene or TNT for example. It is much less common in aliphatic chemistry, although compounds such as nitromethane, CH_3NO_2, do exist. The reason for this is the ➤delocalization stabilization of the nitro group when it interacts with the aromatic ring.

nitrous acid (nitric(III) acid) HNO_2 or HONO An oxoacid of nitrogen with oxidation number +3, hence its alternative name. Being in the middle of nitrogen's range of oxidation numbers, it can act either as a reductant, being oxidized to nitric acid, or an oxidant, being reduced to ammonia. Reaction with alkalis forms ➤nitrite ion.

nitrous oxide Alternative name for 'laughing gas' or ➤dinitrogen oxide, N_2O.

nm Symbol for the unit ➤nanometre.

NMR Abbr. for ➤nuclear magnetic resonance.

No Symbol for the element ➤nobelium.

NO$_x$ An intentionally vague formula that could stand for nitrogen oxide, NO, or nitrogen dioxide, NO_2, or a mixture of the two gases. The level of the air pollutant NO_x is being reduced by the incorporation of ➤catalytic converters in the exhaust systems of petrol engines.

nobelium Symbol No. The element with atomic number 102 and most stable isotope 259, which is a member of the ➤transuranium elements. Its discovery was claimed by a Scandinavian group which named it after Alfred Nobel, the founder of the ➤Nobel prizes. Their claim was in error but the subsequent true discoverers retained the name. The chemistry of nobelium is dominated by its radioactivity.

Nobel prize Any of the six annual awards, named after Alfred Bernhard Nobel (1833–96), made since 1900. The present categories are chemistry, physics, physiology/medicine, literature, economics and peace. Nobel prizes are the most prestigious awards in academic study, along with the Copley Medal of the ➤Royal Society.

noble gases (inert gases, rare gases) Very unreactive monatomic elements constituting Group 18 of the ➤periodic table. The first to be discovered was ➤argon. Nitrogen from the air had a mass very slightly different from nitrogen produced chemically, as the former was contaminated with about 1% argon. This discovery uniquely attracted both the chemistry and physics Nobel prizes in the same year (1904). The first three noble gases (helium, neon and argon) all fail to react with any substance. The last three, and especially ➤xenon, do react, albeit with great reluctance, so the older name of 'inert gases' had to be abandoned.

noble metal A metal that reacts only with difficulty with water and oxygen in particular. Common examples include gold and the ➤platinum metals.

node 1 (Phys.) A point of constant zero displacement in an oscillating system, for example a ➤wave. Compare ➤antinode.
　2 (Astron.) Either of the two points at which an ➤orbit intersects the ➤ecliptic or other reference plane.
　3 (Biol.) The point, often swollen, at which a leaf joins a stem.

node of Ranvier ➤neurone. Named after Louis-Antoine Ranvier (1835–1922).

noise Any undesirable disturbance, usually random, in an electronic, electrical or communication system which interferes with its intended operation. Some noise inevitably arises in almost any system through thermal fluctuations, but it can be made more serious by external disturbances, such as unshielded electromagnetic radiation. It may reach a sufficiently high level to cause spurious errors, thereby corrupting some of the information contained in the signal being propagated. ⯈signal-to-noise ratio; white noise.

Nomex Tradename for an ⯈aramid fibre that is connected in the ⯈meta position. Its particular claim to fame is its ability to retreat from a flame, leading to its use in flame-resistant clothing.

nonane C_9H_{20} The straight-chain alkane with nine carbon atoms, a colourless liquid.

nonaqueous solvent Any solvent other than water. Important examples include ⯈dimethylformamide, ⯈dimethylsulfoxide, ⯈ether, ⯈propanone, ⯈ethanol and liquid ⯈ammonia.

noncompetitive inhibition The reduction of the activity of an ⯈enzyme by a substance that bears no structural relationship to the ⯈substrate. Noncompetitive inhibition occurs by a variety of different mechanisms but usually involves the binding of an inhibitor (e.g. a heavy metal such as lead or mercury) at a site remote from the active site, but which produces a conformational change rendering the active site unrecognizable to the substrate. Compare ⯈competitive inhibition.

nonconductor Alternative name for ⯈insulator.

non-Euclidean geometry A geometry in which the ⯈parallel postulate of ⯈Euclidean geometry is replaced by an ⯈axiom stipulating either that no parallels may be drawn to a given line (**elliptic geometry**) or that many parallels can be drawn (**hyperbolic geometry**). The 19th-century discovery that there was nothing contradictory about such geometries was a major intellectual advance. ⯈general relativity.

nonferrous metal A metal that does not contain iron.

nonlinear Not ⯈linear. In particular, the term is applied to systems or components for which the output for the sum of two input signals, A and B, is not the sum of what would be obtained if the two inputs were processed separately: $f(A + B) \neq f(A) + f(B)$, where f represents the effect of the system or component. The absence of this property of ⯈superposition means that the effects of the component on a complex signal (e.g. an amplifier on a voice recording) *cannot* be predicted by breaking the signal up into simpler signals (the individual frequencies making up the voice) and looking at the effect of the system on those separately.

nonmetal There is no simple method for identifying an element as a nonmetal. However, typical properties include an elemental structure that has only a small number of atoms (e.g. S_8 or O_2) and the formation of one or more oxides that are acidic (e.g. P_4O_6 and P_4O_{10}) together with chlorides that are covalent and rapidly hydrolysed by water (e.g. PCl_3). Nonmetals are typically electrical insulators. The nonmetals are to be found above and to the right of the ⯈metalloids in the ⯈periodic table. ⯈Appendix table 5.

non-Newtonian fluid A fluid in which the ➤viscosity is dependent on the velocity gradient through it. An example is a ➤colloid. Compare ➤Newtonian fluid.

nonpolar molecule A molecule without a dipole moment. Symmetric molecules, such as tetrachloromethane or carbon dioxide, are nonpolar. Their ➤intermolecular forces therefore cannot include ➤dipole–dipole forces and so are purely ➤dispersion forces.

nonradiative transition ➤radiative transition.

nonsense codon ➤termination codon.

nonsingular matrix A ➤square matrix with nonzero ➤determinant, or, equivalently, a square matrix which has an inverse (➤inverse matrix).

nonstoichiometric compound A compound that does not have a simple ratio between the number of atoms of each element present. Examples are most common in the ➤transition metals, especially for their hydrides and nitrides; titanium hydride has an approximate formula of $TiH_{1.7}$. This is related to their ability to exist with more than one common ➤oxidation number.

noradrenaline (norepinephrine) A ➤neurotransmitter secreted by neurones of the sympathetic ➤nervous system which induces such events as vasoconstriction, an increase in heart rate and the constriction of sphincters in the gut. It is also secreted as a ➤hormone from the ➤adrenal medulla and has a similar action to ➤adrenaline.

norm A general conception of length in a ➤vector space so that the norm of $x - y$ is a measure of the distance between x and y.

normal **1** A line through a point on a curve that is ➤perpendicular to the tangent to the curve at that point.
  2 A vector perpendicular to the ➤tangent plane of a ➤surface at a given point. If the equation of the surface is $\phi(x, y, z) = 0$, its normal is grad ϕ (➤gradient).

normal distribution The ➤continuous random variable with ➤probability density function

$$f(x) = \frac{1}{\sigma\sqrt{2\pi}}\exp[-\tfrac{1}{2}(x-\mu)^2/\sigma^2].$$

It has ➤mean μ and ➤variance σ^2 and a symmetrical bell-shaped distribution. Some naturally occurring frequency distributions are normal (e.g. errors), but it is the ➤central limit theorem which guarantees the normal distribution a central role in ➤statistics.

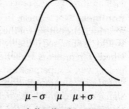

normal distribution The curve has a maximum at $x = \mu$ and points of ➤inflection at $x = \mu \pm \sigma$.

normalization The process of multiplying a set of things by a constant factor to make their sum (or the sum of some related quantity) 1. For example a ➤vector (a, b, c) is normalized in length by dividing by $\sqrt{a^2 + b^2 + c^2}$; a ➤wavefunction $\psi(\mathbf{r})$ is normalized by dividing by $\int |\psi(\mathbf{r})|^2 \, dV$ (where the integral, corresponding to the sum of components for the vector, is over all space and dV is an element of volume).

normal mode (mode) A collective ➤oscillation of a coupled system, where all parts of the system are oscillating with the same frequency (or are stationary ➤nodes). For example, if three masses m are joined by two springs (with spring constant k) as in diagram (a), there are two normal modes. One is with the middle mass stationary and the outer masses oscillating with opposite phase, as in diagram (b); in the other the outer masses are in phase and oscillating together, and the middle mass is out of phase and oscillating with twice the amplitude, as in diagram (c). Molecules have normal modes; the more complex the molecule, the more normal modes it has. Normal modes in solids (where there are of the order of 10^{23} ion cores, and thus a similar number of normal modes) are called ➤phonons. Normal modes can be found by solving an eigenvalue problem (➤eigenvector). In the example above, the force matrix (representing the forces applied by the springs) multiplied by the displacement must be a multiple of the displacement if all the masses are to move together with the same angular frequency ω:

(a)
(b)
(c)
normal mode

$$\begin{bmatrix} k & -k & 0 \\ -k & 2k & -k \\ 0 & -k & k \end{bmatrix} \begin{bmatrix} x_1 \\ x_2 \\ x_3 \end{bmatrix} = m\omega^2 \begin{bmatrix} x_1 \\ x_2 \\ x_3 \end{bmatrix}.$$

northern blotting A technique for transferring ➤RNA fragments separated by gel ➤electrophoresis to a nitrocellulose or other membrane on which they may be detected by a suitable ➤DNA probe. So named since it is the 'opposite' of Southern blotting (➤DNA blotting).

Norton's theorem For any circuit (that consists of components obeying ➤Ohm's law) from which just two terminals emerge, the network between the terminals can be replaced by an ➤equivalent circuit consisting of a current generator in parallel with an internal resistance. Named after Edward Lawry Norton (1898–1983).

notochord A dorsal supporting rod between the ➤nerve cord and the ➤gut present at some stage in the life history of a ➤chordate. In vertebrates it is apparent in the early embryo, but is largely replaced by the development of the vertebral column.

nova (plural **novae)** A sudden brightening of a star by many ➤magnitudes (2) over a time period of about a day. **Fast novae** then decay by several magnitudes over a period of months; **slow novae** may take years to decay but are typically not as bright. All novae are ➤binary stars, consisting of a main-sequence star (➤Hertzsprung–Russell diagram) and a ➤white dwarf. Matter flows from the main-sequence star into an accretion disk around the white dwarf, from which it is slowly transferred to the surface. Nuclear reactions have all but ceased there, but the strong gravitational attraction compresses the hydrogen, resulting eventually in a huge thermonuclear explosion, visible as the nova.

Np Symbol for the element ➤neptunium.

NP-complete A term applied to **NP problems**, problems that apparently cannot be solved exactly by any algorithm whose execution time is a polynomial function of the complexity of the problem, so they become prohibitively slow as the complexity increases. An NP-complete problem is a 'universal' NP problem for which, if a polynomial-time algorithm were to be found, every NP problem would be rendered solvable in polynomial time. An example is the **travelling salesman problem**, to find the shortest route on a given network of towns that visits each town. An important open problem is to prove that NP-complete problems are not solvable in polynomial time.

NPK fertilizer A fertilizer that supplies the soil with the three elements nitrogen (N), phosphorus (P) and potassium (K), which are essential for plants.

n–p–n transistor A ➤bipolar junction transistor consisting of a layer of ➤p-type semiconductor between two layers of ➤n-type semiconductor.

ns Symbol for the unit ➤nanosecond.

n-type semiconductor An ➤extrinsic semiconductor, with an excess of electrons, produced by, for example, doping silicon with a Group 15 element such as arsenic. The electrons are the ➤majority carriers.

nuclear battery A battery often carried by spacecraft to provide energy even when far away from the Sun. In one common type a ➤beta emitter, such as tritium ^{3}H, is used to ionize a gas such as argon to produce a current of about 1 nA.

nuclear cross-section The ➤scattering cross-section for the scattering of particles by an atomic nucleus.

nuclear energy ➤nuclear fission; nuclear fusion.

nuclear fission The process in which a nucleus breaks up. Fission is often initiated by the absorption of a neutron, which causes an unstable nucleus to form and then spontaneously split. This releases the original neutron, plus an additional one or two neutrons, as well as the remaining two (or occasionally more) pieces of the nucleus. In nuclear energy generation, the two most important elements that undergo fission are uranium and plutonium. Furthermore, only some of the isotopes of each are ➤fissile, notably ^{235}U and ^{239}Pu. The resulting ➤chain reaction (2) can be either controlled, as in a nuclear reactor, or uncontrolled, as in a nuclear weapon. The first sustained controlled nuclear fission was achieved by the Italian physicist Enrico Fermi in 1944. ➤➤spallation.

nuclear force ➤strong interaction; weak interaction.

nuclear fuel A fuel used in a ➤nuclear reactor. The two most important ➤fissile nuclides are ^{235}U and ^{239}Pu, but two others can be used, ^{233}U and ^{241}Pu. The fuel needs to be **reprocessed** to extract the fissionable nuclides from the others.

nuclear fusion The process in which two ➤nuclei are made to fuse together to produce a single nucleus. In contrast to ➤nuclear fission, the lightest nuclides are chosen, usually deuterium and tritium, two isotopes of hydrogen. Fusion has been achieved on an uncontrolled scale in the hydrogen bomb (➤nuclear weapon), but so far efforts to duplicate this feat in a controlled way for energy generation have met with only moderate success (➤Joint European Torus). Fusion is the energy source in stars (➤carbon cycle).

nuclear isomer ➤isomer, nuclear.

nuclear magnetic resonance (NMR) An important spectroscopic technique used for analysis in organic chemistry, and medical diagnosis (where it is called ➤magnetic resonance imaging). The technique probes the nucleus of an element. The most commonly observed nucleus is that of hydrogen, because of its ubiquitous presence in organic compounds, and in water for medical studies. Other nuclei are used, such as phosphorus-31 or carbon-13, but **proton magnetic resonance** is dominant. The proton

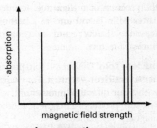

nuclear magnetic resonance
High-resolution NMR spectrum of ethanol.

has spin, and hence a magnetic moment, so in a magnetic field it can take up two orientations, one parallel to the field and one antiparallel. They differ in energy, and in the ground state the lower energy state is occupied. When a photon of exactly the right energy is shone through the sample, resonant absorption occurs from the sample. The area of the electromagnetic spectrum used is the radio frequency, typically around 300 MHz, which requires a strong magnet producing a field of around 10 T. The exact position of each peak, the ➤chemical shift, can identify the chemical environment of each hydrogen atom because changes in the immediate environment of one nucleus due to the electron density cause minute (measured in parts per million) but measurable changes to the resonance frequency. The peak areas identify the number of equivalent nuclei, and thus the structure of the compound can be determined. The NMR spectrum of ethanol is shown in the diagram.

nuclear magneton ➤magneton; Appendix table 2.

nuclear medicine The use of radioactive substances in medicine, typically as tracers or in radiotherapy in cancer treatment.

nuclear physics The physics of the ➤nucleus and nuclear reactions. This contrasts with ➤atomic physics, which is concerned with the interaction of the nucleus with electrons, and ➤high-energy physics, which deals with the subatomic particles that make up the nucleus.

nuclear power The generation of useful energy by ➤nuclear fission or ➤nuclear fusion in a ➤nuclear reactor.

nuclear reaction A reaction in which one nuclide is made into another. The bombarding particle and any particle(s) produced need to be specified. The reaction that led to the discovery of the neutron (^1n) can be written as $^9Be + \alpha \rightarrow {}^1n + {}^{12}C$ or, in a shortened form, $^9Be(\alpha, {}^1n)^{12}C$.

nuclear reactor A plant for generating energy using nuclear reactions, currently only by ➤nuclear fission, but in the future possibly also by ➤nuclear fusion. Nuclear fission reactors come in many different forms. Their common features include

➤nuclear fuel stored in a core, ➤moderators, coolant systems, turbines and containment vessels. A ➤fast reactor has the advantage of generating more fissionable material from a blanket of natural uranium surrounding the core. The hope of clean energy envisioned at the start of the nuclear reactor project has faded as the problem of nuclear fuel reprocessing has been reinforced by the accidents at Three Mile Island and at Chernobyl. Nevertheless, the nuclear industry has a reasonable safety record and countries such as France remain heavily dependent on nuclear reactors.

nuclear spin The overall spin possessed by an atomic nucleus by virtue of the ➤spin of each nucleon within it. For example, the hydrogen nucleus contains one proton with a spin quantum number of $\frac{1}{2}$, which gives rise to a spin of $\frac{1}{2}$ for a hydrogen atom. The spin produces a magnetic moment, which underlies the technique of ➤nuclear magnetic resonance.

nuclear waste (radioactive waste) Radioactive material remaining in a nuclear fission reactor after the ➤nuclear fuel is spent. No satisfactory method for dealing with it is known. Current practice is to separate **low-level waste** from the smaller quantity of highly toxic **high-level waste**, and bury the latter in huge concrete caskets in remote, geologically stable areas.

nuclear weapon (nuclear bomb) The two types of nuclear weapon are the **atomic bomb** and the much more powerful **hydrogen bomb** (or **thermonuclear bomb**). The former, which works by ➤nuclear fission, has been used twice in history, destroying the cities of Hiroshima and Nagasaki in 1945. The latter, which works by ➤nuclear fusion, produced at the height of the Cold War, has never been used in warfare.

nuclease One of a group of enzymes that break nucleic acid strands. DNAase I, DNAase II, micrococcal nucleases and ➤restriction enzymes are examples.

nucleation The process of creating a centre for the formation of a solid from a liquid or a liquid from a gas. Examples include the passage of a charged particle through a supersaturated vapour, as used in the ➤cloud chamber, or the production of vapour trails behind aircraft.

nucleic acid A class of organic acids that are ➤copolymers of ➤nucleotides (➤DNA; RNA). They function in the transmission of hereditary characteristics, protein synthesis and control of cellular activities.

nucleolus (plural nucleoli) A large, diffuse structure inside the cell's ➤nucleus (2) which produces its ➤ribosomes. Newly synthesized ribosomal RNA (rRNA; ➤RNA) is complexed with ribosomal proteins to generate ribosomes in the nucleolus. The nucleolus also contains large loops of DNA whose rRNA genes are transcribed.

nucleon A member of the class of subatomic particles that are present in the atomic nucleus. Thus nucleons consist simply of protons and neutrons.

nucleon number (mass number) The number of nucleons, that is protons and neutrons, in the nucleus of a ➤nuclide, for example 14 in the case of ^{14}C. As carbon has six protons, this nuclide has eight neutrons. Different ➤isotopes have different nucleon numbers.

nucleophile In an organic chemical reaction, the species that donates the electron pair (the term ➤Lewis base is almost synonymous). Examples of nucleophiles include uncharged molecules with lone pairs such as water, H_2O, and ammonia, NH_3, as well as negatively charged species such as iodide ion, I^-, hydroxide ion, OH^-, cyanide ion, CN^-, and alkoxide ion, OR^-. Typically they undergo nucleophilic substitution reactions with halogenoalkanes and nucleophilic addition reactions with carbonyl compounds. Compare ➤electrophile.

nucleophilic substitution A particularly important reaction in ➤aliphatic organic chemistry, enabling substitution of functional groups to take place following attack by a ➤nucleophile. Examples include the conversion of halogenoalkanes into alcohols using hydroxide ion, OH^-, or into nitriles using cyanide ion, CN^-. The comparable reaction in aromatic organic chemistry is less easy as the halogen atom is delocalized with the ring, strengthening the bond. Aromatic nucleophilic substitution only occurs if there is an electron-withdrawing group on the ring.

The mechanisms of nucleophilic substitution are well understood. Two are most important, labelled S_N1 and S_N2 (S stands for 'substitution' and N for 'nucleophilic'). In the former, a tertiary halogenoalkane such as $(CH_3)_3CI$ spontaneously undergoes ➤heterolytic fission to give the carbocation $(CH_3)_3C^+$, which subsequently reacts with a nucleophile in a fast step. The I^- ion is described as a **leaving group**. This is thus a ➤unimolecular reaction, as indicated by the 1 in the name. In contrast, primary halogenoalkanes such as CH_3CH_2Br tend to react by the bimolecular S_N2 route, where the incoming nucleophile attacks at the same time as the leaving group departs, causing ➤Walden inversion at the carbon atom to which the halogen was attached (see the diagram).

nucleophilic substitution showing Walden inversion.

nucleoplasm A granular, amorphous mass in which ➤chromosomes and the ➤nucleoli are embedded.

nucleoprotein Any ➤protein that is complexed with a ➤nucleic acid. ➤Chromosomes consist largely of nucleoprotein in which the DNA is bound to ➤histones. ➤➤chromatin.

nucleoside One of a family of organic molecules consisting of a ➤purine or ➤pyrimidine base covalently bonded to the sugar ribose (a **ribonucleoside**) or deoxyribose (a **deoxyribonucleoside**). Phosphate derivatives of nucleosides (termed ➤nucleotides) are the building units of DNA and RNA, and are important in energy-transfer reactions in cells as well as forming a number of key metabolites. ➤➤ATP.

nucleotide A nucleotide is composed of a ➤nucleoside covalently bonded to one or more phosphate groups. The nucleoside monophosphates are the monomer units of nucleic acids, consisting of a ➤purine or ➤pyrimidine base covalently bonded to a ribose or deoxyribose sugar and one phosphate group. FAD (flavin adenine dinucleotide) and NAD⁺ (nicotinamide adenine dinucleotide) are important coenzymes.

Significant examples of nucleoside di- and triphosphates include ADP (adenosine diphosphate) and ATP (adenosine triphosphate); ATP is the predominant supplier of metabolic energy in living cells.

nucleus (plural **nuclei) 1** (Phys.) The central dense core of an atom, containing a number of protons equal to the element's atomic number, and a number of neutrons equal to the nucleon number minus the atomic number. The diameter of an atom is about one hundred thousand times the diameter of its nucleus. The first **nuclear model** of an atom was Ernest Rutherford's interpretation of the classic ➤Geiger–Marsden experiment. In chemical reactions the nucleus of the atom remains intact, chemical changes occurring by the exchange or sharing of the electrons outside the nucleus, preserving the identity of the nucleus and hence of the element. ➤➤nuclear reaction.

2 (Biol.) A membrane-bound ➤organelle in ➤eukaryotic cells which contains the ➤chromosomes and nuclear cytoplasm or ➤nucleoplasm. The nuclear membrane is a double layer, perforated by pores via which the nucleus can communicate with the cytoplasm. Nuclei vary in size from 10 to 50 μm, but are largest in active secretory or dividing cells.

nuclide An ➤isotope with a specified ➤nucleon number. ➤Radiocarbon dating, for example, uses the nuclide ^{14}C, whereas the standard atom for measuring relative atomic masses is another nuclide of carbon, ^{12}C.

null A deliberately sought zero of a signal of some sort, achieved by varying other parameters of a system. For example, a ➤Wheatstone bridge is used to seek a null current by adjusting one resistance in order to find another.

null hypothesis ➤hypothesis test.

number theory The branch of mathematics concerned with investigating the properties of ➤integers, often using sophisticated techniques of ➤algebra and ➤analysis (2).

numerator ➤fraction.

numerical analysis The branch of mathematics concerned with the construction and implementation of effective ➤algorithms for the numerical solution of otherwise intractable problems. Typical problems are the solution of equations (➤Newton–Raphson method), ➤numerical integration, and step-by-step solution of differential equations (➤Runge–Kutta method).

numerical aperture (NA) In a microscope, the ratio of the radius of the ➤objective to its distance (at its edge) from the object being observed, multiplied by the ➤refractive index of the medium in which the object is situated. The ➤resolving power of a microscope increases with increasing NA.

numerical integration Techniques, such as the ➤trapezium rule and ➤Simpson's rule, which may be used to find the approximate area under a curve. They are useful if either the function involved cannot be integrated easily or the function values are known only at certain points. ➤➤Monte Carlo method.

nutation 1 (Phys.) The oscillation that occurs when a precessing ➤gyroscope is displaced slightly from its equilibrium attitude. Both the angle of inclination and the rate of ➤precession oscillate. An example is the slowly changing orientation of the Earth's poles.

2 (Biol.) A spiral pattern of growth apparent in young shoots of plants and other structures such as the tendrils of climbing plants and the fruiting bodies of fungi.

nutrient A food substance used as a source of energy or of building material during metabolism. ➤➤macronutrient; trace element.

nylon One of a variety of polyamide fibres made by ➤condensation polymerization (➤➤aramid fibre). The most significant is **nylon-6,6**, formed by condensation polymerization of the diamine $H_2N(CH_2)_6NH_2$ with the diacyl chloride $ClOC(CH_2)_4COCl$ to form a polymer with the repeating unit shown in the diagram (note how each half of the repeating unit has six carbon atoms: hence the 6,6 in the name). Another example is **nylon-6** which is not a ➤copolymer, being made by condensation polymerization of the cyclic molecule ➤caprolactam.

$$\text{—NH—(CH}_2)_6\text{—NH—CO—(CH}_2)_4\text{—CO—}$$

nylon-6,6

Nyquist's frequency The minimum sampling frequency for a signal of a ➤bandwidth b required to preserve all the information in the signal after sampling; its value is always $2b$. Named after Harry Nyquist (1889–1976).

O

o- Symbol for the ➤ortho isomer.

O Symbol for the element ➤oxygen.

Ω Symbol for the unit ➤ohm.

objective The lens in a microscope or telescope that is closest to the object being observed.

object-oriented program A style of computer program that encapsulates a ➤subroutine and its associated data as an **object**. The action of the object is well defined, which reduces development time since objects may be reused in different programs. ➤➤C (4).

oblate spheroid ➤spheroid.

oblique Not at right angles.

observable In ➤quantum mechanics, a physical quantity that can be measured, at least hypothetically if not in practice. The position of a particle, its momentum, its energy, its angular momentum and its spin are all observables.

obtuse angle An angle between 90° and 180°.

Occam's razor (Ockham's razor) A principle, originally in philosophy but applicable in science, that when several theories model the available facts adequately, the simplest theory is to be preferred. Named after William of Occam (*c.* 1285 – *c.* 1348).

occlusion The penetration of a gas into the crystal lattice of a metal. Palladium is the element most able to occlude hydrogen.

occultation The temporary blocking of light (or other radiation) from one astronomical body by another. For example, the Moon frequently occults stars as viewed from Earth. ➤➤eclipse.

occupation number The number of particles that occupy a particular ➤quantum state. The occupation number plays an important part in ➤second quantization, where not only are the energies and momenta of particles variables but also the number of particles themselves. For ➤fermions the occupation number of a state can be only 0 or 1; for ➤bosons it is unlimited.

ochre codon The triplet UAA of ➤messenger RNA, one of the three ➤termination codons of the 'universal' genetic code which signifies the end of a protein-coding region and usually causes the ribosome to dissociate from the nascent protein. Mutations that give rise to a termination codon are called 'nonsense' mutations and create truncated proteins by prematurely terminating ➤translation.

octadecanoic acid ➤stearic acid.

octagon A ➤polygon with eight sides.

octahedral In the shape of an octahedron. This shape is very common in chemistry. For example, a complex with six ligands, for example the hexacyanoferrate(III) ion $[Fe(CN)_6]^{3-}$, is octahedral, as is the compound sulfur hexafluoride, SF_6. It is also common in the solid state as the ➤rock-salt structure of NaCl can be thought of as Na^+ ions filling octahedral holes in a close-packed Cl^- array. ➤valence-shell electron-pair repulsion theory.

octahedron A solid that has eight faces. A **regular octahedron** has faces that are equilateral triangles and is one of the five ➤Platonic solids.

octane C_8H_{18} The straight-chain alkane with eight carbon atoms, which is a colourless liquid.

octane number A description of the combustion efficiency of a petrol engine fuel. The scale compares the fuel to a mixture of heptane (which is assigned an octane number of 0) and an isomer of octane, 'iso-octane', 2,2,4-trimethylpentane (see the diagram), which is assigned an octane number of 100. Therefore a 98-octane fuel would burn as efficiently as 98% iso-octane and 2% heptane.

octane number 100-octane fuel (2,2,4-trimethylpentane).

octant One of the eight regions into which space is divided by three mutually perpendicular planes.

octet rule An important simple concept when beginning to understand chemical bonding; the formulae of many binary compounds can be rationalized by assuming that the metal gives away all its outermost electrons to the nonmetal which wants to 'complete its octet', that is to have eight electrons in the outer shell. So whereas lithium chloride is LiCl because lithium can only donate one electron and chlorine only 'needs' one because it already has seven, lithium oxide is Li_2O and lithium nitride is Li_3N because oxygen and nitrogen have only six and five electrons in their outer shell respectively.

It is easy to find examples where this rule fails. Phosphorus forms the expected PCl_3, but it also forms a compound PCl_5, which has five electron pairs around the phosphorus, and is said to have 'expanded its octet'. Instead of simply counting electrons, it is essential to do a proper energy analysis, as in the ➤Born–Haber cycle, to understand compound formation fully.

ocular The eyepiece of a microscope or telescope.

odd function A function f for which $f(-x) = -f(x)$ for all values of x. The graph of an odd function has rotational symmetry about the origin.

odd number An integer that is not divisible by two.

odd parity ➤parity.

oesophagus (US esophagus) ➤gut.

oestrogen (US estrogen) In animals, one of a group of ➤steroids produced by the ➤ovary. The functions of oestrogens include the control of the development of the mammary glands in ➤mammals and of various aspects of the ➤menstrual cycle. Oestrogens usually work in conjunction with other steroids, particularly ➤progesterone. Synthetic oestrogens are used as a component of ➤oral contraceptives.

oestrous cycle A cycle of events in sexually mature female mammals, associated with the release of an ➤egg or eggs from the ➤ovary. Its length varies from, for example, 4 days in hamsters to a year for animals having a distinct breeding season. Usually, females are receptive to males only at the mid-point (**oestrus**) of the cycle, at which an egg is released. ➤➤menstrual cycle.

off-line The state of a piece of equipment that is not currently usable by a computer system, since it is disconnected either physically or logically. It can also refer to a terminal that is not currently connected to its host computer. Compare ➤on-line.

ohm Symbol Ω. The SI unit of electrical ➤resistance. A one-ohm resistor passes a current of one ampere when a potential difference of one volt is applied. Named after Georg Simon Ohm (1789–1854).

Ohm's law The law describing how a typical idealized (**ohmic**) conductor behaves when a voltage is applied; as the voltage V increases, the current I increases in direct proportion, the constant of proportionality being the ➤resistance R:

$$V = IR.$$

If the current increases too much, the energy dissipated in the conductor rises causing its resistance to increase, which in turn causes a breakdown in the simple linear Ohm's law relationship. ➤Semiconductors do not follow Ohm's law.

oil 1 ➤petroleum for crude oil.
  2 ➤lipid for oils and fats.

oil of vitriol An ancient name for ➤sulfuric acid.

oil of wintergreen The ester methyl salicylate, methyl 2-hydroxybenzoate (see the diagram), notable for its attractive smell, along with its ability to reduce the pain of rheumatism.

-ol A suffix indicating the presence of a hydroxyl group. ➤alcohol.

oil of wintergreen

Olbers' paradox The paradox of the black night sky: if the Universe contained an infinite number of stars with a constant number density and uniform brightness, the night sky should be as bright as the surface of the Sun. The number of stars in a shell at distance R would be proportional to R^2, and the intensity of light from each star in that shell proportional to $1/R^2$. Thus each shell out to infinity would continue to add to the intensity of background light and the night sky would be bright, not dark. The paradox is resolved in the ➤big bang theory, for the finite age of the Universe means that it has not had time to fill with light. Named after Heinrich Wilhelm Matthäus Olbers (1758–1840).

olefin Traditional name for an ➤alkene.

oleic acid (*cis*-octadec-9-enoic acid) A long-chain fatty acid (see the diagram), present in oils and fats, notably cows' milk, in larger quantity than any other fatty acid.

$$CH_3(CH_2)_7 \diagdown \atop C=C \diagup (CH_2)_7COOH$$
$$H \diagup \qquad \diagdown H$$

oleic acid

oleum Fuming sulfuric acid, formed by dissolving sulfur trioxide in concentrated sulfuric acid. Its approximate formula is $H_2S_2O_7$. It is a virulent reagent, being a powerful oxidant for example.

olfaction The sense of smell, enabling animals to detect odours.

oligo- A prefix meaning 'few'. In particular it is applied to polymeric molecules to indicate chains of up to about 20 monomeric units, as in **oligomer**, **oligonucleotide**, **oligopeptide** (➤peptide) and ➤oligosaccharide. ➤➤oligotrophic.

Oligocene See table at ➤era.

oligomeric ➤enzyme.

oligosaccharide A carbohydrate molecule comprising a chain of 4–20 monosaccharide sub-units. Oligosaccharides are intermediate digestion products of polysaccharides.

oligotrophic A term used to describe lakes and rivers that have low concentrations of minerals such as nitrates and phosphates, and consequently a low ➤productivity. Upland streams and lakes tend to be oligotrophic. Compare ➤dystrophic; eutrophic; mesotrophic.

olivine A magnesium-rich silicate that also contains iron and manganese, with the approximate formula M_2SiO_4, where M is Mg, Fe or Mn. It is used to make refractory bricks.

omega-minus Symbol Ω^-. A subatomic particle (with mass $1.672\,\text{GeV}/c^2$, charge -1 and spin $\frac{3}{2}$) discovered in 1964, having been predicted to exist by Murray Gell-Mann (b. 1929) on the basis of his ➤quark theory.

omnivore An animal that eats food derived from both plant and animal material.

oncogene A ➤gene involved in the malfunctioning of some cell types that results in the formation of tumours (➤cancer). Oncogenes were recognized after studies on the ➤SV40 virus in rodent cells revealed the virus-induced synthesis of two proteins called small T and large T proteins. In a way that is not fully understood, the accumulation of the large T protein in the nucleus alters the cell's growth regulation mechanisms and causes it to behave like a cancer cell. Although they were first found in viruses, their evolutionary history implies that normal vertebrate cells have genes whose abnormal expression can lead to the proliferation of cancer. Finding out how oncogenes disrupt cell ►metabolism is a major target for cancer research. **Oncoviruses**, usually ►retroviruses, contain viral genes (**v-onc**) that change normal host cells into tumour cells.

one-to-one correspondence A function $f: X \rightarrow Y$ between two ►sets such that every ►element of Y is of the form $f(x)$ for some unique element x of X; an ►inverse function $f^{-1}: Y \rightarrow X$ may then be defined.

on-line The state of a piece of equipment that is connected to a computer system and able to be used by it. It can also refer to someone who is using a terminal to communicate with a computer over a network. Compare ►off-line.

ontogeny The development of an animal throughout its life cycle from ►zygote to adult.

oocyte A cell undergoing ►meiosis that results in the formation of an egg. **Primary oocytes** undergo the first meiotic division; **secondary oocytes** undergo the second division, during ►oogenesis.

oogenesis The production of eggs, usually as a result of ►meiosis followed by a phase of maturation.

oogonium A cell in an animal ovary which gives rise to primary ►oocytes; the term is also applied to the female reproductive organs of some algae and fungi.

opal An amorphous hydrated silica, SiO_2, which shows **opalescence**, a characteristic iridescent colouring, due to interference within internal cavities.

op-amp Abbr. for ►operational amplifier.

opaque Not permitting light (and by extension, any other electromagnetic radiation) to pass. Objects can be opaque in one region of the ►electromagnetic spectrum and transparent in another. Compare ►translucent; transparent.

open circuit The condition in which a pair of terminals (or other points in the circuitry of an electrical device) are not connected by a conducting path, particularly as applied to the output of an electrical device.

open cluster ►cluster.

open-hearth furnace An apparatus formerly used for making steel; it has been superseded by the ►basic oxygen process.

open reading frame A sequence of ►codons, beginning with a ►start codon and not interrupted by a ►termination codon, which is therefore identified as a possible protein-coding sequence. ►►reading frame.

open set A set of points in two or three dimensions, any point of which can be surrounded by a disc or ball which is wholly contained within the set. This definition can be successively generalized to a ►metric space and to a ►topological space.

open system A ►system in which the amount of matter can vary.

operating point The point on a plot of d.c. current against d.c. voltage about which an electronic device such as a transistor operates. ➤bias.

operating system The ➤software in a computer that controls various vital tasks, such as the execution of all other programs and communications with hardware devices, as well as providing useful routines for programs to use. The operating system may be held in ROM; otherwise loading it into RAM is one of the first operations to be performed by a computer when it is switched on. Historically significant examples are Unix and MS-DOS.

operational amplifier (op-amp) A simple ➤amplifier, typically in the form of an ➤integrated circuit, that is used as a component of a more complicated electronic device.

operator **1** (Math.) A function whose ➤domain (2) consists of a set of functions. Examples include integral transforms such as the ➤Fourier and ➤Laplace transforms and differential operators such as the ➤Laplacian.
  2 (Phys.) An abstract object representing an ➤observable in ➤quantum mechanics. The prescription for finding the value of any physical quantity O revolves around finding the corresponding operator O, and solving the ➤eigenfunction equation

$$O|\psi\rangle = O|\psi\rangle,$$

where $|\psi\rangle$ is the ➤quantum state. The operator for the energy is called the ➤Hamiltonian operator.
  3 (Biol.) The site on the DNA to which a ➤repressor protein binds to prevent ➤transcription from initiating at the adjacent ➤promoter (2).

operon A mechanism for the regulation of ➤gene function in ➤bacteria proposed by Jacob and Monod in 1961 which has greatly influenced understanding of molecular biology. The main feature of the mechanism is that it introduced the concept of two types of gene – **structural genes**, responsible for the synthesis of proteins, and **regulatory genes**, which control the expression of the structural genes. The original model referred to the control of the synthesis of the enzyme β-galactosidase, but it appears that a wide range of genes in ➤prokaryotic cells are controlled in more or less the same way. Mechanisms for controlling the regulation of genes in ➤eukaryotic cells are more complex, and there is very little evidence for the existence of operons in such cells. ➤transcription factor.

opiate One of a number of pharmaceutically active compounds derived typically from the latex of the opium poppy (*Papaver somniferum*). Opiates include ➤morphine and diamorphine (heroin). ➤➤endorphins.

opium The dried juice of the unripe seed-pods of the opium poppy. It contains a number of alkaloids, including ➤morphine and codeine. ➤➤narcotic drug.

opposition An occasion on which a planet, as viewed from the Earth, is in the opposite direction to the Sun, and is therefore potentially visible all night. Compare ➤conjunction.

opsin The protein part of a series of light-sensitive pigments in the eye called rhodopsin (➤colour vision). Cloning of the genes for rhodopsin has enabled human

➤colourblindness and the mechanisms of ➤colour vision to be studied in detail. These genes lie mainly on the X chromosome and provide a good example of an inherited character that is sex-linked (➤sex chromosome; ➤sex-linkage).

optical activity Optically active substances rotate the plane of plane-polarized light. For compounds with a single ➤chiral centre, the two **optical isomers** called **enantiomers** are identical in almost every way except for the direction of rotation: one rotates left (➤laevorotatory) whereas the other rotates right (➤dextrorotatory). One other consequence is that the two enantiomers can react differently with another optically active molecule, a fact used to resolve a ➤racemic mixture. Many vital molecules in nature are optically active. ➤Thalidomide caused terrible foetal malformations during the late 1950s. One enantiomer is a ➤teratogen whereas the other would have been innocuous. As another example, one enantiomer of ➤LSD is psychoactive, whereas the other is not. ➤➤(R-S) system.

optical axis The symmetry axis of a linear optical system, running through and perpendicular to the centres of mirrors and lenses.

optical centre The point on a thin lens through which a light ray may pass without deviating.

optical fibre ➤fibre, optical.

optical glass Glass that is used in optical instruments. High-quality glass free from imperfections (such as bubbles and trapped raw material) is required for such applications, and its refractive index must be consistent.

optical isomer ➤optical activity.

optically flat In optics, a surface with imperfections significantly smaller than the wavelength of light to be transmitted or reflected by the surface.

optical path The path taken by light through an optical system. ➤➤Fermat's principle of least time.

optical pumping The injection of energy (by absorption) into a ➤laser to invert the population of the electronic ➤energy levels. It is analogous to the pumping of water from a lower reservoir to a higher reservoir, hence the name.

optical pyrometer An instrument for measuring the temperature of an object by measuring the relative intensity of radiation emitted from it at different frequencies. ➤➤black-body radiation.

optical telescope A ➤telescope that operates in the ➤visible spectrum. Compare ➤radio telescope.

optic axis ➤birefringence.

optics The study of light and its interaction with matter. **Geometrical optics** usually considers the light as rays, and is concerned with the ➤reflection and ➤refraction of those rays at, for example, ➤mirrors and ➤lenses. **Wave optics** considers the light as a wave, and is concerned with the phenomena of ➤diffraction and ➤interference. Although all geometrical optics can be described in terms of wave optics, it is often much easier to consider light as travelling in rays.

optoelectronics The study of the interaction of light with electric fields. Optoelectronics includes the investigation of the optical properties of ➤semiconductors and ➤polymers, the design of solid state ➤lasers and the detection of light by electronic means.

optoisolator A component for transmitting an electrical signal between two parts of a device not physically connected. This ensures that if one part is raised to a high potential with respect to earth, there is no danger of the rest of the device being subjected to the same voltage.

oral contraceptive A pill containing synthetic ➤oestrogens and ➤progesterone used to inhibit ovulation during the ➤menstrual cycle. Oral contraceptives work by preventing the release of an egg from the ovary, yet maintaining the otherwise normal cycle of events in the uterus. Oral contraceptives for men which inhibit the formation of sperm are as yet experimental.

orbit 1 (Astron.) The movement of one body around another under a gravitational force. Since ➤gravity obeys an ➤inverse square law, orbits take the shape of ➤conic sections, such as ellipses (➤Kepler's laws). The eccentricity of an orbit determines whether it is closed or open. If the eccentricity is less than one the orbit is a closed ellipse. If, however, the eccentricity is one or more, the orbit is open, taking the form of a parabola or hyperbola, and the orbiting body has enough energy to escape the gravitational field of the central body.
   2 (Phys.) ➤Bohr model.

orbital A solution of the ➤Schrödinger equation, the name being designed to convey the impression of less certainty than the 'orbit' of the ➤Bohr model. ➤➤atomic orbital; molecular orbital.

orbital angular momentum One of the two sources of angular momentum of an electron in an atom: that produced by the motion of the electron around the nucleus. The other source is ➤spin angular momentum, and the two can be combined in two ways (➤spin–orbit coupling). The angular momentum is quantized; the lowest value for the ➤orbital angular momentum quantum number l is 0, for s orbitals; other important values are 1 for p orbitals, 2 for d orbitals and 3 for f orbitals.

orbital angular momentum quantum number (azimuthal quantum number) Symbol L or l. The ➤quantum number associated with the ➤angular momentum of the spatial ➤wavefunction of one or more electrons, usually in an atom, ion or molecule. Orbital angular momentum is quantized in units $\hbar = h/2\pi$, where h is the ➤Planck constant. Compare ➤spin angular momentum quantum number; ➤magnetic quantum number.

order 1 (Math.) The number of ➤elements in a finite ➤set or ➤group (2).
   2 (Math.) The number of rows in a ➤square matrix.
   3 (Math.) The greatest number of times that a function is differentiated in a ➤differential equation.
   4 (Biol.) A category of biological classification (➤taxonomy) forming subsets of a ➤class and containing related ➤families.
   5 (Phys.) For a ➤phase transition, the lowest order of ➤derivative of the ➤Gibbs energy that shows a ➤discontinuity across the transition. Most common phase

transitions, such as the change of state from a liquid to a gas, are **first order** because the volume (the first differential with respect to pressure) is discontinuous.

order of diffraction The number of nulls in a diffraction pattern between a particular maximum of intensity and the central maximum.

order of magnitude An approximation to a value expressed to the nearest power of 10.

order of reaction The rate of a reaction can usually be expressed as a **rate equation** of the form:

$$\text{rate} = k[A]^m [B]^n,$$

where the order with respect to substance A is the power to which its concentration is raised, that is m. The order with respect to B is n and the overall order $m + n$. Knowing the order of reaction is frequently a vital piece of evidence for finding the ➤mechanism of the reaction.

ordinal number In everyday usage, a whole number denoting the order of terms in a sequence, as in 1st, 2nd, 3rd, ... This concept can be generalized to assign an ordinal number to infinite ➤sets. Compare ➤cardinal number.

ordinary ray A ray, in a crystal exhibiting ➤birefringence, that is parallel to its ➤wave-vector. Compare ➤extraordinary ray.

ordinate Another name for 'y coordinate' (➤Cartesian coordinates).

Ordovician See table at ➤era.

ore A natural mineral from which useful substances can be extracted.

organ A highly organized unit of structure, having a defined function in a multicellular organism and consisting of a range of ➤tissues. Examples of organs are leaves in plants, and the heart, eye and kidney in animals.

organelle A subcellular compartment of ➤eukaryotic cells with a discrete function, usually bound by a membrane. Organelles include ➤mitochondria, ➤chloroplasts and ➤Golgi bodies.

organic chemistry The chemistry of the element carbon, traditionally excluding the oxides and carbonates. Humans are carbon-based lifeforms, and as such we elevate carbon to a unique place among the elements. Of the six million compounds known, five million are organic. The division of all compounds into two classes, organic chemistry and everything else (➤inorganic chemistry), is arbitrary and artificial (➤organometallic chemistry). Carbon is unique for two main reasons. First, long chains can be built up. Although sulfur atoms, for example, also build long chains of up to about 500 000 atoms, these chains cannot be complex as sulfur forms only two bonds. Carbon can form two more bonds having made a chain, hence allowing enormous variation in the side chains. Second, carbon is at the top of Group 14 of the ➤periodic table and so is the smallest atom in the group. Carbon atoms can therefore make stronger bonds than silicon atoms can. The higher bond strength means that carbon compounds are much more stable kinetically than the corresponding silicon compounds. The most important division within organic chemistry is into **aromatic chemistry**, based around rings such as those in benzene, and **aliphatic chemistry**, where molecules do not have benzene rings. Within each of these, further subdivision is possible, depending on the ➤functional group(s) present. Examples of the resulting classes include ➤alcohols, ➤alkanes, ➤alkenes and ➤halogenoalkanes.

organism Any individual living entity, which one can place within a proper biological classification scheme (➤taxonomy).

organometallic chemistry Compounds that contain organic (i.e. carbon-containing) groups attached to metals. Important examples include the antiknock agent ➤tetraethyllead, the ➤Ziegler–Natta catalyst triethylaluminium and the important synthetic intermediates the ➤Grignard reagents, along with more esoteric compounds, such as ➤ferrocene. ➤Vitamin B_{12} can also be so classified. This area of chemistry, which straddles the traditional division into organic and inorganic chemistry, has grown enormously in the last twenty years.

origin That point in a ➤coordinate system all of whose coordinates are zero.

origin of replication A section of a ➤DNA molecule at which replication starts during the copying of DNA, for example before cell division or during laboratory procedures such as the ➤polymerase chain reaction.

ornithine A nonprotein amino acid, particularly important in animals that excrete ➤urea.

orphon An isolated, individual ➤gene related to members of a gene cluster.

orpiment Yellow crystals of arsenic(III) sulfide, As_2S_3.

ortho Symbol o-. The isomer of a disubstituted benzene ring in which the two groups are nearest neighbours; compare ➤meta and ➤para. An example is the arrangement in ➤CS gas. The three isomers sometimes have different properties. For example o-nitrophenol has a lower boiling point than the other two isomers as it has internal ➤hydrogen bonding (see the diagram) rather than hydrogen bonding to neighbouring atoms. Similarly o-benzenedicarboxylic acid dehydrates more easily than the other two isomers.

ortho Intramolecular hydrogen bond in o-nitrophenol.

orthoclase A common aluminosilicate containing potassium, $KAlSi_3O_8$.

orthogonal 1 Describing a set of vectors in which every pair of vectors is mutually perpendicular.
2 Describing a finite or infinite sequence of functions $f_1(x), f_2(x), \ldots, f_n(x)$ in which $\int f_i(x) f_j(x)\, dx = 0$ for distinct suffixes i and j.

orthonormal A set of vectors or functions that are both mutually ➤orthogonal and ➤normalized. Such a set forms a convenient ➤basis for a ➤vector space.

orthophosphoric acid ➤phosphoric acid.

orthorhombic One of the seven crystal systems (➤Bravais lattice).

Os Symbol for the element ➤osmium.

oscillation The periodic (or almost periodic) variation of a physical quantity. It may be in the form of, for example, a ➤vibration (the quantity is ➤displacement) or an a.c. circuit (the quantity is ➤electric current). If the variation is perfectly periodic, it is known as ➤harmonic.

oscillator A system that can exhibit ➤oscillation. In the context of an electrical circuit, an oscillator is a device (often based on a ➤resonant circuit) that takes a d.c. input and provides a single-frequency a.c. output. In ➤quantum mechanics, however, the concept of an oscillator has a fundamental significance (➤second quantization).

oscilloscope (cathode-ray oscilloscope, CRO) A device for displaying an electrical signal as a function of time (or sometimes of a second signal). A traditional oscilloscope uses an ➤electron gun fired at a luminescent screen (a ➤cathode-ray tube). The vertical deflection of the electron beam is governed by the input voltage to be displayed, and the horizontal deflection is governed by a reference voltage of known frequency (the ➤timebase) or the second signal. Modern oscilloscopes tend to use more sophisticated techniques of digital electronics for displaying the signal.

osmiridium An alloy of osmium and iridium that is useful where hardness is a prime requirement, as in fountain pen nibs.

osmium Symbol Os. The element with atomic number 76 and relative atomic mass 190.2, which is in the third row of the ➤transition metals and a member of the ➤platinum metals. It has one important compound with an unusual oxidation number, +8, osmium tetroxide, OsO_4, which is useful for oxidizing double bonds. The metal itself has a very high density, the same to three significant figures as iridium's, and it is a moot point which of these two is the densest element of all.

osmoregulation The ability of animals to regulate their water content and thus the concentration of their body fluids by a variety of physiological mechanisms. In mammals, for example, the ➤kidneys selectively reabsorb different amounts of water from the kidney tubules back into the blood to adjust its concentration. Blood concentration varies according to environmental and physiological conditions. The reabsorption of water according to blood concentration is under the control of the hormone ➤vasopressin, secreted by the ➤pituitary gland. Organisms that live in fresh water have a continual need for osmoregulation since water constantly enters their cells by ➤osmosis.

osmosis The passage of solvent (always water in a biological context), but not solute, through a **selectively permeable membrane** from the more dilute to the more concentrated solution, tending to equalize the concentrations. If the experiment is conducted in a tube, the liquid level on one side will rise and the other fall until a sufficient hydrostatic pressure, the **osmotic pressure**, builds up to prevent further osmosis. A more sophisticated **osmometer** would apply an external pressure just sufficient to prevent any osmosis, hence avoiding complications with dilution effects. Osmosis is vital for many processes in life (➤osmoregulation). ➤➤water potential.

osteoblast A cell responsible for laying down the protein matrix of bone.

osteoclast A cell that destroys bone.

Ostwald process The industrial process used to manufacture nitric acid from ammonia. The ammonia is first oxidized by air over a platinum–rhodium gauze catalyst to nitrogen monoxide. This nitrogen monoxide is then cooled, mixed with air and passed through a water absorption column; the combination of oxidation and absorption produces moderately concentrated (60%) nitric acid. Named after (Friedrich) Wilhelm Ostwald (1853–1932).

Ostwald's dilution law For a weak electrolyte AB at concentration c the ➤equilibrium constant K has the form:

$$K = \alpha^2 c/(1 - \alpha),$$

where α is the ➤degree of ionization.

Ostwald's isolation method A method for determining the ➤order of a reaction by isolating the effect on the rate of each reactant in turn by adding all the others in sufficient excess that the concentration of only the one changes significantly during the reaction.

Otto cycle The cycle of events in an internal combustion engine burning petrol. It consists of four strokes of the piston: an **induction stroke** which draws the fuel–air mixture into the cylinders; a **compression stroke** in which the piston compresses the gas mixture; a **power stroke** in which the mixture combusts after ignition by a spark; and an **exhaust stroke** which pushes the gases out of the cylinder. Named after Nikolaus August Otto (1832–91). ➤➤indicator diagram.

ounce Symbol oz. A unit of mass in Imperial units, equal to one-sixteenth of a ➤pound. 1 oz is equivalent to 28.35 g. ➤➤fl oz.

-ous A suffix labelling acids and ions with an element's lower oxidation number (compare ➤-ic). It was once the common label, but has been superseded in many cases by nomenclature based on specifying oxidation numbers. For example, ferrous was the old name for iron(II).

outbreeding (exogamy) A mechanism in sexual reproduction that promotes the fusion of gametes from unrelated parents. In animals with separate sexes this is common, except in small confined populations. In plants where male and female structures are commonly found on the same individual there are usually mechanisms to promote outbreeding. These include different maturation times of male and female gamete-producing structures and various self-sterility mechanisms. Outbreeding populations are characterized by having a high proportion of genes existing in a ➤heterozygous state. Compare ➤inbreeding.

outer ear ➤ear.

outer planets A collective name for the four major planets farthest from the Sun, beyond the asteroid belt (Jupiter, Saturn, Uranus and Neptune). These are all large and gaseous, as opposed to the inner planets, which are small and rocky.

output impedance The ➤impedance at the output of an electrical device that is internal to the device. It can be measured as the voltage with the output at ➤open circuit, divided by the current when the output terminals are short circuited.

ovary 1 The lower hollow part of the carpel of flowering plants (➤flower) containing the ➤ovules. The carpel may contain a single ovary or a number of ovaries fused together.
    2 The female ➤gonad that is responsible for the production of eggs (**ova**). ➤➤menstrual cycle.

overdamped ➤damped harmonic motion.

overlap ➤molecular orbital.

overtone A musical note of higher frequency than the ➤fundamental note. ➤➤harmonic.

overvoltage (overpotential) An additional voltage over that predicted by thermodynamic arguments (➤standard electrode potential) to be necessary for the discharge of an element during electrolysis, which needs to be applied to overcome a very slow rate of discharge.

oviduct (Fallopian tube) The hollow tube in animals that carries eggs (ova) from the ovary either to the outside of the body or to another part of the reproductive system. In placental mammals, for example, eggs are conveyed to the ➤uterus.

ovulation The process by which the egg (ovum) is released from the ➤ovary. ➤menstrual cycle.

ovule In seed plants, the structure in which the female ➤gametophyte develops to give rise to the female ➤gamete. The ovule is attached to the wall of the ➤ovary by the **funicle**, and is enclosed in the **integuments**, which surround the **micropyle** through which the pollen tube enters the ovule before fertilization. In conifers the ovule is borne on the surface of the scales of the cone; in flowering plants the ovule is enclosed in the ovary. After fertilization the ovule becomes the ➤seed.

ovum (plural **ova)** The unfertilized mature female ➤gamete.

oxalate A salt of oxalic acid (➤ethanedioic acid). Oxalates are poisonous.

oxalic acid Traditional name for ➤ethanedioic acid.

oxidant (oxidizing agent) The species causing oxidation. The oxidant is itself reduced. The quantitative measure of oxidizing strength is the ➤standard electrode potential, a highly positive value indicating a strong oxidant. This classification must be treated with care; for example, hydrogen peroxide can oxidize iodide ions, but is itself oxidized by the stronger oxidant aqueous potassium manganate(VII).
    Common gaseous oxidants include fluorine and oxygen. Common liquid oxidants are concentrated nitric and sulfuric acids. Common oxidants in solution include acidified potassium dichromate(VI) and acidified potassium manganate(VII). A common solid oxidant is manganese(IV) oxide.
    The test for an oxidant is that, when iodide ions are added in acidic solution, the typical brown colour of iodine is seen. This can be quantified by adding excess iodide

ions and measuring the amount of iodine formed by titration with standard aqueous sodium thiosulfate.

oxidation The meaning of oxidation has been sharpened over the last two centuries from meaning combination with oxygen, to the modern meaning of 'loss of electrons'. To see how these concepts are related, consider magnesium oxide, MgO. The compound contains the magnesium ion Mg^{2+} and the oxide ion O^{2-}. In going to the oxide, magnesium has lost electrons via the process:

$$Mg \rightarrow Mg^{2+} + 2e^-.$$

A similar analysis applies even when the oxide is covalently bonded, because the oxidation number rules assign the shared electron pairs to oxygen as it is more electronegative (except in the case of fluorine). The following two simple cases are the most common oxidation reactions. Oxidation occurs when oxygen is added to a substance, as when S becomes SO_2. Oxidation occurs when hydrogen is removed from a substance, as when HCl becomes Cl_2. To deal with the more difficult case of SO_2 going to H_2SO_4, notice that the oxidation number of sulfur in the two compounds is +4 and +6 respectively, so sulfur dioxide has been oxidized when converted to sulfuric acid.

oxidation number (oxidation state) Symbol $Ox(E)$ for element E. An exceptionally useful label for the level of oxidation of an element in a compound that is equal to the number of electrons that need to be added to the element in its combined state to produce the neutral atom. It is a theoretical quantity calculated according to several rules, which should be applied in order.

(i) The sum of the oxidation numbers of all the elements in a species must equal the charge on the species. So in H_2O, if oxygen has $Ox(O) = -2$ then $Ox(H) = +1$, as the species is uncharged. In iodide ion, I^-, $Ox(I) = -1$, as there is only one element and the charge on the species is -1.

(ii) The oxidation number of fluorine in its compounds is always -1. So in the compound SF_6, $Ox(S) = +6$.

(iii) The oxidation number of oxygen in its compounds is usually -2 (except when fluorine is present or in peroxides or superoxides). So in FeO, $Ox(Fe) = +2$, and the name of the compound is iron(II) oxide.

(iv) Assign each shared electron pair to the atom with the greater ➤electronegativity forming the bond. So in ICl_3, Cl is more electronegative and hence $Ox(I) = +3$.

If the oxidation number of an element increases in a reaction, the element is oxidized; if it decreases, the element is reduced. The oxidation number is frequently quoted, in roman numerals, in parentheses after the name of an element, as in the compounds iron(III) oxide or copper(II) sulfate. The oxidation number of an element in a compound can also be indicated by a right superscript in roman numerals, e.g. Fe^{III} or O^{-II}.

oxidative phosphorylation The biochemical process of forming ➤ATP from ADP and inorganic phosphate in the ➤mitochondrion by the transfer of electrons from NADH (➤NAD^+) to oxygen by a sequence of electron carriers. Thirty-two of the thirty-eight ATP molecules formed when glucose is oxidized to carbon dioxide and water are formed during oxidative phosphorylation. An electrochemical potential difference of protons is created across the membranes of the mitochondrion by the ►electron-transport chain. The ►Gibbs energy stored in this electrochemical proton gradient is called **protonmotive force**, and is capable of driving the synthesis of ATP via special channels through which protons can pass. This is the so-called **chemiosmotic coupling hypothesis**, a concept which has been extended to a number of other phenomena involving transport across membranes (►photophosphorylation).

oxide A compound with oxygen. Various classifications are possible. First, **binary oxides** contain only one other element, whereas **mixed oxides**, such as barium titanate, contain at least two others. Second, oxides can be classified according to their acid–base nature. Acidic oxides react with alkalis. Basic oxides react with acids. ►Amphoteric oxides react with both acids and alkalis. Neutral oxides react with neither. Nonmetals tend to form acidic oxides whereas metals tend to form either basic oxides (which may be alkaline) or amphoteric oxides. Metalloids tend to form amphoteric oxides.

oxidizing agent A common alternative name for an ►oxidant.

oxime A derivative of a carbonyl compound in which the carbonyl group C=O is replaced by C=N—OH, formed by reaction with hydroxylamine, NH_2OH.

oxirane ►epoxide.

oxoacid (oxyacid) An acid in which the hydrogen is bonded to *oxygen*. Many examples are known, and care should be taken with formulae as they do not always specify if the acid is an oxoacid. For example, if nitrous acid is written as HNO_2, it is not obviously an oxoacid, so the formula is better expressed as HONO. Similar arguments apply to the chlorine acids such as $HOClO_2$. Sulfuric acid can be more explicitly written as $(HO)_2SO_2$ rather than H_2SO_4 to identify its oxoacid nature.

oxoanion (oxyanion) A negative ion in which the charge is carried predominantly by oxygen. Important examples include the nitrate ion, NO_3^-, and the sulfate ion, SO_4^{2-}.

oxonium ion (hydronium ion, hydroxonium ion) A positive ion in which oxygen is carrying the charge, especially used for the species H_3O^+. This formula provides a more accurate representation of the actual species present in acidic solution than the simpler formula H^+.

oxyacetylene flame The hottest flame achievable in a burner easily transported from one place to another, produced by combustion of acetylene (►ethyne), C_2H_2. It is used in welding.

oxyacid ►oxoacid.

oxyanion ►oxoanion.

oxygen Symbol O. The element with atomic number 8 and relative atomic mass 16.00. It is the element in Group 16 with the lowest atomic number, and is a nonmetal. It is allotropic: one allotrope is ►ozone; the much more common one is dioxygen, O_2, which is a colourless odourless gas.

Oxygen is a very reactive element, reacting often violently with metals such as magnesium, nonmetals such as sulfur, inorganic compounds such as sulfur dioxide and organic compounds such as methane. Burning in oxygen is called ➤combustion, and the majority of mankind's energy needs are met by combustion. Liquid oxygen, which is coloured blue, is particularly violent in its reactions. Combined with a fuel of liquid hydrogen it is used to power the space shuttle.

Oxygen is unusual in being a diradical (i.e. having two unpaired electrons), and the ➤Lewis structure cannot account for this observation. An explanation is provided by molecular orbital theory, which shows that the two unpaired electrons arise because two ➤degenerate orbitals have to be occupied (see the diagram).

Oxygen is essential for life; thus the first question a first-aider must answer is 'Is the casualty breathing?' Humans can do without food for weeks *in extremis*, water for a few days, but oxygen for only a matter of minutes. ➤➤oxygen transport.

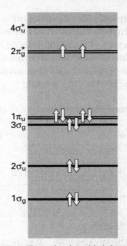

oxygen The molecular orbitals formed by the overlap of the 2s and 2p orbitals on two oxygen atoms, shown with the energy levels increasing upwards. Two oxygen atoms have a total of twelve 2s and 2p electrons. The first ten fill the five lowest orbitals. The final two enter the two $2\pi_g^*$ orbitals. Following ➤Hund's rules, one enters each orbital with parallel spins.

oxygen carrier ➤oxygen transport.

oxygen debt The oxygen required to oxidize the lactic acid that accumulates in active muscles. When the level of activity rises during strenuous exercise, the supply of oxygen to tissues is insufficient to allow complete ➤aerobic respiration. Energy is then made available by the anaerobic process of ➤glycolysis, resulting in the formation of lactate. Lactic acid is toxic and is removed by subsequent oxidation. An obvious manifestation of oxygen debt is the continued heavy breathing by an athlete after exercise.

oxygen transport The conveyance of oxygen around the body of vertebrates and some other animals by an **oxygen carrier**, notably ➤haemoglobin, usually in the red blood cells. Each of haemoglobin's four polypeptide chains contains a prosthetic group, ➤haem, which carries an iron atom. This iron atom combines loosely with one molecule of oxygen. Whether haemoglobin will load or unload oxygen depends on the partial pressure of oxygen in the medium to which haemoglobin is exposed. A high partial pressure results in haemoglobin binding to oxygen; low partial pressure causes haemoglobin to release oxygen.

oxyhaemoglobin HbO_2 ➤Haemoglobin combined with oxygen.

oxytetracycline (Terramycin) $C_{22}H_{24}N_2O_9$ A yellow, crystalline, tetracyclic (four-ringed) broad-spectrum antibiotic. It occurs naturally in fungi of the genus *Streptomyces*, from which it is extracted; it is also produced synthetically. It is used to treat trachoma in developing countries.

oxytocin The ➤peptide hormone composed of nine amino acids synthesized by the ➤hypothalamus and released by the posterior ➤pituitary gland. It acts on the muscles of the uterus causing them to contract, and causes the mammary glands to release milk. Injections of oxytocin are often used to induce labour when pregnancies have passed the 40-week mark.

oz Symbol for the unit ➤ounce.

ozone

ozone (trioxygen) One of the allotropes of oxygen, of formula O_3 (see the diagram). Small-scale preparation of ozone requires an **ozonizer**, which uses a silent electric discharge to provide energy to rip apart an oxygen molecule, the resulting oxygen atom reacting with another molecule to produce O_3. It is an example of a substance that is both harmful and beneficial. It is harmful if released near humans, for example by electrical machines such as laser printers. It is particularly beneficial in the upper atmosphere where the ➤ozone layer protects us. It is also used, particularly in France, to purify water as it is faster acting than chlorine and avoids producing halogenated hydrocarbons.

ozone hole Marked reduction in the level of ozone in the Earth's protective ➤ozone layer, over the north or south pole. The holes are believed to be caused by the release of certain chemicals, most notably the ➤chlorofluorocarbons (CFCs), bans on the use of which have come into force. It is possible that the appearance of ozone holes is partly a natural phenomenon.

ozone layer The layer of ➤ozone in the upper atmosphere that protects life on Earth from much of the harmful effect of ➤ultraviolet radiation from the Sun.

ozonolysis A reaction to identify the position of the double bond in an alkene; the alkene is reacted with ozone, O_3, followed by hydrolysis in acidic solution. This breaks the molecule where the double bond was, creating two ➤carbonyl compounds which are easily identified. The original structure is pieced together from the two fragments.

P

p **1** Symbol for ➤momentum.
   2 Symbol for ➤pressure.

p- Symbol for the ➤para isomer.

P **1** Symbol for the element ➤phosphorus.
   2 Symbol for ➤poise.

P Symbol for ➤power.

π A fundamental mathematical constant defined as the ratio of the circumference of a circle to its diameter. It is a ➤transcendental number with the decimal expansion 3.141 592 653 589 79 . . . More than a trillion decimal places have now been calculated; just 39 places suffice to compute the circumference of the observable ➤Universe accurate to within the radius of a hydrogen atom!

Ψ Symbol for ➤wavefunction.

Ψ particle ➤J/ψ particle.

Pa **1** Symbol for the unit ➤pascal.
   2 Symbol for the element ➤protactinium.

PABA Abbr. for *p*-aminobenzoic acid (see the diagram), a common sun-blocking ingredient in sun-tan lotions.

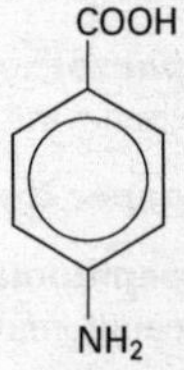

pacemaker ➤heart.

pachytene ➤meiosis.

packing density The number of semiconductor devices per unit area on an ➤integrated circuit.

COOH

NH$_2$

PABA

pairing of electrons Electrons are often found in pairs, with two electrons occupying the same orbital but possessing opposite ➤spins. However, electrons have no innate tendency to pair up. Indeed, pairing is energetically unfavourable. When electrons occupy orbitals of the same energy, they go into different orbitals with their spins parallel (➤Hund's rules; Pauli principle). Normally, they pair because the next available orbital is too far away in energy to be occupied. ➤➤complex.

pair production The formation of an ➤electron and a ➤positron from a ➤photon. In order for both momentum and energy to be conserved in such a process, another particle (typically an atomic nucleus) must be present to take up some of the photon's momentum.

Palaeocene See table at ➤era.

Palaeozoic See table at ➤era.

palladium Symbol Pd. The element with atomic number 46 and relative atomic mass 106.4, which is in the second row of the ➤transition metals and is a member of the ➤platinum metals. The metal itself is used for catalysing hydrogenation reactions, as are its ➤congeners nickel and platinum. Palladium is the best substance for absorbing hydrogen as it can hold 900 times its own volume of the gas at room temperature and pressure (➤occlusion). As with its congeners, its most common oxidation number is +2 as in the oxide PdO, a black solid, and the chloride $PdCl_2$, a dark red solid. Chloride complexes are used in the ➤Wacker process.

palmitic acid A long-chain fatty acid, hexadecanoic acid, $C_{15}H_{31}COOH$, present in ➤lipids.

palynology The study of pollen grains, particularly from fossil deposits.

panchromatic film Photographic film sensitive throughout the visible range.

pancreas A vertebrate organ, lying near the ➤liver, with a wide range of functions. It secretes **pancreatic juice**, containing enzymes that digest ➤lipids, ➤carbohydrates and ➤proteins, into the small intestine. The pancreas has special cells, the ➤islets of Langerhans, which produce the hormones ➤insulin and ➤glucagon, responsible for the regulation of glucose concentration in the blood. ➤➤diabetes.

Paneth mirror An experiment which proved the existence of ➤radicals. Tetra-ethyllead heated in a furnace produced a mirror-like deposit of lead, and butane, C_4H_{10}, could be detected downstream, the result of dimerization of ethyl radicals. Named after Friedrich Adolf Paneth (1887–1958).

Pangea ➤continental drift.

pantothenic acid A ➤vitamin of the B complex acting as a precursor of ➤acetyl coenzyme A.

paper chromatography ➤chromatography.

papillomavirus A DNA-containing ➤virus which infects mammalian cells. Some papillomaviruses cause warts in humans.

papovavirus A group of DNA-containing ➤viruses, including papillomavirus, which are associated with some forms of ➤cancer in mammals.

para Symbol p-. The isomer of a disubstituted benzene ring in which the two groups are opposite each other: examples include ➤PABA and ➤paracetamol. Compare ➤meta and ➤ortho.

parabola A ➤conic with eccentricity 1. With respect to suitable axes, its equation is $y = kx^2$; a general quadratic curve $y = ax^2 + bx + c$ is a parabola since it may be put into this form by a suitable choice of origin. It has one axis of symmetry; rays parallel to this axis are reflected off a **parabolic mirror** through the fixed point (the ➤focus);

this reflection or focal property is used in the design of telescope mirrors and car headlights. The path of a projectile moving freely under gravity is a parabola.

paraboloid A three-dimensional surface which has a parabolic cross-section parallel to two coordinate planes. By a suitable choice of axes, its equation is of the form $x^2/a^2 + y^2/b^2 = cz$ or $x^2/a^2 - y^2/b^2 = cz$; the positive sign giving a bowl-shaped surface and the negative sign a saddle-shaped surface.

paracetamol (acetaminophen) Perhaps the most widely used drug in the world, 4-hydroxy-N-phenylethanamide (see the diagram), being effective at reducing fever and as an analgesic. It has an excellent safety record, but even small overdoses can be dangerous.

paraffin An older name for the ➤kerosene fraction produced during fractional distillation of crude oil. It was once in common use in paraffin lamps and paraffin heaters. The fraction is now most used as jet aviation fuel.

paracetamol

paraffins Traditional name for the ➤alkanes.

paraldehyde A trimer of ➤ethanal (acetaldehyde).

parallax A difference in the perceived position or direction of an object due to a change in the position of an observer. For example, parallax can cause errors of measurement with analogue instruments if the pointer and the scale are separated by a significant distance (see the diagram). Parallax is an important distance-measuring technique in astronomy: for the nearer stars, their distances may be calculated from their small apparent changes in position when viewed from opposite points of the Earth's orbit, six months apart.

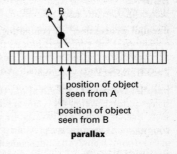

position of object seen from A

position of object seen from B

parallax

parallel 1 (Math.) Describing a set of lines or curves in a plane, or planes or surfaces in space, that never meet, however far they are extended. ➤➤parallel postulate.

2 (Math.) Two ➤vectors are parallel if they have parallel directions; one is then a scalar multiple of the other.

3 (Geog.) A line of equal ➤latitude on the surface of the Earth. The parallels are small circles, with the exception of the ➤equator, which is a ➤great circle.

4 (Phys.) Of two or more electrical components, an arrangement such that each component is connected to all of the others at both ends (see the diagram). The total ➤resistance of a set of resistors R_1, R_2, R_3, ... in a **parallel circuit** is

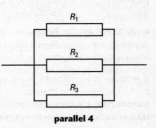

parallel 4

$$\frac{1}{\dfrac{1}{R_1} + \dfrac{1}{R_2} + \dfrac{1}{R_3} + \cdots}.$$

For a set of capacitors in parallel, the total capacitance is simply the sum of the individual capacitances. Compare ➤series (3).

parallelepiped A polyhedron that has six faces, opposite pairs of which are identical ➤parallelograms (see the diagram). A **rectangular parallelepiped** is a ➤cuboid.

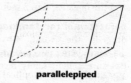

parallelogram A ➤quadrilateral with two pairs of parallel sides.

parallelepiped

parallelogram of vectors, forces, velocities A geometric construction for adding vector quantities, like forces and velocities, in which the diagonal of a parallelogram corresponds to the sum or ➤resultant of the vectors represented by the sides.

parallel postulate An ➤axiom of plane ➤Euclidean geometry: given a line and a point not on the line, there is a line through the point that is parallel to the given line. For a long time after Euclid postulated the axiom (in the third century BC), it was thought to be provable from his other axioms; the realization (in the 19th century) that it was not led to the creation of ➤non-Euclidean geometry.

parallel processing In computing, the simultaneous execution of arithmetic operations. Most computers process their programs sequentially (➤multi-tasking), but parallel computers are capable of using more than one processing unit in order to process several parts of a program at once, in a similar way to the human brain.

parallel-transport A method of extending the concept of 'parallel' into curved spaces and ➤spacetimes. A vector at point A can be moved along a curve in the space such that at each successive point on the curve the vector is parallel to the corresponding vector at the previous point. Only in *flat* spaces will the parallel vector at some point B always be independent of the route taken from A to B. For example, let point A be on the equator at the ➤prime meridian (0°N 0°E) and point B at 0°N 90°E. We can construct the vector at B parallel to a vector pointing east at A. If we move along the equator, the vector still points east at B. However, if we move up the prime meridian to the North Pole, the parallel vector will point down the 90° meridian at the pole, and will remain pointing *south* as we move down to B.

paramagnetism One of the principal types of magnetism, a moderate attraction into a magnetic field, caused by the presence of unpaired electrons. **Paramagnetic** substances have a small positive ➤magnetic susceptibility. Paramagnetism is very uncommon in the gas phase, although two simple molecules, NO and O_2, are paramagnetic, as are ➤radicals. The phenomenon is most common in solutions of many ➤transition metal compounds, where the unpaired electrons arise because several orbitals lie at the same energy level. Strongly paramagnetic species include high-spin d^5 complexes such as $[Mn(H_2O)_6]^{2+}$. ➤➤Pauli paramagnetism.

parameter 1 A constant that characterizes particular cases of a general expression; thus m and c are parameters in the general expression for a straight line, $y = mx + c$.

2 A ➤variable in terms of which the coordinates (➤coordinate system) of points on a curve are expressed. For example, the trajectory of a cricket ball might be described in terms of its position at time t by **parametric equations** such as $x = 10t$, $y = 20t - 5t^2$.

3 A single number such as the ➤mean or ➤standard deviation summarizing some aspect of the distribution of a ➤population (2). By convention, population parameters are denoted by Greek letters and their counterparts for samples by Latin letters.

parasitism ➤symbiosis.

parasympathetic nervous system ➤nervous system.

paraxial ray In an optical system, a ray parallel to the ➤optical axis.

parent nuclide ➤daughter nuclide.

parity (space inversion) Symbol P. A ➤symmetry transformation that transforms a system by mapping every point r into the point $-r$. In ➤quantum mechanics, this corresponds to an observable with a value of either 1 (**even parity**) or −1 (**odd parity**). The operation changes left-handedness into right-handedness (and vice versa), and the indifference of physics to whether, for example, we draw axes as left-handed or right-handed leads to the idea of the conservation of parity. However, in the mid-1950s it was discovered that some processes involving the ➤weak interaction violate the conservation of parity. ➤➤charge–parity–time symmetry; helicity.

parsec Symbol pc. A unit used in astronomy to measure interstellar distances. 1 pc = 3.086×10^{16} m or 3.26 ➤light years. It is the distance at which the Earth's orbit about the Sun would subtend 1 second of arc (the name is a contraction of 'parallax second').

parthenogenesis The production of young from eggs without fertilization. It occurs particularly in animals such as aphids in which males are very rare. In plants, the term **parthenocarpy** is applied to the production of fruits without fertilization. This occurs in, for example, bananas and some varieties of grape. Such fruits do not contain seeds.

partial derivative The ➤derivative of a function of several variables with respect to one of the variables, the others being treated as constants. Notation such as $\partial f/\partial x$ is used to indicate **partial differentiation** with respect to x; **higher partial derivatives** such as $\partial^2 f/\partial x^2$ or $\partial^2 f/\partial x\,\partial y$ arise from successive differentiations.

partial differential equation A ➤differential equation, such as ➤Laplace's equation, involving partial derivatives with respect to one or more variables.

partial differentiation ➤partial derivative.

partial fractions The technique of writing a ➤rational function as a sum of simpler functions, usually as a prelude to integration. For example, $(3x + 1)/(x^2 - 1)$ can be expressed in partial fractions as $2/(x - 1) + 1/(x + 1)$.

partial pressure The pressure exerted by each gas in a mixture of gases. The sum of the partial pressures equals the total pressure of the mixture. ➤Dalton's law.

particle **1** A general term broadly applied to atoms, molecules, ions, or whatever is being considered; 'entity' is another suitable term. ≫elementary particle; subatomic particle.

2 A part of a ≻system that is, for the purposes of the physical model of the system, indivisible. Exactly what can be described as a particle and what as a 'compound body' depends on the system in question. For example, when considering ≻rigid body dynamics in classical mechanics, it is helpful to divide the body into an infinite number of particles (masses of infinitesimal size) and apply ≻Newton's laws of motion to each of those particles. For the purposes of ≻atomic physics and chemistry, electrons, protons and neutrons are particles, whereas in ≻particle physics it is helpful to subdivide protons and neutrons into ≻quarks.

particle accelerator ≻accelerator.

particle in a box A simplistic model of a ≻quantum-mechanical system that consists of a single particle in a cubic region of zero potential (the box), bounded by infinite potential at its walls. The solutions to the ≻Schrödinger equation for such a system are a set of ≻wavefunctions of the form

$$\psi(x, y, z) = \sin\left(\frac{n_x \pi x}{L}\right) \sin\left(\frac{n_y \pi y}{L}\right) \sin\left(\frac{n_z \pi z}{L}\right)$$

where L is the dimension of the box and n_x, n_y, n_z are ≻quantum numbers (positive integers). The energies corresponding to these wavefunctions are given by

$$E = \frac{h^2}{8mL^2}(n_x^2 + n_y^2 + n_z^2),$$

where m is the mass of the particle and h is the ≻Planck constant. Although the system is a simplification, it is a useful starting point for many more sophisticated models of more realistic potentials.

particle physics The physics of ≻elementary particles and ≻subatomic particles. Particle physics deals with the nature of the particles and interactions between them. The experimental study of particle physics is often known as **high-energy physics**.

particular integral (particular solution) A solution to a ≻differential equation which does not involve arbitrary constants (≻integration), usually because these have been determined from ≻initial conditions or ≻boundary conditions.

partition (distribution) When a solute is added to two immiscible solvents (1 and 2) in both of which it is soluble, the solute is partitioned between the two solvents such that the concentration of the solute A in one solvent is directly proportional to its concentration in the other, the constant of proportionality being the **partition coefficient**

$$K_{\text{part}} = [A]_1/[A]_2.$$

partition function Symbol Z. For a thermodynamic system with energy levels E_i, the partition function (a function of T or, for simplicity, of $\beta = 1/kT$, where T is the thermodynamic temperature and k is the ≻Boltzmann constant) is defined as

$$Z(\beta) = \sum_i \exp(-\beta E_i).$$

This is a very important concept in ➤statistical mechanics since this form of Z is recognizable as the normalizing denominator in the calculation of, for example, the mean energy:

$$\bar{E} = \sum_i E_i e^{-\beta E_i} / \sum_i e^{-\beta E_i}.$$

With a little algebra this leads to the conclusion that $\bar{E} = -\partial Z / \partial \beta$ and all the other thermodynamic functions such as ➤entropy and ➤free energy can be expressed in terms of Z and its derivatives. Thus the function $Z(\beta)$ contains *all* the information relevant to the thermodynamics of the system.

parton The generic name given to the hypothetical constituents (such as ➤quarks) of subatomic particles.

pascal Symbol Pa. The SI derived unit of ➤pressure and ➤stress. A pressure of one pascal is produced when a force of one newton is exerted over an area of one square metre: 1 Pa is equivalent to 1 N m^{-2}. Named after Blaise Pascal (1623–62).

Pascal A high-level computer programming language, named after B. Pascal, in honour of the fact that he designed the first successful mechanical computing machine. The language was largely designed to be used for teaching good programming; it is highly structured and well respected.

Pascal's law of fluid pressure The ➤pressure in a fluid exerts a force of constant magnitude on any surface with which it is in contact, regardless of the orientation of the surface. This applies only to ➤static pressure, which acts perpendicular to the surface on which it exerts a force. The ➤dynamic pressure on an object moving through the fluid does depend on the orientation of the object.

Pascal's triangle The triangular pattern of numbers, shown in the diagram, in which every number written in the triangle is the sum of the two above it. The nth row gives the ➤binomial coefficients in the expansion of $(a+b)^n$; for example $(a+b)^3 = 1a^3 + 3a^2b + 3ab^2 + 1b^3$. Named after B. Pascal, although it was known earlier to Chinese mathematicians.

```
        1   1
      1   2   1
    1   3   3   1
  1   4   6   4   1
1   5  10  10   5   1
1  6  15  20  15   6   1
          etc.
```

Pascal's triangle

Paschen–Back effect The limit of the ➤Zeeman effect in very strong magnetic fields. As the splitting caused by the magnetic field (between states with different components m_j of angular momentum in the direction of the field) increases, ➤spin–orbit coupling becomes less dominant until eventually the levels are split according to the sum of the components parallel to the field of both spin (m_s) and orbital (m_l) angular momentum. Named after (Louis Carl Heinrich) Friedrich Paschen (1865–1947) and Ernst Emil Alexander Back (1881–1959).

Paschen series ➤Rydberg–Ritz equation.

passivation (passivity) The formation of a surface coating on some metals, such as iron and aluminium, when reacting with nitric acid, which renders them unreactive.

passive component A component in an electric circuit, such as a ➤resistor, a ➤capacitor or an ➤inductor, that neither introduces power (like an ➤amplifier), nor has a directional effect (like a ➤diode). Compare ➤active component.

passive transport The process in living cells whereby charged ions (such as K^+ or Na^+) cross a membrane from a region of high concentration to one of low concentration (➤diffusion). No energy in the form of ➤ATP is needed for passive transport, unlike ➤active transport.

pasteurization The treatment of milk to kill certain bacterial ➤pathogens, particularly those causing typhoid, brucellosis and tuberculosis, by heating to about $70\,°C$ and rapid cooling to below $10\,°C$. Named after Louis Pasteur (1822–95).

pathogen Any organism (especially a bacterium or fungus) or virus that causes disease.

patina A coating on the surface of a metal. One example is the green colour seen on copper objects (➤verdigris).

Pauli exclusion principle ➤Pauli principle.

Pauli matrices The 2×2 matrices that are the ➤operators (2) for the components of the spin angular momentum of a spin-$\frac{1}{2}$ particle in quantum mechanics:

$$\sigma_x = \begin{bmatrix} 0 & 1 \\ 1 & 0 \end{bmatrix}, \qquad \sigma_y = \begin{bmatrix} 0 & -i \\ i & 0 \end{bmatrix}, \qquad \sigma_z = \begin{bmatrix} 1 & 0 \\ 0 & -1 \end{bmatrix}.$$

A particularly important property is the set of relations:

$$\sigma_x\sigma_y = i\sigma_z,\ \sigma_y\sigma_z = i\sigma_x,\ \sigma_z\sigma_x = i\sigma_y.$$

Named after Wolfgang Pauli (1900–58).

Pauling electronegativity The most common scale of ➤electronegativity. Named after Linus Carl Pauling (1901–94).

Pauli paramagnetism A form of ➤paramagnetism observed in metals. When a magnetic field is applied to a conduction band of free electrons (see the diagram), the energy of the electrons with spins parallel to the magnetic field is reduced, and the energy of the electrons with spins opposite to the magnetic field is increased. Thus the state lowest in energy has a majority of spin-parallel electrons, leading to a magnetic moment in the direction of the field. Compare ➤Van Vleck paramagnetism.

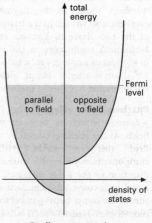

Pauli paramagnetism

Pauli (exclusion) principle One of the most profound ideas in quantum theory, advanced by W. Pauli, which can be stated in a number of forms. Most rigorous is the idea that a many-particle wavefunction must be either symmetric on particle exchange (when the particle will be a ➤boson) or antisymmetric on particle exchange (when the particle will be a ➤fermion). This leads to the more common version of the principle, the **Pauli exclusion principle**, which applies only to fermions such as electrons: electrons, for example, cannot have all four quantum numbers the same; this means that electrons sharing a single orbital must be of opposite spin (i.e. the electrons pair up). ➤identical particles; pairing of electrons.

Pb Symbol for the element ➤lead, from its Latin name *plumbum.*

p-block elements Elements in which the highest-energy occupied orbital is a p orbital. They are the elements in Groups 13–18 of the ➤periodic table.

pBR322 A ➤plasmid commonly used as a ➤cloning vector.

PCB 1 (Phys.) Abbr. for ➤printed circuit board.
 2 (Chem.) Abbr. for polychlorinated biphenyl. PCBs are highly toxic and become concentrated in ➤food chains. They are very resistant to chemical and biological degradation and so persist in the environment.

PCR ➤polymerase chain reaction.

p.d. Abbr. for ➤potential difference.

Pd Symbol for the element ➤palladium.

peak factor The ratio of the maximum absolute value (**peak value**) of a periodically varying quantity to its ➤root mean square value. For example, for a pure sine function of time, the peak factor is $\sqrt{2}$.

pearlite A phase in the ➤iron–carbon phase diagram, so called because of its opalescent appearance. It consists of layers of cementite, Fe_3C, and α-iron; it is soft and malleable.

peat The partially decayed remains of plants deposited particularly in areas of high rainfall and poor drainage where soils are deficient in oxygen. Peat derived from sphagnum moss is used in horticulture as a soil conditioner and the basis of plant composts, and in some places is dried for use as a fuel.

pectin One of a number of substances which are derivatives of carbohydrates and found widely in plants forming gums and mucilages, for example in the ➤middle lamella. The properties of pectin on heating are the basis of the setting process in jam-making.

pectoral girdle (shoulder girdle) The part of the vertebrate skeleton comprising the **scapula** (shoulder blade) and the **clavicle** (collar bone) on each side.

pelagic Describing organisms inhabiting the open seas, rather than the sea bed or coastal zones. Compare ➤benthic.

P element A fragment of DNA used to insert genes into the ➤germ cells of *Drosophila* (fruit flies).

Peltier effect ►thermoelectric effects. Named after Jean Charles Athanase Peltier (1785–1845).

pelvic girdle (pelvis) Three fused bones on each side of the lower part of the ►vertebral column, forming a framework for support and the attachment of the (rear) legs. The three bones are the **ilium** (hip), **ischium** and **pubis**. The girdle is fused at the front at the **pubic symphysis** and joined to the spine at the **sacro-iliac** joint.

pendulum A mechanical system consisting of a ►rigid body free to rotate about a pivot at a point other than its ►centre of mass, in the Earth's ►gravitational field. The equilibrium position of the system is the orientation with the centre of mass vertically below the pivot. If the orientation differs from the equilibrium position, a restoring ►couple acts on the body and causes ►oscillations. If the oscillations are small, they are ►harmonic.

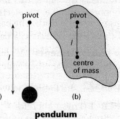

pendulum

A **simple pendulum** (such as a plumb line, or a small ball fastened to a light rod) has almost all of its mass at a single point, and the period of small oscillations is given by:

$$T = 2\pi\sqrt{\frac{l}{g}},$$

where l is the distance between the pivot and the mass, and g is the ►acceleration of free fall (see diagram (a)).

A **compound pendulum** has a more complicated structure, with its mass distributed about the body. The period of small oscillations is:

$$T = 2\pi\sqrt{\frac{k^2 + l^2}{gl}},$$

where l is the distance between the pivot and the centre of mass, and k is the ►radius of gyration of the rigid body about its centre of mass (see diagram (b)).

In neither case does the period depend on the mass of the pendulum. ►►Foucault pendulum.

penicillin A very important class of antibiotics originally extracted from the green mould fungus *Penicillium notatum,* with molecular structures similar to that in the diagram opposite. They work by lysis (breaking) of bacterial cell walls. The discovery by Alexander Fleming is famous; credit also goes to Ernest Chain, Howard Florey and Norman Heatley for their use of penicillin to cure disease.

penis Intromittent male reproductive organ used to insert ►semen into the body of a female, possessed by many animal classes. All mammals and some birds have a penis. Most reptiles have a penis (or a pair of **hemipenes**). Amongst other vertebrates amphibians and fish lack a penis. Some molluscs have a rudimentary penis but it is absent in crustaceans, insects and spiders which use a range of methods to transfer sperm, including the use of jointed appendages.

penicillin V

pentadactyl limb The basic form of tetrapod vertebrate limb, consisting typically of five digits linked to a parallel pair of bones which in turn are linked to a girdle (►pectoral girdle; pelvic girdle) by a single bone to produce a characteristic 5–2–1 structure. The structure is derived from the lobed fins of an ancestral group of fish. The limbs of all vertebrates (other than the fins of modern fish) can be derived from this basis, including the wings of bats and birds, the flippers and flukes of aquatic mammals, and the arms and legs of humans. Thus in the human arm the digits are connected to the paired ►radius (2) and **ulna**, which in turn are joined to the pectoral girdle by the ►humerus.

pentagon A ►polygon that has five sides.

pentahydrate Crystals with five molecules of water of crystallization, a classic example being blue copper(II) sulfate pentahydrate, $CuSO_4 \cdot 5H_2O$.

pentane C_5H_{12} The straight-chain alkane with five carbon atoms, which is a colourless liquid.

pentose A carbohydrate with five carbon atoms; ►ribose is an important example.

penumbra The area surrounding the ►umbra in a shadow produced when an extended source (not a ►point source) of light is obstructed. The penumbra is the area within which the source is partially, but not entirely, obscured. ►►eclipse.

pepsin A ►protease enzyme secreted in the gastric pits of the stomach (►gut) which functions in acidic conditions.

peptidase An ►enzyme that hydrolyses ►peptides (►hydrolysis).

peptide The molecule formed when ►amino acids are joined together via a ►peptide bond. It could be formed from just two or three amino acids (**dipeptides** and **tripeptides**) or several amino acids (**oligopeptides**); a peptide that exceeds 20 amino acids in length is referred to as a **polypeptide**. ►Proteins are natural polypeptides. The ►Merrifield synthesis of polypeptides has increased the ease of making these important compounds in the laboratory.

peptide bond (peptide link) A strong ►chemical bond, like that of a substituted amide (see the diagram), formed as the result of a ►condensation reaction between the carboxyl (—COOH) group of one amino acid and the amino (—NH$_2$) group of another to form a ►peptide. Note that because of partial double-bonding

peptide bond

caused by ➤delocalization, the CO and NH groups are coplanar, making the link flat.

percentage A proportion or ratio expressed as a fraction with the denominator 100. Thus 3/5 as a percentage is 60/100, written 60%. From the Latin *per cent,* in every hundred.

percentile One of the values below which are integral percentages of a given ➤frequency distribution. Thus the 95th percentile of a distribution of heights is 1.88 m if 95% of the heights are below 1.88 m. The ➤median (2) is the 50th percentile; the **upper** and **lower quartiles** are the 75th and 25th percentiles respectively and the difference between them, the **interquartile range**, is sometimes taken as a measure of the spread of a distribution.

perchlorate (chlorate (VII)) A salt of perchloric acid containing the ClO_4^- ion.

perchloric acid (chloric(VII) acid) $HOClO_3$ or $HClO_4$ The oxoacid of chlorine with oxidation number +7. The analogous compounds of bromine and iodine are called **perbromic acid** and **periodic acid** respectively. Perchloric acid is perhaps the strongest common acid and is also a violently strong oxidant. Thus the mixture of perchloric acid and methanol used for etching metal needs to be very carefully made up, as in some proportions the mixture is spontaneously explosive.

perennial A plant that lives for a number of years and is able to survive periods unfavourable for growth. Examples include woody plants such as trees, and herbaceous plants such as tulips which survive the winter as a bulb, the aerial parts being reduced to ground level.

perfect gas equation ➤ideal gas equation.

perfect gas model ➤ideal gas model.

perfect number Any whole number *n* with the property that the sum of all its factors is 2*n*. The first three perfect numbers are 6, 28 and 496. It is not known whether there are infinitely many perfect numbers, nor whether there are any odd perfect numbers.

perianth ➤flower.

pericardial cavity The fluid-filled cavity, bounded by a connective tissue membrane, the **pericardium**, that encloses the ➤heart.

pericarp The ovary wall of a flowering plant after fertilization that develops to form the fruit.

periderm The outer layer of the stem of a woody plant, consisting of the cork (bark), the cork ➤cambium, and underlying cortical cells derived from the cork cambium.

perigee The point in the orbit of the Moon or an artificial satellite orbiting the Earth at which it is closest to the Earth. Compare ➤apogee.

perihelion The point in the orbit of a planet or other body orbiting the Sun at which it is closest to the Sun. Compare ➤aphelion.

perilymph ➤ear.

perimeter The closed curve forming the boundary of a region, or the length of such a curve.

period 1 (Chem.) A horizontal row in the ➤periodic table, sometimes labelled with the principal quantum number of the orbital filled in the s block in the period, such as period 3 for the elements sodium to argon. This consists of elements with successive atomic numbers, which are unlike each other in fundamental ways, and so the connection is less important than the vertical columns, the ➤groups, which contain elements that are similar. The term **first short period** refers to the elements from lithium to neon and the term **second short period** refers to the elements from sodium to argon. ➤➤Appendix table 5.

2 (Phys.) ➤periodic system.
3 (Geol.) ➤era.
4 (Biol.) ➤menstrual cycle.
5 (Math.) ➤periodic function.

periodic function A function f for which there is a constant number p for which $f(x + p) = f(x)$ for all x. The smallest (positive) value of p is called the **period** of f. For example, the sine function has period 2π.

periodic system A system that appears exactly the same after a shift of a variable (usually time) of an integral multiple of the **period** T. Thus the system repeats itself after every time T. Perfect ➤orbits are periodic, as is ➤simple harmonic motion. Perfect crystals are periodic in space (not time).

periodic table A table of the elements listed in order of increasing ➤atomic number (➤Appendix table 5). Arguably the greatest discovery in chemistry occurred when Dmitri Ivanovich Mendeleyev (1834–1907) tried to organize the elements he knew into a coherent pattern that made chemical sense. He listed them in order of increasing mass; similar elements recurred at regular, periodic, intervals. His stroke of genius was to leave gaps, where chemical sense decreed, for undiscovered elements. For example, arsenic did not resemble aluminium, yet it was the next element he knew after zinc. So Mendeleyev concluded that two unknown elements, *eka*-aluminium and *eka*-silicon (*eka* is Sanskrit for 'first'), lay undiscovered. He was uncannily accurate in his predictions for the properties of these unknown elements, so his fame was secured when the elements gallium and germanium were discovered.
 ➤Moseley's law showed that the correct quantity to use for ordering the elements is the atomic number and not the relative atomic mass. Nevertheless, as the two orders are similar, Mendeleyev's table worked very well, despite the odd irregularity such as the fact that argon has a larger mass than potassium.
 The horizontal rows of the periodic table are called ➤periods, and the vertical columns are called ➤groups. The groups are collected into one of four ➤blocks, according to the last ➤orbital to be filled. The **s block** is on the left of the table and the **p block** on the right, with the **d block** in between. The **f block** is often shown at the bottom of the table, essentially to save space across the page.
 There are general trends in the behaviour of the elements down a group and across a period. Groups contain elements which are similar to each other. The formulae of compounds of the elements within a group are usually very similar. For example, the chlorides of the elements in Group 2 have the formulae $BeCl_2$, $MgCl_2$, $CaCl_2$, $SrCl_2$,

BaCl$_2$; their oxides have the formulae BeO, MgO, CaO, SrO, BaO. The changes down a group are often subtle; in general, metallic character increases down a group, with Group 14 showing the largest change in behaviour. At the top of Group 14 is the nonmetal carbon, in the middle is the ➤metalloid germanium, and at the bottom is the metal lead. Trends across a period tend to be much more dramatic. The formulae of the chlorides and oxides across period 3 from sodium to silicon change in a systematic way: NaCl, MgCl$_2$, AlCl$_3$, SiCl$_4$ and Na$_2$O, MgO, Al$_2$O$_3$, SiO$_2$. The bonding in both series of compounds changes from ionic to covalent; NaCl is a solid which dissolves in water whereas SiCl$_4$ is a liquid which hydrolyses violently in water. In general, metallic character decreases across a period. In period 3, the element on the left of the period, sodium, is a reactive metal, whereas the two elements on the right are chlorine, a reactive nonmetal, and argon, a very unreactive nonmetal. Sodium oxide, Na$_2$O, is an alkaline solid whereas the oxide of chlorine with the similar formula, Cl$_2$O, is an acidic gas. The structure of Na$_2$O is well understood using ➤ionic bonding whereas Cl$_2$O is composed of individual covalently bonded molecules.

peripheral A device connected to a computer, usually for the purposes of input or output, or for storage of data, but not an integral part of the computer. Keyboards, monitors, printers and ➤magnetic storage devices are common examples of peripherals.

peripheral nervous system ➤nervous system.

peristalsis A coordinated muscular contraction of longitudinal and circular smooth ➤muscle passing along a duct or other tubular structure in an animal. Examples of peristalsis include the swallowing of food along the oesophagus and the rest of the ➤gut, and the passage of an ➤ovum along an ➤oviduct.

peritoneum The membranous lining of the body cavity that extends around the abdominal organs of vertebrates.

Perkin reaction A ➤base-catalysed condensation reaction. Named after William Henry Perkin (1838–1907).

Permalloy A class of alloys with high ➤magnetic permeability made from iron and nickel, which were used in early computer memories.

permanent hardness ➤hard water.

permanent magnetism ➤magnet, permanent.

permanganate ion Traditional name for ➤manganate(VII) ion, MnO$_4^-$.

permeability 1 The degree to which a solid admits the passage of fluids.
  2 ➤magnetic permeability.

permeability of free space Symbol μ_0; value $4\pi \times 10^{-7}$ H m^{-1}. The ➤magnetic permeability of a vacuum. The permeability of free space has the exact value stated above because of the way the ➤ampere is defined. It is related to the ➤permittivity of free space and the ➤speed of light by $\mu_0 = 1/(\varepsilon_0 c^2)$. ➤➤Appendix table 2.

permeance Unit H. The reciprocal of the ➤reluctance of a circuit.

Permian See table at ➤era.

permittivity Symbol ε; unit F m^{-1}. The ratio of the ➤electric displacement D to ➤electric field strength E in a material. It is often expressed as the dimensionless **relative permittivity** ε_r in terms of the ➤permittivity of free space, with $\varepsilon = \varepsilon_r \varepsilon_0$. For a nonmagnetic material, it equals the square of the ➤refractive index.

permittivity of free space Symbol ε_0; value 8.854×10^{-12} F m^{-1}. The ➤permittivity of a vacuum. It is closely related to the ➤permeability of free space. ➤➤Appendix table 2.

permutation 1 A selection of objects from a given set in which the order of selection matters. The number of permutations of r objects chosen from n is $n!/(n - r)!$, sometimes written ^nP_r. Compare ➤combination.

2 A rearrangement of a set of objects. The set of $n!$ permutations of n objects forms a permutation ➤group (2).

perovskite structure An important crystal structure named after the mineral perovskite, calcium titanate, $CaTiO_3$. It consists of an array of oxide anions with two types of metal ion in the positions shown in the diagram, and hence is a common structure for ➤mixed oxides.

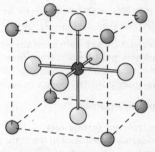

peroxide A compound containing the peroxide ion, $O_2{}^{2-}$. The most common peroxides are those of the alkali metals, such as sodium peroxide, Na_2O_2, which is the dominant product made during the oxidation of sodium in air. The parent acid is ➤hydrogen peroxide, H_2O_2. Compare ➤superoxide.

perovskite structure The central Ti^{4+} ion is surrounded by six O^{2-} ions; the eight ions at the corners of the cube are Ca^{2+} ions.

perpendicular (orthogonal, normal) Pairs of lines or planes are perpendicular if they meet at ➤right angles.

Perspex UK tradename for ➤poly(methyl methacrylate).

perturbation The small difference between a chosen system and a second, standard system. Examples in ➤classical physics are the small gravitational forces exerted by the planets on one another which cause their orbits to depart from perfect ellipses (the standard systems). In ➤quantum mechanics, the effect of a small magnetic or electric field on an atom might be treated as a perturbation on the standard system of a simple atom; the ➤Zeeman effect and ➤Stark effect are usually treated in this way. In ➤particle physics, all interactions between particles are ultimately treated as perturbations on the noninteracting state. ➤➤perturbation theory.

perturbation theory A method used mainly in ➤quantum mechanics for solving differential equations, particularly the Schrödinger equation, in the presence of a small ➤perturbation. **Time-independent perturbation theory** looks at the variation in the ➤quantum states and energy levels with a small permanent change in the ➤Hamiltonian. **Time-dependent perturbation theory** looks at the variation of the quantum states with time when a small time-varying perturbation is applied to the Hamiltonian.

PES Abbr. for ➤photoelectron spectroscopy.

peta- A prefix in SI units denoting 10^{15} times the base unit.

petal ➤flower.

petiole ➤leaf.

PETN Abbr. for pentaerythritol tetranitrate, an extremely powerful explosive, often mixed with TNT. Rather surprisingly, its other main use in tablet form is to relieve the pain of *angina pectoris*.

Petri dish A shallow circular or square plastic or glass dish with an overlapping lid used particularly for growing bacteria or fungi under laboratory conditions. Named after Julius Richard Petri (1852–1921).

petrochemical A chemical ultimately derived from petroleum. Many of our drugs, solvents and polymers come from this one raw material. It was Mendeleyev, the discoverer of the ➤periodic table, who first warned mankind in the late 1870s not to squander this vital resource by burning it.

petrol ➤gasoline.

petroleum (crude oil) A black viscous liquid which assumed extraordinary economic importance in the 20th century, becoming known as 'black gold'. It is a mixture of hundreds of different organic compounds, predominantly the alkanes, from which a vast range of organic chemicals can be made. Primary separation is by ➤fractional distillation, because the fractions have different boiling ranges, to give several fractions similar to those listed in the table. Secondary separation is done by ➤cracking and/or ➤reforming at the oil refinery.

Fraction	Percentage by mass of crude oil	Boiling range/°C	Number of carbon atoms in main components	Uses
Refinery gas	1–2	<20	1–4	Fuel (mainly in the refinery), petrochemicals
Gasoline/ naphtha	15–25	20–175	5–10	Petrol ('gasoline') petrochemicals via cracking
Kerosene	10–15	175–250	11–14	Aviation fuel, domestic heating
Gas oil	15–25	250–400	15–25	Diesel fuel, industrial heating
Residue	40–50	>400	>25	Power station fuel, asphalt for roads

petroleum ether ('pet ether') A mixture of volatile hydrocarbons, mainly alkanes such as pentane and hexane, which is not an ➤ether at all.

pewter An alloy of lead and tin, in the ratio 1:2.

Pfund series ➤Rydberg–Ritz equation. Named after (August) Herman Pfund (1879–1949).

pH A measure of the acidity of a solution: it is defined quantitatively by the formula

$$pH = -\log_{10}\{[H^+ (aq)]/mol\ dm^{-3}\}.$$

As the ➤ionic product for water is $1 \times 10^{-14}\ mol^2\ dm^{-6}$ at 25 °C, and neutral water has equal concentrations of hydrogen and hydroxide ions, both have a value of 1×10^{-7} mol dm^{-3}, and hence neutral pH is $-\log_{10}(1 \times 10^{-7})$, that is 7.0. Acidic solutions have a pH lower than this and, for a 1 mol dm^{-3} ➤strong acid, pH = $-\log_{10} 1$ = 0.0. This is not a lower bound, however. Basic solutions have a pH higher than 7. For a 1 mol dm^{-3} ➤strong base, such as NaOH(aq), [OH$^-$] = 1 mol dm^{-3}, and using the ionic product means that [H$^+$] = 1×10^{-14} mol dm^{-3} and hence pH = 14.0.

Ph A common shorthand for the phenyl group, —C$_6$H$_5$.

phage ➤bacteriophage.

phagocyte A cell that can ingest particles such as microorganisms or organic debris. Many free-living ➤protoctistans are phagocytes, as are certain cells of the mammalian ➤immune system, such as ➤macrophages.

phagocytosis The process by which solid particles, including other cells, can be engulfed by a cell. It involves the invagination of (formation of a hollow in) the cell surface to surround the particle and form a ➤vacuole. Enzymes are then released into the vacuole from ➤lysosomes to digest the contents. Compare ➤pinocytosis.

pharmacology The study of compounds that produce physiological effects in the body, particularly those of medical significance.

pharynx The anterior part of the ➤gut in animals, which in the ➤chordates is perforated by gill slits during at least the early stages of embryonic development.

phase **1** (Chem.) Any part of a system that is uniform chemically and has a recognizable boundary to another part. Thus we speak of the solid, liquid and gas phases of a pure substance. The term can be extended to mixtures such as ➤alloys.
  2 (Math.) For a quantity x that varies with t according to $x = A \sin(\omega t + \phi)$, the value of the quantity $\omega t + \phi$. Phase is usually expressed as a fraction of a full cycle, in ➤radians (which are effectively dimensionless), or in ➤degrees. The term **phase angle** usually refers to a phase difference between two quantities.

phase-contrast microscopy A technique in microscopy used to view transparent objects with a different ➤refractive index from the surrounding medium. The ➤phase shift induced by the difference in refractive index causes ➤interference which can be observed in the intensity of the transmitted light. Differences in the refractive index of the boundary layers of cells and ➤organelles allow detail in living cells to be viewed under the microscope.

phase diagram A diagram showing the conditions, usually of temperature and pressure, under which various phases are thermodynamically stable. A single-component phase diagram (diagram (a), opposite) has lines that show various phase boundaries – solid/liquid (melting line), liquid/gas (boiling line) and solid/gas (sublimation line) and sometimes boundaries between different solid phases, as for the allotropes of sulfur. Two-component phase diagrams become more complicated; a simple one is shown in diagram (b) (⮞Raoult's law). The concept of a phase diagram can be generalized to show the phase of a system as a function of any two thermodynamic quantities. For example, the phase diagram of a superconductor (⮞superconductivity) can be constructed in terms of the temperature and the magnetic field.

phase difference The difference in the ⮞phase (2) of two periodic quantities. If two quantities a and b varying sinusoidally with t have phase difference δ, $a = \sin(\omega t)$ and $b = \sin(\omega t + \delta)$, then $a + b = 2 \sin(\omega t + \frac{1}{2}\delta) \cos(\frac{1}{2}\delta)$. So the sum is sinusoidal with amplitude $\cos(\frac{1}{2}\delta)$. If the phase difference is zero, the amplitude is a maximum and the quantities are said to be **in phase**. If the phase difference is π (or 180°) then the amplitude is zero and the quantities are said to be **out of phase**. Any multiples of 2π (or 360°) in the phase difference can be ignored because this represents a whole period. ⮞interference.

phase modulation ⮞modulation.

phase of the Moon The apparent shape of the Moon as viewed from the Earth. The Moon produces no light of its own: it is visible from the Earth because it reflects light from the Sun, so only the hemisphere facing the Sun can be seen. When the Moon is on the same side of the Earth as the Sun, little of this hemisphere is visible and the phase is **new Moon**. As the Moon moves around the Earth, more of this hemisphere becomes visible and the phase goes through a **waxing crescent** (one-quarter visible), **first quarter** (half visible) and **waxing gibbous** (three-quarters visible) to **full Moon** when the Moon is on the opposite side from the Sun. It then passes through **waning gibbous**, **third quarter** and **waning crescent** back to a new Moon. The complete cycle is a **lunation**; the period is a lunar ⮞month.

phase rule A rule relating the number of ⮞phases that can coexist under particular conditions. The number F of degrees of freedom (conditions such as pressure and temperature) is given by the equation

$$F = C - P + 2,$$

where C is the number of components and P the number of phases. Applied to a single component ($C = 1$), $F = 3 - P$. So a single phase ($P = 1$), such as solid, can exist at various temperatures and pressures ($F = 2$), whereas a phase equilibrium with the liquid ($P = 2$) demands that $F = 1$; that is, once the pressure is fixed, so is the temperature. All three phases ($P = 3$) are present at the ⮞triple point, at which both the pressure and temperature are fixed ($F = 0$); the triple point of water is used as a fixed point on the ⮞kelvin scale.

phase shift A change in the ⮞phase (2) of a periodic quantity, usually a ⮞wave. A phase shift can be caused, for example, by the passage of a wave through a medium

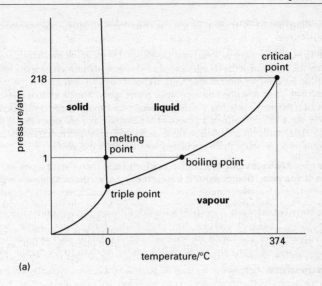

(a)

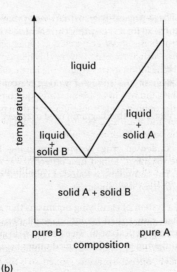

(b)

phase diagram for (a) water and (b) a two-component system.

with a different ➤refractive index from the surrounding medium, or by reflection at boundaries.

phase space A concept of ➤classical physics and ➤statistical mechanics used to visualize the variation of the coordinates of a ➤system with time. Phase space has one dimension for each position or momentum coordinate of a system, and the state of the system is then described by a point in phase space. Thus a system consisting merely of a single particle has a 6-dimensional phase space: three components of position, three of momentum; a system of N particles has a $6N$-dimensional phase space. ➤Hamilton's equations allow the trajectory (position as a function of time) of any point in phase space to be calculated. ➤➤Liouville's theorem.

phase speed (phase velocity) The rate at which a point of constant ➤phase (2) in a ➤wave (e.g. a peak, trough or null) travels through a medium. Compare ➤group speed.

phase-transfer catalysis A means of achieving some organic reactions more easily. For example, halogenoalkanes are immiscible with water while several ➤nucleophiles they react with are water-soluble. A catalyst, such as a quaternary ammonium salt, can carry the nucleophile from the aqueous to the organic phase.

phase transition A change of ➤phase undergone by a system, typically a chemical substance, when its thermodynamic state is altered. For example, melting, boiling and sublimation are all phase changes.

phasor A radial line on an ➤Argand diagram that rotates about the origin with time. Phasors are particularly useful for representing currents and voltages in a.c. electrical circuits.

Phe Abbr. for ➤phenylalanine.

phellem The cork layer of the stems of woody plants, forming part of the ➤periderm.

phenacetin White crystals (see the diagram), used to relieve pain and as an antipyretic.

phenanthrene A condensed ring system containing three benzene rings (see the diagram), joined in a different way from those in ➤anthracene. The compound is found as colourless scales and is used in the synthesis of dyes and drugs.

phenetic Describing a system of classifying organisms that relies on observable characters rather than evolutionary relationships. Phenetic classification is of practical use in, for example, field guides which enable plants and animals to be identified under field conditions. Compare ➤phyletic.

OCH$_2$CH$_3$

NHCOCH$_3$
phenacetin

phenanthrene

phenol C_6H_5OH The usual name for hydroxybenzene. When pure, the crystals are colourless, but they usually darken with age as complex oxidation reactions occur. There are several manufacturing processes, but the ➤cumene process is popular, as the coproduct, propanone, is required in roughly similar quantities. Phenol is acidic; indeed, an old name was carbolic acid. A similar but milder antiseptic, TCP (➤2,4,6-trichlorophe-nol), has proved much more useful. Phenol reacts very quickly during electrophilic aromatic substitution; bromine water, for example, decolorizing rapidly and forming 2,4,6-tribromophenol.

phenol-formaldehyde resins (phenol-methanal resins) Important polymeric substances formed by condensation polymerization of phenol and formaldehyde (➤condensation polymer). The resulting resin is cross-linked in three dimensions, creating a very tough polymer.

phenolphthalein An extremely common acid–base indicator, with structure shown in the diagram. It is useful for strong acid/strong base ➤titrations and also for weak acid/strong base titrations, as its pK_{in} is around 9. It is colourless below pH 8 and pink above pH 10.

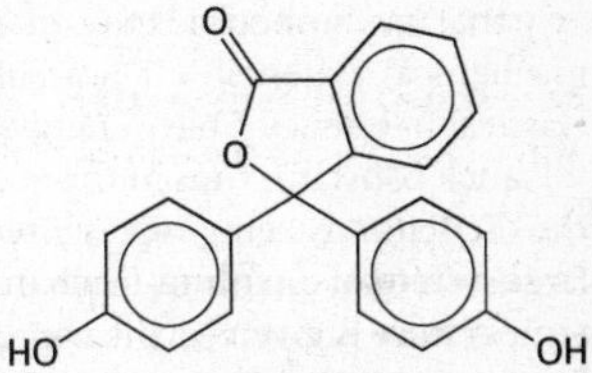

phenotype The characteristics, both externally visible and physiological, of an organism determined by its ➤genes or modified by the environment. Compare ➤genotype.

phenolphthalein

phenylalanine (Phe) A common ➤amino acid. ➤➤Appendix table 7.

phenylamine The preferred, but less commonly used, name for➤aniline, $C_6H_5NH_2$.

phenylethanone $C_6H_5COCH_3$ The systematic name for acetophenone, a solvent used in perfumes.

phenylethene (styrene) $C_6H_5CH{=}CH_2$ A colourless liquid hydrocarbon which is the monomer of ➤ polystyrene.

phenyl group The group formed from the benzene ring by removal of one hydrogen atom, that is —C_6H_5.

phenylhydrazine $C_6H_5NHNH_2$ A reagent used to detect a ➤carbonyl group, with which it undergoes condensation reactions. The substituted form, DNP (➤2,4-dinitrophenylhydrazine), is more useful as its condensation products are more easily crystallized.

phenylmagnesium halides ➤Grignard reaction.

pheromone A chemical substance used as a specific signal by some organisms to communicate with one another. Many insects and some mammals use pheromones to mark out territories and to attract mates. They are usually volatile organic molecules which are effective at very low concentrations (of the order of 1 ppm in many instances).

phloem The part of the ➤vascular system of plants that conducts materials, such as sugar manufactured during ➤photosynthesis, around the plant. The mechanism of phloem transport is the subject of much debate but it appears to be an active process requiring energy to transport materials from a source such as a leaf or a depleting storage organ to a sink such as a developing storage organ, a growing shoot apex or a developing fruit. Phloem consists of **sieve tubes** characterized by thin cellulose walls perforated at the ends to form **sieve plates**. Sieve tubes lack a nucleus, and lie alongside companion cells that possess a nucleus.

pH meter An instrument for measuring pH, consisting of two electrodes to dip into the solution, one sensitive to the aqueous hydrogen ion concentration (frequently a glass electrode) and a reference electrode (often a calomel electrode), together with a meter to measure the e.m.f.

phonon The ➤quantum of lattice vibrations (sound waves) in a solid, in the same way that the ➤photon is the quantum of the electromagnetic wave. The quantum-mechanical states of a ➤normal mode of a crystal lattice (an ➤oscillator) with classical frequency f have energies $(n + \frac{1}{2})hf$, where n is a positive integer and h is the ➤Planck constant; thus in the quantum-mechanical scheme it is natural to think of the oscillator as being represented by a number of phonons of energy hf. Like any large system of quantum-mechanical states, the occupation of levels as a function of temperature is given by the appropriate quantum statistics, in this case the ➤Bose–Einstein statistics. Phonons can interact with electrons, neutrons and photons, and thus their spectrum can be measured by scattering experiments.

phosgene $COCl_2$ A dangerously poisonous gas used to make ➤polycarbonates.

phosphate Any of the salts of phosphoric acid, most usually containing the ion PO_4^{3-}, which are particularly important in their own right. They are frequently used in fertilizers to provide phosphorus for the soil, but overuse can make groundwater ➤eutrophic. ➤➤nucleotide.

phosphate coating (phosphate dip) A protective coating given to metals, such as iron, by immersion in a bath of phosphoric acid, which increases the resistance to rusting.

phosphide A binary compound with phosphorus. A common example is calcium phosphide, Ca_3P_2.

phosphine PH_3 A hydride of phosphorus, which is a flammable gas. Like its nitrogen analogue, ammonia, it is a base and a reductant. It is a weaker base than ammonia, the only stable phosphonium halide at room temperature being PH_4I. On the other hand it is a stronger reductant, reducing copper(II) ions to the metal; ammonia only forms a copper(II) complex.

phospholipid One of a group of triesters of glycerol similar to an ordinary ➤lipid except that one of the organic ester groups has been replaced by the inorganic phosphate ester (see the diagram). The properties of such a molecule are biologically significant since the phosphate group is ➤hydrophilic and the lipid component is ➤hydrophobic. This gives the molecule polarity when in an aqueous medium. Phospholipids form a major component of ➤plasma membranes.

$RCOOCH_2$
|
$R'COOCH$
|
$CH_2OPO_3^{2-}$

phospholipid

phosphonic acid ➤phosphorous acid.

phosphorescence A form of ➤luminescence characterized experimentally by a delay between the removal of the source of illumination and the cessation of emission from the phosphorescent object (a feature put to good use in safety clothing worn by motorway workers and others). Theoretically, the critical difference between phosphorescence and ➤fluorescence is that a triplet state (where the electrons have been unpaired and now have parallel spins) is formed during irradiation, the forbidden nature of a triplet–singlet **intersystem crossing** explaining the delay referred to above. **Phosphors** are used for coating the inside of TV screens, and TV images are visible because of phosphorescence.

phosphoric acid The name for the oxoacids of phosphorus with oxidation number +5. Several exist including **orthophosphoric acid**, H_3PO_4, the most common, and **metaphosphoric acid**, HPO_3. Orthophosphoric acid is a weak tribasic acid. The buffer system $H_2PO_4^-/HPO_4^{2-}$ is one of the two main buffers in biochemistry around neutral pH. Orthophosphoric acid is a much weaker oxidant than sulfuric acid, and can be used to make hydrogen iodide when heated with potassium iodide, under which conditions sulfuric acid would cause oxidation to iodine. It forms a viscous solution and has a boiling point second only to sulfuric acid among acids, both properties stemming from ➤hydrogen bonding. It is used as a catalyst in the ➤direct hydration of ethene.

phosphorous acid (phosphonic acid) H_3PO_3 The oxoacid of phosphorus with the oxidation number +3. Its formula is better written as $(HO)_2HPO$, which explains why it is only a dibasic acid, the third hydrogen being bonded to phosphorus and not oxygen.

phosphorus Symbol P. The element with atomic number 15 and relative atomic mass 30.97, which is in Group 15 of the periodic table directly below nitrogen. Its name means 'light-bringer' since it burns vigorously with a bright white light in oxygen to produce P_4O_{10}. Much of its chemistry shows a superficial resemblance to nitrogen's; it is a nonmetal with acidic oxides and covalent chlorides which easily hydrolyse and a gaseous hydride of formula PH_3. However, significant differences exist because of the availability of d orbitals in period 3 and the larger size of the atom relative to nitrogen. So, whereas nitrogen is a gas containing N_2 molecules, phosphorus is an allotropic solid, white phosphorus containing P_4 tetrahedra (see the diagram) and red phosphorus having a polymeric structure based on P—P single bonds. The structures and formulae of the oxides are also very different. Nitrogen forms gaseous NO and NO_2 as its two most important oxides, while phosphorus forms the solids P_4O_6 and P_4O_{10}. The trichlorides, NCl_3 and PCl_3, are quite similar, both being liquids, although the latter is much more stable and important. Phosphorus can also form the solid PCl_5, by utilizing d orbitals to expand its octet (➤octet rule). The sulfides of phosphorus are more numerous and important than those of nitrogen, P_4S_3 being used in matches. Phosphorus is important in many ways: along with nitrogen and potassium it is one of the three vital elements fertilizers need to provide (➤phosphates are the usual source). It is also heavily involved in life processes, in ➤NADP$^+$ and ➤ATP for example; indeed, it was first isolated in 1669 by Hennig Brand from urine.

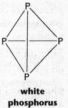

white phosphorus

phosphorus pentachloride (phosphorus(v) chloride) PCl_5 A white solid used as a ➤chlorinating agent in organic chemistry. Its gaseous structure is trigonal bipyramidal, as predicted by ➤valence-shell electron-pair repulsion theory. The solid, however, is composed of the ions PCl_4^+ and PCl_6^-.

phosphorus pentoxide (phosphorus(v) oxide) P_4O_{10} (see the diagram) An important oxide of phosphorus. It is a white solid which is an exceptionally strong dehydrating agent, capable even of dehydrating concentrated sulfuric acid, and is used to dehydrate amides $RCONH_2$ to nitriles RCN.

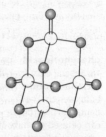

phosphorus pentoxide
Structure of P_4O_{10} in the gas phase.

phosphorylase A class of enzymes responsible for adding phosphate groups to organic molecules. This is frequently necessary for subsequent cellular reactions to take place; for example, ➤glucose must be phosphorylated before it can enter ➤glycolysis during ➤respiration.

photocell A cell that utilizes the ➤photoelectric effect.

photochemistry The study of reactions that occur when light is shone on the reactants. Examples include the photochemical ➤chlorination of alkanes such as methane and the explosive reaction of hydrogen and oxygen in sunlight. The light provides energy to break bonds, which often initiates a ➤chain reaction.

photochromic glass Glass that darkens when light falls on it. The time taken for darkening can be as short as 15 seconds. When removed from the light source, the glass becomes lighter again.

photoconductive effect The increase in the conductivity of a semiconductor when exposed to light. The light excites electrons from the ➤valence band of the material into the ➤conduction band. A ➤photon of visible light (or near infrared) has an energy comparable to that of the ➤energy gap of a typical semiconductor.

photodetector An instrument for detecting low intensities of electromagnetic radiation; that is, relatively small numbers of photons.

photodiode A semiconductor device for measuring light intensity. It is based on the ➤photoconductive effect.

photodisintegration Nuclear decay that occurs when a photon is incident on a nucleus.

photoelasticity The change in the optical properties of a transparent material (such as a polyethylene sheet) when ➤strain is applied. The material shows ➤birefringence with one optic axis parallel to the strain. The resulting strain contours can be viewed by placing it between crossed pieces of ➤Polaroid. The strain built into a toughened glass windscreen can be observed as a distortion by using a pair of Polaroid sunglasses and the natural polarization of sunlight.

photoelectric effect The ejection of electrons from a solid (particularly a metal) when light (or other ➤electromagnetic radiation) is incident on its surface. The maximum energy of the **photoelectrons** that are emitted depends only on the frequency of the light, not on the intensity. No emission occurs below a certain threshold frequency. This was simply explained using Planck's equation, $E = hf$ (where h is the ➤Planck constant): below the threshold frequency the photon cannot carry enough energy to overcome the attraction of the electron to the metal. When the energy is high enough, the kinetic energy the electrons have equals the energy of the photon minus the ➤work function.

photoelectron spectroscopy (PES) An investigative technique based on the ➤photoelectric effect and used primarily to probe the structure and chemical nature of atoms and molecules, particularly at surfaces and interfaces. In **X-ray photoelectron spectroscopy** a monochromatic source of ➤X-rays is used to find the characteristic ➤binding energies of electrons in particular ➤core levels, and thus to identify the presence (and sometimes the relative concentration) of particular elements. **Ultraviolet photoelectron spectroscopy** may be used to investigate the ➤band structure of conductors, along with the detailed electronic energy levels of molecules.

photoemission The emission of photons by the ➤photoelectric effect.

photography The recording of an image on a chemically sensitized surface by the action of radiant energy, usually light. Photography was discovered by chance by Heinrich Schultze in 1727, and was developed rapidly in the 1830s, notably by William Henry Fox Talbot. Silver halides were the basis of most photography because, when light hits a **grain**, metallic silver is formed. A **latent image** is thus built up as silver ions are reduced to silver atoms. Grains containing as few as four silver atoms are sensitive to further reduction during **development**, using benzene-1,4-diol (➤dihydroxybenzene) as the developer. In the **fixing** stage unreacted silver halides are washed off, typically with 'hypo', sodium thiosulfate, leaving a **negative** image of the subject. Film for colour photography consists of three layers, each sensitive to one of the ➤primary colours. In **digital photography** the film is replaced by a ➤CCD; images can be stored on various types of memory card, which can also usually be read by computer.

photoionization The ionization of an atom or molecule by interaction with light.

photolithography ➤lithography.

photoluminescence ➤Luminescence in which the excitation of the atom or molecule is caused by light or other electromagnetic radiation.

photolysis The breaking of a bond by impact of a photon, most usually from visible light. ➤flash photolysis.

photometer An instrument for measuring the intensity of light. Many modern photometers are based on the ➤photoelectric effect or the ➤photoconductive effect.

photomultiplier A device for detecting low intensities of light, sometimes even individual photons. The ➤photoelectric effect produces an electron for each incoming photon, and the electron is then detected by an ➤electron multiplier.

photon The ➤elementary particle that corresponds to an electromagnetic wave (➤wave–particle duality). It is the massless ➤gauge boson for the ➤electromagnetic interaction, and therefore travels at the ➤speed of light in a vacuum. In the same way that ➤normal modes of a crystal lattice can be considered as excitations called ➤phonons, electromagnetic waves are also normal modes (but without a medium) whose energy is in units of hf, where h is the ➤Planck constant and f is the frequency of the oscillation. Thus a photon carries an energy hf. It also carries a momentum $\hbar k$, where k is the ➤wave-vector of the travelling wave. Photons can interact with other particles, particularly electrons, and this interaction is responsible for all the observable phenomena associated with electromagnetic radiation.

photoperiodism The response of certain organisms to changes in the relationship between the length of light and dark periods (the **photoperiod**), usually over a 24-hour day. Many species of flowering plant regulate their times of flowering by being able to detect changes in photoperiod. Such species fall into two broad categories. **Short-day plants** flower in response to longer dark phases and shorter light phases, for example as nights get longer after the summer equinox; **long-day plants** respond to longer light phases and shorter dark phases. Breeding and migratory behaviour in many species of birds and other animals is also under **photoperiodic control**.

photophosphorylation The production of ➤ATP during the ➤light reaction of photosynthesis. ➤➤electron-transport chain; oxidative phosphorylation.

photosphere ➤Sun.

photosynthesis The fundamental biological process by which green plants are able to harness light energy from the Sun and convert it into chemical energy stored in the bonds of organic molecules such as ➤carbohydrates. The process has two main phases: the light-dependent ➤light reaction responsible for the initial capture of energy, and the light-independent ➤dark reaction in which this energy is transferred to carbohydrate using carbon dioxide as a raw material.

photosystem A complex system of membranes and pigments located on the thylakoid membranes of ➤chloroplasts, responsible for trapping light energy in ➤photosynthesis. The reaction centres of these systems involve a number of chlorophyll molecules, accessory pigments, polypeptides and a variety of redox-active prosthetic groups that together are termed antenna complexes. In eukaryotic chloroplasts there are two such photosystems. **Photosystem I** (PSI) is primarily concerned with the capture of light energy resulting in the donation of a pair of electrons from chlorophyll a via a series of intermediates to produce the reduced form of ➤NADP$^+$ (NADPH), which in turn can generate ATP via a membrane-bound electron-transport system. PSI primarily absorbs light of wavelength 700 nm. **Photosystem II** (PSII) captures light energy at 680 nm and brings about the ➤photolysis of water, generating molecular oxygen. In addition, the protons and electrons produced by this process are used to generate ATP. PSII occurs before PSI in the photosynthetic pathway (the numbering reflects the order of discovery). In general terms photosystems can be thought of as light-collecting funnels which direct energy to the specific reaction centres that drive the process of photosynthesis.

phototaxis A response to a directional light stimulus that involves movement of a *whole* organism. An example is the movement of motile cells of certain ➤protoctistans such as *Euglena* towards a light source. Compare ➤phototropism.

phototransistor A semiconductor device with two ➤heterojunctions that is sensitive to light. ➤➤photoconductive effect; photodiode.

phototropism A ➤tropism in which a plant organ responds to a directional source of light. A typical example is the growth of young seedlings towards light. Compare ➤phototaxis.

phthalic acid The common name for *o*-benzenedicarboxylic acid (see the diagram), which exists as colourless crystals.

phthalic anhydride The anhydride of ➤phthalic acid; the ortho relationship of the carboxylic acid groups is convenient for easy dehydration. It is an important intermediate in organic synthesis, used to produce ➤alkyd resins, polymer ➤plasticizers, the indicator ➤phenolphthalein and ➤anthraquinone dyes.

phthalic acid

phthalimide ➤Gabriel synthesis.

phycobilin ➤accessory pigment.

phyletic Describing a system of classifying organisms that reflects their evolutionary relationships. Compare ➤phenetic.

phylum (plural phyla) A major category in the classification of organisms (➤taxonomy). The five ➤kingdoms are divided into phyla, which in turn each contain a number of ➤classes. Important phyla include the arthropods (containing the insects) and the ➤chordates.

physical change A change that does not alter the chemical nature of the substance (compare ➤chemical change). Physical changes are usually easy to reverse and do not involve large energy changes. Examples include melting and boiling as well as the mixing of gases.

physical chemistry The study of the underlying physical causes of the processes in chemistry. It provides the basic language for the study of both inorganic and organic chemistry.

physical map (molecular map) A map that shows the position of ➤genes or ➤markers on a section of DNA, with distances given in base pairs (➤base pairing). A DNA sequence is the most detailed form of physical map. Compare ➤genetic map. ➤➤human genome project; restriction map.

physics The study of ➤systems and their interactions with one another. Physics attempts to characterize these interactions in terms of simple laws. Some of the major subdivisions of the subject are nuclear physics, particle physics, astrophysics, cosmology, solid state physics and low-temperature physics.

physiological saline ➤Ringer's solution.

physiology The study of the functioning of systems that occur in living organisms. Such processes include directly or indirectly homeostatic (➤homeostasis) mechanisms such as the maintenance of water content by the kidney in animals and the growth responses of plants to light.

physisorption ➤adsorption.

phytochrome A pale blue protein pigment in plants that is involved in a number of photomorphogenetic effects. Phytochrome exists in two interconvertible forms which absorb either red or far-red light. The far-red-absorbing form is slowly converted to the red-absorbing form in the dark or on exposure to far-red light. In turn, the red-absorbing form converts to the far-red-absorbing form on exposure to red light. This interconversion provides a mechanism which allows a plant that responds to changes in daylength to measure the relative lengths of light and dark that trigger events such as flowering. Phytochrome operates at the genetic level by switching on genes responsible for initiating morphogenetic events.

phytohormone ➤plant growth substance.

pi ➤π.

pi acid (π acid) A ➤Lewis acid, that is an electron-pair acceptor, in which the electron pair forms a pi bond. An example is the carbonyl group, which forms stable complexes with a number of transition metals. ➤➤spectrochemical series.

pi base (π base) A ➤Lewis base, that is an electron-pair donor, in which the electron pair forms a pi bond. An example is the iodide ion. ➤➤spectrochemical series.

pi bond (π bond) A type of bond in which orbitals overlap sideways such that the resultant molecular orbital, when observed along the internuclear axis, has similar symmetry properties to an atomic p orbital (the Greek letter π echoing the p). Examples are common among the elements of period 2, especially carbon and oxygen; the ➤carbonyl group frequently encountered in organic chemistry being a combination of a sigma bond and a pi bond (see the diagram). Although the pi bond has two regions of high electron density, it is one orbital. Compare ➤sigma bond.

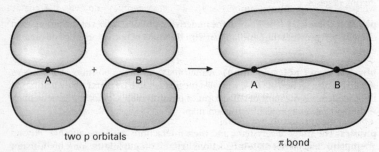

two p orbitals   π bond

pi bond Overlap of two orbitals sideways (with their wavefunctions in phase) forms a pi bond.

pico- Symbol p. A prefix in SI units denoting 10^{-12} of a base unit. For example, 1 pm is equivalent to 10^{-12} m.

picofarad Symbol pF. A useful unit for measuring capacitance: 1 pF is equivalent to 10^{-12} F.

pi complex (π complex) A ➤complex in which the bond formed by the electron pair has the symmetry of a ➤pi bond. A pi complex has been suggested as the first species formed in ➤electrophilic substitution with a number of electrophiles, such as bromine.

picric acid A common name for 2,4,6-trinitrophenol (see the diagram). It exists as explosive yellow crystals.

pie chart A diagram in which sectors of a circle are used to represent the categories of a set of data with the angle of each sector proportional to the number of items in that category. See, for example, the pie chart for the uses of ➤ammonia.

picric acid

piezoelectric effect The production of a ➤dipole (and therefore a potential difference) in a crystal when a strain is introduced, or vice versa. **Piezoelectric crystals** are used to detect or induce very small movements, for example in a ➤scanning tunnelling microscope mapping the surface of a solid. They are also used in ➤microphones and in ➤crystal oscillators.

pig iron Iron that has come from the ➤blast furnace and contains too much carbon (typically 4% by mass) to be of use, being brittle. It is named after the moulds, called 'pigs', used to cool the molten metal. It can be converted into steel in the ➤basic oxygen process. ➤➤cast iron.

pi meson ➤pion.

pinch-off The coming together of the ➤depletion layers of the two junctions in a ➤field-effect transistor. This leaves no conductive path between the source and the drain.

pineal gland A small mass of nerve tissue in the upper part of the midbrain of vertebrates (➤brain). In amphibians and some reptiles, notably the tuatara (*Sphenodon*) of New Zealand, it forms a rudimentary third eye. In mammals it has a glandular function, producing ➤melatonin.

pinna ➤ear.

pinocytosis The ability of some cells to ingest liquids by forming invaginations (hollows) at the cell surface which surround and engulf drops of liquid from an external liquid medium to form ➤vacuoles in the cell. Many ➤protoctistans can obtain nutrients in this way. Compare ➤phagocytosis.

Pinopsida (Gymnospermae) A phylum of trees and shrubs in which the sex organs are borne in cones; ➤seeds are produced on the surface of the cone scales rather than being enclosed in an ovary, as in the ➤Magnoliopsida ('gymnosperm' means 'naked seed'). Typical members of the phylum have needle-like leaves and most are evergreens; examples are pines, junipers, spruces and firs.

pint An Imperial unit of volume. Its value is different in the UK, where it is largely obsolescent, from its value in the US: 6 US pints ≈ 5 UK pints. 1 UK pint = 0.568 26 dm^3; 1 US pint = 0.473 18 dm^3.

pion (pi meson, π meson) Symbol π. A ➤meson, composed of a ➤quark (up or down) and an antiquark (up or down). $π^0$ has a rest mass of 135 MeV/c^2, and zero charge and spin. (There are also $π^+$ and $π^-$ versions, with charges +1 and −1 respectively, and a rest mass of approximately 140 MeV/c^2.) The study of pions was instrumental in the elucidation of the ➤strong interaction.

pipette An instrument for delivering a small volume of liquid, often 25 cm^3, typically during titrations. It has a line marked on the side and, when filled so that the bottom of the meniscus is on the line, an exact volume can be delivered.

Pirani gauge A pressure gauge for measuring the pressure of a medium vacuum (typically 0.1 to 100 Pa). It works by measuring the rate of dissipation of heat in a wire. Named after Marcello Stefano Pirani (1880–1968).

pistil ➤flower.

pitch A quantity related very closely to the ➤frequency of a sound wave. It is a subjective measure based on perception by the human ear.

Pitot tube A device for measuring the speed of an object relative to a fluid, for example the speed of an aircraft relative to the airmass. It senses the dynamic pressure of the fluid flowing into a narrow tube in the direction of motion. Named after Henri Pitot (1695–1771).

pituitary gland A small gland connected to the hypothalamus at the base of the vertebrate ➤brain. It is divided into anterior and posterior lobes. The anterior lobe produces a range of hormones, including growth hormone and the trophic hormones – ➤thyroid stimulating hormone, ➤adrenocorticotrophic hormone and the gonadotrophins (➤gonad). The posterior lobe releases ➤oxytocin and ➤vasopressin. Together with the hypothalamus, the pituitary is the centre of many mechanisms contributing to ➤homeostasis. The pituitary is often referred to as the 'master gland' since its secretions control the functioning of a wide range of other glands.

pixel Acronym for *picture element;* any of the minute dots making up an image on a display screen. Each pixel is assigned a brightness and colour, and the number of pixels determines the screen resolution.

p*K* value A number bearing the same relationship to an ➤equilibrium constant that ➤pH does to the hydrogen ion concentration. For example, there is a p*K* value related to the ➤acid ionization constant:

$$pK_a = -\log_{10} (K_a /\text{mol dm}^{-3}).$$

Other varieties occur; for example, pK_b for ➤bases and pK_w for the ➤ionic product for water.

placenta 1 A composite organ, derived partly from cells of the embryo and partly from maternal cells, which nourishes the developing ➤foetus in the ➤uterus. The placenta connects the uterus wall to the foetus via the **umbilical cord**, which contains blood vessels transporting nutrients to the foetus and removing waste. The placenta also acts as an ➤endocrine gland, producing the hormones ➤oestrogen and ➤progesterone, and in humans placental lactogen and chorionic gonadotrophin (➤gonad). Importantly, the placenta allows materials to be exchanged between mother and foetus without the mixing of maternal and foetal blood. Some antibodies and toxins such as alcohol and nicotine also pass across the placenta. **Placental mammals** (Eutheria) comprise the major group of modern ➤mammals, numbering about 4000 species, in which the young develop in the uterus and are nourished via the placenta.

2 Part of the ➤ovary wall to which the ➤ovules are attached.

planar complex ➤square-planar complex.

Planck constant Symbol h. The fundamental constant involved in the ➤wave–particle duality of light; its value is 6.626×10^{-34} J s. The constant is best known in **Planck's equation**, $E = hf$, relating the energy E of a photon to its frequency f. More useful in many contexts in quantum mechanics is the quantity $h/2\pi$, written as $\hbar$ and called the **rationalized Planck constant** or, more colloquially, 'h-cross' or 'h-bar'. Named after Max Karl Ernst Ludwig Planck (1858–1947). ➤➤Appendix table 2.

Planck's law of radiation The energy density ε of ➤black-body radiation as a function of frequency f and thermodynamic temperature T is:

$$\frac{d\varepsilon}{df} = \frac{8\pi h f^3}{c^3} \frac{1}{e^{hf/kT}-1},$$

where h is the ➤Planck constant, c is the speed of light in a vacuum and k is the ➤Boltzmann constant. At low frequencies the energy density is proportional to f^2 (the Rayleigh–Jeans law: ➤black-body radiation) and when integrated over all frequencies the total energy is proportional to T^4 (Stefan's law: ➤Stefan–Boltzmann constant). The law can be derived by applying ➤Bose–Einstein statistics to electromagnetic oscillations (quantized as photons) in a three-dimensional cavity.

Planck units Units of length and time expressed as combinations of the fundamental constants G, h-bar $(\hbar)$(➤Planck constant) and c. The **Planck length** is $\sqrt{(G\hbar/c^3)}$ (about 10^{-35} m), and the Planck time $\sqrt{(G\hbar/c^5)}$ (about 10^{-43} s). These are the scales at which quantum mechanics is necessary to model the gravitational interaction.

plane A flat surface in three dimensions such that all points on the line joining any pair of points also belong to the surface. In ➤Cartesian coordinates, a plane has equation $ax + by + cz = d$, where $a\mathbf{i} + b\mathbf{j} + c\mathbf{k}$ is a vector normal to the plane.

plane of symmetry A plane through a geometrical figure through which a reflection produces an identical figure.

plane-polarized A state of ➤polarization of light in which the ➤electric field strength vector is fixed in only one plane. Compare ➤circularly polarized; elliptically polarized. ➤➤polarizer.

planet A major body orbiting the Sun (➤Appendix table 4). In 2006 the International Astronomical Union added two further criteria: the body should be nearly spherical and should have 'cleared its neighbourhood'. If a body fails the final criterion, it is now considered to be a **dwarf planet**: the best example is Pluto. This redefinition means that there are only *eight* planets. The discovery (in 1992) of **exoplanets**, bodies orbiting another star, opened a new era in astronomy: over 60 were discovered in 2007 alone. Direct visual images of exoplanets were first obtained in 2008.

planetary nebula ➤nebula.

plane wave A ➤travelling wave in which surfaces of constant ➤phase (2) are planes perpendicular to the direction of motion.

plankton Microscopic plants (**phytoplankton**) and animals (**zooplankton**), and the spores and larvae of macroscopic organisms, that float passively in the oceans and freshwater lakes and rivers. The planktonic community is very diverse, and it serves as the basis of most aquatic ➤food chains.

plano- In the shape of a plane. For example, a plano-convex lens has one face that is in a plane and one that is convex.

plant A member of the ➤kingdom Plantae: ➤eukaryotic multicellular organisms containing ➤chlorophyll and obtaining energy by ➤photosynthesis. There is debate as to whether some algal groups, generally considered to be protoctistans, should be included in this kingdom.

plant growth substance (phytohormone) One of a number of substances produced by plant cells that direct and control developmental processes in plants. Plant growth substances include ➤auxins, ➤cytokinins and ➤gibberellins.

plaque **1** A clear zone in a colony of ➤bacteria on an ➤agar plate that indicates where cells have been killed by infection with a ➤bacteriophage.

2 A layer of organic material over the surface of a tooth that harbours the ➤bacteria responsible for tooth decay.

plasma **1** (Phys.) A ➤state of matter in which almost all the atoms present are fully ionized. It therefore consists of electrons and atomic nuclei moving relatively freely. Plasmas are used in experiments on nuclear fusion. Plasmas occur naturally in the Universe; an example is the ➤solar wind.

2 (Biol.) ➤blood.

plasma cell An ➤antibody-producing cell of the ➤immune system.

plasma frequency The frequency in a conductor at which a collective oscillation of the electrons relative to the ion cores can take place. A very simplistic model of such an oscillation has the electrons (number per unit volume n, charge e and mass m) moving as a block of negative charge, and the electrostatic attraction between this and the block of positive charge of the ion cores causes harmonic oscillations at a frequency of $\sqrt{ne^2/4\pi^2\varepsilon_0 m}$. ➤➤plasmon.

plasma membrane (cell surface membrane) The boundary layer of a ➤cell. Models for the structure of ➤membranes in cells have developed in line with the technology for investigating them. The current model of the plasma membrane (see the diagram) is of a **fluid mosaic** composed of a bilayer of ➤phospholipids (with cholesterol in some types) oriented with their ➤hydrophilic heads towards the membrane's surface and their ➤hydrophobic tails towards the membrane's interior, together with a variety of ➤proteins. Some animal cells may have an additional outer layer of carbohydrates, termed a **glycocalyx**. Plasma membranes are dynamic structures, serving not only as an envelope for the cell but also as a selective barrier for regulating the passage of substances into and out of the cell.

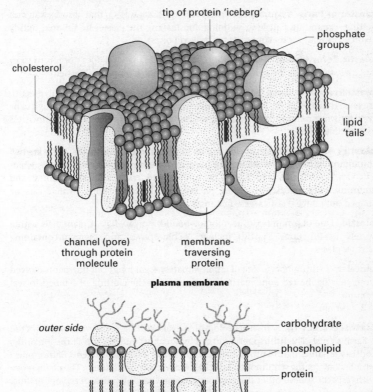

tip of protein 'iceberg'

phosphate groups

cholesterol

lipid 'tails'

channel (pore) through protein molecule

membrane-traversing protein

plasma membrane

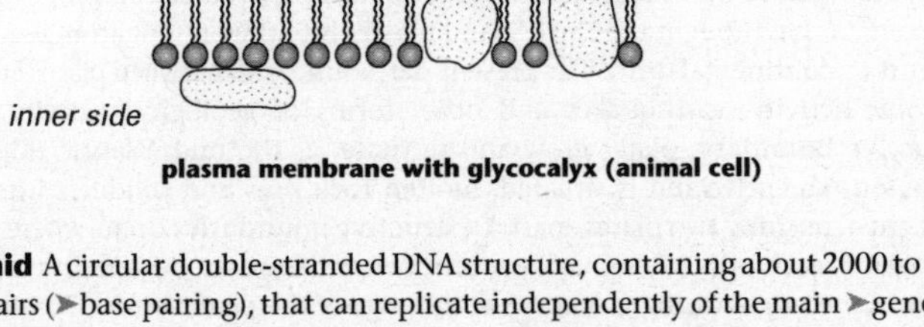

outer side

carbohydrate

phospholipid

protein

inner side

plasma membrane with glycocalyx (animal cell)

plasmid A circular double-stranded DNA structure, containing about 2000 to 10 000 base pairs (➤base pairing), that can replicate independently of the main ➤genome in a cell. Plasmids are widely used in ➤genetic engineering as ➤cloning vectors to insert DNA into recipient cells, and are found naturally in bacteria.

plasmolysis The reduction in volume of the ➤protoplast of a plant cell through loss of water, which results in the plasma membrane shrinking away from the cell wall. Plasmolysis is artificially induced by placing cells in a ➤hypertonic solution, and probably does not occur under natural conditions. ➤➤osmosis; water potential.

plasmon A collective ➤oscillation of electrons in a solid. Like the ➤phonon and the ➤photon, the plasmon is a quantum-mechanical particle corresponding to this oscillation. ➤➤plasma frequency.

plaster of Paris A hydrate of calcium sulfate, $2CaSO_4 \cdot H_2O$, that absorbs water on setting, expanding a little in so doing and making the plaster fit a mould tightly (which is useful for making casts for broken bones).

plastic deformation A deformation in which the material does not return to its original state when the deforming stress is removed. Compare ➤elasticity.

plasticizer A material, such as dioctyl phthalate, added to some polymers to increase their flexibility. For example, uPVC, unplasticized PVC, is rigid enough to be used for window frames but would be useless for making the soles of shoes since it is not flexible enough.

plastics Polymeric materials that can be moulded into shape easily. There are two important classes, **thermosoftening plastics**, for example the polyalkenes such as PVC and polythene, which can be melted and reshaped several times, and **thermosetting plastics**, such as phenol-formaldehyde resins, which can be shaped only once. Plastics were first made in the 1920s.

plastid One of a number of membrane-bound ➤organelles in plant cells with a variety of functions. Plastids include ➤chloroplasts and pigment-containing chromoplasts.

platelet **1** (thrombocyte) Small (approximately 4 μm long) cell fragments derived from cells in the red bone marrow which assist in the clotting of ➤blood to seal wounds.
2 ➤ceramic.

plate tectonics The branch of geology that deals with large-scale movements of the ➤Earth's crust. The **lithosphere**, formed by the crust and the top of the ➤mantle, consists of large, tightly fitting plates (eight major and about twenty minor ones) which 'float' on the semi-molten layer of the mantle beneath them. The plates move with respect to one another at several cm per year. Over the course of geological time, this movement has brought the Earth's land masses into the present arrangement of continents (➤continental drift). The present-day boundaries between plates are sites of volcanic activity, earthquakes and other forms of geological activity. At a **constructive boundary**, plates grow and separate: at the mid-Atlantic ridge, for example, on which Iceland is situated, molten rock rises and solidifies into new oceanic crust, pushing two plates apart. **Destructive boundaries** occur where plates move towards each other; one may be forced below the other, descending into the mantle and melting, causing volcanoes at the edge of the plate above. This process, called **subduction**, has produced, for example, the Andes mountains along the west coast of South America. At a **transform boundary** two plates are sliding past each other; here, earthquakes are frequent, as, for example, along the San Andreas fault in California.

platform The combination of the ➤hardware (particularly ➤CPU) and ➤operating system in a computer. Typically, ➤high-level languages are platform-independent, so are transportable between computer systems, whereas object code (➤source code) is platform-dependent.

platforming A contraction of 'platinum ➤reforming'.

plating ➤electrolysis.

platinum Symbol Pt. The element with atomic number 78 and relative atomic mass 195.1, which is in the third row of the ➤transition metals and is the archetypal ➤platinum metal. It is a very unreactive element, useful therefore in specialist tasks where even gold would be too reactive, as well as for jewellery. It is used as a catalyst, for example in the ➤Ostwald process, and together with rhodium is often used in catalytic converters.

Platinum shows one dominant oxidation number, as for its ➤congeners nickel and palladium, namely +2 as in the important anticancer drug *cis*-platin [$Pt(NH_3)_2Cl_2$]. Other oxidation numbers include +4 in the black solid PtO_2 and +6 in the red solid PtF_6, a violent oxidant important historically for its role in the discovery of the first compounds of ➤xenon.

platinum black A platinum surface coated in a thin film of finely divided platinum that is good at catalysing the equilibrium between hydrogen gas and its ions.

platinum metal Platinum itself, or one of the five elements near it in the periodic table (ruthenium, osmium, rhodium, iridium, palladium) that resemble platinum in being very unreactive metals.

Platonic solid One of the five regular polyhedra (see the diagram overleaf): tetrahedron, cube, octahedron, dodecahedron and icosahedron. Named after Plato (*c.* 428–348 BC).

Platyhelminthes A major animal ➤phylum containing the flatworms and flukes. They have a flattened, unsegmented body with a single opening to the gut (where present). There are many free-living flatworms found in aquatic habitats and in the soil, but many of the group are significant as parasites (➤symbiosis). These include tapeworms and liver flukes.

Pleistocene See table at ➤era.

Plexiglass US tradename for ➤poly(methyl methacrylate).

Pliocene See table at ➤era.

plumbago ➤graphite.

plumbate An oxoanion containing lead.

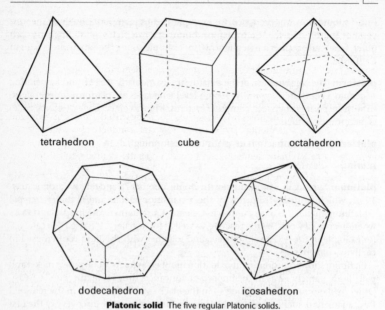

Platonic solid The five regular Platonic solids.

plumule The part of a seedling (epicotyl) developing above the ➤cotyledons in a germinating seed that gives rise to the shoot system. ➤radicle.

Pluto Traditionally, of the nine planets, the one that is, on average, farthest from the Sun (39.5 AU); its radius of only 1200 km kept it undiscovered until 1930. Since 2006, however, its status has been downgraded to that of dwarf planet, meaning that the term 'planet' now strictly applies to just eight bodies. Its 248-year orbit is highly eccentric, taking it within the orbit of Neptune at certain times. Pluto has a rotational period of 6.39 days. Its main satellite, **Charon**, has a radius of 593 km and an orbital period also of 6.39 days, so that **Charon** is always on the same side of Pluto. ➤Appendix table 4.

plutonium Symbol Pu. The element with atomic number 94 and most stable isotope 244, which is a ➤transuranium element most famous for having two fissile nuclides, including ^{239}Pu. It is an exceptionally dangerous substance because of its radioactivity, which dominates its chemistry. Nevertheless such is the significance of the element, to nuclear fuel reprocessing for example, that a great deal is known of its chemistry. In aqueous solution, four oxidation numbers (+3, +4, +5, +6) can be formed easily. The most stable oxide is PuO_2, a yellow-brown solid with the ➤fluorite structure.

Pm Symbol for the element ➤promethium.

PMR Abbr. for proton magnetic resonance (➤NMR).

PNG Abbr. for portable network graphics. A high compression file format for graphics that has nearly as high compression as ➤JPEG with lower loss of quality and much better portability across platforms.

pnicogens An uncommon generic name for the elements in Group 15 of the periodic table, the group headed by nitrogen.

p–n junction A junction formed between layers of ➤p-type semiconductor and ➤n-type semiconductor. The holes from the p-type combine with the electrons from the n-type to form a ➤depletion layer which does not conduct. The width of the depletion layer may depend on applied external potentials (➤bias), and this is the principle by which many semiconductor devices operate. ➤➤diode; transistor.

p–n–p transistor A junction transistor that consists of a layer of ➤n-type semiconductor sandwiched between two layers of ➤p-type semiconductor. ➤➤bipolar junction transistor.

Po Symbol for the element ➤polonium.

poikilothermy (ectothermy, cold-bloodedness) The inability of an animal to exert fine control over its internal temperature by physiological means. The term 'cold-blooded' is somewhat inaccurate since many reptiles, for example, can regulate their temperature within limits by behavioural means (basking in the sun or seeking shade, as appropriate). Compare ➤homoiothermy.

point A dimensionless geometric element that is fully specified by a single assignment of coordinates.

point defect A defect in a crystal that is restricted to a point. Examples are ➤vacancies (Schottky defects) and interstitial ➤defects.

point group 1 The ➤symmetry group of rotations of a crystal ➤lattice; such a rotation transforms any ➤lattice vector into another lattice vector. There are seven different point groups possible for a ➤Bravais lattice (the crystal systems) and 32 point groups for lattices with more than one species in the unit cell.
    2 The group describing all the symmetry elements that a molecule possesses. Knowledge of the point group is essential for interpreting the details of the vibrational spectrum of the molecule and for providing a detailed analysis of the bonding within the molecule.

point source The origin of a ➤wave considered as a single point. In practice, all disturbances that produce waves are extended in space, but treating them as a point may be a good approximation for waves observed a long way from their source. Point sources produce ➤spherical waves whose intensity obeys the ➤inverse square law.

poise Symbol P. The ➤c.g.s. unit of dynamic ➤viscosity. 1 P is equivalent to 1 g cm s^{-1}. Named after Jean Louis Poiseuille (1799–1869).

Poiseuille flow The flow of a ➤viscous liquid in a circular pipe such that the speed of flow of an element of the fluid varies quadratically with the distance from the centre of the pipe. This results in a flow rate proportional to the fourth power of the radius of the pipe.

Poisson distribution A discrete random variable X taking values 0, 1, 2, . . ., with probabilities $P(X = r) = \mathrm{e}^{-\lambda}\lambda^r/r!$. It arises in a variety of contexts: for example, the number of times a randomly occurring event (such as radioactive decay) occurs in a given time interval. The Poisson distribution has mean λ and variance λ and, for large λ, may be approximated by a ➤normal distribution having these parameters. In turn, under certain conditions, it may be used to approximate the ➤binomial distribution. Named after Siméon-Denis Poisson (1781–1840).

Poisson equation The partial differential equation $\nabla^2\phi(\boldsymbol{r}) = f(\boldsymbol{r})$, where ∇^2 is the ➤Laplacian and f is a given function. This is the form of the equation relating the electrostatic potential to the charge density ρ, with $f(\boldsymbol{r}) = -\rho(\boldsymbol{r})/\varepsilon_0$, and the gravitational potential to the mass density, with $f(\boldsymbol{r}) = 4\pi G\rho(\boldsymbol{r})$.

Poisson's ratio In an elastic material, the negative of the ratio of the strain along an axis perpendicular to the stress to the strain along an axis parallel to the stress. A material will tend to bulge out when squeezed, and the ratio is a measure of that tendency. Most materials behave in this way (positive Poisson's ratio) but a cork has a negative Poisson's ratio (its radius contracts when it is squeezed along its length) which makes it easy to push into the neck of a bottle.

polar bond An electron pair (a ➤covalent bond) that is shared unequally: the atom with the higher ➤electronegativity has a greater share of the electron pair and is charged slightly negatively; the less electronegative atom becomes slightly positively charged. A diatomic molecule with a polar bond will always be a ➤polar molecule.

polar coordinates A system for specifying the location of a point P in a plane by means of its distance r from a fixed origin O and the angle θ that OP makes with a fixed line. Polar coordinates (r, θ) are related to ➤Cartesian coordinates by $x = r \cos \theta$, $y = r \sin \theta$; some curves have polar equations (relating r and θ) which are simpler than their Cartesian equations (relating x and y). Polar coordinates are particularly useful in problems involving a degree of rotational symmetry.

polarimeter An instrument for measuring the ➤optical activity of a sample. White light is polarized and passed through the sample, and the rotation of the plane of polarization is examined using an analyser. The analyser is a second polarizing filter, originally set up perpendicular to the plane of polarization of the light so that, before the optically active sample is introduced into the beam, a negligible amount of light passes through the apparatus.

polarity The distinction between poles and the quantities associated with them, whether electrical poles (positive or negative) or ➤magnetic poles (2) (north or south). Thus to swap the leads on a battery powering a circuit would be to change the polarity.

polarizability A measurement of the ease of distortion of a molecule's electron density. When a molecule is placed in an electric field of magnitude E, it has a dipole moment d approximated by $d = \mu + \alpha E$, where μ is the permanent electric dipole moment and α is the polarizability. Polarizability is crucial for the ➤dispersion force between two molecules, which is directly proportional to the product of their polarizabilities.

polarization The direction and time dependence of the electric field strength of electromagnetic radiation, typically light. The direction is measured relative to an arbitrary zero. If the direction stays fixed, the light is ➤plane-polarized. If it rotates (always about the direction of travel of the wave) but remains of a constant magnitude, it is ➤circularly polarized. If the magnitude also varies then it is ➤elliptically polarized. ➤➤polarizer.

polarization, electric ➤electric polarization.

polarizer A device that allows the passage of light with electric field strength in only one direction, rendering it ➤plane-polarized. ➤➤Polaroid.

polarizing angle ➤Brewster angle.

polar molecule Conventionally used to describe a molecule with a nonzero permanent electric dipole moment. So homonuclear diatomic molecules, such as H_2 and Cl_2, are nonpolar whereas heteronuclear diatomic molecules, such as HCl, are polar. Whether or not a polyatomic molecule is polar overall depends on its symmetry. ➤➤dipole–dipole force.

polarography An electrochemical method of analysis using a **dropping mercury cathode** which records the current–voltage plot in a circuit and shows a peak for each species that can be reduced in the solution at a characteristic voltage, known as the **half-wave potential**.

Polaroid Tradename for a semi-transparent film which, because of the ➤anisotropic arrangement of its polymeric molecules, polarizes light. Being a plastic, Polaroid provides a convenient way of polarizing large areas, and is used, for example, in sunglasses, visors, camera filters and car headlights. Since reflected light is partially polarized, Polaroid reduces glare.

polaron An ➤excitation in an ➤ionic solid. When an electron is introduced into the ➤conduction band, it may be energetically favourable for the lattice to distort, creating a bound state for the electron. The combination of electron and lattice distortion is a very mobile entity and is known as a polaron.

polar solvent A solvent composed of polar molecules, for example water. Polar solvents are able to dissolve ionic substances as they can provide ➤solvation energy to compensate for the loss of ➤lattice energy. The solvation energy is due to an ion–dipole electrostatic attraction.

pole 1 (Math.) A ➤complex function has a pole of order n at $z = a$ if $\lim_{z \to a}(z - a)^n f(z)$ exists and is nonzero. The function thus has a ➤singularity at $z = a$, but one which is 'no worse' than $1/(z - a)^n$.

2 (Phys.) A point on a source of electric or magnetic field from which the field appears to originate. ➤➤magnetic pole.

3 (Phys.) In electronics, one of two points on an electronic device from or to which current flows, for example the positive and negative terminals of a ➤cell (2).

4 (Geog.) ➤magnetic pole (2).

5 (Geog.) Either of the two points where the Earth's axis of rotation intersects its surface (the north pole and south pole).

6 (Astron.) ➤celestial poles.

pollen The microspores of seed plants, containing the microgametophyte (the structure that generates the male nucleus, the ➤gamete). Pollen is produced in the microsporangia (anthers of ➤flowering plants), and released when ripe to be carried by wind or animals to the stigma (or female cone in conifers) during **pollination**. Under appropriate conditions, the pollen grain germinates to produce a pollen tube which grows down the style, carrying the male gamete to the female gamete in the ➤ovule.

pollution Any deleterious effect on the natural environment caused by the release of any substance. Water can be polluted by untreated sewage or overuse of fertilizers (➤eutrophic). Air pollution is commonly caused by oxides of nitrogen (collectively known as ➤NO_x), sulfur dioxide, soot and other particulate matter. ➤Ozone is another cause of pollution at ground level, although its presence in the stratosphere is beneficial (➤ozone layer; ➤➤ozone hole).

polonium Symbol Po. The element with atomic number 84 and most stable isotope 209, which is at the bottom of Group 16 in the periodic table and as such has compounds that resemble those of oxygen; its hydride is H_2Po, for example, and **polonides** containing the Po^{2-} ion exist. The chemistry of the element is dominated by its radioactivity, the feature that made Marie Curie search for it; she named it after her native country, Poland. It is the only element which crystallizes with the ➤primitive cubic structure. It is much more metallic than any other element in Group 16.

poly- A common prefix found in a variety of names and almost always meaning 'many'.

polyacrylonitrile A polymer of propenenitrile, $H_2C=CHCN$, which is an important fibre.

polyalkene A polymer made from monomers that are alkenes. The table shows how similar all of the monomers are, the double bond being characteristically broken in the polymerization process to produce a long-chain alkane-like molecule which is unreactive. This unreactivity of polyalkenes is responsible for concern over their indiscriminate use, as they are nonbiodegradable. ➤➤syndiotactic polymer.

Polyalkene	Structure of monomer
Polyethylene	$H_2C=CH_2$
Polypropylene	$H_2C=CHCH_3$
Polystyrene	$H_2C=CHC_6H_5$
PVC	$H_2C=CHCl$
PTFE	$F_2C=CF_2$

polyamide A polymer that contains a substituted amide link (➤nylon).

polyatomic molecule A molecule having many atoms; 'many' can be as few as three as in the triatomic CO_2, or hundreds of thousands as is common in polymers.

polybasic acid An acid containing more than one replaceable hydrogen atom, such as the dibasic sulfuric acid, H_2SO_4.

polycarbonate A polymer based on the repeating unit COORO, often made from ➤phosgene and a dihydroxy compound. Being very hard, polycarbonates are used for bullet-proof windows, riot shields, safety helmets, compact discs and car bumpers. Some are optically clear enough to be used for spectacles, where they are at least ten times more resistant to breakage than conventional glass.

polychlorinated biphenyl ➤PCB (2).

polychromatic Describing a source of light or electromagnetic radiation that consists of a combination of frequencies. Compare ➤monochromatic.

polycrystalline Describing a solid that is composed of many small crystals called **grains**, separated by **grain boundaries**. Most metals are polycrystalline.

polydentate ligand A ligand that can donate many electron pairs to the central metal ion. A classic example is ➤EDTA, which can bond at six sites. A simpler example is the oxalate ion, $(CO_2^-)_2$, which can bond at two sites.

polyester A polymer with repeating units containing ester groups, formed from a dicarboxylic acid reacting with a dihydroxy compound. A good example is ➤Terylene.

polyethylene (polythene) The polymer of ethene, whose systematic name is **poly(ethene)**, which was the first polyalkene to be discovered and remains among the most useful polymers known. It was discovered by chance in 1933 by two ICI chemists (Fawcett and Gibson) when a leak in their apparatus allowed a tiny quantity of oxygen to catalyse the radical polymerization of ethene. The conditions required for the reaction were extreme; producing 1500 atm pressure in particular is very expensive. In 1953 a much cheaper route was found using the ➤Ziegler–Natta catalysts, which also produced a denser material. Later versions of these catalysts were tailored to produce stereoregular polymers (➤stereospecific reaction).

polygon A closed ➤plane figure bounded by a number of straight line segments meeting in pairs of ➤vertices. A **regular polygon** has sides that are equal in length and interior angles that are all equal. Polygons are generally named according to their number of sides, as in hexagon and octagon. The sum of the exterior angles of a polygon is 360°.

polyhedron A solid figure bounded by a number of flat ➤polygonal faces that meet in pairs along common edges, which in turn meet at corners or vertices. **Euler's formula** $V + F = E + 2$ connects the number of vertices (V), faces (F) and edges (E) of a convex polyhedron. Polyhedra are generally named according to their number of faces, as in ➤tetrahedron. A **regular polyhedron** has faces that are identical regular polygons meeting at constant ➤dihedral angles. ➤➤Platonic solid.

polyhydric alcohol An alcohol containing more than one hydroxyl group. The two most important examples are ➤ethane-1,2-diol and ➤glycerol, which have two and three hydroxyl groups respectively.

polymer ➤polymerization.

polymerase A class of ➤enzymes responsible for extending the length of strands of ➤nucleic acids. Most polymerase enzymes require a template from which the required sequence is read. **DNA polymerases** catalyse the synthesis of DNA; most types use DNA as a template (➤DNA replication), but in ➤retroviruses there is a DNA polymerase, called ➤reverse transcriptase, that makes a DNA copy of the viral RNA.

RNA polymerases catalyse the synthesis of RNA from DNA (➤transcription) or RNA templates.

polymerase chain reaction (PCR) A technique (see the diagram) enabling multiple copies to be made of specific sections of ➤DNA molecules. It allows isolation and amplification of such sections from large heterogeneous mixtures of DNA and has many diagnostic applications, for example in detecting genetic mutations and viral infections including HIV, in ➤human leucocyte antigen system tissue typing prior to transplant operations, and in the forensic identification of an individual from blood or semen samples. The technique has revolutionized many areas of molecular biology.

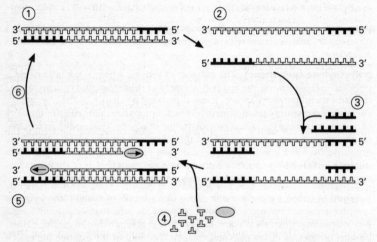

polymerase chain reaction The reaction starts with (1) a double-stranded DNA fragment with known end sequences, which is to be copied. (2) The two DNA strands are separated by heating to 95 °C. (3) Two primers are added that have complementary base pairs to the end sequences in the DNA. Cooling to about 50 °C allows the primer to bond to each strand. (4) DNA nucleoside triphosphates and the heat-stable *Taq* polymerase become involved, and the temperature is raised to 72 °C. (5) New DNA fragments are produced by the nucleotides bonding to the primers, the single strands acting as templates. (6) The process is repeated with the two new double strands. Each cycle of heating and cooling doubles the number of copies of the DNA template, so that after 20 cycles there are over a million copies of the original DNA. (The diagram is indicative only: the DNA fragments are very long, and the primers may be 15–20 bases long.)

polymerization An important process in which many (hence 'poly') small molecules called **monomers** join together to make one gigantic molecule, the **polymer**. Many different classifications are significant; one is based on the mechanism of their formation. **Chain polymerization** produces a polymer of high molar mass almost immediately; the reaction mixture contains only monomer, high molar mass polymer and a few growing chains. Common chain polymers include ➤addition polymers, such as the ➤polyalkenes (see table at polyalkene). **Step polymerization** produces polymers with a wide range of molar masses, the average molar mass rising with longer reaction times. Thermosetting ➤plastics are necessarily step polymers. Common step polymers include ➤condensation polymers, such as polyesters and polyamides.

Other classifications include **synthetic polymers** such as nylon and **natural polymers** such as rubber (➤➤polynucleotide, protein). The polymers have much larger molar masses than the monomers, typically by factors ranging from thousands to hundreds of thousands. ➤➤copolymers; radical polymerization.

poly(methyl methacrylate) The polymer of ➤methyl 2-methylpropenoate; it is sold under the tradenames of **Perspex** (UK) and **Plexiglass** (US).

polymorphism 1 (Chem.) The existence of more than one solid phase for a substance. If the substance is an element, such as sulfur, the term 'allotropy' can be used as well. For a compound, such as calcium carbonate, with polymorphs calcite and aragonite, no such alternative exists.

2 (Biol.) The simultaneous existence in a population of ➤genomes showing genetic variations.

polynomial A polynomial (of one variable) is an algebraic ➤expression of the form $a_n x^n + a_{n-1} x^{n-1} + \cdots + a_0$ where x is a variable and the a_i are constants (usually real or complex numbers). The highest power n is called the **degree**. A polynomial equation $a_n x^n + a_{n-1} x^{n-1} + \cdots + a_0 = 0$ has at most n distinct ➤roots; for $n \leq 4$ these are given by formulae involving roots. There are no such formulae for $n > 4$, but in any case alternative methods such as ➤iteration may be more convenient for finding the roots. ➤➤fundamental theorem of algebra.

polynucleotide A ➤polymer of ➤nucleotides. ➤➤DNA; RNA.

polypeptide A long (in excess of 20) chain of ➤amino acids joined together by ➤peptide bonds.

polyphyletic Describing the origin of a group of organisms from more than one ancestral group. The insectivorous plants are an example. Although they are often grouped together for convenience, they are not a natural group, and insectivory has arisen several times during evolution in different groups of plants.

polyploidy An increase in the number of chromosomes, most commonly in plants. It usually involves the replication of a complete set of chromosomes due to a malfunction in ➤mitosis that converts, for example, a ➤diploid ($2n$) cell into a tetraploid ($4n$) cell. This may occur naturally (**autopolyploidy**), or it can be induced artificially by ➤mutagens such as colchicine. The doubling of chromosomes in this way may allow an otherwise sterile hybrid between two species to become fertile since it makes possible complete pairing of chromosomes during ➤meiosis. Such **allopolyploid** hybrids have been important in the evolution of the plants, since, in effect, they constitute an immediate new ➤species. Allopolyploids are also important in the production of many species of horticultural and agricultural significance, including wheat, potatoes and sugar cane. Polyploidy in animals appears to be restricted to a few species of fish and amphibians, although the extra chromosome found in certain genetic disorders including ➤Down's syndrome, more correctly referred to as **polysomy**, is sometimes included in the definition.

polypropylene A polymer of propylene (propene); see table at ➤polyalkene.

polysaccharide A polymer of ➤monosaccharide molecules. Polysaccharides include ➤starch, ➤cellulose, ➤inulin, ➤glycogen and ➤agarose. Their exact structure depends on the nature of the monomer and how the monomers are bonded together. ➤➤sugar.

polystyrene A polymer of styrene (phenylethene); see table at ➤polyalkene.

polysulfane One of a series of compounds of formula H_2S_n, which consist of a chain of sulfur atoms terminated by a hydrogen at each end. They are thermally unstable yellow oils (except for H_2S_2, which is colourless).

polysulfide An anion that contains more than one sulfur atom joined together. The most important examples are the radical anions S_2^- and S_3^- responsible for the colour of lapis lazuli. Iron pyrites, FeS_2, is another example.

polytetrafluoroethene ➤PTFE.

polythene ➤polyethylene.

polyunsaturated Describing fatty acids that contain *more than one* double bond (this is a rare exception to the meaning 'many' for 'poly'). Examples include ➤linoleic acid and ➤linolenic acid.

polyurethane A polymer made from a low-molar-mass polymer with two hydroxyl end groups and a diisocyanate, giving a repeating unit of the form in the diagram. Their main use is in making foams; poisonous fumes are created if they are combusted.

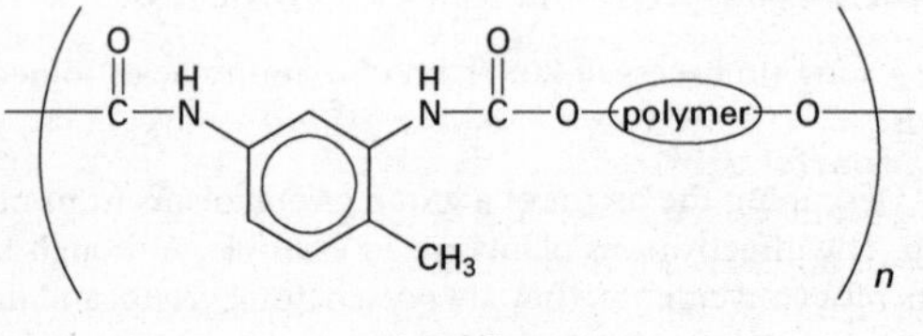

polyurethane

polyvinyl chloride ➤PVC.

population **1** (Biol.) A group of organisms of the same species, usually forming a distinct breeding unit, existing in a particular habitat at a given time.

2 (Math.) The actual or hypothetical collection of all values of a measure of some characteristic of a group of individuals from which a ➤random sample is taken.

3 (Astron.) A classification of stars in our Galaxy (and others), depending on their position. The majority of stars make up **Population I** and are found in the galactic disk; the remainder make up **Population II** and are found in a halo surrounding the disk on each side, and in its central bulge.

population genetics The study of the frequency and patterns of ➤alleles within a ➤population rather than (as in classical genetics) within the parents and offspring of a particular family.

population inversion A condition in, for example, a ➤laser, in which there are more electrons in a higher energy state than in a lower state, which is the reverse of the natural arrangement. This population inversion may be achieved by stimulation with light.

population parameter ➤confidence interval.

p orbitals A set of ➤atomic orbitals and therefore a set of solutions of the ➤Schrödinger equation. The p orbitals are filled across the **p block** of the periodic table and many p-block elements are of considerable importance. This set has an orbital angular momentum quantum number, l, of 1. The number in the set is always given by $2l + 1$, so is 3 in this case. Their shape is shown in the diagram, and they all have the same energy in an isolated atom. They point along the three axes and are labelled p_x, p_y, p_z.

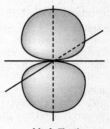

p orbitals The three orbitals lie along the x, y, z axes.

Porifera The animal ➤phylum containing the sponges.

porphyrin One of a family of structures comprising four pyrrole rings in a planar cyclic arrangement (see the diagram), complexed with a metal ion, commonly iron, magnesium or copper. Porphyrins function as ➤prosthetic groups for a variety of ➤proteins and ➤enzymes associated with redox reactions in cells. The porphyrin group is the basis for the involvement of such molecules in redox reactions. The haem groups of ➤haemoglobin are porphyrins.

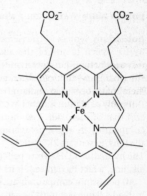

porphyrin Ring as typified by haem.

portal system (portal vein) A vein carrying blood from the capillary system of one organ directly to that of another, without the blood passing into the venous system to be carried via the heart to the second organ. Portal systems are found between the gut and the liver and in the complex formed by the ➤pituitary gland and the ➤hypothalamus.

Portland cement ➤cement.

position vector A ➤vector specifying the position of a point relative to an origin by means of its coordinates (➤coordinate system).

positive 1 Describing a ➤real number that is greater than ➤zero.
  2 ➤electric charge.

positive feedback ➤feedback.

positron Symbol e⁺. The ➤antiparticle of the electron. It has the same mass as the electron but opposite charge and magnetic moment. Its discovery by Carl Anderson (1905–91) in 1932 vindicated the ➤Dirac equation for the electron.

positronium An atomic species consisting of an electron and a positron. The particles readily combine and annihilate each other (releasing photons).

postactinide elements The elements beyond lawrencium, element number 103. Elements with atomic numbers from 104 to 116 have been synthesized (not always in order; 109 was found before 108) together with 118. Their names used to cause controversy (➤rutherfordium). The names now accepted are rutherfordium Rf (104), dubnium Db (105), seaborgium Sg (106), bohrium Bh (107), hassium Hs (108), meitnerium Mt (109), darmstadtium Ds (110) and roentgenium Rg (111). Many have been discovered in minute quantities, as low as a few atoms in some cases. No uses have been found for any of these elements as yet.

posterior Describing the position of a structure in the body of an animal that lies away from the head or behind the surface that is at the front during locomotion. In flowers it refers to a structure that is nearest the main axis of the stem. Compare ➤anterior.

postfix notation ➤reverse Polish notation.

potash alum A common ➤alum of formula $K_2SO_4 \cdot Al_2(SO_4)_3 \cdot 24H_2O$.

potassium Symbol K. The element with atomic number 19 and relative atomic mass 39.10, which is one of the alkali metals. It is the most reactive of the metals commonly found in the laboratory. One example of its impressive reactivity is that the hydrogen produced when potassium reacts with water catches fire (hence the metal must be stored under oil).

Its only oxidation number in compounds is +1, so the formula of the chloride is KCl and that of the oxide K_2O, both of which are white solids. Its other two oxygen compounds are K_2O_2, potassium peroxide, and KO_2, potassium superoxide, where the ions present are $2K^+$ and O_2^{2-} for the peroxide and K^+ and O_2^- for the superoxide. The hydroxide KOH (white pellets normally seen as a colourless aqueous solution) is alkaline, a fact recognized in its traditional name of **caustic potash**.

All potassium compounds are soluble in water, so no precipitation test exists and instead a flame test must be carried out: a lilac flame confirms potassium is present. The potassium ion K^+ is also important in nature as the K^+/Na^+ ion balance controls nerve action. Potassium is one of the three main elements that plants need (➤NPK fertilizers).

There are many important compounds of potassium, a short selection of which follows: carnallite $KCl \cdot MgCl_2 \cdot 6H_2O$, a common ore; the cyanide KCN, notorious as a poison; the iodide KI, used to test for ➤oxidants; the nitrate KNO_3, used in gunpowder; the sulfate K_2SO_4, used as a fertilizer; and the thiocyanate KSCN, used to test for iron(III) ions. All of these compounds are white solids.

potassium dichromate(VI) (potassium dichromate) $K_2Cr_2O_7$ An important oxidant used throughout chemistry, almost invariably acidified, the colour change from orange (due to the $Cr_2O_7^{2-}$ ion) to green (due to the Cr^{3+} ion) being a useful colour test for ethanol for example.

potassium manganate(VII) (potassium permanganate) $KMnO_4$ An important oxidant used throughout chemistry, very often acidified. In inorganic chemistry it is the most useful titrant for reductants as its deep purple colour is decolourized on reduction, the retention of a purple colour signalling the ➤end-point. In organic chemistry it can oxidize alcohols such as ethanol to ethanal and thence to ethanoic acid.

potential 1 (Phys.) ➤electric potential.

2 (Math.) Any scalar function ϕ of position whose gradient f is a given vector field: $f = -\nabla\phi$ or in one dimension (coordinate x) $f = -\mathrm{d}\phi/\mathrm{d}x$. For an electric field this is the electric potential, but other potentials (for a magnetic field, a gravitational field and a fluid velocity field) can be constructed provided the field is ➤conservative. It is usually easier to solve an equation for a scalar potential (such as the ➤Poisson equation or ➤Laplace's equation) than to solve the corresponding vector equation. Potentials have an arbitrary zero: we can always add a constant (**gauge**) to ϕ without changing the value of its gradient. ➤➤potential energy.

potential barrier A region of high ➤potential (2) that must be crossed by a particle moving from one side to the other. A **nuclear barrier** is such a region, around an atomic ➤nucleus. A ball rolling up a hill and over to the other side is a simple example: if it has insufficient total energy it will not reach the top of the hill and cross the gravitational potential barrier; if it has sufficient energy it will reach the other side. ➤➤tunnel effect.

potential difference (p.d., voltage) The difference in ➤electric potential from one point to another. In a conductor in an electric circuit, a difference in electric potential causes an electric current to flow in accordance with ➤Ohm's law. Potential difference is measured in volts.

potential divider (voltage divider) A construction in an electric circuit that outputs a fixed fraction of an input potential difference. In the diagram, the output V_{AB} of the potential divider is a fraction $R_1/(R_1 + R_2)$ of the input V_{AC}.

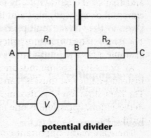

potential energy Symbol V or U. Energy stored in a ➤system by virtue of the state of the system. The energy that an object (close to the Earth's surface) has by virtue of its height above the surface is **gravitational potential energy**. The energy of a stationary charged particle in an electric field is

potential divider

potential energy: if it moves along the direction of electric field lines, it can turn potential energy into ➤kinetic energy as it moves. The zero of potential energy is usually arbitrary; only *changes* in potential energy are meaningful.

potential flow Fluid flow that can be described by a velocity field with a scalar ➤potential (2). Such flow has zero ➤vorticity.

potential well A region of low ➤potential (2) surrounded (in any number of dimensions) by regions of high potential. In ➤classical physics a particle with insufficient energy cannot escape from the potential well (whether electric or gravitational), but in ➤quantum mechanics there is a nonzero probability of finding the particle outside the constraints of the well. In the limit of the high potential tending to infinity, the particle in a potential well becomes a ➤particle in a box.

potentiometer A simple device, based on the ➤potential divider, used in an electric circuit to produce a variable ➤potential difference. The most common type of potentiometer is based on a coil over which a contact moves.

potentiometric titration A ➤titration carried out by monitoring the potential difference between an inert electrode sitting in the solution being titrated and a reference electrode. A significant change indicates the ➤end-point and this proves useful for coloured solutions when indicators are difficult to use.

pound Symbol lb. An Imperial unit of mass dating from Roman times (lb comes from the Latin *libra*). Its definitions in the UK, where it is obsolescent, and the US differ by a fraction of a milligram. In the UK 1 lb is defined as 0.453 592 37 kg exactly.

pound per square inch Symbol psi. An obsolescent Imperial unit of pressure, still used in some nonscientific applications such as measuring tyre pressures. Standard atmospheric pressure is 14.7 psi. 1 psi is equivalent to 6900 Pa.

powder photography An ➤X-ray diffraction technique that uses a powdered sample and photographic film. The characteristic diffraction angles, each corresponding to a ➤reciprocal lattice vector, appear as concentric rings on the photograph.

power 1 (Phys.) Symbol P; unit ➤watt, W. The rate of transfer of energy between one system and another. For example, the rate at which a car's engine turns the chemical energy of the fuel into mechanical energy to drive the car is the power of the engine (a 1.8 litre petrol engine typically produces 90 kW); the rate at which a light bulb turns electrical energy into heat and light is the power (consumption) of the light bulb, given by the voltage multiplied by the current. ➤efficiency.
   2 (Math.) The number of times a number or expression is multiplied by itself. For example, x^4 is the fourth power of x.

power amplifier An ➤amplifier designed to drive large currents as the last stage of an amplification process. It might be used for reproducing sound feeding loudspeakers, or in a radio transmitter feeding the aerial.

power factor ➤loss factor.

power of a lens The reciprocal of the ➤focal length of a ➤lens, taken as positive for a converging lens and negative for a diverging lens ➤dioptre.

power series An ➤infinite series of the form $\sum_{n=0}^{\infty} a_n x^n$ for complex x and constants a_n.

poxvirus One of a group of DNA-containing ➤viruses that cause disease in animals. Examples include myxomatosis and smallpox.

Poynting vector The ➤vector that describes the rate and direction of transfer of energy per unit area of an ➤electromagnetic wave. The Poynting vector is given by $E \times H$, where E is the ➤electric field strength and H the ➤magnetic field strength. Named after John Henry Poynting (1852–1914).

ppm Abbr. for parts per million.

Pr 1 Symbol for the element ➤praseodymium.
  2 Abbr. for the ➤propyl group, —C_3H_7.

praseodymium Symbol Pr. The element with atomic number 59 and relative atomic mass 140.9, which is one of the ➤lanthanides. The name refers to 'twin' and ➤neodymium is indeed very closely similar. Being a typical lanthanide, its dominant oxidation number is +3, as in the chloride $PrCl_3$, which exists as green needles.

pre-amplifier An ➤amplifier designed to feed the input of a further stage of amplification. Typically, it is a small device placed close to the source so as to minimize interference.

Precambrian See table at ➤era.

precession The change of direction of the ➤angular momentum of a rotating body (typically a ➤gyroscope) caused by a couple acting about an axis perpendicular to the rotation axis. If the couple remains perpendicular to the angular momentum as it changes direction, then the angular momentum traces out a circle and the motion is periodic.

precession of the equinoxes The ➤precession of the axis of rotation of the Earth, and therefore of the celestial poles and equinoxes, relative to the ➤fixed stars. Its cause is the small couple exerted on the Earth's ➤equatorial bulge by the Sun and the Moon (see the diagram). The period is about 26 000 years.

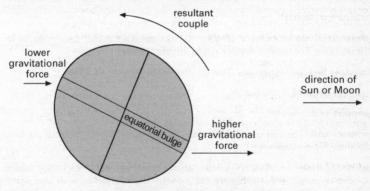

precession of the equinoxes

precipitate A solid produced from a solution when the ➤solubility product of a substance is exceeded. The most common precipitates are those used to test for ions in aqueous solution, including the rust-brown $Fe(OH)_3$ formed from iron(III) ions and hydroxide ions or the white AgCl formed from chloride ions and silver ions; these are examples of ➤ionic association.

precipitation The process by which a ➤precipitate is formed.

precision The number of significant figures quoted in a measurement. This needs to be distinguished from ➤accuracy, which concerns how close a measurement is to the true answer. To state that π was 3.452 546 would be precise and significantly inaccurate, while 3.1 would not be very precise but quite accurate.

preon A hypothetical particle that makes up ➤quarks and ➤leptons. There is at present little evidence for the existence of preons: quarks and leptons appear to be fundamental.

presbyopia The decreasing ability to focus on near objects by human eyes over the age of about forty. Distant vision is less affected, the defect being caused by a loss of elasticity in the lens, preventing it becoming a strong ➤converging lens. The defect is corrected by using external converging lenses, as in 'reading glasses'.

pressure Symbol p; unit ➤pascal, Pa. The force per unit area applied to a surface, usually by a ➤fluid. For example, a gas applies a pressure to the walls of its container. Any fluid of density ρ in a gravitational field of strength g (like the Earth's) exerts a pressure (equally in all directions) of ρgh at a depth h. Thus divers feel the pressure on their bodies exerted by the column of water above them; atmospheric pressure is caused by the weight of atmosphere above.

pressure gauge A device for measuring ➤pressure. Because a wide range of pressures can be achieved in the laboratory (from about 10^{-9} to 10^7 Pa) there are a large number of different types of gauge. For very low pressures of gas, an ➤ionization gauge is often used. For pressures around 10^5 Pa (about atmospheric pressure) a ➤barometer is used.

pressure potential ➤water potential.

pressurized water reactor (PWR) A ➤nuclear reactor that uses water as its ➤moderator. The water is pressurized to prevent it from boiling.

Pribnow box ➤Hogness box. Named after David Pribnow (b. 1948).

primary alcohol ➤alcohol.

primary cell ➤cell (2).

primary coil The coil of a ➤transformer to which the input voltage is applied. Compare ➤secondary coil.

primary colour One of a set of colours, suitable combinations of which can produce all other colours. Artists and printers use pigments that subtract a colour from white light; these **subtractive primaries** are cyan, magenta and yellow. Scientists (and TV sets) work with **additive primaries**: red, green and blue (hence 'RGB' monitors). Yellow is best thought of as white light with blue removed (i.e. yellow is minus-blue). Similarly, magenta is minus-green and cyan is minus-red. ➤➤secondary colour.

primary standard A definition of a fundamental ➤unit of measure. The primary standards for the ➤metre, ➤second, ➤ampere and ➤kelvin are now all definitions in terms of the properties of materials, but the ➤kilogram is still defined in terms of a material primary standard, a block made of platinum and iridium.

primary structure The sequence of ➤amino acids in a ➤peptide or ➤protein, or the sequence of ➤nitrogenous bases in a ➤nucleic acid. ➤protein structure.

primary wave (P wave) The ➤seismic wave that acts in the direction of motion of the disturbance (it is ➤longitudinal). Compare ➤secondary wave.

primate A member of the Primates, the order of ➤mammals that includes the lemurs, monkeys, apes and humans. Primate characteristics include a thumb that is opposable, that is can make contact with other digits on the same limb, a large brain and binocular vision.

prime meridian The meridian that is used as the arbitrary zero of ➤longitude, passing through Greenwich in London.

prime number A whole number greater than 1 that has precisely two factors: itself and 1. There are an infinite number of prime numbers (the largest known by 2008 was $2^{43112609} -1$ ➤**Mersenne prime**) and the **prime number theorem** states that the number of prime numbers less than x is approximately $x/\ln x$. There are many unsolved problems concerning primes, such as **Goldbach's conjecture**: is every even number greater than 2 the sum of two primes?

primer ➤polymerase chain reaction.

primitive cell A shaped volume in a crystal ➤lattice which, when replicated by translation by every lattice vector, exactly fills the lattice, without overlap or voids. Thus a unit cell contains exactly one lattice point (unless its boundary passes through a lattice point). ➤➤Wigner–Seitz cell.

primitive cubic (simple cubic) A simple lattice structure in which the particles are at the corners of a cube. ➤Bravais lattice.

primitive vector One of the vectors that characterizes a ➤Bravais lattice. The entire lattice can be constructed by taking all possible linear combinations of the primitive vectors.

principal axis 1 In a ➤rigid body, the necessarily ➤orthogonal axes with respect to which the ➤moment of inertia tensor is diagonal. They can frequently be determined by symmetry. They are important because analysis of a problem about these three axes allows the (in general very complicated) motion of a rigid body to be simplified into three simple rotations.
  2 optic axis (➤birefringence).

principal focus For a ➤lens or curved mirror, the ➤focus of a beam of light that is incident along or closely parallel to its axis.

principal quantum number Symbol n. The ➤quantum number of a single-electron state of an atom that describes the ➤shell corresponding to the ➤wavefunction. Wavefunctions with increasing n have successively greater mean distances from the nucleus. For potassium, in period 4 of the periodic table, the outermost electron is in the 4s orbital, with principal quantum number 4.

principle of least action ➤Hamilton's principle.

printed circuit An electrical circuit consisting of conductive strips etched or printed onto an insulating ➤substrate (2). The conducting strips are of copper, while the substrate is often a glass/epoxy laminate. A **printed circuit board** (PCB) is a printed circuit on a flat board, with connections for electronic components and integrated circuit chips. PCBs can be single-sided, double-sided or multilayered, so there can be several layers of printed circuit on the same PCB.

prion (proteinaceous infectious particle) A fragment of ➤protein, possibly originating from a ➤virus and apparently with some viral properties, that acts as a disease-causing agent. It is implicated in various neurological disorders in animals, including sheep (e.g. scrapie), cattle (➤BSE) and humans (➤Creutzfeldt–Jakob disease). Recent research interest is centred on a ➤gene for a brain protein, PrP (prion related protein), variants of which are thought to be associated with the development of disease. Evidence suggests that prion diseases can be either infectious or inherited.

prism A ➤polyhedron formed by joining by straight lines corresponding points in two ➤congruent polygonal regions identically situated in parallel ➤planes. The volume of a prism is Ah, where A is the area of each polygon and h is the perpendicular distance between them. Prisms are named according to the shape of their polygonal faces; thus the familiar optical prism is a **triangular prism**.

Pro Abbr. for ➤proline.

probability A measure of the degree of certainty about the occurrence of some event on a scale from 0 (impossible) to 1 (certain). Probabilities may be assigned by symmetry (e.g. the probability of getting a head on tossing a coin is $\frac{1}{2}$), on the basis of long-term behaviour (e.g. life expectancy) or subjectively (e.g. betting). ➤➤Bayesian statistics and probability.

The probability of an event A is denoted $P(A)$ or $\mathrm{Pr}(A)$ and ➤set notation is useful to describe compound events such as 'either A or B' ($A \cup B$) or 'both A and B' ($A \cap B$) or 'not A' (A'). Apart from being a well-developed branch of mathematics and the foundation of statistics and its applications, probability theory is an essential tool in both classical and quantum physics, for example in statistical thermodynamics and in the quantum theory concept of a ➤wavefunction.

probability density function ➤continuous random variable.

probability distribution The distribution of the values of a ➤random variable given either by listing the probabilities (discrete case) or by specifying the probability density function (continuous case).

probe (gene probe) ➤DNA probe.

procaryotic ➤prokaryotic.

procedure ➤subroutine.

producer An organism that fixes energy by ➤photosynthesis to provide a basic food source for all ➤food chains. Compare ➤consumer.

producer gas A mixture consisting mainly of nitrogen, carbon monoxide and hydrogen made by passing air plus a little steam over red hot carbon.

product **1** (Math.) The result of multiplying two or more numbers, polynomials, matrices, etc. The symbol $\prod$ denotes a product thus $\prod_{i=1}^{n} a_i$ means $a_1 a_2 \ldots a_n$.

2 (Chem.) In any chemical reaction, a substance produced during the reaction. Note that all reactions are equilibria (➤equilibrium chemical) and that some reactants will be left, if only in trace quantities usually. In a chemical equation, the products are written on the right-hand side.

productivity The rate at which an individual ➤producer or all the producers in a given ➤ecosystem fix energy. Some of this energy is used to maintain the producer itself or stored as standing crop (of timber, for example), while some is consumed by animals as food. Productivity is measured in joules per square metre per year.

product rule The formula

$$\frac{d}{dx}(uv) = u\frac{dv}{dx} + v\frac{du}{dx}$$

for the ➤differentiation of the product uv of two function u and v.

progesterone A ➤steroid hormone secreted by the corpus luteum of the mammalian ➤ovary (2) and the ➤placenta, associated with the control of the ➤menstrual cycle and maintenance of the ➤uterus wall during pregnancy.

program The set of instructions a computer follows in order to achieve some desired purpose. There are many programming languages, each with a different syntax and set of instructions, in which programs may be written. The basic requirement is that the language is defined without ambiguity. It can be shown that computer languages, once above a very basic level, are all capable of the same operations (➤Turing machine), but the ease with which programs may be written, tested and changed is highly dependent on the language. There are a great many programming languages, some of which were designed for specific purposes; ➤BASIC, ➤C (4), ➤Fortran, ➤Pascal and ➤LISP are particularly well known.

progressive wave ➤travelling wave.

projection **1** (Geog.) ➤map projection.
    2 (Chem.) ➤Fischer projection; Newman projection.

projective geometry A branch of geometry concerned with those properties of geometric figures that are unaffected by projections, that is, ➤transformations (2) such as those used in perspective drawing.

prokaryotic (procaryotic) Describing a type of cell (a **prokaryote**) characterized by the lack of a distinct membrane-bound ➤nucleus (2) and ➤organelles such as ➤mitochondria. Prokaryotic cells comprise the ➤kingdom Monera, which includes the ➤bacteria. Compare ➤eukaryotic.

prolactin A gonadotrophic hormone (➤gonad) secreted by the ➤pituitary gland that promotes the secretion of ➤progesterone by the ovary and is involved in ➤lactation.

prolate spheroid A ➤spheroid whose two equal-length axes are shorter than its third axis. Compare ➤oblate spheroid.

proline (Pro) A common ➤amino acid. ➤➤Appendix table 7.

PROM ➤EPROM.

promethium Symbol Pm. The element with atomic number 61 and most stable isotope 145. Its main claim to fame is that it was the last element in the periodic table to be discovered (in 1945) among those up to uranium, the heaviest naturally occurring element. It is a typical ➤lanthanide, with most common oxidation number +3. However, its chemistry is unimportant as its abundance in nature is some 10^{15} times less than its immediate lanthanide neighbours.

prominence A cloud of material extending from the ➤Sun's chromosphere into the corona, visible as bright arches on the Sun's limb or silhouetted as dark **filaments** against the bright photosphere.

promoter **1** (Chem.) A substance that can be added to a catalyst to speed up a reaction still further. The promoter on its own does not catalyse the reaction. An example comes from the ➤Haber–Bosch process in which the iron catalyst is promoted by the oxides of potassium, calcium and aluminium.
  2 (Biol.) A sequence of ➤nucleotides on a molecule of DNA that is required to initiate ➤transcription by RNA polymerase.

proof by contradiction A mathematical technique of proof, also known by its Latin tag *reductio ad absurdum*, in which a contradiction is shown to follow if the conclusion of a proposed theorem is not true.

proof by induction ➤induction.

proof spirit A term describing a mixture of ethanol and water of a particular strength. In the olden days, proofing a spirit meant checking whether gunpowder would ignite with the spirit present: too much water and the fire goes out. This technique has given way to measuring the density, a rather safer process for customs officers. In the UK, proof spirit is 49.2% by mass of ethanol. US usage rounds this to 50% ethanol.

propagation vector ➤wave-vector.

propanal CH_3CH_2CHO The aldehyde with three carbon atoms, which exists as a colourless liquid. It can be reduced to propan-1-ol and oxidized to propanoic acid.

propane C_3H_8 The alkane with three carbon atoms, which exists as a colourless gas. In its liquid state, it is much used as a fuel.

propane-1,2,3-triol IUPAC name for ➤glycerol.

propanoic acid (propionic acid) CH_3CH_2COOH The carboxylic acid with three carbon atoms, which exists as a colourless liquid.

propan-1-ol $CH_3CH_2CH_2OH$ One of the two isomers possible when a hydroxyl group is substituted for a hydrogen in ➤propane, which exists as a colourless liquid. On oxidation, it forms ➤propanal and then ➤propanoic acid, being a primary ➤alcohol.

propan-2-ol (isopropanol) $CH_3CH(OH)CH_3$ One of the two isomers possible when a hydroxyl group is substituted for a hydrogen in ➤propane, which exists as a colourless liquid. On oxidation, it forms ►propanone, being a secondary alcohol. Propan-2-ol is much used as a solvent, for example in stylus and CD cleaners.

propanone (acetone) CH_3COCH_3 The archetypal ►ketone which is important more for its solvent properties than its chemical reactivity, and which exists as a colourless liquid. It has a large dipole moment, mainly due to the carbonyl group. However, it is not a hydrogen donor and is described as a ►dipolar aprotic solvent. Propanone is thus a useful ingredient in solvent mixtures for ►chromatography.

Chemically it can be reduced, by ►lithium tetrahydridoaluminate for example, to propan-2-ol. It resists oxidation well. It undergoes nucleophilic addition, and most especially condensation reactions, as with DNP (►2,4-dinitrophenylhydrazine). The DNP derivative is an orange solid with a sharp melting point, which makes it useful for characterization.

propene (propylene) $CH_3CH{=}CH_2$ The alkene with three carbon atoms, which is a colourless gas. It is the monomer from which polypropylene is made (see table at ►polyalkene). Many other products too are manufactured from propene.

proper motion The annual angular velocity of a star across the ►celestial sphere. Compare ►radial velocity.

proper time In ►special relativity, the time between two events as measured by an observer present at *both* events (i.e. whose ►world line passes through both points in spacetime).

prophage A ►phage ►genome integrated as part of a bacterial ►chromosome.

prophase ►meiosis; mitosis.

propionic acid Traditional name for ►propanoic acid, deriving from the fact that it was the first (lowest M_r) acid that could form fats (*pion* is the Greek for 'fat').

proportional ►variation.

proportional counter A particle counter in which the size of the output pulse is proportional to the number of particles causing the pulse.

propylene An obsolete name for propene, $CH_3CH{=}CH_2$. Its polymer is still called ►polypropylene.

propyl group The group —C_3H_7, which has two isomeric forms: the *n*-propyl group, —$CH_2CH_2CH_3$; and the **isopropyl group**, —$CH(CH_3)_2$.

prosencephalon ►brain.

prostaglandin One of a number of cyclic derivatives of ►fatty acids containing 20 carbon atoms, particularly arachidonic acid, which have a wide range of regulatory metabolic functions. They affect smooth (involuntary) ►muscle and have a number of important therapeutic applications in the treatment of, for example, high blood pressure, coronary heart disease and anaphylactic shock.

prostate gland A golf-ball-sized gland surrounding the ►urethra at the base of the bladder in male mammals. Its alkaline secretions contain ►prostaglandins, which contribute to the volume of the ➤semen and activate sperm to become motile in the vagina.

prosthetic group A group other than a ➤polypeptide that is bound to a ➤protein to form a complex. The ➤haem groups of haemoglobin are prosthetic groups. Prosthetic groups differ from other ➤cofactors in that they are more closely bound, by ➤coordinate bonding, to the protein.

protactinium Symbol Pa. The element with atomic number 91 and most stable isotope 231 often found associated with the next element in the table, uranium. Its chemistry is dominated by its radioactivity, the feature that had attracted its discoverers, Otto Hahn (1879–1968) and Lise Meitner (1878–1968).

protease An ➤enzyme capable of hydrolysing proteins (➤hydrolysis).

protecting group In attempting to synthesize a protein, for example, care must be taken that only the groups that need to react do so; others that might be affected need to be protected before reaction and then deprotected afterwards. A prime example of a protecting group concerns the carbonyl group, which is protected by conversion into a cyclic ➤ketal (see the diagram), by reaction with ➤ethane-1,2-diol.

ketone   excess ethane-1,2-diol   cyclic ketal

protecting group A ketone can be protected by formation of a cyclic ketal by reaction with excess ethane-1,2-diol and a trace of acid; the ketone can be regenerated by hydrolysis with aqueous acid.

protein A class of ➤polypeptide molecule with a wide variety of types and functions. Proteins are important components of animal diets and provide a source of ➤amino acids for building new proteins. Proteins are also important constituents of ➤plasma membranes; all ➤enzymes and some ➤hormones are proteins. Many ➤seeds (e.g. beans) store proteins, making them important constituents of vegetarian diets. ➤➤protein structure; protein synthesis.

protein blotting ➤western blotting.

protein sequencing The ability to determine the linear sequence of ➤amino acids in a protein chain. Protein sequence data have been used to determine evolutionary relationships between organisms and to investigate the functioning of ➤genes. ➤➤Sanger's reagent.

protein structure The molecular organization of a protein. Protein structure is organized at four levels. The properties of a protein ultimately depend on its **primary structure**, the sequence of ➤amino acids in the protein. The chain starts with an amino acid with a free amino group (the **N terminus**) and ends with one with a free carboxyl group (the **C terminus**). The distribution in the chain of amino acids with charged side-groups causes it to be coiled or folded into areas of helices (➤alpha helix) and pleats (➤beta pleated sheet), to give an energetically stable structure. Such helices and sheets form the **secondary structure**. The overall three-dimensional shape of the polypeptide chain is called the **tertiary structure**. Where the final functional protein consists of more than one polypeptide chain, as in ➤haemoglobin, the spatial arrangement of the individual subunits and any ➤prosthetic groups present is referred to as the **quaternary structure**.

protein synthesis The process by which the information coded in the sequence of ➤nitrogenous bases in DNA is translated into functional proteins in cells. The expression of the information encoded in DNA occurs in two main phases: the copying of a DNA sequence to make RNA (➤transcription), which is then processed (➤splicing) to give ➤messenger RNA, followed by the reading of the base sequence of this messenger RNA by ➤ribosomes to make proteins (➤translation). The process requires energy in the form of ➤ATP, a supply of amino acids, ➤transfer RNA and a variety of ➤enzymes. Many antibiotics work by selectively inhibiting bacterial but not mammalian protein synthesis.

Proterozoic See table at ➤era.

protium The isotope of hydrogen with nucleon number 1: its nucleus contains one proton and no neutrons. Compare ➤deuterium.

protoctistan A member of the ➤kingdom Protoctista (Protista), which contains all ➤eukaryotic single-celled organisms, including the group formerly known as Protozoa and the single-celled algae.

protomer One of a number of ➤polypeptide units that comprise some enzymes consisting of more than one subunit. ➤➤isoenzyme.

proton A subatomic particle with a charge opposite in sign and exactly equal in magnitude to the ➤electronic charge, a mass of 938.3 MeV/c^2 and a spin of $\frac{1}{2}$. It is a ➤nucleon, and feels the ➤strong interaction, which binds it with the other protons and ➤neutrons in the nucleus. It is one of the two most important particles in chemistry (the other being the ➤electron). The number of protons in a nucleus, the atomic number, is the essential characteristic that distinguishes one element from another. The ➤periodic table is a list of the elements in order of their number of protons. The proton's ➤quark constitution of uud explains why it is close in mass but not identical to the neutron, of quark constitution udd. The proton appears to be indefinitely stable, but the lower limit for its longevity is a current research topic of significance in cosmology.

protonation The process during which a proton, equivalent to the hydrogen ion, H$^+$, is added to a molecule. It is often a preliminary step in an acid-catalysed reaction, as, for example, in the substitution of ROH to form RCl; the protonated alcohol ROH$_2^+$ has to be formed before the nucleophile Cl$^-$ can successfully substitute.

proton (magnetic) resonance (PMR) A variant of ➤nuclear magnetic resonance that tunes into the proton's frequency.

proton microscope A microscope similar in principle to the ➤electron microscope but using a beam of protons in place of electrons.

protonmotive force ➤oxidative phosphorylation.

proton number Alternative but uncommon name for ➤atomic number.

proton–proton reaction A sequence of nuclear reactions that forms the basis of energy generation in stars up to about the mass of the Sun. The net result is the formation of one ^{4}He nucleus from four protons; the intermediate products are ^{2}H (deuterium) and ^{3}He. ⫸carbon cycle.

proton synchrotron An ⫸accelerator for protons. Unlike a conventional (electron) ⫸synchrotron, the proton synchrotron must have a very large radius; the 450 GeV Super Proton Synchrotron at ⫸CERN has a diameter of more than 2 km.

protoplasm An archaic term applied to the fundamental material comprising the structure of cells. It is now more commonly referred to as ⫸cytoplasm and ⫸nucleoplasm.

protoplast The metabolically active part of a cell, particularly a plant or bacterial cell, lying within the cell wall. In certain biotechnological techniques the cell wall is removed to expose protoplasts which can then be manipulated (fused together, for example, to form hybrid cells) in ways impossible with the walls intact.

protostar The precursor of a ⫸star. Protostars form when condensation takes place in clouds of material in the ⫸interstellar medium. ⫸stellar evolution.

Protozoa Formerly the ⫸phylum containing single-celled ⫸eukaryotic ⫸heterotrophs. These organisms are now included in the ⫸protoctistans.

provirus A stage in the infective cycle of a ⫸retrovirus in which RNA is converted into DNA before being incorporated into the host DNA.

Prussian blue A famous blue pigment made by reacting either hexacyanoferrate(II) ions with iron(III) ions or hexacyanoferrate(III) ions with iron(II) ions; *both* oxidation numbers must be present. Despite having been used for centuries, amongst other things for blueprints, its structure was not finally elucidated until the late 1970s. The structure involves iron ions at the corners of a cube, with cyanide ions acting as bridges; the deep colour is due to charge transfer across the bridges from Fe^{2+} to Fe^{3+}.

prussic acid An old name for ⫸hydrocyanic acid, HCN, named because of the vital presence of cyanide in ⫸Prussian blue.

pseudogene A copy of a ⫸gene that exists in addition to the corresponding normal functional genes. Pseudogenes are not expressed because they lack parts of their coding sequences or have alterations in their regulatory sequences. There may be between one and one thousand pseudogenes for any particular gene; they form part of the ⫸repeated DNA in the genomes. ⫸intron.

pseudohalogen A species that produces compounds similar to the ⫸halogens but without any connection to any element in Group 17 of the ⫸periodic table. Perhaps the best example is CN, which forms a dimer $(CN)_2$, a soluble sodium salt NaCN, an insoluble silver salt AgCN and an oxoanion CNO^-.

pseudo-order An ⫸order of reaction lower than the true order because of a large excess of one substance. Hydrolysis reactions often appear to be zero order with respect to water as it is in vast excess.

pseudovector ⫸axial vector.

psi Abbr. for ➤pound per square inch.

Pt Symbol for the element ➤platinum.

PTFE Abbr. for poly(tetrafluoroethene), the polymer made from C_2F_4 (see table at ➤polyalkene). It is used on nonstick frying pans. Its exceptionally low coefficient of friction has been described by saying it is like 'wet ice on wet ice'.

p-type semiconductor A semiconductor in which holes are the ➤majority carrier. Compare ➤n-type semiconductor.

Pu Symbol for the element ➤plutonium.

pubis ➤pelvic girdle.

pulmonary Of or pertaining to the ➤lungs. Thus the **pulmonary artery** carries blood to the lungs from the heart, and blood is returned to the heart via the **pulmonary vein**.

pulsar A rotating ➤neutron star that sends out pulses of electromagnetic radiation at regular intervals (typically of the order of seconds, though it can be much less: 1.6 ms is the current record).

pulse An increase, and subsequent decrease, in the magnitude of a signal or waveform to a level higher than its mean.

pulse modulation A form of ➤modulation in which the signal is encoded on the carrier in ➤digital form.

punctuated equilibrium ➤evolution.

pupa (chrysalis) The stage in the life cycle of insects showing complete ➤metamorphosis between the larva and the adult, during which larval tissues are transformed into a fully functioning sexually mature adult.

pupil ➤eye.

purine One of two classes of ➤nitrogenous base found in ➤nucleic acids which contains two heterocyclic aromatic rings (see the diagram). ➤Adenine and ➤guanine are purines and also give rise to a number of other biologically important compounds. Compare ➤pyrimidine. ➤➤ATP; DNA; RNA.

purine   adenine   guanine

purine The two most important purines are adenine and guanine.

Purkinje fibres Modified muscle fibres in the ➤heart. Named after Jan Evangelista Purkinje (1787–1869).

putrescine An aptly named diamine, 1,4-diaminobutane $H_2N(CH_2)_4NH_2$, found together with **cadaverine** in rotting flesh.

PVC Abbr. for poly(vinyl chloride), the polymer made from the monomer vinyl chloride (chloroethene); see table at ➤polyalkene. The unplasticized form, **uPVC**, is a rigid material used for making long-lasting window frames, whereas the addition of ➤plasticizers makes the polymer more pliable and this more common form is found in water pipes, 'vinyl' records and clothing.

P wave Abbr. for ➤primary wave.

PWR Abbr. for ➤pressurized water reactor.

pyramid A ➤polyhedron formed by joining by straight lines all points in a plane ➤polygonal region (the **base**) to a common point (the **apex**) not in the plane of the base (see the diagram). The volume of a pyramid is $\frac{1}{3}Ah$ where A is the area of the base and h is the perpendicular distance from the base to the apex. The ancient Egyptian pyramids are square-based pyramids; a ➤tetrahedron is a triangular-based pyramid.

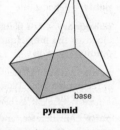

apex

base

pyramid

pyramid of numbers The numerical relation between different types of organism in a ➤food chain. Typically, organisms at a particular trophic level in the chain are more numerous than those in the level immediately above. The concept has only limited value, however, and it is preferable to express the various levels in terms of ➤biomass or ➤productivity.

pyranose A structure present in many sugars, comprising a ring with six members, five carbons and one oxygen atom (see the diagram). Compare ➤furanose.

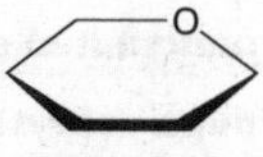

pyranose

Pyrex The trade name for a ➤borosilicate glass with a particularly low coefficient of thermal expansion and good resistance to the attack of acids and alkalis, and therefore much used in laboratory glassware.

pyridine A heterocyclic aromatic hydrocarbon (see the diagram), related to benzene by replacing one C—H group by a nitrogen atom. It is a colourless liquid used as a solvent; it is different from benzene, another good solvent, in that the nitrogen atom confers basicity on the molecule.

pyridine

pyridoxine ➤vitamin B_6.

pyrimidine One of two classes of ➤nitrogenous base found in ➤nucleic acids which contains a single heterocyclic aromatic ring (see the diagram opposite). ➤Cytosine, ➤thymine and ➤uracil are pyrimidines and also give rise to a number of other biologically important compounds. Compare ➤purine. ➤➤DNA; RNA.

pyrites (iron pyrites, pyrite) An ore of iron of formula FeS_2 which contains Fe^{2+} and S_2^{2-} ions. Its yellow colour and lustre led to its alternative name of 'fool's gold'.

pyrogallol The traditional name for 1,2,3-trihydroxybenzene, a white solid, a solution of which is used as a photographic developer.

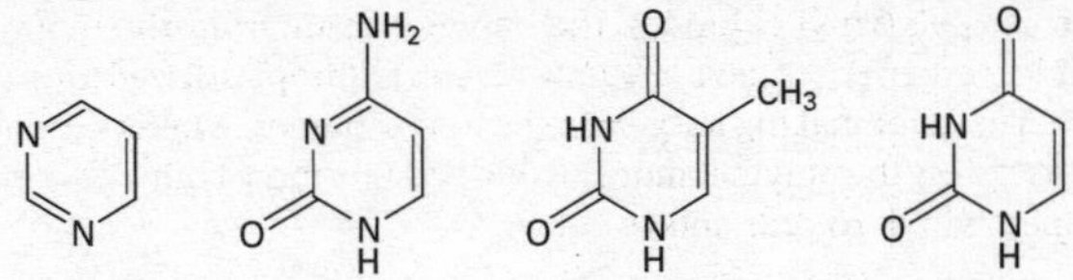

pyrimidine The three most important pyrimidines are cytosine, thymine and uracil.

pyrolusite An ore of manganese, one of the polymorphs of manganese(IV) oxide, MnO_2.

pyrolysis Breaking of a molecule by heating.

pyrometer An instrument for measuring the temperature of a very hot body. Some work by contact as high temperature ➤thermometers and ➤thermocouples; others, like the ➤optical pyrometer, by detecting radiation.

pyrophoric Spontaneously inflammable on contact with air or oxygen. The property usually arises because a metal is very finely divided, speeding up the rate of combustion greatly.

pyroxenes An important class of silicate minerals.

pyrrole A colourless oil with the structure shown in the diagram. Derivatives of pyrrole occur naturally (➤porphyrin).

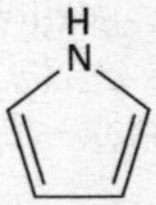

pyruvate A salt of pyruvic acid. Pyruvate is the end product in ➤glycolysis. Two ➤ATP molecules are generated in the conversion of glucose to pyruvate.

pyruvic acid $CH_3COCOOH$ The traditional name for 2-oxopropanoic acid, a colourless liquid, which as pyruvate is the product of ➤glycolysis and a precursor for the ➤Krebs cycle.

pyrrole

Pythagoras' theorem The relationship $a^2 = b^2 + c^2$ which holds between the lengths of the sides of any ➤right-angled triangle, where a is the length of the ➤hypotenuse and b and c are the lengths of the other two sides. Named after Pythagoras (6th century BC).

Q

q Symbol for ➤electric charge.

ℚ Symbol for the set of ➤rational numbers, chosen because a rational number is a ➤quotient of ➤integers.

Q_{10} (temperature coefficient) The rate of a process at a thermodynamic temperature $(T + 10\,K)$ expressed as a multiple of the rate of the process at T. For most biochemical (enzyme-medicated) reactions, Q_{10} is about 2, whereas for most physical processes (such as diffusion) it is about 1.5.

QCD Abbr. for ➤quantum chromodynamics.

QED Abbr. for ➤quantum electrodynamics.

Q factor (quality factor) In the forced ➤oscillation of a system with light ➤damping, described by the equation:

$$a\frac{d^2x}{dt^2}(t) + b\frac{dx}{dt}(t) + cx(t) = f_0\cos\omega t,$$

with solution $x(t) = x_0\cos(\omega t + \phi)$ (x_0 being a function of ω), the quality factor is defined as:

$$Q = \sqrt{ac}/b.$$

Here Q represents three important attributes of the system, as long as Q is much greater than 1.

(i) It is the ratio of the resonant frequency ω_0 to the width of the ➤resonance peak, $\Delta\omega$.
(ii) It is the ratio of the amplitude of the oscillation at resonance, $\omega = \omega_0$, to the steady state response as $\omega \to 0$.
(iii) It is the number of periods of oscillation for a decay of the amplitude to $1/e$ of the original amplitude if the oscillation is free ($f_0 = 0$).

Q-switched laser A laser in which the resonant cavity itself is suddenly improved in resonant quality (by the Q factor). The result is a rapid intense flash of light, delivered in around 10 ns. This is very useful for ➤flash photolysis experiments on very fast reactions.

quadrant **1** A quarter of a circle.
2 One of the four regions into which two ➤perpendicular lines divide a plane.

quadrat A device used in ➤ecology investigations to sample the size of a population of plants or sessile animals. The most commonly used quadrat is one metre square, but other sizes may be used depending on the organisms to be sampled. For example, in woodlands a 10 m × 10 m quadrat might be used to estimate numbers of trees, whereas barnacles on a seashore might be counted using a 25 cm × 25 cm quadrat.

quadratic A ➤polynomial of degree 2. The **quadratic equation** $ax^2 + bx + c = 0$ has two ➤roots given by the **quadratic formula**

$$\frac{-b \pm \sqrt{b^2-4ac}}{2a}.$$

quadrature 1 An old-fashioned term for the process of finding or approximating the area under a curve.

2 A description of ➤periodic quantities with a ➤phase difference of $\frac{1}{2}\pi$ (90°) between them. For example, $\sin \omega t$ and $\cos \omega t$ are in quadrature.

quadric A surface such as an ➤ellipsoid, ➤hyperboloid or ➤paraboloid described by a ➤polynomial equation of degree 2.

quadrilateral A four-sided ➤polygon.

quadrupole The result of bringing together (from a distance a) two equal and opposite ➤dipoles (with dipole moment of magnitude p) while pa remains constant.

qualitative analysis A determination of the elements present in a sample, without regard to how much of each is present. For solutions it is usual to identify metal ions by precipitating out the metal hydroxides: see the table for a simple scheme for the eight most important ions. More complicated **qual schemes** exist and use in addition the precipitation of sulfides, first in alkaline and then in acidic conditions. Modern qualitative analysis depends on spectroscopy, usually ➤atomic absorption spectroscopy, which can analyse a sample for up to 20 elements simultaneously at concentrations as low as 1 part per billion.

Observation on adding dilute NaOH (aq)	Ions that may be present
White precipitate, insoluble in an excess of aqueous sodium hydroxide	Mg^{2+} Ca^{2+}
White precipitate, soluble in an excess of aqueous sodium hydroxide	Al^{3+} Pb^{2+} Zn^{2+}
Coloured precipitate, insoluble in an excess of aqueous sodium hydroxide	Fe^{2+} (green) Fe^{3+} (red-brown) Cu^{2+} (pale blue)

quanta The plural of ➤quantum.

quantitative analysis An analytical scheme that requires measuring exactly how much of each element is present. This can be done in many ways. For example, excess of a reagent known to give a highly insoluble salt can be added and the precipitate filtered off and weighed. Alternatively the amount of some ions in solution can be measured by ➤titration.

quantization **1** The process of turning a model of a system in ➤classical physics into a model in ➤quantum mechanics. In particular, it applies to the replacement of coordinates (x_i) and momenta (p_i) by the corresponding ➤operators (2), along with the use of the ➤commutator (2) $[\hat{x}_i, \hat{p}_j] = i\hbar\delta_{ij}$ where δ_{ij} is the ➤Kronecker delta. ➤➤second quantization.

2 The method of turning an analogue signal into digital form.

quantum (plural quanta) A discrete amount that is the minimum by which a physical quantity can increase or decrease in a ➤quantum-mechanical system. It almost invariably involves the ➤Planck constant, h. Thus if a simple ➤oscillator of (classical) frequency f has energy levels $hf(n+\frac{1}{2})$, where n is an integer, the quantum of energy is hf. Similarly, the quantum of ➤orbital angular momentum is $h/2\pi$.

quantum chromodynamics (QCD) The ➤quantum field theory describing the interactions of ➤quarks under the ➤strong interaction. Like QED, QCD is a ➤gauge field theory; its ➤symmetry group is ➤SU(3), the elements of which are 3×3 matrices that act on the ➤colour charge of the quarks (hence the 'chromo' in the name).

quantum electrodynamics (QED) The ➤quantum field theory describing the behaviour of charged particles under the ➤electromagnetic interaction. QED is a ➤gauge field theory; its ➤symmetry group is U(1), the group of complex numbers with magnitude 1 under multiplication (the wavefunction of a particle may be multiplied by a complex number without changing the physics of the problem). As with any other such symmetry, this leads to conservation laws, in this case of electric charge. QED is arguably the most successful current theory of matter, providing a detailed and rigorously accurate description of all aspects of the electromagnetic interaction. For example, the calculated value of the ➤gyromagnetic ratio of the electron agrees with experimental results to eight decimal places.

quantum electronics The study and design of electronic devices that rely on ➤quantum mechanics for their operation.

quantum field theory A type of theory of particle physics (➤quantum electrodynamics and ➤quantum chromodynamics are examples) in which all entities are treated as ➤fields. Once the idea of an electron having a ➤wavefunction rather than a fixed location is accepted, together with the idea of ➤wave–particle duality of the electromagnetic field and the photon, the concept of all particles as fields (and their interactions as the interactions between the fields) follows.

quantum gravity The treatment of ➤gravity by ➤quantum mechanics. Since the gravitational interaction is so weak, quantum gravity would be observed only at scales of the order of the Planck length (➤Planck units). Thus most gravitational effects can be dealt with quite adequately using classical physics (plus ➤general relativity), but quantum gravity will be essential for understanding➤ black holes and the ➤big bang. ➤➤supergravity.

quantum jump A transition of a ➤quantum-mechanical system from one state to another, usually under an external influence (➤perturbation). Because the energy in atoms and molecules is quantized, an electron can only ever be in one of the quantum states. When it moves to another quantum state a quantum jump occurs, usually accompanied by the absorption or emission of a photon of energy exactly matching the ➤energy gap. ➤electronic transition.

quantum-mechanical system A ➤system for which ➤quantum mechanics is necessary or desirable for analysing its behaviour.

quantum mechanics The study of systems that are on such a small scale that the ideas of ➤quantum state and ➤quantization must be applied. There are two fundamental differences between quantum mechanics and ➤classical physics (e.g. ➤Hamilton's equations). First, the parameters associated with a quantum system (e.g. the position and momentum of an electron) are directly affected by the act of measuring them. Second, the parameters are not precise values but are instead expressed in terms of probability distributions (e.g. the squared modulus of the ➤wavefunction is the probability distribution for position).

quantum number In general, a number corresponding to an operator $\hat{A}$ for a quantum state $|\phi\rangle$ that is an eigenstate of $\hat{A}$. For electrons in atoms, four quantum numbers are required to define the state: the ➤principal quantum number, ➤orbital angular momentum quantum number, ➤magnetic quantum number and spin quantum number (➤spin). These numbers, invariably integral or half-integral, can be used to calculate the energy of the electron.

quantum state Symbol $|\psi\rangle$, $|\phi\rangle$ or similar. A description of all that is known about a quantum-mechanical system. **Stationary states** are based on solutions to the ➤Schrödinger equation, the eigenvalue problem $H|\phi\rangle = E|\phi\rangle$, where H is the ➤Hamiltonian of the system. Stationary states are characterized by an energy (the eigenvalue above) and other ➤quantum numbers. The state of a system can always be expressed in terms of (complex) combinations of these stationary states (➤superposition). All the states together comprise a complex ➤vector space. ➤Operators (2), of which the Hamiltonian H is an example, act on this state space as ➤matrices act on vectors, changing one state vector into another. When an ➤observable is measured, the result is an eigenvalue a_i of the ➤Hermitian operator $\hat{A}$ for that observable, and the state of the system after measurement is the eigenstate $|\phi_i\rangle$ corresponding to it. A state is an abstract concept and is more fundamental to quantum mechanics than the concept of a ➤wavefunction, which is a *representation* of a state in terms of the possible outcomes of an experiment to measure the position of a particle.

quantum statistics The generic name given to the ➤Bose–Einstein statistics and ➤Fermi–Dirac statistics for the occupation of a set of energy levels by ➤bosons and ➤fermions respectively.

quantum theory The theory underpinning ➤quantum mechanics. Before the 20th century, it had been believed that energy could be given to a particle in arbitrary amounts. Between 1900 and 1925 physicists discovered that energy was quantized: only multiples of a certain quantity of energy could be exchanged. The idea first arose in 1900 when Max Planck explained black-body radiation, via the pivotal Planck's equation $E = hf$ (➤Planck constant). This thermodynamic context was extended in 1905 when Albert Einstein used the idea of energy quanta to explain the ➤photoelectric effect. When Niels Bohr used the same idea of energy quantization to advance a tentative model for the structure of the atom in 1913, it was clear that quantum theory was all-pervasive. The second phase of development came between 1924 and 1927. The ➤Bohr model for the atom was deeply unsatisfactory because the assumptions made were untenable. An electron orbiting a nucleus would have to be accelerating (like any object in circular motion); an accelerating charge would emit electromagnetic radiation and thus lose energy. The Bohr atom would rapidly self-destruct as the orbiting electron spiralled into the nucleus. The first step towards the **new quantum theory** was ➤de Broglie's equation. Erwin Schrödinger then advanced the ➤Schrödinger equation, which was to prove a decisive step forward, as the solutions to the equation show that energy is quantized. The remaining stage was the recognition that the Schrödinger equation is not correct from a relativistic viewpoint, as it treats space and time differently. **Relativistic quantum theory** was therefore developed, first by Paul Dirac; its latest versions are termed ➤quantum field theories.

quantum yield The number of molecules of product made per photon absorbed in a photochemical reaction. In the photochemical chlorination of methane, for example, the number of molecules of chloromethane produced per photon absorbed is typically around 1000, demonstrating that the mechanism is a ➤chain reaction.

quark An elementary particle that is a constituent of a subatomic particle, such as the proton. The theory of quarks was advanced by Murray Gell-Mann (b. 1929) in 1955. Gell-Mann got the name from James Joyce's book, *Finnegan's Wake*, and pronounced it to rhyme with 'pork', not 'park'. All particles known before 1974 could be explained in terms of just three quarks (up (u), down (d) and strange (s), the

quark	symbol	approx. mass	charge/e
up	u	5 MeV/c^2	$+\frac{2}{3}$
down	d	10 MeV/c^2	$-\frac{1}{3}$
strange	s	100 MeV/c^2	$-\frac{1}{3}$
charm	c	1.5 GeV/c^2	$+\frac{2}{3}$
bottom	b	4.7 GeV/c^2	$-\frac{1}{3}$
top	t	30 GeV/c^2	$+\frac{2}{3}$

names of which should be treated simply as labels) combined in a couple of different ways. ➤Hadrons are composed of three quarks, the proton being uud and the neutron udd; ➤mesons are composed of a quark and an antiquark. All particles had a logical explanation and all expected combinations existed, with one exception. Gell-Mann's detailed prediction of the nature of the ➤omega-minus particle, subsequently discovered in 1964, vindicated the quark model. In 1974 a new particle was discovered called the ➤J/Ψ particle. Theoreticians quickly explained that this was a meson formed from a new, fourth quark named charm (c). There were thus four quarks – u, d, s and c – plus four ➤leptons (with no internal structure) the ➤electron, the ➤muon and their related ➤neutrinos. A few years later, evidence was found for another quark, called bottom (b), and another lepton, called the ➤tauon. A sixth quark, called the ➤top (t) quark, was discovered in 1995. There are good theoretical reasons for expecting a sixth lepton, the tauon neutrino; experimental proof is currently limited.

quark confinement A theory developed to explain why, despite the success of the ➤quark theory in explaining the nature of the elementary particles, no free quark has ever been seen. It argues that quarks can exist only within the matter that is made of them.

quart An Imperial unit of volume, obsolete in the UK, equal to two ➤pints; it derives its name from quarter-gallon. It is still commonly used in the US, where it equals $0.946\ 36\ dm^3$. 1 UK quart equals $1.1365\ dm^3$; 6 US quarts $\approx$ 5 UK quarts.

quarter-wave plate A plate of material exhibiting ➤birefringence designed to introduce a ➤phase shift of $\frac{1}{2}\pi$ between a component of the ➤electric field strength E of light in two perpendicular directions. Thus if ➤plane-polarized light passes through a quarter-wave plate it can either be unaffected (E parallel to one of the axes) or produce circularly polarized light (E at 45° to one of the axes) or elliptically polarized light (for any other polarization).

quartile ➤percentile.

quartz The polymorph of silicon dioxide, SiO_2, which is stable at room temperature.

quartz crystal oscillator A very stable ➤oscillator that uses a piezoelectric crystal to produce a very sharp ➤resonance. Such devices are used for time-keeping in clocks, watches and timers for video recorders. ➤➤piezoelectric effect.

quasar (QSO) An astronomical object that appears star-like on optical images, but which has a very high ➤redshift; the conclusion is that quasars are among the Universe's most distant objects. The name is a contraction of quasi-stellar object. Quasars may be powered by the accretion of matter onto massive ➤black holes at the centres of distant galaxies.

quasistatic change A change that is sufficiently slow that it may be considered as a sequence of equilibrium positions. This is a particularly useful concept in thermo-dynamics.

Quaternary See table at ➤era.

quaternary salt A salt of the form $R_4N^+X^-$ with four alkyl groups attached to nitrogen. An important example is the neurotransmitter ➤acetylcholine.

quaternary structure ➤protein structure.

quaternion A four-dimensional number system consisting of entities of the form $a_0 + a_1i + a_2j + a_3k$, where the a_r terms are real numbers and i, j, k are multiplied according to rules such as $i^2 = j^2 = k^2 = -1$, $ij = k$ and $ji = -k$. Quaternions may be used to represent three-dimensional rotations and, in the 19th century, were used for purposes similar to those of ➤vector analysis. The set of all quaternions is denoted by $\mathbb{H}$ in honour of their inventor, William Rowan Hamilton (1805–65).

quench 1 To cool very rapidly. Quenching can change the properties of materials; it hardens steel, for example.
 2 A capacitor and/or resistor placed in parallel with a circuit with a high ➤inductance to avoid the build-up of very high voltages (and the possibility of sparking) when the current ceases.

quick-freeze drying ➤lyophilization.

quicklime (lime) Traditional name for calcium oxide, CaO.

quicksilver An old name for the element ➤mercury.

quiet Sun The ➤Sun during those periods of the ➤solar cycle when there is little activity in the form of, for example, ➤sunspots or ➤solar flares.

quinine An alkaloid present in cinchona bark used to combat malaria as it can kill the malarial parasite after it has entered the red blood cells. Indian tonic water contains quinine.

quinone One of a group of small fat-soluble ➤coenzymes that function as electron carriers in ➤redox reactions in the ➤electron-transport chain. Because they are so widespread, they are usually referred to as **ubiquinones**. A common ubiquinone is ➤coenzyme Q, which occurs in several forms in ➤mitochondria and ➤microorganisms.

quotient The result of dividing two numbers or algebraic expressions. ➤➤division.

quotient rule The formula

$$\frac{d}{dx}\frac{u}{v} = \frac{1}{v^2}\left(v\frac{du}{dx} - u\frac{dv}{dx}\right)$$

for the ➤differentiation of the quotient u/v of two functions u and v.

R Symbol for an ➤alkyl group (from the word 'residue').

R **1** Symbol for the ➤gas constant.
2 Symbol for ➤resistance.

(*R*)– ➤(*R–S*) system.

ℝ Symbol for the set of ➤real numbers.

ρ **1** Symbol for ➤density.
2 Symbol for ➤charge density.
3 Symbol for ➤resistivity.

Ra Symbol for the element ➤radium.

race A group of organisms within a ➤species that are genetically or physiologically distinct from other members of the species. In anthropology the term is used to describe distinct human types such as Caucasian, Negroid and Mongoloid. **Land races** are cultivated varieties that have been produced locally by informal artificial selection by farmers (sometimes living only a few kilometres apart) of crop plants showing characters that make them suited to particular growing conditions. There are estimated to be over 120 000 land races of rice.

racemic mixture (racemate) A mixture containing equal quantities of both enantiomers (➤optical activity), of a molecule with one ➤chiral centre, that has no effect on plane-polarized light. As the two enantiomers rotate the plane equally in opposite directions, their rotations cancel.

This racemic mixture can be separated, or resolved, into the two enantiomers by several methods. The first successful **resolution** occurred when Louis Pasteur noticed that the (+)- and (−)-enantiomers of sodium ammonium tartrate had different shaped crystals when seen under a microscope; they could be picked out of the mixture. More sophisticated measures are usually needed, involving reaction with an optically active molecule which gives isomers that have sufficiently different melting points to be separable. A common compound used to resolve acids is the optically active base ➤strychnine. ➤➤chromatography.

rad Symbol for ➤radian.

radar A technique for finding the bearing and distance of a remote object by measuring the total time required for a ➤radio frequency beam to travel to the object and back; the bearing is simply the direction from which the reflection is received.

The term is an acronym for radio direction and ranging. Radar is used extensively to determine the positions of aircraft and ships, usually with a view to collision avoidance. It is also used to detect precipitation since raindrops reflect in the radio frequency very well. This basic technique is known as **primary radar**. In **secondary radar**, the object actively responds to the incident beam. For example, for air traffic control many aircraft carry transponders that return an identification code and the aircraft's altitude when interrogated by a ground-based radar.

radial Directed towards or away from some centre. Compare ➤tangential.

radial velocity (line-of-sight velocity) The component of the velocity of a celestial body along the line of sight to Earth, corrected for the Earth's orbital motion. It is measured in terms of the Doppler shift (➤Doppler effect) of lines in the body's spectrum. Compare ➤proper motion. ➤➤redshift.

radian A measure of ➤angle defined so that, in a circle of radius r, an ➤arc of length r ➤subtends an angle of 1 radian at the centre of the circle. It follows that 1 radian is $180/\pi \approx 57.296°$. Radians are of fundamental importance for ➤trigonometric functions; formulae like $d(\sin x)/dx = \cos x$ are valid only for x in radians, as are approximations such as $\tan x \approx \sin x \approx x$. The name 'radian' is a relatively recent invention, dating from 1873.

radiance Symbol L_e; unit $W\,m^{-2}$. The ➤radiant intensity emitted from a source per unit surface area. ➤Luminance is the corresponding quantity restricted to light.

radiant flux Symbol ϕ_e; unit W. The total power (over all frequencies of the electromagnetic spectrum) emitted or received by a body or surface. ➤Luminous flux is the corresponding quantity restricted to light.

radiant intensity Symbol I_e; unit W. The power (over all frequencies of the electromagnetic spectrum) emitted by a radiating point source per unit solid angle. ➤Luminous intensity is the corresponding quantity restricted to light.

radiation The process of the emission of particles (➤radioactivity) or ➤electromagnetic waves, or the particles or waves themselves. ➤➤black-body radiation; heat radiation.

radiation belts ➤Van Allen Belts.

radiation hazard ➤hazchem symbols.

radiation pressure The pressure exerted on a surface reflecting or absorbing radiation. The effect can be considered either within ➤classical physics (the incident electric field induces currents on which the magnetic field applies a force) or within ➤quantum mechanics (the incident photons transfer their momentum to the surface).

radiation temperature ➤effective temperature.

radiation units ➤becquerel; gray; sievert.

radiative capture The acquisition of a neutron by an atomic nucleus: the neutron collides with the nucleus to form an ➤excited state; the excited state decays by emitting electromagnetic radiation (a ➤gamma ray); and the neutron has insufficient energy to escape.

radiative collision Any collision in which some energy is transformed into electromagnetic radiation.

radiative transition The emission of a photon when an excited particle loses energy, as opposed to a **nonradiative transition**, caused by collision with and transfer of energy to another particle. Radiative transitions occur, for example, in atomic emission spectroscopy.

radical In mathematics, another name for ➤root.

radical (free radical) A species containing at least one unpaired electron. Perhaps the most common examples are halogen atoms created during ➤chlorination of organic compounds. During the reaction, organic radicals (such as methyl radical $CH_3^•$) are formed. The dot is a reminder that the species is a radical. Other gaseous radicals include nitrogen oxide, NO, which has an *odd* total number of electrons, and, more surprisingly, ➤oxygen, O_2. ➤➤chain reaction.

radical chain reaction ➤chain reaction.

radical polymerization Polymerization that is initiated by radicals, after which the carbon backbone grows via a ➤chain reaction. The radicals are often produced by thermal decomposition, for example of benzoyl peroxide.

radicle The part of the plant embryo that gives rise to the primary ➤root (2). During germination it usually emerges from the seed before the ➤plumule.

radio The transmission and reception of electromagnetic radiation of ➤radio frequency. Its primary application is for communication. A signal is broadcast on a radio-frequency ➤carrier wave by ➤modulation, and is demodulated at the receiver.

radio- Prefix denoting a relationship either to ➤radio or to ➤radioactivity.

radioactive decay ➤radioactivity.

radioactive waste ➤nuclear waste.

radioactivity The emission of particles or electromagnetic energy from the nucleus of an atom. It was discovered by Henri Becquerel in 1896, and many new discoveries, such as the elements radium and polonium, quickly followed. Three types of **radioactive decay** occur: ➤alpha (α) decay, ➤beta (β) decay and ➤gamma (γ) decay. Radioactive decay is a first-order process (➤order of reaction); as such the decay is exponential, so a ➤half-life can be tabulated for each radioisotope. When the decay product itself decays, and so on, the sequence of nuclides is known as a **radioactive series**. There are several such series in which a succession of decays occurs until a stable nuclide is formed, typically one of the isotopes of lead.

radio astronomy The study of radio-frequency emissions from astronomical objects, such as stars or ➤radio galaxies. ➤Radio telescopes are used for the observations, which extend to the sub-millimetre range (➤microwaves) as well as radio frequencies as normally defined. A low ➤signal-to-noise ratio is inevitable in such observations, and substantial effort goes into the processing of experimental results.

radiocarbon dating A technique for dating archaeological artefacts of organic origin (e.g. wood) that uses the fact that radioactive decay is an exponential process with a known ➤half-life: 5730 years for the nuclide ^{14}C, **radiocarbon**. The ratio of radiocarbon present in an artefact is measured using a mass spectrometer. Assuming that the ^{14}C/^{12}C ratio in nature has fluctuated little over the last few millennia, and that a living organism exchanges its carbon often enough to preserve the ^{14}C/^{12}C ratio, the time since death of the organism can be evaluated. This assumption has been shown to be slightly incorrect by comparison with ➤dendrochronology, but the radiocarbon date is close enough for useful estimates to be obtained.

radio frequency (r.f.) The range of frequencies of electromagnetic radiation below 300 GHz (wavelength greater than 1 mm). ➤frequency band.

radio-frequency heating ➤induction heating.

radio galaxy A galaxy that emits strongly at ➤radio frequencies. Radio galaxies are typically giant ➤elliptical galaxies.

radio interferometer ➤radio telescope.

radioisotope An isotope that is radioactive. Radioisotopes can be used for ➤labelling atoms so that their progress through a reaction pathway can be followed. High levels of ^{90}Sr in milk caused concern during the early 1960s and resulted in a successful Test Ban Treaty in 1963 banning the explosion of nuclear weapons in the atmosphere.

radioluminescence ➤Fluorescence that occurs after radioactive decay (➤radioactivity).

radiolysis The breakdown of a chemical substance caused by the impact of radiation.

radionuclide A ➤radioisotope with a specified ➤nucleon number.

radio receiver A device for receiving and demodulating a ➤radio signal. Typically, the receiver turns the signal into sound. ➤➤heterodyne receiver; superheterodyne receiver.

radio source Any astronomical object that emits radio-frequency radiation. They include ➤radio galaxies and ➤quasars.

radio telegraphy Communication using ➤radio transmissions that does not carry a voice as a signal, but uses some encoding format for messages.

radio telephony Communication using ➤radio transmissions in which a voice is carried as a signal by ➤modulation. The high-frequency band and higher bands are used for this purpose.

radio telescope An instrument designed to detect and observe radio sources. Since radio frequencies correspond to large wavelengths, large reflectors in the form of ➤paraboloids are used as aerials to achieve good ➤resolution. Alternatively, an array of aerials can be used (a **radio interferometer**), joined to the same receiver in such a way that the phase of the signal is preserved, and the image can then be reconstructed.

radiotherapy The use of a ➤radioisotope in the treatment of ➤cancer.

radium Symbol Ra. The element with atomic number 88 and most stable isotope 226, which is the last member of Group 2 of the ➤periodic table. Like the other members of the group, its dominant oxidation number is +2 as in its most famous compound radium sulfate, $RaSO_4$. This can be coprecipitated with barium sulfate, the small difference in solubility being utilized during fractional crystallization by Pierre and Marie Curie in their separation of radium from the ore **pitchblende**. Radium's dominant characteristic is its radioactivity, as its name implies. This can be put to good use as the radiation can destroy cancers in **radium therapy**.

radius (plural **radii) 1** (Math.) The line segment joining the centre of a circle to any point on it, or the length of such a line.
   2 (Biol.) One of the two bones of the mammalian forearm. In humans it is the bone on the side of the thumb when the palm is turned upwards.
   3 (Phys., Chem.) The 'size' of an atom or ion. The radius is commonly tabulated, but the values need to be treated with care. For example, an atom is not a rigid sphere and can respond slightly to the presence of neighbouring atoms, so atomic radius should be likened to a tennis ball's radius rather than that of a cricket ball. The ➤electron density of an atom or ion falls off exponentially, and it is impossible to specify where the electron density stops.

radius of curvature ➤curvature.

radius of gyration Symbol k or R_0. For a ➤rigid body of mass M, rotating about a ➤principal axis, the radius at which a point mass M would have to be placed to have the same ➤moment of inertia, I. Thus the radius of gyration, given by $k = \sqrt{(I/M)}$, depends on the body's shape, not its density.

radius vector The vector from an origin to a point on a curve.

radix The ➤base (2) of a number system.

radon (radium emanation) Symbol Rn. The element with atomic number 86 and most stable isotope 222, which is the noble gas with the highest mass. This radioactive gas is emitted from the decay of radioactive elements such as radium and uranium. It accumulates in certain rocks, particularly the granite of south-west England. As with xenon, a few compounds are known such as the fluoride RnF_2, but its chemistry is dominated by its radioactivity.

raffinose A trisaccharide composed of glucose, fructose and galactose.

rainbow An arc of spectral colours in the sky, seen when sunlight illuminates small raindrops. The light undergoes internal reflection within the droplet, and refraction at the surface of the droplet splits the colours (see the diagram overleaf). If the light is reflected twice within the droplet, the light forms a **secondary rainbow** with the order of colours reversed.

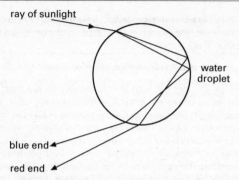

ray of sunlight

water droplet

blue end

red end

rainbow Formation of a primary rainbow by one internal reflection.

r.a.m. Abbr. for ➤relative atomic mass.

RAM Acronym for random access memory. A type of computer memory used for temporary storage of programs or data; the contents of the memory are lost if the power is turned off. Data in RAM can be accessed in any order and the length of time taken to access a single item of data, for reading or writing, is independent of previous accesses, hence the label 'random access'. Compare ➤ROM.

Raman effect A scattering process in which the absorption of a photon is followed by the emission of another photon. Usually viewed at right angles to the incident beam, this scattering (the **Raman spectrum**) gives information about molecular vibrations, since the frequency of the two photons differs by a vibrational frequency of the molecule investigated. The Raman spectrum complements the data available from infrared spectroscopy. **Laser Raman spectroscopy** is particularly useful because of the high energy and tunability of a laser. Named after Chandrasekhara Venkata Raman (1888–1970).

random number A number chosen in such a way that the overall frequency distribution of such numbers is predictable, but each individual number is independent of those that have gone before and is therefore not predictable. Producing random numbers is impossible in practice. When produced by a computer algorithm, they necessarily follow from their predecessors; such **pseudorandom numbers** are adequate for most practical purposes.

random sample A sample from a ➤population (2) chosen freely so that every element has an equal chance of being selected. In any practical situation, considerable care must be taken over sampling strategies to try to ensure a representative sample free from bias.

random variable A function assigning a number to each outcome of a set of possible outcomes. Random variables may be **discrete**, as in the score on a die, or **continuous**, as in the height of a teenager.

random walk A path consisting of individual steps, each of which arises from some random choice which is independent of previous steps. ➤Brownian motion is a classic example of a random walk.

Raney nickel A highly porous form of nickel, made by heating a nickel/aluminium alloy with aqueous sodium hydroxide, useful for catalysing hydrogenation reactions. Named after Murray Raney (1885–1966).

range 1 The set of all values taken by a ➤function.
 2 The difference between the largest and the smallest items in a ➤random sample, which serves as a crude measure of the spread of values in the sample.

rank 1 The maximum number of linearly independent rows in a ➤matrix.
 2 ➤tensor.

Raoult's law The vapour pressure above a mixture of two liquids varies linearly with the composition of the mixture. Two very similar liquids, such as benzene and methylbenzene or pentane and hexane, have a vapour pressure variation (see the diagram) given by

$$p = x_A p_A^* + x_B p_B^*,$$

where x_A and x_B are the ➤mole fractions of the two components A and B, and p_A^* and p_B^* are the vapour pressures of the pure liquids. For many mixtures, the forces within the liquids are very different and deviations occur from Raoult's law (➤azeotrope). Named after François Marie Raoult (1830–1901).

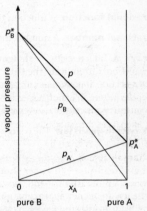

Raoult's law for constant temperature.

rare earth Alternative name for a ➤lanthanide. They are actually not all that 'rare', as most exceed iodine's abundance in nature.

rare gas An obsolete name for a ➤noble gas. 'Rare' is a particularly inappropriate adjective for argon, which constitutes 1% of air, and is therefore thirty times more abundant than carbon dioxide.

Ras One of a number of ➤genes responsible for several types of human cancer. The term is derived from 'rat sarcoma', since the genes also occur in rodents. The gene H-*Ras*, for example, is located on chromosome 11 and causes bladder cancer, while K-*Ras* is on chromosome 12 and causes cancer of the lung and colon. ➤➤oncogene.

rate constant (rate coefficient, velocity constant) When the rate of a reaction can be expressed as a **rate equation** of the form

$$\text{rate} = k[A]^m[B]^n,$$

the rate constant k is the constant of proportionality. Note that the dimensions of k depend on the ➤order of reaction. ➤➤Arrhenius equation.

rate-limiting step (rate-determining step) In a multistep reaction, the slowest step; it will be this slow step that limits the overall rate of the reaction. For an example, see ➤mechanism.

rate of change A derivative (➤differentiation), especially with respect to time. For example, velocity is the rate of change of displacement with respect to time.

rate of reaction The rate of change of concentration as a function of time:

rate = d[P]/dt,

where [P] is the concentration of a product P and t is the time. Studying the rate of a reaction is an important step in understanding its mechanism. ➤chemical kinetics.

ratio The dimensionless ➤quotient of two quantities written $a{:}b$ or a/b so as to highlight their relative sizes.

rational function A function that is the ➤quotient of two ➤polynomials.

rational number A number that is the ➤quotient of two ➤integers.

ray A line (usually a straight line) constructed perpendicular to successive wavefronts to indicate the path of travel of light. Rays undergo ➤refraction and ➤reflection at boundaries in a straightforward way, so they are useful in geometrical optics where it is not necessary to resort to a full wave model of the light. Ray constructions cannot, however, explain ➤diffraction.

Rayleigh–Jeans law The energy density ε of ➤black-body radiation as a function of frequency f is given by

dε/d$f = 8\pi f^2 kT/c^3$

where k is the ➤Boltzmann constant and T the ➤thermodynamic temperature. It is the low-frequency approximation to ➤Planck's law of radiation. Named after John William Strutt, Lord Rayleigh (1842–1919) and James Hopwood Jeans (1877–1946).

Rayleigh scattering ➤Scattering of light (or other electromagnetic radiation) from particles that are smaller than the ➤wavelength of the scattered radiation. The electric field may therefore be treated as uniform across the particle for the purposes of calculating the results of scattering. The scattering intensity is proportional to the fourth power of the frequency, so blue light scatters about ten times as much as red. This makes the sky (scattering light towards us from small particles) blue, but sunsets (where relatively unscattered light is observed) red. Named after Lord Rayleigh. ➤Tyndall effect.

rayon An artificial cellulose fibre made from either ➤viscose or ➤cellulose ethanoate.

ray tracing **1** (Comput.) A method for producing highly realistic computer images. The success of ray tracing lies in the fact that it simulates the way images find their way into the eye in nature, though often in reverse: the paths of imaginary light rays are followed from the viewer's eye, through the computer screen and into the scene to be drawn. However, this method requires intensive calculations so it is notoriously slow.

2 (Phys.) In ➤geometric optics, the analysis of an optical system by constructing the paths of individual rays through it.

Rb Symbol for the element ➤rubidium.

rDNA Abbr. for ribosomal DNA, that is the DNA encoding ribosomal ➤RNA.

Re Symbol for the element ➤rhenium.

Re Symbol for ➤Reynolds number.

reactance Symbol X. The ➤imaginary part of an ➤impedance, such as capacitance or inductance. A purely **reactive load** has no resistance, but it acts as a capacitor or an inductor.

reactant One of the species present at the start of a chemical reaction. In chemical equations the reactant(s) is/are always written on the left-hand side of the equation.

reaction The force exerted by body B on body A in response to a force applied by body A on body B (as in the third of ➤Newton's laws of motion). In particular, if an object exerts a force on a mechanical component (e.g. a beam or surface) that supports it, the component exerts an equal and opposite force, known as the reaction, on the object.

reaction, chemical A reaction in which ➤chemical change takes place.

reaction quotient For a reaction such as $a\mathrm{A} + b\mathrm{B} \rightleftharpoons c\mathrm{C} + d\mathrm{D}$, the reaction quotient Q is defined as the ratio of the concentrations of the substances present, raised to their ➤stoichiometric coefficients:

$$Q = [\mathrm{C}]^c[\mathrm{D}]^d/[\mathrm{A}]^a[\mathrm{B}]^b.$$

It features in the ➤Nernst equation. When the concentrations are those at ➤equilibrium, the reaction quotient becomes the ➤equilibrium constant.

reactivity series An alternative name for the ➤electrochemical series.

reactor ➤nuclear reactor.

reading frame One of the ways in which a sequence of ➤nucleotides may be grouped into ➤codons. An RNA sequence (or single-stranded DNA sequence) has three reading frames, each offset from the other two by one nucleotide, while a double-stranded DNA sequence has six reading frames (three on each strand). When ➤messenger RNA is translated into protein, the ➤ribosome must read the sequence in the correct reading frame (that which contains the codons encoding the amino acids of the protein chain). This is achieved by the presence of an initiation codon (AUG) that signals a ribosome to start translation, in the frame of the initiation codon. **Frameshift mutations** are deletions or insertions of a number of nucleotides not equal to a multiple of three, since these cause the correct reading frame to change, so further translation of a frameshifted mRNA results in the incorporation of incorrect amino acids, and usually a change in length of the resulting protein. ➤➤open reading frame; protein synthesis.

reagent A substance used to react with another. Sometimes the term is used synonymously with 'reactant', but it can also specify a more general behaviour, as for example when citing ➤lithium tetrahydridoaluminate as a reagent used to reduce compounds containing the carbonyl group.

real axis The x axis in the ➤Argand diagram.

real gas A gas that does not obey the ➤ideal gas equation, because the interaction between the particles of the gas is significant.

real image ➤image.

real number Informally, any ➤rational or ➤irrational number corresponding to a point on a continuous number scale. Any formal definition involves taking the limit of sequences of rational numbers. Every real number has an expansion as an infinite decimal, with rational numbers necessarily having a recurring decimal expansion. ➤➤countable set; ➤➤dense.

real part Symbol Re. The part x in the ➤complex number $z = x + iy$.

real-time processing The process of performing calculations with a computer, on data sampled from a physical system, sufficiently quickly that the results are output in time to have the required effect on the system. A typical application of real-time processing can be found in the systems vital to a 'fly by wire' aeroplane.

rearrangement A reaction in which a molecule retains the same molecular formula but changes structure, so that an isomer of the compound is formed. Examples are widespread, from the isomerization to more highly branched hydrocarbons that occurs in ➤cracking, to specific rearrangements that make useful changes to the functional groups, as in the ➤Beckmann rearrangement.

receptor 1 In the ➤nervous system, any cell or organ that can receive a ➤stimulus and generate a ➤nerve impulse. The eye, ear and nerve endings in the skin are receptors.
  2 In the ➤immune system, a ➤protein molecule on the surface of a cell that enables the cell to recognize ➤antigens.
  3 Any macromolecule to which another molecule binds leading to a physiological response.

recessive Describing an allele that does not alter the ➤phenotype if it is ➤heterozygous with a ➤dominant form of the same gene. For a recessive allele to be expressed in the phenotype, it must be present in the ➤homozygous condition.

reciprocal The reciprocal of a nonzero real or complex number x is the number $1/x$.

reciprocal lattice A ➤Bravais lattice with ➤lattice vectors (**reciprocal lattice vectors**) K formed from another lattice (the **direct lattice**) such that $e^{iK.R} = 1$ where R is any direct lattice vector. The ➤primitive vectors of the reciprocal lattice are formed by constructing a vector perpendicular to each pair of primitive vectors of the direct lattice in turn. Thus there is a reciprocal lattice vector perpendicular to every *plane* of the direct lattice. For example, a ➤body-centred cubic lattice has a reciprocal lattice that is ➤face-centred cubic, and vice versa. ➤➤von Laue condition.

reciprocal proportions, law of (equivalent proportions, law of) The proportions in which two elements separately combine with the same mass of a third element are also the proportions in which the first two elements combine together.

recombinant DNA ➤DNA.

recombination 1 (Biol.) In genetics, the formation of new combinations of ➤genes during ➤gamete formation by ➤meiosis. Recombination is significant since it can increase genetic diversity in a population providing potential for evolution. Recombination can also occur in ➤somatic cells, for example by the action of ➤transposons.

2 (Phys.) The annihilation of an ➤electron–hole pair in a ➤semiconductor. In effect, the electron fills the hole.

recrystallization A very common last step in the synthesis of an organic solid: the last traces of impurities are removed by finding a suitable solvent, dissolving the substance (plus impurities) and then evaporating off most of the solvent. The substance required crystallizes while impurities stay in solution as the solvent can dissolve the small quantity of impurity. The purity is much higher afterwards, although the yield will have fallen as some of the desired substance also remains in solution.

rectangle A ➤quadrilateral with four interior angles of 90°.

rectangular coordinates Another name for ➤Cartesian coordinates.

rectified spirit Alternative name for the azeotropic mixture of ethanol and water (➤azeotrope).

rectifier An electrical device that converts an ➤alternating current to a ➤direct current. It usually employs ➤diodes. ➤➤full-wave rectifier; half-wave rectifier.

rectify To find the length of a curve, usually by ➤integration.

rectilinear In a straight line.

recurrence relation A way of specifying the terms of a sequence by giving the first terms of the sequence and a formula giving the general connection between the terms of the sequence. For example, the ➤Fibonacci numbers may be defined by the recurrence relation $u_1 = 0$, $u_2 = 1$ and $u_{n+1} = u_n + u_{n-1}$.

recurring decimal A number with a decimal expansion that eventually consists of repetitions of a fixed block of digits.

recursive procedure (recursive subroutine) A ➤subroutine in a computer program that calls itself. This allows concise and elegant algorithms to be implemented, but they may be memory intensive.

red blood cell ➤erythrocyte.

red giant A ➤giant star with a surface temperature of around 4000 K. They are over ten times the Sun's mass, and a hundred or more times as luminous. Many red giants are ➤variable stars. ➤➤Hertzsprung–Russell diagram.

red lead An interesting mixed oxide with lead having both its main oxidation numbers; the formula of Pb_3O_4 is best thought of as an oxide ion array with two Pb^{2+} ions for every Pb^{4+} ion. It is a beautiful red solid used to make leaded glass and as an anticorrosive primer for steel.

redox reaction Abbr. for reduction–oxidation reaction, that is, one in which both ➤oxidation and ➤reduction occur. Thus redox reactions involve electron transfer: oxidation is loss of electrons and reduction is gain of electrons. This concept, together with the ➤Lewis theory of acids and bases, covers virtually all chemical reactions. Redox reactions in aqueous solution can be quantified using ➤standard electrode potentials.

redshift **1** A Doppler shift (➤Doppler effect) of light to a lower frequency, characteristic of objects moving away from the observer at high speed. The redshift of well-known lines in atomic spectra (➤spectroscopy) can be used to calculate the speed of recession of distant galaxies. ➤➤Hubble constant.

 2 ➤Einstein shift.

reduced instruction set computer ➤RISC processor.

reduced mass A system consisting of two bodies or particles in motion around each other can be analysed as motion around their centre of gravity, with the mass m of the lighter one converted to a reduced mass $\mu = mM/(m + M)$, where M is the mass of the heavier one. An important application is in the analysis of the hydrogen atom spectrum, where a change of about 1 part in 1000 is made to the mass, producing exact agreement with experiment.

reductant (reducing agent) The substance that performs the reduction in a ➤redox reaction. The reductant is itself oxidized. Common examples of reductants are the gases hydrogen, H_2, and carbon monoxide, CO, the acidified solutions potassium iodide, $KI(aq)$, and tin(II) chloride, $SnCl_2(aq)$, and the solid carbon, C.

reductio ad absurdum A Latin tag for ➤proof by contradiction.

reduction Gain of electrons. The definition of reduction has been steadily refined over the centuries, paralleling the clarification of its opposite, ➤oxidation. Hence when oxygen reacts with metals, it formally gains electrons, to become O^{2-} if the compound is ionic. Having gained electrons, it has been reduced. Other simple rules for recognizing reductions are as follows (all are opposite to the rules for oxidation): gain of hydrogen, as when Cl_2 is reduced to HCl; loss of oxygen, as when Fe_2O_3 is reduced to Fe; and decrease in oxidation number, as when Fe^{3+} is reduced to Fe^{2+}.

reference electrode An electrode against which ➤electrode potentials are measured. The primary reference electrode is the ➤standard hydrogen electrode. In practice, other secondary references such as a ➤calomel electrode are used, as they are less inconvenient to set up.

reference state (of an element) The most stable form of a pure element at the standard pressure of 1 bar. For carbon, for example, the reference state at 298 K is graphite.

refining The term used to describe a variety of processes that occur at an oil refinery, from ➤fractional distillation of the crude oil to ➤cracking and ➤reforming.

reflectance Symbol ρ. The ratio of the flux reflected by a surface to the total incident flux. ➤➤transmittance.

reflecting telescope ➤telescope.

reflection 1 (Phys.) The deflection of a ➤ray of light when it reaches a surface that is the boundary between two different media, for example glass and air. The component of the ray perpendicular to the surface reverses direction; the parallel component is unaffected. The **reflected ray** does not penetrate the second medium (compare ➤refraction). The **angle of reflection** is the angle between the ➤normal (2) to a reflecting surface and the reflected ray. The reflected ray, the normal to the surface and the incident ray all lie in the same plane and the reflected ray makes the same angle with the normal as the incident ray (see the diagram). ➤total internal reflection.

2 (Math.) The geometrical ➤transformation (2) defined relative to a mirror plane that maps every point along the perpendicular to the mirror to a point equidistant from the mirror on the other side of it.

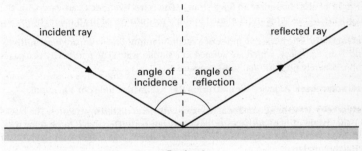

incident ray   reflected ray

angle of | angle of
incidence | reflection

reflection 1

reflection nebula ➤nebula.

reflex An innate and automatic type of nervous reaction characterized by its rapidity. Most reflexes are of survival significance, such as the blinking of the eyelid or the withdrawal of a hand from a hot object, but are also involved with posture, as in the knee-jerk reflex. The rapidity of reflexes arises because the ➤nerve impulses involved pass through a minimum number of nerve cells – usually three (two in the knee jerk), forming a **reflex arc**. The reflex arc links a sensory nerve fibre, via a relay fibre in the spinal cord, to a motor nerve fibre which initiates the response. The action thus does not require conscious thought. ➤nervous system.

reflex angle An angle that is between $180°$ and $360°$.

reflux A term used to describe the orientation of a ➤Liebig condenser when it is arranged so that any vapour that escapes goes vertically upwards where it is then condensed, runs back into the flask and continues to react. For example, when ethanol is oxidized, two possible products can be formed, ethanal and ethanoic acid. As ethanal is itself oxidized to ethanoic acid, the latter can be made as the exclusive product under reflux.

reforming A process in which the molecular structure of a mixture of hydrocarbons is rearranged. Isomers are produced, frequently with more branched chains, which are better fuels. A catalyst is required; if this is platinum the process is nicknamed **platforming**.

refracting telescope ➤telescope.

refraction The change in direction of a ➤ray of light as it crosses the boundary between two media of different ➤refractive index, for example glass and air (see the diagram at ➤Snell's law). The **refracted ray** is the ray that penetrates the second medium (compare ➤reflection). The **angle of refraction** is the angle between the ➤normal (2) to a refracting surface (not the incident ray) and the refracted ray. The refracted ray, the normal to the surface and the incident ray all lie in the same plane, and the angles between them obey Snell's law.

refractive index Symbol n. For a transparent medium, the ratio of the ➤speed of light in a vacuum to the speed of light in that medium. The ratio of the refractive indices of two media with a common boundary determines the angle of ➤refraction of light at that boundary (➤Snell's law). The refractive index can vary with the frequency of the light, and is related to the ➤permittivity of the medium by $n = \sqrt{\varepsilon_r}$.

refractivity The refractive index of a medium minus one (compare ➤permittivity and ➤susceptibility, which are defined in a similar way). The refractivity of a gas is approximately proportional to its pressure.

refractometer A device for measuring the ➤refractive index of a medium.

refractory Describing a material that can withstand high temperatures. The bricks on the inside of blast furnaces or basic oxygen converters have to be refractory. Calcined ➤dolomite, which is a mixture of calcium and magnesium oxides, is a refractory material.

refrigerant A substance that can be used in an apparatus to keep other materials cold. In systems open to the atmosphere, liquid nitrogen (above 77 K) or liquid helium (above 4 K) can be used, as both gases are safe for humans. In systems that are closed (e.g. a refrigerator) more harmful chemicals can be used, such as ammonia or, until recently, ➤CFCs.

regenerator A ➤heat exchanger that reduces the heat loss from a furnace by using hot exhaust gases to preheat the cold air drawn into the furnace.

register A part of the ➤CPU of a computer where data are temporarily held for processing. Registers can have special purposes, such as storing the memory location of the instruction that is currently being processed (**address register**), or they may be general purpose, for use by programs (**data register**).

regression A statistical technique used for prediction in which the expected values of a dependent random variable Y are modelled as a function of one or more **explanatory variables** $x_1, \ldots, x_n$. An important example is **linear regression**, where:

$$E(Y| x_1, \ldots, x_n) = \beta_0 + \beta_1 x_1 + \cdots + \beta_n x_n.$$

relative Indicating a quantity expressed as a ratio to a standard quantity. Compare ➤absolute.

relative aperture The ratio of the effective ➤aperture of a lens or mirror to its ➤focal length.

relative atomic mass (r.a.m., atomic weight) Symbol A_r. The average mass of an atom, of a specified isotopic composition, compared with the mass of one atom of carbon–12 (^{12}C) taken as exactly 12. If no isotopic composition is specified, the natural distribution of isotopes is assumed. This is the quantity needed for calculations involving the masses of substances that react together (➤molar mass).

relative density ➤density.

relative error The error in a given value, expressed as the ratio of the actual error to the correct value.

relative formula mass A more precise, albeit much less commonly used, term for ➤relative molecular mass, as it strictly applies to ionic compounds as well as molecules.

relative humidity The ratio of the mass per unit volume of water vapour to the mass per unit volume of water vapour that would be found in saturated air. Alternatively it may be defined as the ratio of the vapour pressure of water in the atmosphere to the ➤saturated vapour pressure at that temperature; in practice the difference between the two definitions is small. Relative humidity is normally expressed as a percentage. Compare ➤absolute humidity.

relative molecular mass (r.m.m., molecular weight) Symbol M_r. The equivalent for molecules of the ➤relative atomic mass and defined in a similar way. The relative molecular mass is the sum of the relative atomic masses of the constituent elements:

$$M_r(CO_2) = A_r(C) + 2A_r(O) = 12.01 + 2 \times 16.00 = 44.01.$$

Despite being technically incorrect, the term is commonly used for ionic compounds as well, in place of the more precise term ➤relative formula mass.

relative permeability ➤magnetic permeability.

relative permittivity ➤permittivity.

relativistic Pertaining to ➤relativity. The word is often used to describe effects that are not predicted by ➤Galilean transformations or ➤Newtonian mechanics.

relativistic mass In ➤relativity, the ➤energy of a particle divided by c^2 (where c is the ➤speed of light in a vacuum). For a particle at rest it is the ➤rest mass. Modern practice is to use 'natural units' in which $c = 1$ and mass and energy then have the same dimension. The term 'mass' on its own is best used to mean rest mass explicitly. ➤➤energy–momentum.

relativity The area of physics that deals with the behaviour of objects at high speed or in high gravitational fields. ➤general relativity; special relativity.

relay A device that allows one circuit to control another circuit without there being an electrical connection between them. This allows a relatively low-powered circuit to control a circuit with a high current. An **electromechanical relay** uses the magnetic force of a current to close a switch. A **solid state relay** uses ➤semiconductor devices to perform the same function, and has the advantages of being quiet and having no moving parts.

reluctance (magnetic resistance) Symbol R; unit H^{-1}. The ➤magnetomotive force required to drive unit ➤magnetic flux in a magnetic circuit. It is analogous to electrical resistance.

remainder That number or polynomial remaining when one number or polynomial is not exactly divisible by another.

remanence The ➤magnetic flux density B that remains after a ➤hysteresis cycle when the applied ➤magnetic field strength H has been reduced to zero. ➤Ferromagnetic materials tend to be **remanent**, enabling them to be used as permanent magnets.

renewable energy source A source of energy that can be regenerated and hence can last indefinitely. Examples include solar energy, wave energy and wind energy. Compare ➤fossil fuel.

renormalization An important feature of ➤quantum field theory. The electron, photon and electric charge are redefined with quantum corrections to avoid infinite series in calculations of the ➤interactions between them. In essence, it is a mathematical demonstration that particles and their interactions in such a theory are consistent with quantum mechanics.

repeated DNA Repeated sequences of ➤nucleotides found in eukaryotic DNA. The repeats may be joined together 'head to tail' (e.g. in ➤satellite DNA) or may be dispersed through the genome (e.g. ➤pseudogene). Some repeated sequences encode ➤RNA needed in large quantities, but others have no known function and may have been generated by ➤transposon activity. ➤➤selfish gene.

repeating decimal Another name for ➤recurring decimal.

replication ➤DNA replication.

repressor A protein that can bind with DNA to block its ➤transcription. Repressors and the genes that encode them are important in the regulation of gene expression in cells. ➤➤operon.

reproduction The characteristic feature of all organisms that allows them to produce new individuals. ➤asexual reproduction; sexual reproduction.

reptile A member of the vertebrate class **Reptilia**, whose members have a scaly skin and typically lay amniotic eggs with a leathery shell. They are an ancient and very diverse group and dominated the earth's fauna until the end of the Triassic ➤era. Modern reptiles include snakes, lizards, crocodilians, turtles and terrapins.

residue 1 (Math.) A ➤complex number assigned to a ➤pole of a complex function at $z = a$ that equals $\lim_{z \to a}(z - a)\, f(z)$ if this limit exists, and zero otherwise. ➤residue theorem.
   2 (Chem.) ➤filtrate.

residue theorem The integral around a simple closed curve of a ➤function that is ➤analytic except at some ➤poles is equal to $2\pi i$ times the sum of the residues at the poles within the curve. The residue theorem is a powerful theorem because it reduces the evaluation of integrals to the simpler task of evaluating residues.

resin A high molar mass material that can be either natural, such as shellac, or artificially produced, such as ➤phenol–formaldehyde resins.

resistance Symbol R; unit ➤ohm, Ω. The ➤potential difference required across a conductor to drive unit current through it. The resistance of a conductor is proportional to its ➤resistivity and length but inversely proportional to its cross-sectional area; it may also depend on temperature. ➤Ohm's law.

resistance thermometer A thermometer that measures the resistance of a conductor, and uses this to determine its temperature.

resistivity Symbol ρ; unit Ω m. The ➤resistance of a material of unit cross-sectional area and unit length. Thus $R = \rho l / A$, where R is the resistance of a conductor, l its length and A its cross-sectional area. ➤➤conductivity.

resistor An electrical component designed to provide a known ➤resistance.

resolution 1 (Phys.) The ability to distinguish between two sources, either spatially or in a spectrum. Quantitatively it is a measure of the ability of an instrument to do this. The resolution of a spectrometer is the separation in frequency (or energy) of two lines that can just be distinguished from each other. The resolution of a telescope is the angle between two sources that can just be distinguished; typically, the resolution increases with the telescope's aperture.
2 (Math.) The process of writing a given vector as the sum of its component vectors in each of a fixed set of perpendicular directions.
3 (Chem.) ➤racemic mixture.

resolving power Alternative name for ➤resolution.

resonance 1 A large-amplitude oscillation of a system in response to a small driving force (➤forced oscillation). The **resonant frequency** is the frequency at which the ratio of the response to the driving force is maximized, and is close to the ➤natural frequency. The frequency response around resonance is typically in the form of a ➤Lorentzian function. Other phenomena with a similar dependence on frequency or energy (usually resulting from some sort of ➤exponential decay) are also given the name 'resonance', including short-lived states of high-energy particles.
2 (Chem.) An obsolete and unhelpful technique in bonding theory in which the structure of some molecules is a **resonance hybrid** between two or more **canonical structures**. Benzene was regarded as a hybrid of two structures, in both of which there were alternating double and single bonds (➤Kekulé structure). Neither structure was correct, and the resonance hybrid was imagined to be somewhere between the two; even the name was unfortunate as it wrongly implied that the actual molecule 'resonated' between the two structures. The failure of resonance theory lay in its insistence on treating all bonds as two-centre bonds, so ignoring the reality of ➤delocalization.

resonant circuit A circuit that includes inductors and capacitors with the purpose of creating resonance by ➤forced oscillation.

respiration The series of chemical reactions characteristic of organisms that enables them to convert the energy stored in organic food materials into energy which can be used by cells. ➤aerobic respiration; anaerobic respiration.

respiratory chain ➤electron-transport chain.

respiratory quotient (RQ) The ratio of the volume of carbon dioxide produced to oxygen consumed during ➤aerobic respiration. In *in vitro* experiments the RQ can give an indication of the nature of respiratory substrates, but the application of this information to whole organisms is complicated by the fact that most organisms can use a variety of respiratory substrates at any one time.

response ➤nervous system.

rest energy In ➤relativity, the energy of a particle in its own ➤rest frame. It is the ➤rest mass of the particle multiplied by c^2. Like the term ➤relativistic mass, it is not commonly used in modern practice.

rest frame A ➤frame of reference in which a particle remains at rest. ➤➤proper time; rest mass.

resting nucleus ➤interphase.

resting potential The ➤potential difference that exists across the membranes of many cells, particularly nerve cells. It is maintained by an energy-requiring process which actively maintains different concentrations of sodium ions on each side of the membrane. ➤action potential; nerve impulse; sodium pump.

rest mass In relativity, the mass of a particle in its ➤rest frame. It is the magnitude of the ➤energy–momentum four-vector of the particle (divided by c^2):

$$m_0^2 c^4 = E^2 - p^2 c^2,$$

where m_0 is the rest mass, E the total energy of the particle and p its momentum. Crucially, because it is the magnitude of a four-vector, it is ➤invariant under a ➤Lorentz transformation. The photon has a zero rest mass.

restriction enzyme (restriction nuclease, restriction endonuclease) A ➤DNAase obtained from ➤bacteria that cuts DNA molecules into smaller fragments. Over a hundred different restriction enzymes have now been identified and characterized. They cut at specific sites (**restriction sites**) on the DNA determined by particular base sequences and are named after the bacterium from which they were obtained. For example, the enzyme *Eco*RI is obtained from *E. coli* and cuts a DNA strand between the G and the A of the sequence GAATTC. Restriction enzymes are so called because they restrict the development of ➤plaques (2) caused by ➤bacteriophage infection of bacterial cells. Restriction enzymes are extensively and routinely used to map the structure of DNA and in ➤genetic engineering. ➤➤DNA sequencing; human genome project; restriction map.

restriction fragment length polymorphism (RFLP) A ➤polymorphism (2) in the lengths of DNA fragments that result from cutting an individual's DNA at a particular ➤locus (2) with a particular ➤restriction enzyme. The difference in length is a consequence of a difference in the positions of restriction sites for that enzyme, which may vary widely between individuals, even within the same gene. The fragments can be separated by gel ➤electrophoresis, often followed by ➤DNA blotting. The significance of RFLPs is that they greatly increase the number of

➤markers available when performing genetic mapping experiments, and they provide a link between the ➤physical map and ➤genetic map of an organism. RFLPs are also used as diagnostic tools to screen for certain inherited diseases; an RFLP is useful in this respect if it is closely linked to the gene responsible, or if a deletion in the gene is directly responsible for an RFLP.

restriction map The specification, often in the form of a diagram, of the positions of restriction sites for one or more ➤restriction enzymes in a piece of DNA. This is a ➤physical map, since the physical distances between the sites (in base pairs) are given. A restriction map may be found by analysis of fragments obtained by cutting the DNA with restriction enzymes, or may be derived from knowledge of the DNA sequence.

resultant A single vector that is the sum of, and thus equivalent to, two or more given vectors: ➤parallelogram of vectors.

retardation (deceleration) The rate at which speed decreases. For cases in which ➤velocity is considered (as a ➤vector), the term ➤acceleration is preferred for any change of velocity, whether positive or negative.

retina ➤eye.

retinol ➤vitamin A.

retort A glass apparatus with a long stem tapering at the end.

retrograde Of an orbiting body, moving in the opposite sense to the direction of motion of the planets about the Sun.

retrotransposon A segment of ➤DNA that can move from one place to another in a ➤genome, so producing new combinations of ➤genes. The movement of a retrotransposon depends on its ➤transcription, followed by ➤reverse transcription of the resultant ➤RNA. ➤➤transposon.

retrovirus A ➤virus containing ➤RNA that is able to copy the RNA into ➤DNA when it infects a cell by means of the enzyme ➤reverse transcriptase, which is encoded in its ➤genome. A number of cancer-causing viruses and ➤HIV are retroviruses.

reverse bias ➤diode.

reverse genetics The process of using ➤genetic engineering to create ➤mutations in a ➤gene of interest, in order to investigate its function. Typically, a gene may be identified as a candidate for involvement in a particular process as a result of ➤genome sequencing, by comparison with genes of known function. This reverses the order of progress in classical genetics, in which the process starts with a mutant ➤phenotype and proceeds via gene mapping to the identification of the mutant gene responsible for the phenotype.

reverse osmosis The passage of a solvent through a ➤selectively permeable membrane in the opposite direction to that usual in ➤osmosis when a pressure greater than the osmotic pressure is applied to a solution. Reverse osmosis can be used to desalinate water.

reverse Polish notation (postfix notation) A notation for mathematical expressions which removes the need for brackets. Often expressions are converted into reverse Polish by computers before processing since one pass, from left to right, is sufficient to evaluate the expression. When a constant or a variable is encountered in the expression, it is added to the top of a stack (initially empty). When an arithmetical operator is encountered, the last two numbers in the stack are taken off and used as the operands, and the result is put back on the stack. Thus one reverse Polish version of $2^{[6 + (5 - 3) \times 4]}$ is $6\ 5\ 3\ -\ 4 \times\ +\ 2\ \wedge$.

reverse transcriptase An enzyme that copies RNA into DNA. It is obtained from ➤retroviruses and is used by molecular geneticists to synthesize cDNA (➤DNA) from messenger RNA *in vitro*. ➤➤reverse genetics.

reverse transcription The synthesis of ➤DNA from an ➤RNA template, requiring the enzyme ➤reverse transcriptase. Compare ➤transcription.

reversible In thermodynamics, describing a process that can be reversed by moving through the same set of states in the reverse order. ➤quasistatic change; (second law of) thermodynamics.

reversible reaction A reaction with an equilibrium constant (➤equilibrium, chemical) reasonably close to unity. As *all* reactions reach chemical equilibrium, the distinction between a reversible reaction and one that 'goes to completion' is blurred.

Reynolds number Symbol *Re*. A dimensionless number used in ➤fluid mechanics to describe the nature of fluid flow in a particular situation: $Re = \rho v l / \eta$, where ρ is the density, v the flow velocity, l a characteristic dimension such as the diameter of a pipe and η the viscosity. At low *Re* the viscous forces dominate and extend over large distances causing ➤laminar flow; at high *Re* the flow can be unstable and change to ➤turbulent flow. For example, ➤Poiseuille flow in a pipe depends on the Reynolds number being less than about 30. For flow of a stream of fluid around a cylinder, *Re* less than about 10 results in steady flow; for *Re* between 10 and about 40, vortices are produced downstream of the obstruction; as *Re* approaches a few hundred a ➤boundary layer forms, and at *Re* greater than about 10 000, a turbulent boundary layer extends well downstream. Named after Osborne Reynolds (1842–1912).

r.f. Abbr. for ➤radio frequency.

Rf Symbol for the element ➤rutherfordium.

RFLP Abbr. for ➤restriction fragment length polymorphism.

Rg Symbol for the element ➤roentgenium.

Rh 1 (Chem.) Symbol for the element ➤rhodium.
　2 (Biol.) ➤Rhesus factor.

rhenium Symbol Re. The element with atomic number 75 and relative atomic mass 186.2, which is in the third row of the ➤transition metals. Like its ➤congener manganese, its highest oxidation number is +7, the number of electrons in its outer shell. The rhenate(VII) ion, ReO_4^-, is a much weaker oxidant than MnO_4^-. Rhenium also forms ReH_9^{2-}, which is unusual in having a coordination number of 9.

rheostat A variable ➤resistor.

Rhesus factor An ➤antigen on the surface of ➤erythrocytes, first recognized in Rhesus monkeys, which occurs in approximately 85% of the human population. Such people are **Rhesus positive** (Rh+). The Rhesus factor is particularly important during blood transfusions and for women who are **Rhesus negative** (Rh−), but who are pregnant with a Rh+ foetus. If Rh+ blood is transfused into a Rh− patient it will produce anti-Rh antibodies in the blood plasma. If there is a subsequent transfusion of Rh+ blood, there will be an agglutination reaction with the Rh+ antigen. If a Rh− mother carries a Rh+ foetus she may develop anti-Rh antibodies in her blood as a result of seepage of blood across the placenta at birth. This is not of significance to the first child since there is usually no mixing of maternal and foetal blood until birth itself. However, the anti-Rh antibodies will react with a second Rh+ foetus to cause foetal haemolysis in the newborn. This may be averted by injecting a Rh− mother with anti-Rh antibodies before the birth of a Rh+ child or by immediate transfusion of the newborn baby's blood.

rhodium Symbol Rh. The element with atomic number 45 and relative atomic mass 102.9, which is in the second row of the ➤transition metals and is a member of the platinum metals. An increasing use of the metal is as an alloy with platinum in catalytic converters. Another important rhodium-containing catalyst is ➤Wilkinson's catalyst.

rhodopsin ➤colour vision.

rhombencephalon ➤brain.

rhombohedral One of the seven crystal systems. ➤Bravais lattice.

rhombus A ➤parallelogram with all four sides of equal length.

riboflavin ➤vitamin B_2. ➤➤flavin.

ribonuclease (RNAase) An enzyme that hydrolyses bonds between the ➤nucleotides of RNA.

ribonucleic acid ➤RNA.

ribose A sugar containing five carbon atoms (see the diagram), which forms an integral part of the structure of ➤RNA. Its derivative ➤**deoxyribose** is found in DNA.

D-ribose Open-chain and ring structures (two views): the OH on the right-hand carbon in the ring structures is in the ➤beta position.

ribosomal RNA (rRNA) ➤RNA.

ribosome A particle in a cell that assembles ➤proteins by organizing ➤amino acids into particular sequences determined by the ➤nucleotide sequence of ➤messenger RNA. Ribosomes, synthesized in the ➤nucleolus, are composed of RNA and protein and occur in the cytoplasm, usually attached to the rough ➤endoplasmic reticulum, and also in some ➤organelles. Bacteria have smaller ribosomes than eukaryotes.

ribozyme (catalytic RNA) An ➤RNA molecule that has catalytic properties similar to those of ➤enzymes. Specifically, ribozymes are involved in the cleavage of large RNA molecules into smaller fragments to form ➤transfer RNA. Ribozymes were first recognized in *E. coli* in 1981.

Richter scale The scale used to quantify the strength of earthquakes. The strength on this scale is proportional to the logarithm of the amplitude of the movement of the ground. The earthquake that destroyed San Francisco in 1906 registered 8.25 on the Richter scale. Named after Charles Francis Richter (1900–85).

Riemannian geometry A general conception of geometry in which the infinitesimal distance between two points in n–dimensional space is given by an expression of the form $ds^2 = \sum g_{ij}\, dx_i\, dx_j$. Shortest paths (➤geodesics) and curvature can be defined in terms of this ➤metric and there are applications to ➤general relativity. Named after (Georg Friedrich) Bernhard Riemann (1826–66).

Riemann integral A conception of integration in which a function f defined on an ➤interval $[a, b]$ is declared integrable if the limit of the sum of the areas of (thin) rectangles with heights on the graph $y = f(x)$ and bases abutting to fill up $[a, b]$ exists as the maximum width of the rectangles shrinks to zero. All continuous functions are Riemann integrable, but for advanced applications the ➤Lebesgue integral is superior.

right angle An angle of $90°$.

right-angled triangle A ➤triangle with one interior angle equal to a right angle.

right ascension ➤celestial sphere.

rigid body (compound body) A body composed of a large number of particles in different positions, which do not move with respect to each other as the body moves. In mechanics it contrasts with a ➤particle, which is a mass located at a single point. A rigid body has a nonzero ➤moment of inertia.

rigidity modulus ➤shear modulus.

ring closure A reaction in which a cyclic ring is formed. **Robinson annulation**, which is commercially important in the synthesis of the steroid hormone oestrogen, builds a ring in two steps, the first of which is a ➤Michael reaction. The ring closure step involves an internal ➤aldol condensation.

Ringer's solution (physiological saline) A solution of salts of set composition and concentration that mimics the composition of animal body fluids. Ringer's solutions are used to keep excised tissues and organs functioning during experimental investigations. There are different formulations for different animals and tissues. Named after Sydney Ringer (1835–1910).

ring main An arrangement for wiring (as in a domestic electricity supply) in the form of a ring. If the ring is connected to a power supply at one point only, then between that point and any other point on the ring there are two independent electrical paths. Hence, in the event of a fault in the ring, the supply is maintained.

ring opening ➤cyclopropane.

ring strain ➤cyclopropane.

RISC processor Abbr. for reduced instruction set computer processor. A successful modern design of ➤CPU, in which the number of available instructions is reduced to a simple subset of those normally found on complex processors. This reduction, perhaps surprisingly, makes the processor *more* powerful because, with a smaller number of possible instructions, the processor can be designed to execute them much faster. Several of the simple instructions can be combined in order to perform the functions of the complex instructions, but this is not a great disadvantage since they are used much less often.

r.m.m. Abbr. for ➤relative molecular mass.

r.m.s. value Abbr. for ➤root mean square value.

Rn Symbol for the element ➤radon.

RNA (ribonucleic acid) A ➤polynucleotide molecule composed of the bases ➤adenine, ➤guanine, ➤cytosine and ➤uracil and containing the sugar ➤ribose. RNA has a variety of functions. ➤Messenger RNA (mRNA) is copied from DNA in the nucleus and conveys information encoded in its base sequence to instruct ➤ribosomes to make ➤protein; RNA constitutes the ➤genome of some ➤viruses. ➤Transfer RNA (tRNA) is responsible for marshalling amino acids during ➤protein synthesis. **Ribosomal RNA** (rRNA) forms an integral part of the structure of ribosomes. In addition, other smaller RNAs occur in the nucleus where they have a variety of functions, including splicing sections of DNA together and modifying the structure of rRNAs and tRNAs to their functional forms. RNA generally is single-stranded, but in some forms (such as tRNA) helical regions occur with looped regions doubled back on themselves. Compare ➤DNA; ➤➤ribozyme.

RNAase ➤ribonuclease.

RNA polymerase An enzyme that initiates and elongates RNA chains by assembling ➤ribose-containing ➤nucleotides in a sequence determined by a ➤DNA template.

robot A programmable, multifunctional device with mechanical parts, capable of moving itself or manipulating objects, and having sensors for providing information to the controlling computer. The computer, which may be on board the robot, often uses this information to determine the next instructions to send to the robot, hence the robot is controlled by a ➤feedback loop. Robots are increasingly used in industry.

Rochelle salt $Na^+ \,^-O_2CCH(OH)CH(OH)CO_2^- \,K^+$ A convenient name for sodium potassium 2,3-dihydroxybutanedioate (sodium potassium tartrate), named after the port of La Rochelle.

rocket A system of propulsion used for weapons and spacecraft. A rocket carries both fuel and oxidant (as opposed to a missile, which may be powered by a conventional ➤jet engine) and is therefore suitable for use outside the Earth's atmosphere. Chemical reaction between the fuel and oxidant forces a stream of gas out of a nozzle at high velocity, and the law of ➤conservation of momentum means that the momentum of the rocket increases in the opposite direction.

rock-salt structure (halite structure) The structure of sodium chloride, NaCl (see the diagram), which is the most important crystal structure, adopted by hundreds of binary compounds. The structure can be interpreted in a number of different ways. First, each ion is in a ➤face-centred cubic array, so the structure can be described as two interpenetrating face-centred cubic arrays. This highlights the difference from the ➤caesium chloride structure, which consists of two interpenetrating *primitive* cubic arrays. Alternatively, it can be noted that the sodium ions are in all the octahedral holes in a face-centred cubic array of chloride ions. This interpretation contrasts usefully with the ➤fluorite structure and the ➤nickel arsenide structure.

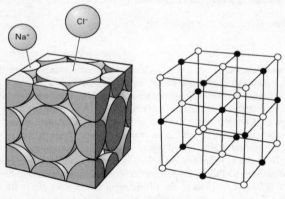

rock-salt structure

rod ➤eye.

rodent A diverse and widespread mammalian order **Rodentia**, which includes mice, rats, squirrels, guinea pigs and beavers. Rodents are typified by chisel-shaped incisor teeth that grow continually, but are worn down by their typical diet of vegetables fibres.

roentgen A unit of the ➤dose of exposure to radiation (X-rays or γ-rays). The SI-derived unit is the coulomb per kilogram, and 1 roentgen = 2.58×10^{-4} C kg^{-1}. Named after Wilhelm Conrad Röntgen (1845–1923). (Note that the unit has no accent.)

roentgenium Symbol: Rg. The element with atomic number 111. Named after W. C. Röntgen.

rolling friction ➤Friction that opposes the motion of a body rolling on a surface. The energy lost to this sort of friction is dissipated chiefly in the cycle of compression and subsequent elastic expansion of the surface as the body passes over it.

ROM Acronym for read only memory. A type of semiconductor computer memory which does not lose its memory when the power is switched off (contrast ➤RAM). The data is fixed into the ROM when it is manufactured and cannot be altered, hence 'read only'. This type of memory is used for storing programs and data which are required when the computer is switched on, typically so that the ➤operating system can be loaded into RAM. On some computers, however, the operating system itself is held in ROM. ➤➤EPROM.

Röntgen rays An old name for ➤X-rays, after their discoverer W. C. Röntgen.

root 1 (Math.) The nth root of a number is that number whose nth power is the given number. For example, the 5th root of 32, written $\sqrt[5]{32}$, is 2 since $2^5 = 32$. The '2nd root' and '3rd root' are always called **square root** and **cube root**, respectively.
  2 (Biol.) Typically the part of a vascular plant that grows beneath the soil and serves to anchor the plant and absorb water and minerals. Root systems are very diverse and include the fibrous roots of, for example, many grasses and the underground taproot storage organs of carrots and parsnips. Some **epiphytic** plants, such as tropical orchids, have **aerial roots** capable of absorbing water from humid atmospheres; others have side or prop roots for support. Roots never bear leaves and typically do not produce buds (though buds may arise on certain root tubers and from groups of cells derived from roots in tissue culture) and the vascular tissue is arranged in a central cylinder rather than in ➤vascular bundles, as they are in stems. The roots of leguminous plants in the Fabaceae family have nodules containing mutualistic (➤symbiosis) ➤bacteria that can fix nitrogen. Compare ➤stem; ➤➤nitrogen cycle.
  3 (Biol.) The origin of a nerve in the central ➤nervous system.

root mean square value (r.m.s. value) For a set of quantities $(x_1, x_2, \ldots, x_n)$ the quantity $\sqrt{\Sigma x_n^2 / n}$.

rotary converter An ➤alternating current motor used to drive a ➤direct current generator.

rotation A geometric ➤transformation (2) in which points are moved through a fixed angle in a plane perpendicular to a fixed axis.

rotational spectrum The rotational energy of a molecule is quantized, and transitions between these ➤quantum states can be induced by radiation of the appropriate frequency to produce a rotational spectrum. A pure rotational spectrum is observed in the ➤microwave region as the ➤energy gaps are small compared with vibrational or electronic energies; **rotational fine structure** can often be observed in ➤infrared spectra and the ➤Raman effect.

rotational symmetry A geometric figure has rotational symmetry if some rotation leaves it identical to itself (although in a new orientation): ➤symmetry transformation.

rotor The moving part of an electric motor or generator. Compare ➤stator.

roughage ➤fibre (3).

rounding error The error made in approximating a number to a fixed number of significant figures or decimal places. Calculations by computer are also subject to rounding errors, as only a fixed number of bytes are available to represent each number.

roundworm ➤nematode.

Royal Society, The The premier science society in Britain, founded in 1662 by the amalgamation of discussion groups in London and Wadham College, Oxford. **FRS** stands for Fellow of the Royal Society.

RQ Abbr. for ➤respiratory quotient.

(R–S) system (Cahn–Ingold–Prelog system) A convention for designating the enantiomers at a ➤chiral centre. The four groups attached to the central carbon atom are first ranked according to **priority**; in simple cases, the priority increases with increasing atomic number of the atom directly attached to carbon. The lowest-priority group is imagined as furthest away from the viewer. The (R) (Latin *rectus*) isomer has the other three groups in order of descending priority arranged in a clockwise manner and the (S) (Latin *sinister*) anticlockwise. In the diagram the arrows are clockwise, so the enantiomer shown is (R)-butan-2-ol.

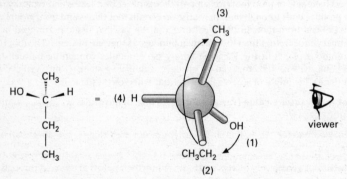

(R–S) system The isomer of butan-2-ol shown is described as (R).

Ru Symbol for the element ➤ruthenium.

rubber The latex of the rubber tree, with well-known elastic properties. In 1826 Michael Faraday worked out that the empirical formula of rubber was C_5H_8 and by 1860 it was known that the molecule ➤isoprene could be produced on heating rubber strongly. Attempts to synthesize rubber failed until the crucial feature of the *cis* orientation around each remaining double bond was incorporated in 1955 using ➤Zeigler–Natta catalysts to produce poly(*cis*-isoprene). ➤Vulcanization was responsible for a huge increase in the applicability of rubber. Several artificial rubbers have been developed, such as ➤SBR, but the natural material is still very useful.

rubella (German measles) A short-lived and contagious viral disease characterized by a rose-coloured rash (hence the name). Other symptoms include fever, sore throat and swelling of the lymph glands behind the ears. Rubella is generally a mild disease most common in young adults, although it may have severe consequences for women in early pregnancy since foetal abnormalities such as heart defects and mental disability can result from the virus crossing the ➤placenta.

rubidium Symbol Rb. The element with atomic number 37 and relative atomic mass 85.47, which is a member of the alkali metals. As such, it is a very powerful reductant, for example reducing water violently. It forms compounds very similar to sodium's and exclusively with oxidation number +1. There is no precipitation test for rubidium as all its compounds are soluble and so a flame test is used; the red flame produced explains the name 'rubidium'.

ruby An impure form of aluminium oxide, much more precious than the pure compound. The impurity that gives its red colour is the chromium(III) ion, Cr^{3+}. Apart from its use in jewellery, a ruby rod was the first material from which a laser was constructed, by Theodore Maiman in 1960.

Runge–Kutta method One of a class of methods for the numerical solution of differential equations by a step-by-step method using ➤Taylor expansions. Named after Carl David Tolmé Runge (1856–1927) and Wilhelm Martin Kutta (1867–1944).

Russell–Saunders coupling ➤spin–orbit coupling. Named after Henry Norris Russell (1877–1957) and Frederick Albert Saunders (1875–1963).

rusting The natural decay process afflicting all iron and steel objects; it is the reconversion of the metal to its very stable hydrated oxide, from which the metal has been extracted only after very considerable energy investment. **Rust** is a hydrated oxide of iron, $Fe_2O_3.xH_2O$, where the x indicates that the extent of hydration is variable (as for ➤bauxite). The conditions necessary for rusting of iron are the presence of air (or oxygen) and water. The extent of rusting increases in an acidic environment, as in cities with high sulfur dioxide levels. The rate of rusting is accelerated by ionic compounds such as common salt, especially at the seaside.

   Formation of the oxide is extremely exothermic ($-1648 \text{ kJ mol}^{-1}$) so in practice it is only possible to reduce the rate of rusting, not prevent it altogether. This is done in a number of ways, of which the most trivial is painting, a useful protection only until the paint is damaged. Better protection is afforded by ➤sacrificial protection. This is the reason behind the success of ➤galvanizing steel and the use of magnesium blocks to protect battleships.

ruthenium Symbol Ru. The element with atomic number 44 and relative atomic mass 101.1, which is in the second row of the ➤transition metals and a member of the ➤platinum metals. It shows a variety of oxidation numbers, from +8 in the yellow solid oxide RuO_4 downwards. Note that its ➤congener iron does not reach such a high oxidation number.

rutherfordium Symbol Rf. The name suggested for the element with atomic number 104, then proposed for the element with atomic number 106 before being *officially* reallocated to element 104. Named after Ernest Rutherford (1871–1937).

Rutherford scattering The elastic ➤scattering of one particle from another by an ➤inverse square law force. Rutherford proposed the ➤Geiger–Marsden experiment, in which alpha particles were observed for the first time being scattered from nuclei of gold.

rutile structure An important crystal structure, named after the ore **rutile** (titanium dioxide, TiO_2).

Rydberg–Ritz equation The equation quantifying the ➤hydrogen atom emission spectrum:

$$\frac{1}{\lambda} = R_H \left(\frac{1}{n_1^2} - \frac{1}{n_2^2} \right),$$

where λ is the wavelength of the emission line, R_H is the **Rydberg constant** (1.097×10^7 m^{-1}; ➤Appendix table 2), n_2 is the ➤principal quantum number of the upper energy level and n_1 the principal quantum number of the lower energy level. The various series in the spectrum are named after their discoverers and the value of n_1 for each series is given in brackets: Lyman (1), Balmer (2), Paschen (3), Brackett (4), Pfund (5) and Humphreys (6). The ➤Balmer series is seen in the visible region, the ➤Lyman series is in the ultraviolet and the others are in the infrared or lower frequency band. Named after Johannes Robert Rydberg (1854–1919) and Walter Ritz (1878–1909).

S

s **1** Symbol for the unit ➤second.
2 (Chem.) Label used in chemical equations to indicate the solid state.

s- Abbr. for ➤secondary.

S **1** Symbol for the element ➤sulfur.
2 Symbol for the unit ➤siemens.

S **1** Symbol for ➤entropy.
2 Symbol for ➤spin angular momentum.

(S)- ➤(R–S) system.

∑ ➤ series.

Sabatier–Senderens process The hydrogenation of an organic compound, typically an unsaturated fat, using hydrogen gas and a nickel catalyst at 150 °C. It is used for producing margarine. Named after Paul Sabatier (1854–1941), who won the Nobel prize in 1912 for the discovery, and Jean Baptiste Senderens (1856–1937).

saccharin One of the most common artificial sweeteners with structure shown in the diagram, being about 400 times sweeter than sugar albeit with a bitter after-taste. Its sweet taste was first discovered accidentally in 1879 by a chemist who forgot to wash his hands before eating.

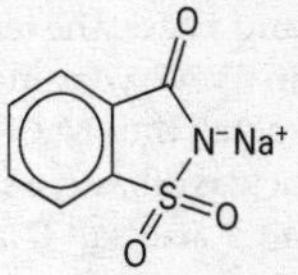

saccharin

sacculus ➤ear.

sacrificial protection A method of protecting iron and steel objects against ➤rusting by alloying with a more reactive element which supplies the electrons required by oxygen, hence becoming the ➤anode, the site of oxidation. The sacrificial element needs renewing occasionally. For example, a block of magnesium is used to protect battleships from corrosion. ➤➤galvanize.

saddle point As its name suggests, a stationary point on a surface that is a maximum in one plane and a minimum in another. In the diagram, P is a saddle point.

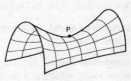

saddle point

sal ammoniac An old name for ammonium chloride, NH_4Cl.

salbutamol (Ventolin) $C_{13}H_{21}O_3N$ White crystals used as a bronchodilating nasal spray to help asthma sufferers.

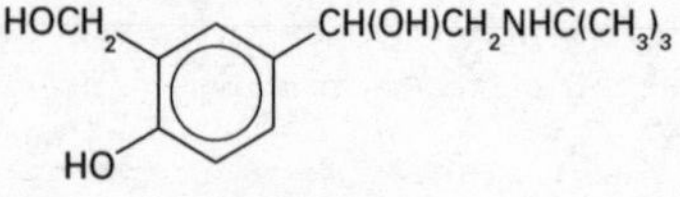

$$HOCH_2 \quad \diagdown \quad CH(OH)CH_2NHC(CH_3)_3$$

$$HO$$

salbutamol

salicylic acid (2-hydroxybenzoic acid) Colourless needles with structure shown in the diagram which are used mainly in the production of aspirin by ➤ethanoylation. The methyl ester of salicylic acid is famous as the attractive smell of ➤oil of wintergreen.

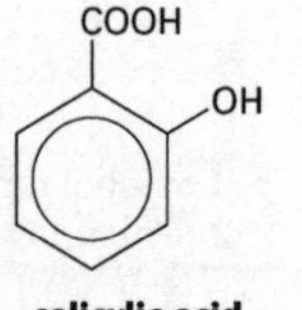

COOH
OH

salicylic acid

saline Containing a ➤salt, as in medical 'saline solution', sterilized 0.9% aqueous sodium chloride, which is ➤isotonic with blood.

saliva A secretion of the ➤salivary glands which in humans consists of an aqueous solution of the protein lubricant mucin and the ➤enzyme salivary amylase. Saliva is mixed with food during chewing (mastication), facilitates swallowing and starts the digestion of starch.

salivary gland Glands linked to the mouth of many land animals that secrete ➤saliva. Humans have three pairs of salivary glands. The salivary glands of the ➤larvae of some insects, notably *Drosophila,* have so-called giant ➤chromosomes which may be utilized to study the action of ➤genes.

Salmonella A ➤genus of rod-shaped Gram-negative bacteria (➤Gram's stain) responsible for some forms of food poisoning.

salt A generic name for the product of the neutralization reaction between an acid and a base; the term is also used for the archetypal example sodium chloride NaCl (as in its everyday meaning). Salts are produced by replacing a hydrogen in an acid with a metal: thus NaCl is produced from HCl. The salt typically has ➤ionic bonding. If the acid is dibasic, a salt exists (e.g. Na_2SO_4 from H_2SO_4) but there is also the possibility of an ➤acid salt ($NaHSO_4$).

salt bridge A typically U-shaped tube containing a salt solution used to achieve electrical connection between two ➤half-cells. A common salt used in the bridge is potassium chloride.

salt cake An industrial name for sodium sulfate decahydrate, $Na_2SO_4 \cdot 10H_2O$.

salting out The reduction in solubility of a substance by the addition of a ➤salt. Soaps can be salted out using sodium chloride.

saltpetre (nitre) Traditional name for the ore potassium nitrate, KNO_3. ➤➤Chile saltpetre.

samarium Symbol Sm. The element with atomic number 62 and relative atomic mass 150.4, which is a member of the ➤lanthanides. As expected therefore its dominant oxidation number is +3 as in the yellow crystals $SmCl_3$. Its most important use is in a ➤ferromagnetic alloy with cobalt ($SmCo_5$), which forms strong permanent magnets.

sample The outcome of a single experiment (usually in the context of a series of repetitions of the same experiment). ➤error of measurement; random sample.

sampling The taking of measurements of some physical quantity, often at regular time intervals. For example, the magnitude of sound waves is sampled at 44.1 kHz for compact disc recordings.

sand A granular mixture dominated by silicon dioxide, SiO_2.

Sandmeyer reaction The conversion of a ➤diazonium salt into the corresponding chloro (or bromo) compound by treatment with copper(I) chloride (bromide). It is the best way of introducing either halogen into an aromatic ring. Named after Traugott Sandmeyer (1854–1922).

sandstone ➤Sedimentary rock consisting of grains of quartz, cemented by various minerals.

sandwich compound The generic name for a class of compounds in which one (or more) metal ion is trapped between two (or more) flat ligands (two ions between three ligands produces a 'double-decker' sandwich). An important early example was ➤ferrocene; another example is ➤dibenzenechromium.

Sanger's reagent The reagent 2,4-dinitrofluorobenzene (see the diagram), used in the identification of the terminal amino acid in a protein. Named after the double Nobel prizewinner in chemistry, Frederick Sanger (b. 1918).

saponification The production of a soap (Latin *sapo*) by the alkaline hydrolysis of a ➤lipid. The soap is the salt of a long-chain fatty acid, such as sodium stearate, $Na^+C_{17}H_{35}CO_2^-$. 'Soft' soaps contain potassium salts.

Sanger's reagent

sapphire A precious stone that is impure aluminium oxide, Al_2O_3, the impurities being responsible for its blue colour. Two Al^{3+} ions are replaced by one Ti^{4+} ion and either a Co^{2+} or an Fe^{2+} ion at a few locations.

saprotroph (saprophyte) An organism (typically a bacterium or fungus) that feeds by secreting digestive ➤enzymes into an organic ➤substratum and absorbing the digestion products directly into its cells.

sarcoma ➤cancer.

sarin ➤nerve gas.

SARS Acronym for Severe Acute Respiratory Syndrome. A SARS epidemic started in Guangdong, China in November 2002. By the time the epidemic was over there were 8000 cases with a mortality rate of 10% (which compares with that for flu of 0.6%).

satellite **1** (Astron.) A smaller body orbiting a larger one, usually a planet. For example, the Moon is the satellite of the Earth. An **artificial satellite** is an unmanned space vehicle designed to orbit a larger body, usually the Earth. Earth satellites are used for remote sensing, navigation and communications.

2 (Phys.) In spectroscopy, a smaller line that appears at an energy or frequency slightly different from the main line.

satellite DNA ➤DNA consisting of repeated sequences of ➤nucleotides that appear to have no clear genetic function. The repeated sequences of satellite DNA are the basis of ➤genetic fingerprinting.

saturated 1 (Phys.) Of air, containing as much water vapour as possible. The water has a ➤saturated vapour pressure equal to the pressure of the air.

  2 (Phys.) ➤saturation.

  3 (Chem.) ➤saturated compound.

  4 (Chem.) ➤saturated solution.

saturated compound A compound that only contains single bonds, as distinct from an **unsaturated compound**, which contains at least one multiple bond. The simplest examples are the alkanes, which contain only single bonds, compared with the alkenes which have carbon–carbon double bonds. A **saturated fat** is one formed from fatty acids, such as stearic acid, with no double bonds.

saturated solution A solution of a substance in which no more of that particular substance can dissolve. ➤➤supersaturated solution.

saturated vapour pressure (SVP) The ➤vapour pressure of a gas coexisting with its corresponding liquid phase. This is the maximum vapour pressure possible for a substance at a specified temperature.

saturation For a ➤ferromagnetic material, a state in which all ➤domains are aligned and the ➤magnetization is therefore at its maximum value.

Saturn The sixth planet from the Sun, at an average distance of 9.54 AU, and with an orbital period of 29.46 years. Its radius is 60 270 km (nearly ten times the Earth's), though its gravitational field is of similar magnitude to the Earth's. Different features on the planet rotate at slightly different rates varying from 10 h 14 min to 10 h 39 min: the best estimate of the rotation period of the planet's core (as of 2007) is 10 h 33 min. Its 1-bar 'surface' magnetic field is similar to that on Earth, but of opposite polarity. However, if approximated as a dipole field created by a current loop with a specified magnetic moment, Saturn's magnetic moment is about 580 times larger than Earth's. In most respects Saturn resembles Jupiter but it is distinguished by its beautiful rings, which consist of billions of tiny particles (diameters from about 1 μm to 1 m or more, made partly of ice) orbiting in very thin layers. It has at least 18 satellites, of which **Titan** is by far the largest (with a radius of 2575 km, only just smaller than Jupiter's largest satellite, Ganymede). ➤➤Appendix table 4.

SAW Abbr. for ➤surface acoustic wave.

sawtooth A waveform which, when plotted against time, looks like the teeth of a saw. Each 'tooth' consists of a linear rise followed by a sharp fall back to the minimum value.

Sb Symbol for the element ➤antimony, from its Latin name *stibium*.

s-block elements The elements in which the outermost orbital being filled is an ➤s orbital. They comprise Groups 1 and 2 of the ➤periodic table.

SBR An artificial rubber made from ➤styrene and ➤butadiene, used, for example, to make car tyres.

Sc Symbol for the element ➤scandium.

scalar A physical quantity with no direction associated with it (compare ➤vector and ➤tensor). Mass is a scalar quantity, as is temperature. The magnitude of a vector is a scalar (e.g. ➤ speed is a scalar, the magnitude of the vector ➤velocity).

scalar field A function that assigns a scalar to each point of a multidimensional region. For example, air temperature and gravitational potential are both scalar fields.

scalar product (dot product) The scalar $|a||b| \cos \theta$, written $a \cdot b$, associated with two vectors a and b, where $|a|$ and $|b|$ are the magnitudes of the two vectors and θ is the angle between them. If $a = a_1 \mathbf{i} + a_2 \mathbf{j} + a_3 \mathbf{k}$ and $b = b_1 \mathbf{i} + b_2 \mathbf{j} + b_3 \mathbf{k}$, then the scalar product is also given by $a_1 b_1 + a_2 b_2 + a_3 b_3$. The scalar product gives an efficient method for finding angles in three dimensions; in particular two vectors are perpendicular if $a \cdot b = 0$. It is also useful in physics: for example, $F \cdot \hat{a}$ gives the component of the vector F in the direction of the unit vector $\hat{a}$.

scalar triple product The scalar $a \cdot (b \times c)$ formed from three three-dimensional vectors a, b, c. If the three vectors are the position vectors of points A, B and C with respect to the origin O, then $|a \cdot (b \times c)|$ is the volume of the ➤parallelepiped, four of whose vertices are OABC.

scaler A counting circuit that outputs a pulse or signal when a designated number of input pulses have been received.

scaling A concept in physics closely related to the idea of ➤dimensional analysis. Scaling is the way that one variable relates to another in similar systems. For example, for similarly shaped objects, the volume and mass scale as the cube of the length, whereas the surface area scales as the square of the length. For their size, smaller animals (ants, for example) appear to be stronger than larger animals (elephants, for example). The mass of an animal scales with the cube of its length, whereas the strength of an animal scales with the cross-sectional area of its body and hence with the square of its length. This seemingly trivial method of reasoning has important applications in particle physics, in which **scale invariance** plays a large part.

scan The act of moving across a surface (or solid angle) and applying the same experimental technique at each point (➤sampling) in order to map out some property of the surface. For example, in scanning the intensities of parts of a visual image, the image is examined in successive scan lines, in a similar way to the raster scan of a television set.

scandium Symbol Sc. The element with the atomic number 21 and relative atomic mass 44.96, which is in Group 3 of the ➤periodic table. It resembles aluminium in its chemistry: for example, its dominant oxidation number is +3, as in the oxide Sc_2O_3 used in electronic components.

scanning electron microscope (SEM) An electron microscope in which a picture of a surface is built up by scanning an electron beam across the surface and detecting the ➤secondary electrons emitted as a result. The image obtained appears similar to the image formed in an optical microscope. Compare ➤transmission electron microscope.

scanning tunnelling microscope (STM) A device for mapping the properties of surfaces on the atomic scale. By moving a tip held at a high potential over the surface and maintaining a constant current (constant because of the ➤tunnel effect) by moving the tip up and down with a ➤feedback loop, it is possible to map the electron density of a surface in remarkable detail, imaging individual atoms.

scatter diagram A plot of points showing the value of two attributes (such as height and weight) for the members of a random sample.

scattering The change in direction of a moving particle or ➤travelling wave when it interacts with other particles. In **elastic scattering**, the energy of the incident particle or frequency of the incident wave is unchanged (though there may be a ➤phase shift). In **inelastic scattering**, some energy is transferred to the scatterer or dissipated as radiation. Scattering can be treated classically, as in ➤Rayleigh scattering and ➤Rutherford scattering. However, a detailed treatment of most scattering requires quantum mechanics. ➤➤Compton effect; Raman effect; Tyndall effect.

scattering cross-section Symbol σ; unit m^2 or (much more conveniently) ➤barn. The number of particles scattered per unit time divided by the incident number per unit flux density. Cross-section can be subdivided, for example into the cross-section for scattering into a particular solid angle (the **differential cross-section**), or with a particular energy loss. For a target of cross-sectional area S in a uniform beam of particles, the cross-section for hitting the target would simply be S, hence the name of the quantity.

SCF theory Abbr. for self-consistent field theory (➤Hartree–Fock theory).

Scheele's green A green pigment, copper(II) hydrogen arsenite ($CuHAsO_3$), named after Carl Wilhelm Scheele (1742–86).

Schiff's reagent A reagent used to test for aldehydes as distinct from ketones. It is made by removing the colour of rosaniline dye with sulfur dioxide. Aliphatic aldehydes and aldose sugars restore the magenta colour. Named after Hugo Josef Schiff (1834–1915).

Schlieren photography A method of photographing ➤convection in fluids by detecting the change in ➤refractive index due to the change of density of the fluid in the convection currents. The currents appear as streaks (German *Schlieren).*

Schottky defect ➤vacancy. Named after Walter Schottky (1886–1976).

Schottky diode A diode based on a metal–semiconductor junction, known as a **Schottky barrier**. The separation of the energy of the bottom of the ➤conduction band and the ➤Fermi energy is increased at the junction, creating a barrier that the carriers must cross, either by having sufficient thermal energy or by the ➤tunnel effect, and the current depends on the ➤bias voltage. Schottky diodes can be used as ➤rectifiers, and switch faster than conventional ➤p–n junction diodes.

Schottky noise (shot noise) Noise in the output of an electronic device that is a consequence of the randomness of a process involving individual electrons or holes.

Schrödinger equation A differential equation for the ➤wavefunction ψ, the most important equation in nonrelativistic quantum mechanics. In its most succinct form the time-dependent equation is

$$H\psi = i\hbar \, d\psi/dt,$$

here $\hbar = h/2\pi$, h being the ➤Planck constant, and i is the square root of minus one. In time-independent form,

$$H\psi = E\psi$$

where E is the energy. More generally, the Schrödinger equation governs not only the wavefunction ψ but the entire ➤quantum state $|\psi\rangle$. The ➤Hamiltonian H is composed of two parts: the quantum-mechanical version of the kinetic energy $(-\hbar^2/2m)\nabla^2$ (➤Laplacian), and the potential energy V. Hence the latter equation can be written as

$$(-\hbar^2/2m)\nabla^2\psi + V\psi = E\psi.$$

It is exactly soluble only for the hydrogen atom, and approximations are needed for other atoms and molecules (➤Hartree–Fock theory). The introduction of this equation by Erwin Schrödinger (1887–1961) in 1926 demonstrated the quantization of atomic energy levels naturally for the first time, as the equation is soluble only for certain discrete values of the energy.

Schwann cell ➤neurone. Named after Theodor Schwann (1810–82).

Schwarzschild radius ➤black hole. Named after Karl Schwarzschild (1873–1916).

science The ongoing search for knowledge about the Universe. Typical of the progression in our knowledge is the level of understanding about the solar system. Nicholas Copernicus speculated that the sun was at the centre of the solar system. This could be proved only after detailed naked-eye observations made by Tycho Brahe were correctly interpreted by Johannes Kepler (➤Kepler's laws). The fundamental principles of gravitational attraction were discovered by Isaac Newton: ➤Newton's law of gravitation together with his laws of motion explain Kepler's laws. The theory Newton introduced was adequate to enable us to put a man on the Moon. Einstein's theory of ➤general relativity introduced further refinements.

scintillation counter A detector for ionizing radiation that works by counting flashes (scintillations) emitted by a suitable material when a ➤gamma ray is incident.

sclera ➤eye.

scleroprotein One of a group of protein derivatives forming tough connective tissues. Scleroproteins include ➤collagen and ➤keratin.

SCP Abbr. for ➤single-cell protein.

SCR Abbr. for ➤silicon-controlled rectifier.

screening The reduction in ➤electric field strength caused by the presence of intervening charge. For example, the electrons occupying inner shells of an atom screen the outer electrons from the nuclear charge. ➤➤effective nuclear charge.

Se Symbol for the element ➤selenium.

seaborgium Symbol Sg. The element with atomic number 106. Named after Glenn Theodore Seaborg (1912–99), the only person so honoured in his lifetime.

search engine A program used to search for the presence of specified terms on the ➤Internet. Over one hundred different search engines can be used to navigate the Internet.

sea water An aqueous solution found in the oceans containing dissolved ions; those present at a concentration of greater than 1000 ppm are Cl^-, Na^+, SO_4^{2-} and Mg^{2+}. Magnesium is extracted from sea water, as is bromine despite its low concentration of just 65 ppm.

sebaceous gland ➤skin.

sec ➤trigonometric functions.

sec- Abbr. for ➤secondary.

secant 1 ➤trigonometric functions.
   2 A line that cuts a curve. Compare ➤tangent (2).

sech ➤hyperbolic functions.

second 1 Symbol s. The SI unit of time. Its definition as 1/31 556 925.974 7 of the tropical year 1900 (now known as the **ephemeris second**) was replaced in 1968 by a new definition as 9 192 631 770 periods of the radiation corresponding to the transition between two ➤hyperfine levels of the ground state of the ^{133}Cs nuclide.
   2 Symbol ″. A unit of angle equal to one-sixtieth of a minute (which is, in turn, one-sixtieth of a ➤degree) so $3600″ = 1°$. One **second of arc** is a very small angular separation in everyday terms: two objects separated by 4 mm subtend an angle about $1″$ at an observer 1 km away. However, in observational astronomy the second becomes a significant unit.

secondary Symbol *sec-* or *s-*. When applied to an alcohol or a halogenoalkane, this indicates that the hydroxyl group or halogen atom is attached to a carbon with two alkyl groups attached. An example of a secondary alcohol is ➤adrenaline. When applied to an amine the meaning is different. A secondary amine has two alkyl groups attached to the *nitrogen* atom. Again, adrenaline is a secondary amine.

secondary cell ➤cell (2).

secondary coil The output coil of a ➤transformer; that is, the coil in which a voltage is generated by induction. Compare ➤primary coil.

secondary colour A colour that is composed of two ➤primary colours.

secondary electron An electron emitted (e.g. in ➤electron spectroscopy or in a ➤photomultiplier) as a result of a collision between more energetic electrons and atoms. Because these are ➤inelastic collisions and can happen in succession, electrons in **secondary emission** usually have a broad range of energies.

secondary structure ➤protein structure.

secondary wave (S wave) A ➤seismic wave that is perpendicular to the direction of propagation (a ➤shear wave). Compare ➤primary wave.

second derivative That function denoted by d^2y/dx^2 or $f''(x)$ that is the ➤derivative of the derivative of a given function $y = f(x)$.

second law of thermodynamics ➤thermodynamics.

second quantization An extension of ➤quantum mechanics which treats interacting particles as excitations of an underlying field. This allows for the treatment of the creation and annihilation of particles. It is fundamental to ➤quantum field theory, and underpins the treatment of ➤phonons, ➤photons and ➤plasmons as particles.

secretin A ➤polypeptide ➤hormone secreted by the gut lining which stimulates the ➤pancreas to produce pancreatic juice and the release of ➤bile from the liver. Its production is stimulated by hydrochloric acid from the stomach, the release of which it inhibits by a ➤feedback mechanism.

section A plane figure that arises as the intersection of a plane with a solid figure. A **cross-section** is a section arising from a plane that is perpendicular to an axis of the solid.

sector That portion of a circle bounded by two radii and an ➤arc (see the diagram).

secular equation An old-fashioned synonym for ➤characteristic equation.

secular variation The variation of ➤magnetic declination with time, as the Earth's magnetic poles move with respect to its rotational axis.

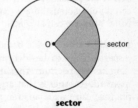

sector

sedimentary rock One of the three main types of rock, formed from pre-existing rocks by the erosion of the latter into particles of varying sizes, which are subsequently transported and deposited as sediments usually in seas and lakes. As these sediments accumulate they form sedimentary rocks. They are classified as **clastic**, including sandstones and shales, **chemical**, including chalk and some limestones, or **organic**, including coal. Many sedimentary rocks contain fossils. Compare ➤igneous rock; metamorphic rock.

Seebeck effect ➤thermoelectric effects. Named after Thomas Johann Seebeck (1770–1831).

seed The structure that develops after ➤fertilization of the ovule in the ➤Magnoliopsida and ➤Pinopsida. In the Magnoliopsida, the seed contains the

---

embryo together with a stored food reserve, the ➤endosperm, which may be partly or wholly transferred into a single or pair of embryo seed leaves (➤cotyledons). The seed is surrounded by the **testa** (usually a tough outer coating), derived from the outer layers of the ovule, and forms inside the ➤pericarp or fruit wall, derived from the ovary. Seeds and fruits provide a means of dispersing plant offspring over a wide area, and may have features such as hairs which facilitate transport by wind, or hooks to attach to the fur of animals. Many are eaten and survive passage through an animal's gut.

seeding Use of a seed crystal to induce a ➤supersaturated solution to crystallize.

seed plants ➤Spermatophyta.

segment That portion of a circle bounded by a ➤chord and an ➤arc (see the diagram).

segregation During ➤meiosis the separation of ➤alleles that occur together in a ➤germ cell into individual ➤gametes.

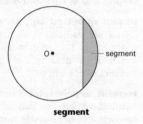

segment

seismic wave A wave of displacement of the material making up the Earth, such as is produced by an earthquake. Seismic waves (measured using a **seismograph** or **seismometer**) are similar to sound waves, but are generally of much lower frequency. ➤primary wave; Richter scale; secondary wave.

seismology The study of ➤earthquakes and ➤seismic waves.

selection pressure The degree to which organisms in a population possessing certain characteristics are either advantaged or disadvantaged by those characteristics. ➤sickle cell anaemia.

selection rule A rule based on symmetry that enables the likely transitions in ➤spectroscopy to be identified. For example, one selection rule is that a pure ➤rotational spectrum is possible only if the molecule has a permanent electric dipole moment. Another is that any electronic transition in an atom caused by light (a photon) causes a change in orbital angular momentum of one unit, because the angular momentum of the photon is one unit. Selection rules depend on specific ➤transitions between quantum states having zero probabilities. If the probabilities are nonzero but very low, selection rules are only approximate.

selectively permeable membrane (semi-permeable membrane, SPM) A membrane that is permeable only to some particles and not to others. A common example is ➤cellulose ethanoate, which allows water to pass but not dissolved ions. The SPM is central to ➤osmosis.

selenium Symbol Se. The element with atomic number 34 and relative atomic mass 78.96, which is in Group 16 below sulfur, which it closely resembles. For example, it forms **selenides** such as lead(II) selenide, PbSe. An important aspect of selenium's chemistry is its ability to react well with heavy metals, as this enables them to be excreted from the body. Selenium is an essential trace element in the human body, the major source of which is garlic. Selenium deficiency leads to **Kashin–Beck syndrome**. The electrical conductivity of selenium is changed when light shines on it, a property put to use in photocopiers. Selenium dioxide, SeO_2, is a white solid used in organic chemistry to oxidize CH_2 groups immediately adjacent to ➤carbonyl groups to form the structure —COCO—. As for sulfur, the range of oxidation numbers is wide, from -2 in the Se^{2-} ions in selenides through $+1$ in the red liquid Se_2Br_2, $+2$ in the red solid Se_4S_4, $+4$ in SeO_2 to $+6$ in the octahedral molecule SeF_6.

self-consistent field (SCF) theory ➤Hartree–Fock theory.

self-exciting generator A ➤generator that uses the output current to power its own electromagnets.

self-inductance ➤inductance.

selfish gene (selfish DNA) The theory that bodies of organisms are vehicles for the survival of sets of ➤genes, which are seen as the basic unit of evolutionary selection. In effect, genes that code for features which produce bodies that can propagate more copies of those genes are regarded as successful. The concept helps to explain the apparently altruistic behaviour of, for example, social insects such as bees. Individual bees may 'sacrifice' themselves by stinging an intruder and dying. This is explained in selfish gene theory by the fact that the sacrifice of one bee might actually enhance the survival of more copies of the same genes in its relatives in the rest of the colony.

SEM Abbr. for ➤scanning electron microscope.

semen A fluid produced by the ➤testis and associated glands (such as the ➤prostate gland) consisting of an alkaline solution of salts and organic compounds such as ➤prostaglandins and containing sperm. Semen provides a medium in which sperm can swim to fuse with an egg cell, particularly in animals with internal fertilization.

semicarbazide $NH_2NHCONH_2$ A colourless crystalline solid used in solution to test for aldehydes and ketones, which form **semicarbazones**, crystalline solids with characteristically sharp melting points useful for identifying the parent aldehyde or ketone.

semiclassical model In solid state physics, a model of a conductor in which the quantum-mechanical nature of the electron states (the ➤band structure) is recognized, but many other aspects of the model are treated as in ➤classical physics, such as the effect of electric and magnetic fields on the movement of electrons.

semiconductor A material whose electrical conductivity *increases* with increasing temperature (this is in contrast to conductors, whose conductivity decreases). The distinction between semiconductors and insulators is less clear, as it depends on the magnitude of the ➤energy gap between the ➤conduction band and the ➤valence band. Insulators have a large gap, whereas semiconductors have a small one. Semiconductors fall into two categories: intrinsic and extrinsic. **Intrinsic semiconductors**, such as pure silicon or germanium, operate by thermal excitation taking an electron from the valence band to the conduction band, the electron excited plus the ➤hole left behind causing conduction. In **extrinsic semiconductors**, added impurities (**dopants**) provide additional energy levels between the valence and conduction bands which allow either electrons (n-type) or holes (p-type) to be the ➤majority carriers. Impurities used include elements from Group 15, such as phosphorus and arsenic, which produce n-type behaviour, and elements from Group 13, such as aluminium and gallium, which produce p-type behaviour. While silicon and germanium are the most important semiconducting materials, compounds such as gallium arsenide (GaAs) are becoming increasingly useful. A third ➤isoelectronic series consists of compounds such as lead(II) telluride (PbTe). The ➤p–n junction assumes central importance in electronic circuits. ➤➤diode; inhomogeneous semiconductor; semimetal; transistor.

semiconductor diode ➤diode.

semiconductor junction ➤p–n junction.

semimetal A conductor similar to a metal but with a carrier concentration several orders of magnitude lower. The ➤valence band and ➤conduction band in a semimetal touch rather than overlap (as in a metal), so there is a very small ➤density of states at the ➤Fermi energy. Graphite, arsenic, antimony and bismuth are semimetals. Compare ➤semiconductor, in which there is an ➤energy gap.

semi-permeable membrane ➤selectively permeable membrane.

sense organ A specialized structure in animals, containing ➤receptors.

sensitivity The ratio of the output of a measuring device to its input.

sensory neurone ➤neurone.

sepal ➤flower.

separating funnel An apparatus used to separate ➤immiscible liquids. The two liquids are shaken together and left for a time to settle. The denser phase can be run off slowly, leaving behind the less dense phase in the funnel.

separation energy The energy input required to remove a nucleon from an atomic nucleus. It is the nuclear equivalent of ➤ionization energy (but compare ➤binding energy).

sequence A succession of numbers or other entities $a_1, a_2, \ldots$ indexed by the natural numbers: sequences are denoted by (a_n). If a_n is expressed as a function of n, it is known as the general term, but some sequences are more conveniently defined by a ➤recurrence relation.

sequestration Complexation of metal ions by enveloping them in a large ligand such as the highly effective ➤EDTA ligand. This is useful in water softening and in analysis.

Ser Abbr. for ➤serine.

series **1** (Math.) The sum of a finite or infinite number of terms of a sequence. Summation notation such as $\sum_{i=1}^{n} a_i$ and $\sum_{i=1}^{\infty} a_i$ is used to denote a series; the latter is also used to denote the value of the sum, if the series converges.

2 (Phys.) ➤spectral series.

3 (Phys.) Of two or more electrical components, an arrangement such that the components are connected in a chain (see the diagram). The total ➤resistance of a set of resistors in series is simply the sum of the individual resistances; likewise a set of ➤impedances in a **series circuit** has a total impedance that is the sum of the individual impedances. However, a set of ➤capacitors C_1, C_2, C_3, . . . in series has a total capacitance of

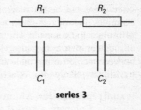

series 3

$$\frac{1}{\dfrac{1}{C_1} + \dfrac{1}{C_2} + \dfrac{1}{C_3} + \cdots}.$$

Compare ➤parallel (4).

serine (Ser) A common ➤amino acid. ➤Appendix table 7.

serology The study of the interactions between ➤antigens and ➤antibodies.

serotonin A derivative of the ➤amino acid tryptophan, secreted by the vertebrate brain, which acts as a ➤neurotransmitter. It appears to act antagonistically to ➤noradrenaline, takes part in a wide range of body functions including blood clotting and blood pressure, and may have a role in the control of moods and certain endogenous rhythms (➤biological clock) associated with the day–night cycle.

Sertoli cell ➤testosterone. Named after Enrico Sertoli (1842–1910).

serum ➤blood.

serum albumin ➤albumin.

sesquicarbonate A compound containing the salt and the ➤acid salt of carbonic acid. Sodium sesquicarbonate is $Na_2CO_3 \cdot NaHCO_3 \cdot 2H_2O$.

sesquioxide An obsolete term denoting an oxide with an element-to-oxygen ratio of 2:3, as in iron(III) oxide, Fe_2O_3.

set Any collection of distinguishable mathematical objects (**elements**) thought of as a whole. Sets are usually specified either by listing all the elements or by describing their properties. The usual notation is to enclose the elements in braces, with a colon as the condition delineator (read as 'such that'). Thus the set of even integers could be described by {n: n is an integer and 2 divides n to give an integer}.

sex chromosome One of a pair of chromosomes in animals that determines sex. There are usually two kinds of sex chromosome, designated X and Y. Human males have an X and a Y, while females have two X chromosomes. ➤Barr body.

sex hormones ➤oestrogen; progesterone; testosterone.

sex-linkage A term applied to certain genes that lie on the X chromosome of animals whose sex is determined by a nonhomologous pair of ➤sex chromosomes. The principle relies on the fact that if a ➤recessive allele is present on the non-homologous part of the X chromosome in a male, it is expressed in the ➤phenotype since there can be no ➤dominant allele present on the Y chromosome. Classic examples include red–green ➤colourblindness and ➤haemophilia in humans, and 'white eye' in *Drosophila*. The term sex-linkage thus refers to the increased frequency of such conditions in males. Colourblind females are rare since they require a cross between a colourblind male and a female ➤carrier, and only half the offspring are, statistically, likely to be affected.

sexual reproduction A form of reproduction in which two ➤gametes fuse to form a ➤zygote. Compare ➤asexual reproduction. ➤➤fertilization; meiosis.

sexual selection A concept in ➤evolution that explains why the two sexes in many animals are ➤dimorphic. Animals are considered to have a preference for mates that show certain 'sex-limited' characters, such as bright plumage and courtship rituals in male birds.

Seyfert galaxy A ➤spiral galaxy that has a bright, active centre that emits strongly. The activity is thought to be from an accretion disk surrounding a massive ➤black hole. Named after Carl Keenan Seyfert (1911–60).

Sg Symbol for the element ➤seaborgium.

shadow An area that is not fully illuminated by a beam of light because of the presence of an obstruction. A shadow from an extended source consists of a dark ➤umbra and a partially illuminated ➤penumbra.

SHE Abbr. for ➤standard hydrogen electrode.

shear The transformation that consists of moving some parts of an object parallel to a fixed plane; the object changes shape (compare a ➤rotation, where one point is fixed and the object does not change shape). Ionic crystals are ➤brittle because, when subjected to shear, ions of the same charge are brought next to each other, which causes repulsion.

shear modulus (rigidity modulus) The ratio of ➤shear stress in a material to the ➤shear strain.

shear strain A measure of the deformation of an object by ➤shear. It is the ratio of the distance d moved parallel to the fixed plane divided by the perpendicular distance D from that plane (see the diagram).

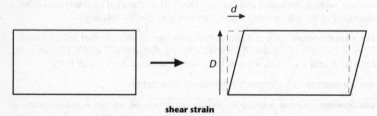

shear strain

shear stress A measure of the forces experienced by an object undergoing ➤shear. It is the force that one plane exerts on a neighbouring plane per unit area of contact.

shear wave A ➤transverse wave in which the varying quantity is ➤shear strain and the driving force is ➤shear stress. A ➤secondary wave is a shear wave.

shell An arrangement of electrons in atoms around the nucleus, each successive shell being further from the nucleus on average. In the obsolete ➤Bohr model, they corresponded to different fixed distances from the nucleus. In ➤quantum mechanics the distance of the electron can be given only as a probability distribution, so the term now refers to all orbitals with the same ➤principal quantum number. The $n = 3$ shell has a 3s orbital, the 3p ➤subshell and the 3d subshell.

shell model A model of the atomic nucleus in which the nucleons occupy shells very much like the electron ➤shells in an atom. ➤➤magic numbers.

SHF Abbr. for ➤super high frequency.

shielding constant ➤effective nuclear charge.

shift reaction A part of the manufacturing process of hydrogen, during which carbon monoxide is reacted with steam to produce hydrogen and carbon dioxide in a shift converter:

$$CO(g) + H_2O(g) \rightarrow H_2(g) + CO_2(g).$$

A catalyst of iron(III) oxide, Fe_2O_3, is used at $400\,°C$.

shift reagent ➤lanthanide shift reagent.

SHM Abbr. for ➤simple harmonic motion.

shock wave A wavefront with a very high pressure gradient associated with a local disturbance, typically an object moving faster than the speed of sound (a **sonic boom**) or an explosion.

shoot The aerial part of a plant (**stem**) arising from the ➤epicotyl in seed plants, bearing leaves and buds, and typically capable of ➤photosynthesis.

short circuit The state of two points in a circuit with a conducting path between them, particularly where the path is unintentional. Compare ➤open circuit.

short-day plant ➤photoperiodism.

short sight ➤myopia.

shoulder girdle ➤pectoral girdle.

shower The production of a number of ➤subatomic particles after a single high-energy particle interacts with matter.

shunt A resistor placed in parallel with an ammeter such that only a known fraction of the current to its terminals passes through the ammeter itself. Thus the shunt increases the range of the ammeter, but decreases its sensitivity.

shuttle vector A fragment of ➤DNA, usually a ➤plasmid, that can replicate in cells from two different species. Some occur naturally, but many are now constructed by ➤genetic engineering techniques, with ➤nucleotide sequences that initiate replication in both ➤yeast and ➤E. coli cells, for example.

Si Symbol for the element ➤silicon.

sickle cell anaemia A genetically determined condition of human blood that produces a characteristic distorted shape of the ➤erythrocytes (hence the 'sickle'). It is caused by a mutation involving a single ➤amino-acid substitution of valine for glutamic acid in position 6 of the β (beta) chain of ➤haemoglobin. Individuals ➤heterozygous for the gene show increased resistance to malaria, although the ➤homozygous ➤recessive condition is lethal. Since the gene for the condition is particularly common in parts of Africa where malaria is endemic, it is often cited as an example of a genetic response in a population to ➤selection pressures of the environment.

side chain An alkyl group that is not part of the longest chain of the molecule. For example, ➤methylpropane has a methyl side chain attached to the main three-carbon chain.

sidereal Pertaining to the stars, particularly with reference to the ➤fixed stars.

sidereal day The time taken for the Earth to rotate once on its axis with respect to the ➤fixed stars. Since the Earth also orbits the Sun once each year, there is one more sidereal day in the year than there are standard days, and the sidereal day is equal to 23 hours, 56 minutes, 4.091 seconds.

sidereal month ➤month.

sidereal period The orbital period of an astronomical body relative to the ➤fixed stars.

sidereal year The orbital period of the Earth relative to the ➤fixed stars (about 365.256 ➤days). This differs by about 20 minutes from the ➤tropical year because of the ➤precession of the equinoxes.

siemens Symbol S. The SI derived unit of ➤conductance. 1 S is equivalent to $1\,\Omega^{-1}$. Named after Ernst Werner Siemens (1816–92).

sievert Symbol Sv. The SI derived unit of equivalent ➤dose. One sievert is the radioactive dose equivalent of $1\,J\,kg^{-1}$. Named after Rolf Maximilian Sievert (1896–1966).

sigma bond (σ bond) A covalent bond that is symmetric about the bond axis (see diagram (a) opposite), the name echoing the symmetrical orbital in atoms, the s orbital. Most covalent bonds are of sigma character (the **sigma framework** of the benzene molecule is shown in diagram (b)); the remaining bonds are usually ➤pi bonds. ➤➤delta bond.

sigma particle A ➤baryon, related to the ➤lambda particle, with one ➤strange quark. Unlike the lambda, it has an isotopic spin equal to one; it can be charged (positive or negative) or uncharged.

signal A variation of some quantity with time in a way designed to convey information of some sort, for example a voice. In a broader sense, 'signal' is often used to mean the useful part of an experimental measurement, in contrast with ➤noise. ➤➤modulation; signal-to-noise ratio.

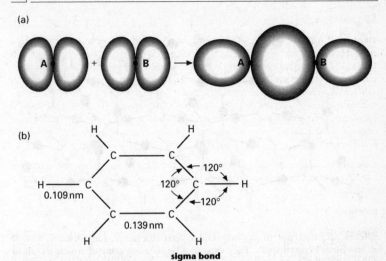

(a)

(b)

sigma bond

signal-to-noise ratio The extent to which the power of a ➤signal is larger than the underlying ➤noise. The ratio is often stated using the logarithmic ➤decibel scale. In many experiments, particularly spectroscopy, there is often a trade-off: increasing the ➤resolution of the detector reduces the signal-to-noise ratio.

significance level ➤hypothesis test.

significant figures A phrase used to indicate the precision with which a given number has been expressed in terms of the number of digits in the decimal expansion of the number that have been retained. For example, in both 106.6 and 0.010 66 the significant digits are 1066 and, to three significant figures, the numbers are 107 and 0.0107. ➤decimal places; precision.

silanes A series of hydrides of silicon with formulae analogous to the ➤alkanes, Si_nH_{2n+2}. Fewer exist than for the alkanes, the largest being Si_8H_{18}, and they are all less stable, reacting rapidly with oxygen.

silica (silicon dioxide) SiO_2 An abundant material on Earth, being the major constituent of sand. It is a white solid, with a high melting point (1710 °C) due to the structure of the solid, which is a giant covalent lattice with four Si—O covalent bonds arranged tetrahedrally (see the diagram overleaf). It forms three polymorphs: ➤quartz, tridymite and cristobalite (all of which can exist in both low- and high-temperature forms). At high temperatures, silicates can be formed by fusion with other substances, as in the formation of ➤slag in a blast furnace. When fused with limestone ($CaCO_3$) and sodium carbonate (Na_2CO_3) at around 1500 °C and then cooled, a ➤glass is formed.

silica gel A common drying agent, produced in an amorphous form by precipitating hydrated silica. Many items such as cameras are protected from dampness using silica gel which is also used in ➤desiccators.

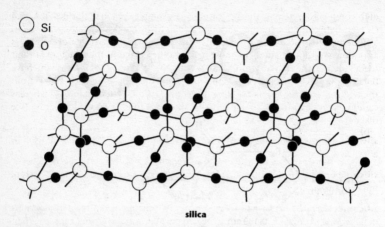

silica

silicate An anion containing silicon and oxygen. Silicates account for nearly 95% of all the Earth's crustal rocks. They exist in a bewildering array of structures, all of which are variations on the theme of SiO_4 tetrahedra. The simplest is the lone ion SiO_4^{4-}, as in zircon, $ZrSiO_4$. Complexity increases with rings, chains, double chains and sheets (see the diagram), and finally fully three-dimensional networks. Examples of silicates abound and include ➤feldspars, ➤micas and ➤zeolites.

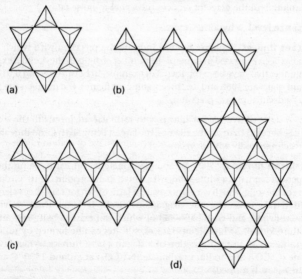

silicate SiO_4 tetrahedra can combine to fom (a) rings, (b) chains, (c) double chains or (d) sheets.

silicon Symbol Si. The element with atomic number 14 and relative atomic mass 28.09, which is in Group 14 of the ➤periodic table, immediately below carbon. Some of its compounds resemble those of carbon; for example the tetrachlorides $SiCl_4$ and CCl_4 are both colourless liquids, although $SiCl_4$ is hydrolysed much faster with water, a property used to produce smokescreens. This theme of increased kinetic reactivity is also shown in the ➤silanes. The most striking difference is in the structure of their oxides, despite the similar formulae SiO_2 and CO_2. Whereas CO_2 is a gas which we breathe out, SiO_2 (➤silica) is a solid we stand on. The difference stems from the relative lack of multiple bonding in Period 3 of the periodic table, so whereas carbon dioxide has the structure $O=C=O$, silica has a single-bonded structure (see diagram).

silicon carbide (carborundum) SiC An exceptionally hard, black solid used in abrasive tools.

silicon chip A single silicon crystal containing an integrated circuit. ➤chip; microelectronics.

silicon-controlled rectifier (SCR) A device consisting of three ➤p–n junctions designed to operate as a solid state ➤relay or switch. A pulse greater than a certain threshold voltage to the ➤gate allows current to flow through its main conduction path, and this can be used with a sinusoidal input to allow through only a certain portion of the input waveform.

silicone (siloxane) A polymer based on the structural motif in which two alkyl groups are attached to a silicon atom which is part of a chain of alternating silicon and oxygen atoms, as in $[—Si(CH_3)_2O—]_n$. They are produced by hydrolysis of the corresponding dichlorides. They form water-repellent oils, waxes and rubbers.

Silurian See table at ➤era.

silver Symbol Ag (from the Latin *argentum*). The element with atomic number 47 and relative atomic mass 107.9, which is at the end of the second row of the d block of the ➤periodic table, below copper and above gold. Like gold, it has been prized for centuries for its reluctance to react with water or oxygen, a fact reflected in its use for coinage and jewellery. It is also used in mirrors. Because of its high electrical conductivity and unreactivity, silver is used in electrical components. The major consumption was probably for the production of photographic film, which used silver halide emulsions. All its major compounds have oxidation number +1 as in AgCl (the most important silver(II) compound is AgF_2, a solid used as a fluorinating agent). Silver ions are used to test for halide ions in aqueous solutions as they precipitate the corresponding silver halide.

silver halide A binary compound between silver and one of the halogens. Silver ions react with aqueous chloride, bromide and iodide ions to form precipitates, which are coloured white, cream and yellow respectively, for example:

$$Ag^+(aq) + Cl^-(aq) \rightarrow AgCl(s).$$

This is useful both in qualitative and quantitative analysis. Silver halides darken on exposure to light and are used in the production of photographic emulsions.

silver nitrate $AgNO_3$ The most important compound of silver. It is a colourless crystalline solid which dissolves to form a colourless solution that is kept in a dark bottle as it is light-sensitive. A chance observation by Heinrich Schultze in 1727 that silver nitrate darkened in sunlight led to the development of photography. In the laboratory, an aqueous solution is used to test for halide ions (see above) and in ➤Tollens' reagent to test for aldehydes.

simian virus 40 ➤SV40.

similar Two geometrical figures that have the same shape but not necessarily the same size; one could be an enlargement of the other.

Simmons–Smith reagent The compound ICH_2ZnI which adds stereospecifically *cis* to alkenes to give cyclopropanes in high yield. Named after two du Pont chemists, Howard Ensign Simmons (1929–97) and Ronald D. Smith (b. 1930).

simple harmonic motion (SHM) The simplest possible type of ➤oscillation: a system consisting of a time-varying quantity $x(t)$ that obeys the differential equation:

$$\frac{d^2x}{dt^2} = -\omega^2 x.$$

The general solution for the equation is $x = A \sin(\omega t + \phi)$, where ω is the ➤angular frequency of the oscillation, ϕ its ➤phase (2) and A its amplitude. For a mechanical system this means that the restoring force is proportional to the displacement. For example, for a mass m on a spring (with ➤spring constant k), if the displacement from the equilibrium position is given by x with (according to the second of ➤Newton's laws of motion):

$$m\frac{d^2x}{dt^2} = -kx,$$

the motion is SHM with $\omega = \sqrt{k/m}$. For any restoring force, if it can be expanded in a ➤power series in x, then for small values of x the motion will be approximately SHM.

Thus SHM is frequently used as a model in any situation involving small oscillations. ➤➤damped harmonic motion.

simple pendulum ➤pendulum.

simplex A communications link that only ever allows data transmission in one direction. Compare ➤duplex.

simplex method The most widely used algorithm for solving ➤linear programming problems. The constraints of the problem define the boundaries of a region, or **simplex**, and the maximum or minimum value sought occurs at a **vertex** of this region. The simplex method checks vertices successively, each ➤iteration homing in on the required optimum value.

Simpson's rule The numerical integration formula

$$\tfrac{1}{3}h[y_0 + y_n + 4(y_1 + y_3 + \cdots + y_{n-1}) + 2(y_2 + y_4 + \cdots + y_{n-2})]$$

that gives the approximate area under a curve in terms of the y coordinates at an odd number of x coordinates that are evenly spaced a distance h apart. Because it is derived by fitting a parabola to successive triples of points, it is usually more accurate than the ➤trapezium rule. Named after Thomas Simpson (1710–61).

simulation The modelling of the processes involved in some system. Computer-based simulations have many applications; simulations of aircraft flight decks are used to train pilots, for example, and the world's weather system is simulated in supercomputers in order to predict the weather.

simultaneous equations A set of two or more equations (often linear) for which values are sought that solve all the equations simultaneously. If the graphs of the equations are drawn, the solution is given at the point(s) at which the graphs all meet.

sin ➤trigonometric functions.

sine ➤trigonometric functions.

sine curve (sine wave, sinusoid) The graph $y = f(x) = a \sin(bx + c)$ of a general sine function is shown in the diagram. The basic sine curve arises in the theory of ➤simple harmonic motion. ➤Fourier analysis extends this to model general periodic phenomena by combining sine curves.

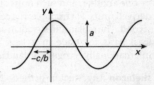

sine curve The graph $y = a \sin(bx + c)$, where a is the amplitude, $2\pi/b$ is the period and the phase angle c controls the x-intercept.

sine rule A formula used in ➤trigonometry to find unknown sides or angles of a triangle. With the usual notation (side of length a opposite angle A, etc.), it states that $a/\sin A = b/\sin B = c/\sin C = 2R$, where R is the radius of the **circumcircle** of the triangle.

single bond A single bond has two electrons, an electron pair, between the atoms joined together; for example the bond in H—Cl is a single bond, as it is in H—H or Cl—Cl. In contrast, the oxygen molecule is bonded by a ➤double bond, O=O.

single-cell protein (SCP) ➤Protein derived from microorganisms, usually ➤yeasts or ➤bacteria grown commercially in large fermentation vessels. Such protein can be processed and variously textured and flavoured to form food for animals and humans.

singlet state ➤triplet.

singularity Either a point at which a function is not ➤continuous, or a point at which a curve has a ➤cusp or crosses itself. For example, $x = 0$ is a singularity for each of $y = 1/x$, $y^2 = x^3$, $y^2 = x^3 + x^2$.

sinh ➤hyperbolic functions.

sinter To make a solid coalesce into a single mass by heating it strongly below its melting point.

sinus An expanded cavity in animals, particularly in the circulatory system, such as the **sinus venosus** in the venous system of sharks, or the respiratory tract, as in the nasal sinuses of land vertebrates.

sinusoidal In the shape of a ➤sine curve.

siphon A simple device for maintaining a flow of fluid from a higher reservoir to a lower reservoir. The fluid flows because the ➤static pressure at point X exceeds the static pressure at Z by ρgh, where ρ is the density of the fluid, g is the acceleration of free fall and h is the vertical distance between X and Z (see the diagram).

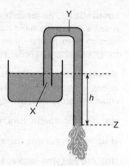

SI units The system of units, known in full as Système International d'Unités, that evolved from the metre, kilogram and second (m.k.s.) system over a number of years. The seven ➤base units are the ampere, candela, kelvin, kilogram, metre, mole and second, as finally agreed upon in 1960. Units for any physical quantity may be derived by multiplying or dividing the base units by one another, and such units are therefore known as **derived units**. For example, the coulomb is one ampere-

siphon

second. The radian and steradian are termed **supplementary units**. SI units form the legal basis for trade in a large number of countries world-wide and are the internationally recognized standards for science. ➤➤Appendix table 1.

skeleton Any structure in the body of an animal that serves to provide support or for the attachment of ➤muscles. In the ➤annelids, support is generated hydrostatically. The skeleton is most developed in the ➤arthropods, which have an exterior ➤exoskeleton containing ➤chitin, and in the ➤vertebrates, which have a sophisticated interior ➤endoskeleton made of bone or cartilage comprising a system of jointed struts and levers. ➤➤pectoral girdle; pelvic girdle; pentadactyl limb; vertebral column.

skin An extensive organ covering most of the body surface in vertebrates. It consists of three layers: an outer **epidermis** produced by the generative **Malpighian layer**, an underlying **dermis**, and an inner **subcutaneous** layer of fat and connective tissues, by which the skin is attached to the underlying structures (see the diagram opposite). The exact structure and function vary considerably. In mammals the epidermis is tough and waterproof, but in amphibians this layer is moist and functions in gas exchange. Hair, feathers, claws, fingernails and the scales of fish are derived from the epidermis. The hairs of mammals grow from **follicles** and are rendered waterproof by secretions from the **sebaceous gland**. Hairs may be raised or lowered by **erector muscles**, trapping a suitable thickness of air to assist in insulation. Mammalian skin is protective and also functions as a sense organ (it contains a large number of different types of nerve ending). In addition, the insulating subcutaneous fat, the ability to trap an insulating layer of air in the hair and the ability to lose water through the **sweat glands** to cool the body make the skin an important organ of ➤homeostasis.

skin effect The inductive effect that causes a large proportion of the current carried in a conductor at high frequency to be carried in the outer part (skin) of a wire. The current varies exponentially with distance x from the surface as $\exp(-x/\delta)$, and the **skin depth** δ is given by $\delta = \sqrt{2/\sigma\mu\omega}$, where σ is the conductivity of the wire, μ is its permeability and ω is the angular frequency of the current. Thus very high frequencies are carried as well by thin tubes as they are by thick wires.

skin friction The drag exerted on a body (e.g. an ➤aerofoil) moving in a fluid by the shear forces close to the surface of the body.

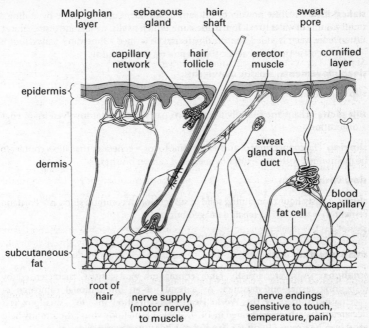

Malpighian layer
sebaceous gland
hair shaft
sweat pore

capillary network
hair follicle
erector muscle
cornified layer

epidermis {

sweat gland and duct

dermis {

blood capillary

fat cell

subcutaneous { fat

root of hair
nerve supply (motor nerve) to muscle
nerve endings (sensitive to touch, temperature, pain)

skin Vertical section through mammalian skin.

skip distance The spread of ranges within which a high-frequency (HF) radio transmission cannot be received because it is beyond the range of the ➤ground wave, but is not far enough for the sky wave (➤ionospheric wave) to be received. This creates zones in which communication is impossible when HF is used between ground stations and aircraft or ships.

skull The ➤superior part of the skeleton of vertebrates. The skull consists of a number of fused bones which provide protection for the brain and sense organs such as eyes, nose and ears, and support and attachment for the jaws. ➤➤vertebral column.

sky wave ➤ionospheric wave.

slag A by-product of the refining of metals, most importantly in the reduction of iron in a blast furnace. It is formed because the iron ore has impurities such as silicon dioxide, SiO_2, which can react with the calcium oxide produced by thermal decomposition of the limestone added to the furnace. This high-temperature reaction produces the slag calcium silicate:

$$CaO(s) + SiO_2(s) \rightarrow CaSiO_3(l).$$

The molten slag is less dense than molten iron and floats on the surface from where it can be tapped off. Blast furnace slag can be used in manufacturing cement and for breeze blocks and road foundations.

slaked lime A white powder of calcium hydroxide, $Ca(OH)_2$, made by adding a small volume of water to calcium oxide, lime, which swells, steams and disintegrates. When more water is added, the solution formed is termed ➤limewater. Slaked lime is used to increase the pH of acidic soils, a process called **liming**.

sleep movements ➤nastic movements.

slip ➤dislocation.

slip plane The plane that divides the two parts of a crystal involved in an edge ➤dislocation.

slip ring The copper rings of an ➤electric motor or ➤generator that allow contact to be maintained with the ➤rotor by means of carbon brushes.

slow virus ➤prion.

slurry A pasty liquid containing solid in suspension. A common slurry is ➤Portland cement in water used to repair damaged brickwork.

Sm Symbol for the element ➤samarium.

small intestine ➤gut.

smallpox A severe, usually fatal, contagious viral disease characterized by disfiguring rashes and pustules. As a result of an international campaign of vaccination started by the World Health Organization in 1967, smallpox was declared to have been eradicated in 1979. Stocks of smallpox virus are currently held for research purposes in only a handful of laboratories world-wide.

smectic liquid crystal ➤liquid crystal.

smog Smoke-filled fog that pollutes the air over cities. The London smogs were notorious until the Clean Air Act of 1956. **Photochemical smog**, a brown haze containing a complex cocktail of molecules such as nitrogen dioxide created by reactions initiated by the absorption of sunlight, remains a problem in some cities, such as Los Angeles.

smoke A ➤suspension of a solid in a gas, for example carbon in air produced by the combustion of organic substances.

Sn Symbol for the element ➤tin, from the Latin *stannum*.

Snell's law The chief law governing the refraction of light (see the diagram). The angle of refraction θ_r is related to the angle of incidence θ_i by:

$$\sin \theta_r = \frac{n_i}{n_r} \sin \theta_i,$$

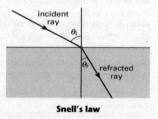

Snell's law

where n_r is the refractive index of the medium in which the refracted ray is travelling, and n_i is the refractive index of the medium in which the incident ray is travelling. Named after Willebrord van Roijen Snell (1591–1626).

snRNP (pronounced '**snurp**'**)** Abbr. for small nuclear ribonucleoprotein particle. One of a number of small particles which catalyse ➤splicing, composed of protein and snRNAs (small nuclear ➤RNAs).

snurp ➤snRNP.

soap ➤saponification.

soda-ash Industrial name for anhydrous sodium carbonate, Na_2CO_3.

soda-lime A solid mixture of sodium and calcium hydroxides, $NaOH/Ca(OH)_2$, which being strongly alkaline can absorb carbon dioxide gas. It is widely used in anaesthesia to remove exhaled carbon dioxide.

sodamide (sodium amide) $NaNH_2$ A compound produced by heating sodium with gaseous ammonia, which is used in organic chemistry as a reductant.

sodium Symbol Na (from the Latin *natrium*). The element with the atomic number 11 and relative atomic mass 22.99, which is in Group 1 of the ➤periodic table. Along with potassium, it is a typical element of Group 1, a vigorously reacting metal which forms ionic compounds with oxidation number +1 by losing an electron to water or a nonmetal, for example. Oxygen reacts to form both the oxide Na_2O and the peroxide Na_2O_2 (which is bonded $2Na^+O_2^{2-}$). As sodium reacts with oxygen and water, it must be stored under oil. Sodium metal is formed by electrolysis in a ➤Downs cell. It has a low melting point of 98 °C and liquid sodium is used as a coolant in nuclear reactors.

The sodium ion Na^+ is vital for the action of the ➤nervous system and is present in a huge range of important compounds, only a small selection of which follow. ➤Sodamide and ➤sodium tetrahydridoborate are strong reductants. ➤Sodium carbonate and ➤sodium hydroxide are vital alkalis in industry. Sodium hydrogencarbonate (bicarbonate) is **baking soda**. Sodium nitrate is ➤Chile saltpetre. Sodium silicate is **water glass**. Sodium sulfate decahydrate is **Glauber's salt**. ➤Sodium thiosulfate is an important reagent for iodine titrations. Sodium benzoate, $Na^+C_6H_5CO_2^-$, is used as a corrosion inhibitor and as a treatment for gout.

sodium bismuthate $NaBiO_3$ An off-white solid which is a very strong oxidant, used primarily to oxidize manganese(II) ions to purple manganate(VII) ions, a useful colour test for manganese.

sodium borohydride ➤sodium tetrahydridoborate.

sodium carbonate Na_2CO_3 An important alkali in industry, manufactured by the ➤Solvay process or increasingly from ➤trona. In solution, the carbonate ion extracts a proton from water:

$$CO_3^{2-}(aq) + H_2O(l) \rightleftharpoons HCO_3^-(aq) + OH^-(aq).$$

The hydroxide ions make the solution alkaline; it is a weak alkali as the ionization is incomplete. The anhydrous salt is called soda-ash whereas the decahydrate is called ➤washing soda. Over half of the production of sodium carbonate is used in the manufacture of ➤glass.

sodium chloride (common salt, rock-salt) NaCl The archetypal ionic compound (➤ionic bonding). It occurs naturally as rock-salt; the ➤rock-salt structure is the most important crystal structure in nature. It is manufactured by evaporation of sea water or brine. The many uses to which it is put include flavouring and preserving food, lowering the temperature of water on roads to prevent formation of ice and as a raw material for the production of sodium hydroxide, chlorine and sodium carbonate.

sodium cyclamate ➤sweetener.

sodium hydroxide (caustic soda) NaOH The single most important alkali both in industry and in the laboratory. It is manufactured by a variety of electrolytic processes (➤chlor–alkali industry). Sodium hydroxide is normally produced as white pellets which dissolve very exothermically in water, producing aqueous hydroxide ion, OH^-, which is the characteristic species in an ➤alkali. The solution neutralizes acids, hydrochloric acid forming sodium chloride:

$$NaOH(aq) + HCl(aq) \rightarrow NaCl(aq) + H_2O(l).$$

sodium pump A process that uses energy from ➤ATP to create concentration gradients of sodium ions across ➤plasma membranes, particularly those of ➤nerve cells in which a resting potential is produced. Sodium pumps are implicated in the transport of a wide range of substances across membranes.

sodium spectrum The four series of lines in the sodium emission spectrum, which were historically important in establishing the structure of atoms. The brightest feature is a doublet at 589 nm, but the two lines are so close together that they are hard to resolve. The **principal** ➤spectral series arises from transitions from np (where n is the ➤ principal quantum number of the upper state) to 3s. The **sharp** and **diffuse** series have the same series limit, arising from a common lower state: the transitions are ns to 3p and nd to 3p. The **fundamental** series has transitions from nf to 3d. The strange notation for ➤atomic orbitals comes from the initial letters of the four series: sharp, principal, diffuse and fundamental.

sodium sulfite Na_2SO_3 A colourless crystalline solid often used as an antioxidant food additive (coded as E221). It should be avoided by asthmatics as it can deoxygenate blood.

sodium tetrahydridoborate (sodium borohydride) $NaBH_4$ A strong reductant for aldehydes and ketones that converts them to the corresponding alcohols. Its advantage over ➤lithium tetrahydridoaluminate is that, being less reactive, it can be used in aqueous solution. An example of its industrial use is for the reduction of nickel(II) chloride to produce a nickel surface on a plastic for high-quality vinyl record masters.

sodium thiosulfate $Na_2S_2O_3$ A colourless crystalline solid used in aqueous solution to test for iodine, as it reacts quantitatively to form iodide ions and tetrathionate ions:

$$I_2(aq) + 2S_2O_3^{2-}(aq) \rightarrow 2I^-(aq) + S_4O_6^{2-}(aq).$$

A common test for an oxidant is therefore to add iodide ions, and then titrate the liberated iodine with a ➤standard solution of sodium thiosulfate. It is also used in photographer's ➤'hypo'.

soft acids and bases ➤hard and soft acids and bases.

soft radiation An imprecise name given to ➤ionizing radiation of low frequency which tends to be less penetrating than radiation of higher frequency. Compare ➤hard radiation.

software A generic term referring to any computer program. Software is an abstraction without physical reality (contrast ➤hardware), but it may be stored in a permanent physical state, such as ➤ROM, in which case it is often known as firmware.

soft water ➤hard water.

soil The upper part of the surface of the Earth's crust that provides a medium for plant growth. Soil is a complex and dynamic assemblage of components: rock particles of varying size derived by weathering from the parent rock, ➤humus, minerals in solution in water, air, and organisms including ➤bacteria, ➤arthropods and worms. Its formation, exact composition, and chemical and physical properties depend on the interaction of geological, climatic and biological factors. Basic soils tend to form over chalk and limestone; nutrient-poor acidic soils form in areas of high rainfall, which leaches minerals from the upper layers. Soils are classified according to their physical and chemical properties. **Loam soils** are the most fertile, and constitute most of the productive agricultural soils of the world.

sol ➤colloid.

solar cell A device generating electric power from sunlight used, for example, on calculators and space vehicles. It typically consists of a large flat ➤p–n junction which is exposed to sunlight. This produces ➤electron–hole pairs in each half of the junction, and these flow across the junction, generating a current.

solar constant The power of radiation received from the Sun per unit area (perpendicular to the direction of incidence) at the top of the Earth's atmosphere (at a distance of one ➤astronomical unit). It is the ➤radiant flux density of the Sun at the Earth, and has an average value of about 1370 W m^{-2}; this leads to an estimate of 4×10^{26} W for the total power output from the Sun.

solar cycle The periodic variation in the number of ➤sunspots visible on the Sun's disc, and in the level of other solar activity (such as ➤solar flares). The period is about 11 years.

solar day The time between successive passages of the Sun across the ➤celestial meridian.

solar energy The energy radiated by the Sun (➤solar constant). The energy comes from nuclear fusion reactions in the Sun (➤carbon cycle). Ultimately, most energy used on the Earth derives from solar energy. ➤Fossil fuel is the result of plants storing energy after ➤photosynthesis. Hydroelectric power is available because heat from the Sun evaporates water which subsequently falls as rain on high ground. Even wind power is available only because temperature differences on the surface of the Earth lead to pressure differences and therefore air currents. ➤➤solar power.

solar flare A violent release of matter and energy from the surface of the Sun, caused by instabilities in the magnetic field between ➤sunspots.

solar power Power produced directly from the Sun's radiation. This can be achieved either by the use of ➤solar cells, or by using the Sun's radiation to heat water, which is subsequently passed to a ➤heat exchanger. Europe's first solar power tower, with an array of moveable mirrors directing sunlight onto a receiver on top of a 115 m tower, was opened outside Seville, Spain in 2007.

Solar System The Sun and the bodies that orbit it: the planets and their ➤satellites, ➤asteroids, ➤comets, ➤meteoroids and other interplanetary material.

solar units A system of units, most importantly mass, related to the properties of the Sun, used in astrophysics. One **solar mass** is about 2×10^{30} kg, one **solar radius** about 7×10^5 km and one **solar luminosity** unit about 4×10^{26} W.

solar wind A ➤plasma of very low density that is emitted from the Sun.

solder A mixture of tin and lead, and sometimes a little antimony, used to join two pieces of metal together.

solenoid A long, tight coil of current-carrying wire designed to produce a magnetic field within its loops. Solenoids are used in ➤relays and switches. The magnitude of the ➤magnetic field strength at the centre of a solenoid is $H = nI$, where I is the current and n is the number of turns per unit length.

solid A ➤state of matter in which the particles have very little if any motion away from their lattice sites, simply vibrating about their mean position. Solids have a definite shape and generally a high density, and can be ➤crystalline or ➤amorphous. When heated, a solid melts at a characteristic temperature, its ➤melting point, to form a liquid.

solid angle The three-dimensional conical surface formed by all rays from a given point that extend through a given closed curve (➤steradian).

solid of revolution A solid figure formed by rotating a plane curve through $360°$ about an axis. The volume of the solid of revolution formed by rotating the area under a curve between $x = a$ and $x = b$ about the x axis is $\pi \int_a^b y^2 \, dx$.

solid state physics The physics of solids, particularly their electronic, optical and magnetic properties. ➤Semiconductors and their electronic properties are an important area of solid state physics.

solidus In a solid–liquid ➤phase diagram, a line indicating the commencement of melting or the completion of solidification of a particular phase. Compare ➤liquidus.

solstice The two points on the ➤ecliptic farthest from the ➤celestial equator. Like the term ➤equinox, 'solstice' is also used for the times at which the Sun appears at one of the solstices, which correspond to the longest (**summer solstice**) and shortest (**winter solstice**) days of the year.

solubility The extent to which one substance dissolves in another. The normal solubility records the maximum mass of a solid that can dissolve in a specified mass of water to form a ➤saturated solution. When the solubility is exceeded, any more solid will appear as a precipitate. The solubility is temperature-dependent, often strongly so. To prepare crystals from a solution, some of the water is evaporated off, causing a very concentrated solution. On cooling, the solubility is exceeded at some intermediate temperature so that when room temperature is reached crystals have formed.

solubility product Symbol K_{sp}. An equilibrium constant (➤equilibrium, chemical) that describes ionic concentrations for solids that have low solubility in water. The ions formed in solution reach an equilibrium concentration such that, for a binary solid A^+B^-:

$$[A^+(aq)][B^-(aq)] = K_{sp}.$$

When this equilibrium concentration has been reached, any further addition of solid causes precipitation. The solubilities of metal hydroxides are low, as are those of metal sulfides, which is the basis of ➤qualitative analysis.

soluble Capable of dissolving in a solvent, which can be taken to be water unless stated otherwise.

soluble RNA (sRNA) An old name for ➤transfer RNA.

solute ➤solution.

solution The homogeneous mixture of a **solute**, which can be a solid or a liquid or a gas, dissolved in a **solvent**. The most important solvent is water and **aqueous solutions** abound (➤solubility; solubility product). For covalent compounds, such as organic solids, good solvents are those with high ➤dispersion forces, such as tetrachloromethane, ethoxyethane, benzene, methylbenzene, etc. Two other important solvents are ethanol and propanone. The solute can be separated from the solution usually by evaporation of the solvent. Solvents do not have to be liquid: a **solid solution** can exist.

solvation The process by which a substance dissolves in a solvent. When the solvent is water, the more specific term ➤hydration can be used.

Solvay process (ammonia–soda process) The industrial manufacture of sodium carbonate. Carbon dioxide is passed up a tower down which ammoniacal brine is flowing:

$$CO_2(g) + NH_3(aq) + NaCl(aq) + H_2O(l) \rightarrow NaHCO_3(s) + NH_4Cl(aq).$$

The sodium hydrogencarbonate, insoluble under these conditions, precipitates out. After filtering, this solid is heated in rotating calciners where decomposition to the carbonate occurs:

$$2NaHCO_3(s) \rightarrow Na_2CO_3(s) + H_2O(g) + CO_2(g).$$

Named after Ernest Solvay (1838–1922), who used some of the profit from the process to fund the **Solvay Congresses**, which were instrumental in solving the structure of atoms.

solvent ➤solution.

solvent extraction A method of separation whereby a solid is extracted from a solution by adding a second solvent immiscible with the first. Most of the solid is extracted from the first solvent because of its greater solubility in the second. Penicillin is extracted from aqueous solution using trichloromethane.

somatic cell Any cell of the body, except a ➤germ cell.

somatotrophin ➤growth hormone.

sonar A technique, similar in principle to ➤radar, for finding the distance and direction of a remote object in water by transmitting sound waves and detecting reflections from it. The time of travel is then used to estimate the distance to the object. The name is an acronym for sound navigation ranging.

sonic boom The audible manifestation of the ➤shock wave produced by a body moving at ➤supersonic speed through the air.

s orbital An atomic orbital and therefore a solution of the ➤Schrödinger equation. The s orbital is filled across the **s block** of the periodic table. The orbital has an orbital angular momentum quantum number, l, of 0. The number of orbitals with any given value of l is always $2l + 1$, so is only 1 in this case. The shape of an s orbital is spherical, drawn in 2D as a circle. Compare ➤p orbitals, ➤d orbitals.

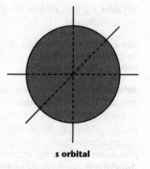

s orbital

sorption pump A ➤vacuum pump (for producing a low vacuum) that works by ➤adsorption of gas onto a very large surface area of a solid (usually a honeycomb structure) that is cooled to a low temperature. Because the pump uses no oil, a very clean vacuum is produced.

sound A pressure ➤wave with a frequency detectable by the human ear (approximately 20 Hz to 20 kHz). The wave is ➤longitudinal and requires a medium, such as air, for its transmission. The speed of sound is equal to $\sqrt{E/\rho}$ where E is the ➤elastic modulus or ➤bulk modulus of the medium, and ρ is its density. For a gas, since both E and ρ are proportional to pressure, the speed is simply proportional to the square root of the thermodynamic temperature. ➤➤acoustics; decibel.

sound pressure level ➤decibel.

source 1 The object or point from which a ➤wave or similar phenomenon originates.
　2 ➤ field-effect transistor.

source code In computing, the high-level program that is presented to a ➤compiler or ➤assembler in order to produce **object code**. Once linked with standard library routines, the object code can be run as a program.

Southern blotting ➤DNA blotting. Named after Edwin Mellor Southern (b. 1938).

Soxhlet extractor An apparatus for extracting a solute by continuous passage of a boiling solvent. It is named after Franz Soxhlet (1848–1913), a German food analyst.

space 1 Any region outside the Earth's atmosphere with a very low particle density.
  2 Collectively, the three dimensions that manifest themselves as distances as opposed to the fourth dimension, time. In ➤Newtonian mechanics space and time are treated distinctly, but in relativity they are treated as a single entity, ➤spacetime.

space-charge In a semiconductor, a region in which there is a net unbalanced charge. For example, at a ➤depletion layer where the electrons and holes have recombined, there is a deficit of electrons in the ➤n–type material (so there exists a net positive charge) and a deficit of holes in the ➤p–type material (so there exists a net negative charge).

spacecraft Any artificial object designed to travel outside the Earth's atmosphere, such as a ➤space probe, artificial ➤satellite or manned craft.

space group The ➤symmetry group of a crystal that includes ➤translations (2). ➤➤Bravais lattice.

space inversion ➤parity.

space-like Describing an interval in ➤Minkowski spacetime that is positive (i.e. $\Delta x^2 + \Delta y^2 + \Delta z^2 > c^2\Delta t^2$).

space probe An unmanned spacecraft designed to travel great distances to investigate other astronomical bodies or conditions in space.

spacetime The combination of space and time as coordinate systems. This allows an event to be specified by a single point in spacetime, which defines where it happens and when it happens. ➤➤Minkowski spacetime.

spallation The splitting of the nucleus of an atom into at least three parts. Compare ➤nuclear fission.

spark The visible result of ➤breakdown of air. The air in a path joining high potential to low potential is ionized, causing excitations which appear as bluish light. Spectacular examples can be found in high-voltage equipment such as the ➤Van de Graaff generator.

spark photography A method of photography used, for example, in ballistics, in which the illumination comes from a spark or flash whose duration controls the exposure (rather than the timing of the camera shutter).

Spearman's rank correlation coefficient The ➤correlation coefficient between two data sets allocated ranks 1, 2,. . ., n by some criterion. Because of the special form of the data, the shortcut formula

$$1 - 6\Sigma d^2/n(n^2 - 1)$$

is used where the ds are the differences between individual rankings. Named after Charles Edward Spearman (1863–1945).

special function Any one of the nonelementary functions deemed worthy of a special name. Examples include ➤Bessel functions and the ➤gamma function. Such functions often arise in physics when solving ➤differential equations in various coordinate systems. ➤➤elementary function.

special (theory of) relativity The area of physics dealing with high speeds (a significant fraction of the ➤speed of light). In special relativity the ➤Galilean transformation (used in classical physics for transforming coordinate systems between moving observers) is replaced by the ➤Lorentz transformation. A crucial feature is that the scale of time is different for observers moving at different speeds, which gives rise to the phenomena of ➤time dilation and ➤length contraction.

special unitary group The group of all ➤unitary matrices of a given order with ➤determinant +1, for example ➤SU(3).

speciation ➤evolution.

species 1 (Biol.) A population of individuals that share a high degree of common characters and interbreed freely to form fertile offspring. There may be one or many species within a ➤genus. However, the definition is not straightforward since under certain circumstances some species can interbreed to form fertile hybrids (➤hybridization (2)). Since evolution is a continuous process, the boundaries between species should be expected not to be absolute.

2 (Chem.) A general term that can be used to mean a set of chemically identical atoms, molecules, ions, etc.

specific When applied to most physical quantities, per unit mass or, exceptionally, per unit amount of substance (➤mole). For example, specific heat capacity is the heat capacity of a substance per unit mass, usually measured in $J \, K^{-1} \, kg^{-1}$. Use of specific quantities relates an ➤extensive property of a system to an ➤intensive property of the material making up the system.

specific gravity An obsolete name for relative ➤density.

specific heat capacity (specific heat) The ➤heat capacity per unit mass.

specific latent heat The ➤latent heat per unit mass.

specific resistance An obsolete term for ➤resistivity.

speckle interferometer A technique used in astronomy in which a large number of short-exposure images are taken and then combined by computer. This compensates for the turbulence in the Earth's atmosphere.

spectator ion An ion that remains in solution after a reaction between other ions has taken place. For example, in the reaction between aqueous silver nitrate and an aqueous sodium halide that precipitates a ➤silver halide, the nitrate and sodium ions are spectator ions.

spectra The plural of ➤spectrum.

spectral Relating to a ➤spectrum.

spectral line A narrow peak of intensity in a ➤spectrum.

spectral series A component of an emission spectrum consisting of a series of spectral lines resulting from transitions from different initial states to the same final state (➤Rydberg–Ritz equation). Examples are the ➤Lyman series and ➤Balmer series. ➤➤sodium spectrum.

spectral type (spectral class) A measure of the ➤effective temperature of a star. The relative intensity of spectral lines is used to determine the temperature of the emitter (➤Planck's law of radiation). The ➤Harvard classification is the standard for spectral types.

spectrochemical series A series listing ligands by the magnitude of the d-orbital splitting (➤ligand field theory) they produce. The series can be broadly divided into ➤pi bases, such as Cl^- and Br^-, which produce the smallest splitting, ligands forming no ➤pi bonds, such as H_2O and NH_3, and finally ➤pi acids, which cause the largest splitting, especially the ligands CN^- and CO.

spectroheliogram An image of the Sun obtained by photography using a very narrow filter permitting only a single spectral line to pass (typically calcium at 393 nm or hydrogen at 656 nm).

spectrometer (spectroscope) Any instrument used to detect the intensity of emission or absorption (of electromagnetic radiation or particles) as a function of frequency or energy. There are a vast number of different types of spectrometer which employ a wide range of techniques to detect photons, electrons and other particles. In a typical **double-beam absorption spectrometer** a monochromator is placed in front of a source of radiation in order to extract a single frequency. This monochromatic beam is then passed alternately through the sample and a reference. The difference in intensity of radiation detected by photomultipliers is interpreted as the absorption spectrum.

spectrophotometer A ➤spectrometer used with visible light. An atomic absorption spectrophotometer can analyse for up to twenty elements simultaneously at concentrations as low as one part per billion.

spectroscope ➤spectrometer.

spectroscopic binary A ➤binary star whose components are detectable by the Doppler shifts (➤Doppler effect) of the radiation coming from them as they alternately approach and recede from the Earth in the course of their orbits around each other.

spectroscopy The study of the interaction of electromagnetic radiation with matter (using a ➤spectrometer). In ➤emission spectroscopy the radiation is emitted (as in the ➤hydrogen atom emission spectrum or the ➤sodium spectrum), whereas in ➤absorption spectroscopy the radiation is absorbed. Atoms produce **line spectra**, whereas molecules produce **band spectra**. The various ranges of the ➤electromagnetic spectrum are probed as follows, in order of increasing frequency. In the radio region, ➤nuclear magnetic resonance provides information, especially on protons, and hence on the structure of organic compounds in particular. In ➤microwave spectroscopy, pure rotational transitions of molecules can cause absorption, and ➤electron spin resonance can be used to look at ➤radicals. ➤Infrared spectroscopy provides a detailed analysis of the vibrations molecules and crystals can undergo; characteristic absorptions can be used to identify specific structural features, especially in organic compounds. In the visible and ultraviolet, electronic transitions are observed. Gamma-ray spectroscopy is a more specialized technique (➤Mössbauer effect). Spectroscopy has a large number of applications, from probing the chemical composition, physical structure and electronic structure of materials in the laboratory to the elucidation of the composition and temperature of distant stars. ➤➤Auger effect; photoelectron spectroscopy; Raman effect.

spectrum (plural **spectra**) A measurement of the intensity of a quantity as a function of frequency or energy. It is often obtained as the experimental result of ➤spectroscopy, but is also used in applications where an intensity as a function of time is broken into its frequency components.

specular reflection ➤Reflection in which the angle of incidence of a ray is exactly equal to the angle of reflection. This can occur when light reflects from a flat surface, but the term is also used to describe, for example, collisions between gas molecules and the walls of a container in which no energy is lost.

speed The magnitude of ➤velocity. Speed, like the magnitudes of all vector quantities, is a ➤scalar. The speed of a body is the distance travelled by the body per unit time.

speed of light (in a vacuum) Symbol c. A quantity which, according to ➤special relativity, is the same in any ➤frame of reference, regardless of the motion of the observer (➤Lorentz transformation). The ➤metre is defined in such a way as to make the speed of light in a vacuum equal to exactly 299 792 458 m s^{-1}.

spermatogenesis The stages by which ➤germ cells (primary **spermatocytes**) in the ➤testis divide by ➤meiosis and mature to become functional sperm.

Spermatophyta (seed plants) The vascular plants that reproduce by ➤seeds, in some classifications regarded as a ➤phylum. The group includes the ➤Magnoliopsida and the ➤Pinopsida.

spermatozoon (plural **spermatozoa**) A motile male ➤gamete (sperm).

sphalerite An ore of zinc of approximate formula ZnS, which may also contain some iron.

sphere A surface in three-dimensional space, every point of which is a fixed distance from a fixed point. The fixed distance r is the **radius** of the sphere and the fixed point is its **centre**. With the centre as origin, the equation of the sphere is $x^2 + y^2 + z^2 = r^2$. Its surface area is $4\pi r^2$ and the volume it encloses is $\frac{4}{3}\pi r^3$. Of all closed surfaces with a given area, the sphere encloses the maximum volume.

spherical aberration ➤aberration.

spherical harmonic The angular portion $Y_{lm}(\theta,\phi)$ of the solution to ➤Laplace's equation in ➤spherical polar coordinates given by

$$Y_{lm}(\theta,\phi) = N_{lm} \, P_l{}^m(\cos\theta)\exp(im\phi)$$

where

$$P_l{}^m(x) = (1 - x^2)^{m/2} \, d^m/dx^m \, P_l(x);$$

$P_l(x)$ is the Legendre polynomial and $P_l{}^m(x)$ is the associated Legendre polynomial, N_{lm} is a normalizing constant.

spherical polar coordinates A system for specifying the location of a point in space by means of its distance r from a fixed origin and two angles θ and ϕ which may be visualized as 'colatitude' and 'longitude' (see the diagram). Spherical polar coordinates are related to ➤Cartesian coordinates by:

$$x = r \sin \theta \cos \phi,$$
$$y = r \sin \theta \sin \phi,$$
$$z = r \cos \theta,$$

and they are particularly useful in problems involving a degree of spherical symmetry.

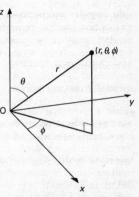

spherical polar coordinates

spherical wave A wave whose ➤wavefronts are spherical. Such a wave emanates from a ➤point source.

spheroid The surface formed by rotating an ellipse about one of its axes of symmetry. An **oblate spheroid** (like the shape of the Earth) is formed by rotation about the minor axis; a **prolate spheroid** (like the shape of a rugby ball) is formed by rotation about the major axis.

sphincter A circular tract of ➤muscle encircling an opening of a tubular structure in animals. Examples include the anal sphincter, the **pyloric sphincter** of the stomach, and others in the urinary and male reproductive system.

spin Symbol S. The intrinsic ➤angular momentum associated with a subatomic particle, which can be loosely associated with a picture of the particle spinning. Spin is inherently quantum mechanical in its nature. It can be pictured as a vector quantity; thus there is a total spin and a component of spin in a specified direction. The total spin has a **spin quantum number** (symbol s) with a value equal to an integer for a ➤boson, and a half-integer for a ➤fermion (the unit being $\hbar = h/2\pi$, where h is the ➤Planck constant), and the word 'spin' is often used to mean this quantum number. However care should be taken as the relationship between the **spin angular momentum** and the spin quantum number is not trivial: $|S| = \hbar \sqrt{s(s+1)}$, where S is the spin angular momentum and s is the spin quantum number (➤Pauli principle; spin–orbit coupling; Stern–Gerlach experiment). Spin is the basis of ➤nuclear magnetic resonance and ➤electron spin resonance.

spinal column ➤vertebral column.

spinal cord The extension of the ➤brain forming part of the ➤nervous system of vertebrates which runs along the ➤dorsal part of the body, surrounded by the ➤vertebral column. The spinal cord is a major tract of nervous tissue carrying ➤nerve impulses to and from the brain and along ➤reflex arcs. The cell bodies of most of the

➤neurones in the spinal cord lie in the central region and form the **grey matter**. The ➤axons form the outer zone, and appear as a distinct zone of **white matter**.

spin angular momentum quantum number Symbol S or s. The ➤angular momentum quantum number associated with the ➤spin of one or more electrons, usually in an atom, ion or molecule. Spin angular momentum is quantized in units of $h/4\pi$, where h is the ➤Planck constant. Compare ➤orbital angular momentum quantum number.

spindle ➤meiosis; mitosis.

spine 1 ➤vertebral column.
   2 The leaves of plants such as cacti, modified and reduced into robust narrow pointed structures which deter grazing by animals and reduce water loss.

spinel A mixed oxide of magnesium and aluminium, $MgAl_2O_4$, which gives its name to an important crystal structure of form AB_2O_4. The oxide ion array is close-packed and the metal ions fit into the holes in this array. In a **normal spinel** the A^{2+} ions, such as Mg^{2+}, are in the tetrahedral holes whereas the B^{3+} ions, such as Al^{3+}, are in the octahedral holes. ➤inverse spinel.

spin glass An alloy consisting mostly of a nonmagnetic material (e.g. gold) with a small proportion of magnetic material (e.g. iron). The magnetic moments (due to the spin of the unpaired electrons) on the magnetic atoms, instead of being in a regular crystalline array as they would be in the pure magnetic material, are scattered randomly through the system, and there is no ordered magnetic structure (in the same way that glass has no crystal structure).

spinneret The silk-spinning organ of certain animals such as spiders. This natural technique is mimicked industrially, using a thimble with a flat base containing holes through which a solution containing, for example, rayon is forced in order to produce fibres.

spin–orbit coupling An interaction between the ➤orbital angular momentum and spin angular momentum (➤spin) of an electron (or set of electrons) in an atom as a result of the magnetic moment that each creates. The energy difference that spin–orbit coupling creates between states with different quantum numbers lies behind ➤Hund's rules. If spin–orbit coupling is small compared with the interactions between electrons (as is typical for lighter atoms), it is helpful to add the spins together to give a total S, and orbital angular momenta together to give a total L, and then use spin–orbit coupling to derive the small correction to the energy. This is called **Russell–Saunders coupling** or LS **coupling**. If the spin–orbit coupling is more significant (as is typical for heavier atoms), it may first be necessary to add spin s and orbital angular momentum l for each individual electron to give an overall angular momentum j, and then add the j values for the electrons together as a perturbation (jj **coupling**).

spin quantization Although all electrons have a ➤spin quantum number of $\frac{1}{2}$, when in an atom the electron's spin can point in one of two directions ($m_s = +\frac{1}{2}$ or $m_s = -\frac{1}{2}$, conventionally labelled 'up' and 'down'). ➤➤pairing of electrons.

spiracle The modified ➤gill slit forming a round hole posterior to the gills of a shark or ray through which water passes from the pharynx. The term also applies to the paired pores on the body segments of insects by which the respiratory system connects with the exterior.

spiral Any one of a class of curves with the whirling shape suggested by their name. Examples include the **Archimedean spiral** with polar equation $r = a\theta$ and the **logarithmic spiral** with polar equation $r = ae^{\theta}$, which occurs in nature in the shape of the nautilus shell. ➤➤polar coordinates.

spiral galaxy A ➤galaxy with apparent spiral arms. Most spirals are **normal spirals**, with arms emerging from a central spheroidal bulge, but some 20% are **barred spirals** in which the arms emerge from a straight bar.

spleen A mass of lymphoid tissue forming an organ, lying below the stomach of vertebrates, that acts as a reserve store of ➤lymphocytes and ➤plasma cells. It is also partly responsible for the breakdown and recycling of ➤erythrocytes, and is involved in regulating the number of erythrocytes in the circulation.

splicing The process by which ➤RNA molecules, formed by ➤transcription in the nucleus, have ➤introns removed. It is achieved by cutting RNA between the junction of each intron and ➤exon, and joining the two flanking exons to generate a continuous stretch of RNA containing only ➤nucleotide sequences that code for ➤protein. ➤➤snRNP.

SPM Abbr. for ➤selectively permeable membrane.

sponge ➤Porifera.

spontaneous emission Emission of a photon (from an excited atom, ion, molecule or solid) that is not related to the absorption of a photon of the same energy. Compare ➤stimulated emission.

spontaneous fission ➤Nuclear fission that takes place with neither energy input nor collisions with incident particles.

spontaneous reaction A reaction that has a natural tendency to occur; the ➤standard Gibbs energy change must be negative.

spore A reproductive cell formed by ➤meiosis that can develop into another individual or another multicellular reproductive structure without undergoing ➤fertilization. Spores are produced in plants, and are especially obvious in mosses and ferns.

sporophyte The ➤ diploid phase of the life cycle of a plant which typically gives rise to ➤spores by ➤meiosis.

spreadsheet A computer program, or corresponding data file, that can perform mathematical functions on rows and columns of figures. It removes the need for tedious calculations on long lists of figures, as the spreadsheet will be automatically updated if any of the numbers are changed, so that different scenarios may be easily examined.

spring balance A simple instrument for measuring force. It uses the extension of a spring (usually assumed to obey ➤Hooke's law) to indicate the force applied to it.

spring constant The ratio of the force exerted by a spring to its extension. For a material obeying ➤Hooke's law, the spring constant is indeed constant, but it may vary with extension in more complex systems.

SPS Abbr. for the Super Proton Synchrotron particle accelerator at ➤CERN, used in the discovery of the ➤gauge bosons.

sputtering The bombardment of a surface by atoms or ions, either with a view to depositing a layer on the surface (➤epitaxy) or to clean it at an atomic level (argon ions are typically used for this purpose).

square **1** A ➤quadrilateral with sides of equal length and with all interior angles equal to 90°.
 2 The result, or operation, of multiplying a number or expression by itself.

square matrix A ➤matrix with the same number of rows and columns.

square-planar complex A ➤complex with four ➤ligands arranged in the geometrical shape of a square. The most important examples are complexes of the elements of Group 10 of the ➤periodic table (nickel, palladium and platinum) in oxidation state +2. The drug *cis*-platin (➤*cis-trans* isomerism) is an important anticancer treatment.

square root ➤root (1).

square wave A periodic ➤waveform that switches sharply from one constant value to another and back again (see the diagram).

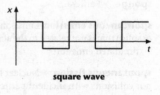

square wave

SQUID Acronym for superconducting quantum interference device, a device for detecting very small magnetic fields based on a Josephson junction (➤Josephson effect).

Sr Symbol for the element ➤strontium.

stability The property of a ➤system of being either ➤stable or ➤unstable. A particular feature that produces unstable behaviour is often known as an **instability**.

stability constant Symbol K_{stab}. An equilibrium constant describing the stability of a ➤complex. The **stepwise stability constant** describes the step-by-step replacement by one ligand at a time, whereas the overall stability constant describes the complete process: the overall constant is the product of each of the stepwise constants.

stable Describing a ➤system in a state in which a small change induces a restoring force to oppose the change and return the system to its initial state. For example, a ball resting in the valley between two hills is a stable system. If the ball were to roll from equilibrium to a higher position, its weight would pull it back towards the equilibrium. Compare ➤unstable.

staining A technique in microscopy in which dyes are used to increase the contrast or to identify the specific chemical composition of a specimen. Stains are very diverse, and may be applied to living cells (**vital staining**) or may require the killing and fixation (stabilization) of a specimen. In **counter-staining**, two or more dyes are used to produce a contrast between cells of different types.

stainless steel ➤Steel that resists corrosion well and hence is suitable for making cutlery, for example. A typical stainless steel contains 18% chromium and 8% nickel by mass.

stamen ➤flower.

standard atmosphere ➤international standard atmosphere.

standard deviation 1 A measure of the spread of a set of numbers $x_1, x_2, \ldots, x_n$ about their ➤mean $\bar{x}$ given by $\sqrt{\Sigma(x_i - \bar{x})^2/n}$. If the numbers are a ➤sample from a ➤population (2), the unbiased or **sample standard deviation** with a divisor of $n - 1$ instead of n is sometimes used.
2 The standard deviation of a ➤frequency distribution is given by $\sqrt{\Sigma f_i(x_i - \bar{x})^2/\Sigma f_i}$, where f_i is the class frequency and x_i is the mid-class value.
3 The square root of the ➤variance of a ➤random variable.

standard electrode potential The e.m.f. developed when a **standard half-cell**, a ➤half-cell in which the concentration is 1 mol dm^{-3}, is connected to a ➤standard hydrogen electrode, which is assigned an arbitrary value of zero. The standard electrode potential enables quantitative calculations of ➤redox reactions in aqueous solution to be performed. The half-cell with the *lower* standard electrode potential causes reduction ('the lower reduces'). So any metal with a negative standard electrode potential, such as sodium or calcium, can reduce water to hydrogen gas. A metal with a positive standard electrode potential, such as copper, cannot perform this reduction. ➤➤electrochemical series.
Standard electrode potentials can also be used for nonmetals. The value for iodine is the least positive for the halogens, showing that the ➤iodide ion is the best reductant in the group. Conversely, the standard electrode potential for fluorine is the most positive, reflecting the fact that ➤fluorine is the strongest oxidant in the group.
When the concentration varies from the standard value of 1 mol dm^{-3}, the value for the electrode potential is slightly different (➤Nernst equation).

standard enthalpy change (standard reaction enthalpy) Symbol $\Delta H^{\ominus}$; unit kJ mol^{-1}. The ➤enthalpy change per mole for conversion of reactants in their ➤standard states into products in their standard states, at a specified temperature (often, but not necessarily, 298 K).

standard enthalpy change of combustion (standard combustion enthalpy) The ➤standard enthalpy change for the complete combustion of a substance in oxygen.

standard enthalpy change of formation (standard formation enthalpy) The ➤standard enthalpy change for the formation of a substance from its elements in their ➤reference states.

standard form (scientific notation, exponential notation) A method of writing all numbers as the product of a decimal number between one and ten and a power of ten. It is especially convenient for expressing very large and very small numbers and is used in all scientific calculators.

standard Gibbs energy change (standard reaction Gibbs energy) Symbol $\Delta G^{\ominus}$; unit kJ mol^{-1}. The ►Gibbs energy change per mole for conversion of reactants in their ►standard states into products in their standard states, at a specified temperature (often, but not necessarily, 298 K). It is important to realize two points. First, the standard Gibbs energy change is the Gibbs energy change that would accompany the *complete* conversion of reactants into products, even if in practice the reaction stops at an intermediate composition, as for the synthesis of ammonia by the ►Haber–Bosch process. Second, the standard Gibbs energy change can be used to predict equilibrium positions at any temperature, not just 298 K, although most published data are for this temperature. For a chemical reaction, $\Delta G^{\ominus}$ is related to the equilibrium constant K (►equilibrium, chemical) for the reaction by the equation

$$\Delta G^{\ominus} = -RT \ln K,$$

where R is the ►gas constant and T is the thermodynamic temperature. This equation provides the link between thermodynamics and the experimentally measurable quantities.

standard Gibbs energy change of formation (standard formation Gibbs energy) The ►standard Gibbs energy change for the formation of a substance from its elements in their ►reference states.

standard hydrogen electrode (SHE, standard hydrogen half-cell) All measurements on cells can only detect the *difference* between two electrode potentials, never the actual value of one of them alone. The standard chosen is the SHE, a half-cell consisting of a solution containing hydrogen ions at unit concentration (1 mol dm^{-3}, i.e. pH = 0) into which is placed a platinum electrode over which hydrogen gas is passed at a pressure of 1 bar. The SHE is defined to have a ►standard electrode potential of 0 V.

standard model The model, central to particle physics, that describes the ►strong interaction in terms of ►quarks and a conserved ►colour charge with an ►SU(3) symmetry group, and the ►electroweak interaction in terms of quarks, ►leptons and the ►Higgs boson, a conserved electric charge and an SU(2) × U(1) symmetry group.

standard reduction potential Alternative name for the ►standard electrode potential.

standard solution A solution of known concentration.

standard state (of a substance) The pure substance at the standard pressure of 1 bar. ►reference state.

standard temperature and pressure (STP) An arbitrary standard for measuring, for example, gas volumes, where the temperature is defined as 0 °C (273.15 K) and the pressure is defined as 1 atm (101 325 Pa). The ➤molar volume of an ➤ideal gas at STP is 22.4 dm^3 mol^{-1}.

standing wave (stationary wave) A ➤wave that is confined to a fixed region rather than moving steadily though a medium. A standing wave is governed by boundary conditions: for example, for a standing wave on a string, there may be fixed points at which the displacement of the string is constrained to be zero, or points where the lateral force is constrained to be zero (a free end).

stannane The hydride of tin, SnH$_4$. There is just one hydride of tin compared with at least 75 known ➤alkanes and eight ➤silanes.

stapes ➤ear.

staphylococcus A genus of spherical Gram-positive (➤Gram's stain) ➤bacteria, typically occurring in clusters, many of which are implicated in human diseases. Staphylococcal infections include boils and abscesses, and one form of food poisoning. MRSA (methicillin-resistant *Staphylococcus aureus*) is an extreme ➤pathogen which is resistant to a wide spectrum of ➤antibiotics.

star A large, gaseous, self-luminous astronomical body which radiates light and other electromagnetic radiation when energy is produced by ➤nuclear fusion reactions in its core. The Sun is a star; stars are grouped into ➤clusters and ➤galaxies. ➤binary star; dwarf; giant; Hertzsprung–Russell diagram; neutron star; stellar evolution; supergiant; variable star.

starch A ➤polysaccharide comprising α-(1-4)-linked glucose molecules (see the diagram). Between 10 and 30% of natural starch is made of **amylose**, an unbranched chain with M_r between 4×10^3 and 1.5×10^5. Starch also contains **amylopectin**, which is heavily branched. Compare ➤glycogen. ➤➤cellulose.

starch

Stark effect The splitting of degenerate energy levels (➤degenerate states) when an electric field is applied to an atom. Named after Johannes Stark (1874–1957).

start codon The ➤codon ATG (in ➤DNA) or AUG (in ➤RNA) that signifies the beginning of an ➤open reading frame. In the universal ➤genetic code, this codon represents the amino acid methionine, known in this instance as the **initiating methionine**. The start codon signifies to ➤ribosomes the position at which ➤translation of ➤messenger RNA should start.

state A description of the attributes of a ➤system under certain conditions. ➤quantum state; states of matter.

state function A thermodynamic function whose value depends only on the state of the system, not on the route by which the state was reached. Examples include ➤enthalpy, ➤entropy and ➤Gibbs energy.

states of matter The three states – solid, liquid and gas – in which matter commonly exists. ➤➤change of state.

state symbols Symbols that label a spectroscopic state (energy level) by its ➤multiplicity, its angular momentum and, occasionally, its symmetry.

static 1 Not moving; the opposite of ➤dynamic.
2 Disturbance of a radio signal by natural atmospheric phenomena, for example lightning.

static electricity Electricity associated with charges at rest. ➤electrostatics.

static friction ➤Friction that acts between a body and a surface when the body is not moving. Static friction originates from the microscopic irregularity of surfaces. Compare ➤dynamic friction.

static pressure The pressure exerted by a fluid on a body moving freely with the fluid. It is also the pressure exerted on a surface that is parallel to the direction of motion of the fluid. Compare ➤dynamic pressure.

statics The branch of mathematics concerned with the ➤equilibrium of ➤rigid bodies, when the vector sum both of the forces acting and of their moments about any point are zero.

stationary orbit ➤geostationary orbit.

stationary phase ➤chromatography.

stationary point A point at which the ➤derivative of a function is zero: a stationary point gives rise to a ➤maximum, ➤minimum or point of ➤inflection. For more than one variable, a stationary point occurs at a point where all partial derivatives of a function are zero. ➤➤partial derivative.

stationary state ➤quantum state.

stationary-state approximation ➤steady-state approximation.

stationary wave ➤standing wave.

statistic A single representative number such as the ➤mean or ➤standard deviation calculated from the values of a sample and summarizing some aspect of the data.

statistical inference The branch of statistics dealing with inferences about a ➤population (2) on the basis of the evidence of a ➤sample. Statistical inference hinges on the surprising fact that, because of the ➤central limit theorem, the sampling distribution of a parameter may be known even if knowledge of the population distribution is very limited.

statistical mechanics The area of physics that relates the ➤microscopic ➤dynamics of a system to its ➤macroscopic thermodynamic properties. The properties (e.g. pressure) of a system with a very large number of weakly interacting particles (as in a gas) are determined by averages over that very large number of particles. This statistical averaging allows macroscopic properties to be predicted to within very narrow probability distributions, which for most purposes can be treated as exact values for the property. The dynamics of the system is described in terms of ➤microstates, which are either points in phase space (in a classical model) or quantum states (in a quantum-mechanical model). A large number of microstates may correspond to a small range of macroscopic parameters, the ➤macrostate. The role of energy in statistical mechanics is pivotal, as is temperature, and the ➤Boltzmann distribution plays a key role in determining probability distributions. The macroscopic concept of ➤entropy can be related to the degree of uncertainty about the microstate of the system.

statistics The branch of mathematics concerned with the collection, presentation and interpretation of data usually in the context of using (known) ➤sample data to make inferences about an (unknown) ➤population (2).

stator The nonmoving part of an electric motor or generator. Compare ➤rotor.

steady state A description of a system whose overall attributes are not changing with time, even though ➤dynamic equilibrium may be established.

steady-state approximation (stationary-state approximation) In reaction kinetics, an approximation that simplifies prediction of the form of rate equations. It assumes that the concentration of all intermediates is small as they are likely to be significantly more reactive than the reactants themselves.

steady-state theory A theory in which the Universe appears the same from all positions at all times. It relies on the production of new matter in the spaces between galaxies, and is at odds with the ➤big bang theory, which has effectively superseded it.

steam Water in the vapour phase.

steam distillation A separation procedure useful for temperature-sensitive organic compounds immiscible with water, especially ➤aniline. Steam is passed through the mixture causing boiling at a lower temperature than normal as water's vapour pressure adds to that of the compound being distilled. The resulting immiscible mixture is separated with a ➤separating funnel.

steam engine A machine in which combustion occurs outside the working chamber. Steam generated on combustion is used to force down a cylinder; the reciprocating motion of a piston is turned into rotary motion by a flywheel. Compare ➤internal combustion engine.

steam point The temperature at which the ➤saturated vapour pressure of water is exactly 1 atm. The Celsius scale is defined so as to make the steam point exactly $100\,^{\circ}\text{C}$.

stearic acid (octadecanoic acid) $C_{17}H_{35}COOH$ An important saturated fatty acid which exists as colourless crystals. ➤➤saponification.

steel Iron with a carefully controlled quantity of carbon, typically 0.2% by mass for **mild steel** to 2.0% for **high-carbon steel**. (The ➤iron–carbon phase diagram is extremely complex.) It is made by the ➤basic oxygen process, which has largely superseded the older ➤Bessemer process. Further elements can be added to make **alloy steels**, such as ➤stainless steel or ➤Hadfield steel.

Stefan–Boltzmann constant Symbol σ. The constant relating the power per unit area M_e emitted by a radiating black body (➤black-body radiation) to the thermodynamic temperature T of the body by means of **Stefan's law**: $M_e = \sigma T^4$. The Stefan–Boltzmann constant can be related to the fundamental constants using quantum mechanics:

$$\sigma = \frac{2\pi^5 k^4}{15h^3c^2},$$

where k is the ➤Boltzmann constant, h the ➤Planck constant and c the ➤speed of light in a vacuum. Named after Josef Stefan (1835–93) and Ludwig Edward Boltzmann (1844–1906). ➤➤Appendix table 2.

stellar Pertaining to a ➤star.

stellar evolution The formation and development of a star. A star similar to the Sun probably starts off as a rotating cloud of gas and dust which subsequently collapses under gravity to form a **protostar** within a dust shell extending to perhaps 10 000 solar radii. As the contraction continues, the temperature rises and ➤nuclear fusion reactions begin in the core (➤➤carbon cycle). The star joins the main sequence of the ➤Hertzsprung–Russell diagram. Its subsequent fate depends on its size. Stars initially between about 0.8 and 10 solar masses (➤solar units) evolve through red ➤giant and ➤supergiant stages, shed their outer layers as a planetary ➤nebula to become a ➤white dwarf, and slowly cool. (No star below 0.8 solar masses has had time to evolve off the main sequence.) Stars initially between about 10 and 50 solar masses become supergiants which then suffer a catastrophic core collapse and explode as a ➤supernova. The remaining mass exceeds the ➤Chandrasekhar limit for the size of a white dwarf, and collapses into a ➤neutron star. It is rare for stars to form with an initial mass above 50 solar masses. Such stars evolve very quickly to the supernova stage, after which they collapse to form ➤black holes.

stem ➤shoot.

stem cell An undifferentiated cell, particularly in an embryo, capable of unlimited division and of producing daughter cells that develop into different cell types, which in turn give rise to different tissues and organs. Stem cells of more limited ability for growth and differentiation are also found in adults. Stem cells are currently the focus of intense research, since in principle they could be used to grow tissues and organs as replacements for those damaged by disease or trauma. In 2008 a patient's own stem cells were used to reduce the chance of rejection of a transplanted section of ➤trachea.

steradian A measure of ➤solid angle defined so that the surface of a sphere of unit radius subtends a solid angle of 4π steradians at its centre; see the diagram.

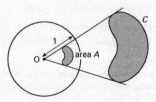

steradian The curve C subtends a solid angle of A steradians.

stereochemistry The study of the arrangement in space of the atoms in a molecule. ➤conformation; stereoisomerism.

stereographic projection (stereogram) A conformal azimuthal projection (➤map projection) used on maps and for crystal structures. ➤conformal transformation.

stereoisomerism The existence of isomers that differ in the spatial arrangement of the groups, as in ➤*cis–trans* isomerism or ➤optical activity.

stereoregular polymer A polymer with a particular arrangement of the groups in space, for example a ➤syndiotactic polymer. ➤Ziegler–Natta catalysts.

stereospecific reaction A reaction that results predominantly in only one of a variety of possible stereoisomers (e.g. ➤inversion of configuration).

steric hindrance A slowing in the rate of reaction of a molecule because of the presence of a large bulky group. For example, the ➤Friedel–Crafts reaction of $C_6H_5C(CH_3)_3$ with $(CH_3)_3CCl$ gives almost exclusively the ➤para isomer as the bulky groups would be too close together in the ➤ortho isomer.

Stern–Gerlach experiment An important experiment, completed in 1922, that confirmed the **space quantization** of angular momentum, and also confirmed the existence of electron ➤spin (though this was not recognized at the time). A beam of silver atoms deflected by an inhomogeneous magnetic field followed just one of two paths, depending on their ➤spin quantum number, rather than being distributed in a ➤Gaussian distribution centred on the original direction of the beam. Named after Otto Stern (1888–1969) and Walther Gerlach (1889–1979).

steroid One of a group of biologically active compounds derived from ➤cholesterol. Steroids have a four-ring structure consisting of one cyclopentane and three six-membered rings (see the diagram), and have an important role as ➤hormones. The sex hormones ➤testosterone, ➤progesterone and ➤oestrogen are steroids, as are the ➤corticosteroids, which are synthesized in the ➤adrenal cortex. ➤anabolic steroid.

steroid The steroid testosterone.

stibnite (antimony glance) The only important ore of antimony, antimony(III) sulfide, Sb_2S_3.

sticky ends Single-stranded extensions of ➤nucleotides left after double-stranded ➤DNA has been cut by certain types of ➤restriction enzyme. Sticky ends are significant since mixtures of DNA fragments from different sources cut with the same restriction enzyme can join together by complementary ➤base pairing of these extensions. This is the basis for making recombinant DNA in ➤genetic engineering.

stigma ➤flower.

still An apparatus used for ➤distillation.

stimulated emission Emission of radiation caused by incident radiation. ➤laser; compare ➤spontaneous emission.

stimulus (plural stimuli) Any event or factor that evokes a response by an organism. Stimuli to which animals respond include sound, light, touch and odours. Plants respond to stimuli such as gravity, directional light, touch and the day–night cycle.

Stirling's formula (Stirling's approximation) The approximation to $n!$ (➤factorial) given by the formula:

$$n! \approx \sqrt{2\pi n}(n/e)^n.$$

It becomes an increasingly better approximation as n increases and is an important tool in statistical mechanics. Taking natural logarithms of this equation gives an even more approximate equation that is occasionally called Stirling's approximation:

$$\ln n! \approx n \ln n - n.$$

Named after James Stirling (1692–1770).

STM Abbr. for ➤scanning tunnelling microscope.

stochastic process A process that develops with time in a statistically random way, such as a ➤random walk. The prices of stocks and shares are often modelled as stochastic processes.

stoichiometric In a ratio that corresponds to that given in the formula for the compound or the equation for the reaction. The **stoichiometric coefficient** is positive for products and negative for reactants. Hence for the reaction of nitrogen and hydrogen to form ammonia, the stoichiometric coefficients are +2 for NH_3, −1 for N_2 and −3 for H_2:

$$N_2(g) + 3H_2(g) \rightleftharpoons 2NH_3(g).$$

Compare ➤nonstoichiometric compounds.

stokes Symbol St. The c.g.s. unit of ➤kinematic viscosity. 1 St is equivalent to $1 \text{ cm}^2 \text{ s}^{-1}$. Named after George Gabriel Stokes (1819–1903).

Stokes's law The resistance felt by a sphere of radius r moving through a fluid at speed v is $6\pi\eta rv$, where η is the ➤viscosity of the fluid. It is valid only for a ➤Reynolds number less than about 0.2.

Stokes's theorem The identity

$$\iint_S \text{curl } \boldsymbol{F} \cdot \boldsymbol{n} \, dS = \int_C \boldsymbol{F} \cdot \boldsymbol{t} \, ds$$

that relates the ➤surface integral of the ➤normal (2) component of the ➤curl of a ➤vector field F over a surface S to the ➤line integral of its tangential component $\boldsymbol{F} \cdot \boldsymbol{t}$ over the curve C bounding its surface.

stoma (plural stomata) A pore on the surface of a vascular plant that facilitates ➤gas exchange between cells in leaves and stems and the external atmosphere. Stomata consist of a pair of special cells (**guard cells**) bounding a central hole, the size of which can be regulated by movements of the guard cells caused by changes in ➤turgor. The opening and closing of stomata to allow gas exchange yet prevent excessive water loss are controlled by a wide range of environmental and ➤endogenous factors.

stomach ➤gut.

stop A means of restricting the light entering an optical system.

stop codon ➤termination codon.

stopping power A property of a material that measures its ability to slow a moving particle. It is the rate of loss of kinetic energy per unit distance, sometimes measured per unit mass density (**mass stopping power**) and sometimes per unit number density of atoms in the material (**atomic stopping power**).

STP ➤standard temperature and pressure.

straight-chain isomer An ➤alkane with all the carbon atoms in a continuous chain, whose names form the basis of the ➤IUPAC nomenclature. Traditional nomenclature used the notation *n*-alkane to distinguish the straight-chain isomer.

strain The extension of an object per unit original length when it is subjected to a ➤stress. Strain is a dimensionless quantity. ➤➤shear strain.

strain gauge A device for measuring strain, usually by measuring the change in resistance or capacitance of an electrical component attached to the object under strain. For very small strains piezoelectric crystals are used (➤piezoelectric effect).

strain hardening ➤work hardening.

strange attractor The ➤fractal ➤trajectory in ➤phase space followed by chaotic nonlinear dynamic systems for some values of initial ➤parameters (2) as time tends to infinity. ➤➤chaos theory.

strangeness A ➤quantum number associated with subatomic particles that is conserved in the strong and electromagnetic interactions, but not in the weak interaction. Its existence was postulated to explain the fact that some particles were observed to have much longer lifetimes than expected. Strangeness is associated with the presence of one or more ➤strange quarks in the particle: the strange quark has a strangeness value of -1, and the strange antiquark a value of $+1$.

strange quark A ➤flavour of ➤quark. The strange quark was the third of the six flavours of quark of the ➤standard model to be discovered, and the first of them that is not commonly found in matter (unlike the ➤up quark and ➤down quark). The ➤kaon was the first particle now known to contain a strange quark to be discovered (in 1947). A ➤quantum number, **strangeness**, is associated with strange particles.

stratopause The boundary in the atmosphere between the ➤stratosphere and the ➤mesosphere, delineated by a change from rising to falling temperature with increasing altitude.

stratosphere The layer of the ➤atmosphere above the ➤troposphere. Its base is at 11 km altitude in the ➤international standard atmosphere, and is characterized by a slow rise in temperature with altitude. This creates stability, and means that weather (in the form of clouds) is almost entirely confined to the troposphere below.

streamline A line through a fluid in the direction of the velocity of the fluid at each point. In an incompressible fluid such lines are continuous, and are analogous to magnetic field lines for describing a magnetic field.

streptococcus A genus of spherical, Gram-positive (➤Gram's stain), aerobic bacteria occurring as pairs or chains of cells. Some species (in particular *Streptococcus pyogenes*) are pathogenic in humans and cause throat infections, scarlet fever and some forms of pneumonia. Nonpathogenic streptococci are used in the manufacture of dairy products such as cheese and buttermilk.

stress Unit Pa or $N m^{-2}$. The force applied to or within a body per unit cross-sectional area. A stress applied to a body normally causes a ➤strain. ➤➤shear stress; Young's modulus.

stress–energy tensor In relativity, the second-rank symmetric ➤tensor that describes the density and flux of energy and momentum. ➤➤Einstein field equation.

string theory A theory of elementary particles which treats them as tiny one-dimensional strings rather than zero-dimensional objects. States of these strings correspond to standing waves, and frequencies correspond to energies and therefore masses, which serve to distinguish different types of particle.

stroboscope (strobe) A flashing light whose frequency can be adjusted to make a periodic motion (such as a rotating object) appear stationary.

stroma ➤chloroplast.

strong acid An acid that is essentially completely ionized in aqueous solution, such as hydrochloric or nitric acid. Compare ➤weak acids and bases.

strong base (strong alkali) A base that is essentially completely ionized in aqueous solution, such as sodium hydroxide or potassium hydroxide. Compare ➤weak acids and bases.

strong electrolyte An electrolyte that is effectively completely ionized in aqueous solution.

strong interaction One of the four ➤fundamental interactions of nature. The strong interaction is felt by ➤hadrons, not by ➤leptons. It has been called the **strong nuclear force** because it is the attractive force that holds together protons and neutrons in atomic nuclei. It is a very short-range force, which dominates the other interactions only at distances of about 10^{-15} m or less; it conserves ➤baryon number and ➤isotopic spin, and is carried by the ➤gluon. ➤➤colour charge.

strontium Symbol Sr. The element with atomic number 38 and relative atomic mass 87.62 (named after the village Strontian in Scotland), which is in Group 2 of the ➤periodic table, below calcium. Chemically, strontium resembles calcium in having just one dominant oxidation number, +2, as in its carbonate ore strontianite, $SrCO_3$, or its sulfate ore celestine, $SrSO_4$. Strontium salts are used to produce a crimson colour in fireworks or flares. ➤➤fall-out.

structural formula The formula that specifies the molecule's connectivity (i.e. exactly which atoms are bonded to which other ones), thus providing more information than the molecular formula can give. The **condensed structural formula** conveys the information in its most concise form (e.g. propane can be written $CH_3CH_2CH_3$), whereas the **displayed structural formula** fills in all the bonds too (see the diagram).

(a)   (b)

structural formula (a) A structural formula for propane showing the three-dimensional arrangement around each carbon atom. (b) A simpler structural formula for propane showing which atoms are connected together.

structural isomer ➤isomer.

structure factor The factor that modifies the ➤diffraction pattern of a crystal structure because of a ➤basis (2) within the unit cell.

strychnine The principal alkaloid (see the diagram), notoriously poisonous, from the aptly named *Nux vomica* tree. It is one of the most important optically active bases for resolving a ➤racemic mixture of optically active acids.

style ➤flower.

styptic Capable of staunching the flow of blood.

styrene The more common name for ➤phenylethene.   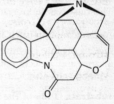 ➤➤polystyrene.

strychnine

styrene–butadiene rubber ➤SBR.

SU(3) The ➤special unitary group of unitary 3×3 matrices with determinant +1. It is particularly important in particle physics as the group for ➤symmetry transformations of the ➤colour of ➤quarks.

subatomic particle A particle smaller in size than an atom. The three most important are the ➤electron, ➤neutron and ➤proton. The electron is an ➤elementary particle. The neutron and the proton are composed of ➤quarks. ➤➤particle physics.

subcutaneous Under the ➤skin.

sublimate A solid formed by ➤sublimation.

sublimation The process in which a solid turns to a gas without first forming a liquid (or vice versa). Carbon dioxide sublimes at pressures below 5.1 atm, the pressure of the ➤triple point. Iodine is often collected as a sublimate.

submillimetre The frequency band (strictly microwave rather than radio frequency) covering wavelengths from 1 mm down to about 0.1 mm (frequencies

300 to 3000 GHz). Several important molecular emissions are found in this range, and the submillimetre region is of interest in radioastronomy.

subroutine (procedure) A part of a computer program that performs some well-defined task, and may be called upon at many points during the program. This saves duplication of the instructions in the procedure, and aids structured programming by allowing a problem easily to be broken into a series of smaller, self-contained sections.

subset A ➤set formed by taking some, but not necessarily all, of the ➤elements (2) of a given set. The notation $B \subseteq A$ means that B is a subset of A.

subshell A subdivision of a ➤shell in which the ➤orbitals all have the same ➤orbital angular momentum quantum number, hence there is a p subshell and a d subshell in addition to the s orbital for elements in period 3.

subsonic Moving at less than the speed of sound. Compare ➤supersonic.

substituent Any atom or group that replaces a hydrogen atom in an organic compound.

substitution reaction A reaction in which one group is substituted for another. This can occur in inorganic chemistry in **ligand substitution reactions**, such as:

$$[Ni(H_2O)_6]^{2+}(aq) + 6NH_3(aq) \rightarrow [Ni(NH_3)_6]^{2+}(aq) + 6H_2O(l),$$

and in organic chemistry where both ➤nucleophilic substitution of halogeno-alkanes and ➤electrophilic substitution of aromatic hydrocarbons are important synthetic processes.

substrate 1 (Biol.) The substance on which an ➤enzyme acts.
   2 (Phys.) A material, typically a crystal or a semiconductor, on which a second material is to be deposited or adsorbed.

substratum The underlying surface, such as the sea bed, over which organisms move or to which they are attached.

subtend The angle subtended by two points A and B at the ends of a line segment or curve from a point P is the angle APB.

subtraction The mathematical operation in which the difference of two numbers, vectors, matrices, polynomials, etc., is found. It is best thought of as the inverse of ➤addition.

succinic acid ➤butanedioic acid.

succinic anhydride The anhydride of succinic acid (see the diagram); it forms colourless needles, and is used in the manufacture of ➤alkyd resins.

succus entericus A copious secretion (between two and three litres per day in humans) of the glandular walls of the small intestine (➤gut). It contains a variety of enzymes, including ➤lipases, ➤carbohydrases and ➤proteases, and is responsible for the majority of digestion of food.

succinic anhydride

sucrase The enzyme that catalyses the conversion of sucrose into ➤glucose and ➤fructose.

sucrose (cane sugar, beet sugar, household 'sugar') A ➤disaccharide consisting of ➤glucose and ➤fructose combined in a particular manner, the glucose being in a six-membered ➤pyranose ring whereas the fructose is in a five-membered ➤furanose ring (see the diagram). It is a nonreducing sugar. When hydrolysed, the products rotate the plane of plane-polarized light in the opposite direction from sucrose itself (➤invert sugar).

sucrose

sufficient condition A condition that is sufficient for the truth of a statement in the sense that the statement may be derived from it. Thus for a triangle to be an ➤isosceles triangle, it is sufficient for it to be ➤equilateral.

sugar A ➤carbohydrate with a definite, fixed molar mass; monosaccharides and disaccharides are examples. ➤Glucose and ➤sucrose are examples of sugars. ➤Polysaccharides, such as starch, which can have a variable molar mass, are carbohydrates but are not sugars. For **reducing sugar**, ➤Benedict's solution.

sulfa drugs (sulfonamide drugs) Drugs related to ➤sulfanilamide and sharing its ability to treat bacterial diseases. The drugs work by inhibiting the growth of bacteria, allowing the host's natural defences to overcome them more quickly. Examples are **sulfacetamide**, used to treat conjunctivitis, **sulfathiazole**, effective against pneumococci and staphylococci, and **sulfapyridine**. They are generally made by

sulfa drugs (a) Sulfacetamide; (b) sulfathiazole; (c) sulfapyridine.

treating ethanoylated *p*-aminobenzenesulfonyl chloride with an amine, and then removing the protecting ethanoyl group by hydrolysis (see the diagram overleaf).

sulfamic acid NH_2SO_2OH A colourless crystalline solid, normally existing as a ►zwitterion. It is a strong acid used in a variety of processes with products ranging from sweeteners to weed-killers.

sulfane ►polysulfane.

sulfanilamide The typical ►sulfa drug, used as a treatment for bacterial infection since it was shown to be the active ingredient in 'prontosil', which first proved effective against bacteria in 1934. Sulfanilamide can treat streptococcal, gonococcal and meningococcal infections, although occasionally with unwelcome side effects such as rashes.

sulfanilic acid (*p*-aminobenzenesulfonic acid) $H_2NC_6H_4SO_2OH$ Colourless crystals used to make both ►diazo dyes and ►sulfa drugs.

sulfate A compound that contains the **sulfate ion**, SO_4^{2-}, the oxoanion containing sulfur with its highest oxidation number of +6, hence the alternative name of **sulfate(VI) ion**. The test for the sulfate ion is to add barium nitrate acidified with nitric acid (the purpose of which is to destroy any carbonate or sulfite ions, which give the same test result) and observe a white precipitate of barium sulfate:

$$Ba^{2+}(aq) + SO_4^{2-}(aq) \rightarrow BaSO_4(s).$$

Important sulfate ores include ►anhydrite, $CaSO_4$, and ►Epsom salts, $MgSO_4 \cdot 7H_2O$.

sulfide The **sulfide ion** S^{2-} is present in binary ionic sulfides such as sodium sulfide, Na_2S. Other metallic sulfides have more covalent character, which is one reason why they are often very insoluble in water (►solubility product). The parent acid is ►hydrogen sulfide. Important sulfide ores include zinc blende, ZnS, and cinnabar, HgS.

sulfite ion (sulfate(IV) ion) SO_3^{2-} The oxoanion containing sulfur with oxidation number +4, an intermediate value in sulfur's range of oxidation numbers. Hence sulfites can act as oxidants or reductants under particular conditions. They are most commonly used as antioxidants (►sodium sulfite).

sulfonamide drugs ►sulfa drugs.

sulfonation The introduction of a sulfonic acid group, —SO_2OH, into a molecule, most commonly performed in aromatic organic chemistry. The archetypal molecule produced is benzenesulfonic acid, $C_6H_5SO_2OH$, made by reacting concentrated sulfuric acid with benzene, the mechanism for which is electrophilic substitution with SO_3 as the electrophile.

sulfoxide A compound of the general form RSOR′, the most important example of which is dimethylsulfoxide, a useful solvent.

sulfur (traditional UK **sulphur)** Symbol S. The element with atomic number 16 and relative atomic mass 32.07, which is in Group 16 of the ►periodic table, beneath oxygen. It exhibits ►allotropy, the stable allotrope at room temperature being

sulfur S_8 ring.

rhombic sulfur, which forms a yellow solid. On controlled heating just above the transition temperature of 96 °C, the other allotrope, **monoclinic sulfur**, can be formed. Both are based on S_8 crown-shaped rings (see the diagram opposite), but the packing is less ordered in monoclinic sulfur. The liquid's behaviour on heating is unusual: very suddenly at around 160 °C the viscosity increases greatly. The S_8 rings break and long-chain polymeric sulfur molecules can form. The polymers reach a maximum of 250 000 atoms at around 180 °C, thereafter becoming shorter, causing the viscosity to decrease gradually. At 444 °C, the liquid boils. When poured into water, **plastic sulfur** results. The element is extracted in the ➤Frasch process and 90% of the production goes to the manufacture of ➤sulfuric acid, although it is also important as a fungicide and in the➤vulcanization of rubber.

The element shows a wide range of oxidation numbers, as is common with nonmetals. First, in combination with metals it shows negative oxidation numbers, commonly -2 as in the ➤sulfides. With oxygen, two important compounds are formed, SO_2 and SO_3, with oxidation numbers +4 and +6. These oxidation numbers are mirrored in the oxoacids, ➤sulfurous acid and ➤sulfuric acid, as well as their salts, the ➤sulfites and ➤sulfates. With the halogens a range of oxidation numbers is shown, from +1 in the toxic golden-yellow liquid S_2Cl_2, used in rubber vulcanization despite its revolting smell, through +2 in the red liquid SCl_2 and +4 in the colourless gas SF_4 to +6 in sulfur hexafluoride SF_6. Because of its exceptional stability, SF_6 is used as a gaseous insulator in transformers and electrical switch gear.

sulfur dichloride dioxide (sulfuryl chloride) SCl_2O_2 A colourless liquid used as a ➤chlorinating agent.

sulfur dichloride oxide (thionyl chloride) SCl_2O A colourless liquid used as a ➤chlorinating agent.

sulfur dioxide SO_2 A colourless, choking gas produced on a very large scale from a variety of sources, some natural (such as volcanoes) and some artificial. The latter include emission during the ➤contact process for the synthesis of sulfuric acid and as a by-product of the combustion of fossil fuels. As sulfur dioxide contributes to ➤acid rain, efforts have been taken to reduce the levels of the gas and desulfurizing plants are installed in modern fossil fuel stations, sometimes using calcium hydroxide to react with the gas stream. In the laboratory the test for sulfur dioxide is that acidified potassium dichromate(VI) on filter paper held in the gas turns from orange to green, as sulfur dioxide is a reductant. ➤➤hydrogen sulfide.

sulfuric acid (sulfuric(VI) acid) H_2SO_4 or more specifically $(HO)_2SO_2$. The oxoacid of sulfur with oxidation number +6, hence its alternative name. It is manufactured in greater quantity than any other chemical, in terms of the mass produced. Currently the major manufacturing process is the ➤contact process; older methods include the ➤lead-chamber process. During the 19th century it became clear that the consumption of sulfuric acid was a good indicator of a nation's prosperity, so ubiquitous are its uses. Concentrated sulfuric acid, a colourless viscous liquid, is not only an acid, forming both salts (the ➤sulfates) and ➤acid salts, but also a powerful oxidant, oxidizing, for example, copper to copper(II) sulfate:

$$Cu(s) + 2H_2SO_4(l) \rightarrow CuSO_4(aq) + SO_2(g) + 2H_2O(l).$$

It is also an excellent drying agent and dehydrating agent (➤dehydration), as well as a component of a ➤nitrating mixture. It can be used to produce hydrogen chloride from chlorides and nitric acid from nitrates as it is the most involatile acid (b.p. 338 °C) due to very strong ➤hydrogen bonding, which also explains its viscous nature.

sulfurous acid (sulfuric(IV) acid) Approximate formula H_2SO_3 The acid, containing sulfur with its oxidation number of +4, made by dissolving sulfur dioxide in water:

$$SO_2(g) + H_2O(l) \rightarrow H_2SO_3(aq).$$

This reaction is a major contributor to ➤acid rain. The salts of sulfurous acid are the ➤sulfites.

sulfur oxoacid By far the most important ➤oxoacid of sulfur is ➤sulfuric acid. At least a dozen other ones occur, however, such as ➤sulfurous acid and ➤oleum.

sulfur trioxide SO_3 A white solid produced, for example, during the ➤contact process manufacturing sulfuric acid. Its structure is interesting as it is polymorphic. One form consists of a ring of three molecules, $(SO_3)_3$, while another form has chains of molecules. In the gas phase the molecule is trigonal planar in structure as expected from ➤valence-shell electron-pair repulsion theory.

sulfuryl chloride ➤sulfur dichloride dioxide.

sulphur The traditional UK form of ➤sulfur: hence other UK forms are **sulphates**, **sulphides**, etc., rather than sulfates, sulfides, etc.

summation check A method for detecting some errors in a set of data, used for example in bar codes. The sum of the individual items of data is found, and the total taken modulo (➤modular arithmetic) some number (typically the ➤word length of the data), and stored along with the data. In order to check the data at some later date, the same sum is calculated and the resulting **check-sum** is compared with that stored. If they differ, the data have changed in some way, due perhaps to an error in storage, retrieval or transmission.

summation notation Notation such as $\Sigma_{i=1}^{n} a_i$ used as shorthand for the sum $a_1 + a_2 + \ldots + a_n$.

Sun The star closest to the Earth and about which the Earth orbits; it is the central body of the Solar System. Its radius is approximately 700 000 km (over a hundred times the Earth's) and its mass is approximately 2×10^{30} kg. Energy is generated in its core by nuclear fusion, primarily via the ➤proton–proton reaction. Outside the core, energy is transferred by radiation through the **radiative zone** and then by convection cells through the **convective zone**. The outer visible layer is the **photosphere** ('sphere of light'), just a few hundred kilometres thick, from which the energy is radiated into space. Outside this is the **chromosphere**, a few thousand kilometres thick, and the highly rarefied **corona**, extending outwards for several million kilometres. The temperature at the Sun's core is about 15 million K; at the photosphere, about 5800 K; and in the corona, 1–2 million K. ➤➤eclipse; prominence; solar constant; solar cycle; solar flare; sunspot. ➤➤Appendix table 4.

sunspot A dark patch on the surface of the ➤Sun's photosphere. Sunspots tend to appear in pairs, corresponding to a loop of magnetic field lines leaving the surface and re-entering, and their number fluctuates with the ➤solar cycle.

superacid A very powerfully acidic solvent produced, for example, by mixing hydrofluoric acid with antimony(v) fluoride, $HF-SbF_5$, which enables certain cationic species to be isolated for much longer than in other media. The structure of the intermediate cation formed in electrophilic aromatic substitution was confirmed by ^{13}C NMR in a superacid solvent.

supercomputer A very powerful computer with fast processors, typically capable of several billion ➤flops, and large amounts of main memory. They are used for predicting weather patterns and for other computationally intensive tasks.

superconducting quantum interference device ➤SQUID.

superconductivity A property exhibited by certain materials (**superconductors**) at low temperatures in which electricity is conducted without resistance. A superconductor also shows perfect ➤diamagnetism (it excludes magnetic fields: ➤Meissner effect) and appears as if it has an ➤energy gap in its one-electron states around the ➤Fermi energy. Superconductivity is observed below some critical temperature (e.g. 7.2 K in lead) and critical magnetic field in the majority of nonmagnetic ➤transition metals, ➤lanthanides and ➤actinides. The microscopic theory (**BCS theory**) of superconductivity is complicated and revolves around the formation of bound pairs of electrons (**Cooper pairs**) at energies close to the Fermi energy. While this low-temperature superconductivity has been known since 1912, **high-temperature superconductors** that superconduct at temperatures up to about 100 K were not discovered until the 1980s. They are typically ceramics containing lanthanides; the mechanism of superconductivity in these materials appears to be different from the standard BCS model. The record transition temperature as of 2007 was 138 K.

supercooling When a solution cools quickly, the temperature may fall below its freezing point without solid being formed because of a kinetic delay in the crystallization process. This forms a ➤supersaturated solution. Once freezing does occur, the temperature rises again to the freezing point. Clouds and fog at temperatures below 0 °C often contain supercooled water droplets which form ice only when some substance is introduced to allow nucleation (such as dust, a car windscreen or an aircraft's wings).

superfluid A fluid that flows without resistance. The phenomenon is observed only in very special circumstances; the best-known example is helium−4 at temperatures less than 2.2 K.

supergiant A very large, extremely luminous star, 25 to as much as 1000 times the Sun's size. Supergiants are found at the top of the ➤Hertzsprung–Russell diagram. ➤➤stellar evolution.

supergravity A quantum-mechanical theory of gravity, which is based on ➤supersymmetry. It incorporates the **gravitino**, a hypothetical supersymmetric partner to the ➤graviton. ➤➤quantum gravity.

superheated Describing a liquid that has been heated above its boiling point. The most familiar example is superheated steam.

superheterodyne (superhet) receiver By far the most common type of radio receiver; it mixes the incoming signal with a locally generated waveform. The result, the **intermediate frequency** signal, is amplified and passed on to an audiofrequency amplifier. The advantage of this technique is that it avoids the need for a radiofrequency amplifier, performing the amplification at the much lower intermediate frequency.

super high frequency (SHF) The ➤frequency band from 3 to 30 GHz.

superior Describing a structure that lies above (or ➤anterior to) a reference point in an organism. Thus the superior vena cava (anterior vena cava) is the main vein conducting blood to the heart from the head and forelimbs. Compare ➤inferior.

supernatant liquid The liquid lying above a ➤precipitate.

supernova (plural **supernovae)** A huge explosion marking the final stage of ➤stellar evolution. The most recent visible to the naked eye, designated Supernova 1987A, took about 100 days to reach an absolute visual ➤magnitude (2) of −15, some 10 000 times brighter than the brightest star. One type of supernova occurs when the core of a ➤supergiant or supermassive star has burnt all its nuclear fuel and can no longer resist the gravitational attraction to its centre by ➤radiation pressure. It then undergoes rapid **gravitational collapse** to a ➤neutron star or ➤black hole, releasing huge amounts of gravitational energy, as much energy as an entire galaxy radiates in a year. Another type results from the transfer of material onto the ➤white dwarf component of a ➤binary star, as in a ➤nova, but the extra material takes the white dwarf above the maximum stable mass (the ➤Chandrasekhar limit) and it explodes.

superoxide A compound containing the **superoxide ion**, O_2^-, formed when very reactive metallic elements combine with oxygen. The most important superoxide is potassium superoxide, KO_2, which is the dominant product when potassium is heated in air. Potassium superoxide is used to purify air in submarines, not only by removing carbon dioxide but also by regenerating oxygen:

$$4KO_2(s) + 2CO_2(g) \rightarrow 2K_2CO_3(s) + 3O_2(g).$$

As a ➤radical, the superoxide ion is particularly toxic to living tissues because it disrupts a number of biological processes. It can deactivate ribonucleotide reductase (an enzyme needed to form deoxyribonucleotides from ribonucleotides) and other enzymes, especially those containing iron and sulfur, critical in many biochemical pathways. Such damage is held to be a factor in determining an individual's lifespan. The enzyme **superoxide dismutase** is needed to eliminate superoxide ions by conversion into hydrogen peroxide and oxygen:

$$2O_2^- + 2H^+ \rightarrow H_2O_2 + O_2.$$

superphosphate of lime A fertilizer mixture consisting mainly of calcium dihydrogenphosphate and calcium sulfate.

superposition The addition of two functions, usually of position, to create a resultant. If the functions are solutions of a ➤linear equation then the resultant will also be a solution of the equation. In particular, the effect of two different sources of light can be calculated by adding the amplitudes of the waves from the individual sources at each point. ⋙interference.

supersaturated solution A solution that contains more dissolved solid than a ⋗saturated solution at a given temperature. This means that the supersaturated solution has a thermodynamic tendency to precipitate. The rate may be sufficiently slow that it remains metastable for some time. ⋙cloud chamber; supercooling.

supersonic Travelling at greater than the speed of sound. ⋙hypersonic; Mach number.

supersymmetry A theory of particle physics in which ⋗fermions and ⋗bosons are related. Every fermion has a **superpartner sfermion** (e.g. slepton, squark) of integral spin, and every boson has a superpartner bosino (e.g. photino, gluino, wino, zino) of half-integral spin, related by a supersymmetry. These superpartners appear to be so massive that they cannot be detected; as of 2008, there is no direct experimental evidence that supersymmetry is indeed a real symmetry of nature.

supplementary angles Two angles that add up to $180°$.

supplementary unit ⋗SI units.

surd A numerical expression involving ⋗irrational ⋗roots of whole numbers, such as $\sqrt{7}$ and $2 + \sqrt{3}$.

surface A two-dimensional geometric figure, often specified by equations of the form $g(x, y, z) = 0$ or $z = f(x, y)$.

surface acoustic wave (SAW) A wave of ⋗ultrasound passed along the surface of a solid. **Acoustic wave devices** use these waves for signal processing.

surface integral Any integral of a ⋗scalar field taken over a two-dimensional surface. When ⋗parameters are used to define the surface, a surface integral may be evaluated as a ⋗double integral.

surface of revolution A surface formed by rotating a curve through $360°$ about an axis. The area of the surface of revolution formed by rotating the arc of a curve between $x = a$ and $x = b$ about the x axis is given by:

$$2\pi \int_a^b y\sqrt{1 + (dy/dx)^2}\, dx.$$

surface tension symbol γ; unit $N\,m^{-1}$. The energy required to increase the surface of a liquid by unit area. The tension arises as a result of the attractive forces between the molecules of the liquid.

surgical spirit Essentially ⋗methylated spirit together with a few other ingredients including castor oil.

susceptance The ⋗imaginary part of the ⋗admittance of a circuit.

susceptibility ⋗electric susceptibility; magnetic susceptibility.

suspension When a solid is present in a solution but has not had time to settle to the bottom of the tube, the mixture is said to be a suspension. Given time, the solid will gradually fall to the bottom under gravity. The terms 'suspension' and 'precipitate' are often used interchangeably.

SV40 (simian virus 40) A ➤papovavirus first isolated from monkeys but which can also infect humans. Its study has led to a wider understanding of ➤introns and ➤splicing, and of the structural complexities of the mechanisms governing ➤gene expression in ➤eukaryotic cells.

SVP Abbr. for ➤saturated vapour pressure.

S wave ➤secondary wave.

sweat gland ➤skin.

sweetener A substance that can add a sweet taste to food. The most important artificial sweeteners are aspartame, cyclamates and ➤saccharin.

sylvine (sylvite) An ore of potassium containing mainly potassium chloride, KCl.

symbiosis Two (or occasionally more) different organisms consistently living together, usually in close physical and physiological contact (literally, 'living together'). Three main kinds of **symbiotic** relationship are recognized: **commensalism**, an often rather loose association in which one of the organisms benefits with no apparent advantage or disadvantage to the other (e.g. the behaviour of cattle egrets which loosely associate themselves with herds of game animals and feed off the small animals they disturb); **mutualism**, in which both organisms benefit (e.g. the fungus–alga relationship in a lichen); and **parasitism**, in which one organism (the **parasite**) lives in or on the body of another (the **host**) and obtains nutrition and other benefit from it. A parasitic relationship is usually to the detriment of the host (➤malaria), but some host species have evolved remarkable tolerance to high levels of parasitic infection (e.g. sleeping sickness in African antelopes). The boundaries between these three kinds of symbiosis are not always clear because it can be difficult to establish advantage or disadvantage in the relationship.

symbol, chemical A 'fully laden' symbol such as $^{16}O_2^{2-}$ can be explained as follows. The left superscript is the ➤nucleon number of the nuclide of the element. There are two oxygen atoms in the ion, which is indicated by the right subscript. The charge of -2 is indicated by the right superscript. Sometimes a left subscript is used to give the atomic number, but this is superfluous, given the symbol of the element.

symbolic logic A formal system of logic, using symbols and well-defined rules for their manipulation, in order to establish logical deductions. ➤➤Boolean algebra.

symmetric function A function of several variables that takes identical values for all ➤permutations (2) of its variables.

symmetric matrix A ➤square matrix that remains unchanged when its rows and columns are interchanged (➤transpose matrix).

symmetry element One of the elements of symmetry that a molecule can possess, which include a ➤centre of symmetry, a ➤rotation axis, ➤improper rotation axis,

➤reflection (2) plane, etc. The symmetry elements that a molecule possesses are encapsulated in its ➤point group (2).

symmetry group A ➤group with elements that are transformations that leave a given object unchanged, with the binary operation given by forming composite transformations.

symmetry transformation A geometrical ➤transformation (2) that leaves an object unchanged. For example, a reflection about a horizontal axis is a symmetry transformation for the letter E, while a rotation through 180° is a symmetry transformation for the letter S. ➤➤symmetry group.

sympathetic nervous system ➤nervous system.

sympatric Describing two or more species or varieties within a species whose geographical range overlaps to at least some degree. The term is often used to describe speciation events (➤evolution), but there is considerable debate as to whether sympatric speciation has occurred or can occur. Compare ➤allopatric.

symphysis A partly fused joint in the ➤skeleton of a vertebrate formed by bones bound together with cartilage and fibrous connective tissue at which a limited degree of movement is possible. The classic example is the **pubic symphysis**, joining the pubis on each side of the ➤pelvic girdle together at the front. The joint can ease to give increased room in the pelvic girdle during childbirth.

synapse The junction between two ➤neurones. ➤Neurotransmitter substances allow presynaptic neurones to trigger impulses in postsynaptic cells. Whether or not this happens is a major factor determining control and coordination by the ➤nervous system.

synchrocyclotron A particle ➤accelerator based on the ➤cyclotron, in which the frequency of the accelerating electric field is decreased as the energy of the particles increases so that the particles can remain in phase with the field at ➤relativistic speeds.

synchronous Describing two or more (usually periodic) events that happen at the same time, often by design.

synchronous orbit ➤geostationary orbit.

synchrotron A particle ➤accelerator that uses a ring of powerful magnets and a ➤radio-frequency electric field to accelerate the particles to ➤relativistic speeds in a loop. Typically the energy of electrons would rise to a thousand times their ➤rest mass, thus their speed is within about 1% of the speed of light. Usually the particles are electrons (though proton synchrotrons also exist) which radiate (because they are accelerating charges). Modern electron synchrotrons are designed as sources of this **synchrotron radiation**, which is intense, forward focused and highly polarized. It covers an extremely broad spectrum extending to frequencies in the hard X-ray region of the electromagnetic spectrum, including some frequency ranges (e.g. around a photon energy of 100 eV) that are difficult to produce in a standard laboratory. Thus synchrotron radiation is extremely effective for ➤spectroscopy. A very powerful new machine, named Diamond Light Source, which is the largest UK-funded scientific facility built in the last 40 years, began operation on the Harwell site in Oxfordshire in 2008.

syndiotactic polymer (syntactic polymer) A particular arrangement of the side chains off a polymer chain where the groups are alternately in front of and then behind the polymer chain (see the diagram). This arrangement can be tailor-made using ➤Ziegler–Natta catalysts. Compare ➤atactic polymer; ➤isotactic polymer.

syndiotactic polymer In isotactic polypropylene (a), the methyl groups all lie on one side of the chain; in syndiotactic polypropylene (b), the groups lie alternately on opposite sides of the chain; in atactic polypropylene (c), the arrangement of the groups is random.

syngamy ➤fertilization.

synodic period The orbital period of an astronomical body as defined by successive ➤conjunctions or ➤oppositions. Compare ➤sidereal period.

synoptic chart A weather chart showing ➤isobars, and often reports from weather stations. The spacing and direction of the isobars may be used to deduce the wind velocity. ➤➤geostrophic flow.

synovial fluid A viscous fluid consisting of a solution of ➤glycoproteins which lubricates joints in vertebrate skeletons. It is contained in the joint by the **synovial membrane**.

synthesis 1 The formation of a compound from its elements, as distinct from any other route, as in the ammonia synthesis:

$$N_2(g) + 3H_2(g) \rightleftharpoons 2NH_3(g).$$

The ➤Haber–Bosch process is therefore a synthesis, whereas the laboratory preparation which involves heating an ammonium salt with an alkali is not.

2 In a more general sense, used typically in organic chemistry, the **synthetic route** outlines how a chemical is made from others in a logical sequence of steps.

synthesis gas A 2:1 mixture of hydrogen and carbon monoxide produced from steam and methane, used to synthesize methanol and hence other organic molecules.

system That part of the world to which a particular physical model is applied. Thus in thermodynamics a system might be a well-insulated cylinder with a piston containing gas; in astronomy, a certain set of planets or stars; in quantum mechanics, the electrons making up an atom, together with the potential due to the nucleus and any externally applied fields. As the model becomes more sophisticated, more components are added to the system to simulate reality more closely. In the examples above, this might be (respectively) the heat loss through the cylinder walls bringing the surroundings into the system; the effect of more distant planets or stars on the chosen set; or the effect of neighbouring atoms or molecules.

systematics ➤taxonomy.

Système International ➤SI units.

systole ➤heart.

T

t Symbol for the unit ➤tonne.

t Symbol for ➤time.

$t_{\frac{1}{2}}$ Symbol for ➤half-life.

t- Symbol for ➤tertiary.

T 1 Symbol for ➤tritium.
    2 Symbol for the unit ➤tesla.
    3 Symbol for the prefix ➤tera-.
    4 Symbol for the pyrimidine base ➤thymine.

T **1** Symbol for ➤thermodynamic temperature.
    2 Symbol for ➤kinetic energy.

θ Symbol often used for ➤angle.

2,4,5-T Abbr. for 2,4,5-trichlorophenoxyethanoic acid (see the diagram), a selective herbicide. It and ➤2,4-D (which differs only in having one less chlorine on the benzene ring) are among the most-used herbicides. Unless great care is taken in its manufacture, ➤dioxin is also created and this is a potent human ➤teratogen.

Ta Symbol for the element ➤tantalum.

tabun A tasteless, odourless ➤nerve gas. It is an organic phosphate which acts rapidly, causing respiratory and cardiac paralysis. The antidote is ➤atropine.

tachometer An instrument for measuring the speed of rotation of an engine.

tachyon A hypothetical particle that can travel faster than the ➤speed of light.

TAI Abbr. for ➤International Atomic Time (from its name in French).

tail, cometary Material ejected from the nucleus of a ➤comet in the direction away from the Sun. The straight **gas tail** (also called the **ion tail** or **plasma tail**) consists of ionized gas carried away by the solar wind. The curved **dust tail** contains dust particles which may spread around the comet's orbit and give rise to a meteor shower.

talc (talcum, French chalk) A hydrated magnesium silicate which is notable as the mineral with the lowest rating on the ➤Mohs scale of hardness.

tan ➤trigonometric functions.

tangent 1 ➤trigonometric functions.

2 A line touching a curve. For some curves, tangents may be found by a geometrical construction; thus the tangent to a circle is perpendicular to the radius at the point of contact. In general, differential calculus is required: the gradient of the tangent at a point is the value of dy/dx there.

tangential Directed along a ➤tangent (2).

tangent plane A plane touching a surface. If the equation of the surface is $f(x, y, z) = 0$, the ➤normal (2) vector to the tangent plane at a point is given by grad f (➤gradient).

tanh ➤hyperbolic functions.

tannic acid Approximate formula $C_{76}H_{52}O_{46}$. A yellow-white powder with an astringent taste, often identified as a constituent of wines stored in oak casks.

tannin Any of a group of astringent compounds used to cure leather. Tannins are esters of gallic acid, derived from a number of plants, and bind to the proteins in leather rendering them resistant to fungal attack.

tantalum Symbol Ta. The element with atomic number 73 and relative atomic mass 180.9, which is at the bottom of the group of d-block elements headed by vanadium. As with vanadium, the highest oxidation number is +5, in which all the outer electrons participate in bonding. Examples include the white solid Ta_2O_5 and the yellow solid $TaCl_5$. The lower halides are interesting as they are not as simple as their stoichiometry suggests, being based on metal–metal clusters such as $[Ta_6Cl_{12}]^{2+}$. Tantalum is used in surgical instruments as it can be sterilized by heating without loss of hardness and it is unreactive with body fluids. It is also used as a catalyst for producing artificial diamonds.

tar The product of destructive distillation of a range of carbon-containing substances such as coal, coke, peat and wood. This nonaqueous liquid mixture contains many substances which are very useful. **Coal tar** was a major early source of ➤phenol, for example.

tartaric acid (2,3-dihydroxybutanedioic acid) $(CH(OH)COOH)_2$ An acid manufactured from wine lees. The molecule has two identical halves, each having a ➤chiral carbon atom. This gives rise to three optical isomers, rather than the four that might be expected (➤*meso* isomer). The main use of tartaric acid is as an ingredient in fizzy drinks.

tartrate A salt of tartaric acid. Probably the most well known is ➤Rochelle salt.

TATA box ➤Hogness box.

tauon (tau particle) The most massive ➤lepton (rest mass 1.78 GeV/c^2, more than three thousand times the mass of the electron). It has a lifetime of about 3×10^{-13} s. ➤➤standard model.

tautomerism A form of dynamic isomerism in which two ➤isomers can be readily interconverted, as in ➤keto-enol tautomerism.

taxis A response to a directional ➤stimulus in which the whole organism (or cell in unicellular forms) moves. The swimming of a sperm towards an egg and the response of certain photosynthetic ➤protoctistans, such as *Euglena*, to light are examples. The stimulus may be a chemical (**chemotaxis**), light (**phototaxis**) or gravity (**geotaxis**), for example.

taxonomy The branch of biology concerned with the classification of organisms. Organisms were formerly grouped into various taxa (singular **taxon**) on the basis of observable anatomical and morphological structures, but nowadays molecular and genetic information is used increasingly, and classification systems are continually being revised as more information becomes available. **Cladistics** is an approach to classification based on evolutionary relationships and the assumption that organisms that exhibit ➤homologous structures are derived from a common ancestor and are therefore related by genealogy. An organism is classified according to a hierarchical system as follows:

kingdom, phylum, class, order, family, genus, species.

Intermediate categories such as subphylum or superorder may be used in particularly large groups such as flowering plants, in which the term 'division' traditionally replaced phylum. The purpose of taxonomy is basically twofold: to provide a means of ready identification of organisms, and to infer and illustrate evolutionary relationships between them. Classification and the naming of organisms are closely related. All organisms have a binomial Latinized name based on the ➤Linnaean system. ➤➤Appendix table 8.

Taylor expansion The representation of a function $f(x)$ near a point a by the **Taylor series** $\sum_{n=0}^{\infty} f^{(n)}(a)(x-a)^n/n!$, where $f^{(n)}(a)$ denotes the value of the nth ➤derivative of $f(x)$ at $x = a$. Named after Brook Taylor (1685–1731).

Tay–Sachs disease (TSD) A fatal, inherited ➤recessive condition in children resulting from the inability to produce hexosaminidase A (Hex-A). In the absence of Hex-A, a lipid, GM2 ganglioside, accumulates in nerve cells, particularly in the brain. A TSD baby appears normal at birth but progressive deterioration in mental and physical ability results in death, usually before the age of five. To date there is no effective cure or treatment. The disease is of interest to population geneticists since the proportion of carriers, who can now be identified by DNA screening, varies in different ethnic groups. For example approximately 1 in 27 European Jews are carriers of TSD, whereas the proportion in the general population is about 1 in 250. Named after Warren Tay (1843–1927) and Bernard Sachs (1858–1944).

Tb Symbol for the element ➤terbium.

TB (tuberculosis) An acute and chronic infectious disease caused by the bacillus *Mycobacterium tuberculosis*, usually found in the lungs. The name is descriptive of characteristic cyst-like structures called **tubercles**, in which the bacteria are located. The disease is also often found in domesticated animals such as cattle, chickens and pigs.

Tc Symbol for the element ➤technetium.

TCA cycle Abbr. for tricarboxylic acid cycle. ➤Krebs cycle.

T cell (T lymphocyte) A lymphocyte of the ➤immune system which matures in the ➤thymus gland. There are two main classes. ➤Helper T cells have highly variable surface receptors composed of two ➤polypeptide chains. An ➤antigen binding to a receptor causes proliferation of a ➤clone of cells with the same receptor. ➤Killer T cells interact directly with other cells carrying foreign antigens and bring about the death of the cell. Some types of T cell are also involved in ➤autoimmune diseases.

TCP Abbr. for ➤2,4,6-trichlorophenol.

t-distribution The ➤probability distribution of the ➤means of ➤random samples from a population having a ➤normal distribution with unknown ➤variance. It is also known as **Student's t-distribution** after the pseudonym adopted by its discoverer, W. S. Gosset, at the insistence of the brewery for which he worked.

Te Symbol for the element ➤tellurium.

tear gas A gas such as ➤CS gas which is ➤lachrymatory (causes tears).

technetium Symbol Tc. The element with atomic number 43 and most stable isotope 98, which is in the second row of the d-block elements in the periodic table. Its name is derived from the Greek for 'artificial', which is appropriate as it does not occur naturally. All its isotopes are radioactive, the nuclide ^{99}Tc having clinical use for scanning bones and organs.

Teflon Tradename for polytetrafluoroethene (➤PTFE), used in nonstick frying pans.

TEL ➤tetraethyllead.

telecommunications Any means of transmitting information to a remote site. Sometimes there is no physical link between transmitter and receiver, as in ➤radio or ➤television; otherwise, ➤transmission lines are used.

telemetry The transmission of measurements over a ➤telecommunications link for analysis. Data transmission from space vehicles is a well-established example, but telemetry is also used, for example, in Formula 1 motor racing to allow mechanics to monitor the condition of a car while it is in motion.

telephoto lens A camera lens designed to allow detailed photography of distant objects. It uses a lens system very similar to that in a refracting ➤telescope.

telescope An optical instrument for viewing distant objects in detail. Rays from an object are brought to a focus where they are magnified by an eyepiece, consisting of one or more lenses in combination. Different methods are used to focus the rays (see the diagram overleaf). **Refracting telescopes** use lenses; they include the **Galilean telescope**, now used only in the form of opera-glasses, and the **Keplerian telescope**, used for terrestrial and small astronomical instruments. **Reflecting telescopes** use mirrors; the main types are the **Newtonian telescope** and the **Cassegrain telescope**. **Catadioptric telescopes** use a mirror–lens combination; examples are the **Schmidt telescope** and **Schmidt–Cassegrain telescope**. All major astronomical instruments use mirrors since large lenses distort under their own weight. Telescopes can also be designed for wavelengths outside the optical range; see, for example, ➤radio telescope.

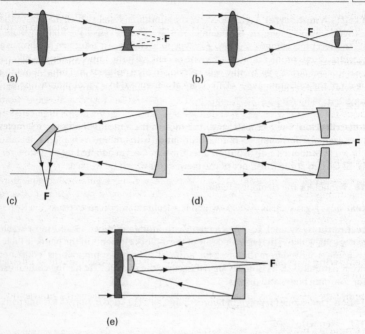

telescope Schematic representations of light paths in types of telescope: (a) Galilean, with diverging eyepiece; (b) Keplerian, with converging eyepiece; (c) Newtonian, with paraboloidal primary mirror and flat secondary mirror; (d) Cassegrain, with hole in paraboloidal primary mirror and hyperboloidal secondary; (e) Schmidt–Cassegrain, with spheroidal primary and secondary mirrors, and a specially figured lens called a corrector plate. F is the focus in each case; eyepiece lenses are omitted from (c), (d) and (e).

television (TV) Transmission of moving pictures by telecommunications. The images are converted by a television camera into electrical signals for transmission, and are decoded at the remote site and displayed on a monitor. Conventional broadcast **terrestrial television** uses UHF radio waves (about 500 to 900 MHz) for transmission. **Satellite television** broadcasts via artificial satellites on microwave frequencies. **Cable television** sends signals over a network. **Digital television** was launched in 1998.

telluride A binary compound in which the more electronegative element is ➤tellurium.

tellurium Symbol Te. The element with atomic number 52 and relative atomic mass 127.6, which is in Group 16 of the ➤periodic table below selenium (their names are connected as selenium comes from 'Moon' and tellurium from 'Earth'). It is best classified as a ➤metalloid, the grey allotrope having a low electrical conductivity. Cationic clusters, such as the bright-red species Te_4^{2+}, also exist. Its chemistry resembles that of sulfur, with a similar range of oxidation numbers: −2, +2, +4 and +6 being most common. One infamous compound is hydrogen telluride, a gas with at least as revolting a smell as that of hydrogen sulfide, but which suffers from the disadvantage that it is absorbed by the skin causing the unpleasant odour to persist longer.

telomere Classically, a term used to describe the ends of eukaryotic chromosomes which seem never to get incorporated into the body of the chromosome, or to bind closely with their homologue during ➤meiosis. Telomeres are now known to contain repeated sequences of DNA which seem to protect the chromosome from degradation – a process which appears to be related to ageing and is implicated in some cancers. An enzyme, telomerase reverse transcriptase, rebuilds the telomeres after each cell division.

telophase ➤meiosis; mitosis.

temperature A measure of how hot a system is. ➤Empirical temperature determines the direction of heat flow from one body to another (from hotter to colder). ➤Thermodynamic temperature incorporates ➤absolute zero as the lowest temperature available. ➤➤Celsius scale; Fahrenheit scale; Kelvin scale.

temperature coefficient ➤Q_{10}.

tempering The process in which a quenched steel is reheated to relieve internal stresses.

template A DNA or RNA strand that is used to direct the sequence of ➤nucleotides in a new strand of DNA or RNA. ➤➤transcription.

temporary hardness ➤hard water.

temporary magnetism Magnetism in a material that exists only when an external magnetic field is applied. There is no ➤remanence, unlike in a permanent ➤magnet.

tendon A tough inelastic tract of connective tissue connecting muscles to bone, composed mainly of ➤collagen fibres.

tensile stress The stress in a material that is being stretched. It is the tension per unit cross-sectional area. Compare ➤compressive stress.

tensiometer A device for measuring ➤tension or ➤surface tension.

tension An internal force in a body that is being stretched. For any plane that divides the body in two, there is a force applied to one part of the material that is in the direction of the other.

tensor A generalization of the concept of a ➤vector. In general the **rank** of a tensor is the number of indices required to specify a component; vectors are first-rank tensors. Tensors are encountered frequently where two vectors are related by a more complex dependence than a simple scalar multiplication and the related quantities are not parallel. For example, the inertia tensor I relates angular momentum L to angular velocity ω as $L = I\omega$. The dielectric tensor ε relates electric field strength to electric displacement as $D = \varepsilon E$. These **second-rank tensors** can be expressed as square matrices (with two indices), just as vectors can be expressed as row or column matrices (with one index).

tera- Abbr. T. An SI prefix meaning 10^{12} times the base unit.

teratogen A substance capable of damaging the human foetus. The most notorious chemical teratogen is ►thalidomide, which causes birth defects including incompletely developed limbs. The most common viral teratogen is ►rubella (German measles).

terbium Symbol Tb. The element with atomic number 65 and relative atomic mass 158.9, which is a ►lanthanide. As expected, its dominant oxidation number is +3, as in the pale pink Tb^{3+} ion and the white solid oxide Tb_2O_3. There is another oxidation number of +4 because Tb(IV), as in the dark solid TbO_2, enjoys the extra stability of a half-filled subshell (f^7 here).

terephthalic acid (1,4-benzenedicarboxylic acid) A benzene ring with two carboxylic acid groups arranged in the para position (see the diagram), which exists as colourless needles used in the manufacture of ►Terylene.

COOH

COOH
terephthalic acid

terminal 1 (Comput.) A workstation that is connected to a computer network. The terminal may be **dumb**, in which case it acts merely as an ►I/O unit, depending for processing entirely upon the computer to which it is connected. Alternatively, it may be **intelligent**, in which case it has at least some processing power of its own.

2 (Phys.) A conductor, at the input or output of an electrical device, to which a conducting connection can easily be made. Terminals usually come in multiples, for example positive and negative for ►direct current, or live, neutral and earth for mains alternating current.

terminal speed The limiting speed of a falling body. This is achieved when the ►drag force (which increases with velocity) equals its weight.

termination codon (nonsense codon, stop codon) A ►codon (TAA, TGA or TAG in DNA, or UAA, UGA or UAG in RNA) that signals the end of an ►open reading frame which is undergoing ►translation. ►►genetic code.

terminator 1 (Phys.) A device connected to the end of a ►transmission line to prevent reflections of the signal at the end. ►►matched termination.

2 (Astron.) The line separating night and day on a planet or satellite.

ternary compound A compound consisting of three elements, for example ►spinel, $MgAl_2O_4$.

terpene Any member of an important group of naturally occurring compounds including the carotenoid pigments of plants (►carotene) and the steroid ►hormones of animals.

Terramycin Tradename for the antibiotic ►oxytetracycline.

terrestrial magnetism ►geomagnetism.

terrestrial planets ►inner planets.

tert- A prefix meaning ►tertiary, as in *tert*-butyl, the alkyl group $(CH_3)_3C$—.

tertiary When applied to a ➤halogenoalkane or an ➤alcohol, designating a structure in which the halogen or oxygen atom is attached to a carbon that has three other alkyl groups attached, for example $(CH_3)_3COH$. Tertiary alcohols are much harder to oxidize than primary or secondary alcohols as they lack a hydrogen on the carbon attached to the —OH group. When applied to amines, confusingly, it means that the *nitrogen* atom has three alkyl groups attached, for example $(CH_3)_3N$.

Tertiary See table at ➤era.

tertiary structure ➤protein structure.

Terylene (US Dacron) Tradename for a polyester made from ➤terephthalic acid and ethylene glycol (now called ➤ethane-l,2-diol). It is stronger than cotton and more resistant to moisture, hence its use in shirts.

tesla Symbol T. The SI derived unit of ➤magnetic flux density, named after Nikola Tesla (1856–1943).

Tesla coil A device based on an induction coil for producing high voltages at high frequencies.

test cross ➤back cross.

testis The male reproductive organ of animals, producing ➤spermatozoa. In vertebrates the testis also produces the male ➤sex hormones, particularly ➤testosterone, and a variety of secretions contributing to ➤semen.

testosterone A ➤steroid hormone produced by the **Sertoli cells** of the ➤testis. It is responsible for the maintenance of sperm production and also for the appearance of secondary sexual characteristics in males at puberty. See diagram at ➤steroid.

tetanus **1** A disease ('lock jaw') caused by the bacterium *Clostridium tetani*, commonly found associated with soil and dirt. The bacterium is anaerobic and reproduces in wounds where the blood supply may be compromised, releasing a powerful toxin which causes muscle spasms – notably in the jaw muscles. Progressive and severe convulsions can result in death. Prevention of tetanus is by immunization – a series of two injections is given in childhood followed by a booster at 5–10-year intervals.

2 Sustained muscle contraction due to nervous stimulation at a rate that does not allow for relaxation.

tetraammine- A prefix meaning that a complex contains four ammonia ligands: for example, the tetraamminezinc ion is $[Zn(NH_3)_4]^{2+}$.

tetraaqua- (tetraaquo-) A prefix meaning that a complex contains four water ligands: for example, the tetraaquazinc ion is $[Zn(H_2O)_4]^{2+}$.

tetrachloromethane (carbon tetrachloride) CCl_4 A colourless volatile liquid; an important organic solvent which is effective at dissolving organic compounds due to the high ➤dispersion forces in the molecule. Its use as a solvent has waned since the discovery that it is strongly carcinogenic.

tetracyanoethene (tetracyanoethylene, TCNE) $C(CN)_2{=}C(CN)_2$ A very powerful dienophile in the ➤Diels–Alder reaction.

tetrad The term applied to the four ➤haploid cells produced as a result of ➤meiosis, particularly while they are still attached to one another.

tetraethyllead (lead tetraethyl, TEL) $Pb(CH_2CH_3)_4$ The classic ➤anti-knock agent for petrol engines which enables the use of higher compression ratios without the problem of pre-ignition. Its use has been discontinued because of concerns over the toxicity of ➤lead.

tetrafluoroethene $CF_2{=}CF_2$ A colourless gas that is the monomer of polytetra-fluoroethene (➤PTFE).

tetragonal One of the seven ➤crystal systems. A tetragonal lattice is like a stretched cubic lattice. There are two ➤Bravais lattices corresponding to this: **simple tetragonal** and **centred tetragonal**.

tetrahedral A shape resembling a tetrahedron, which is very common throughout chemistry. In the gas phase, this is the usual structure adopted by molecules with four bonding pairs of electrons (➤valence-shell electron-pair repulsion theory), such as methane, tetrachloromethane, etc. The **tetrahedral angle** between the bonds is 109° 28′. In the solid state **tetrahedral holes** are important for many structures including the ➤fluorite structure. ➤➤spinel.

tetrahedron A ➤pyramid that has four triangular ➤faces. A **regular tetrahedron** has faces that are ➤equilateral triangles and is one of the five ➤Platonic solids.

tetrahydrofuran (THF) A colourless liquid (see the diagram) extensively used as a solvent both for simple molecules and for plastics.

tetramethylsilane (TMS) $Si(CH_3)_4$ The standard for comparison of chemical shifts in ➤nuclear magnetic resonance.

tetrahydrofuran

tetrathionate ion $S_4O_6{}^{2-}$ An ion formed when ➤iodine reacts with thiosulfate ions. The name arises because there are four sulfur atoms in the ion. (The stronger oxidant chlorine forms sulfate ions instead.)

Th Symbol for the element ➤thorium.

thalidomide A notorious drug (see the diagram for its structure) used to treat morning sickness before it was appreciated that it was a ➤teratogen. It has, perhaps surprisingly, continued to be used to treat multiple myelomas in developing countries. ➤➤optical activity.

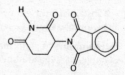

thalidomide

thallium Symbol Tl. The element with atomic number 81 and relative atomic mass 204.4, which is at the bottom of Group 13 of the ➤periodic table. The preceding element in the period is mercury and the following element is lead, so it is perhaps not surprising that its compounds are also cumulative poisons. Like aluminium, its most important oxidation number is +3, as in the dark solid Tl_2O_3. However, because of the ➤inert pair effect, there is another important oxidation number, +1, as in the strong alkali TlOH. Indeed the compound TlI_3 which appears to be thallium(III) iodide is in fact thallium(I) triiodide, $Tl^+I_3{}^-$.

Thallophyta An archaic term for the lower plants, undifferentiated into root and shoot. It included types now placed in the various groups of algae, the taxonomic status of which is a matter of debate.

thallus The body of plants such as multicellular algae and some bryophytes (particularly the liverworts), which is not differentiated into distinct root and shoot systems.

T helper cell An alternative name for ➤helper T cell.

theodolite An instrument used in surveying to measure angles. It is in essence a telescope with a set of fine scales against which its orientation can be measured with great accuracy.

theorem A significant mathematical result in which the conclusion has been logically deduced from the premises.

thermal conductivity Symbol κ; unit $W\ K^{-1}\ m^{-1}$. The rate of conduction of heat through a material over unit length per unit area per unit temperature gradient. ➤➤Biot–Fourier equation.

thermal cracking ➤cracking.

thermal emission (thermionic emission) The emission of electrons from a hot metal surface. As the temperature of a metal rises, more electrons occupy energy levels above the Fermi energy (➤Fermi–Dirac statistics), and some of them will have enough energy (➤work function) to escape from the metal and become free electrons.

thermal equilibrium A state in which there is no net flow of heat between the parts of a system. ➤thermodynamics (zeroth law).

thermal imaging The detection and imaging of objects by means of the ➤infrared radiation emitted by the objects. The spectrum of radiation emitted by a body depends on its temperature (➤Planck's law of radiation), so hotter objects can be detected against cooler ones by comparing the radiation emitted at two or more frequencies.

thermal neutron A neutron whose energy is of the order of kT, where k is the ➤Boltzmann constant and T is the ➤thermodynamic temperature.

thermal reactor A ➤nuclear reactor in which thermal neutrons are responsible for the ➤nuclear fission process. Most thermal reactors use uranium fuel rods embedded in a ➤moderator. Compare ➤fast reactor.

thermion An electron or ion ejected by ➤thermal emission from a metal surface.

thermionic emission ➤thermal emission.

thermionic valve An arrangement of electrodes within an evacuated tube in which current is carried by ➤thermions. Such valves were the forerunners of ➤transistors for controlling and amplifying current.

thermistor A ➤semiconductor device whose resistance is very sensitive to temperature. Typically thermistors are used in situations where temperature is regulated by a ➤feedback loop.

thermit reaction A spectacularly violent reaction in which aluminium is used to reduce iron(III) oxide to metallic iron:

$$2Al(s) + Fe_2O_3(s) \rightarrow Al_2O_3(s) + 2Fe(l).$$

The reaction is so exothermic that the iron melts and this can be used to fill moulds around the ends of sections of railway track to weld them together. The mixture is safe at room temperature, as the rate of reaction is then very low, so the reaction is initiated by a magnesium fuse.

thermoammeter ➤ammeter.

thermochemistry The application of ➤thermodynamics to chemistry. It deals mostly with the measurement and interpretation of ➤enthalpy changes.

thermocouple A device for measuring temperature, based on ➤thermoelectric effects. It measures the difference in temperature between two junctions, each composed of two different metals. Since the ➤junction potential depends on temperature, a difference in temperature will produce a potential difference. A variety of types are available for different temperature ranges and sensitivities.

thermodynamics The scientific study of the transformation of energy by heat and work. There are four laws of thermodynamics; the first and second laws are particularly important.

The **zeroth law** of thermodynamics states that if ➤systems A and B are in thermal equilibrium, and B and C are in thermal equilibrium, then A and C are also in thermal equilibrium. This law is tacitly assumed in every measurement of ➤temperature.

The **first law** of thermodynamics states that the total energy of a closed system remains constant; when energy is transferred to a system, the change in the internal energy of the system equals the heat added to the system plus the work done on the system. This is effectively the law of ➤conservation of energy applied to the system. However, the law of conservation of energy can be applied to individual collisions between molecules, whereas thermodynamics should not be applied on the molecular scale. ➤➤enthalpy; Hess's law.

The **second law** of thermodynamics can be stated in a number of ways. The simplest is that heat will not pass spontaneously from a colder to a hotter body. This qualitative statement can be understood quantitatively by introducing the concept of entropy. In terms of ➤entropy, the second law states that the total entropy of a closed system cannot decrease. It is the second law of thermodynamics that explains why chemical reactions occur. ➤standard Gibbs energy change; van't Hoff isotherm.

The **third law** of thermodynamics states that the entropy of a perfect crystalline substance at ➤absolute zero is zero.

thermodynamic temperature (absolute temperature) Symbol T. A measure of temperature that builds on ➤empirical temperature (θ) to define a scale. Whereas any monotonically increasing function of θ would serve as a scale of empirical temperature, thermodynamic temperature T is defined in terms of the limiting behaviour of a gas at low pressure: T is the limit of (pV_m/R) as the pressure p tends to zero, where V_m is its ➤molar volume and R is the ➤gas constant. This does not enforce a particular choice of units, and although ➤kelvin is almost universally used, thermodynamic temperature should not be referred to as 'temperature in kelvin'.

thermoelectric effects The collective name for several different interrelations of thermal and electrical behaviour. The **Seebeck effect** is the production of a current when the junctions of a circuit made of two different metals are at different temperatures; it forms the basis of the ➤thermocouple. The **Peltier effect** is the converse: the flow of current in such a circuit causes a temperature difference. The **Thomson effect** (or **Kelvin effect**) is the production of a potential difference between two points in the same material at different temperatures.

thermoluminescence ➤Luminescence that is the result of heating. It arises from the excitation of electrons from states localized around ➤defects.

thermolysis (pyrolysis) Chemical decomposition of a compound induced by heating it.

thermometer An instrument for measuring temperature. Thermometers often rely on the thermal expansivity of liquids. Typically mercury (for temperatures greater than room temperature) or ethanol (for room temperature and lower) is enclosed in a bulb with a thin calibrated tube attached. The expansion of the liquid with increasing temperature forces the liquid up the tube.

thermonuclear bomb Alternative name for a hydrogen bomb (➤nuclear weapon).

thermopile An instrument for detecting heat radiation, consisting of a number of ➤thermocouples in series.

thermoplastics Alternative name for thermosoftening ➤plastics.

thermosetting plastics ➤plastics.

thermosoftening plastics ➤plastics.

thermosphere The layer of the ➤atmosphere above the ➤mesosphere, from approximately 100 km. The thermosphere, like the ➤stratosphere, is characterized by a temperature that increases with altitude, in this case to several thousand kelvin. Molecules tend to atomize and ionize in the thermosphere (➤ionosphere).

thermostat A device designed to regulate a temperature. It consists of a temperature-sensing device like a ➤bimetallic strip which switches a heating or cooling system on or off as appropriate.

THF ➤tetrahydrofuran.

thiamine (thiamin) Vitamin B_1, a white crystalline solid. Absence of thiamine from the diet leads to the disease beri-beri. Its pyrophosphate derivative is a coenzyme in a range of processes including the ➤Krebs cycle and the conversion of alanine to acetyl coenzyme A. See table at ➤vitamin.

thigmotropism (haptotropism) A ➤tropism in which the stimulus is a directional touch. A classic example is the curling response of a tendril of, for example, a pea plant around a support which is in contact with it.

thin-layer chromatography (TLC) ➤chromatography.

thio- A common prefix meaning that the compound contains sulfur, often in place of oxygen. For example, the thiosulfate ion is $S_2O_3^{2-}$ whereas the sulfate ion is SO_4^{2-}.

thiocyanate ion SCN$^-$ The ion present in, for example, potassium thiocyanate used to test for iron(III) ions, which give a very intense blood-red colour due to the complex $[Fe(SCN)(H_2O)_5]^{2+}$.

thiols Systematic name for ➤mercaptans.

thionyl chloride ➤sulfur dichloride oxide.

thiopentone (sodium 5-ethyl-5-(1-methylbutyl)-2-thiobarbiturate) A yellowish powder used to produce general anaesthesia.

thiophene A five-membered heterocyclic aromatic hydrocarbon (see the diagram), which is a colourless liquid smelling like benzene and is used as a solvent.

thiophene

thiosulfate ion $S_2O_3^{2-}$ The ion used to test for ➤iodine. It is also useful for complexing silver ions as $[Ag(S_2O_3)_2]^{3-}$ is especially stable (➤➤hypo). When acid is added to thiosulfate ions, a yellow precipitate of sulfur gradually forms and sulfur dioxide gas is evolved.

thistle funnel A glass funnel with a long thin tube topped by a thistle-shaped head.

thixotropic Capable of becoming fluid on agitation and then coagulating again. This property is very useful for paints.

Thomson effect ➤thermoelectric effects. Named after William Thomson, Baron Kelvin (1824–1907).

Thomson scattering The treatment of scattering of light by electrons in classical physics. The oscillating electric field of the light causes the electrons to oscillate and re-radiate electromagnetic radiation of a lower frequency in all directions. Named after Joseph John Thomson (1856–1940).

thorax The middle of the three divisions of the body of an ➤insect, bearing the legs and wings and joining the head to the **abdomen**. The term also refers to the chest region of vertebrates; in vertebrates the region below the ribcage is called the abdomen.

thorium Symbol Th. The element with atomic number 90 and relative atomic mass 232.0, which is one of the ➤actinides. It is a grey radioactive metallic solid. Its most common oxidation number is +4 as in the white solid oxide **thoria**, ThO_2, used in the ➤Fischer–Tropsch synthesis. The **thorium series** of radioactive decay starts with the nuclide ^{232}Th and ends with the nuclide ^{208}Pb.

thorium emanation (thoron) An obsolete name for the gas ^{220}Rn produced by alpha decay of thorium.

Thr Abbr. for ➤threonine.

three-centre bond An electron pair that bonds three atoms together. The simplest example occurs in the ion H_3^+. Each of the three hydrogen atoms provides one electron, the positive charge signifies that one electron has been lost; the two remaining electrons hold the ion together. Another example is the pair of three-centre bonds in diborane (see the diagram). ➤delocalization.

three-centre bond The bridging hydrogen atoms are held together by a pair of electrons joining both of the two boron atoms and the hydrogen atom.

three-phase power A power supply with three alternating voltages, each 120° out of phase with the others. It is convenient and efficient to generate three phases in this way rather than just one.

threonine (Thr) A common ➤amino acid. ➤Appendix table 7.

threose A tetrose ➤monosaccharide (see the diagram), with the two —OH groups situated on opposite sides. Compare ➤erythrose.

threshold In general, a value of a quantity beyond which the behaviour of a system changes abruptly in some way. For example, in the ➤photoelectric effect the threshold frequency corresponds to the energy of photons falling on a metal surface that is just enough for the electrons of highest energy in the solid to escape (➤work function). ➤➤nerve impulse.

D-**threose**

thrombosis ➤blood clot.

thrust A force produced, usually as a means of propulsion, by pushing fluid in the opposite direction to the required force. Jet engines produce thrust, as do propellers.

thulium Symbol Tm. The element with atomic number 69 and relative atomic mass 168.9, which is a ➤lanthanide. Typically, its dominant oxidation number is +3, as in the pale green Tm^{3+} ion. It is used as a portable source of X-radiation on bombardment with neutrons.

thymine Symbol T. A ➤pyrimidine base (see the diagram) present in DNA and tRNA. Other forms of RNA use ➤uracil in place of thymine (the two nitrogenous bases differ only by the methyl group at carbon atom 5). Its ➤base pairing with adenine occurs by mutual ➤hydrogen bonding. ➤➤genetic code.

thymine

thymol A disubstituted phenol (see the diagram overleaf), which exists as colourless crystals used in antiseptic mouth washes and to make a range of indicators such as thymolphthalein or bromothymol blue.

thymus gland An organ in vertebrates, located in the pharyngeal or thoracic region, containing primary lymphoid tissue. The thymus has an important function as part of the ➤immune system and is responsible for the maturation of ➤T cells. The

thymol

major function appears to be the conditioning of T cells to be able to distinguish between 'self' and 'non-self' ➤antigens.

thyristor A ➤silicon-controlled rectifier in which a pulse is used to switch current on and off.

thyroid gland An ➤endocrine gland in vertebrates, located in the neck region, which produces a variety of thyroid ➤hormones including ➤thyroxine. Iodine is required in the diet for the correct functioning of the gland, and an inadequate supply of this element results in enlargement of the gland (**goitre**).

thyroid stimulating hormone (TSH) A protein ➤hormone secreted by the anterior ➤pituitary gland which stimulates the production of ➤thyroxine and other **thyroid hormones**. The control of thyroid function by TSH is by a ➤feedback mechanism, and is an important component of ➤homeostasis.

thyroxine A ➤hormone secreted by the ➤thyroid gland associated with the regulation of the ➤basal metabolic rate and certain developmental processes (e.g. the metamorphosis of amphibian tadpoles).

Ti Symbol for the element ➤titanium.

tibia One of the pair of bones in the distal part of the rear leg of limbed vertebrates. In humans the tibia is the larger of the pair (alongside the **fibula**) and forms the shin bone.

tide A periodic rise or fall of the level of the oceans caused by the gravitational effect of the Moon and Sun. The Moon exerts its greatest force on the side of the Earth nearest the Moon, and the least force on the opposite side. Subtracting the force exerted at the centre of the Earth from the force exerted at each point on the surface shows that there is a net force acting towards the line joining the centres of the Earth and the Moon, as shown in diagram (a) opposite. As the Earth rotates below these two tidal bulges, coastal regions experience two high and two low tides each day. The Sun raises tides in the same way, but only about one-third as strongly as the Moon. When the Sun and Moon are at right angles, their effects are subtractive and they produce the lower **neap tides**, as shown in diagram (b). When they are aligned, their effects are additive and the higher **spring tides** result.

TIFF Acronym for Tagged Image File Format. It is a digital image format using much lower compression than a JPEG, meaning that the data can be recovered more accurately, at the expense of a larger file size.

tight-binding approximation ➤energy band.

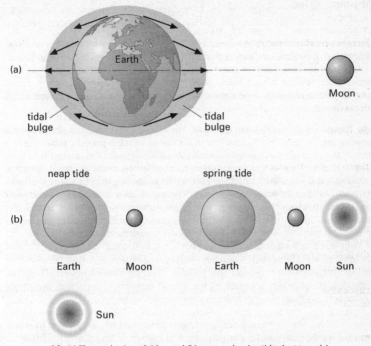

tide (a) The mechanism of tides, and (b) neap and spring tides (not to scale).

time Symbol *t*; unit ➤second, s. One of the four coordinates of spacetime that define events. An event occurs at a particular position and a particular time. Time appears to differ from space, at least in the human perception of it, in that the world moves steadily forward in time. ➤charge–parity–time symmetry; relativity.

timebase A signal, usually with a ➤sawtooth waveform, used, for example, to display another signal on an oscilloscope by varying the horizontal deflection of the electron beam.

time dilation The phenomenon in which time appears to run more slowly for fast-moving objects than for stationary observers; it is a consequence of ➤special relativity. A ➤proper time interval τ, for an observer moving with the object, appears as $\gamma\tau$ to a stationary observer, where:

$$\gamma = \frac{1}{\sqrt{1 - v^2/c^2}},$$

v being the speed of the object and *c* the speed of light.

time-lapse photography A technique in which photographs of an object are taken at regular intervals and then played back in rapid succession, with the effect of making time appear to run much faster. Time-lapse methods are particularly impressive when used, for example, to observe the development of plants and clouds.

time reversal symmetry A ➤symmetry transformation in which time is reversed, that is $t \rightarrow -t$. The laws of classical physics are invariant under such a transformation, and this is also the case for the ➤strong interaction and the ➤electromagnetic interaction. For the ➤weak interaction, however, the symmetry group is much smaller (➤charge–parity–time symmetry), and time is not necessarily reversible for all reactions.

tin Symbol Sn (from the Latin *stannum*). The element with atomic number 50 and relative atomic mass 118.7, which is in Group 14 of the ➤periodic table. Tin is a metal, the normal silvery coloured solid having a distorted close-packed structure. Below 13.5 °C, there is a slow transition to another less dense allotrope, grey tin, which has the ➤diamond structure. This causes crumbling over time; tin structures such as organ-pipes suffer '**tin plague**'. The metal itself is used in a number of important alloys, such as ➤bronze (with copper), pewter (with lead and antimony) and solder (with lead), in addition to its major use as a coating for steel objects which provides protection from attack notably by the organic acids in food.

Tin's more important oxidation number is +4, as is usual in Group 14. A useful compound with the oxidation number +4 is tin(IV) sulfide, SnS_2, which because of its golden-yellow colour is used to give a gilded effect on wooden objects. The other oxidation number (+2) is caused by the ➤inert pair effect. Tin(II) ions are strong reductants. For example, they reduce chromate(VI) ions to green chromium(III) ions, whereas lead(II) ions simply form a yellow precipitate of lead(II) chromate(VI). Tin has two chlorides: tin(II) chloride, $SnCl_2$, a largely ionic solid which dissolves in water, and tin(IV) chloride, $SnCl_4$, a largely covalent colourless liquid which hydrolyses rapidly. The more common oxide is SnO_2, which occurs as the ore ➤cassiterite or **tinstone**. Organometallic tin compounds include halogenated compounds such as $(CH_3)_2SnCl_2$, used as stabilizers to prevent discolouration of PVC, and $(CH_3)_3SnCl$, used as a pesticide.

tincture of iodine A solution of iodine and potassium iodide in ethanol used to clean wounds.

tissue An assemblage of cells of the same type, function and origin in a multicellular organism. Tissues are organized together into ➤organs.

tissue culture A range of techniques for maintaining actively growing cells isolated from a multicellular organism in sterile conditions. Cells may be maintained either in liquid medium or on ➤agar plates, supplied with a range of appropriate nutrients and an energy source. Tissue culture techniques form the basis of many applications of ➤biotechnology and ➤genetic engineering.

Titan ➤Saturn.

titania (titanium dioxide, titanium(IV) oxide) TiO_2 The main ore of titanium (➤rutile structure).

titanium Symbol Ti. The element with atomic number 22 and relative atomic mass 47.87, which is in the first row of the ➤transition metals. The element itself, manufactured in the ➤Kroll process, is a strong metal with a much lower density than iron and a very high melting point, favoured in the production of aircraft and of valves for a high-revving engine. Its most common oxidation number is +4, as in rutile, TiO_2, widely used as a white pigment in paints (and horseradish), and in titanium(IV) chloride, $TiCl_4$, used as a ➤Ziegler–Natta catalyst. Ease of interconversion between different oxidation numbers is the key to the usefulness of transition metals as catalysts, and titanium(III) chloride, $TiCl_3$, can also be used as a Ziegler–Natta catalyst. The hexaaquatitanium(III) ion, $[Ti(H_2O)_6]^{3+}$, is purple, the colour being due to a ➤d–d transition in this d^1 complex. Other important compounds include titanium carbide, TiC, which is used in tool tips as it is very resistant to chemical attack, and barium titanate, $BaTiO_3$, which is piezoelectric (➤piezoelectric effect) and hence used in transducers.

titration A technique for quantitative measurement of the concentration of a solution. Although the most common form of titration is that between an acid and a base, **redox titrations** between oxidants and reductants can be performed, as for iodine–thiosulfate. **Precipitation titrations**, as for silver and chloride ions, and **complexometric titrations**, often using ➤EDTA, are other rarer examples. The traditional **acid–base titration** is performed by measuring a fixed volume of one reactant using a ➤pipette. Then an indicator is added and the second reactant is run in from a ➤burette until the indicator changes colour, when the volume of the second reactant is noted. From these volumes and the concentration of one of the two solutions, the other concentration can be worked out. While this works well when either a strong acid or strong base (or both) is involved (see the diagram overleaf), the ➤end-point for a weak acid and a weak base cannot be found using an indicator. In this case, a **conductometric titration** is used.

titre The volume determined in a ➤titration.

T killer cell An alternative name for ➤killer T cell.

Tl Symbol for the element ➤thallium.

TLC Abbr. for thin-layer ➤chromatography.

T lymphocyte ➤T cell.

Tm Symbol for the element ➤thulium.

TMS Abbr. for ➤tetramethylsilane.

TNT Abbr. for trinitrotoluene (➤2,4,6-trinitromethylbenzene).

tobacco mosaic virus (TMV) A rod-shaped ➤RNA-containing ➤virus which causes deformation and patchy bleaching of leaves of plants, particularly tobacco. It is transmitted by insects such as aphids, and was the first virus to be isolated and characterized, by Wendell Stanley in the 1930s.

tocopherol One of a closely related series of fat-soluble compounds similar to α-tocopherol (**vitamin E**, see the diagram overleaf), found in vegetable lipids. One action is to prevent the oxidation of unsaturated lipids; and in rats a deficiency is linked to infertility.

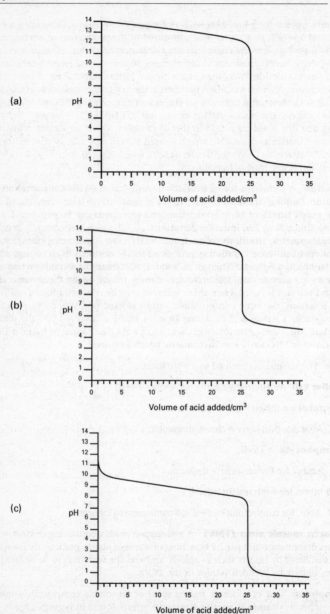

titration pH curves for the progress of acid–base titrations for (a) strong acid and strong base, (b) weak acid and strong base, (c) strong acid and weak base.

α-**tocopherol**

Tollens' reagent (ammoniacal silver nitrate) A solution, named after Bernhard Christian Gottfried Tollens (1841–1918), used to distinguish between ➤aldehydes and ➤ketones. The reagent is made up by adding a small volume of aqueous sodium hydroxide to aqueous silver nitrate and then adding enough aqueous ammonia to redissolve the precipitate originally formed. This solution contains the complex ion $[Ag(NH_3)_2]^+$ which is reduced by aldehydes to silver, forming a **silver mirror** on the side of the tube.

toluene A very common name for ➤methylbenzene.

tomography A high-resolution but low-dose ➤X-ray ➤scanning technique which allows detailed images of the internal structure of the body to be constructed. Commonly linked to a computer to provide CT (**computerized tomography**) images, it is a powerful diagnostic tool, for example, for identifying cancers or finding blood clots.

ton An Imperial unit of mass. In UK usage it is equivalent to 2240 lb; in the US it is equivalent to 2000 lb. 1 UK ton (sometimes called a **long ton**) is about 1.016 ➤tonnes; 1 US ton (a **short ton**) is about 0.907 tonnes.

tongue A muscular organ in the floor of the mouth of vertebrates associated with the chewing (mastication) and swallowing of food. The tongue is unique in being the only organ in the body to contain voluntary ➤muscle which does not connect bone to bone. In humans the tongue has numerous taste buds which are sensitive to flavours in food and is also used to help form sounds during speech.

tonne (metric ton) An SI unit of mass, equivalent to a thousand kilograms. It is often pronounced 'tunny' to distinguish it from the ➤ton.

tooth One of a variety of hard structures in the mouth of vertebrates, used primarily for biting and processing food before swallowing. In mammals and reptiles the teeth are confined to the **jaws**, but in amphibians and fish they may occur across the palate. Modern birds do not have teeth.

topaz An aluminosilicate which contains fluorine, $Al_2SiO_4(F,OH)_2$. The yellow variety is a gemstone.

topological space A ➤set, together with a designated collection of open subsets satisfying certain ➤axioms, which serves as the most general framework within which questions of continuity (➤continuous) can be considered.

topology A major branch of mathematics concerned with the classification and characterization of those properties of geometric figures and more general sets of points, ➤topological spaces, that are unaffected by continuous transformations.

top quark (truth quark) The quark in the ➤standard model that has the highest mass; it is the partner of the ➤bottom quark. It was the last of the six to be found: its discovery was announced in 1995, though evidence for its existence had come from statistical analysis of experiments carried out in previous years.

toroidal In the shape of a ➤torus.

torque ➤moment of force.

torr A unit of pressure equal to 1/760 ➤atmospheres (2), or about 133 Pa. The torr was originally the pressure exerted by a 1 mm column of mercury, written as 1 mm Hg. Named after Evangelista Torricelli (1608–47).

Torricellian vacuum The name given to the vacuum above a column of mercury in a sealed tube (typically in a ➤barometer).

torsion The twisting of an object produced by a torque (➤moment of force) or ➤couple when one part of it is held in place. Torsion corresponds to rotation as ➤shear corresponds to linear displacement.

torsion balance A device for measuring a torque (➤moment of force), and hence also a small force on an arm of known length, by measuring the ➤torsion of a wire.

torus The surface formed by rotating a circle through 360° about an axis that does not cut the circle (see the diagram). A bicycle inner tube has the shape of a torus.

torus

tosyl chloride A more memorable name for p-methylbenzenesulfonyl (p-toluenesulfonyl) chloride p-$H_3CC_6H_4SO_2Cl$. Tosylates are very useful synthetic intermediates as the tosylate ion is a good leaving group (➤nucleophilic substitution).

total angular momentum quantum number Symbol J or j. The ➤angular momentum quantum number associated with the total of ➤orbital angular momentum (L) and ➤spin angular momentum (S) of one or more electrons, usually in an atom, ion or molecule. Since its constituent parts are both vectors, J can take on a range of values from $L + S$ to $L - S$.

total differential ➤differential.

total eclipse ➤eclipse.

total internal reflection The ➤reflection of a ray from a boundary with an optically less dense medium at an angle that permits no refracted ray. The angle of reflection must exceed the critical angle θ_c where $\sin \theta_c = n_1/n_2$, n_1 being the ➤refractive index of the less dense medium and n_2 the refractive index of the denser medium. ➤refraction.

totality The total obscuration of the Sun by the Moon during a solar ➤eclipse or the time during which the moon is entirely within the Earth's shadow during a lunar eclipse.

toxin Any poisonous substance produced by an organism. The production of such substances is usually either defensive and deters predators, or assists in the capture of prey (as in snake venoms).

trace The sum of the diagonal elements of a ➤square matrix; it is also equal to the sum of its eigenvalues (➤eigenvector).

trace element (essential element, micronutrient) An ➤element required in minute amounts (usually <1 ppm) for normal growth of organisms. Both plants and animals require trace elements, which must be supplied from the soil or from the diet respectively. Trace elements are usually ➤cofactors of enzymes essential for metabolism, and include boron, cobalt, manganese, zinc and molybdenum.

tracer An ➤isotope added to enable the course of an element's progress in a reaction to be investigated. The isotope can be radioactive or nonradioactive (➤labelling).

trachea (plural **tracheae)** The windpipe, joining the pharyngeal cavity to the lungs in air-breathing vertebrates. The trachea is reinforced and prevented from collapse by rings of cartilage. Tracheae also form part of the ➤gas-exchange system of insects being segmentally arranged tubes accessing the external air via ➤spiracles.

tracheid ➤xylem.

Tracheophyta An obsolete term used to group together plants with a ➤vascular system. Ferns, gymnosperms and flowering plants were included.

trajectory The position of an object as a function of time. The position is typically in space (e.g. the trajectory of an artillery shell) but can also be in ➤phase space or similar domains where coordinates vary as a function of time.

trans- ➤*cis–trans* isomerism.

transactinide elements ➤postactinide elements.

transcendental number A real or complex number such as e or π that is not an ➤algebraic number; it is thus necessarily an ➤irrational number.

transcription The copying of a ➤DNA chain into a strand of ➤RNA by the action of RNA polymerase. Transcription is an essential part of the process by which the information encoded in the DNA sequences of ➤genes is converted into functional molecules in cells. Transcription takes place in the nucleus, and the messenger RNA molecules so formed pass out via nuclear pores into the cytoplasm where they attach to ➤ribosomes. Other types of RNA that do not encode protein (e.g. ribosomal RNA and transfer RNA) are also produced by transcription from DNA. Compare ➤reverse transcription. ➤➤protein synthesis; translation.

transcription factor A ➤protein that facilitates the ➤transcription of a specific gene or set of genes by ➤RNA polymerase.

transcription unit All the segments of ➤DNA associated with a ➤gene that are copied into ➤RNA during ➤transcription.

transducer   A device that measures a physical quantity (such as pressure, temperature or power dissipation) and transmits the measurement in another form, usually as a small electrical voltage or current.

transduction   The introduction of foreign ➤DNA into a bacterial cell by a ➤bacteriophage.

transfer function   ➤frequency response.

transfer RNA (tRNA, soluble RNA, sRNA)   A small ➤RNA molecule with an affinity for a specific amino acid. tRNAs are responsible for the marshalling of amino acids into the correct position on a ➤ribosome during ➤protein synthesis. Part of the molecule has a set of three ➤nucleotide bases (an ➤anticodon) that undergo complementary ➤base pairing with the ➤codons on a ➤messenger RNA molecule.

transformation 1 (Math.) A change in a mathematical expression to an equivalent form, such as that caused by a substitution of one set of variables for another.
  2 (Math.) Another name for a function, most commonly used in the context of **geometric transformations** (where functions such as rotations are most naturally described geometrically) and **linear transformations** (such as those defined by matrices). ➤Lorentz transformation.
  3 (Biol.) A permanent inheritable change in the ➤DNA content of a cell brought about either by insertion of foreign DNA into the cell or by ➤mutation.

transformer   A device for changing an alternating voltage from one value to another which works by the principle of electromagnetic ➤induction. An alternating current in the **primary coil** creates an alternating magnetic field, which induces an alternating current in the **secondary coil**. The ratio of the voltage in the secondary to the voltage in the primary is approximately equal to the ratio of the number of turns of wire in the secondary coil to the number of turns in the primary. To ensure efficiency, the core of the transformer is often made from ➤laminated iron to provide a high permeability and minimize ➤eddy currents.

transgene   A foreign gene that has been artificially introduced into the genome of an organism. A transgene present in ➤germ cells is inherited by offspring; a transgene present only in ➤somatic cells is not passed on to the ➤gametes.

transgenic   Describing an organism containing a ➤transgene. A transgenic organism is described as **chimeric** if only some of its cells contain the transgene, or it may be fully transgenic if it was derived from a parent in which the germline contained the transgene. ➤➤genetic engineering.

transient   The response of an oscillating system to a short input (or change in input) that is not sustained. ➤➤Q factor; resonance.

transistor   A ➤semiconductor device with three electrodes, which has almost completely superseded the thermionic valve as a circuit-controlling device. In general, a small voltage or current applied to one pair of electrodes is used to adjust another current flowing between a different pair. There are two major categories: the ➤bipolar junction transistor; and the ➤field-effect transistor. Transistors can be incorporated in ➤integrated circuits, where they act as switches or amplifiers. Their increasing miniaturization has been the driving force behind advances in the processing speeds of digital computers.

transistor–transistor logic ➤TTL.

transition The change of a ➤quantum-mechanical system from one state to another as a result of interaction with an external influence. For example, a transition of a molecule from one state to another may be caused by electromagnetic radiation (such as light).

transition metal (transition element) This name is sometimes used injudiciously to mean the same as ➤d-block element; the important distinction is that the transition metals are d-block elements that have at least one stable ion with a *partially filled* d subshell. Both the first and last group of the d block (scandium and zinc in the first row) do not form such an ion and so are excluded from the transition metals. The five main transition metal characteristics are: variable oxidation numbers; catalytic activity; complex formation; formation of coloured ions; and paramagnetism. These are highlighted under the individual transition metals, such as ➤iron. ➤➤Appendix table 5.

transition state The highest energy state through which a reaction passes on the way from reactants to products. In the diagram at ➤intermediate, the distinction is made between the two transition states and the intermediate formed during the reaction: the latter is, at least in principle, isolable, whereas the transition states are not. ➤activation energy.

transition temperature The temperature at which a transition occurs between two phases. For example, the transition temperature between rhombic and monoclinic sulfur is $96\,^{\circ}C$.

translation 1 (Biol.) The process by which the ➤nucleotide base sequence of ➤messenger RNA is converted by ➤ribosomes into a sequence of ➤amino acids to form a ➤polypeptide. ➤➤genetic code; protein synthesis.
2 (Math.) A transformation that consists of a shift of spatial coordinates. Translations feature in the symmetry space group of lattices.

translocation The movement over long distances inside a plant of materials manufactured by metabolism. For example, ➤carbohydrates are translocated from a source (the leaves) in the form of sucrose to a sink (the roots). ➤➤phloem.

translucent Permitting light to pass, but in such a way that images cannot be clearly distinguished (because of scattering by the translucent material). Compare ➤opaque; transparent.

transmission coefficient At a boundary or surface, the ratio of the amplitude of a transmitted wave to the amplitude of the incident wave.

transmission electron microscope (TEM) The simplest form of electron microscope; it passes a beam of electrons through a very thin sample and detects the electrons that are transmitted through it without scattering. Compare ➤scanning electron microscope.

transmission line A pair of conductors (e.g. a coaxial cable) for carrying an electrical signal over significant distances. Each section of a transmission line has a resistance R, a capacitance C and an inductance L. The circuit equations for this arrangement reduce to the ➤wave equation for the voltage and current as a function of position, with an ➤impedance per unit length of approximately $\sqrt{(L/C)}$, typically about 70 Ω for a simple coaxial cable. ➤➤matched termination.

transmittance The ratio of the intensity of a transmitted wave to the intensity of the incident wave. Compare ➤reflectance; ➤➤transmission coefficient.

transmitter 1 (Phys.) A device for transmitting a signal, typically by means of a ➤radio wave. It usually consists of a means of generating the ➤carrier wave and a means of introducing the signal by ➤modulation.
  2 (Biol.) ➤neurotransmitter.

transparent Allowing light to pass without significant ➤scattering. Objects can be clearly viewed through a transparent medium. Glass, although transparent to visible light, absorbs in the infrared. Compare ➤opaque; translucent.

transpiration The loss of water by evaporation from the aerial parts of a plant. This is an entirely physical process, and is influenced by factors such as temperature and humidity. Plants can regulate the degree of transpiration by opening or closing pores (stomata; ➤stoma) on their leaves.

transplant ➤graft.

transponder ➤radar.

transport number The fraction of the current through an electrolyte carried by a particular ion.

transport phenomena The set of phenomena associated with the motion of particles, for example either gas molecules or electrons in a conductor. They include electrical and ➤thermal conductivity and ➤thermoelectric effects.

transpose matrix That ➤matrix formed by interchanging the rows and columns of a given matrix.

transposon (transposable element, mobile element, 'jumping gene') A DNA segment, usually between 1000 and 10 000 base pairs long, that can move from one position to another in the same ➤genome. Transposons cause ➤mutations of various kinds and have important applications in ➤genetic engineering. Several different kinds of transposon exist, characterized by their mechanisms of movement. Their existence was first proposed by Barbara McClintock in the 1940s.

transuranium elements (transuranic elements) The elements that follow uranium, all of which are radioactive and have been made artificially. The first to be discovered were neptunium and plutonium in the aftermath of the first nuclear explosion. The person most responsible for investigating the transuranium elements was Glenn T. Seaborg. The last of these elements to have any significant uses is element 98, californium. Element 106 is now named seaborgium. ➤➤Appendix table 5.

transverse Perpendicular to the direction of motion of some object or wave.

transverse electric (TE) and transverse magnetic (TM) fields ➤wave-guide.

transverse wave A wave that features a quantity whose direction is ➤transverse. For example, the electric and magnetic fields in an ➤electromagnetic wave are both vectors perpendicular to the direction of motion (➤Poynting vector), so such a wave is transverse. Compare ➤longitudinal wave.

trapezium (plural trapezia) A ➤quadrilateral with one pair of ➤parallel sides. If these sides have lengths a and b and are a perpendicular distance h apart, the area of the trapezium is $\frac{1}{2}h(a + b)$.

trapezium rule The ➤numerical integration formula:

$$\tfrac{1}{2}h[y_0 + y_n + 2(y_1 + \cdots + y_{n-1})],$$

which gives the approximate area under a curve in terms of the y coordinates at x coordinates that are equally spaced, distance h apart. It is derived by approximating the area under the curve by a sequence of abutting trapezia.

travelling wave (progressive wave) A wave that moves in one direction rather than being confined by its boundary conditions. In one dimension a travelling wave of the form $y(x, t) = f(x - ct) + g(x + ct)$ cannot be separated into the form $y(x, t) = A(x)B(t)$, as can a ➤standing wave.

triacylglycerol (triglyceride) The generic name for the common structure of ➤lipids in which the glycerol molecule is acylated three times (see the diagram).

CH_2OCOR
|
$CHOCOR'$
|
CH_2OCOR''

triacylglycerol

triad A collection of three closely related elements, such as chlorine, bromine and iodine. Identifying triads was an early step in the development of the ➤periodic table.

triangle A polygon with three sides. The area of a triangle is $\frac{1}{2}bh$, where b is the length of one side and h is the perpendicular distance from the other vertex to that side. Unknown sides and angles of triangles may be found by ➤trigonometry. The sum of the angles in a triangle is $180°$.

Triassic See table at ➤era.

triatomic molecule A molecule containing three atoms, such as ozone, O_3, or carbon dioxide, CO_2.

tribasic acid (triprotic acid) An acid that can donate up to three protons to a base, such as phosphoric acid, H_3PO_4. Phosphoric acid can therefore react with sodium hydroxide to form three different species, NaH_2PO_4, Na_2HPO_4 and Na_3PO_4. ➤buffer solution.

triboelectricity Electricity produced by ➤friction.

tribology The study of ➤friction.

triboluminescence Emission of light caused by crushing crystals.

tribromomethane IUPAC name for bromoform (➤bromine).

tricarboxylic acid cycle (TCA cycle) ➤Krebs cycle.

1,1,1-trichloroethane (TCE) CCl_3CH_3 A colourless liquid halogenoalkane formerly much used as a solvent, for example for liquid paper.

trichloroethene (trichloroethylene, Trilene) $CCl_2{=}CHCl$ A colourless liquid halogenoalkane much used as a solvent, especially in metal degreasing and dry cleaning.

trichloromethane IUPAC name for ➤chloroform.

2,4,6-trichlorophenol (TCP) A halogenated phenol (see the diagram) which is the most well-known antiseptic. This compound illustrates the principle that chemists can turn two harmful chemicals (chlorine and phenol cause severe skin acne) into a compound that is actually beneficial (when applied to cuts and grazes).

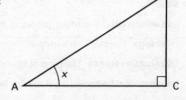

2,4,6-trichlorophenol

trichome ➤hair (2).

triclinic The ➤Bravais lattice (and ➤crystal system) of minimum symmetry. There is no special relationship between the ➤primitive vectors of the lattice.

tricuspid valve ➤heart.

tridymite A polymorph of ➤silica stable from $867\,°C$ to $1470\,°C$.

triethylaluminium (aluminium triethyl) $Al(CH_2CH_3)_3$ The most common organoaluminium compound used as a ➤Ziegler–Natta catalyst.

trifluoroethanoic acid (trifluoroacetic acid) CF_3COOH A colourless liquid which is very acidic, useful, for example, for cleaving esters in the synthesis of peptides.

triglyceride ➤triacylglycerol.

trigonal An alternative name for ➤rhombohedral.

trigonal bipyramidal The most common shape for molecules with five electron pairs around the central atom (➤valence-shell electron-pair repulsion theory). It is the gas-phase structure of PCl_5, for example. The structure is not symmetrical as there are three different bond angles: $90°$, $120°$ and $180°$.

trigonometric functions (trigonometric ratios) Six functions – sine, cosine, tangent, cosecant, secant and cotangent (abbreviated sin, cos, tan, cosec, sec and cot) – that were first defined in terms of the ratios of lengths in a right-angled triangle:

$$\sin x = \frac{BC}{AB}, \quad \operatorname{cosec} x = \frac{1}{\sin x}$$
$$\cos x = \frac{AC}{AB}, \quad \sec x = \frac{1}{\cos x}$$
$$\tan x = \frac{BC}{AC}, \quad \cot x = \frac{1}{\tan x}$$

Since sin, cos, tan may be regarded as known functions, these ratios may be used to calculate unknown sides and angles. Trigonometric functions satisfy a host of ➤identities such as $\sin^2 x + \cos^2 x = 1$, the ➤addition formulae and the ➤double-angle formulae.

The derivatives of the trigonometric functions may be expressed in terms of trigonometric functions. When x is in radians, $\sin x$ and $\cos x$ have the ➤Maclaurin expansions

$$\sin x = \sum_{n=0}^{\infty} (-1)^n x^{2n+1}/(2n+1)!, \quad \cos x = \sum_{n=0}^{\infty} (-1)^n x^{2n}/(2n)!.$$

Trigonometric functions are used extensively in the analysis of wave motions.

trigonometry The branch of mathematics concerned with the calculation of unknown sides and angles in triangles, using ➤trigonometric functions and results such as the ➤sine rule and the ➤cosine rule.

trihydric alcohol (triol) An alcohol containing three hydroxyl groups. The most important example is ➤glycerol.

triiodomethane IUPAC name for ➤iodoform.

triiron tetroxide Another name for magnetite (➤lodestone).

trilobite A member of the extinct arthropod class **Trilobita**, common as fossils in rocks of Cambrian and Silurian age. Trilobites typically possessed flattened oval bodies and showed clear specialization of segments in different regions of the body. Most were small, around 5–10 cm in length. They were essentially bottom dwellers and are thought to have occupied a variety of different feeding niches, including detritus feeding.

2,4,6-trinitromethylbenzene (2,4,6-trinitro-toluene, TNT) IUPAC name for the most famous high explosive, with the structure shown in the diagram. It exists as yellow leaflets.

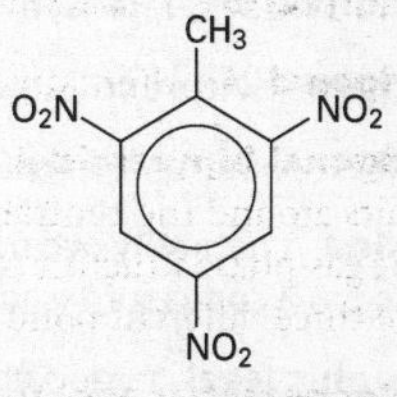

2,4,6-trinitrophenol ➤picric acid.

trinitrotoluene (TNT) ➤2,4,6-trinitromethyl-benzene.

2,4,6-trinitromethylbenzene

triode A ➤thermionic valve with three electrodes; the current between the collector and the emitter can be varied by changing the voltage of the **control grid** between the two. They have been almost entirely replaced by ➤transistors in modern electronics.

trioxygen IUPAC name for ➤ozone.

triphenylmethyl radical $(C_6H_5)_3C^{\bullet}$ A radical which is unusually stable because of extensive ➤delocalization.

triphenylphosphine $(C_6H_5)_3P$ A particularly useful ligand for stabilizing lower oxidation numbers for the ➤d-block elements.

triple bond A bond between two atoms that has three shared electron pairs. The most famous examples are the triple bonds in the N_2 molecule and in ➤ethyne.

triple integral A multiple integral involving three variables, typically written as

$$\iiint f(x, y, z) \, \mathrm{d}x \, \mathrm{d}y \, \mathrm{d}z.$$

triple point A point on a ➤phase diagram where three phases coexist in equilibrium. The triple point of water occurs at 273.16 K (0.01 °C) and 611 Pa. The temperature at water's triple point defines the second fixed point on the ➤Kelvin scale. If the triple point for a substance lies above atmospheric pressure, the substance can undergo ➤sublimation, as does carbon dioxide, for example. ➤➤phase rule.

triple product ➤scalar triple product; vector triple product.

triplet **1** (Phys.) A set of three ➤degenerate quantum states. In particular, for two ➤identical particles with ➤spin $\frac{1}{2}$, there are four states possible for the overall system. One of these (the **singlet state**) is antisymmetric in spin (so the spatial wavefunction must be symmetric under exchange of particles because the particles are ➤fermions). Three states (the triplet) are symmetric in spin (so the spatial wavefunction must be antisymmetric).
 2 (Biol.) ➤genetic code.

trisomy The presence in some or all cells in an otherwise ➤diploid organism of three copies of a particular chromosome, for example human chromosome 21 in **trisomy 21** (➤ Down's syndrome). ➤➤polyploidy.

tritium Symbol T. The third isotope of hydrogen, ^3H; it has one proton and two neutrons. It is radioactive with a ➤half-life of 12.3 years. ➤➤nuclear fusion.

Triton ➤Neptune.

tRNA Abbr. for ➤transfer RNA.

trona A naturally occurring mineral, $Na_2CO_3 \cdot NaHCO_3 \cdot 2H_2O$, which is increasingly used as a source of ➤sodium carbonate.

trophic level ➤food chain.

tropical year The time between successive apparent passages of the Sun through the Earth's equatorial plane at the vernal ➤equinox.

tropics The zone on the surface of the Earth between latitudes 23.5°N (the **tropic of Cancer**) and 23.5°S (the **tropic of Capricorn**). At some stage of the year, the plane of the ➤ecliptic passes through every point in the tropics.

tropism A response to a directional **stimulus** in plants that involves growth. The bending of a seedling towards light (**phototropism**) and the response of roots to gravity (**gravitropism**) are examples.

tropopause The boundary between the ➤troposphere and the ➤stratosphere in the Earth's atmosphere. The tropopause is characterized by a change in the temperature profile of the atmosphere: the temperature fall with altitude halts, and instead the temperature remains steady or rises with altitude. In the ➤international standard atmosphere the tropopause occurs at an altitude of 11 km, but the actual level varies with latitude (higher at the equator and lower at the poles) and the season. The level of the tropopause is of interest to pilots of jet aircraft because the ➤jet stream tends to be strongest just below the tropopause.

troposphere The layer of the ➤atmosphere closest to the surface of the Earth. Its upper boundary is the ➤tropopause, and it is the layer in which almost all weather occurs.

tropylium ion The traditional name for the cycloheptatrienyl ion, $C_7H_7^+$, which has six electrons delocalized across seven atoms causing it to be unusually stable.

Trouton's rule A rough rule of thumb that states that the ➤latent heat of vaporization divided by the boiling point (in kelvin) is approximately constant for a number of liquids. This is because the standard entropy of vaporization is itself roughly constant, being dominated by the large entropy of the gas. However, the rule notably fails for liquids that have significant ➤hydrogen bonding, such as ethanol and water. Named after Frederick Thomas Trouton (1863–1922).

Troy weight An obsolete system of weights used for precious stones and metals. The name comes from Troyes, France. 12 troy ounces = 1 pound.

Trp Abbr. for ➤tryptophan.

truncation error The error involved in using an approximate value in a calculation, for example in using just the first one hundred terms to approximate the sum of an infinite series.

truth table A table that shows all the possible results of a logical operation. For example, consider the AND operator and its truth table. For a two-input AND gate, its output will only be 1 if both inputs are 1; otherwise it will be 0.

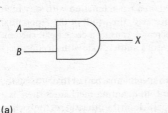

A	B	X
0	0	0
0	1	0
1	0	0
1	1	1

(a)   (b)

truth table (a) A two-input AND gate and (b) its truth table.

trypsin A ➤protease enzyme secreted in vertebrates by the ➤pancreas. Trypsin is produced in an inactive form (**trypsinogen**) which is activated by **enterokinase**, an enzyme secreted by the wall of the intestine.

tryptophan (Trp) A common ➤amino acid. ➤➤Appendix table 7.

TSH Abbr. for ➤ thyroid stimulating hormone.

tsunami A series of water waves formed when a large body of water, such as an ocean, is rapidly disturbed, for example by an underwater earthquake. When the waves arrive at shallower depth, they slow down and so form huge destructive walls of water which crash down on beaches. The Boxing Day 2004 Tsunami caused the largest recorded death toll from such an event with at least 250,000 lives lost in South East Asia.

TTL Abbr. for transistor–transistor logic. A class of logic gates constructed from bipolar ➤transistors. TTL circuits use 0 V to represent ➤binary 0, and +5 V to represent binary 1.

tuberculosis ➤TB.

tumour ➤cancer.

tuned circuit A ➤resonant circuit with a component (usually a variable capacitor) that allows the resonant frequency to be varied. Such circuits are used in ➤radio receivers.

tungsten Symbol W (from the German name *Wolfram*). The element with atomic number 74 and relative atomic mass 183.8, which is in the d block of the periodic table, below chromium. The metal itself is famous for its exceptionally high melting point (3410 °C), hence its use in electric bulb filaments. Steel alloys containing around 8% tungsten are used where great hardness is required. Tungsten shows a wide range of oxidation numbers up to +6, as for chromium. The yellow solid oxide WO_3 is an example, but much more extensive are the range of tungstates based on WO_6 octahedra sharing corners or edges. Reducing **tungstates** with sodium produces a series of tungsten bronzes of approximate formula $NaWO_3$ (in which tungsten has an average oxidation number of +5). All chlorides from WCl_6 to WCl_2 are known, the last two containing metal–metal clusters: WCl_2 is better written $[W_6Cl_8]^{4+}$ $4Cl^-$. An important compound of tungsten is tungsten carbide, WC, an exceptionally hard substance used in cutting tools.

tunnel diode (Esaki diode) A ➤diode based on a ➤p–n junction with very high levels of ➤doping. As the forward ➤bias is increased, the width of the ➤potential barrier between the p- and n-type materials increases, causing a decrease in current. Thus there is a section of the characteristic curve of a tunnel diode with an effectively negative resistance. Compare ➤Gunn diode.

tunnel effect The passage of a particle through a ➤potential barrier that, classically, it would not have enough energy to surmount. In quantum mechanics there is a small amplitude of the ➤wavefunction for the particle on the other side of the barrier, and the particle appears to be able to tunnel through the barrier.

turbine A part of a machine or engine in which the flow of fluid against rotating, angled ➤aerofoils is used to do work. In a generator based on a water turbine, for example, the rotation is used to generate electricity directly. In a turbine engine such as a jet engine, some of the energy is used in compressing the mixture for combustion. This is in contrast to a piston engine, in which the energy for the compression in one cylinder comes from the combustion in other cylinders.

turbocharging A method of increasing the power of an engine by using the exhaust gases to spin a **turbocharger** and thereby force the fuel/air mixture into the cylinders. Two problems are that there is a **turbo lag** between applying the throttle and getting the power increase, and that the turbocharger runs very hot.

turbulent flow Fluid flow with a high ➤Reynolds number, in which the velocity of the fluid varies rapidly with both position and time. Eddies form and transfer momentum rapidly from one part of the fluid to another. Compare ➤laminar flow.

turgor A condition in a plant cell in which the cell has taken up water so that the plasma membrane exerts an equal and opposite force to that exerted by the surrounding cell wall. Turgor is important in support, and plants which lose turgor start to wilt. ➤water potential.

Turing machine A theoretical computing machine proposed by Alan Mathison Turing (1912–54) in 1936, before digital computers existed. His machine can be in a number of discrete states S, and it operates on an infinitely long tape (corresponding in modern terms to its memory). The tape is divided into cells, each of which may contain one of a discrete number of symbols X. The machine acts on a single cell at a time, and can move left and right along the tape. The program consists of a set of conditions, of the following form: if the machine is in state S_i and the symbol in the current cell is X_m, then change the symbol to X_n, move left/right one cell (or stay still), and go into state S_j. It turns out that this machine has the necessary complexity to perform any ➤computable operation (i.e. one that can be carried out by performing a suitable algorithm). In fact, Turing used his idea in an innovative way to show that some problems are not computable, which at the time was a major unsolved problem in mathematics.

Turing test A test proposed by Alan Turing to establish whether a computer can think. The test involves a human interrogator who may ask any questions of two hidden interviewees, by typed communication. One of these is human, the other a computer. The aim of the interrogator is to establish which of the interviewees is human. The aim of the interviewees is to convince the interrogator of their humanity. If the computer is consistently as able as the human at convincing the interrogator, it passes the test. Turing proposed this as an operational test of computer intelligence.

turning point A ➤stationary point on a curve that is either a ➤maximum or a ➤minimum.

turpentine A variable mixture of several hydrocarbons, notably α-pinene, used as a solvent, especially for paint.

twilight The period before sunrise or after sunset during which the light scattered from the atmosphere illuminates the surface of the Earth.

twisted pair A pair of insulated conductors twisted together so as to minimize the interference induced by alternating magnetic fields.

two's complement A method of representing negative binary numbers in computers in which binary strings of N digits represent integers in the range -2^{N-1} to $2^{N-1} - 1$: binary numbers with left-most digit 1 are considered negative and those with 0 are positive. To represent a negative integer in two's complement form, one starts from the binary form of the ➤absolute value of the integer, converts all zeros to ones and vice versa to form the **one's complement**, and then adds one.

tympanum ➤ear.

Tyndall effect The scattering of light by small particles, such as dust in a light beam. Whereas a true solution does not scatter light, a colloidal one will, this being one of the best distinguishing features of a ➤colloid. Named after John Tyndall (1820–93). ➤➤Rayleigh scattering.

typhoid An acute infectious disease caused by the bacillus *Salmonella typhi*. Typhoid is transmitted in water and food contaminated by faeces of infected persons. Symptoms include high fever, nausea and diarrhoea, and, if untreated, the disease may progress to pneumonia and internal haemorrhage. Control is by improved sanitary conditions, immunization and the use of the antibiotic **chloramphenicol**.

Tyr Abbr. for ➤tyrosine.

tyrosine (Tyr) A common ➤amino acid. ➤➤Appendix table 7.

U **1** Symbol for the element ➤uranium.
   2 Symbol for the pyrimidine base ➤uracil.

U **1** Symbol for ➤internal energy.
   2 Symbol for ➤potential energy.

ubiquinone ➤quinone.

UHF Abbr. for ➤ultra high frequency.

UHT Abbr. for ultra-heat-treated. Milk subjected to a temperature of about 135 °C for 2 s will keep for months in sterile containers.

ulna ➤pentadactyl limb.

ultracentrifuge A high-speed ➤centrifuge, developed by Theodor Svedberg in 1923, which, under appropriate conditions, will sediment and separate subcellular particles and ➤organelles and some macromolecules such as ➤proteins and ➤nucleic acids. It can also sediment colloidal sols (➤colloid).

ultra high frequency (UHF) The ➤frequency band between 300 MHz and 3 GHz. It is used for communications, including television transmissions, and for ➤radar.

ultramarine A sulfur-containing sodium aluminosilicate of blue colour used as a gemstone.

ultramicroscope A device for viewing very small particles by observing the light scattered by them from an intense beam. ➤➤Rayleigh scattering.

ultrasound Pressure waves of higher frequency than ➤sound (i.e. greater than about 20 kHz). **Ultrasonics**, the study and application of ultrasound, is used where light is unable to penetrate but X-rays would be damaging, for example in the imaging of foetuses.

ultrastructure The structure of cells and cell inclusions as imaged by the ➤electron microscope.

ultraviolet (UV) A region of the ➤electromagnetic spectrum that lies beyond the violet end of the visible region. The **near UV** extends from roughly 400 to 280 nm, being subdivided into **UVA** (400–320 nm) and **UVB** (320–280 nm); the **far UV** is roughly from 280 to 200 nm, and the **vacuum UV** is below 200 nm. Ultraviolet photons carry more energy than those in the visible, and so can cause harmful effects such as skin cancers on long exposure, so the reduction in the protection afforded by the ➤ozone layer is a cause of concern. ➤spectroscopy.

umbilical cord ➤placenta.

umbra The dark central region of a shadow cast by an object interrupting an extended light source. It is surrounded by the ➤penumbra. ➤➤eclipse.

uncertainty principle ➤Heisenberg uncertainty principle.

uncountable set A ➤set, such as the ➤real numbers, that is not ➤countable.

undecane The ➤alkane with eleven carbon atoms, $C_{11}H_{24}$, which exists as a colourless liquid.

undulipodium (plural **undulipodia)** An extension protruding through the plasma membrane of ➤eukaryotic cells. **Flagella** and **cilia** are about 100 μm and 10 μm in length respectively, and 300 nm in diameter. All undulipodia have a pair of central ➤microtubules surrounded by nine paired microtubules. They are ➤organelles that can undulate using energy from ➤ATP. They are associated with cell locomotion, for example in the swimming of ➤spermatozoa and many ➤protoctistans, and in creating currents past cells, for example to move an egg along the lining of the oviduct.

uniaxial crystal ➤birefringence.

unicellular Consisting of a single cell.

unified field theory In theoretical physics, a theory that attempts to explain all the known ➤fundamental interactions in a unified way in terms of ➤quantum field theory.

uniform The same at all positions. 'Uniform' is used for ➤scalar and ➤vector fields in the same way that 'constant' would be used for a quantity that does not vary with time.

unimolecular reaction A reaction in which the ➤transition state contains only one species. A good example is the S_N1 reaction discussed under ➤nucleophilic substitution.

union That ➤set consisting of those ➤elements that belong to at least one of a given family of sets. The union of two sets A and B is denoted $A \cup B$.

unit A physical quantity of a standard size, for example the ➤kilogram, defined in such a way that other physical quantities may be expressed in multiples of it. It is fundamental to physics that a physical law never depends on a particular choice of units. Units are always arbitrary, though some standards are more useful and practical than others. ➤SI units; Appendix table 1.

unitary matrix A ➤square matrix U with complex number elements that is equal to the ➤inverse of the ➤transpose of the matrix whose elements are the ➤complex conjugates of the elements of U.

unitary symmetry ➤SU(3).

unit cell The fundamental volume from which a lattice is constructed. A **primitive unit cell** is the three-dimensional shape that, when replicated and translated by every lattice vector of a ➤Bravais lattice, exactly fills space. ➤➤Wigner–Seitz cell.

unit vector A ➤vector of ➤magnitude 1. The notation $\hat{n}$ is used to denote a unit vector in the direction of n.

universal gas constant ➤gas constant.

universal gravitational constant ➤Newton's law of (universal) gravitation.

universal indicator A mixture of ➤indicators that changes through a range of colours as the pH goes from about 4 to about 10.

universal motor An electric motor, containing a ➤commutator, that is designed to run on either ➤alternating current or ➤direct current.

universal set In a given context, such as a ➤Venn diagram, a specific ➤set large enough to include all the elements relevant to the context.

Universal Time (UT) The scale that is used to define time on Earth, regardless of location. It is essentially the same as **Greenwich Mean Time** (GMT) defined in terms of the transit of certain stars across the prime ➤meridian at Greenwich in London. Corrections are applied to UT to take account of irregularities in the Earth's rotation, and the resulting time scale is designated **UT1**. When atomic clocks (measuring ➤International Atomic Time, TAI) became more accurate than the motion of the Earth as a measure of time, it became apparent that the length of the year varied by almost a second. Broadcast time signals are therefore now based on **Coordinated Universal Time** (UTC, from its abbreviation in French), which is defined to be TAI plus or minus a whole number of seconds, which may be changed by the addition or subtraction of **leap seconds**. Such a scale is accurate to the standard of the atomic clocks, but never varies by more than 1 second from UT1.

Universe Everything that exists as matter, and the space in which it is found. ➤big bang.

UNIX An operating system introduced in 1971 for computer systems which can support more than one user at the same time, on different terminals.

unnil- A prefix used in the now-outdated name for each of the six elements following element 103 (lawrencium), the ➤postactinides. It stands for 1 followed by 0, as in unnilquadium, element 104.

unsaturated The opposite of ➤saturated in all its meanings.

unstable The opposite of ➤stable; that characteristic of a system which, after a small displacement from ➤equilibrium, tends to move away from that equilibrium.

up quark One of the six ➤quarks of the ➤standard model. It is a constituent of both the proton and the neutron.

upsilon particle The first particle containing the ➤bottom quark to be observed. The upsilon is a ➤meson consisting of a bottom quark and its antiquark. It has a mass of approximately 9.5 GeV/c^2 and is extremely short-lived.

uPVC Unplasticized ➤PVC.

uracil Symbol U. A ➤pyrimidine base forming a ➤nucleotide, important as an integral part of the structure of ➤RNA but *not* ➤DNA. Specifically, when DNA is being copied into RNA during ➤transcription, uracil replaces the base ➤thymine.

uranium Symbol U. The element with atomic number 92 and relative atomic mass 238.0, which is one of the ➤actinides. It is the last naturally occurring element in the periodic table, later ones being called transuranium elements. Its most famous property relates to two of its isotopes, ^{235}U and ^{233}U, which can undergo ➤nuclear fission, the basis of nuclear power generation as well as nuclear weaponry. The fissile isotope ^{235}U is only present in 0.7% of the natural element and uranium has to be **enriched** for fission to take place. The separation from the much more abundant ^{238}U can be done in several ways, most notably by gaseous diffusion of the compound uranium(VI) fluoride, UF_6, although laser isotope separation is becoming more prevalent. Uranium is found with a variety of oxidation numbers up to +6; for example all four chlorides from UCl_3 to UCl_6 can be made. A large variety of oxygen compounds are also known, of which the most interesting is the **uranyl ion**, UO_2^{2+}. Uranyl nitrate, which exists as yellow crystals, is used as an indicator and in ceramics and glasses.

uranium(VI) fluoride (uranium hexafluoride) UF_6 A white solid that sublimes easily to a violently corrosive gas used to separate the ➤isotopes of uranium by gaseous diffusion.

Uranus The seventh planet from the Sun, orbiting in 84.01 years at an average distance of 19.2 AU. Its equatorial radius is 25 560 km, about four times that of the Earth. Like Saturn, Uranus has rings, but optically they are almost invisible. Particularly unusual is its axial inclination of 98°, putting its rotational axis very close to its orbital plane. There is very little structure discernible in the atmosphere (83% hydrogen, 15% helium, 2% methane). Uranus has 17 known satellites; the largest two are Titania and Oberon, both with radii approaching 800 km. ➤➤Appendix table 4.

urea (carbamide) $CO(NH_2)_2$ The main excretory product of many animals, particularly ➤mammals and fish of the shark and ray family. It is produced in the **ornithine (urea) cycle** in the liver by deamination of excess ➤amino acids and is also a breakdown product of ➤purines and ➤pyrimidines. ➤➤Wöhler's synthesis.

ureter In mammals, the tube connecting the ➤kidney to the bladder, along which urine is transported.

urethra In mammals, the tube connecting the bladder to the exterior of the body, responsible for the transport of ➤urine. In males the urethra extends through the penis and also carries ➤semen.

uric acid A nitrogenous derivative of ➤purines. Uric acid is an important excretory product, particularly in reptiles, birds and insects, where water conservation is important. Overproduction of uric acid in humans, and its deposition in joints, is the cause of gout.

uridine A ➤nucleoside consisting of the base ➤uracil covalently bonded to the sugar ➤ribose.

urine An aqueous solution of mineral salts and nitrogenous excretory products formed by the ➤kidney. The exact composition of urine varies even within a single animal, depending on factors such as the external environment, diet and state of health.

URL Abbr. for uniform resource locator. The ➤www uses URLs to address ➤Internet resources such as Web pages, as, for example, in:

http://www.tonbridge-school.co.uk

USB Abbr. for universal serial bus. Computers come equipped with multiple USB connectors which enable a huge range of peripherals to be connected, from printers and cameras to **USB disks** for storing copies of files.

US Customary System The system of units traditionally used in the USA. It is based on the UK system of ➤Imperial units, but with some variations: the ➤gallon and ➤ton, for example, are defined differently.

UT Abbr. for ➤Universal Time.

UTC ➤Universal Time.

uterus The part of the female reproductive tract in placental mammals in which the ➤foetus develops. It is connected to the exterior via the ➤vagina. For some other animals the term is also used to describe a part of the female tract that can temporarily hold eggs (as in tapeworms). ➤➤menstrual cycle.

utriculus ➤ear.

UV Abbr. for ➤ultraviolet.

V

v Symbol for ➤velocity.

V **1** Symbol for the element ➤vanadium.
  2 Symbol for the unit ➤volt.

V **1** Symbol for ➤volume.
  2 Symbol for ➤potential difference (voltage).
  3 Symbol for ➤potential energy.

VA Symbol for ➤volt-ampere.

vacancy (Schottky defect) A site in a crystal lattice (that would normally be occupied by an atom or ion) where the atom or ion is missing.

vaccine A substance used clinically for the immunization of the recipient organism against a ➤pathogen. Vaccines are usually preparations of cell-free products of, or attenuated (less ➤virulent) forms of, disease-causing cells or ➤viruses. Specifically, the word vaccine refers to the Vaccinia virus that causes cowpox. In 1796 Edward Jenner used the liquid from cowpox vesicles to immunize people against smallpox. The cowpox ➤antigen is sufficiently similar to that of smallpox, so that ➤antibodies raised against cowpox are effective against smallpox. ➤➤immune system; immunization.

vacuole One of a number of different types of membrane-bound ➤organelles in cells. Plant cells are typically characterized by having a large vacuole that may occupy over 90% of the cell's volume and contributes to the cell's ➤turgor. The vacuole is also a site for the storage of various ➤metabolites or pigments (as in beetroot cells). Vacuoles may be present in animal cells where they have a number of storage or metabolic functions including the 'food vacuoles' of ➤phagocytes.

vacuum (plural **vacua)** A volume of space that contains no particles of any sort. In experimental physics a vacuum is normally taken to mean a volume of space where the density of gas present is extremely low. Vacua are normally measured according to the pressure of gas remaining. Pressures of 10^{-7} Pa are routinely achievable in a laboratory.

vacuum distillation Distillation under reduced pressure. For example, the residue from normal fractional distillation of petroleum constitutes about half the original mass, and contains hydrocarbons with more than about twenty-five carbon atoms. Those with less than about thirty carbon atoms are useful as lubricating oils and can be further separated by vacuum distillation, where the lower boiling points at reduced pressure allow a reasonable separation.

vacuum evaporation A technique used to deposit thin layers (sometimes with a thickness of only a few atomic radii) of an **evaporant** material on a ➤substrate (2). Typically, both evaporant and substrate are metals or semiconductors. In a high or ultra-high vacuum chamber, the evaporant is heated and sublimes; the resulting stream of atomic gas is directed towards the substrate and collides with its surface, forming a thin layer. The process is carried out in a vacuum to avoid collisions between the evaporant and the residual gas in the chamber. ➤➤molecular beam epitaxy.

vacuum pump A pump used to create a low pressure. A backing pump, such as a rotary pump, reduces the pressure to below 10 Pa. Further pressure reduction requires a pump such as a ➤diffusion pump. ➤➤ion pump; sorption pump.

vacuum state A quantum state for which no particles are present in a system. It is not normally necessary to include such a state unless ➤second quantization is used.

vacuum ultraviolet (VUV) The region of the ➤electromagnetic spectrum with frequency between the far ➤ultraviolet and the longest-wavelength X-rays (approximately 200 to 100 nm). In this region of the spectrum light is strongly attenuated by absorption in air, and most experiments must be carried out in a ➤vacuum, hence the name.

vagina The duct of the female reproductive organs connecting the ➤uterus with the exterior. It receives the ➤penis during copulation and forms the birth canal through which the fully formed ➤foetus is born.

Val Abbr. for ➤valine.

valence ➤valency.

valence band The ➤energy band in a semiconductor that is immediately below the Fermi energy (see the diagram overleaf). The valence band is almost completely occupied, though at nonzero temperature a small number of electrons will occupy states in the ➤conduction band, leaving ➤holes in the valence band (➤Fermi–Dirac statistics). In metals the valence band is less relevant, as the Fermi energy falls within the conduction band, but the label is still used to describe bands for which the electronic ➤wavefunction spreads over a large number of atomic centres.

valence bond theory A method of applying quantum theory to the calculation of ➤chemical bonding. The more common method is ➤molecular orbital theory. When both theories are applied in their most rigorous forms, the initially significant differences vanish.

valence electrons The electrons in the outermost ➤shell of an atom that are responsible for the bonding between two atoms.

valence-shell electron-pair repulsion theory (VSEPR theory) A theory explaining the shape of simple molecules. First, a ➤Lewis structure is drawn to show the electron pairs in the valence shell (all other electrons are ignored as they are held too close to their parent atoms). These electron pairs come in two categories:

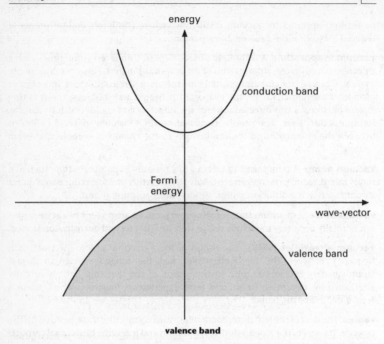

valence band

bond pairs and ➤lone pairs, the latter being attached to only one atom. These electron pairs are then moved as far apart as possible, to account for the mutual repulsion between like charges. Lone pairs repel a little more than bond pairs. Three electron pairs form a trigonal planar shape, four a tetrahedral shape, five a trigonal bipyramidal shape, and six an octahedral shape (see the diagram opposite). The ➤methane molecule has four bond pairs and no lone pairs, so the shape is predicted to be tetrahedral, as observed experimentally. The water molecule too has four pairs, two bond pairs and two lone pairs. Its angular shape is also based on a tetrahedral distribution of the electron pairs, with the angle (104.5°) between the two bonds being a little less than the tetrahedral angle (109° 28′), because of the greater lone pair/lone pair repulsions.

valency (valence) The number of atoms of hydrogen with which one atom of an element will combine or which one atom will replace. Thus magnesium with a valency of 2 will combine with chlorine, which has a valency of 1, to form magnesium chloride of formula $MgCl_2$. The term is now used much less often; instead, the specification of the ➤oxidation number of each element also helpfully designates which atom has a greater share of the bonding electrons.

valine (Val) A common ➤amino acid. ➤➤Appendix table 7.

Valium Tradename for ➤diazepam.

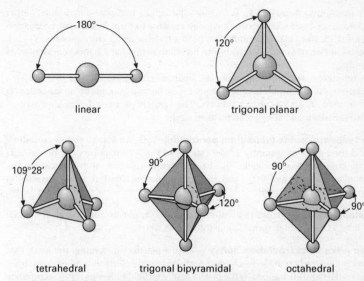

valence-shell electron-pair repulsion theory (VSEPR theory) The energetically most favourable arrangements for electron pairs.

valve 1 (Biol.) One of two halves of some fruits, particularly of the cabbage family (Brassicaceae), or of the shell of some ➤molluscs such as mussels (bivalves), or of the external wall of a ➤diatom.

    2 (Biol.) A fold or flap of tissue that promotes the one-way flow of fluid in a duct or blood vessel. ➤➤heart.

    3 (Phys.) ➤thermionic valve.

vanadate An oxoanion containing the element vanadium. The most common are those with oxidation number +5. **Metavanadates** contain the VO_3^- ion and **ortho-vanadates** the VO_4^{3-} ion. There are a large number of different species that are present in aqueous solution, varying in concentration according to the pH of the solution.

vanadium Symbol V. The element with atomic number 23 and relative atomic mass 50.94, which is in Group 5 of the ➤periodic table. The pure element is a silvery metal, with a body-centred cubic crystal structure. Vanadium shows the usual features associated with ➤transition metals: it has several common oxidation numbers and forms coloured ions. This can be demonstrated by reducing a solution of ammonium metavanadate with zinc amalgam and acid. The yellow colour is due to the dioxovanadium(v) ion, VO_2^+. There follows a series of colours which explains why the element was named after Vanadis, the Scandinavian goddess of beauty. The blue colour is due to the oxovanadium(IV) ion, VO^{2+}, the green colour to the vanadium(III) ion, V^{3+}, and finally, after a long time, a violet colour can be seen which is due to the vanadium(II) ion, V^{2+}.

Vanadium(v) fluoride, VF_5, is a white solid rapidly hydrolysed by water. Neither chlorine nor bromine is a strong enough oxidant to form halides with oxidation number +5. In oxidation number +4, both a fluoride and a chloride exist: VF_4 is a green solid whereas VCl_4 is a red-brown liquid, showing that it is more covalent in its bonding.

Complexes also exist, especially the hydroxo complexes such as $[VO_2(OH)_2]^-$. Catalysts often incorporate vanadium, the most famous example being vanadium(v) oxide used in the ➤contact process. The element is most frequently used as **ferrovanadium**, an alloying element in steels.

vanadium(v) oxide (vanadium pentoxide) V_2O_5 An orange-yellow crystalline solid used most importantly as the catalyst in the ➤contact process. It can react with both acids and alkalis. In strongly acidic solution, it is present as the dioxovanadium(v) ion, VO_2^+, and, in strongly alkaline solution, as the vanadate(v) ion, VO_3^-.

vanadyl ion The oxovanadium(IV) ion, VO^{2+}, found in, for example, vanadyl sulfate, $VOSO_4$, which forms a blue solution in water.

Van Allen Belts (radiation belts) Belts of ➤plasma surrounding the Earth that emit intense radiation. Some charged particles, mostly electrons and protons, from the ➤solar wind are trapped by the Earth's magnetic field as they pass and accumulate in two doughnut-shaped zones. The movement of the particles within the zones causes them to emit electromagnetic radiation. Named after James Alfred Van Allen (1914–2006). ➤➤aurora; magnetosphere.

Van de Graaff generator An ➤electrostatic generator in which charge is carried on an insulated belt from an external source to an isolated metal sphere. It is thus possible to produce ➤potential differences of millions of volts, limited only by the effectiveness of the insulation. Named after Robert Jemison Van de Graaff (1901–67).

van der Waals equation An ➤equation of state which describes real gases more accurately than the ➤ideal gas equation does, but at the expense of introducing two constants to take into account the forces between the molecules, a for the attractive forces and b for the repulsive forces:

$$\left(p + \frac{an^2}{V^2}\right)(V - nb) = nRT,$$

Where p, V, n, R and T have the same meaning as in the ideal gas equation. Named after Johannes Diderik van der Waals (1837–1923).

van der Waals forces ➤Intermolecular forces which account for the formation of condensed phases, and that explain the attractive term in the ➤van der Waals equation. There are two main types: ➤dipole–dipole forces and ➤dispersion forces, of which the latter usually dominate. (A third type, **Debye forces**, exists between the permanent dipole of one molecule and the temporary dipole of another.) The only other important intermolecular force, ➤hydrogen bonding, is not usually considered to be a van der Waals force.

vanillin (4-hydroxy-3-methoxybenzaldehyde) The molecule (see the diagram) responsible for the smell of vanilla, and a contributor to the smell of some perfumes. It is a white crystalline solid.

vanillin

van't Hoff isochore A fairly uncommon name for the equation describing the temperature variation of the ➤equilibrium constant K at constant pressure:

$$d \ln K/dT = \Delta H^{\ominus}/RT^2.$$

Thus exothermic reactions (those for which $\Delta H^{\ominus}$ is negative) respond to increasing temperature by having a reduced equilibrium constant. This quantifies the qualitative shift predicted by ➤Le Chatelier's principle. Named after Jacobus Henricus van't Hoff (1852–1911).

van't Hoff isotherm A rather uncommon name for the equation relating the ➤standard Gibbs energy change $\Delta G^{\ominus}$ for a chemical reaction to the equilibrium constant K for the reaction:

$$\Delta G^{\ominus} = -RT \ln K.$$

Van Vleck paramagnetism A form of ➤paramagnetism arising from the presence of unpaired electrons in an unfilled atomic ➤energy level. Compare ➤Pauli paramagnetism. Named after John Hasbrouck Van Vleck (1899–1980).

vapour A term used almost synonymously with 'gas'; strictly, a vapour must be able to be turned into a liquid by compression alone. The term should therefore be used only below the substance's ➤critical temperature.

vapour density The ratio between the density of a specific volume of a particular ➤vapour and the density of the same volume of hydrogen.

vapour pressure The ➤partial pressure of a vapour. If the liquid and gas are in ➤dynamic equilibrium, the resulting vapour pressure of the gas is known as the **saturated vapour pressure** (SVP), which depends on the temperature of the system.

variable A symbol such as x, y, z for an arbitrary member of a given ➤set. In particular, when defining the function $y = f(x)$, x is the **independent variable** and y is the **dependent variable**.

Variable Number Tandem Repeats ➤genetic fingerprinting.

variable region, variable segment ➤V segment.

variable star A star that varies in brightness. The variation can result from the orbital motion of a ➤binary star, as the two components periodically eclipse each other as seen from Earth – an **eclipsing binary** – or from actual physical changes taking place within a single star. The best-known of this latter type are the ➤Cepheid variable stars, whose luminosity varies with a well-defined period as they pulsate. Many ➤red giant stars are variable, with less well-defined periods of typically several months. The most dramatic types of variable star are ➤novae and, in particular, ➤supernovae.

variance The square of the ➤standard deviation, usually denoted by σ^2. The variance of a ➤discrete random variable X is $\Sigma_x (x - \mu)^2 P(X = x)$, where μ is its ➤mean and $P(X = x)$ is the ➤probability that X takes the value x; the variance of a ➤continuous random variable is $\int(x - \mu)^2 f(x)\,\mathrm{d}x$, where μ is its mean and $f(x)$ is its ➤probability density function.

variation 1 (Math.) A way of describing certain relationships between two variables. To say that y is **directly proportional** to x means that $y = kx$ for some constant k. To say that y is **inversely proportional** to x means that $y = k/x$ (or $xy = k$) for some constant k. ➤Hooke's law, the ➤ideal gas equation and ➤Newton's law of gravitation are often cast in these forms.

   2 (Biol.) The range of differences between individuals in a population. Variation may be determined by genetic factors, as for coat colour in dogs, or by environmental factors – the yield of a genetically identical crop may depend on the nutrients available locally in the soil. Variation is commonly further classified according to whether it is **continuous** between two extremes in a population (as for height in humans) or **discontinuous** in which the population can be divided into two or more distinct 'morphs' or forms (as for blood groups in humans). Genetically determined variation is fundamental to the process of ➤evolution.

variation, magnetic ➤magnetic declination.

variational principle A way of expressing fundamental laws of physics, particularly mechanics, in terms of the form of function that minimizes or maximizes an integral rather than in terms of differential equations. ➤➤calculus of variations; Hamilton's principle.

variometer A form of variable ➤inductor that consists of two coils; the coupling between the coils is varied by moving or rotating one of them, varying the self-inductance (➤inductance) of the component as a whole.

varistor A form of ➤resistor with a resistance that varies with applied voltage. Varistors are used as voltage limiters.

varve dating A method of dating fossils that relies on counting **varves**, which are laminated clay-based structures, the layers being alternately light- and dark-coloured. The light-coloured layers are deposited in winter, and the darker ones in summer, when the warmer streams are carrying more dissolved material.

vascular system A transport system in organisms. Examples are the blood system of animals, and the xylem and phloem (**vascular bundles**) of some plants.

vas deferens The duct carrying spermatozoa from the epididymis of the ➤testis to the outside via the ➤urethra.

vas efferens The duct carrying sperm from the sperm-forming tubules in the ➤testis to the epididymis.

vasoconstriction, vasodilation The ability of blood vessels, particularly capillaries, respectively to decrease or increase in diameter under nervous or hormonal control. This ability is significant in ➤homoiothermic animals. When an animal is at risk of overheating, vasodilation (or vasodilatation) of capillaries in the

➤skin allows more blood to flow to the surface of the body, where heat can be lost by ➤heat radiation. When an animal is cold, vasoconstriction prevents blood from flowing near the surface, and heat is conserved deeper in the body.

vasopressin (antidiuretic hormone, ADH) A ➤hormone, secreted by the posterior ➤pituitary gland, associated with the regulation of the concentration of body fluids. Specifically, vasopressin controls the amount of fluid reabsorbed from the tubules of the ➤kidney into the surrounding blood capillaries. A deficiency of vasopressin causes the condition *diabetes insipidus* (➤diabetes), in which large quantities of dilute urine are produced.

vat dye ➤dye.

VDU Abbr. for visual display unit.

vector **1** (Math.) A physical quantity that has magnitude and direction, and combines according to the **parallelogram law** (so that the combined effect of two vectors is given by the diagonal of the ➤parallelogram with sides representing the two vectors). Examples of vectors are ➤force, ➤dipole moment and ➤angular momentum. A vector can be defined in any number of dimensions; for example, in ➤special relativity it is useful to define vectors in four-dimensional spacetime. Any vector can be written as an array of ➤components that expresses the vector in terms of a linear combination of ➤unit vectors. Vectors are conventionally denoted by boldface letters in print. A vector can be viewed as a first-rank ➤tensor.

2 (Math.) An ➤element (2) of a ➤vector space.

3 (Biol.) An organism that carries a parasite (➤symbiosis) from one host to another, thus spreading the infection. Many vectors are themselves hosts to the parasite, which may complete part of its life cycle in the vector's body. Examples of vectors are the *Anopheles* mosquito, which carries the agent causing ➤malaria, and the tsetse fly, which carries the agent causing sleeping sickness. Virtually all blood- and sap-feeding animals are potential vectors; aphids transmit many viral diseases of plants.

4 (Biol.) An agent that can be used to carry a particular fragment of ➤DNA into a recipient cell. Originally, ➤bacteriophages were used to bring about the ➤transduction of bacterial cells, but recent techniques rely on the use of ➤plasmids. **Cloning vectors** are those used to produce multiple copies of a piece of inserted DNA, the choice of vector being determined by the size of the DNA in question. Small DNA fragments (less than 15 000 bases long) can be cloned in the bacteriophage lambda or plasmids, but larger fragments are cloned in artificially generated **cosmids**. For cloning fragments over 100 000 bases long, a specifically constructed ➤yeast artificial chromosome is used to introduce the DNA into yeast cells. ➤**Expression vectors** are those in which the fragment of DNA of interest is not only cloned but also expressed, so that the ➤gene product can be generated in the recipient cell. Vectors are designed and constructed using ➤genetic engineering techniques for specific applications so that recipient cells that have taken up the vector can be readily identified. Such techniques include ➤antibiotic resistance or the production of a characteristic colour. ➤➤lambda phage; shuttle vector.

vector analysis The application of the methods of ➤calculus to ➤scalar and ➤vector fields involving operators such as div (➤divergence), grad (➤gradient (3)) and ➤curl, and tools such as the ➤divergence theorem and ➤Stokes's theorem.

vector field A function that assigns to each point of a multidimensional region a multidimensional vector. This definition generalizes the familiar ➤fields met in physics where the regions and vectors involved are three-dimensional. ➤➤divergence; irrotational.

vector potential, magnetic ➤magnetic potential.

vector product (cross product) A ➤vector formed from two vectors a and b, denoted by $a \times b$, with magnitude $|a||b| \sin \theta$ (where θ is the angle between a and b). It is perpendicular to the plane that contains a and b and points in the direction in which a right-handed corkscrew moving from a to b would travel. If $a = a_1 \mathbf{i} + a_2 \mathbf{j} + a_3 \mathbf{k}$ and $b = b_1 \mathbf{i} + b_2 \mathbf{j} + b_3 \mathbf{k}$, then $a \times b = (a_2 b_3 - a_3 b_2)\mathbf{i} + (a_3 b_1 - a_1 b_3)\mathbf{j} + (a_1 b_2 - a_2 b_1)\mathbf{k}$. The vector product finds many applications in mechanics and electromagnetic theory. ➤➤Poynting vector.

vector space An abstract mathematical structure consisting of a ➤commutative ➤group of vectors and a suitable notion of multiplication by scalars (usually real or complex numbers). Vector spaces are characterized by their dimension, which may be infinite. ➤➤Hilbert space.

vector triple product The vector $a \times (b \times c)$ formed from three three-dimensional vectors a, b and c. It is also equal to $(a \cdot c)b - (a \cdot b)c$.

vein A blood vessel carrying blood from capillaries towards the heart. Veins are characterized by having relatively thin walls in relation to their cross-sectional area, containing smooth ➤muscle and connective tissue, and typically have ➤valves (2) to promote the one-way flow of blood. Veins carry blood at low pressure. In mammals and birds all veins carry deoxygenated blood, except the **pulmonary vein** which carries oxygen-rich blood from the lungs to the heart (compare ➤artery). The term is also applied to the support and conducting structures in plant leaves and the toughened support tubes of insect wings.

velocity Symbol v. For a body with a time-dependent displacement (with an arbitrary origin) $r(t)$, the rate of change of r, dr/dt. Velocity, like displacement, has a ➤magnitude and a direction and is therefore a ➤vector. The magnitude of v is known as ➤speed.

velocity, relative The rate of change of the ➤displacement vector between two bodies. This is equal to the difference between the velocities of the two bodies, expressed as a vector.

velocity constant ➤rate constant.

velocity modulation The bunching of electrons in a beam that is subjected to a ➤radio-frequency electromagnetic field.➤klystron.

velocity potential In fluid mechanics, for ➤irrotational velocity fields, a quantity analogous to the ➤electric potential arising in electromagnetism: $v(r) = \nabla\phi(r)$. The sign of the potential is the opposite of that conventionally adopted for ➤ electric potential.

velocity ratio (distance ratio) In a mechanical device, the distance through which the effort must act divided by the corresponding distance moved by the load. ➤efficiency; mechanical advantage.

vena cava (plural **venae cavae)** ➤heart.

venation The pattern, often characteristic, of the arrangement of veins on the leaves of plants.

Venn diagram A diagrammatic representation of ➤sets in which the ➤elements (2) of a set are allocated to a region (such as a circle) so that overlap of the regions corresponds to ➤intersection of the sets (see the diagram). Named after John Venn (1834–1923).

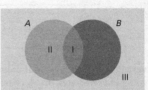

Venn diagram If *A* represents, say, left-handed people, and *B* represents men, then region I represents left-handed men, region II left-handed women and region III right-handed women.

ventral Describing the position of a surface, structure or organ lying on the underside of an organism. Compare ➤dorsal.

ventricle 1 One of a number of cavities in the ➤brain of vertebrates.

 2 A muscular chamber in the lower part of the ➤heart in vertebrates.

Venturi tube A tube whose cross-sectional area reduces in a chosen section, causing a fluid that passes through the tube to increase its velocity at the narrow section. ➤Bernoulli's theorem shows that the pressure will fall in the section where the velocity is higher. Venturi tubes are most commonly used for measuring the speed of fluid flow. Named after Giovanni Battista Venturi (1746–1822).

Venus The second planet from the Sun, orbiting in 224.7 days at an average distance of 0.72 AU. Its radius of 6052 km is slightly less than that of Earth, its mean density is similar, and thus its gravitational field strength at the surface is similar. However, its atmosphere is 96% carbon dioxide with a surface pressure about a hundred times the Earth's. Such a high concentration of greenhouse gas (➤greenhouse effect) is responsible for the very high surface temperature of 740 K. Venus is covered by an unbroken layer of cloud, containing a sulfuric acid haze about 50 km above the surface. Its rotational period is 243 days. ➤Appendix table 4.

verdigris The green covering on copper or bronze objects that develops over time on exposure to the atmosphere. It is normally a basic copper(II) carbonate, $Cu(OH)_2 \cdot CuCO_3$, but near the sea it will be a mixture with basic copper(II) chloride, $Cu(OH)_2 \cdot CuCl_2$.

vernier An auxiliary scale on an instrument used for determining a fractional distance between two divisions on a linear or angular measuring scale. One end of the vernier scale coincides with the reference mark against which the measurement is read off (see the diagram overleaf). The divisions on the vernier are slightly smaller than those on the main scale, and the fraction of a division on the measuring scale is related to the number of divisions on the vernier at which the marks on the measuring scale and vernier scale coincide. Named after Pierre Vernier (1580–1637).

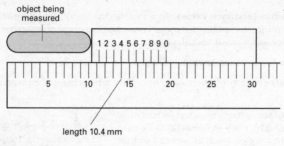

object being measured

1 2 3 4 5 6 7 8 9 0

5   10   15   20   25   30

length 10.4 mm

vernier

vertebral column (spinal column) The major supporting structure forming the **spine** of vertebrates, and a characteristic feature of this subphylum. The vertebral column consists of a number of individual **vertebrae**, each formed from a number of fused bones, between which there is limited movement. In quadrupeds and humans different regions of the vertebral column are recognized. The **cervical** region (7 vertebrae in humans) forms the neck, the **thoracic** region (12 vertebrae) forms the chest region and allows for the attachment of ribs, the **lumbar** region (5 vertebrae) forms the lower spine and provides support and flexibility, the **sacral** region (5 fused vertebrae) attaches the column to the ➤pelvic girdle at the sacro-iliac joint, and the **caudal** region (4 reduced vertebrae) forms a short 'tail', the coccyx.

vertebrate A member of the subphylum **Vertebrata** of the ➤chordates characterized by a dorsal ➤vertebral column or backbone. Vertebrates comprise the cartilaginous and bony fishes, amphibians, reptiles, birds and mammals.

vertex (plural **vertices)** A point at which the edges of a ➤polygon or ➤polyhedron meet.

vertical In the direction of the local ➤gravitational field.

very high frequency (VHF) The ➤frequency band of the electromagnetic spectrum between 30 and 300 MHz. Radio waves between 88 and 108 MHz are used for high-fidelity music and voice radio broadcasts; 108 to 118 MHz is used for aircraft navigation, and 118 to 137 MHz for aircraft communications.

very large scale integration (VLSI) A measure of the detail of an ➤integrated circuit, VLSI refers to a component capable of storing from 16 kilobits to 1 megabit (or the equivalent number of transistors if the component is not a memory circuit).

very low frequency (VLF) The ➤frequency band of the electromagnetic spectrum between 3 and 30 kHz. VLF radio signals are used for long-range aeronautical and marine navigation, but such systems are being made obsolete with the development of the ➤global positioning system.

vesicle Any small membrane-bound sac within the ➤cytoplasm of cells.

vessel 1 A specialized tubular cell in the ➤xylem responsible for the transport of water and minerals in vascular plants.

2 Any of a number of tubular structures that transport fluids in the body, for example ➤blood or ➤lymph vessels.

vestigial organ Any structure within an organism that has become reduced in both size and function during evolutionary time. Vestigial organs are useful in evolutionary studies since they can be used to infer relationships with ancestral types in which the organ was functional. Thus the presence of a vestigial ➤pelvic girdle in the skeletons of some snakes, such as boas, implies a link with an ancestral limbed form. The human appendix is another example.

VHF Abbr. for ➤very high frequency.

vibration A periodic motion about equilibrium (an ➤oscillation), usually mechanical in nature. ➤➤normal modes; simple harmonic motion.

vibrational energy The energy of an ➤oscillator, such as a diatomic molecule, that is due to the vibration itself. ➤➤harmonic oscillator.

vibrational spectroscopy A method of measuring the ways in which a molecule can vibrate. The experiments are performed most commonly in the ➤infrared region of the electromagnetic spectrum. Infrared radiation is passed through the sample and the absorption detected. The infrared radiation is absorbed at certain characteristic frequencies, those corresponding to the ➤normal modes of vibration of the molecule, since the vibrations are quantized. There is almost always a rotational ➤fine structure which complicates the ➤lineshape, but this can be analysed fully only for very simple molecules. An alternative method utilizes the ➤Raman effect. The information provided proves to be complementary for centrosymmetric molecules as all vibrations that are not observable in the infrared are observable in the Raman.

vic- A prefix, short for **vicinal**, still occasionally used to indicate an isomer in which the functional groups are on adjacent carbon atoms, as opposed to the same one in the ➤*gem-* isomer (see the diagram).

vic- (a) *vic-* and (b) *gem-* isomers of a dihalide.

villus (plural **villi)** A finger-like projection of the wall or surface of an organ involved in the absorption or exchange of materials. Villi have extensive internal networks of ➤capillaries and are found, for example, in the walls of the ➤gut and in the ➤placenta.

vinyl The traditional name for the group $CH_2\!\!=\!\!CH\!\!-\!\!$. Common uses are in vinyl chloride and vinyl alcohol.

vinyl acetate (ethenyl ethanoate) $CH_2\!\!=\!\!CHOCOCH_3$ A colourless liquid used as an intermediate in the preparation of poly(vinyl alcohol).

vinyl alcohol The traditional name for the molecule $CH_2\!\!=\!\!CHOH$, which is the ➤enol form of ➤ethanal, CH_3CHO.

vinyl chloride ➤chloroethene.

vinylidene chloride Traditional name for 1,1-dichloroethene, $CH_2=CCl_2$. It is a colourless liquid which can be polymerized to a thermoplastic material used for mouldings and fibres.

virial expansion An ➤equation of state for a real gas in which the pressure is treated as a power series expansion, of either the pressure or the inverse of the volume. The first term in the expansion comes from the ➤ideal gas equation; the second term uses the **second virial coefficient** B, and so on:

$$pV_m = RT + Bp + Cp^2 + \cdots$$

It provides a better description of real gases, and the second virial coefficient can be roughly calculated from the intermolecular forces of the particular gas.

virion A single, fully assembled ➤virus particle.

virtual image ➤image.

virtual particle In particle physics, a particle that exists for a short time and only participates in the interaction between two or more real particles. The concept is essential to ➤quantum field theory. The virtual particles are part of the interaction process and cannot exist in isolation. For example, a virtual photon (the ➤gauge boson of the electromagnetic interaction) participates in the interaction between two electrons.

virulent Capable of producing disease or severe pathological symptoms.

virus A noncellular infective agent consisting of a ➤genome contained within a protein coat and, in some complex forms, a surrounding envelope of a fat and protein. The genome can be either ➤DNA or ➤RNA in single- or double-stranded form. Viruses range in size from 20 to 400 nm, and form particles which may crystallize and are resistant to treatment with ➤antibiotics. They are generally considered to be non-living, but are clearly biological in nature and are thought by some to have originated from ➤plasmid-like fragments of ➤nucleic acids.
   Viruses can replicate only by invading a living cell (the host) and taking over the host cell's metabolic processes in order to synthesize more virus particles. Most viruses therefore produce pathological symptoms, since virus infection leads to severe cellular dysfunction, and some cause severe and ➤virulent disease. Examples of viruses include poliomyelitis, measles, ➤smallpox, ➤HIV, various cancer-inducing tumour viruses (➤SV40) and ➤bacteriophages. ➤➤retrovirus.

viscometer A device for measuring the ➤viscosity of a fluid. One common type measures the torque required to achieve a relative rotation of two coaxial cylinders when the space between them is filled with the fluid under test.

viscose A viscous liquid made from ➤cellulose by the addition of carbon disulfide and sodium hydroxide. The resulting cellulose xanthate can be forced through holes into an acid bath that decomposes it into cellulose, leaving a cellulose fibre known as ➤rayon.

viscosity Symbol η; unit Pa s. For two layers of viscous fluid moving relative to each other, the force that each exerts on the other, per unit area of contact, and per unit velocity gradient. With a fluid moving parallel to the y axis, if the velocity v_y varies with the coordinate along the x axis, then the shear force F_y is given by:

$$F_y = \eta A \ dv_y/dx.$$

where η is the viscosity and A is the area of fluid in contact. In fluid mechanics it is often also useful to define the **kinematic viscosity** v of a fluid as η/ρ, where ρ is the density of the fluid. The kinematic viscosity has unit $m^2 \ s^{-1}$.

viscous Describing fluids that are resistant to ➤shear. Viscous fluids tend to flow slowly through pipes and channels (the rate of flow is inversely proportional to the ➤viscosity) and are difficult to pour. Examples of viscous fluids are glycerol, molten glass and ➤viscose.

visible spectrum That part of the spectrum of electromagnetic radiation that is visible to the human eye, called 'light' or 'visible light', with wavelengths from about 380 to 780 nm (frequency 7.9 to 3.8×10^{14} Hz) (➤electromagnetic spectrum).

visual binary A ➤binary star whose components can be seen individually in visible light (through an optical telescope), as opposed to, for example, a ➤spectroscopic binary. This requires the angular separation to exceed about 0.5 ➤seconds of arc.

visual purple ➤colour vision.

vitamin An organic compound required in minute amounts by an animal but which it usually cannot synthesize. Most vitamins are key ➤cofactors for ➤enzymes that are important in ➤metabolism and must be obtained in the diet. If insufficient of a particular vitamin is available a deficiency disease results, the classic example being scurvy caused by a lack of vitamin C. Vitamins are usually classified into **fat-soluble** and **water-soluble**. A balanced human diet contains sufficient amounts of all the normally recognized vitamins. Nutritional information on food packaging commonly gives the vitamin contents expressed as a percentage of the **RDA** (recommended daily amount) for an adult human (see the table overleaf). Vitamins for one species may not be essential in another.

vitreous Having a glassy appearance, as in vitreous silica.

vitreous humour ➤eye.

vitriol ➤blue vitriol; green vitriol; oil of vitriol.

VLF Abbr. for ➤very low frequency.

VLSI Abbr. for ➤very large scale integration.

vocal cord One of a pair of membranous folds in the ➤larynx of air-breathing vertebrates. When vibrated by air passing over them from the lungs, they produce sound. The pitch and quality of the sound are controlled by muscles in the larynx which operate cartilaginous flaps. In humans this control is highly developed and makes speech possible.

volatile Describing a substance that can easily become a vapour.

volt Symbol V. The SI unit of ➤electric potential and ➤potential difference. Work of 1 joule must be done on a positive charge of 1 coulomb to raise the potential by 1 volt.

Name	Chemical name or derivation	Symptom of deficiency	Sources
Water-soluble			
B vitamins (vitamin B complex)			
Thiamin (B₁)	Thiamin pyrophosphate	Beri beri, paralysis	Wholemeal grain, liver, yeast, green vegetables
Riboflavin (B₂)	Flavin mononucleotide, flavin adenine dinucleotide	Skin lesions	As above
Niacin	Nicotinic acid	Pellagra	Liver, fish, cereals, legumes, nuts
Pyridoxine (B₆)	Constituent of pyridoxal phosphate	Skin disorders, anaemia, kidney stones	Whole-grain cereal, bananas, liver
Cyanocobalamin (B₁₂)	Nucleotide cobalt derivative	Pernicious anaemia	Liver, eggs, milk, fish
Folic acid	Pteroylglutamic acid	Anaemia (rare)	Liver, legumes, yeast, whole-grain cereals
Pantothenic acid	Component of acetyl CoA (Krebs cycle)	Rare or unclear	Meat, vegetables
Biotin	Coenzyme for various enzymes	Rare	Widespread; synthesized by intestinal bacteria
Vitamin C	Ascorbic acid	Scurvy	Fresh fruit and vegetables
Fat-soluble			
Vitamin A	Retinol	Night-blindness, xerophthalmia	Fruit, green vegetables
Vitamin D	Group of two vitamins derived from steroids	Rickets, lack of absorption of Ca in ileum	Fish liver oils; some synthesis in the skin
Vitamin E	Tocopherol	Infertility in rats, liver dysfunction (role in humans is unclear)	Vegetable oils, whole-grain cereals
Vitamin K	Isoprene derivative	Blood-clotting disorders	Fish oils, green vegetables

In an electric circuit, a power of 1 watt is dissipated when a current of 1 ampere passes through a potential difference of 1 volt. Named after Alessandro Giuseppe Antonio Anastasio Volta (1745–1827).

voltage For a ➤direct current circuit, the term is used synonymously with ➤potential difference. For an ➤alternating current circuit, it is a measure of the amplitude of a time-varying potential difference, such as its ➤root mean square value.

voltage divider ➤potential divider.

voltage drop The ➤potential difference across a component in an electric circuit, or along a power line.

voltage gain The ratio of the output voltage to the input voltage in an ➤amplifier.

voltameter (coulometer) An electrolytic cell used to measure the charge flowing in a circuit. A **silver voltameter** consists of two platinum electrodes dipping into aqueous silver nitrate. The mass of silver deposited at the cathode is weighed and this can be related to the charge passed using ➤Faraday's laws. Two other common examples are the **Hofmann voltameter**, which uses dilute sulfuric acid, and the **copper voltameter**, which uses copper(II) sulfate.

volt-ampere Symbol VA. A unit of ➤power for ➤alternating current circuits. A ➤root mean square current of 1 A multiplied by a root mean square voltage of 1 V gives a power, through a resistor, of 1 VA.

voltmeter A device for measuring ➤potential difference (voltage). A variety of such instruments are available for both direct and alternating current voltages, though analogue, moving-coil voltmeters are increasingly being replaced by digital, solid state instruments.

volume **1** Unit m^3. A physical quantity that measures the extent of a three-dimensional space; the volume of a body is the space that it occupies.
  2 A measure of the loudness of audio signals. ➤decibel.

volumetric analysis The measurement of the concentration of a solution by ➤titration with an appropriate reagent.

voluntary muscle ➤muscle.

von Laue condition An equation relating the ➤wave-vector k of the incident wave and the wave-vector k' of the scattered wave in ➤X-ray diffraction. The von Laue condition is:

$$k' = k + K,$$

where K is a ➤reciprocal lattice vector of the diffracting crystal. Named after Max Theodor Felix von Laue (1879–1960).

vortex (plural **vortices)** A region of a fluid with high ➤vorticity; a swirl of fluid rotating about a line. Vortices are generated by, for example, aircraft wings.

vorticity Symbol w. A vector field in fluid dynamics that describes the tendency of the ➤velocity field to form closed loops. It is defined as $w = \nabla \times v$.

VRML Abbr. for virtual reality markup language. A set of definitions that allows computer users to 'surf the Net' in three dimensions. ➤HTML.

V segment (variable segment) One of an estimated 300 DNA fragments coding for the variable regions (**V regions**) of ➤immunoglobulins, the polypeptides that form the ➤antibodies produced by ➤lymphocytes. The associations that are possible between these and other antibody-producing components enable the ➤immune system to produce antibodies in response to a seemingly unlimited variety of ➤antigens. It is estimated that lymphocytes are probably capable of producing of the order of 10^{10} different antibodies. ➤C segment; J segment.

VSEPR theory Abbr. for ➤valence-shell electron-pair repulsion theory.

vulcanization The treatment of ➤rubber with, most commonly, sulfur. The sulfur produces cross-links between some of the double bonds in the molecule. The resulting material is much more elastic and less sticky than the original unvulcanized rubber. This treatment, first introduced by Charles Goodyear in 1839 (but patented by Thomas Hancock in 1843, one month before Goodyear), expanded the usefulness of rubber greatly.

VUV Abbr. for ➤vacuum ultraviolet.

VX ➤nerve gas.

W **1** Symbol for the element ➤tungsten, from its German name *Wolfram*.
2 Symbol for the unit ➤watt.

Wacker process The industrial aerial oxidation of ethene, C_2H_4, to ethanal, CH_3CHO, using a catalyst of palladium(II) chloride, $PdCl_2$, and copper(II) chloride, $CuCl_2$. Developed by J. Smidt and colleagues at Wacker Chemie in 1959.

Walden inversion The inversion that occurs when a nucleophile reacts in an S_N2 ➤nucleophilic substitution mechanism. If the carbon atom concerned is ➤chiral, this process will invert the configuration (➤(*R–S*) system), rather like an umbrella flip. Named after Paul Walden (1863–1957).

warm-bloodedness ➤homoiothermy.

washing soda $Na_2CO_3 \cdot 10H_2O$ The common name for the decahydrate (specifically) of sodium carbonate, so named after its major use. Coloured bath salts are a very expensive way of buying this chemical.

water H_2O One of the most important molecules on Earth: humans are about 65% water (and jellyfish 98%). The isolated molecule (see the diagram) has two hydrogen atoms bonded to one oxygen atom, the bond angle being 104.5° (➤valence-shell electron-pair repulsion theory).

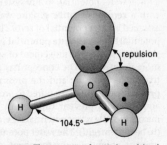

The water molecule has a permanent dipole; hence the molecules are held together by ➤dipole–dipole forces as well as ➤dispersion forces. There is also ➤hydrogen bonding between the molecules. The ➤intermolecular forces are so much larger than usual for a molecule of its mass that water is a liquid at room temperature. Both water's melting point (0 °C, 273 K) and its boiling point (100 °C, 373 K) are anomalously high. Another

water The structure of water is explained by the presence of two lone pairs on the oxygen atom; these two lone pairs and two bonding pairs form a tetrahedral arrangement, which accounts for the V-shaped molecule.

consequence of the hydrogen bonding between water molecules is that liquid water has a maximum density at 4 °C, because the solid (➤ice) has an open structure created by the ordered hydrogen bonds. This means that ponds freeze from the top down, insulating the lower layers, and enabling aquatic life to survive below.

Water reacts in many different ways. First, consider its acid–base nature. It is neutral, able to respond to the presence of Brønsted acids by accepting a proton and to Brønsted bases by donating one (➤ionic product of water). The lone pairs on water make it a ➤Lewis base and so it bonds with Lewis acids, such as metal ions, to give aqua complexes such as $[Fe(H_2O)_6]^{3+}$. Water causes ➤hydrolysis of covalent halides, phosphorus pentachloride giving phosphoric acid, for example. Hydrolysis is also common in organic chemistry: halogenoalkanes such as $(CH_3)_3CCl$ hydrolyse to alcohols such as $(CH_3)_3COH$.

Second, consider its redox nature. Strong reductants, such as reactive metals, can reduce it to hydrogen gas. Oxidants are rarely able to oxidize it to oxygen, although this does occur with fluorine.

water cycle ➤hydrological cycle.

water gas A mixture of approximately 50% hydrogen and 40% carbon monoxide with the remainder made up of methane, carbon dioxide and nitrogen, used as a fuel. The conversion of the carbon monoxide into carbon dioxide, the **water gas shift reaction**, is an important step in the preparation of hydrogen for the ➤Haber–Bosch process.

water of crystallization A definite quantity of water that is chemically combined in a crystal when it is formed. The bond that holds such water molecules in place can be seen as a coordinate bond from the water. Examples occur in ➤washing soda, $Na_2CO_3 \cdot 10H_2O$, and ➤blue vitriol, $CuSO_4 \cdot 5H_2O$.

water potential Symbol ψ; unit Pa. A measure of the tendency of a system or cell (in biological contexts) to give up water to another system. Water flows from a higher to a lower water potential, so any system with a positive ψ will pass water into a system with a zero or negative ψ. Pure water at ➤standard temperature and pressure is assigned a water potential of zero. The presence of a solute in water lowers the water potential to give a **solute potential** ψ_s, which is therefore always negative. ➤Osmosis should be seen as the movement of water from an area of less negative ψ_s to an area of more negative ψ_s. The concept has particular significance for plant cells, in which hydrostatic pressure and the pressures associated with the cell wall generate the **pressure potential** ψ_p required to produce support for structures such as young stems and leaves to prevent them from wilting. When the pressure potential is at a maximum, the water potential is zero and the cell is said to have maximum ➤turgor (to be fully **turgid**). The water potential ψ is related to the solute potential ψ_s and the pressure potential ψ_p by the equation

$$\psi = \psi_s + \psi_p.$$

water vapour Water in the form of ➤vapour present in the Earth's atmosphere in amounts that vary, from place to place and from season to season, between zero and about 5% by mass. ➤➤absolute humidity; dew point; relative humidity.

watt Symbol W. The SI derived unit of ➤power. One watt is the release or transfer of one joule per second. Named after James Watt (1736–1819).

wattmeter An instrument for measuring power in an electrical circuit.

wave A moving disturbance. A wave differs from an ➤oscillation in that the time-varying quantity also depends on position. Usually the nature of the disturbance is directional, and the result is a ➤longitudinal wave if this direction is the same as the direction of motion of the wave, or a ➤transverse wave if it is perpendicular to the direction of motion. Most waves travel through a medium, the properties of which determine the speed of the wave. The disturbance may be a vibration of the medium (e.g. a ➤sound wave in a solid where the atoms vibrate as the wave passes) or an oscillation of some other property of the medium (e.g. a water wave, where the height of the water is the oscillating quantity). In the case of an ➤electromagnetic wave, the disturbances are the electric and magnetic field strengths. ➤➤standing wave; travelling wave.

waveband A range of frequencies in the electromagnetic spectrum (➤frequency band).

wave energy The generation of power from the energy of water waves. Such a source of energy is environmentally clean, but technically very difficult to achieve, and very few economically viable generators of wave power have been produced.

wave equation The common mathematical link in all wave motion. In one dimension:

$$\frac{\partial^2 \psi}{\partial x^2} = \frac{1}{c^2} \frac{\partial^2 \psi}{\partial t^2},$$

where ψ is the time- and position-dependent quantity, x is the position coordinate, t is time and c is the wave speed, which depends on the medium. In three dimensions, the wave equation becomes:

$$\nabla^2 \psi = \frac{1}{c^2} \frac{\partial^2 \psi}{\partial t^2},$$

(➤Laplacian). The ➤boundary conditions of the wave equation can usually be specified as $\psi = 0$ at a chosen point (for a **standing wave**) or continuity of ψ between media (for a **travelling wave**). For one solution of the wave equation, see ➤wave motion.

waveform The time-dependence of a signal (usually electrical), as would be pictured on an ➤oscilloscope. This is a misnomer, as no wave motion need be involved.

waveform generator A device for generating electrical signals of variable frequency.

wavefront A surface (line) of constant phase of a wave in three- (two-) dimensional space. The concept of a wavefront is used in ➤Huygens' construction for determining wave propagation. For convenience, the surface is usually the position of a peak or trough of the wave (an antinode) and this, for a mechanical wave, is an observable feature whose motion can be followed.

wavefunction In quantum mechanics, the complex scalar quantity $\psi(r)$, which is the solution of the Schrödinger equation for a particle. The square of the modulus of the wavefunction, $|\psi(r)|^2$, is the ➤probability density function for finding the particle at displacement **r**. The wavefunction is a representation of the spatial part of its ➤quantum state.

waveguide A device for directing a wave (usually electromagnetic) in a particular direction. Without a waveguide, a wave that originates from a point source would fall off in intensity as the inverse square of the distance travelled. With a waveguide the intensity can be preserved (with a little ➤attenuation). Typically a waveguide for an electromagnetic wave consists of a metal tube of rectangular cross-section with dimensions similar to the wavelength. The boundary condition imposed by the walls means that the wave has either a **transverse electric** (TE) field with a magnetic field in the direction of propagation, or vice versa (a **transverse magnetic** (TM) field).

wavelength Symbol λ. The distance over which a simple wave is periodic. It is the distance between successive peaks (or troughs) of a wave. The key relationship is $f\lambda = c$, where f is the frequency of the wave and c is its ➤phase speed.

wave mechanics The study of solutions of the ➤Schrödinger equation in spatial coordinates. For a free particle such as an electron in a constant potential, one solution for the ➤wavefunction is simply a ➤plane wave. For more complicated potentials, the plane-wave solutions may be combined to deduce an overall solution.

wave motion The time development of a wave according to the ➤wave equation. The general solution to the one-dimensional wave equation with a constant value for the wave speed c is:

$$\psi(x,\ t) = f(x - ct) + g(x + ct),$$

where f and g are arbitrary functions. The first term is a wave travelling forwards (towards positive x), the second a wave travelling backwards.

wavenumber Symbol $\tilde{v}$ in a vacuum, σ in a medium. A quantity used primarily in spectroscopy, defined as the reciprocal of the ➤wavelength of a wave (usually electromagnetic). It is often measured in cm^{-1}, even though this is not an SI unit. Particularly in physics, it has become common practice to refer to the magnitude k of the ➤wave-vector of a wave as its wavenumber, but this differs from σ by a factor of 2π (as $k = 2\pi/\lambda$).

wave packet An arbitrary function of $x - ct$ or $x + ct$ that is a valid solution to the ➤wave equation (➤➤wave motion) and can propagate as a wave. If the ➤phase speed c depends on frequency, then the wave packet is no longer an exact solution of the wave equation, and it will change shape as it moves (➤dispersion).

wave–particle duality The concept that objects such as electrons are at the same time both particles and waves. It is one of the fundamental principles of quantum mechanics, first appreciated by Albert Einstein in 1905. He examined the ➤photoelectric effect and concluded that the energy in light is carried by particles in 'packets' (or ➤quanta). The complementary idea, that any particle has associated with it a wavelength, is predicted by ➤de Broglie's equation. Quantum mechanics treats waves and particles on the same footing, describing the ➤wavefunction of a particle but also allowing waves to be quantized.

wave theory of light ➤electromagnetic wave.

wave-vector (propagation vector) Symbol k. A vector quantity that describes not only the magnitude of a wave but also its direction. The magnitude of the wave-vector is $2\pi/\lambda$, where λ is the wavelength of the wave. Compare ➤wavenumber; ➤➤wave motion.

wax A ➤lipid made from the esters of fatty acids with monohydric fatty alcohols. A simple example is melissyl palmitate, $C_{30}H_{61}OCOC_{15}H_{31}$, present in beeswax. Waxes are water-repellent and are used to coat surfaces. Synthetic waxes include the ➤silicones.

W boson (W particle) A ➤gauge boson of the ➤weak interaction. It is a very massive (80.4 GeV/c^2), charged (positively or negatively) elementary particle, predicted from the theory of the ➤electroweak interaction before it was discovered (in 1983). ➤➤Z boson.

weak acids and bases ➤Brønsted acids and bases that are *not* completely ionized. So 1 mol dm^{-3} ethanoic acid, the classic weak acid, produces a little less than 1% hydrogen ions. (Similarly the classic **weak base**, aqueous ammonia, has about the same percentage of hydroxide ions at the same concentration.) The quantitative extent of ionization depends on the ➤acid ionization constant or the ➤base ionization constant. Compare ➤strong acid; strong base.

weak interaction (weak nuclear force) A ➤fundamental interaction that is associated with the atomic nucleus. It is responsible for ➤beta decay, and ranks in strength and range between the ➤strong interaction and the ➤electromagnetic interaction. ➤➤electroweak theory; W boson.

weakly interacting massive particle ➤WIMP (2).

weber Symbol Wb. The SI derived unit of ➤magnetic flux. 1 Wb is the magnetic flux of a magnetic flux density of 1 T crossing a surface of area 1 m^2. Named after Wilhelm Eduard Weber (1804–91).

weight Unit ➤newton, N. The ➤force on a body due to ➤gravity (usually the gravitational field of the Earth), equal in magnitude to its mass m multiplied by the gravitational field strength g, directed towards the centre of the Earth. The word 'weight' is also used colloquially as a synonym for the mass of a body, but the distinction is vital because g, far from being a fundamental constant, is not even constant over the surface of the Earth.

weightless Describing a body in an environment where no ➤reaction to ➤weight is felt. This may be because a gravitational force is not acting on the body, but more commonly it arises because the force acts unopposed and causes an ➤acceleration. For example, an astronaut in orbit in a space station is weightless.

western blotting (protein blotting) A technique for identifying the presence of a specific ➤protein in a mixture. The mixture is separated by gel ➤electrophoresis, and the separated bands are transferred (by 'blotting') onto a polymer membrane. A readily detectable ➤antibody is then added and if the protein in question is present, this will bind to it, so identifying it. It is named by analogy with ➤DNA blotting.

Weston cell A widely used standard ➤cell (2) of e.m.f. 1.02 V, which uses an electrolyte of saturated cadmium sulfate, $CdSO_4$. Named after Edward Weston (1850–1936).

wet bulb thermometer A thermometer with a mechanism for keeping its bulb wet; traditionally the bulb is enclosed in moist muslin. In atmospheres of less than 100% ➤relative humidity, the evaporation of the water cools the bulb and the thermometer reads less than a corresponding **dry bulb thermometer**. The relative humidity and ➤dew point can be calculated from the pair of wet and dry bulb temperatures.

Wheatstone bridge A four-arm electrical circuit for measuring resistance (see the diagram). The value of the resistance R_2 is varied until the potential difference across the galvanometer is zero, under which circumstance the value of the unknown resistance R_1 can be found from:

$$\frac{R_1}{R_2} = \frac{R_3}{R_4}.$$

The technique relies only on being able to measure a null (zero current or voltage), and hence it does not depend on the calibration of the measuring device, only on the resistors. Named after Charles Wheatstone (1802–75).

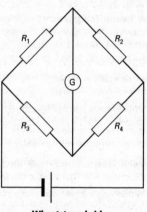

Wheatstone bridge

white blood cell ➤leucocyte.

white dwarf A very small star, typically the size of the Earth, that is the end product of ➤stellar evolution for stars up to about 10 solar masses. They are dense stars made up of mainly carbon and oxygen. There is a maximum possible size of a white dwarf (the ➤Chandrasekhar limit) above which the ➤degeneracy pressure of the electrons in the core of the star is not enough to compensate for the gravitational compression. ➤➤supergiant.

white matter ➤brain; spinal cord.

white noise Noise in a communications system or electrical device that covers a broad range of frequencies. ➤➤Johnson noise.

white spirit (turpentine substitute) A mixture of hydrocarbons, typically boiling over the range 130 to 210 °C, used as a solvent, in paint thinners for example.

Wiedemann–Franz law An empirical (and approximate) law stating that the ratio of the ➤thermal conductivity κ to ➤electrical conductivity σ of a metal is proportional to its thermodynamic temperature T:

$$\frac{\kappa}{\sigma} = \frac{\pi^2}{3}\left(\frac{k}{e}\right)^2 T,$$

where k is the Boltzmann constant and e the elementary charge. Moreover, the ratio is independent of the metal chosen, being a fundamental property only of the electrons that carry the current and heat. The approximation is accurate to about 10% for a wide variety of common metals over a temperature range of 0 to 100 °C. Named after Gustav Heinrich Wiedemann (1826–99) and Rudolf Franz (1827–1902).

Wien bridge A four-arm electrical circuit designed to measure either capacitance or frequency. Compare ➤Wheatstone bridge. Named after Wilhelm Carl Werner Otto Fritz Franz Wien (1864–1928).

Wien's (displacement) law ➤black-body radiation.

Wigner–Seitz cell A specific choice of primitive ➤unit cell for a ➤Bravais lattice. It is constructed by selecting a lattice point and taking the volume closer to that lattice point than to any other lattice point. Named after Eugene Paul Wigner (1902–95) and Frederick Seitz (1911–2008).

Wilkinson's catalyst An important ➤homogeneous catalyst for hydrogenation, consisting of a rhodium complex $[Rh(P(C_6H_5)_3)_3Cl]$. Named after Geoffrey Wilkinson (1921–96).

Williamson's ether synthesis Probably the most widely applicable synthesis for ethers which uses the sodium salt of an alcohol to attack a halogenoalkane:

$$RHal + NaOR´ \rightarrow ROR´ + NaHal.$$

Named after Alexander William Williamson (1824–1904), who had to overcome adversity, having lost an arm and an eye in childhood.

WIMP **1** (Comput.) Acronym for windows, icons, menus, pointers. A graphical user interface for computers that uses a device such as a 'mouse', which the user moves around on a table top to control an on-screen pointer. This pointer can be moved to point at ➤icons. Buttons on the mouse may be pressed to cause relevant functions to be executed. The windows may be moved around and resized on the imaginary desktop that the screen represents. Menus can be called up in their own windows, with lists of available functions.

  2 (Phys.) Acronym for weakly interacting massive particle, a hypothetical subatomic particle suggested to explain the existence of ➤dark matter. Because it interacts weakly, it is argued, it is not readily observed.

Wimshurst machine An electrostatic generator that collects charge on a pair of counter-rotating discs. Named after James Wimshurst (1832–1903).

wind Large-scale air currents in the atmosphere caused by pressure differences from place to place. Wind direction is normally referred to by the direction, relative to north, from which it is blowing (thus a westerly wind blows towards the east). At altitudes of about 300 m and higher, wind direction is normally parallel to the ➤isobars at that level, and wind speed is determined by a balance between the pressure gradient and the ➤Coriolis force due to the rotation of the Earth. At lower levels the wind is usually at lower speed and in a slightly different direction because of the drag caused by the surface.

window **1** (Phys.) In various contexts, a transparent section in an otherwise opaque domain. This opaque domain may be spatial ('window' in its common usage), it may be a spectrum (ranges of the spectrum of the Earth's atmosphere that allow radiation to penetrate are 'windows') or it may even be temporal (a time period during which a spacecraft may be launched to achieve a particular objective is a **launch window**).

2 ➤WIMP (1).

Wöhler's synthesis A historically significant step in the development of organic chemistry when in 1828 Friedrich Wöhler (1800–82) made urea $CO(NH_2)_2$ from ammonium isocyanate NH_4NCO. This was at the time supposed to be impossible as the former compound was regarded as 'organic' and the latter compound 'inorganic'.

wood A plant tissue comprising the ➤xylem. Wood forms the greater part of the bulk of trees (both stem and roots) and, depending on the species it is derived from, has a wide range of properties which have been exploited by humans for diverse purposes, including house and boat building and making musical instruments, furniture and cricket bats. **Wood pulp** is used to make paper, and cellulose ethanoate derivatives of pulp are used to make some forms of plastic and fibres used in fabrics. ➤➤lignin.

Woodward–Hoffmann rules (rules of orbital symmetry) Rules to enable prediction of the course of cycloaddition reactions such as the ➤Diels–Alder reaction. The electronic structures of the ➤frontier orbitals, the HOMO and LUMO, and their possible overlaps can explain the stereochemistry of the product. Named after Robert Burns Woodward (1917–79), doyen of 20th-century synthetic organic chemists, and Roald Hoffmann (b. 1937).

word In computing, the unit of data that a processor can deal with in one cycle of operation. Early microprocessors had 8-bit word sizes but 32-bit microprocessors are now usual, and 64-bit ones will soon be common. The larger the word size, the faster the operation of the microprocessor, since entire complex instructions can be contained within a single word and executed in one cycle.

word processor A computer program that allows the creation, editing and formatting of text. The layout of the text and style of the letters can be finely controlled, graphics can usually be incorporated, and most word processors have a facility for checking spelling or even grammar.

work Symbol W; unit ➤joule, J. A transfer of energy as a consequence of a force acting through a distance. If a force F acts through a displacement d then the work done is the ➤scalar product:

$$W = F \cdot d.$$

For rotational motion, work can be expressed in terms of a torque of magnitude G moving though an angle θ ($W = G\theta$).

work function The minimum energy required for an electron at the ➤Fermi energy of a conductor to escape from the conductor. It is the difference in energy between the vacuum level and the Fermi level. ➤➤photoelectric effect.

work hardening (strain hardening) The hardening of a material that is strained to beyond its ➤elastic limit.

world line In relativity, the line joining the coordinates of an object showing its position with time as a parameter.

W particle ➤W boson.

wrought iron A form of nearly pure iron that is very tough.

wurtzite structure An important crystal structure (see the diagram) named after the ore zinc sulfide, ZnS. The structure has the sulfur atoms in an array with hexagonal ➤close packing, with the zinc atoms in half of the tetrahedral holes (compare ➤zinc blende structure). Other compounds with this structure include zinc oxide, ZnO, beryllium oxide, BeO, and aluminium nitride, AlN.

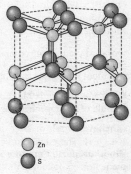

Zn

S

wurtzite structure

Wurtz synthesis A synthesis for ➤alkanes which uses sodium in dry ether plus a halogenoalkane, such as chloroethane, CH_3CH_2Cl. Simplistically, a radical such as $CH_3CH_2^{\bullet}$ is created and this can dimerize to give butane, C_4H_{10}. In practice, yields are usually very low due to side reactions. Named after Charles-Adolphe Wurtz (1817– 84).

www Abbr. for World Wide Web. A subset of the ➤Internet that links vast information resources containing text, graphics, sound and even video. These sites for multimedia are platform-independent, meaning that in theory any computer can view the information. **Web pages** are the electronic medium on which the data are presented and a **Web browser** is the piece of software used to view them. The milestone of one million Web sites was passed in 1997. ➤➤hypertext; HTML; HTTP; Java; URL.

X

x Symbol for ➤mole fraction.

X A common symbol for a ➤halogen in an organic compound: RX is the general formula of a halogenoalkane.

xanthate A salt or ester of the unstable acids ROCSSH, the sulfur analogue of ROCOOH. The esters can be made from an alcohol using sodium hydroxide and carbon disulfide. **Cellulose xanthate** is used in the ➤viscose process for making rayon.

xanthophyll A carotenoid plant pigment (➤carotene) responsible for yellow and brown colours in petals and leaves.

X chromosome ➤sex chromosome.

Xe Symbol for the element ➤xenon.

xenon Symbol Xe. The element with atomic number 54 and relative atomic mass 131.3, which is a colourless odourless gas present at approximately 1 ppm in the atmosphere. It is a member of the noble gases and, as such, is highly unreactive. The name for the group was, however, altered from 'inert' gases to 'noble' gases because in 1962 Neil Bartlett managed to get xenon to react (➤dioxygenyl ion). Subsequently a fairly rich xenon chemistry has been discovered, including several fluorides such as XeF_2, XeF_4 and XeF_6, a couple of oxides such as XeO_3, as well as other compounds such as $XeOF_4$ and $XeCl_2$.

XPS Abbr. for X-ray ➤photoelectron spectroscopy.

X-ray Electromagnetic radiation of a wavelength shorter than about 100 nm (frequency greater than 3×10^{15} Hz). Radiation of wavelength less than 10^{-11} m is often distinguished as a ➤gamma ray, reflecting the fact that X-rays are created by processes outside the ➤nucleus, whereas gamma rays are produced from inside the nucleus. X-rays were discovered in 1895 by Wilhelm Röntgen; they were emitted from a solid target bombarded with cathode rays (electrons). X-rays can be hazardous because of their interactions with the human body, but they are commonly used for diagnosis in medicine, and to detect metallic objects.

X-ray absorption spectroscopy A form of spectroscopy in which a beam of X-rays of variable frequency from a ➤synchrotron radiation source is directed at a sample and the absorption of the X-rays as a function of frequency is measured. Steps or 'edges' in the spectrum occur at the frequencies that correspond to transitions between atomic levels. An analysis of the ➤fine structure of the edges allows the distances between the absorbing atom and its neighbours to be calculated, and this can be used to find the structure of the sample. This latter technique is known as **extended X-ray absorption fine structure** (EXAFS).

X-ray crystallography The use of ➤X-ray diffraction to investigate the structure of crystals by identifying the possible angles at which X-rays can be diffracted and the corresponding intensities using a diffractometer. Structures are determined by using ➤Bragg's law, in addition to other mathematical methods. Significant applications of the technique have seen the identification of the simple ➤rock-salt structure by Lawrence Bragg, the vitamin B_{12} structure by Dorothy Hodgkin, and the structure of the haemoglobin molecule by Max Perutz. A clear X-ray diffraction pattern for ➤DNA proved to be the crucial evidence in favour of the Watson–Crick model of the double helix. This technique used to require painstaking work over many months or years to elucidate complicated structures. After the introduction (by Jerome Karle and Herbert A. Hauptman in the 1950s) of sophisticated algorithms based on probability theory, in addition to the advent of dramatically faster computers, the technique has become much more routine. By 2001 more than 250,000 small-molecule structures were available on the Cambridge Structural Database with a further 150,000 being added in the following 5 years.

X-ray diffraction The diffraction of an X-ray beam, passed through a crystal, in certain directions (determined by the ➤von Laue condition or ➤Bragg's law) in which the path difference between the beams scattered by adjacent atoms differs by a whole number of wavelengths. This allows both the structure and the lattice spacing to be determined. The necessary experiments can be carried out with a crystal of fixed orientation, or by observing the diffraction of an X-ray beam from a powdered sample which contains pieces of the crystal in all orientations (➤powder photography). The latter technique determines the lattice spacing directly and allows the structure to be deduced.

X-ray emission spectroscopy A technique for determining the chemical composition of a sample by bombarding it with X-rays (or electrons) and examining the ➤X-ray spectrum it then emits.

X-ray fluorescence A technique, usually used to complement ➤scanning electron microscopy, in which the chemical composition of material below the surface of a sample is determined by analysing the X-rays emitted by ➤fluorescence in response to an incident X-ray beam.

X-ray lithography A variant of ➤lithography that takes advantage of the short wavelength of X-rays: in the production of a ➤semiconductor device, an X-ray beam can be used to create a pattern on the ➤substrate (2) surface.

X-ray photoelectron spectroscopy (XPS) ➤photoelectron spectroscopy.

X-ray source 1 (Phys.) A device for generating X-rays in the laboratory. Electrons produced by thermal emission from a filament are accelerated through a potential difference of several thousand volts into a metal ➤anode (typically of tungsten), which must be cooled. The wavelengths of the X-rays produced by this bombardment are characteristic of the element used for the anode (➤X-ray spectrum).

2 (Astron.) Any astronomical object that emits X-rays.

X-ray spectrum The ➤spectrum of X-rays emitted by a material. Each element, for example in an ➤X-ray source, emits X-rays of characteristic frequencies that form **lines** in the spectrum. The energies corresponding to these frequencies are the differences between atomic ➤energy levels and represent transitions from one electronic state to another. ➤➤Moseley's law.

xylem Part of the ➤vascular system of plants responsible for the transport of water, dissolved minerals and some organic substances. Xylem consists of two distinct cell types, both of which lack living cell contents: **vessels**, which are open-ended tubes with lignified walls of various patterns; and **tracheids**, which are smaller closed cells with tapering ends. The composition of the xylem defines the main difference between hardwoods and softwoods. Vessels are formed only in hardwoods derived from trees in the ➤Magnoliopsida, whereas tracheids are found in both, and exclusively form the xylem of softwoods derived from conifers.

xylene Traditional name for dimethylbenzene, $C_6H_4(CH_3)_2$, which has three positional isomers (➤ortho, meta and para). All are colourless liquids boiling close to $140\,°C$. Industrially xylene is formed as a mixture of isomers by reforming of naphthalene and used as a solvent and petrol additive. The most important isomer, o-xylene, is used to make ➤phthalic anhydride.

Y

Y Symbol for the element ➤yttrium.

YAC Acronym for ➤yeast artificial chromosome.

YAG Acronym for yttrium aluminium garnet. ➤Neodymium in YAG is an important laser system.

Yagi aerial A directional ➤aerial commonly used, among many other applications, for television reception. It consists of one or two simple dipole aerials together with a **reflector** aerial and a number of shorter parallel **director** aerials. The directors focus the signal onto the dipole. Named after Hidetsugu Yagi (1886–1976). Yagi aerials were seen on the nosecones of many WWII aircraft, such as the Junkers Ju 88.

yard Symbol yd. An Imperial unit of distance equal to 3 feet. A statute ➤mile is 1760 yd.

Yb Symbol for the element ➤ytterbium.

Y chromosome ➤sex chromosome.

year A unit of time approximately equal to the orbital period of the Earth. ➤Gregorian calendar; sidereal year; tropical year.

yeast A member of a group of widespread unicellular ➤fungi of the phylum Ascomycota. Yeasts are economically important in the production of alcohol by ➤fermentation and in baking. There is a very large number of 'domesticated' varieties of yeast, mostly of the one species *Saccharomyces cerevisiae*, but many occur naturally, on the skins of fruit, for example. Some yeasts are ➤pathogens and cause diseases such as thrush in humans. *S. cerevisiae* is also an important laboratory organism, and is the focus of much genetic research using recombinant ➤DNA and DNA cloning techniques. Yeast extracts are also used in the food industry as sources of ➤vitamins, and mass culture of yeast cells can yield ➤single-cell protein.

yeast artificial chromosome (YAC) A system used for producing large segments of ➤DNA (more than 200 000 bases). ➤Genetic engineering techniques have been used to construct DNA molecules that contain a yeast ➤centromere and several yeast genes. Segments of DNA can be inserted into such molecules and, when introduced into yeast cells, they behave like ➤chromosomes and divide in step with the cell. YACs have become important host ➤vectors (4) for cloning large regions of ➤genomes, and were extensively used in the ➤human genome project.

yield point The ➤stress above which a material, often a metal, will begin to flow rather than simply deform.

yoke A component of high ➤magnetic permeability used to complete a ➤magnetic circuit.

Young's fringes The interference pattern produced when coherent light passes through a pair of narrow slits. The intensity of the light, of wavelength λ, is given by:

$$I(\theta) \propto \cos^2\left(\frac{\pi D \sin \theta}{\lambda}\right),$$

for an angle θ, the angle to the normal to the plane of the slits, separated by a distance D. Thus maximum intensity occurs when $n\lambda = D \sin \theta$, where n is any integer. Named after Thomas Young (1773–1829).

Young's modulus Symbol E. The constant of proportionality in the relationship between ➤stress and ➤strain for an elastic solid: $\sigma = E\eta$, where σ is the compressive or tensile stress and η is the strain in the direction of that stress.

ytterbium Symbol Yb. The element with atomic number 70 and relative atomic mass 173.0, which is one of the ➤lanthanides. (The reason that its name sounds so similar to those of yttrium, terbium and erbium is that all four were discovered in the same small area around the town of Ytterby in Sweden.) It shows the typical properties of lanthanides, with its chemistry dominated by compounds with oxidation number +3, such as the colourless Yb^{3+} ion.

yttrium Symbol Y. The element with atomic number 39 and relative atomic mass 88.91, which is the Group 3 element immediately below scandium. Its dominant oxidation number is +3, as in the colourless ion Y^{3+} and the oxide Y_2O_3, which is used when combined with europium as a red phosphor in TV sets.

Z

Z   **1** Symbol for ➤atomic number.
    2 Symbol for ➤partition function.
    3 Symbol for ➤impedance.

ℤ   Symbol for the set of all ➤integers, derived from the German word *Zahlen* for 'numbers'.

Z boson (Z particle)   A ➤gauge boson of the ➤weak interaction. It is a very massive (91.2 GeV/c^2), uncharged particle. ➤➤W boson.

Zeeman effect   The splitting of atomic ➤energy levels in the presence of an external magnetic field. The degeneracy of orbitals with the same ➤orbital angular momentum quantum number is lifted, and the energies of orbitals with different ➤magnetic quantum numbers are split apart, the splitting being proportional to the magnetic field strength. Named after Pieter Zeeman (1865–1943) who shared the 1902 Nobel prize with his tutor Hendrik Lorentz, who had predicted that the effect should occur. ➤➤Paschen–Back effect.

Zeise's salt   The first ➤pi complex discovered (see the diagram) in 1825. Named after William Christopher Zeise (1789–1847).

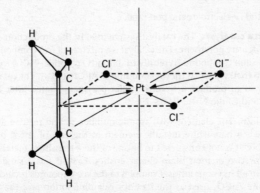

Zeise's salt This platinum complex has a square planar shape, with three of the four vertices filled by chloride ligands. The fourth vertex is occupied by an ethene molecule attached by its ➤pi bond.

Zener breakdown The sudden increase in current observed when the reverse bias of a heavily doped ➤p–n junction allows electrons to tunnel from the valence band of the p-type material to the conduction band of the n-type. Named after Clarence Melvin Zener (1905–93). Compare ➤avalanche breakdown.

Zener diode A type of ➤diode, used as a voltage regulator, that uses ➤Zener breakdown to ensure a well-defined maximum voltage (reverse bias) across it. The diode consists of a ➤p–n junction with a high ➤doping concentration.

zenith The point on the ➤celestial sphere directly above the observer. Compare ➤nadir.

zeolite One of a group of complex aluminosilicates which have an intricate system of holes of molecular dimensions into which other molecules can fit. For example, water may be incorporated and then released on heating: hence the name, which comes from the Greek for 'boiling stone'. Zeolites can be used as molecular sieves and in ➤ion exchange. ➤➤clathrate.

zero **1** The name given to the number 0 (nought).
 2 A zero of a function $f(x)$ is a real or complex number a for which $f(a) = 0$.

zero matrix A ➤matrix all of whose entries are zero.

zero-order reaction A reaction in which the rate is independent of the concentration of a specified substance. For example, in the iodination of propanone, CH_3COCH_3, the rate is independent of the concentration of iodine which proves that its ➤mechanism must be a multistep process with iodine reacting in a second, fast step.

zero-point energy In ➤quantum mechanics, the nonzero energy present in the ➤ground state of a system. For a simple harmonic oscillator of frequency f, the zero-point energy is $\frac{1}{2}hf$. In ➤classical physics oscillators can have zero energy, so this is a feature peculiar to quantum mechanics. All ➤bound states have a zero-point energy.

zero vector A vector with zero magnitude.

zeta potential ➤electrokinetic potential.

Ziegler–Natta catalysts The catalyst system used in the production of polymers such as high-density polythene. The catalyst is a mixture of two compounds. One of them is a titanium compound (typically titanium(IV) chloride, $TiCl_4$) together with an aluminium trialkyl such as triethylaluminium, $Al(CH_2CH_3)_3$. The catalysts can be subtly modified to produce a ➤stereoregular polymer. Named after Karl Ziegler (1898–1973) and Giulio Natta (1903–79).

zinc Symbol Zn. The element with atomic number 30 and relative atomic mass 65.39, which is a blue-white metallic element at the end of the d block of the ➤periodic table. It is not considered to be one of the ➤transition metals, because it does not show two of their main characteristics. First, it has a single oxidation number (+2) in all its compounds. Examples of these compounds include the white solid carbonate $ZnCO_3$ used as the base for calamine lotion and the white solid chloride $ZnCl_2$ used as a wood preservative and in batteries. Second, it does not form any coloured ions in aqueous solution, the aqueous ion being colourless Zn^{2+}(aq). It does resemble the transition metals in that it forms complexes such as an ammine complex $[Zn(NH_3)_4]^{2+}$ and hydroxo complexes such as tetrahydroxozincate $[Zn(OH)_4]^{2-}$. The formation of this hydroxo complex makes its hydroxide ➤amphoteric, reacting with excess aqueous sodium hydroxide. The element tends to occur as the sulfide rather than the oxide, two common ores being zinc blende, ZnS, and sphalerite, which also contains iron. The ore is roasted in air to form the oxide from which the element is won by reduction using carbon. One of its main uses is to ➤galvanize iron and steel objects; zinc chromate is incorporated in rust-inhibiting paints. In addition it is used in zinc–acid batteries and in the alloys ➤solder and ➤brass.

zinc blende structure An important crystal structure (see the diagram) shown by zinc blende, a sulfide ore of zinc, ZnS. The structure has the sulfur atoms in an array with cubic ➤close packing, with the zinc atoms in half of the tetrahedral holes. This is closely related to the ➤wurtzite structure, except that the sulfur atom array is the other close-packed form. Examples of compounds with the zinc blende structure include silicon carbide, SiC, and boron nitride, BN. ➤➤fluorite structure.

zinc blende structure

zinc oxide ZnO A solid which is unusual in that its colour is white when cold and yellow when hot. This behaviour is caused by the loss of a small quantity of oxygen on heating. On cooling the reverse happens as oxygen is gained from the atmosphere. It is used medicinally as a disinfectant and as a white pigment called Chinese white.

zinc sulfide ZnS A polymorphic solid, two polymorphs of which have important structures (➤wurtzite structure; zinc blende structure). The solid is used as a phosphorescent material.

zirconium Symbol Zr. The element with atomic number 40 and relative atomic mass 91.22, which is in the second row of the ➤transition metals directly below titanium. Like titanium, it shows variable oxidation numbers, with +4 as the dominant one, as in the aqueous ion Zr^{4+}(aq). The oxide ZrO_2 is one ore of zirconium called baddeleyite. The more important ore is **zircon**, $ZrSiO_4$, which is a gemstone.

Z line A transverse line in striated ➤muscle.

Zn Symbol for the element ➤zinc.

zodiac Twelve ancient constellations on the ➤ecliptic. Viewed from the Earth, the Sun appears to pass through each of these once per year.

zone The part of the surface of a sphere cut off by two ➤parallel planes. If the sphere has radius r and if d is the distance between the planes, the area of the zone is $2\pi rd$; this was Archimedes' favourite theorem.

zone refining A purification technique in which a heating coil is passed slowly from one end of a solid (such as a metal, an alloy or a ➤metalloid) to the other. The solid melts locally, and the impurities dissolve in the molten zone. When the heating coil moves to another part of the solid, the impurities remain in the molten zone, and accumulate at one end, which is then discarded. The technique is used to produce ultra-pure silicon and germanium for semiconductor use (which are often then intentionally made impure again by ➤doping).

zoology The scientific study of animals. Zoology is now a very wide, multidisciplinary field and this is reflected in the fact that most academic institutions now commonly subsume 'zoology' into animal or biological sciences.

zooplankton ➤Plankton consisting of microscopic animals, or the larvae or other juvenile forms of larger organisms such as crustaceans, fish and molluscs.

Z particle ➤Z boson.

Zr Symbol for the element ➤zirconium.

zwitterion An ➤ion that can exhibit both acidic and basic properties, depending on ➤pH. ➤Amino acids provide good examples (see the diagram).

$$H_3\overset{+}{N}-\overset{\overset{\displaystyle R}{|}}{\underset{\underset{\displaystyle H}{|}}{C}}-CO_2^-$$

zygote The cell resulting from the fusion of two ➤gametes during ➤sexual reproduction.

zwitterion of an amino acid.

zygotene ➤meiosis.

zymase A complex of enzymes, found in yeast, which plays a role in ➤fermentation.

Appendix

1. SI Units

Base units

Physical quantity	Name	Symbol
Length	metre	m
Mass	kilogram	kg
Time	second	s
Electric current	ampere	A
Thermodynamic temperature	kelvin	K
Luminous intensity	candela	cd
Amount of substance	mole	mol

Derived units

Physical quantity	Name	Symbol	Expression in terms of other SI units
Frequency	hertz	Hz	s^{-1}
Force	newton	N	$kg\ m\ s^{-2}$
Pressure	pascal	Pa	$N\ m^{-2}$
Energy	joule	J	$N\ m$
Power	watt	W	$J\ s^{-1}$
Electric charge	coulomb	C	$A\ s$
Electric potential	volt	V	$J\ C^{-1}$
Capacitance	farad	F	$C\ V^{-1}$
Resistance	ohm	Ω	$V\ A^{-1}$
Conductance	siemens	S	Ω^{-1}
Magnetic flux density	tesla	T	$V\ s\ m^{-2}$
Magnetic flux	weber	Wb	$V\ s$
Inductance	henry	H	$V\ s\ A^{-1}$
Activity	becquerel	Bq	s^{-1}

Decimal multiples and submultiples of units

Submultiple	Prefix	Symbol	Multiple	Prefix	Symbol
10^{-1}	deci	d	10	deca	da
10^{-2}	centi	c	10^{2}	hecto	h
10^{-3}	milli	m	10^{3}	kilo	k
10^{-6}	micro	μ	10^{6}	mega	M
10^{-9}	nano	n	10^{9}	giga	G
10^{-12}	pico	p	10^{12}	tera	T
10^{-15}	femto	f	10^{15}	peta	P
10^{-18}	atto	a	10^{18}	exa	E

2. Fundamental Constants

Quantity	Symbol	Value
Speed of light in vacuum	c	$299\ 792\ 458$ m s^{-1} (defined)
Planck constant	h	6.6261×10^{-34} J s
	$\hbar = h/2\pi$	1.0546×10^{-34} J s
Boltzmann constant	k	1.3807×10^{-23} J K^{-1}
Permittivity of free space	ε_0	8.8542×10^{-12} F m^{-1}
Mass of electron	m_e	9.1094×10^{-31} kg
Mass of proton	m_p	1.6726×10^{-27} kg
Elementary charge	e	1.6022×10^{-19} C
Gravitational constant	G	6.6726×10^{-11} m^3 kg^{-1} s^{-2}
Avogadro constant	L	6.0221×10^{23} mol^{-1}
Gas constant	$R = Lk$	8.3145 J K^{-1} mol^{-1}
Faraday constant	$F = Le$	9.6485×10^4 C mol^{-1}
Permeability of free space	$\mu_0 = 1/\varepsilon_0 c^2$	$4\pi \times 10^{-7}$ H m^{-1} (defined)
Bohr magneton	$\mu_B = e\hbar/2m_e$	9.2740×10^{-24} J T^{-1}
Nuclear magneton	$\mu_N = e\hbar/2m_p$	5.0508×10^{-27} J T^{-1}
Stefan–Boltzmann constant	$\sigma = 2\pi^5 k^4/15h^3 c^2$	5.6704×10^{-8} W m^{-2} K^{-4}
Fine-structure constant	$\alpha = e^2/2\varepsilon_0 hc$	7.2974×10^{-3}
Bohr radius	$a_0 = \varepsilon_0 h^2/\pi m_e e^2$	5.2918×10^{-11} m
Hartree energy	$E_H = \hbar^2/m_e a_0^2$	4.3597×10^{-18} J
Rydberg constant	$R_\infty = E_H/2hc$	1.0974×10^7 m^{-1}

3. Derivatives and Integrals

y	$\dfrac{\mathrm{d}y}{\mathrm{d}x}$	$\int y\,\mathrm{d}x$
x^n	nx^{n-1}	$\dfrac{1}{n+1}x^{n+1}$
$\dfrac{1}{x}$	$-\dfrac{1}{x^2}$	$\ln x$
e^{ax}	$a\mathrm{e}^{ax}$	$\dfrac{1}{a}\mathrm{e}^{ax}$
$\ln x$	$\dfrac{1}{x}$	$x\,(\ln x - 1)$
$\cos ax$	$-a\sin ax$	$\dfrac{1}{a}\sin ax$
$\sin ax$	$a\cos ax$	$-\dfrac{1}{a}\cos ax$
$\tan ax$	$a\sec^2 ax$	$-\dfrac{1}{a}\ln\cos ax$
$\cot x$	$-\operatorname{cosec}^2 x$	$\ln\sin x$
$\sec x$	$\tan x\sec x$	$\ln(\sec x + \tan x)$
$\operatorname{cosec} x$	$-\cot x\operatorname{cosec} x$	$\ln(\operatorname{cosec} x - \cot x)$
$\sin^{-1}\dfrac{x}{a}$	$\dfrac{1}{\sqrt{a^2 - x^2}}$	$x\sin^{-1}\dfrac{x}{a} + \sqrt{a^2 - x^2}$
$\cos^{-1}\dfrac{x}{a}$	$\dfrac{-1}{\sqrt{a^2 - x^2}}$	$x\cos^{-1}\dfrac{x}{a} + \sqrt{a^2 - x^2}$
$\tan^{-1}\dfrac{x}{a}$	$\dfrac{a}{a^2 + x^2}$	$x\tan^{-1}\dfrac{x}{a} - \dfrac{a}{2}\ln(a^2 + x^2)$

4. The Solar System

Planet	Diameter/km	Distance from Sun/10^6 km	Period of revolution	Period of rotation	Eccentricity of orbit	Inclination of orbit/deg	Number of Satellites
Mercury	4 879	57.91	87.97 d	58 d 15 h 30 min	0.21	7.0	0
Venus	12 104	108.21	224.70 d	243 d 0 h 30 min R	0.01	3.4	0
Earth	12 756	149.60	365.26 d	23 h 56 min +	0.02	0.0	1
Mars	6 794	227.94	686.98 d	24 h 37 min	0.09	1.9	2
Jupiter	142 984	778.33	11.862 y	9 h 51 min	0.05	1.3	16
Saturn	120 540	1426.99	29.457 y	10 h 33 min	0.06	2.5	18*
Uranus	51 118	2869.55	84.01 y	17 h 14 min	0.05	0.8	17
Neptune	49 528	4496.64	164.80 y	16 h 7 min	0.01	1.8	8
Sun	1 392 500	—		25 d 9 h 7 min	—	—	—
Moon	3 476	0.3844†	27 d 7 h 44 min	27 d 7 h 44 min	0.05†	5.2†	—

R Retrograde motion.

* Other satellites have been reported but not confirmed.

† Relative to Earth.

+ Sidereal day.

N.B. The status of Pluto was downgraded to dwarf planet in 2006.

5. The Periodic Table The ➤ atomic number for each element is shown as a left subscript (➤ symbol, chemical). The number below each symbol is the element's ➤ relative atomic mass (to four significant figures); if all its isotopes are radioactive the ➤ nucleon number of the most stable nuclide is given instead in brackets.

The bold numbers across the top are those of the ➤ groups; the numbers on the extreme left are those of the ➤ periods. The letters s, p, d and f identify the ➤ blocks.

Period	1	2	3	4	5	6	7	8	9	10	11	12	13	14	15	16	17	18
1	$_1$H 1.008																	$_2$He 4.003
2	$_3$Li 6.941	$_4$Be 9.012											$_5$B 10.81	$_6$C 12.01	$_7$N 14.01	$_8$O 16.00	$_9$F 19.00	$_{10}$Ne 20.18
3	$_{11}$Na 22.99	$_{12}$Mg 24.31											$_{13}$Al 26.98	$_{14}$Si 28.09	$_{15}$P 30.97	$_{16}$S 32.07	$_{17}$Cl 35.45	$_{18}$Ar 39.95
4	$_{19}$K 39.10	$_{20}$Ca 40.08	$_{21}$Sc 44.96	$_{22}$Ti 47.87	$_{23}$V 50.94	$_{24}$Cr 52.00	$_{25}$Mn 54.94	$_{26}$Fe 55.85	$_{27}$Co 58.93	$_{28}$Ni 58.69	$_{29}$Cu 63.55	$_{30}$Zn 65.38	$_{31}$Ga 69.72	$_{32}$Ge 72.64	$_{33}$As 74.92	$_{34}$Se 78.96	$_{35}$Br 79.90	$_{36}$Kr 83.80
5	$_{37}$Rb 85.47	$_{38}$Sr 87.62	$_{39}$Y 88.91	$_{40}$Zr 91.22	$_{41}$Nb 92.91	$_{42}$Mo 95.96	$_{43}$Tc (98)	$_{44}$Ru 101.1	$_{45}$Rh 102.9	$_{46}$Pd 106.4	$_{47}$Ag 107.9	$_{48}$Cd 112.4	$_{49}$In 114.8	$_{50}$Sn 118.7	$_{51}$Sb 121.8	$_{52}$Te 127.6	$_{53}$I 126.9	$_{54}$Xe 131.3
6	$_{55}$Cs 132.9	$_{56}$Ba 137.3	$_{57}$La 138.9	$_{72}$Hf 178.5	$_{73}$Ta 180.9	$_{74}$W 183.8	$_{75}$Re 186.2	$_{76}$Os 190.2	$_{77}$Ir 192.2	$_{78}$Pt 195.1	$_{79}$Au 197.0	$_{80}$Hg 200.6	$_{81}$Tl 204.4	$_{82}$Pb 207.2	$_{83}$Bi 209.0	$_{84}$Po (209)	$_{85}$At (210)	$_{86}$Rn (222)
7	$_{87}$Fr (223)	$_{88}$Ra (226)	$_{89}$Ac (227)	$_{104}$Rf	$_{105}$Db	$_{106}$Sg	$_{107}$Bh	$_{108}$Hs	$_{109}$Mt	$_{110}$Ds	$_{111}$Rg							

Lanthanides

$_{58}$Ce 140.1	$_{59}$Pr 140.9	$_{60}$Nd 144.2	$_{61}$Pm (145)	$_{62}$Sm 150.4	$_{63}$Eu 152.0	$_{64}$Gd 157.3	$_{65}$Tb 158.9	$_{66}$Dy 162.5	$_{67}$Ho 164.9	$_{68}$Er 167.3	$_{69}$Tm 168.9	$_{70}$Yb 173.1	$_{71}$Lu 175.0

Actinides

$_{90}$Th 232.0	$_{91}$Pa 231.0	$_{92}$U 238.0	$_{93}$Np (237)	$_{94}$Pu (244)	$_{95}$Am (243)	$_{96}$Cm (247)	$_{97}$Bk (247)	$_{98}$Cf (251)	$_{99}$Es (252)	$_{100}$Fm (257)	$_{101}$Md (258)	$_{102}$No (259)	$_{103}$Lr (262)

6. The Elements

Element	Symbol	Atomic number	Density* /g cm^{-3}	Melting point/°C	Boiling point/°C
Actinium	Ac	89	10.06	1050	3200
Aluminium	Al	13	2.70	660	2470
Americium	Am	95	13.67	990	2600
Antimony	Sb	51	6.69	631	1640
Argon	Ar	18	1.66	−189	−186
Arsenic	As	33	5.78	613s	
Astatine	At	85		300	350
Barium	Ba	56	3.59	710	1640
Berkelium	Bk	97	14.79	986	
Beryllium	Be	4	1.85	1285	2470
Bismuth	Bi	83	9.75	271	1560
Boron	B	5	2.47	2030	3700
Bromine	Br	35	3.12	−7	59
Cadmium	Cd	48	8.65	321	770
Caesium	Cs	55	1.90	29	686
Calcium	Ca	20	1.53	840	1490
Californium	Cf	98			
Carbon	C	6	2.27	3700s	
Cerium	Ce	58	6.71	800	3430
Chlorine	Cl	17	2.03	−101	−34
Chromium	Cr	24	7.19	1860	2600
Cobalt	Co	27	8.80	1494	2900
Copper	Cu	29	8.93	1085	2580
Curium	Cm	96	13.30	1340	
Dysprosium	Dy	66	8.53	1410	2600
Einsteinium	Es	99			
Erbium	Er	68	9.04	1520	2860
Europium	Eu	63	5.25	820	1450
Fermium	Fm	100			
Fluorine	F	9	1.14*	−220	−188
Francium	Fr	87		30	650
Gadolinium	Gd	64	7.87	1310	3000
Gallium	Ga	31	5.91	30	2400
Germanium	Ge	32	5.32	959	2850
Gold	Au	79	19.28	1064	2850
Hafnium	Hf	72	13.28	2230	5300
Helium	He	2	0.12*		−269

6. The Elements *(continued)*

Element	Symbol	Atomic number	Density* /g cm^{-3}	Melting point/°C	Boiling point/°C
Holmium	Ho	67	8.80	1470	2700
Hydrogen	H	1	0.076	−259	−253
Indium	In	49	7.29	157	2050
Iodine	I	53	4.95	114	184
Iridium	Ir	77	22.55	2447	4550
Iron	Fe	26	7.87	1540	2760
Krypton	Kr	36	2.82	−157	−153
Lanthanum	La	57	6.17	920	3450
Lawrencium	Lr	103			
Lead	Pb	82	11.34	328	1760
Lithium	Li	3	0.53	180	1360
Lutetium	Lu	71	9.84	1700	3400
Magnesium	Mg	12	1.74	650	1100
Manganese	Mn	25	7.47	1250	2120
Mendelevium	Md	101			
Mercury	Hg	80	13.55	−39	357
Molybdenum	Mo	42	10.22	2620	4830
Neodymium	Nd	60	7.00	1024	3100
Neon	Ne	10	1.44	−249	−246
Neptunium	Np	93	20.45	640	
Nickel	Ni	28	8.91	1455	2730
Niobium	Nb	41	8.58	2425	5000
Nitrogen	N	7	1.04	−210	−196
Nobelium	No	102			
Osmium	Os	76	22.58	3030	5000
Oxygen	O	8	2.00	−219	−183
Palladium	Pd	46	12.00	1554	3000
Phosphorus	P	15	1.82	44	280
Platinum	Pt	78	21.45	1772	3720
Plutonium	Pu	94	19.81	640	3200
Polonium	Po	84	9.40	254	960
Potassium	K	19	0.86	63	777
Praseodymium	Pr	59	6.78	935	3000
Promethium	Pm	61	7.22	1168	3300
Protactinium	Pa	91	15.37	1200	4000
Radium	Ra	88	5.00	700	1140
Radon	Rn	86	4.40*	−71	−62

6. The Elements *(continued)*

Element	Symbol	Atomic number	Density* /g cm^{-3}	Melting point/°C	Boiling point/°C
Rhenium	Re	75	21.02	3180	5600
Rhodium	Rh	45	12.42	1963	3700
Rubidium	Rb	37	1.53	39	705
Ruthenium	Ru	44	12.36	2310	4100
Samarium	Sm	62	7.54	1060	1800
Scandium	Sc	21	2.99	1540	2800
Selenium	Se	34	4.81	220	685
Silicon	Si	14	2.33	1410	2355
Silver	Ag	47	10.50	962	2160
Sodium	Na	11	0.97	98	900
Strontium	Sr	38	2.58	770	1380
Sulfur	S	16	2.09	115	445
Tantalum	Ta	73	16.67	3000	5400
Technetium	Tc	43	11.50	2200	4600
Tellurium	Te	52	6.25	450	990
Terbium	Tb	65	8.27	1360	3100
Thallium	Tl	81	11.87	304	1460
Thorium	Th	90	11.73	1700	4500
Thulium	Tm	69	9.33	1550	2000
Tin	Sn	50	7.29	232	2270
Titanium	Ti	22	4.51	1670	3300
Tungsten	W	74	19.25	3387	5650
Uranium	U	92	19.05	1135	4000
Vanadium	V	23	6.09	1920	3400
Xenon	Xe	54	3.56	−112	−108
Ytterbium	Yb	70	6.97	824	1500
Yttrium	Y	39	4.48	1510	3300
Zinc	Zn	30	7.14	420	913
Zirconium	Zr	40	6.51	1850	4400

For some of the radioactive elements, data are not available.
* Densities for the liquid state; all others are for the solid state.
s Sublimes.

7. Amino Acids

Name	Abbreviation	Symbol	Structure
Aliphatic amino acids			
Glycine	Gly	G	$H_2N-\overset{\displaystyle H}{\underset{\displaystyle H}{C}}-COOH$
Alanine	Ala	A	$H_2N-\overset{\displaystyle H}{\underset{\displaystyle CH_3}{C}}-COOH$
Valine	Val	V	$H_2N-C-COOH$, with H above and $HC-CH_3$ and CH_3 below
Leucine	Leu	L	$H_2N-C-COOH$, with H above and CH_2, $HC-CH_3$, CH_3 below
Isoleucine	Ile	I	$H_2N-C-COOH$, with H above and $HC-CH_3$, CH_2, CH_3 below

7. Amino Acids *(continued)*

Name	Abbreviation	Symbol	Structure

Hydroxy amino acids

Serine — Ser — S

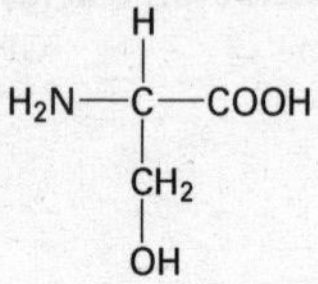

$$H_2N-\overset{\overset{\displaystyle H}{|}}{\underset{\underset{\displaystyle OH}{|}}{\underset{\underset{\displaystyle CH_2}{|}}{C}}}-COOH$$

Threonine — Thr — T

$$H_2N-\overset{\overset{\displaystyle H}{|}}{\underset{\underset{\displaystyle CH_3}{|}}{\underset{\underset{\displaystyle HC-OH}{|}}{C}}}-COOH$$

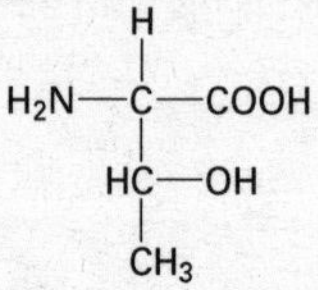

Dicarboxylic amino acids and their amides

Aspartic acid — Asp — D

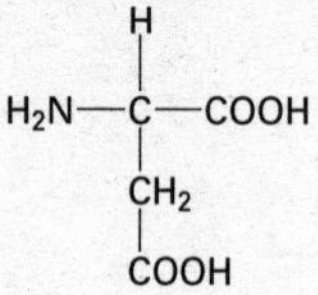

$$H_2N-\overset{\overset{\displaystyle H}{|}}{\underset{\underset{\displaystyle COOH}{|}}{\underset{\underset{\displaystyle CH_2}{|}}{C}}}-COOH$$

Asparagine — Asn — N

$$H_2N-\overset{\overset{\displaystyle H}{|}}{\underset{\underset{\displaystyle CONH_2}{|}}{\underset{\underset{\displaystyle CH_2}{|}}{C}}}-COOH$$

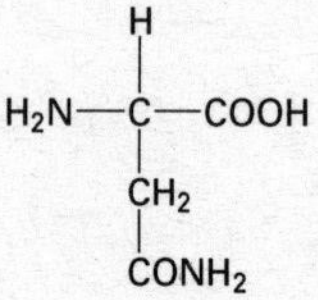

Glutamic acid — Glu — E

$$H_2N-\overset{\overset{\displaystyle H}{|}}{\underset{\underset{\underset{\displaystyle COOH}{|}}{\underset{\displaystyle CH_2}{|}}}{\underset{\underset{\displaystyle CH_2}{|}}{C}}}-COOH$$

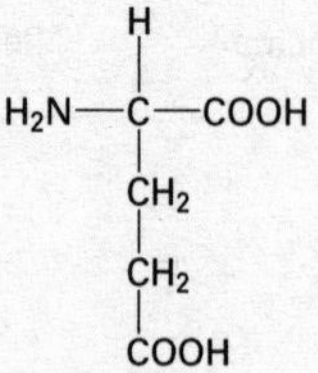

7. Amino Acids *(continued)*

Name	Abbreviation	Symbol	Structure
Glutamine	Gln	Q	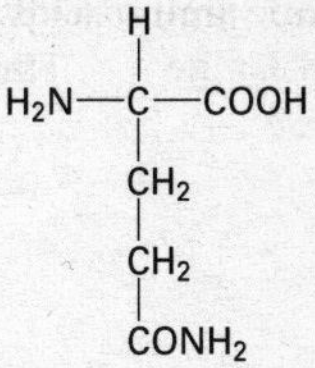

$$H_2N-\overset{\overset{\displaystyle H}{|}}{C}-COOH$$
$$|$$
$$CH_2$$
$$|$$
$$CH_2$$
$$|$$
$$CONH_2$$

Amino acids with basic side chains

Name	Abbreviation	Symbol	Structure
Lysine	Lys	K	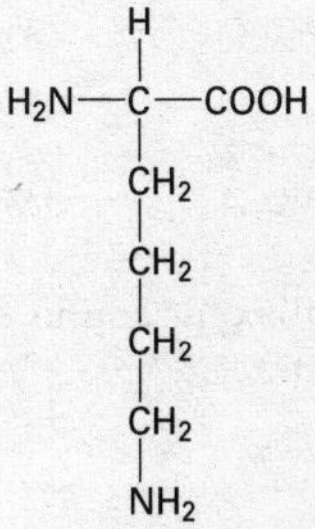

$$H_2N-\overset{\overset{\displaystyle H}{|}}{C}-COOH$$
$$|$$
$$CH_2$$
$$|$$
$$CH_2$$
$$|$$
$$CH_2$$
$$|$$
$$CH_2$$
$$|$$
$$NH_2$$

| Histidine | His | H 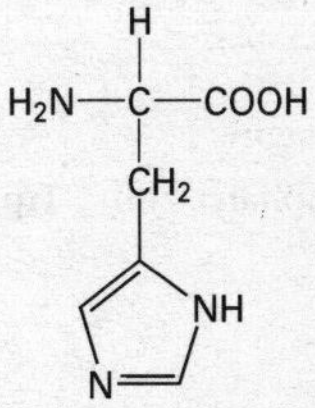 |

$$H_2N-\overset{\overset{\displaystyle H}{|}}{C}-COOH$$
$$|$$
$$CH_2$$

(imidazole ring with NH and N)

| Arginine | Arg | R 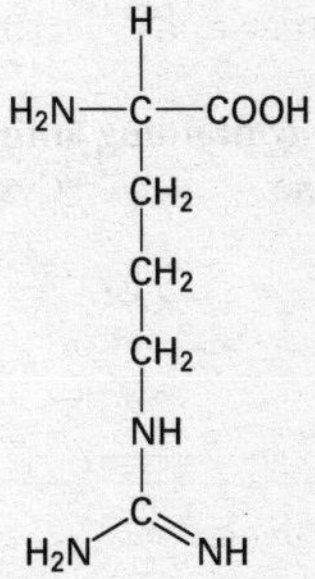 |

$$H_2N-\overset{\overset{\displaystyle H}{|}}{C}-COOH$$
$$|$$
$$CH_2$$
$$|$$
$$CH_2$$
$$|$$
$$CH_2$$
$$|$$
$$NH$$
$$|$$
$$H_2N-C=NH$$

7. Amino Acids *(continued)*

Name	Abbreviation	Symbol	Structure
Aromatic amino acids			
Phenylalanine	Phe	F	
Tyrosine	Tyr	Y	
Tryptophan	Trp	W	
Sulfur-containing amino acids			
Cysteine	Cys	C	

7. Amino Acids *(continued)*

Name	Abbreviation	Symbol	Structure
Methionine	Met	M	

Imino acid

Proline	Pro	P	

8. Outline Classification of Living Organisms

There is no one universally accepted system for the classification of organisms, and a variety of different approaches will be met in textbooks. The differences lie largely in the nomenclature and status of the phyla, particularly in the plant and fungal kingdoms. The reader should thus expect to encounter (particularly) different nuances of word ending (e.g. Ascomycota might appear as Ascomycotina in some schemes). The academic level and understanding of students also has a bearing on this; thus it would be appropriate for primary school children to think of seashore 'algae' as 'plants', but by Advanced level such organisms should be discussed as protoctistans.

The generally accepted five kingdoms with their major phyla are given below. Latin names of examples of organisms in each phylum are appended in italics in brackets.

We have attempted to adopt a common approach using mainly the terminology suggested in *Biological Nomenclature* (1997), Institute of Biology, London. For the plants the terminology of hierarchies is based on that adopted by C. A. Stace (1991) *New Flora of the British Isles,* CUP, Cambridge, with terms in brackets cross-referencing to *Biological Nomenclature.*

(1) **Prokaryota** (formerly Monera) Currently under review and regarded by some authorities as containing organisms sufficiently diverse to warrant separating the group into other kingdoms. Unicellular prokaryotic cells including the Bacteria and Cyanobacteria (*Salmonella, Nostoc*).

(2) **Protoctista** Very diverse – primarily single-celled eukaryotic organisms, but also held to include the multicellular groups of algae.

Phyla: Rhizopoda – locomotion by amoeboid movement (*Amoeba*)
Zoomastigina – locomotion by one or more flagella (*Trypanosoma*)
Apicomplexa – mostly parasites (*Plasmodium*)
Ciliophora – ciliates (*Paramecium*)
Euglenophyta – photosynthetic flagellates (but some heterotrophs) (*Euglena*)
Oomycota – included in the Fungi by some, but having a stage in the life cycle with flagella (*Phytophthora*)
Chlorophyta (green algae) – photosynthetic with chlorophyll *a* and *b* (*Ulva*)
Rhodophyta (red algae) – photosynthetic with chlorophyll *a* and *d* (*Porphyra*)
Phaeophyta (brown algae) – photosynthetic with chlorophyll *a* and *c* (*Fucus*)
Bacillariophyta (diatoms) – photosynthetic with cells with two 'valves' (*Diatoma*)

(3) **Fungi** Eukaryotic heterotrophs (chemoorganotrophs), feeding by absorption, with cells rarely containing cellulose, spores without flagella.

Phyla: Zygomycota – sexual reproduction producing a zygospore. Hyphae
      without cross-walls (*Mucor*)
      Ascomycota – sexual reproduction producing a sac-like ascus (*Sordaria*)
      Basidiomycota – sexual reproduction producing spores on a basidium.
      Mushrooms and toadstools (*Agaricus*)
      Deuteromycota – fungi without clear sexual stages (*Aspergillus*)

(4) **Plantae** (Plants) Photosynthetic, multicellular, eukaryotic with cellulose cell walls.

Phyla: Bryopsida (Bryophyta) – plants with a conspicuous gametophyte generation
      with true roots absent and no highly differentiated vascular tissue. Mosses
      and liverworts (*Polytrichum, Pellia*)

All phyla following have a conspicuous sporophyte with a clearly organized vascular system with xylem and phloem, and are collectively regarded as tracheophytes (formerly the phylum Tracheophyta).

      Lycopodiopsida (Lycopodophyta) – small spirally arranged leaves –
      sporangia in rather lax cones – clubmosses (*Lycopodium*)
      Equisetopsida (Sphenophyta) – leaves reduced to scales – distinct cone
      (*Equisetum*)
      Pteropsida (Filicinophyta) – leaves typically frond-like, unfurling distinctly
      as 'fiddle heads', sporangia in clusters (sori) – ferns (*Dryopteris*)
      Pinopsida (Coniferophyta) – woody trees and shrubs with cones (*Pinus*)
      Magnoliopsida (Angiospermophyta) – plants with true flowers
      Class: Magnoliidae (Dicotyledoneae) – leaves with net veins, embryos with
            two cotyledons, floral parts variously arranged singly or in 2's, 4's or
            5's (*Helianthus, Ranunculus*)
            Liliidae (Monocotyledoneae) – leaves with parallel veins, embryos
            with one cotyledon, floral parts generally in 3's or 6's (*Avena, Lilium*)

(5) **Animalia** (Animals) Heterotrophic, generally feeding via a mouth, with nervous systems.

Phyla: Cnidaria (Coelenterata) – animals with two layers of cells, radially symmetrical
      with tentacles and nematocysts (*Hydra, Actinia*)
      Platyhelminthes – unsegmented flat bodies, with mouth but no anus
      (*Fasciola, Taenia*)
      Nematoda – unsegmented bodies with mouth and anus ('round worms')
      (*Caenorhabditis*)
      Annelida – segmented worms (*Nereis, Lumbricus*)
      Mollusca – unsegmented with muscular 'foot' – slugs, snails, mussels,
      octopus
      Arthropoda – exoskeleton, segmented bodies with jointed appendages

–numerically the largest group on the planet. Includes insects, centipedes and millipedes, crustaceans and arachnids (spiders and scorpions)

Echinodermata – marine, pentaradiate symmetry with water vascular system and tube feet. Starfish, brittle stars, sea cucumbers and sea urchins

Chordata – bodies with a notochord and gill slits at at least some stage of their development, a dorsal nerve cord and a post anal tail.

There are a number of nonvertebrate classes in the Chordata that have larvae possessing notochords or a very reduced notochord in the adult. These include the sea squirts and the proboscis worms. The vertebrate (regarded as the sub-phylum Vertebrata) classes are as follows:

Class: Chondrichthyes – fish with cartilaginous skeletons and separate gill slits – sharks and rays (*Scyliorhinus*)

Osteichthyes – bony fish with fins with supporting rays, gills with single opening covered by the operculum (*Gadus*)

Amphibia – typically terrestrial quadrupeds with soft skins and gilled, aquatic larvae. Frogs, toads, salamanders and newts (*Rana, Triturus, Salamandra*)

Reptilia – scaly skinned, eggs with leathery shells. Snakes, lizards, tortoises and turtles (*Lacerta, Natrix*)

Aves – birds. Endothermic animals with wings and feathers and typically capable of flight. Eggs with hard, calcareous shells (*Columba*)

Mammalia – mammals. Endothermic animals with hair and mammary glands. Young are suckled on milk. A very diverse group including whales, bats, bears, cats and dogs, elephants, antelope and cattle, and humans (*Canis, Rattus, Homo*)

PENGUIN REFERENCE LIBRARY

THE PENGUIN DICTIONARY OF ARCHITECTURE & LANDSCAPE ARCHITECTURE

EDITED BY JOHN FLEMING, HUGH HONOUR & NIKOLAUS PEVSNER

'Immensely useful, succinct and judicious … rich in accurate fact and accumulated wisdom' *The Times Literary Supplement*

Need a handbook to your architect's business or studies? Struggling with your latest DIY project? Interested in the art of architecture? This classic work, now in its fifth edition, covers every aspect of architecture and landscape architecture. Ranging from ancient times to contemporary trends, it adopts a truly international perspective, focussing on countries and cultures such as *Coptic*, *Tibet* and *De Stijl*.

- Provides biographies of architects and landscape architects, ancient and modern, from *Le Corbusier* to *Frank Lloyd Wright* to *Sir Christopher Wren*

- Surveys architecture and landscape architecture by country

- Discusses the history of urban design, including parks and motorways

- Covers structural and technical innovations and new terminology

- Includes clear and detailed drawings

ONLY PENGUIN GIVES YOU MORE

PENGUIN REFERENCE LIBRARY

THE PENGUIN DICTIONARY OF GEOLOGY

EDITED BY PHILIP KEAREY

'User friendly ... concise but informative'
The Times Higher Education Supplement

Siderophile, *interfluve*, *charnockite*: these are the terms that define the earth that we live on. The study of earth sciences is a complex one, full of specialist terms, making an authoritative dictionary a crucial purchase. *The Penguin Dictionary of Geology* is the most extensive and authoritative guide to earth sciences available. Covering over 7,700 key terms in meticulous detail, it explains not only the core vocabulary of 'the study of the solid earth' but also incorporates the related disciplines of astronomy, biology, environmental geoscience, chemistry and physics. The result is an up-to-date dictionary that will be a must-buy for student, researcher or keen amateur geologist.

- Gives comprehensive coverage of the core subject areas in sharp, concise definitions, from *labile* to *birnessite* to *ventifact*

- Provides helpful tables showing geological ages of the earth and SI conversion units

- Includes an extensive easy-to-use topic-based bibliography for further reference

ONLY PENGUIN GIVES YOU MORE

PENGUIN REFERENCE LIBRARY

THE PENGUIN DICTIONARY OF ACCOUNTING

EDITED BY CHRISTOPHER NOBES

You're an accountant needing *golden handcuffs*, you're faced with a jargon-laden business exam, you're the proud owner of an incomprehensible tax form. Where do you turn? *The Penguin Dictionary of Accounting* is your answer. Demystifying the most complex terms in simple, easy-to-understand language, this is the ideal reference book for anyone who reads, speaks or writes accountancy – whether student, businessperson, investor, preparer of financial statements, auditor or accounts manager. Compiled by Christopher Nobes, PricewaterhouseCoopers Professor of Accounting at Reading University, this fully up-to-date edition is your first stop in unravelling the ambiguities of accounting.

Contains information in the fields of *tax*, *auditing* and *management accounting*

Includes relevant terms relating to finance and law, from *depreciation* to the *ratchet effect* via *hyperinflation*

Analyses all the terms used in International Financial Reporting Standards

Caters for accountants worldwide, making clear where terminology differs across countries

ONLY PENGUIN GIVES YOU MORE

PENGUIN REFERENCE LIBRARY

THE PENGUIN DICTIONARY OF MATHEMATICS

EDITED BY DAVID NELSON

'Clear, authoritative, up-to-date … an absolutely superb reference book – a treasure trove of impeccably presented information' *Mathematical Gazette*

There are some who say that the world is a mass of mathematical equations. Whether true or not, they certainly have a point: mathematics is important. *The Penguin Dictionary of Mathematics* is the definitive handbook to the study, taking in all branches of pure and applied mathematics, from *algebra* to *mechanics* and from *number theory* to *statistics*. Written for students and teachers of all levels, it is also a useful and versatile source book for economists, business people, engineers, technicians and scientists of all kinds.

- Defines over 3,200 terms, supported by dozens of explanatory diagrams
- Useful for anyone from GCSE standard upwards
- Gives extensive entries on such topics as *chaos*, *fractals* and *graph theory*
- Includes biographies of over 200 key figures in mathematics, from *Einstein* to *Pythagoras* to *Descartes*
- Provides comprehensive coverage of subjects taught in school and college

ONLY PENGUIN GIVES YOU MORE

PENGUIN REFERENCE LIBRARY

THE PENGUIN DICTIONARY OF CHEMISTRY

EDITED BY DAVID W. A. SHARP

'A quick and easy-to-use resource…ideal for those who are starting to study the subject' *The Times Higher Education Supplement*

Saliginen, Hemiketals. Chitinase. It's what the world is made of, but how much do you know about it? *The Penguin Dictionary of Chemistry* is your guide. Crucial for students and employees of all sciences at any level, this bestselling text (over 150,000 sold) is both accessible and detailed, offering clear definitions alongside chemical symbols and diagrams. Comprehensive, illustrated and cross-referenced with the *Penguin Dictionary of Biology* and *Physics*, the latest edition also takes account of recent developments in this key subject area.

- Explains chemical terms from all branches of the subject, from *AAS* to *zwitterions*

- Gives concise entries on elements (*neon*), compounds (*ethene*) and other substances (*heptenophos*)

- Provides succinct accounts of important chemical operations (*hydrations, entrainment*), including industrial processes (*polarography, urea adduction*)

ONLY PENGUIN GIVES YOU MORE

PENGUIN REFERENCE LIBRARY

THE PENGUIN DICTIONARY OF BIOLOGY
EDITED BY M. THAIN & M. HICKMAN

'A marvellous compendium: accurate, clear and complete'
Matt Ridley, author of *Genome*

Worried about your *maternal effect* or *biological clock*? Need to know a *rhizoid* from a *rhizome*? Think you're going to fail your *zoology* or *botany* exam? *The Penguin Dictionary of Biology* is your saviour, defining some 6000 terms relating to this rich, complex and constantly expanding subject – from *amino acids*, *bacteria* and the *cell cycle* to *X-ray diffraction*, *Y chromosome* and *zygotes*. Long established as the definitive single-volume source, this dictionary has sold over 200,000 copies and is extensively updated for its eleventh edition.

- Contains over 400 new entries to take account of the latest thinking on *genetics*, *human physiology*, *disease* and *cell biology*
- Superbly complemented by *The Penguin Dictionary of Human Biology*, due in Autumn 2007
- Ideal for students, teachers, professionals and amateur biologists
- Extensively illustrated throughout, with charts demonstrating *vertebrate spermatogenesis*, *primate evolution* and much more

ONLY PENGUIN GIVES YOU MORE

PENGUIN SUBJECT DICTIONARIES

Penguin's Subject Dictionaries aim to provide two things: authoritative complimentary reference texts for the academic market (primarily A level and undergraduate studies) *and* clear, exciting and approachable reference books for general readers on subjects outside the core curriculum.

Academic & Professional

ACCOUNTING
ARCHEOLOGY
ARCHITECTURE
BUILDING
BUSINESS
CLASSICAL MYTHOLOGY
CRITICAL THEORY
ECONOMICS
INTERNATIONAL RELATIONS
LATIN
LITERARY TERMS & THEORY
MARKETING (forthcoming)
MEDIA STUDIES
MODERN HISTORY
PENGUIN HUMAN BIOLOGY (forthcoming)
PHILOSOPHY
PSYCHOLOGY
SOCIOLOGY

Scientific, Technical and Medical

BIOLOGY
CHEMISTRY
CIVIL ENGINEERING
COMPUTING
ELECTRONICS
GEOGRAPHY
GEOLOGY
MATHEMATICS
PHYSICAL GEOGRAPHY
PHYSICS
PSYCHOANALYSIS
SCIENCE
STATISTICS

English Words & Language

CLICHÉS
ENGLISH IDIOMS
PENGUIN ENGLISH GRAMMAR
PENGUIN RHYMING DICTIONARY
PROVERBS
SYNONYMS & ANTONYMS
SYNONYMS & RELATED WORDS
ROGET'S THESAURUS
THE COMPLETE PLAIN WORDS
THE PENGUIN A–Z THESAURUS
THE PENGUIN GUIDE TO PLAIN ENGLISH
THE PENGUIN GUIDE TO PUNCTUATION
THE PENGUIN WRITER'S MANUAL
USAGE AND ABUSAGE

Religion

BIBLE
ISLAM (forthcoming)
JUDAISM (forthcoming)
LIVING RELIGIONS
RELIGIONS
SAINTS
WHO'S WHO IN THE AGE OF JESUS

General Interest

BOOK OF FACTS
FIRST NAMES
MUSIC
OPERA
SURNAMES (forthcoming)
SYMBOLS
THEATRE

Penguin Reference – making knowledge everybody's property